Advanced Mathematics

A UNIFIED COURSE: BOOK TWO

L. K. Turner
F. J. Budden
D. Knighton

CONTENTS OF BOOK ONE

LONGMAN

ADVANCED MATHEMATICS

A UNIFIED COURSE: BOOK TWO

L. K. Turner
F. J. Budden
D. Knighton

Longman Group Limited
London
Associated companies, branches and representatives throughout the world

First published 1975
Third impression 1977
ISBN 0 582 35240 1

Printed in Great Britain by
William Clowes & Sons, Limited,
London, Colchester and Beccles

Acknowledgements

We are grateful to the following for permission to reproduce copyright material which appears in Advanced Mathematics Book 1 and Advanced Mathematics Book 2: The Associated Examining Board for the General Certificate of Education for questions from 'A' Level Mathematics Papers:—1970: Statistics Paper No. 2, 1970: Statistics Paper 10 No. 2, 1972: Applied Mathematics Paper 3 No. 8 and 1972: Modern Syllabus Paper 6 No. 10; Joint Matriculation Board for questions set by the Board in previous examinations in Mathematics; Oxford Delegacy of Local Examinations for questions from 1972 'A' Level Mathematics II Nos. 21 and 22 and Oxford Scholarship Examinations 1969 Mathematics III No. D.14 and 1968 Natural Science I (Mathematics) No. 17; University of Cambridge Local Examinations Syndicate for questions from Entrance & Schools Examination Papers:—1969 Maths III 4A and Further Maths 2A, 1970 Maths III 3A and Further Maths IIB and for questions from four 'A' Level Mathematics Papers 1971 and questions from four 'A' Level Mathematics Papers 1972: University of London for four questions from London 'GCE' Mathematics Examinations Papers 1968 and 1969 and the Welsh Joint Education Committee for questions from July 1972 Paper:—Mathematics A2 and Special Paper No. 11.

Finally, particular thanks are expressed to the Oxford and Cambridge Schools Examinations Board for permission to use a wide selection of questions, both from their own papers and from those of the S.M.P. and M.E.I. examinations.

Contents

Sections marked * may be delayed until a second reading.

Preface

The reader of Book 1 has already been introduced to simple functions, to their derivatives and integrals, to elementary sequences and series, to the foundations of probability and statistics, to vectors and mechanics. Our present task is to build upon this base.

Whatever we might have achieved has been due to joint authorship, chapters 10, 13 and 14 having been written by F. J. Budden and chapter 12 by D. Knighton, whilst L. K. Turner has been responsible for the remaining chapters and for general editorship. As in Book 1, the treatment within each chapter follows a natural progression and the individual teacher or reader can exercise very great freedom in deciding how the chapters themselves are to be combined.

Finally, it is a pleasure to thank Mr Michael Spincer and all his colleagues at Longmans for their constant support and consideration throughout our work, Oxford Illustrators for the quality of their diagrams and the staff of William Clowes for the customary excellence of their printing.

L.K.T.
F.J.B.
D.K.

Notation

Vectors

$\boldsymbol{AB}$, $\boldsymbol{CD}$, ...	displacement vectors
$\boldsymbol{a}$, $\boldsymbol{b}$, ...	free vectors, or position vectors, with moduli a, b, c, ...
$\hat{\boldsymbol{a}}$, $\hat{\boldsymbol{b}}$, ...	unit vectors in the directions $\boldsymbol{a}$, $\boldsymbol{b}$, ...
$\boldsymbol{a} + \boldsymbol{b}$, $\boldsymbol{a} - \boldsymbol{b}$	vector sum and difference
$\boldsymbol{i}, \boldsymbol{j}, \boldsymbol{k}$	unit vectors, with right-hand set of mutually perpendicular axes
$\boldsymbol{r} = x\boldsymbol{i} + y\boldsymbol{j} + z\boldsymbol{k}$	vector $\boldsymbol{r}$ with components x, y, z
$\boldsymbol{v} = \frac{\mathrm{d}\boldsymbol{r}}{\mathrm{d}t} = \dot{\boldsymbol{r}}$	velocity
$\boldsymbol{a} = \frac{\mathrm{d}\boldsymbol{v}}{\mathrm{d}t} = \ddot{\boldsymbol{r}}$	acceleration
$\boldsymbol{v}_{\mathrm{P}}(\mathrm{Q}) = \frac{\mathrm{d}}{\mathrm{d}t}(\boldsymbol{PQ})$	velocity of Q relative to P
$\boldsymbol{a}_{\mathrm{P}}(\mathrm{Q}) = \frac{\mathrm{d}}{\mathrm{d}t}\boldsymbol{v}_{\mathrm{P}}(\mathrm{Q})$	acceleration of Q relative to P
$\omega = \dot{\theta}$	angular velocity
$\dot{\omega} = \ddot{\theta}$	angular acceleration
$\boldsymbol{r} = \frac{\alpha\boldsymbol{a} + \beta\boldsymbol{b}}{\alpha + \beta}$	$\boldsymbol{r}$ divides $\boldsymbol{a}$, $\boldsymbol{b}$ in the ratio $\beta:\alpha$
$\boldsymbol{g} = \frac{1}{n}\sum_{1}^{n}\boldsymbol{r}_i$	centroid of $\boldsymbol{r}_i (i = 1, \ldots, n)$

$\bar{\boldsymbol{r}} = \dfrac{\sum_1^n m_i \boldsymbol{r}_i}{\sum_1^n m_i}$	centre of mass of m_i at $\boldsymbol{r}_i (i = 1, \ldots, n)$
$\boldsymbol{a}.\boldsymbol{b} = ab \cos\theta$ $= a_1b_1 + a_2b_2 + a_3b_3$	scalar product of $\boldsymbol{a}$, $\boldsymbol{b}$
$\boldsymbol{a} \times \boldsymbol{b} = (ab \sin\theta)\hat{\boldsymbol{n}}$ $= (a_2b_3 - a_3b_2)\boldsymbol{i} + (a_3b_1 - a_1b_3)\boldsymbol{j} + (a_1b_2 - a_2b_1)\boldsymbol{k}$	vector product of $\boldsymbol{a}$, $\boldsymbol{b}$

Mechanics

$\boldsymbol{F}$	force $\boldsymbol{F}$
$\boldsymbol{R} = \boldsymbol{P} + \boldsymbol{Q}$	resultant force $\boldsymbol{R} = \boldsymbol{P} + \boldsymbol{Q}$
$\boldsymbol{g}$	acceleration due to gravity ($g \approx 9.8$ m s^{-2} on Earth)
$\boldsymbol{F}, \boldsymbol{N}, \boldsymbol{R}$	frictional component ($\boldsymbol{F}$) and normal component ($\boldsymbol{N}$) of resultant reaction ($\boldsymbol{R}$)
μ	coefficient of friction $\left(\dfrac{F}{N} \leqslant \mu\right)$
λ	angle of friction $(= \tan^{-1}\mu)$
$\boldsymbol{I} = \boldsymbol{F}t$	impulse of constant force $\boldsymbol{F}$ in time t
$\boldsymbol{I} = \int_{t_1}^{t_2} \boldsymbol{F}\,\mathrm{d}t$	impulse of force $\boldsymbol{F}$ in interval (t_1, t_2)
$m\boldsymbol{v}$	momentum of mass m moving with velocity $\boldsymbol{v}$
e	coefficient of restitution (or *resilience*)
$W = \boldsymbol{F}.\boldsymbol{s}$	work done by constant force $\boldsymbol{F}$ in displacement $\boldsymbol{s}$
$W = \int \boldsymbol{F}.\mathrm{d}\boldsymbol{s}$	work done by variable force $\boldsymbol{F}$
$\frac{1}{2}m\boldsymbol{v}^2$	kinetic energy of mass m moving with velocity $\boldsymbol{v}$
$P = \dfrac{\mathrm{d}W}{\mathrm{d}t} = \boldsymbol{F}.\boldsymbol{v}$	power of force $\boldsymbol{F}$ (= rate of doing work)
$k \ (T = kx)$	stiffness of a spring
$\lambda \left(T = \dfrac{\lambda x}{l}\right)$	elastic modulus of spring or string of natural length l

$\ddot{x} = -\omega^2 x$ $\dot{x}^2 = \omega^2(a^2 - x^2)$ $x = a \sin(\omega t + \varepsilon)$	standard equations of simple harmonic motion with amplitude a and period $2\pi/\omega$
$\boldsymbol{G} = \boldsymbol{r} \times \boldsymbol{F}$	moment about O of a force $\boldsymbol{F}$ acting at point $\boldsymbol{r}$
$I = \sum mr^2$	moment of inertia
$G(= 6.67 \times 10^{-11}$ in SI)	constant of gravitation

Logarithmic, exponential, and hyperbolic functions

$\ln x = \log_e x = \int_1^x \frac{dt}{t} \quad (x > 0)$	natural (or Naperian) logarithmic function
$e\ (\approx 2.718)$	base of natural logarithms ($\ln e = 1$)
e^x (or $\exp x$)	exponential function
$\sinh x = \frac{1}{2}(e^x - e^{-x})$	hyperbolic sine
$\cosh x = \frac{1}{2}(e^x + e^{-x})$	hyperbolic cosine
$\sinh^{-1} x = \ln [x + \sqrt{(x^2 + 1)}]$ $\cosh^{-1} x = \ln [x + \sqrt{(x^2 - 1)}]$	inverse hyperbolic functions

Coefficients in binomial series

$$\binom{n}{r} = \frac{n(n-1)(n-2)\ldots(n-r+1)}{r!} \quad (n \in R, r \in Z^+)$$

$$= {}^nC_r \quad (n \in Z^+)$$

Complex numbers

$z = x + iy$ $= r(\cos\theta + i\sin\theta) = r \operatorname{cis}\theta$	complex number z
$R(z) = x$	real part of z
$I(z) = iy$	imaginary part of z
$\lvert z\rvert = r = \sqrt{(x^2 + y^2)}$	modulus of z
$\arg z = \theta\ (-\pi < \theta \leqslant \pi)$	argument of z
$z^* = x - iy = r \operatorname{cis}(-\theta)$	complex conjugate of z

Probability and statistics

$\mu = E[x] = \sum p_r x_r$	mean or expected value of random variable x, distributed with probability p
$\sigma^2 = V[x] = E[(x - \mu)^2]$ $= E[x^2] - \mu^2$ $= \sum p_r x_r^2 - \mu^2$	variance of random variable x
σ	standard deviation
$G(t) = \sum p_r t^r$	probability generator

$\mu = G'(1)$
$\sigma^2 = G''(1) + G'(1) - [G'(1)]^2$

$\mu = np$ $\sigma^2 = npq$	mean and variance of binomial distribution
$\Phi(x)$	cumulative probability, or distribution, function
$\phi(x) = \Phi'(x)$	probability density function
$\Rightarrow \Phi(X) = \int_{-\infty}^{X} \phi(x)\,\mathrm{d}x$	
$\mu = \int x\phi(x)\,\mathrm{d}x$ $\sigma^2 = \int (x-\mu)^2\phi(x)\,\mathrm{d}x$ $= \int x^2\phi(x)\,\mathrm{d}x - \mu^2$	mean and variance of continuous distribution with p.d.f. $\phi(x)$
$\phi(x) = \frac{1}{\sqrt{2\pi}\,\sigma} e^{-(x-\mu)^2/2\sigma^2}$	normal distribution with mean μ and standard deviation σ
$\phi(x) = \frac{1}{\sqrt{2\pi}} e^{-x^2/2}$	normal distribution with mean 0 and standard deviation 1
$p_r = \frac{\mu^r}{r!} e^{-\mu}$	Poisson distribution with mean μ
$E[m] = \mu$ $V[m] = \frac{\sigma^2}{n}$ $\frac{\sigma}{\sqrt{n}}$	expected value, variance and standard deviation of distribution of mean m of sample size n (drawn from population with mean μ and standard deviation σ)
$E[s^2] = \frac{n-1}{n}\sigma^2$	expected value of variance s^2 of sample size n
$\Rightarrow \hat{\sigma} = \sqrt{\left(\frac{n}{n-1}\right)}\, s$	unbiassed estimate for standard deviation of a population when sample size n has standard deviation s

For a bivariate distribution (x_i, y_i):

$\bar{x} = \frac{\sum x_i}{n}, \quad \bar{y} = \frac{\sum y_i}{n}$	means of x, y
$s_x^2 = \frac{\sum (x_i - \bar{x})^2}{n} = \frac{\sum x_i^2}{n} - \bar{x}^2$	variance of x

$s_y^2 = \dfrac{\sum (y_i - \overline{y})^2}{n} = \dfrac{\sum y_i^2}{n} - \overline{y}^2$	variance of y
$s_{xy} = \dfrac{\sum (x_i - \overline{x})(y_i - \overline{y})}{n}$ $= \dfrac{\sum x_i y_i}{n} - \overline{x}\,\overline{y}$	covariance of x, y
$\dfrac{s_{xy}}{s_x^2}$	regression coefficient of y on x
$\dfrac{s_{xy}}{s_y^2}$	regression coefficient of x on y
$r = \dfrac{s_{xy}}{s_x s_y}$	correlation coefficient
$\rho = 1 - \dfrac{6 \sum d_i^2}{n(n^2 - 1)}$	Spearman's rank correlation coefficient
$\tau = \dfrac{\delta}{\frac{1}{2}n(n - 1)}$	Kendall's rank correlation coefficient

Matrices and determinants

$\boldsymbol{A}, \boldsymbol{B}, \boldsymbol{C}, \ldots$	matrices
det $\boldsymbol{A}$ or $\lvert\boldsymbol{A}\rvert$	determinant of square matrix $\boldsymbol{A}$
$\boldsymbol{A}', \boldsymbol{B}', \boldsymbol{C}', \ldots$	transposed matrices
$\boldsymbol{A}^*, \boldsymbol{B}^*, \boldsymbol{C}^*, \ldots$	adjoint matrices
$\boldsymbol{A}^{-1}, \boldsymbol{B}^{-1}, \boldsymbol{C}^{-1}, \ldots$	inverse matrices
A_{ij}	cofactors of a_{ij} in det $\boldsymbol{A}$
$\boldsymbol{AB}$	product matrix

Sets

Z	integers
Q	rational numbers
R	real numbers
C	complex numbers
Z^+, Q^+, R^+	positive integers etc.
Z^*, Q^*, R^*	Z, Q, R, but omitting zero

Groups

$G, *$	group formed by elements of G under binary operation $*$
C_n	cyclic group of order n
D_n	dihedral group of order $2n$ (group of reflections and rotations of regular n-sided polygon)

S_n	symmetric (permutation) group of order n
$\cong$	is isomorphic to

Boolean Algebra

$A + B \ (= A \cup B)$	union of sets A, B
$AB \ (= A \cap B)$	intersection of sets A, B
$\subset$, $\supset$	inclusion ($A \subset B$ or $A \supset B$)
$\varnothing$	null (or empty) set
$\mathscr{E}$	universal set
A'	complementary set

Throughout the book, and in obedience to the recommendations of the British Standards Institution and the Royal Society (though contrary to the preference of F.J.B.), the aligned dot denotes a decimal point and not multiplication: so that 2.3, for instance, means $2\frac{3}{10}$ rather than 6.

9

Further calculus

9.1 The logarithmic function, ln *x*

When we first considered integration, we found that

$$\int t^n \, \mathrm{d}t = \frac{t^{n+1}}{n+1} + c, \quad \text{provided that } n \neq -1.$$

When $n = -1$, however, this expression contains the meaningless term $t^0/0$. Nevertheless, the problem of finding $\int t^{-1} \, \mathrm{d}t$ is certainly not meaningless, and will arise whenever (which is frequently) we need to know the area beneath a section of the curve $f(t) = 1/t$. But the reader will probably find that, despite all his efforts, he fails to discover a function whose derivative is $1/t$, and must therefore return to a more careful examination of the area which this integral represents.

We start by considering the area beneath the curve $f(t) = 1/t$ between $t = 1$ and $t = x$, which we call $l(x)$.

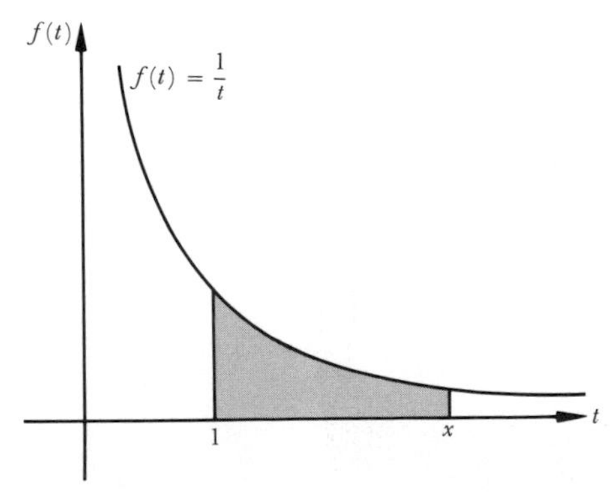

So $l(x) = \int_1^x \frac{1}{t}\,dt$, which is defined for all $x > 0$.

Clearly, if $x > 1$, $l(x) > 0$;

if $0 < x < 1$, $l(x) < 0$;

and $l(1) = 0$.

We can now estimate $l(x)$ for any positive value of x by means of counting squares or by Simpson's rule.

Exercise 9.1a

1 Count squares or use Simpson's rule (with four intervals in each case) to estimate the values of:
(*i*) $l(2)$; (*ii*) $l(3)$; (*iii*) $l(6)$; (*iv*) $l(9)$; (*v*) $l(12)$; (*vi*) $l(\frac{1}{2})$; (*vii*) $l(\frac{1}{3})$.

Properties of $l(x)$

After this last exercise, the reader may have suspected that

$l(6) = l(2) + l(3)$, $\quad l(12) = l(2) + l(6)$,
$l(3^2) = 2l(3)$, $\quad l(2^{-1}) = -l(2)$.

More generally, we shall now prove that
(*i*) $l(xy) = l(x) + l(y)$,
and (*ii*) $l(x^n) = nl(x)$.

(*i*) By definition,

$$l(x) + l(y) = \int_1^x \frac{1}{t}\,dt + \int_1^y \frac{1}{t}\,dt,$$

and we can transform the second integral by changing the variable.

Regarding x as constant, we let $u = xt$ so that $du = x\,dt$.

Also $t = 1 \Rightarrow u = x$, and $t = y \Rightarrow u = xy$.

$$\text{So} \quad \int_1^y \frac{1}{t}\,dt = \int_x^{xy} \frac{1}{u}\,du$$

$$= \int_x^{xy} \frac{1}{t}\,dt$$

(as the letters t and u are themselves irrelevant).

$$\text{Hence} \quad l(x) + l(y) = \int_1^x \frac{1}{t}\,dt + \int_x^{xy} \frac{1}{t}\,dt$$

$$= \int_1^{xy} \frac{1}{t}\,dt$$

$$\Rightarrow \quad l(x) + l(y) = l(xy).$$

This substitution can be illustrated by considering the shaded area between $t = 1$ and $t = y$ and squeezing it, like toothpaste, into the area between $t = x$ and $t = xy$: horizontal dimensions are stretched by a factor x and vertical dimensions compressed by a factor $1/x$, so that areas remain unchanged:

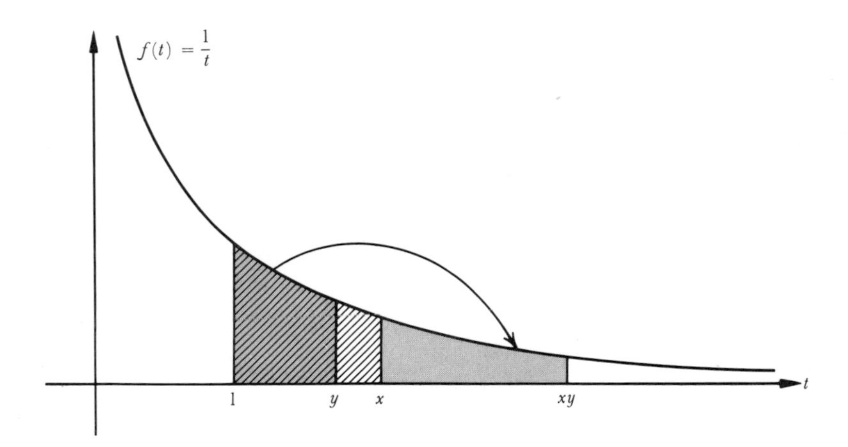

Hence $l(x) + l(y) = l(xy)$.

(*ii*) Similarly, $l(x^n) = \displaystyle\int_1^{x^n} \frac{1}{t}\,\mathrm{d}t$;

and if we now let $t = u^n$, it follows that $\mathrm{d}t = nu^{n-1}\,\mathrm{d}u$.

Also $t = 1 \Rightarrow u = 1$ and $t = x^n \Rightarrow u = x$.

So $l(x^n) = \displaystyle\int_1^{x} \frac{nu^{n-1}}{u^n}\,\mathrm{d}u = n\int_1^{x} \frac{1}{u}\,\mathrm{d}u$

$\Rightarrow l(x^n) = nl(x)$.

Hence
$$\boxed{\begin{aligned} l(xy) &= l(x) + l(y) \\ l(x^n) &= nl(x) \end{aligned}}$$

Now as these are also properties of logarithms, the question immediately arises whether $l(x)$ itself is in fact a logarithm. But first we must introduce the crucial number e such that

$l(e) = 1$,

i.e., such that $t = e$ is the ordinate up to which there is unit area beneath the given curve:

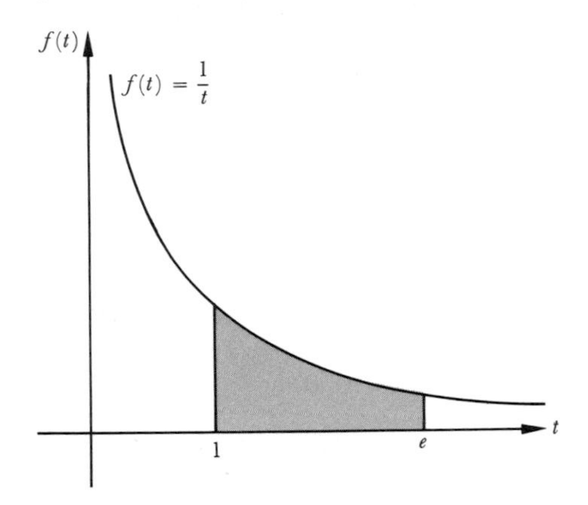

By counting squares, we can show that $e \approx 2.72$, and we shall soon be able to calculate it to a much greater degree of accuracy.

Using the above results, we see that

$$l(e^{l(x)}) = l(x)\,l(e) = l(x).$$

So the areas under the curve to the value $e^{l(x)}$ and to the value x are equal. These must therefore be the same, and $e^{l(x)} = x$

$$\Rightarrow \quad \boxed{l(x) = \log_e x}$$

$\log_e x$ is usually called a natural, or Naperian, logarithm (after John Napier, 1550–1617) and is sometimes denoted without a base, simply as $\log x$; but more customarily as $\ln x$. We shall also usually write

$$\int \frac{1}{t}\,\mathrm{d}t \text{ and } \int \frac{1}{x}\,\mathrm{d}x \quad \text{as} \quad \int \frac{\mathrm{d}t}{t} \text{ and } \int \frac{\mathrm{d}x}{x}.$$

So, in summary, if $x > 0$:

$$\boxed{\begin{aligned} &\int_1^x \frac{\mathrm{d}t}{t} = \log_e x = \ln x \\ &\int \frac{\mathrm{d}x}{x} = \ln x + c \qquad \frac{\mathrm{d}}{\mathrm{d}x}(\ln x) = \frac{1}{x} \end{aligned}}$$

and the relationship between

$$\frac{1}{x} \quad \text{and} \quad \ln x$$

can be conveniently illustrated by their graphs:

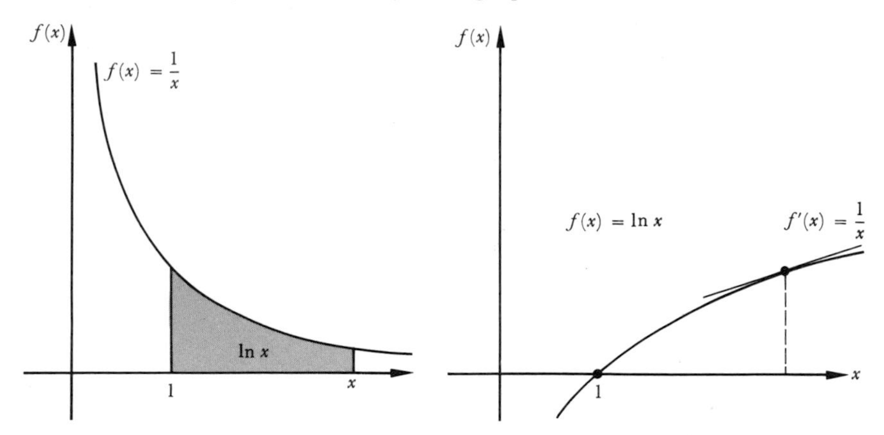

In particular, we notice that

when $x \leqslant 0$, $\quad \ln x$ is not defined;

as $x \to +\infty$, $\quad \ln x \to +\infty$;

as $x \to 0+$, $\quad \ln x \to -\infty$.

We are now able to find other areas bounded by such graphs.

Example 1

Investigate the areas represented by

$(i)\ \displaystyle\int_2^5 \frac{\mathrm{d}x}{x}$; $\quad (ii)\ \displaystyle\int_{-3}^{-2} \frac{\mathrm{d}x}{x}$; $\quad (iii)\ \displaystyle\int_{-1}^{2} \frac{\mathrm{d}x}{x}$.

$$(i)\quad \int_2^5 \frac{\mathrm{d}x}{x} = [\ln x]_2^5$$
$$= \ln 5 - \ln 2$$
$$= \ln \tfrac{5}{2}$$
$$= \ln 2.5 \approx 0.92.$$

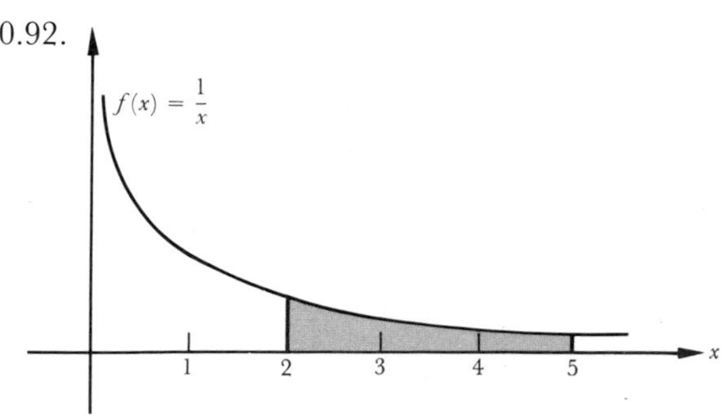

$(ii)\ \displaystyle\int_{-3}^{-2} \frac{\mathrm{d}x}{x}$ represents an area below the x-axis, and our natural approach would be similar:

$$\int_{-3}^{-2} \frac{\mathrm{d}x}{x} = [\ln x]_{-3}^{-2}$$
$$= \ln(-2) - \ln(-3)$$
$$= \ln \frac{-2}{-3} = \ln \tfrac{2}{3} \approx -0.41.$$

But the reader will rightly object that $\ln(-2) - \ln(-3)$ is completely meaningless, since neither $\ln(-2)$ nor $\ln(-3)$ has been defined. We can, however, easily side-step this obstacle by using the symmetry of the graph about the origin, observing that

$$\int_{-3}^{-2} \frac{\mathrm{d}x}{x} = -\int_{2}^{3} \frac{\mathrm{d}x}{x} = -[\ln x]_{2}^{3}$$
$$= -(\ln 3 - \ln 2)$$
$$= -\ln 1.5$$
$$\approx -0.41.$$

(*iii*) $\displaystyle\int_{-1}^{2} \frac{\mathrm{d}x}{x}$.

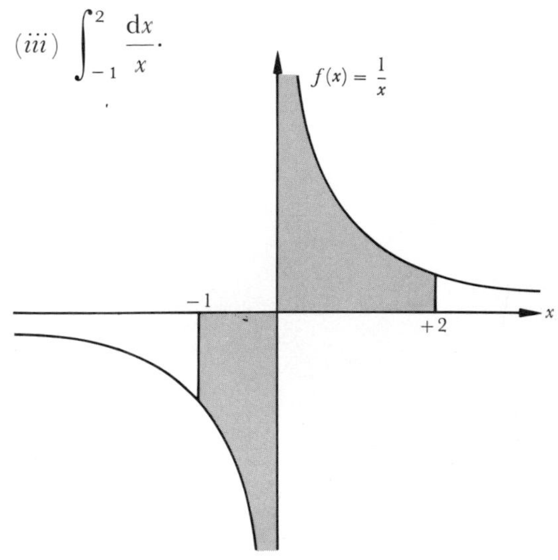

We have already seen that $\ln x \to -\infty$ as $x \to 0+$, so the area beneath the curve between $x = 0$ and $x = 2$ is infinite; and similarly the area to the left of $x = 0$ is infinite.

So $\int_{-1}^{2} \mathrm{d}x/x$ is composed of two parts, both of which are infinite. It is therefore meaningless and without value.

Example 2

Differentiate:

(*i*) $\ln 5x^3$; (*ii*) $\ln \dfrac{x^2+1}{x}$; (*iii*) $\log_{10} x$.

(*i*) $\ln 5x^3 = \ln 5 + \ln x^3 = \ln 5 + 3 \ln x$

$$\Rightarrow \quad \frac{\mathrm{d}}{\mathrm{d}x} \ln 5x^3 = \frac{\mathrm{d}}{\mathrm{d}x}(\ln 5 + 3 \ln x) = \frac{3}{x}.$$

Or $\dfrac{\mathrm{d}}{\mathrm{d}x}(\ln 5x^3) = \dfrac{1}{5x^3} \times 15x^2 = \dfrac{3}{x}.$

(*ii*)
$$\frac{\mathrm{d}}{\mathrm{d}x}\left(\ln \frac{x^2+1}{x}\right) = \frac{\mathrm{d}}{\mathrm{d}x}[\ln (x^2+1) - \ln x]$$
$$= \frac{1}{x^2+1} \times 2x - \frac{1}{x}$$
$$= \frac{x^2-1}{x(x^2+1)}$$

Or
$$\frac{\mathrm{d}}{\mathrm{d}x}\left(\ln \frac{x^2+1}{x}\right) = \frac{x}{x^2+1} \times \frac{x \times 2x - (x^2+1)}{x^2}$$
$$= \frac{x^2-1}{x(x^2+1)}.$$

(*iii*) Let $y = \log_{10} x$.

Then $x = 10^y$

$$\Rightarrow \quad \ln x = y \ln 10$$

$$\Rightarrow \quad y = \frac{\ln x}{\ln 10}$$

$$\Rightarrow \quad \frac{\mathrm{d}y}{\mathrm{d}x} = \frac{1}{(\ln 10)x} \approx \frac{1}{(2.302\,6)x}.$$

So $\dfrac{\mathrm{d}}{\mathrm{d}x}(\log_{10} x) \approx \dfrac{0.434\,3}{x}.$

Example 3

(*i*) $\displaystyle\int \frac{x^2}{x^3+1}\,\mathrm{d}x$; (*ii*) $\displaystyle\int \frac{f'(x)}{f(x)}\,\mathrm{d}x$; (*iii*) $\displaystyle\int_0^{\pi/4} \tan x \,\mathrm{d}x$.

(*i*) Let $I = \displaystyle\int \frac{x^2\,\mathrm{d}x}{x^3+1}.$

Now $u = x^3 + 1 \Rightarrow \mathrm{d}u = 3x^2\,\mathrm{d}x.$

So $$I = \int \frac{\frac{1}{3}\,\mathrm{d}u}{u} = \tfrac{1}{3}\ln u + c$$

$$\Rightarrow \quad \int \frac{x^2\,\mathrm{d}x}{x^3 + 1} = \tfrac{1}{3}\ln(x^3 + 1) + c.$$

(*ii*) Let $I = \displaystyle\int \frac{f'(x)}{f(x)}\,\mathrm{d}x.$

Here again we let $u = f(x) \Rightarrow \mathrm{d}u = f'(x)\,\mathrm{d}x.$

So $$I = \int \frac{\mathrm{d}u}{u} = \ln u + c$$
$$= \ln f(x) + c,$$

so that we are now able to integrate any quotient whose numerator is the derivative of its denominator.

(*iii*) $I = \displaystyle\int_0^{\pi/4} \tan x\,\mathrm{d}x.$

This can be expressed as

$$I = \int_0^{\pi/4} \frac{\sin x}{\cos x}\,\mathrm{d}x = -\int_0^{\pi/4} \frac{-\sin x}{\cos x}\,\mathrm{d}x.$$

So, using (*ii*) above,

$$I = [-\ln \cos x]_0^{\pi/4}$$
$$= -\left(\ln \frac{1}{\sqrt{2}} - \ln 1\right) = \tfrac{1}{2}\ln 2.$$

Hence $$\int_0^{\pi/4} \tan x\,\mathrm{d}x = \tfrac{1}{2}\ln 2.$$

Exercise 9.1b

1 Given that $\ln 2 = 0.693\,15$, $\ln 3 = 1.098\,61$, $\ln 5 = 1.609\,44$, calculate to 5 places of decimals:

(*i*) $\ln 4$; (*ii*) $\ln 6$; (*iii*) $\ln 10$; (*iv*) $\ln 12$; (*v*) $\ln \frac{1}{2}$;
(*vi*) $\ln \frac{1}{6}$; (*vii*) $\ln \frac{1}{12}$; (*viii*) $\ln 10^6$; (*ix*) $\ln \sqrt{2}$; (*x*) $\ln \sqrt[3]{100}$.

2 Evaluate:

(*i*) $\ln e^3$; (*ii*) $\ln 1$; (*iii*) $\ln \dfrac{1}{e}$; (*iv*) $\ln \sqrt{e}$.

3 Solve the equations:
(*i*) $\ln x = 2$; (*ii*) $\ln x = 100$; (*iii*) $\ln x = \frac{1}{10}$; (*iv*) $\ln x = -2$.

4 Differentiate with respect to x:

(*i*) $\ln 2x$; (*ii*) $\ln x^3$; (*iii*) $\ln 3x^4$;

(*iv*) $\ln \dfrac{x}{3}$; (*v*) $\ln \dfrac{1}{x}$; (*vi*) $\ln \sqrt{x}$;

(*vii*) $\ln (x^2 + 1)$; (*viii*) $\ln \dfrac{x+1}{x}$ (*ix*) $\ln \sqrt{(x^2 - 1)}$;

(*x*) $\ln x\sqrt{(x + 1)}$; (*xi*) $x \ln x$; (*xii*) $\dfrac{\ln x}{x}$.

5 Differentiate with respect to x:

(*i*) $\ln \sin x$; (*ii*) $\ln \cos x$; (*iii*) $\ln \sec x$;
(*iv*) $\ln (\sec x + \tan x)$; (*v*) $\ln (3 \sin x)$; (*vi*) $\ln (\sin 3x)$;
(*vii*) $\ln (\sin^2 x)$; (*viii*) $\ln (4 \sin^2 x)$; (*ix*) $(\ln x)^2$;
(*x*) $\ln (\ln x)$.

6 Find the following integrals:

(*i*) $\displaystyle\int \frac{dx}{3x}$; (*ii*) $\displaystyle\int \frac{dx}{x+2}$; (*iii*) $\displaystyle\int \frac{dx}{2x+3}$;

(*iv*) $\displaystyle\int \frac{x+1}{x}\,dx$; (*v*) $\displaystyle\int \frac{dx}{1-x}$; (*vi*) $\displaystyle\int \frac{2x\,dx}{x^2+1}$;

(*vii*) $\displaystyle\int \frac{x^2\,dx}{1-x^3}$; (*viii*) $\displaystyle\int \frac{x-1}{x^2-2x+3}$ (*ix*) $\displaystyle\int \frac{x^2+1}{x^3+3x}\,dx$;

(*x*) $\displaystyle\int \cot x\,dx$; (*xi*) $\displaystyle\int \tan 3x\,dx$; (*xii*) $\displaystyle\int \cot \tfrac{1}{2}x\,dx$;

(*xiii*) $\displaystyle\int \frac{\sin x + \cos x}{\sin x - \cos x}\,dx$; (*xiv*) $\displaystyle\int \frac{\sec^2 x}{1 + \tan x}\,dx$.

7 Evaluate (or, if this is impossible, illustrate):

(*i*) $\displaystyle\int_2^5 \frac{dx}{x}$; (*ii*) $\displaystyle\int_4^6 \frac{dx}{x+1}$; (*iii*) $\displaystyle\int_2^3 \frac{dx}{2x+3}$;

(*iv*) $\displaystyle\int_1^3 \frac{dx}{x-4}$; (*v*) $\displaystyle\int_1^3 \frac{dx}{x-2}$; (*vi*) $\displaystyle\int_2^3 \frac{x\,dx}{x^2-1}$;

(*vii*) $\displaystyle\int_2^4 \frac{dx}{3-2x}$; (*viii*) $\displaystyle\int_0^1 \frac{dx}{1-2x}$; (*ix*) $\displaystyle\int_{\frac{1}{6}\pi}^{\frac{1}{2}\pi} \cot \theta\,d\theta$;

(*x*) $\displaystyle\int_0^{\frac{1}{4}\pi} \frac{\cos \theta - \sin \theta}{\cos \theta + \sin \theta}\,d\theta$.

8 Given that $\ln 2 \approx 0.693\,15$, use differentiation to estimate (to 5 places of decimals):
(*i*) $\ln 2.001$; (*ii*) $\ln 1.999\,9$.

9 Find the volume of the solids of revolution formed when the area
(*i*) between $y = 1/\sqrt{x}$, $x = 1$, $x = 2$, and $y = 0$ is rotated about the x-axis;

(*ii*) between $y = 1/x^2$, $y = 1$, $y = 4$ is rotated about the y-axis.

10 By using the Newton–Raphson method, or any other suitable iterative method, solve the equation

$$x + \ln x \doteq 3,$$

given that the root is close to 2. Obtain the root correct to two decimal places. (J.M.B.)

11 (*i*) Using the Newton–Raphson method or otherwise, solve the equation

$$\ln x = \tfrac{1}{5}x,$$

giving your answers to four significant figures.
(*ii*) Find the gradient of the straight line through the origin which touches the graph of $\ln x$, and hence show that

$$\ln x = \alpha x$$

has a solution whenever $\alpha \leqslant 1/e$.

12 Boyle's law states that, at constant temperature, the pressure p and volume V of a gas are such that $pV = c$, where c is a constant. If the work done by the gas when expanding is $\int p \, \mathrm{d}V$, show that this is constant $(= c \ln 2)$ whenever it doubles its volume.

13 If an animal grows without changing its shape then its length x, surface area S and volume V are such that

$$S = ax^2, \quad V = bx^3 \quad \text{(where } a, b \text{ are constants)}.$$

Hence show by differentiation: (*a*) without use of logarithms; (*b*) after first taking logarithms, that slight increases δx, δS and δV are related by

$$\frac{\delta S}{S} = 2\frac{\delta x}{x}, \qquad \frac{\delta V}{V} = 3\frac{\delta x}{x},$$

i.e., that *fractional* increases of S (and V) are twice (and thrice) the fractional increase of x.

9.2 The exponential function, e^x

We now proceed to investigate the function $f(x) = e^x$. This is clearly a power function, and we start by finding its derivative.

Now $\quad y = e^x$

$\Rightarrow \quad x = \ln y$

$\Rightarrow \quad \dfrac{\mathrm{d}x}{\mathrm{d}y} = \dfrac{1}{y}$

$\Rightarrow \quad \dfrac{\mathrm{d}y}{\mathrm{d}x} = y = e^x.$

So $\dfrac{d}{dx}(e^x) = e^x$

and we immediately see that e^x is the particular power function whose rate of increase is always equal to its value:

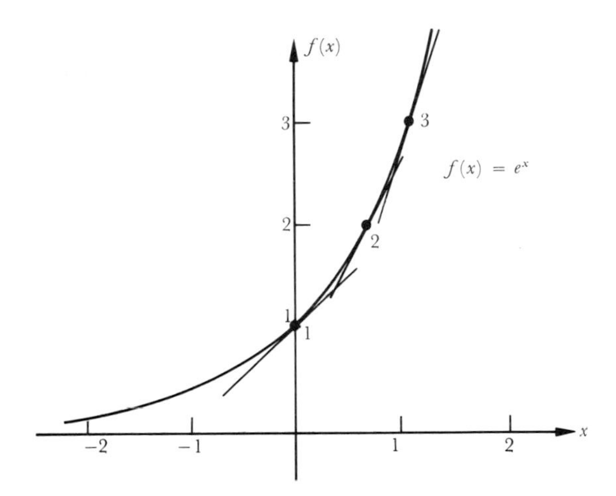

It was mentioned earlier that power functions are sometimes referred to as exponential functions. Because of the supreme importance of e^x it is usually called *the* exponential function, and is sometimes written exp x.

Similarly we see that

$$\frac{d}{dx}(e^{-x}) = e^{-x} \times -1 = -e^{-x},$$

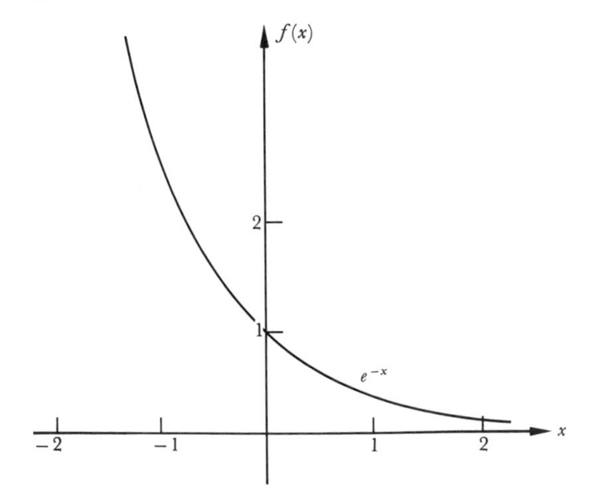

so that e^{-x} is the corresponding function of exponential decay.

These two functions, and variants of them, occur whenever the rate of growth of a quantity, or its rate of decay, is determined by its present amount. So, for instance, the spread of an infection in the first stage of an epidemic is likely to involve a positive exponential function, whilst the decay of a radioactive element is governed by a negative exponential.

Lastly, just as $\dfrac{d}{dx}(e^x) = e^x$

so $\int e^x \, dx = e^x + c$

Example 1

Differentiate: (*i*) e^{x^2}; (*ii*) 10^x.

(*i*) $\dfrac{d}{dx}(e^{x^2}) = e^{x^2} \times 2x = 2x\, e^{x^2}$.

(*ii*)
$$\frac{d}{dx}(10^x) = \frac{d}{dx}(e^{\ln 10})^x$$
$$= \frac{d}{dx}(e^{x \ln 10})$$
$$= e^{x \ln 10} \times \ln 10$$
$$= 10^x \ln 10.$$

Alternatively, $y = 10^x$

$\Rightarrow \quad \ln y = \ln 10^x = x \ln 10$

$\Rightarrow \quad \dfrac{1}{y}\dfrac{dy}{dx} = \ln 10$

$\Rightarrow \quad \dfrac{dy}{dx} = y \ln 10$

$\Rightarrow \quad \dfrac{d}{dx}(10^x) = 10^x \ln 10.$

Example 2

Find the integrals (*i*) $\displaystyle\int_0^1 e^{2x} \, dx$; (*ii*) $\displaystyle\int 2^x \, dx$.

(*i*) Since $\dfrac{d}{dx}(e^{2x}) = 2\, e^{2x}$,

it immediately follows that

$$\int_0^1 e^{2x} \, dx = \left[\tfrac{1}{2} e^{2x}\right]_0^1 = \tfrac{1}{2}(e^2 - 1).$$

(*ii*) As in Example 1, we can easily show that

$$\frac{d}{dx}(2^x) = 2^x \ln 2.$$

So $\displaystyle\int 2^x \, dx = \frac{2^x}{\ln 2} + c.$

Example 3

When a charged condenser is discharged through a particular circuit containing an inductance and resistance, the current x at time t obeys the equation

$$\frac{d^2x}{dt^2} + 2\frac{dx}{dt} + 5x = 0.$$

Show that $x = e^{-t} \cos 2t$ satisfies this equation and sketch the graph of the function.

If $\qquad x = e^{-t} \cos 2t,$

then $\qquad \dfrac{dx}{dt} = e^{-t} \times -2 \sin 2t - e^{-t} \cos 2t$

$$= e^{-t}(-\cos 2t - 2 \sin 2t),$$

and $\qquad \dfrac{d^2x}{dt^2} = e^{-t}(2 \sin 2t - 4 \cos 2t) - e^{-t}(-\cos 2t - 2 \sin 2t)$

$$= e^{-t}(-3 \cos 2t + 4 \sin 2t).$$

So we verify that

$$\frac{d^2x}{dt^2} + 2\frac{dx}{dt} + 5x$$
$$= e^{-t}\{(-3 \cos 2t + 4 \sin 2t) + 2(-\cos 2t - 2 \sin 2t) + 5 \cos 2t\}$$
$$= 0.$$

Now $x = e^{-t} \cos 2t$ is the product of an exponential decay and a cosine function:

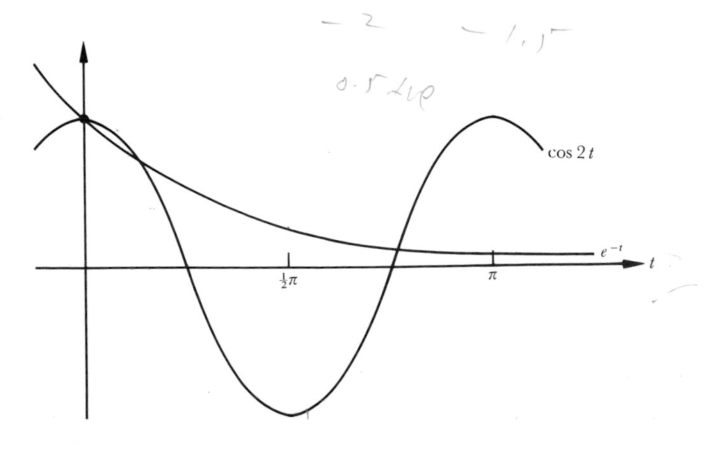

Combining these, we obtain the graph of $x = e^{-t}\cos 2t$ and we readily see that it represents a decaying oscillation.

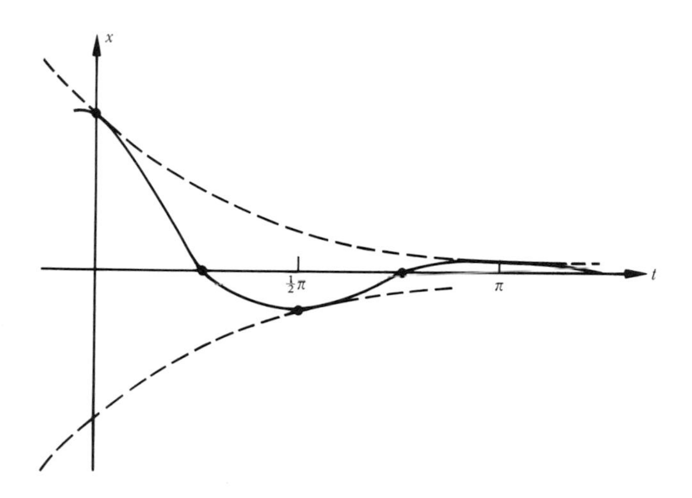

Example 4

A bathful of water is at a temperature of 80 °C in a large room whose temperature is 20 °C. The water cools at a rate which is proportional to its excess temperature and takes 5 min to reach 70 °C.

Find the temperature of the water after t min.

If the *excess* temperature is θ °C, the rate of cooling is $-\dfrac{d\theta}{dt}$.

So $\quad -\dfrac{d\theta}{dt} = k\theta,\quad$ where k is a constant,

$$\Rightarrow \quad \frac{dt}{d\theta} = -\frac{1}{k\theta}$$

$$\Rightarrow \quad t = -\frac{1}{k}\ln\theta + c.$$

But, when $t = 0$, $\theta = 60$.

So $\quad 0 = -\dfrac{1}{k}\ln 60 + c \quad \Rightarrow \quad c = \dfrac{1}{k}\ln 60$

$$\Rightarrow \quad t = -\frac{1}{k}\ln\theta + \frac{1}{k}\ln 60$$

$$= -\frac{1}{k}\ln\frac{\theta}{60}$$

$$\Rightarrow \quad \ln\frac{\theta}{60} = -kt.$$

So far we have only used the general 'law of cooling' and the 'initial' condition that $\theta = 60$ when $t = 0$. But we also know that when $t = 5$, $\theta = 50$.

$$\text{So} \quad \ln \tfrac{50}{60} = -k \times 5$$
$$\Rightarrow \quad k = -\tfrac{1}{5} \ln \tfrac{5}{6} = \tfrac{1}{5} \ln \tfrac{6}{5} \approx 0.036\,5.$$
$$\text{So} \quad \ln \frac{\theta}{60} = -0.0365t$$
$$\Rightarrow \quad \frac{\theta}{60} = e^{-0.0365t}$$
$$\Rightarrow \quad \theta = 60\, e^{-0.0365t}.$$

So the temperature of the bath after t minutes will be

$$20 + 60\, e^{-0.0365t}.$$

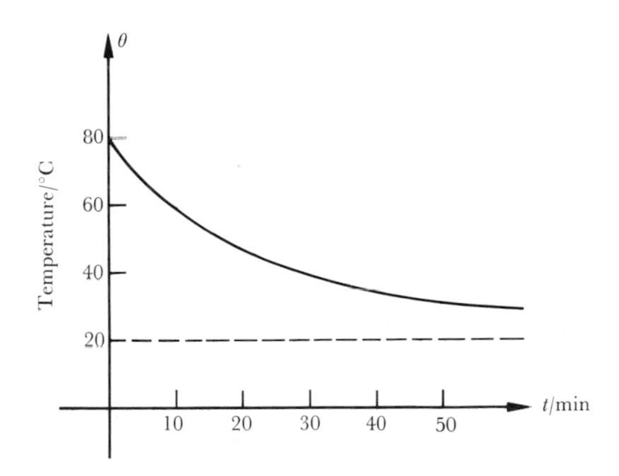

We can now use this function to calculate the temperature of the bath at any subsequent time. For example:

if $t = 10$, temperature $= 20 + 60\, e^{-0.365}$
$= 20 + 60 \times 0.694$
$= 20 + 41.6 = 61.6\ ^\circ\text{C}$;

if $t = 20$, temperature $= 20 + 60\, e^{-0.73}$
$= 20 + 60 \times 0.482$
$= 20 + 28.9 = 48.9\ ^\circ\text{C}$.

Exercise 9.2

1 Use tables to evaluate e^x when $x = 10, 100, \frac{1}{10}, -10$.

2 Find values of x such that
(*i*) $e^x = 10, \quad 1\,000, \quad 10^6$; (*ii*) $e^{-x} = \frac{1}{100}, \quad 10^{-4}, \quad 10^{-6}$.

3 Differentiate with respect to x:

(*i*) e^{3x}; (*ii*) $e^{-\frac{1}{2}x}$; (*iii*) $x\,e^{-x}$;

(*iv*) $x^2\,e^x$; (*v*) $e^{-\frac{1}{2}x^2}$; (*vi*) $\dfrac{e^x}{x}$;

(*vii*) $e^{-x}\sin x$; (*viii*) $e^{2x}\cos 3x$; (*ix*) 3^x; (*x*) x^x.

4 Find the following integrals:

(*i*) $\displaystyle\int e^{-2x}\,dx$; (*ii*) $\displaystyle\int e^{\frac{1}{3}x}\,dx$; (*iii*) $\displaystyle\int x\,e^{x^2}\,dx$;

(*iv*) $\displaystyle\int e^{\sin x}\cos x\,dx$; (*v*) $\displaystyle\int x\,e^{-\frac{1}{2}x^2}\,dx$; (*vi*) $\displaystyle\int 10^x\,dx$.

5 Evaluate the following integrals:

(*i*) $\displaystyle\int_0^2 e^{-\frac{1}{2}x}\,dx$; (*ii*) $\displaystyle\int_1^3 e^{2x}\,dx$; (*iii*) $\displaystyle\int_0^1 x\,e^{-\frac{1}{2}x^2}\,dx$; (*iv*) $\displaystyle\int_0^2 10^x\,dx$.

6 Investigate the stationary values and sketch the graphs of the following functions:

(*i*) $x\,e^{-x}$; (*ii*) $x^2\,e^{-x}$; (*iii*) $x^n\,e^{-x}$; (*iv*) e^{-x^2}.

7 Find $\displaystyle\int_0^X e^{-x}\,dx$.

By letting $X \to +\infty$, investigate the area which lies between the curve $y = e^{-x}$ and the two axes (for convenience called $\int_0^\infty e^{-x}\,dx$).

8 After time t the displacement x of a heavily damped pendulum is given by $x = 2\,e^{-t}\sin t$.

(*i*) For what values of t is the displacement zero?
(*ii*) For what values of t is the displacement greatest?
(*iii*) Sketch the graph of x against t.
(*iv*) Prove that

$$\frac{d^2x}{dt^2} + 2\frac{dx}{dt} + 2x = 0.$$

9 A biologist knows that a certain organism, of which there is initially 2 g, is growing *continuously* at a rate of $\frac{1}{10}$ g/day for each gram of its mass. If its mass after t days is m g,

(*i*) show that $\dfrac{dm}{dt} = \frac{1}{10}m$;

(*ii*) deduce that $m = 2\,e^{t/10}$;

(*iii*) by what percentage does it increase each day?

(*iv*) how long does it take to double its mass?

10 A sample of radium loses mass at a rate which is proportional to the amount present.
(*i*) If its mass is m after time t, show that

$$\frac{dm}{dt} = -km;$$

(*ii*) if its initial mass (when $t = 0$) is m_0, deduce that

$$m = m_0 e^{-kt};$$

(*iii*) if its mass is halved in 1 600 years, calculate k;
(*iv*) calculate the percentage rate of decrease per century.

11 A man with toothache leaves home at noon for the dentist, whose surgery is 1 km away. He starts by running at 10 km h^{-1}, but gradually slows down so that his speed is always proportional to his distance from the surgery.
(*i*) How far has he gone in the first 5 min?
(*ii*) How long will he take to get within 100 m?
(*iii*) When will he arrive?

9.3 Interlude: infinite series

In Chapter 6 we investigated a number of finite series and also mentioned the 'sum to infinity' of a geometric progression.

It was proved, for instance, that if S_n is the sum of the first n terms of the series

$$1 + \tfrac{1}{2} + \tfrac{1}{4} + \tfrac{1}{8} + \cdots,$$

then $$S_n = \frac{1 - (\frac{1}{2})^n}{1 - \frac{1}{2}} = 2 - \frac{1}{2^{n-1}}.$$

Now as $n \to \infty$, $$\frac{1}{2^{n-1}} \to 0.$$

So $S_n \to 2$,

and we say that the *infinite series*

$$1 + \tfrac{1}{2} + \tfrac{1}{4} + \tfrac{1}{8} + \tfrac{1}{16} + \cdots$$

is *convergent*, and that its *sum* (or *sum to infinity*) is 2.

More generally, we can consider the infinite series

$$u_1 + u_2 + u_3 + \cdots$$

and suppose that the sum of its first n terms is S_n, so that

$$S_n = u_1 + u_2 + u_3 + \cdots + u_n.$$

If $S_n \to S$ as $n \to \infty$, we say that the infinite series is *convergent*, and that S is its *sum*.

It is beyond our present scope to investigate the convergence of series, but we can consider some elementary examples:

Example 1

$1 + \frac{1}{2} + \frac{1}{3} + \frac{1}{4} + \frac{1}{5} + \cdots.$

Certainly the terms of this series become very small. If we look at the successive values of S_n, we obtain (approximately):

1, 1.500, 1.833, 2.083, 2.283, ...

and before long it seems that the values of S_n are having such a struggle to increase that they must necessarily tend to a limit, which will be the sum of the infinite series. But a little more thought will cause us to write the series as

$$\begin{aligned} &1 + \tfrac{1}{2} + (\tfrac{1}{3} + \tfrac{1}{4}) + (\tfrac{1}{5} + \tfrac{1}{6} + \tfrac{1}{7} + \tfrac{1}{8}) + (\tfrac{1}{9} + \tfrac{1}{10} + \tfrac{1}{11} + \cdots + \tfrac{1}{16}) + \cdots \\ &> 1 + \tfrac{1}{2} + 2 \times \tfrac{1}{4} + 4 \times \tfrac{1}{8} + 8 \times \tfrac{1}{16} + \cdots \\ &> 1 + \tfrac{1}{2} + \tfrac{1}{2} + \tfrac{1}{2} + \tfrac{1}{2} + \cdots \end{aligned}$$

Now this series, although growing very slowly, is completely unlimited. So, contrary to our expectations, the original series

$1 + \frac{1}{2} + \frac{1}{3} + \frac{1}{4} + \cdots$

is also unlimited and does not converge (even though, as we shall later prove, the sum of its first million terms is less than 15).

Example 2

The series

$$\frac{1}{1^2} + \frac{1}{2^2} + \frac{1}{3^2} + \frac{1}{4^2} + \frac{1}{5^2} + \frac{1}{6^2} + \frac{1}{7^2} + \frac{1}{8^2} + \cdots$$

can be written as

$$\begin{aligned} &\frac{1}{1^2} + \left(\frac{1}{2^2} + \frac{1}{3^2}\right) + \left(\frac{1}{4^2} + \frac{1}{5^2} + \frac{1}{6^2} + \frac{1}{7^2}\right) + \left(\frac{1}{8^2} + \cdots + \frac{1}{15^2}\right) + \cdots \\ &< 1 + 2 \times \frac{1}{2^2} + 4 \times \frac{1}{4^2} + 8 \times \frac{1}{8^2} + \cdots \\ &< 1 + \tfrac{1}{2} + \tfrac{1}{4} + \tfrac{1}{8} + \cdots = 2 \end{aligned}$$

In this case, therefore, the series

$$1 + \frac{1}{2^2} + \frac{1}{3^2} + \frac{1}{4^2} + \cdots$$

does converge to a sum less than 2 (and actually, though we cannot prove it here, to $\frac{1}{6}\pi^2$).

Example 3

As a warning that we must treat infinite series with a certain respect, we now consider the series

$$1 - \tfrac{1}{2} + \tfrac{1}{3} - \tfrac{1}{4} + \tfrac{1}{5} + \cdots.$$

It can be shown that this series is convergent and also, perhaps surprisingly, that its sum is $\ln 2 \approx 0.693$. So it would appear to be an innocent operation to rearrange its terms as follows:

$$\begin{aligned}
&1 - \tfrac{1}{2} + \tfrac{1}{3} - \tfrac{1}{4} + \tfrac{1}{5} - \tfrac{1}{6} + \tfrac{1}{7} - \tfrac{1}{8} + \cdots \\
&= 1 - \tfrac{1}{2} - \tfrac{1}{4} + \tfrac{1}{3} - \tfrac{1}{6} - \tfrac{1}{8} + \tfrac{1}{5} - \tfrac{1}{10} - \tfrac{1}{12} + \cdots \\
&= (1 - \tfrac{1}{2}) - \tfrac{1}{4} + (\tfrac{1}{3} - \tfrac{1}{6}) - \tfrac{1}{8} + (\tfrac{1}{5} - \tfrac{1}{10}) - \cdots \\
&= \quad \tfrac{1}{2} \quad - \tfrac{1}{4} + \quad \tfrac{1}{6} \quad - \tfrac{1}{8} + \quad \tfrac{1}{10} \quad - \cdots \\
&= \tfrac{1}{2}(1 - \tfrac{1}{2} + \tfrac{1}{3} - \tfrac{1}{4} + \tfrac{1}{5} - \cdots).
\end{aligned}$$

But the reader will immediately see that we have 'proved' that

$$\ln 2 = \tfrac{1}{2} \ln 2.$$

This paradox, and others like it, show that the theory of infinite series must be developed with very great care: though we cannot attempt this here, it forms a major branch of modern mathematics.

Example 4

We now return to the geometric series

$$1 + x + x^2 + x^3 + \cdots.$$

The reader will recall that

$$S_n = \frac{1 - x^n}{1 - x} = \frac{1}{1 - x} - \frac{x^n}{1 - x};$$

and that if $|x| < 1$, $x^n \to 0$ as $n \to \infty$.

So if $|x| < 1$ the series is convergent to the sum $\dfrac{1}{1 - x}$.

This is a particular case of a *power series*, whose terms are successive powers of x and which converges for a set of values of x called its *domain of convergence*.

The general power series can be written

$$a_0 + a_1 x + a_2 x^2 + a_3 x^3 + \cdots$$

where $a_0, a_1, a_2, \ldots$ are particular constants, and we should expect such a series, if convergent, to have a sum that is a function of x. It will, for instance, be shown in the next section that

$$x - \tfrac{1}{2}x^2 + \tfrac{1}{3}x^3 - \tfrac{1}{4}x^4 + \cdots$$

is convergent to the sum $\ln(1 + x)$, provided that $1 < x \leqslant 1$;

and that

$$1 + x + \frac{x^2}{2!} + \frac{x^3}{3!} + \frac{x^4}{4!} + \cdots$$

is convergent to the sum e^x for all values of x.

9.4 Maclaurin series

We shall begin by reversing the question of the last section and ask whether, if we are given a particular function $f(x)$, we can discover a convergent power series of which it is the sum. For this purpose we shall have to make a number of assumptions.

Firstly we shall restrict ourselves to functions which are differentiable not only once, but any number of times, i.e. whose derived curves are all continuous and smooth.

Secondly, we shall assume the *existence* of such a series for a certain domain of x, and our problem will be simply to *find* the series, granted that it exists. This may seem rather strange but, throughout mathematics, proving such existence is frequently a major problem, and the only statement we shall be able to make here is that, *if* a particular function is expressible as a power series, then this series must be such and such.

Lastly, we assume the fact that if this series exists and is convergent for a certain domain of x, then, within this domain, it is differentiable 'term by term'. By this we mean simply that the derivative of its sum is equal to the sum of its derivatives, even though there is an infinite series of them. In other words, that just as

$$\frac{d}{dx}(u + v) = \frac{du}{dx} + \frac{dv}{dx}$$

and $$\frac{d}{dx}(u + v + w) = \frac{du}{dx} + \frac{dv}{dx} + \frac{dw}{dx}$$

and so on for any *finite* series, so also it is true for our *infinite* series that

$$\frac{d}{dx}(u + v + w + \cdots) = \frac{du}{dx} + \frac{dv}{dx} + \frac{dw}{dx} + \cdots.$$

Let us suppose that $f(x)$ is a function which, throughout a certain domain including $x = 0$,

(*i*) is differentiable any number of times, and

(*ii*) is the sum of a convergent power series.

Let this series be

$$f(x) = a_0 + a_1x + a_2x^2 + a_3x^3 + a_4x^4 + \cdots.$$

Differentiating term by term and putting $x = 0$, we obtain successively:

$$\begin{aligned}
f(x) &= a_0 + a_1x + a_2x^2 + a_3x^3 + a_4x^4 + \cdots &&\Rightarrow f(0) = a_0 \\
f'(x) &= a_1 + 2a_2x + 3a_3x^2 + 4a_4x^3 + \cdots &&\Rightarrow f'(0) = a_1 \\
f''(x) &= 2a_2 + 6a_3x + 12a_4x^2 + \cdots &&\Rightarrow f''(0) = 2!a_2 \\
f'''(x) &= 6a_3 + 24a_4x + \cdots &&\Rightarrow f'''(0) = 3!a_3 \\
f''''(x) &= 24a_4 + \cdots &&\Rightarrow f''''(0) = 4!a_4, \text{ etc.}
\end{aligned}$$

So $a_0 = f(0), \quad a_1 = f'(0), \quad a_2 = \dfrac{f''(0)}{2!}, \quad a_3 = \dfrac{f'''(0)}{3!},$

and similarly $a_n = \dfrac{f^{(n)}(0)}{n!}$.

Hence $$f(x) = f(0) + f'(0)x + \frac{f''(0)}{2!}x^2 + \cdots + \frac{f^{(n)}(0)}{n!}x^n + \cdots$$

This is known as Maclaurin's series (after the Scots mathematician Colin Maclaurin, 1698–1746), which we can now illustrate by three very important examples:

Exponential series

If $\quad f(x) = e^x$,
then $\; f'(x) = f''(x) = \cdots = f^{(n)}(x) = e^x$.
So $\; f(0) = f'(0) = \cdots = f^{(n)}(0) = 1$.

Hence $$e^x = 1 + x + \frac{x^2}{2!} + \frac{x^3}{3!} + \frac{x^4}{4!} + \cdots + \frac{x^n}{n!} + \cdots$$

It can be proved that this series is valid for all values of x; as, indeed, we would expect, since however large x might be (say $x = 1\ 000$), the great size of x^n is eventually overcome by the even greater size of $n!$.

We can now use this series to calculate e to any required degree of accuracy. For, putting $x = 1$,

$$e = 1 + \frac{1}{1!} + \frac{1}{2!} + \frac{1}{3!} + \frac{1}{4!} + \frac{1}{5!} + \cdots$$

and the calculation can be set out to seven places of decimals:

1		1.000 000 0
$\frac{1}{1!}$	2)	1.000 000 0
$\frac{1}{2!}$	3)	0.500 000 0
$\frac{1}{3!}$	4)	0.166 666 7
$\frac{1}{4!}$	5)	0.041 666 7
$\frac{1}{5!}$	6)	0.008 333 3
$\frac{1}{6!}$	7)	0.001 388 9
$\frac{1}{7!}$	8)	0.000 198 4
$\frac{1}{8!}$	9)	0.000 024 8
$\frac{1}{9!}$	10)	0.000 002 5
$\frac{1}{10!}$		0.000 000 2
		2.718 281 5

We see that this series is rapidly convergent and that $e \approx 2.718\,281\,5$.

The careful reader will note that this value of e is subject to two sources of error:

(*i*) because individual terms have been calculated only to seven places of decimals.

When, for instance, we state that $\frac{1}{3!} = 0.166\,666\,7$,

we are asserting that it lies between 0.166 666 65 and 0.166 666 75 and so is subject to a maximum error of 5 in its eighth decimal place, i.e., 5×10^{-8}. As there are eight such terms which have been corrected in this way, the maximum error from this source is $40 \times 10^{-8} = 4 \times 10^{-7}$. This is known as the *rounding error*.

(*ii*) because the series was cut short at $1/10!$.

The remainder after this term is clearly

$$\frac{1}{11!} + \frac{1}{12!} + \frac{1}{13!} + \cdots$$

$$= \frac{1}{11!}\left(1 + \frac{1}{12} + \frac{1}{12 \times 13} + \cdots\right)$$

$$< \frac{1}{11!}\left(1 + \frac{1}{12} + \left(\frac{1}{12}\right)^2 + \cdots\right)$$

$$< \frac{1}{11!} \times \frac{1}{1 - \frac{1}{12}} = \frac{12}{11 \times 11!}.$$

Now $\dfrac{12}{11 \times 11!} < 2.7 \times 10^{-8}$.

So the maximum error from this source is less than 2.7×10^{-8}. This is known as the *truncation error*.

We therefore see that the total error is less than the sum of these, and so is less than $4.27 \times 10^{-7} < 5 \times 10^{-7}$.

Hence e lies in the range $2.718\,281\,5 \pm 5 \times 10^{-7}$.

$\Rightarrow \quad 2.718\,281 < e < 2.718\,282$

$\Rightarrow \quad e = 2.718\,28$, correct to five places of decimals

(in fact $e = 2.718\,281\,8$ to seven places).

Exercise 9.4a

1 Evaluate the following functions at intervals of 0.5 from $x = -1$ to $x = +2$:
(*i*) 1; (*ii*) $1 + x$; (*iii*) $1 + x + \frac{1}{2}x^2$;
(*iv*) $1 + x + \frac{1}{2}x^2 + \frac{1}{6}x^3$; (*v*) $1 + x + \frac{1}{2}x^2 + \frac{1}{6}x^3 + \frac{1}{24}x^4$;
(*vi*) e^x.

Hence, on a single accurate figure, plot the graphs of these functions from $x = -1$ to $x = +2$.

2 Find the power series, as far as the x^4 term, of:
(*i*) e^{-x}; (*ii*) e^{2x}; (*iii*) $\frac{1}{2}(e^x + e^{-x})$; (*iv*) $\frac{1}{2}(e^x - e^{-x})$.

3 Use the exponential series to calculate to five decimal places:
(*i*) $\sqrt{e}$; (*ii*) $\dfrac{1}{e}$; (*iii*) $e^{0.1}$.

Logarithmic series

If $\ln x = a_0 + a_1x + a_2x^2 + a_3x^3 + \cdots$

it would immediately follow that

$$a_0 = \ln 0.$$

But $\ln 0$ does not exist, so we cannot express $\ln x$ as a power series.

We can, however, consider the function $\ln(1 + x)$.

Now $f(x) = \ln(1 + x) \quad \Rightarrow \quad f(0) = \ln 1 = 0$

$$\Rightarrow \quad f'(x) = \frac{1}{1+x} \qquad \Rightarrow \quad f'(0) = 1$$

$$\Rightarrow \quad f''(x) = -\frac{1}{(1+x)^2} \qquad \Rightarrow \quad f''(0) = -1$$

$$\Rightarrow \quad f'''(x) = +\frac{2}{(1+x)^3} \qquad \Rightarrow \quad f'''(0) = +2$$

$$\Rightarrow \quad f^n(x) = (-1)^{n-1}\frac{(n-1)!}{(1+x)^n} \qquad \Rightarrow \quad f^n(0) = (-1)^{n-1}(n-1)!$$

So, with the usual assumptions for Maclaurin's series,

$$\ln(1+x) \equiv 0 + 1x + \frac{-1}{2!}x^2 + \frac{2}{3!}x^3 + \cdots + \frac{(-1)^{n-1}(n-1)!}{n!}x^n + \cdots$$

$$\Rightarrow \quad \boxed{\ln(1+x) = x - \tfrac{1}{2}x^2 + \tfrac{1}{3}x^3 - \tfrac{1}{4}x^4 + \cdots + \frac{(-1)^{n-1}}{n}x^n \cdots} \qquad (1)$$

and it can be shown that in this case the domain of convergence is $-1 < x \leqslant 1$.

Putting $-x$ for x, we obtain

$$\ln(1-x) = -x - \tfrac{1}{2}x^2 - \tfrac{1}{3}x^3 - \tfrac{1}{4}x^4 - \cdots \qquad (2)$$

whose domain of convergence is $-1 < -x \leqslant 1$, i.e., $-1 \leqslant x < 1$.

If we now subtract these two results, we obtain

$$\ln(1+x) - \ln(1-x) = 2(x + \tfrac{1}{3}x^3 + \tfrac{1}{5}x^5 + \cdots)$$

$$\Rightarrow \quad \tfrac{1}{2}\ln\frac{1+x}{1-x} = x + \tfrac{1}{3}x^3 + \tfrac{1}{5}x^5 + \cdots \qquad (3)$$

which is convergent provided both the original series are convergent, i.e., if

$-1 < x < 1$.

It is instructive to use these series to calculate ln 2.

If we put $x = 1$ in (1), we obtain

$$\ln 2 = 1 - \tfrac{1}{2} + \tfrac{1}{3} - \tfrac{1}{4} + \cdots.$$

But the millionth term is $-1/10^6$ and so is still affecting the sixth decimal place. The series, therefore, converges so slowly that it is not very useful for the calculation of ln 2.

If, on the other hand, we use (3), we see that to calculate ln 2 it will be necessary to let

$$\frac{1+x}{1-x} = 2 \quad \Rightarrow \quad x = \tfrac{1}{3}.$$

Putting $x = \frac{1}{3}$, we obtain

$$\tfrac{1}{2}\ln 2 = \tfrac{1}{3} + \tfrac{1}{3}(\tfrac{1}{3})^3 + \tfrac{1}{5}(\tfrac{1}{3})^5 + \cdots$$

$$\Rightarrow \quad \ln 2 = 2[\tfrac{1}{3} + \tfrac{1}{3}(\tfrac{1}{3})^3 + \tfrac{1}{5}(\tfrac{1}{3})^5 + \cdots]$$

and it is clear that this series converges much more rapidly. How many terms of this series need to be taken in order to determine the sixth decimal place?

Exercise 9.4b

1 Evaluate the following functions at intervals of 0.5 from $x = -1$ to $x = +1$:
(*i*) x; (*ii*) $x - \frac{1}{2}x^2$; (*iii*) $x - \frac{1}{2}x^2 + \frac{1}{3}x^3$;
(*iv*) $x - \frac{1}{2}x^2 + \frac{1}{3}x^3 - \frac{1}{4}x^4$; (*v*) $\ln(1 + x)$.
Hence graph these functions on a single diagram from $x = -1$ to $+1$.

2 Find the power series, as far as the x^4 term of:

(*i*) $\ln\left(1 + \dfrac{x}{2}\right)$; (*ii*) $\ln(1 + 2x)$; (*iii*) $\ln \dfrac{1}{1 - x^2}$.

3 Use the series for $\frac{1}{2}\ln \dfrac{1 + x}{1 - x}$ to evaluate, to three decimal places:

(*i*) $\ln 3$; (*ii*) $\ln \frac{5}{3}$.
Hence calculate $\ln 5$.

Binomial series

We now turn to the function

$$f(x) = (1 + x)^n \quad (n \in R).$$

$$\begin{aligned} \text{Now}\quad & f(x) = (1 + x)^n && \Rightarrow f(0) = 1 \\ \Rightarrow\quad & f'(x) = n(1 + x)^{n-1} && \Rightarrow f'(0) = n \\ \Rightarrow\quad & f''(x) = n(n - 1)(1 + x)^{n-2} && \Rightarrow f''(0) = n(n - 1). \end{aligned}$$

More generally,

$$f^{(r)}(x) = n(n - 1)\ldots(n - r + 1)(1 + x)^{n-r}$$
$$\Rightarrow \quad f^{(r)}(0) = n(n - 1)\ldots(n - r + 1).$$

So, using the Maclaurin series, we obtain the *binomial series*:

$$(1 + x)^n = 1 + nx + \frac{n(n - 1)}{2!}x^2 + \cdots + \frac{n(n - 1)(n - 2)\ldots(n - r + 1)}{r!}x^r + \cdots$$

and it can be shown that the series is convergent, provided $|x| < 1$.

If $n \in Z^+$, we notice that the series terminates and reduces to the binomial theorem.

For instance, if $n = 4$,

$$\begin{aligned}(1+x)^4 &= 1 + 4x + \frac{4\times 3}{2!}x^2 + \frac{4\times 3\times 2}{3!}x^3 + \frac{4\times 3\times 2\times 1}{4!}x^4 \\ &\quad + \frac{4\times 3\times 2\times 1\times 0}{5!}x^5 \\ &\quad + \frac{4\times 3\times 2\times 1\times 0\times -1}{6!}x^6 + \cdots \\ &= 1 + 4x + 6x^2 + 4x^3 + x^4.\end{aligned}$$

More generally, if $n \in Z^+$,

$$\frac{n(n-1)(n-2)\ldots(n-r+1)}{r!} = {}^nC_r \quad \text{if } r \leqslant n,$$

$$\text{and} \quad 0 \quad \text{if } r > n$$

so that the binomial series simply becomes the binomial theorem.

If $n \notin Z^+$ it is impossible to speak of ${}^nC_r$, as ${}^{-3}C_2$, ${}^{\frac{1}{2}}C_3$ etc. have no meaning. Nevertheless, we find it convenient to have an abbreviation for the coefficients of the binomial series, and write

$$\frac{n(n-1)(n-2)(n-3)\ldots(n-r+1)}{r!} \quad \text{as} \quad \binom{n}{r}.$$

So $$\binom{\frac{1}{2}}{2} = \frac{\frac{1}{2}(\frac{1}{2}-1)}{2!} = -\tfrac{1}{8},$$

$$\binom{-2}{3} = \frac{(-2)(-2-1)(-2-2)}{3!} = -4,$$

whilst $$\binom{4}{2} = \frac{4(4-1)}{2!} = 6 = {}^4C_2.$$

With this notation we can write the binomial series as

$$\boxed{(1+x)^n = 1 + \binom{n}{1}x + \binom{n}{2}x^2 + \cdots + \binom{n}{r}x^r + \cdots}$$

Example 1

Find a series for $\left(1 - \dfrac{x}{2}\right)^{-2}$.

$$\begin{aligned}\left(1-\frac{x}{2}\right)^{-2} &= 1 + (-2)\left(-\frac{x}{2}\right) + \frac{(-2)(-3)}{2!}\left(-\frac{x}{2}\right)^2 \\ &\quad + \frac{(-2)(-3)(-4)}{3!}\left(-\frac{x}{2}\right)^3 + \cdots \\ &= 1 + x + \tfrac{3}{4}x^2 + \tfrac{1}{2}x^3 + \cdots\end{aligned}$$

which is convergent if $\left|-\frac{x}{2}\right| < 1 \quad \Leftrightarrow \quad |x| < 2.$

Example 2

Use the binomial series to estimate $\sqrt[3]{999}$ to seven places of decimals.

$$\sqrt[3]{999} = (1\,000 - 1)^{1/3}$$
$$= 10(1 - 0.001)^{1/3}.$$

Now $(1 - 0.001)^{1/3} = 1 + \frac{1}{3}(-0.001) + \dfrac{\frac{1}{3} \times -\frac{2}{3}}{2!}(-0.001)^2$

$$+ \frac{\frac{1}{3} \times -\frac{2}{3} \times -\frac{5}{3}}{3!}(-0.001)^3 + \frac{\frac{1}{3} \times -\frac{2}{3} \times -\frac{5}{3} \times -\frac{8}{3}}{4!}(-0.001)^4 + \cdots$$

The first three terms are

$$\begin{array}{rl} & +\ 1.000\,000\,000\,0 \\ & -\ 0.000\,333\,333\,3 \\ & -\ 0.000\,000\,111\,1 \\ \hline = & +\ 0.999\,666\,555\,6 \end{array}$$

It is easy to see that the maximum rounding error is 10^{-10}. Also the remainder is

$$-\frac{5}{81}(0.001)^3 - \frac{10}{243}(0.001)^4 - \cdots,$$

so the truncation error is

$$\frac{5}{81}(0.001)^3 + \frac{10}{243}(0.001)^4 + \cdots$$
$$< \frac{5}{81}(0.001)^3\,[1 + (0.001) + (0.001)^2 + \cdots]$$
$$< \frac{5}{81} \times 10^{-9} \times \frac{1\,000}{999} < 10^{-10}.$$

Hence the total error $< 2 \times 10^{-10}$

and $\sqrt[3]{999} = 9.996\,665\,556$

with a maximum possible error of 2×10^{-9}.

So $\sqrt[3]{999}$ lies between $9.996\,665\,554$
and $9.996\,665\,558$.

Hence $\sqrt[3]{999} = 9.996\,665\,6$ to seven decimal places.

Exercise 9.4c

1 Find the first four terms of the power series of the following functions, in each case stating their domain of convergence:

(*i*) $\sqrt{(1+x)}$; (*ii*) $\dfrac{1}{1+x}$; (*iii*) $(1+x)^{-3}$;

(*iv*) $(1-2x)^{-1/2}$; (*v*) $(2+x)^{-1}$; (*vi*) $\sqrt{(4-2x)}$.

2 Evaluate, to five places of decimals:

(*i*) $\sqrt{1.01}$; (*ii*) $\sqrt[4]{0.99}$; (*iii*) $\sqrt{4.1}$; (*iv*) $\sqrt[3]{27.27}$; (*v*) $\dfrac{1}{(2.01)^2}$.

3 Obtain binomial series for $1/(1-x)^2$ and $1/(1-x)^3$, and check by means of differentiating the series for $1/(1-x)$.

Hence sum the series:

$1 + 2(\tfrac{1}{2}) + 3(\tfrac{1}{2})^2 + 4(\tfrac{1}{2})^3 + \cdots$.

4 (*i*) Using $\sqrt{(x^2+1)} = x(1+1/x^2)^{1/2}$, show that if x is large:

$\sqrt{(x^2+1)} \approx x + \dfrac{1}{2x} - \dfrac{1}{8x^3} + \cdots$ (known as an *asymptotic* series).

(*ii*) Find the first three terms of similar asymptotic series for $1/(x-1)^2$ and $\sqrt[3]{(x^3+2)}$.

Taylor series

The Maclaurin series enable us to find the value of a function near $x = 0$ in terms of its value and those of its derivatives at $x = 0$. If we now seek to investigate its value in the neighbourhood of another point $x = a$, we can let the point be $x = a + h$ and write

$$f(a+h) \equiv F(h) \quad \Rightarrow \quad f(a) = F(0)$$

so that

$$f'(a+h) \equiv F'(h) \quad \Rightarrow \quad f'(a) = F'(0)$$

$$f''(a+h) \equiv F''(h) \quad \Rightarrow \quad f''(a) = F''(0).$$

But $F(h) = F(0) + hF'(0) + \dfrac{h^2}{2!}F''(0) + \cdots$

$$\Rightarrow \quad \boxed{f(a+h) = f(a) + hf'(a) + \frac{h^2}{2!}f''(a) + \cdots}$$

which is known as the Taylor series (after Brook Taylor, 1685–1731), and which enables us to find the value of the function $f(x)$ *near* $x = a$ in terms of its value and those of its derivatives at $x = a$.

In particular, if h is very small,

$f(a+h) \approx f(a) + hf'(a)$

provides a linear approximation to the function, already familiar in the form

$$f'(a) \approx \frac{f(a+h) - f(a)}{h}.$$

Exercise 9.4d

1 Find, as far as the x^4 term, Maclaurin series for
(*i*) $\sin x$; (*ii*) $\cos x$; (*iii*) $\tan x$; (*iv*) $\sec x$; (*v*) $e^x \sin x$; (*vi*) $\ln \sec x$.

2 Use Maclaurin series to calculate to five decimal places, $\tan 0.1$, $\tan 1°$, $\tan 1'$.

3 Find the first four terms of the Taylor series for $\tan(\pi/4 + h)$ and hence calculate, to five decimal places, $\tan 46°$ and $\tan 45°\ 1'$.

4 Obtain the Maclaurin series for $\tan^{-1} x$ and check by integrating the binomial series for $1/(1 + x^2)$. Hence
(*i*) obtain Gregory's series:

$$\frac{\pi}{4} = 1 - \tfrac{1}{3} + \tfrac{1}{5} - \tfrac{1}{7} + \cdots;$$

(*ii*) use the result (see Exercise 5.13a, no. 3) that

$$\frac{\pi}{4} = \tan^{-1} \tfrac{1}{2} + \tan \tfrac{1}{3}$$

to show that

$$\frac{\pi}{4} = [\tfrac{1}{2} - \tfrac{1}{3}(\tfrac{1}{2})^3 + \tfrac{1}{5}(\tfrac{1}{2})^5 - \cdots] + [\tfrac{1}{3} - \tfrac{1}{3}(\tfrac{1}{3})^3 + \tfrac{1}{5}(\tfrac{1}{3})^5 - \cdots].$$

9.5 Integration by parts: reduction formulae

We now return to the problem of integration. As was observed in Chapter 4, a result about differentiation can often be transformed into an equally useful one about integration, and we saw how the chain rule for derivatives led to the method of substitution for finding integrals. We now proceed to look again, in just this way, at the formula for the derivative of a product.

If u and v are two functions of x, we know that

$$\frac{d}{dx}(uv) = u\frac{dv}{dx} + \frac{du}{dx}v$$

So

$$uv = \int u\frac{dv}{dx}\,dx + \int \frac{du}{dx}v\,dx$$

$$\Rightarrow \quad \int u\frac{dv}{dx}\,dx = uv - \int \frac{du}{dx}v\,dx.$$

This is the rule for *integrating by parts* and is useful whenever the product $\frac{du}{dx}v$ is easier to integrate than $u\frac{dv}{dx}$.

If we now replace u and v by U and V, where

$$U = u \quad \text{and} \quad V = \frac{dv}{dx},$$

the result can be written as

† $$\boxed{\int UV\,dx = U\int V\,dx - \int\left(\frac{dU}{dx}\int V\,dx\right)dx}$$

Example 1

$$\int x\cos x\,dx = x\sin x - \int 1\sin x\,dx$$
$$= x\sin x + \cos x + c.$$

Example 2

$$\int_0^1 x^2 e^x\,dx = [x^2 e^x]_0^1 - \int_0^1 2x\,e^x\,dx$$
$$= e - 2\left\{[x\,e^x]_0^1 - \int_0^1 1\,e^x\,dx\right\}$$
$$= e - 2\{e - (e - 1)\}$$
$$= e - 2.$$

Example 3

$$\int \ln x\,dx = \int 1 \times \ln x\,dx$$
$$= x\ln x - \int x\frac{1}{x}\,dx$$
$$= x\ln x - x + c.$$

Example 4

$$\int e^{-x}\sin x\,dx = (-e^{-x})\sin x - \int(-e^{-x})\cos x\,dx$$
$$= -e^{-x}\sin x + \int e^{-x}\cos x\,dx$$

† Conveniently remembered as

$$\int UV = U\int V - \int\left(U'\int V\right).$$

$$= -e^{-x} \sin x + (-e^{-x}) \cos x - \int (-e^{-x})(-\sin x)\, dx$$

$$= -e^{-x} (\sin x + \cos x) - \int e^{-x} \sin x\, dx + c$$

$$\Rightarrow \quad 2 \int e^{-x} \sin x\, dx = -e^{-x} (\sin x + \cos x) + c$$

$$\Rightarrow \quad \int e^{-x} \sin x\, dx = -\tfrac{1}{2} e^{-x} (\sin x + \cos x) + c'.$$

Exercise 9.5a

Find the following integrals:

1 (*i*) $\int x\, e^x\, dx$; (*ii*) $\int x \sin x\, dx$; (*iii*) $\int x\, e^{2x}\, dx$;

(*iv*) $\int x \ln x\, dx$; (*v*) $\int x \cos \frac{x}{2}\, dx$; (*vi*) $\int \frac{\ln x}{x^2}\, dx$.

2 (*i*) $\int x^2 e^{-x}\, dx$; (*ii*) $\int x^2 \sin x\, dx$;

(*iii*) $\int \theta \sec^2 \theta\, d\theta$; (*iv*) $\int u 10^u\, du$.

3 (*i*) $\int e^x \sin x\, dx$; (*ii*) $\int e^x \cos x\, dx$;

(*iii*) $\int e^{-x} \cos x\, dx$; (*iv*) $\int e^x \sin 2x\, dx$.

4 (*i*) $\int (\ln x)^2$; (*ii*) $\int \tan^{-1} x\, dx$;

(*iii*) $\int \sin^{-1} u\, du$; (*iv*) $\int x \tan^{-1} x\, dx$.

5 Evaluate:

(*i*) $\int_0^1 x\, e^{-x}\, dx$; (*ii*) $\int_0^{\frac{1}{2}\pi} x \sin 2x\, dx$;

(*iii*) $\int_0^{\pi} t^2 \cos \tfrac{1}{2}t\, dt$; (*iv*) $\int_1^2 u^2 \ln u\, du$.

6 A mathematical model is constructed to describe the situation in a factory. The number of workers w which an employer will require is a function of the wage rate £r per week for each worker where $w = 100\, e^{-0.04r}$. If the union wishes to maximize the total pay of its members required by the employer, what rate will it ask for? (Assume that all workers belong to the union.) (S.M.P.)

Reduction formulae

It is sometimes possible to find a formula which will reduce a given integral to a simpler one of the same type, and then to find the required integral by successive applications of this process.

Example 5

If $I_n = \int x^n e^{-x}\,dx$, evaluate I_3.

$$\text{Now}\quad I_n = x^n(-e^{-x}) - \int (nx^{n-1})(-e^{-x})\,dx$$

$$= -x^n e^{-x} + n\int x^{n-1} e^{-x}\,dx$$

$$\Rightarrow \quad I_n = -x^n e^{-x} + nI_{n-1}.$$

$$\begin{aligned}\text{Hence}\quad I_3 &= -x^3 e^{-x} + 3I_2\\ I_2 &= -x^2 e^{-x} + 2I_1\\ I_1 &= -x\,e^{-x} + I_0.\end{aligned}$$

$$\text{But}\quad I_0 = -e^{-x} + c.$$

$$\begin{aligned}\text{So}\quad I_3 &= -x^3 e^{-x} + 3(-x^2 e^{-x} + 2I_1)\\ &= -(x^3 + 3x^2)\,e^{-x} + 6(-x\,e^{-x} + I_0)\\ &= -(x^3 + 3x^2 + 6x)\,e^{-x} + 6I_0\\ &= -(x^3 + 3x^2 + 6x + 6)\,e^{-x} + c.\end{aligned}$$

Example 6

$$I_n = \int_0^{\frac{1}{2}\pi} \sin^n x\,dx$$

$$\begin{aligned}I_n = \int_0^{\frac{1}{2}\pi} \sin^n x\,dx &= \int_0^{\frac{1}{2}\pi} \sin^{n-1} x \times \sin x\,dx\\ &= \left[\sin^{n-1} x\,(-\cos x)\right]_0^{\frac{1}{2}\pi}\\ &\qquad - \int_0^{\frac{1}{2}\pi} (n-1)\sin^{n-2} x \times \cos x\,(-\cos x)\,dx\\ &= 0 + (n-1)\int_0^{\frac{1}{2}\pi} \sin^{n-2} x \cos^2 x\,dx\\ &= (n-1)\int_0^{\frac{1}{2}\pi} \sin^{n-2} x\,(1 - \sin^2 x)\,dx.\end{aligned}$$

$$\text{So}\quad I_n = (n-1)\left\{\int_0^{\frac{1}{2}\pi} \sin^{n-2} x\,dx - \int_0^{\frac{1}{2}\pi} \sin^n x\,dx\right\}$$

$$\Rightarrow \quad I_n = (n-1)(I_{n-2} - I_n)$$
$$\Rightarrow \quad nI_n = (n-1)I_{n-2}$$
$$\Rightarrow \quad I_n = \frac{n-1}{n} I_{n-2}$$

and this reduction formula enables us to calculate I_n in terms of I_{n-2}.

But we know that

$$I_1 = \int_0^{\frac{1}{2}\pi} \sin x \,dx = [-\cos x]_0^{\frac{1}{2}\pi} = 1$$

and $$I_0 = \int_0^{\frac{1}{2}\pi} 1 \,dx = [x]_0^{\frac{1}{2}\pi} = \tfrac{1}{2}\pi.$$

So it follows, for instance, that

$$I_5 = \tfrac{4}{5}I_3 = \tfrac{4}{5} \times \tfrac{2}{3}I_1 = \tfrac{4}{5} \times \tfrac{2}{3} \times 1 = \tfrac{1}{15}$$

and $$I_6 = \tfrac{5}{6}I_4 = \tfrac{5}{6} \times \tfrac{3}{4}I_2 = \tfrac{5}{6} \times \tfrac{3}{4} \times \tfrac{1}{2}I_0$$
$$= \tfrac{5}{6} \times \tfrac{3}{4} \times \tfrac{1}{2} \times \tfrac{1}{2}\pi = \tfrac{5\pi}{32}.$$

Similarly, if $J_n = \int_0^{\frac{1}{2}\pi} \cos^n x \,dx$, we can put

$$u = \tfrac{1}{2}\pi - x$$
$$\Rightarrow \quad du = -dx.$$

So $$J_n = \int_{\frac{1}{2}\pi}^{0} \cos^n (\tfrac{1}{2}\pi - u) \times -du$$
$$= \int_0^{\frac{1}{2}\pi} \sin^n u \,du = I_n.$$

Summarising, if $I_n = \int_0^{\frac{1}{2}\pi} \sin^n x \,dx, \qquad J_n = \int_0^{\frac{1}{2}\pi} \cos^n x \,dx$

then $$\boxed{I_n = \frac{n-1}{n} I_{n-2}, \qquad J_n = \frac{n-1}{n} J_{n-2}}$$

Exercise 9.5b

1 Evaluate:

(*i*) $\int_0^{\frac{1}{2}\pi} \sin^4 \theta \,d\theta$; (*ii*) $\int_0^{\frac{1}{2}\pi} \cos^3 \theta \,d\theta$; (*iii*) $\int_0^{\frac{1}{2}\pi} \cos^6 \theta \,d\theta$;

(*iv*) $\int_0^{\frac{1}{2}\pi} \sin^7 \theta \,d\theta$; (*v*) $\int_0^{\frac{1}{2}\pi} \cos^8 \theta \,d\theta$.

2 Evaluate:

(*i*) $\int_0^{\pi} \sin^5 \theta \, d\theta$; (*ii*) $\int_{-\frac{1}{2}\pi}^{\frac{1}{2}\pi} \cos^3 \theta \, d\theta$; (*iii*) $\int_0^{\pi} \cos^3 \theta \, d\theta$;

(*iv*) $\int_{-\frac{1}{2}\pi}^{\frac{1}{2}\pi} \sin^6 \theta \, d\theta$; (*v*) $\int_{-\frac{1}{2}\pi}^{\frac{1}{2}\pi} \sin^7 \theta \, d\theta$.

3 If $I_n = \int (\ln x)^n \, dx$, show that $I_n = x(\ln x)^n - nI_{n-1}$.

Hence find I_3.

4 Use $\tan^n \theta = \tan^{n-2} \theta (\sec^2 \theta - 1)$ to find a reduction formula for

$I_n = \int \tan^n \theta \, d\theta$, and hence find I_4 and I_6.

5 If $I_n = \int \sec^n \theta \, d\theta$, prove that

$(n + 1) I_{n+2} = \tan \theta \sec^n \theta + nI_n$.

9.6 Interlude: partial fractions

At an early age we learnt such simplifications as

$$\frac{2}{3} - \frac{1}{5} = \frac{10 - 3}{15} = \frac{7}{15};$$

and, rather later, when we learned that

$$\frac{1}{x - 1} - \frac{1}{x + 1} = \frac{(x + 1) - (x - 1)}{(x - 1)(x + 1)} = \frac{2}{x^2 - 1},$$

we had little doubt that this too represented a simplification, as $2/(x^2 - 1)$ is generally more useful and more convenient than $1/(x - 1) - 1/(x + 1)$.

In the next section we shall frequently be needing to find such integrals as

$$\int \frac{1}{x^2 - 1} \, dx$$

for which there is no immediately obvious answer. But a little meditation might lead us to reverse the above process and say that

$$\frac{1}{x^2 - 1} = \frac{\frac{1}{2}}{x - 1} - \frac{\frac{1}{2}}{x + 1}$$

$$\Rightarrow \int \frac{dx}{x^2 - 1} = \int \left(\frac{\frac{1}{2}}{x - 1} - \frac{\frac{1}{2}}{x + 1} \right) dx$$

$$= \tfrac{1}{2} \ln (x - 1) - \tfrac{1}{2} \ln (x + 1) + c$$

$$= \tfrac{1}{2} \ln \frac{x - 1}{x + 1} + c.$$

Obviously we were extremely fortunate to know that

$$\frac{1}{x^2 - 1} = \frac{\frac{1}{2}}{x - 1} - \frac{\frac{1}{2}}{x + 1},$$

and our next task will be to find a process whereby similar rational functions (i.e. the quotients of two polynomials) can be expressed in terms of such partial fractions.

The full theory of partial fractions is beyond our present scope, but we can illustrate the procedure by a number of examples:

Example 1

Express in partial fractions

$$\frac{x + 1}{(x - 1)(x - 2)}.$$

Hopefully, we write

$$\frac{x + 1}{(x - 1)(x - 2)} \equiv \frac{A}{x - 1} + \frac{B}{x - 2},$$

where A, B are constants which are to be determined

$$\Rightarrow \qquad x + 1 \equiv A(x - 2) + B(x - 1).$$

Putting $x = 2$, $3 = A \times 0 + B \times 1 \quad \Rightarrow \quad B = 3.$

Putting $x = 1$, $2 = A \times -1 + B \times 0 \quad \Rightarrow \quad A = -2.$

So the required partial fractions would be

$$\frac{-2}{x - 1} + \frac{3}{x - 2},$$

and it is quickly verified that these can be combined to give the original expression.

Example 2

We now seek to put into partial fractions the more complicated expression

$$\frac{2x^3 - 3x^2 - x + 5}{2x^2 + x - 1}.$$

First of all we observe that $2x^3 - 3x^2 - x + 5$ can be divided by $2x^2 + x - 1$ to give a quotient $x - 2$ and remainder $2x + 3$.

So $2x^3 - 3x^2 - x + 5 \equiv (2x^2 + x - 1)(x - 2) + (2x + 3)$

$$\Rightarrow \quad \frac{2x^3 - 3x^2 - x + 5}{2x^2 + x - 1} \equiv x - 2 + \frac{2x + 3}{2x^2 + x - 1}.$$

Now $$\frac{2x + 3}{2x^2 + x - 1} \equiv \frac{2x + 3}{(2x - 1)(x + 1)},$$

so we write,

$$\frac{2x + 3}{(2x - 1)(x + 1)} \equiv \frac{A}{2x - 1} + \frac{B}{x + 1}$$

$$\Rightarrow \qquad 2x + 3 \equiv A(x + 1) + B(2x - 1).$$

Putting $x = -1$, $\quad 1 = A \times 0 - 3B \quad \Rightarrow \quad B = -\frac{1}{3}$.

Putting $x = \frac{1}{2}$, $\quad 4 = A \times \frac{3}{2} \quad \Rightarrow \quad A = \frac{8}{3}$.

So (as can easily be checked),

$$\frac{2x + 3}{2x^2 + x - 1} \equiv \frac{\frac{8}{3}}{2x - 1} - \frac{\frac{1}{3}}{x + 1}$$

and $$\frac{2x^3 - 3x^2 - x + 5}{2x^2 + x - 1} \equiv x - 2 + \frac{\frac{8}{3}}{2x - 1} - \frac{\frac{1}{3}}{x + 1}.$$

In this case the key to our success lay in first dividing the numerator by the denominator, and wherever possible (i.e., unless the degree of the numerator is already less than that of the denominator) this must always be the initial step.

Example 3

We now take a function whose denominator contains a repeated factor:

$$f(x) = \frac{x^2 + 5x + 9}{(x + 3)(x + 2)^2}.$$

Again, we first try the simple partial fractions:

$$\frac{x^2 + 5x + 9}{(x + 3)(x + 2)^2} \equiv \frac{A}{x + 3} + \frac{B}{(x + 2)^2} \tag{1}$$

$$\Rightarrow \quad x^2 + 5x + 9 \equiv A(x + 2)^2 + B(x + 3).$$

Equating coefficients of x^2, x, and the constant terms, we see that it would be necessary to have

$$\left.\begin{aligned} 1 &= A \\ 5 &= 4A + B \\ 9 &= 4A + 3B \end{aligned}\right\} \text{ which are clearly inconsistent.}$$

So $f(x)$ *cannot* be expressed in terms of partial fractions (1).

For our next attempt, therefore, above the quadratic denominator $(x + 2)^2$ we place a term $Bx + C$, so that

$$\frac{x^2 + 5x + 9}{(x + 3)(x + 2)^2} \equiv \frac{A}{x + 3} + \frac{Bx + C}{(x + 2)^2}$$

$$\Rightarrow \quad x^2 + 5x + 9 \equiv A(x+2)^2 + (Bx + C)(x+3).$$

Putting $x = -3$: $\quad 3 = A \quad \Rightarrow \quad A = 3.$

Equating coefficient x^2: $\quad 1 = A + B \quad \Rightarrow \quad B = -2.$

Equating constants: $\quad 9 = 4A + 3C \quad \Rightarrow \quad C = -1.$

Hence (as can easily be checked) the required partial fractions are

$$\frac{x^2 + 5x + 9}{(x+3)(x+2)^2} \equiv \frac{3}{x+3} - \frac{2x+1}{(x+2)^2}.$$

Sometimes it is convenient to split this last fraction still further, by saying that

$$\frac{2x+1}{(x+2)^2} \equiv \frac{2(x+2) - 3}{(x+2)^2} = \frac{2}{x+2} - \frac{3}{(x+2)^2}$$

so that $$\frac{x^2 + 5x + 9}{(x+3)(x+2)^2} \equiv \frac{3}{x+3} - \frac{2}{x+2} + \frac{3}{(x+2)^2}.$$

Example 4

Finally, we take a rational function whose denominator contains a quadratic term which cannot be factorised:

$$f(x) = \frac{x-3}{(x-1)(x^2+1)}.$$

As it is impossible to express $x^2 + 1$ in real factors, we follow the precedent of the last example and write

$$\frac{x-3}{(x-1)(x^2+1)} \equiv \frac{A}{x-1} + \frac{Bx + C}{x^2+1}$$

$$\Rightarrow \qquad x - 3 \equiv A(x^2+1) + (Bx + C)(x-1)$$

$x = 1$: $\quad \Rightarrow \quad -2 = 2A \quad \Rightarrow \quad \mathrm{A} = -1$

x^2 term: $\quad \Rightarrow \quad 0 = A + B \quad \Rightarrow \quad B = +1$

constants: $\quad \Rightarrow \quad -3 = A - C \quad \Rightarrow \quad C = +2.$

So it appears that

$$\frac{x-3}{(x-1)(x^2+1)} \equiv \frac{-1}{x-1} + \frac{x+2}{x^2+1},$$

which can quickly be verified.

Partial fractions are principally used in integration, but it is also interesting to see how they can help us to find the sums of certain series.

If, for instance, we consider the series

$$\frac{1}{1 \times 3} + \frac{1}{3 \times 5} + \frac{1}{5 \times 7} + \cdots,$$

we see that its rth term is $\dfrac{1}{(2r - 1)(2r + 1)}$.

This can easily be expressed in partial fractions as

$$\frac{\frac{1}{2}}{2r - 1} - \frac{\frac{1}{2}}{2r + 1}$$

$$= \frac{1}{2}\left(\frac{1}{2r - 1} - \frac{1}{2r + 1}\right).$$

Hence the sum of the first n terms is

$$S_n = \frac{1}{1 \times 3} + \frac{1}{3 \times 5} + \frac{1}{5 \times 7} + \cdots + \frac{1}{(2n - 1)(2n + 1)}$$

$$= \frac{1}{2}\left\{\left(\frac{1}{1} - \frac{1}{3}\right) + \left(\frac{1}{3} - \frac{1}{5}\right) + \left(\frac{1}{5} - \frac{1}{7}\right) + \cdots + \left(\frac{1}{2n - 1} - \frac{1}{2n + 1}\right)\right\}$$

$$= \frac{1}{2}\left\{1 - \frac{1}{2n + 1}\right\} = \frac{n}{(2n + 1)}.$$

Also, as $n \to \infty$, $S_n \to \frac{1}{2}$.

So the sum of the infinite series is $\frac{1}{2}$.

Exercise 9.6a

1 Express in partial fractions

(*i*) $\dfrac{x + 1}{(x - 1)(x - 2)}$; (*ii*) $\dfrac{x}{x^2 - 4}$; (*iii*) $\dfrac{1}{x(2x + 1)}$;

(*iv*) $\dfrac{x^2}{(x - 1)(x - 2)(x - 3)}$; (*v*) $\dfrac{x - 1}{x^2 + x}$; (*vi*) $\dfrac{x^2 + 1}{x^2 - 2x}$;

(*vii*) $\dfrac{x^2}{x^2 - 4}$; (*viii*) $\dfrac{x^3}{x^2 - 1}$.

2 Express in partial fractions

(*i*) $\dfrac{1}{x(x - 1)^2}$; (*ii*) $\dfrac{x - 1}{x^2(x + 1)}$;

(*iii*) $\dfrac{x}{(x^2 - 1)(x + 1)}$; (*iv*) $\dfrac{x^3}{(x - 1)(x^2 - 1)}$.

3 Express in partial fractions

(*i*) $\dfrac{1}{x(x^2+1)}$; (*ii*) $\dfrac{1}{x^4-1}$; (*iii*) $\dfrac{1}{x^3-1}$; (*iv*) $\dfrac{x}{x^3+1}$.

4 Find the first four terms of the power series for $\dfrac{1}{(1+x)(1+2x)}$ by first expressing it

(*i*) in partial fractions;
(*ii*) as $[1+(3x+2x^2)]^{-1}$;
(*iii*) as $(1+x)^{-1}(1+2x)^{-1}$.

5 Find the rth term, the sum of the first n terms and the sum to infinity of the series

(*i*) $\dfrac{1}{1\times 2}+\dfrac{1}{2\times 3}+\dfrac{1}{3\times 4}+\cdots$;

(*ii*) $\dfrac{1}{1\times 3}+\dfrac{1}{2\times 4}+\dfrac{1}{3\times 5}+\cdots$;

(*iii*) $\dfrac{1}{1\times 2\times 3}+\dfrac{1}{2\times 3\times 4}+\dfrac{1}{3\times 4\times 5}+\cdots$;

Use of partial fractions in integration

In the next section we shall return to the general problem of integration, but we can already see the use of partial fractions when we need to integrate the quotient of two polynomials (or *rational function*, as it is usually called).

Example 5

$$I=\int\frac{x+1}{x^2-4}\,\mathrm{d}x.$$

It is quickly seen that

$$\frac{x+1}{x^2-4}\equiv\frac{\frac{3}{4}}{x-2}+\frac{\frac{1}{4}}{x+2}.$$

So
$$I=\int\left(\frac{\frac{3}{4}}{x-2}+\frac{\frac{1}{4}}{x+2}\right)\mathrm{d}x$$
$$=\tfrac{3}{4}\ln(x-2)+\tfrac{1}{4}\ln(x+2)+c.$$

Example 6

$$I=\int\frac{x^2}{(x-1)(x+1)^2}\,\mathrm{d}x.$$

We can find the full partial fractions in such a case, with a repeated factor, by letting

$$\frac{x^2}{(x-1)(x+1)^2} \equiv \frac{A}{x-1} + \frac{B}{x+1} + \frac{C}{(x+1)^2}$$

$$\Rightarrow \qquad x^2 \equiv A(x+1)^2 + B(x-1)(x+1) + C(x-1).$$

Putting $x = 1$, $\quad 1 = 4A \quad \Rightarrow \quad A = \frac{1}{4}$.

Putting $x = -1$, $\quad 1 = -2C \quad \Rightarrow \quad C = -\frac{1}{2}$.

Equating coefficient x^2, $\quad 1 = A + B \quad \Rightarrow \quad B = \frac{3}{4}$.

$$\text{So} \quad I = \int \left(\frac{\frac{1}{4}}{x-1} + \frac{\frac{3}{4}}{x+1} - \frac{\frac{1}{2}}{(x+1)^2} \right) \mathrm{d}x$$

$$= \tfrac{1}{4} \ln (x-1) + \tfrac{3}{4} \ln (x+1) + \frac{1}{2(x+1)} + c.$$

Example 7

$$I = \int \frac{x^2 \, \mathrm{d}x}{(x-1)(x^2+1)}.$$

We can express this in partial fractions as

$$\frac{x^2}{(x-1)(x^2+1)} \equiv \frac{A}{x-1} + \frac{Bx+C}{x^2+1}.$$

$$\text{So} \qquad x^2 \equiv A(x^2+1) + (Bx+C)(x-1).$$

Putting $x = 1$, $\quad 1 = 2A \quad \Rightarrow \quad A = \frac{1}{2}$.

Equating coefficient x^2, $\quad 1 = A + B \quad \Rightarrow \quad B = \frac{1}{2}$.

Equating constant term, $\quad 0 = A - C \quad \Rightarrow \quad C = \frac{1}{2}$.

$$\text{So} \quad \frac{x^2}{(x-1)(x^2+1)} \equiv \frac{\frac{1}{2}}{x-1} + \frac{\frac{1}{2}x + \frac{1}{2}}{x^2+1}.$$

$$\text{Hence} \quad I = \tfrac{1}{2} \int \frac{\mathrm{d}x}{x-1} + \tfrac{1}{2} \int \frac{x}{x^2+1} \mathrm{d}x + \tfrac{1}{2} \int \frac{\mathrm{d}x}{x^2+1}$$

$$= \tfrac{1}{2} \ln (x-1) + \tfrac{1}{4} \ln (x^2+1) + \tfrac{1}{2} \tan^{-1} x + c.$$

In this last example we encountered a quadratic term $x^2 + 1$ which cannot be split into real linear factors (and which is therefore said to be *irreducible*). As such terms, and their corresponding partial fractions, occur quite frequently, our next example will show how they can best be treated:

Example 8

$$(i) \quad \int \frac{2x-6}{x^2-6x+25} \mathrm{d}x; \qquad (ii) \quad \int \frac{1}{x^2-6x+25} \mathrm{d}x;$$

(*iii*) $\int \frac{x}{x^2 - 6x + 25}\,dx.$

In (*i*) we immediately see that the numerator is the derivative of the denominator.

So $\int \frac{2x - 6}{x^2 - 6x + 25}\,dx = \ln(x^2 - 6x + 25) + c.$

In (*ii*) we first write

$$\int \frac{1}{x^2 - 6x + 25}\,dx = \int \frac{1}{(x - 3)^2 + 16}\,dx.$$

Putting $x - 3 = 4\tan\theta \Rightarrow dx = 4\sec^2\theta\,d\theta$ we obtain

$$\int \frac{1}{x^2 - 6x + 25}\,dx = \int \frac{4\sec^2\theta\,d\theta}{16\tan^2\theta + 16}$$

$$= \tfrac{1}{4}\int d\theta = \tfrac{1}{4}\theta + c.$$

But $x - 3 = 4\tan\theta$

$$\Rightarrow \quad \tan\theta = \frac{x - 3}{4} \quad \Rightarrow \quad \theta = \tan^{-1}\frac{x - 3}{4}.$$

So $\int \frac{1}{x^2 - 6x + 25}\,dx = \frac{1}{4}\tan^{-1}\frac{x - 3}{4} + c.$

In (*iii*) we express the required integral as a linear combination of the two previous parts:

$$\int \frac{x}{x^2 - 6x + 25}\,dx = \int \frac{\frac{1}{2}(2x - 6)}{x^2 - 6x + 25}\,dx + \int \frac{3}{x^2 - 6x + 25}\,dx$$

$$= \tfrac{1}{2}\ln(x^2 - 6x + 25) + \tfrac{3}{4}\tan^{-1}\frac{x - 3}{4} + c.$$

Example 9

$$I = \int \frac{x}{x^3 + 1}\,dx.$$

As $x^3 + 1 \equiv (x + 1)(x^2 - x + 1),$

we can express $x/(x^3 + 1)$ in partial fractions:

$$\frac{x}{x^3 + 1} \equiv \frac{x}{(x + 1)(x^2 - x + 1)} \equiv \frac{A}{x + 1} + \frac{Bx + C}{x^2 - x + 1}.$$

So $x \equiv A(x^2 - x + 1) + (Bx + C)(x + 1).$

Putting $x = -1$, $\quad -1 = A(1 + 1 + 1) \Rightarrow A = -\frac{1}{3}$.

Equating coefficient x^2, $\quad 0 = A + B \Rightarrow B = +\frac{1}{3}$.

Equating constants, $\quad 0 = A + C \Rightarrow C = +\frac{1}{3}$.

So $\dfrac{x}{x^3 + 1} \equiv \dfrac{-\frac{1}{3}}{x + 1} + \dfrac{\frac{1}{3}x + \frac{1}{3}}{x^2 - x + 1}$

$$\Rightarrow \qquad I \equiv -\tfrac{1}{3}\int \frac{1}{x + 1}\,\mathrm{d}x + \tfrac{1}{3}\int \frac{x + 1}{x^2 - x + 1}\,\mathrm{d}x.$$

Now $\displaystyle\int \frac{1}{x + 1}\,\mathrm{d}x = \ln(x + 1)$

and
$$\int \frac{x + 1}{x^2 - x + 1}\,\mathrm{d}x = \int \left(\frac{\frac{1}{2}(2x - 1)}{x^2 - x + 1} + \frac{\frac{3}{2}}{x^2 - x + 1}\right)\mathrm{d}x$$
$$= \tfrac{1}{2}\ln(x^2 - x + 1) + \tfrac{3}{2}\int \frac{\mathrm{d}x}{x^2 - x + 1}.$$

But $\displaystyle\int \frac{\mathrm{d}x}{x^2 - x + 1} = \int \frac{\mathrm{d}x}{(x - \frac{1}{2})^2 + \frac{3}{4}}$

and we now let $x - \frac{1}{2} = \dfrac{\sqrt{3}}{2}\tan\theta \Rightarrow \mathrm{d}x = \dfrac{\sqrt{3}}{2}\sec^2\theta\,\mathrm{d}\theta$.

Hence
$$\int \frac{\mathrm{d}x}{x^2 - x + 1} = \int \frac{\frac{1}{2}\sqrt{3}\sec^2\theta\,\mathrm{d}\theta}{\frac{3}{4}\tan^2\theta + \frac{3}{4}}$$
$$= \frac{2}{\sqrt{3}}\int \frac{\sec^2\theta\,\mathrm{d}\theta}{\tan^2\theta + 1} = \frac{2}{\sqrt{3}}\int \mathrm{d}\theta = \frac{2}{\sqrt{3}}\theta.$$

But $x - \frac{1}{2} = \dfrac{\sqrt{3}}{2}\tan\theta \Rightarrow \dfrac{2x - 1}{\sqrt{3}} = \tan\theta \Rightarrow \theta = \tan^{-1}\dfrac{2x - 1}{\sqrt{3}}$.

So
$$\int \frac{\mathrm{d}x}{x^2 - x + 1} = \frac{2}{\sqrt{3}}\tan^{-1}\frac{2x - 1}{\sqrt{3}}$$

and
$$\int \frac{x\,\mathrm{d}x}{x^3 + 1} = -\tfrac{1}{3}\ln(x + 1) + \tfrac{1}{3}\Big[\tfrac{1}{2}\ln(x^2 - x + 1)$$
$$+ \tfrac{3}{2} \times \frac{2}{\sqrt{3}}\tan^{-1}\frac{2x + 1}{\sqrt{3}}\Big] + c$$
$$= \tfrac{1}{6}\ln\frac{x^2 - x + 1}{(x + 1)^2} + \frac{1}{\sqrt{3}}\tan^{-1}\frac{2x - 1}{\sqrt{3}} + c.$$

Exercise 9.6b

Find the following integrals:

1 (*i*) $\displaystyle\int \frac{\mathrm{d}x}{x^2 - 1}$; (*ii*) $\displaystyle\int \frac{\mathrm{d}x}{x^2 + x}$; (*iii*) $\displaystyle\int \frac{\mathrm{d}x}{2x - x^2}$;

(*iv*) $\int \frac{x^2}{x^2 - 4}\,dx$; (*v*) $\int \frac{dx}{4x^2 - 1}$; (*vi*) $\int \frac{dx}{2x^2 + x - 1}$.

2 (*i*) $\int \frac{dx}{(x + 1)^2}$; (*ii*) $\int \frac{dx}{x^2(x + 2)}$;

(*iii*) $\int \frac{x + 1}{x(x - 1)^2}\,dx$; (*iv*) $\int \frac{x^2\,dx}{(x - 1)^2}$.

3 (*i*) $\int \frac{2x\,dx}{x^2 + 9}$ $\int \frac{dx}{x^2 + 9}$ $\int \frac{x + 3}{x^2 + 9}\,dx$;

(*ii*) $\int \frac{8x\,dx}{4x^2 + 9}$ $\int \frac{dx}{4x^2 + 9}$ $\int \frac{2x - 3}{4x^2 + 9}\,dx$;

(*iii*) $\int \frac{2x - 6}{x^2 - 6x + 10}\,dx$ $\int \frac{dx}{x^2 - 6x + 10}$ $\int \frac{x\,dx}{x^2 - 6x + 10}$.

4 (*i*) $\int \frac{x\,dx}{x - 1}$; (*ii*) $\int \frac{x^2 + 1}{x^2 - 1}\,dx$; (*iii*) $\int \frac{x^3 + 1}{x^2 - 1}\,dx$;

(*iv*) $\int \frac{x\,dx}{(x - 1)^2}$; (*v*) $\int \frac{x^2\,dx}{x^3 + 1}$; (*vi*) $\int \frac{x\,dx}{x^3 - 1}$.

9·7 General integration

In the last section we learned to integrate rational functions by means of partial fractions. We now seek to draw together other methods we have employed for finding integrals.

Faced with a particular function whose integral is required, our primary need is for a very thorough knowledge of differentiation so that we can try to find another function, or 'primitive', whose derivative is the given function. This may not always be possible: for instance, until we invented the function $\ln x$ it was impossible to integrate even such a simple function as $1/x$, and we shall frequently encounter similar examples. Nevertheless, our first step must be to re-state the standard derivatives and the integrals to which they give rise.

$$\frac{d}{dx}(x^n) = nx^{n-1} \qquad \int x^n\,dx = \frac{x^{n+1}}{n + 1} + c \quad (n \neq -1)$$

$$\frac{d}{dx}(\ln x) = \frac{1}{x} \qquad \int \frac{dx}{x} = \ln x + c$$

$$\frac{d}{dx}(e^x) = e^x \qquad \int e^x\,dx = e^x + c$$

$$\frac{d}{dx}(\sin x) = \cos x \qquad \int \cos x\,dx = \sin x + c$$

$$\frac{d}{dx}(\cos x) = -\sin x \qquad \int \sin x \, dx = -\cos x + c$$

$$\frac{d}{dx}(\tan x) = \sec^2 x \qquad \int \sec^2 x \, dx = \tan x + c$$

$$\frac{d}{dx}(\sin^{-1} x) = \frac{1}{\sqrt{(1 - x^2)}} \qquad \int \frac{dx}{\sqrt{(1 - x^2)}} = \sin^{-1} x + c$$

$$\frac{d}{dx}(\tan^{-1} x) = \frac{1}{1 + x^2} \qquad \int \frac{dx}{1 + x^2} = \tan^{-1} x + c$$

These standard integrals frequently need to be used in conjunction with general methods such as that of substitution or integration by parts and, although these have been investigated in previous sections, considerable practice is necessary in order to sense what is the most appropriate method for each particular example. We have already stressed the crucial importance of integration in finding areas and volumes, and it will soon be seen that it is similarly central in statistics, mechanics, and throughout all branches of applied mathematics.

Example 1

$$I = \int \sqrt{(a^2 - x^2)} \, dx.$$

Here a substitution converts the integral into another which is more tractable,

for $x = a \sin \theta \Rightarrow dx = a \cos \theta \, d\theta$.

$$\begin{aligned} \text{So} \quad I &= \int \sqrt{(a^2 - a^2 \sin^2 \theta)} \, a \cos \theta \, d\theta \\ &= a^2 \int \cos^2 \theta \, d\theta \\ &= \tfrac{1}{2}a^2 \int (1 + \cos 2\theta) \, d\theta \\ &= \tfrac{1}{2}a^2(\theta + \tfrac{1}{2} \sin 2\theta) + c \\ &= \tfrac{1}{2}a^2(\theta + \sin \theta \cos \theta) + c. \end{aligned}$$

Finally, however, it is necessary to revert to the original variable x.

$$\text{So} \quad \int \sqrt{(a^2 - x^2)} \, dx = \tfrac{1}{2}a^2 \sin^{-1} x + \tfrac{1}{2}x\sqrt{(a^2 - x^2)} + c.$$

Example 2

$$I = \int \tan^{-1} x \, dx.$$

Because $\tan^{-1} x$ is a function which we can easily differentiate, there is an immediate attraction in using the method of integration by parts:

$$I = \int 1 \tan^{-1} x \, dx$$

$$= x \tan^{-1} x - \int x \frac{1}{1 + x^2} \, dx.$$

So $$I = x \tan^{-1} x - \tfrac{1}{2} \int \frac{2x}{1 + x^2} \, dx$$

$$= x \tan^{-1} x - \tfrac{1}{2} \ln (1 + x^2) + c.$$

Example 3

$$I = \int \operatorname{cosec} x \, dx.$$

It may be recalled that $\sin x$, $\cos x$, and $\tan x$ are easily expressible in terms of $\tan \frac{1}{2}x$:

$$t = \tan \tfrac{1}{2}x$$

$$\Rightarrow \quad \sin x = \frac{2t}{1 + t^2}, \qquad \cos x = \frac{1 - t^2}{1 + t^2}, \qquad \tan x = \frac{2t}{1 - t^2}.$$

So the substitution

$$t = \tan \tfrac{1}{2}x$$

is sometimes particularly convenient for simplifying trigonometric integrals.

Now $$t = \tan \frac{x}{2}$$

$$\Rightarrow \quad dt = \tfrac{1}{2} \sec^2 \tfrac{1}{2}x \, dx = \tfrac{1}{2}(1 + t^2) \, dx$$

$$\Rightarrow \quad dx = \frac{2 \, dt}{1 + t^2}.$$

So $$I = \int \frac{dx}{\sin x}$$

$$= \int \frac{2 \, dt/(1 + t^2)}{2t/(1 + t^2)} = \int \frac{dt}{t} = \ln t + c$$

$$= \ln \tan \tfrac{1}{2}x + c.$$

Hence $$\int \operatorname{cosec} x \, dx = \ln \tan \tfrac{1}{2}x + c$$

and it is quickly verified that

$$\frac{d}{dx}(\ln \tan \tfrac{1}{2}x) = \frac{1}{\tan \frac{1}{2}x}(\sec^2 \tfrac{1}{2}x) \times \tfrac{1}{2}$$

$$= \frac{1}{2 \sin \frac{1}{2}x \cos \frac{1}{2}x} = \frac{1}{\sin x} = \operatorname{cosec} x.$$

To complete the catalogue of the integrals of elementary trigonometric functions, we recall that

$$\int \tan x \, dx = \int \frac{\sin x}{\cos x} \, dx$$

$$= -\log \cos x + c$$

$$= \log \sec x + c$$

and $$\int \sec x \, dx = \int \frac{\sec x (\sec x + \tan x)}{\sec x + \tan x} \, dx$$

$$= \int \frac{\sec x \tan x + \sec^2 x}{\sec x + \tan x} \, dx$$

$$= \ln (\sec x + \tan x) + c.$$

Finally, we give two examples which are quickly dealt with by a preliminary rearrangement:

Example 4

$$I = \int \tan^4 x \, dx.$$

Now $$I = \int (\tan^2 x)(\tan^2 x) \, dx$$

$$= \int \tan^2 x (\sec^2 x - 1) \, dx$$

$$= \int (\tan^2 x \sec^2 x - \sec^2 x + 1) \, dx$$

$$= \tfrac{1}{3} \tan^3 x - \tan x + x + c.$$

Example 5

$$I = \int \frac{\sin x}{\sin x + \cos x} \, dx.$$

We first notice that we could easily find

$$\int \frac{\sin x + \cos x}{\sin x + \cos x} \, dx = \int dx = x$$

and $$\int \frac{\cos x - \sin x}{\sin x + \cos x} \, dx = \ln (\sin x + \cos x).$$

So $\displaystyle\int \frac{\sin x}{\sin x + \cos x}\,dx = \frac{1}{2}\int \frac{\sin x + \cos x}{\sin x + \cos x}\,dx - \frac{1}{2}\int \frac{\cos x - \sin x}{\sin x + \cos x}\,dx$

$$= \tfrac{1}{2}x - \tfrac{1}{2}\ln(\sin x + \cos x) + c.$$

Exercise 9.7

Find the following integrals:

1 (*i*) $\int 5x^4\,dx$; (*ii*) $\int t^6\,dt$; (*iii*) $\int \frac{3\,du}{u^2}$;

(*iv*) $\int \frac{dx}{2\sqrt{x}}$; (*v*) $\int \sqrt{v}\,dv$; (*vi*) $\int (x+1)^4\,dx$;

(*vii*) $\int \frac{du}{(2u+1)^3}$; (*viii*) $\int \frac{dv}{\sqrt{(3v-2)}}$; (*ix*) $\int x(x^2-1)^5\,dx$;

(*x*) $\int \frac{x^2\,dx}{(1-x^3)^2}$; (*xi*) $\int x\sqrt{(x^2+1)}\,dx$; (*xii*) $\int \frac{x\,dx}{\sqrt{(1-x^2)}}$.

2 (*i*) $\int \frac{2}{x}\,dx$; (*ii*) $\int \frac{3}{x-3}\,dx$; (*iii*) $\int \frac{du}{1-u}$;

(*iv*) $\int \frac{v\,dv}{v^2-1}$; (*v*) $\int \frac{x^2\,dx}{1-x^3}$; (*vi*) $\int \frac{2x+1}{x^2+x}\,dx$;

(*vii*) $\int e^{2x}\,dx$; (*viii*) $\int 2\,e^{3x}\,dx$; (*ix*) $\int x\,e^{-x^2}\,dx$;

(*x*) $\int \frac{e^u}{1+e^u}\,du$; (*xi*) $\int 10^t\,dt$; (*xii*) $\int \frac{\ln x}{x}\,dx$.

3 (*i*) $\int \cos 2x\,dx$; (*ii*) $\int \sin \tfrac{1}{2}x\,dx$; (*iii*) $\int \sin^2\theta\cos\theta\,d\theta$;

(*iv*) $\int \cos^2\theta\sin\theta\,d\theta$; (*v*) $\int \cos^3\theta\,d\theta$; (*vi*) $\int \sin^3\theta\,d\theta$;

(*vii*) $\int \sin x\cos x\,dx$; (*viii*) $\int \sin^2 \tfrac{1}{2}x\,dx$; (*ix*) $\int \cos^2 x\,dx$;

(*x*) $\int \frac{\cos x}{\sqrt{(\sin x)}}\,dx$; (*xi*) $\int \frac{\cos\theta}{\sin\theta}\,d\theta$; (*xii*) $\int \frac{\sin\theta}{\cos\theta}\,d\theta$.

4 (*i*) $\int \sec^2 x\,dx$; (*ii*) $\int \tan^2 x\,dx$;

(*iii*) $\int \sec\theta\tan\theta\,d\theta$; (*iv*) $\int \operatorname{cosec}\theta\cot\theta\,d\theta$;

(*v*) $\int \operatorname{cosec}^2\theta\,d\theta$; (*vi*) $\int \cot^2\theta\,d\theta$;

(*vii*) $\int \tan\theta \sec^2\theta \, d\theta$; (*viii*) $\int \sec^3\theta \tan\theta \, d\theta$;

(*ix*) $\int \tan 2x \, dx$; (*x*) $\int \tan^3 x \, dx$;

(*xi*) $\int \cot \tfrac{1}{2}x \, dx$; (*xii*) $\int \cot^3 x \, dx$.

5 (*i*) $\int \frac{dx}{\sqrt{(4 - x^2)}}$ $(x = 2\sin\theta)$;

(*ii*) $\int \frac{dx}{\sqrt{(9 - 4x^2)}}$ $(2x = 3\sin\theta)$;

(*iii*) $\int \frac{dx}{1 + 4x^2}$ $(2x = \tan\theta)$;

(*iv*) $\int \frac{dx}{4 + 9x^2}$ $(3x = 2\tan\theta)$;

(*v*) $\int \frac{du}{\sqrt{(u^2 + 1)}}$ $(u = \tan\theta)$;

(*vi*) $\int \frac{du}{\sqrt{(u^2 - 1)}}$ $(u = \sec\theta)$;

(*vii*) $\int \sqrt{(1 - x^2)} \, dx$ $(x = \sin\theta)$;

(*viii*) $\int x\sqrt{(1 - x^2)} \, dx$;

(*ix*) $\int \frac{dx}{4x^2 + 9}$; (*x*) $\int \frac{x \, dx}{4x^2 + 9}$;

6 (*i*) $\int x \cos x \, dx$; (*ii*) $\int x \, e^{-x} \, dx$; (*iii*) $\int t \sin 3t \, dt$;

(*iv*) $\int x^2 \sin x \, dx$; (*v*) $\int \sin^{-1} x \, dx$; (*vi*) $\int \ln x \, dx$;

(*vii*) $\int \theta \sec^2\theta \, d\theta$; (*viii*) $\int x^2 e^{2x} \, dx$; (*ix*) $\int e^{-t} \sin t \, dt$;

(*x*) $\int e^{2t} \cos 3t \, dt$; (*xi*) $\int x^2 \ln x \, dx$; (*xii*) $\int x \, 10^x \, dx$.

7 Use the substitution $t = \tan \tfrac{1}{2}x$ to find:

(*i*) $\int_0^{\frac{1}{2}\pi} \frac{dx}{1 + \sin x}$; (*ii*) $\int_0^{\frac{1}{2}\pi} \frac{dx}{1 + \sin x + \cos x}$; (*iii*) $\int_0^{\frac{1}{2}\pi} \frac{dx}{\sin x + \cos x}$.

8 Show that

$$\int_0^a f(x)\,dx = \int_0^a f(a-x)\,dx$$

and hence (using no. 7) evaluate:

(*i*) $\displaystyle\int_0^{\pi} \frac{x\,dx}{1+\sin x}$; (*ii*) $\displaystyle\int_0^{\frac{1}{2}\pi} \frac{x\,dx}{1+\sin x+\cos x}$; (*iii*) $\displaystyle\int_0^{\frac{1}{2}\pi} \frac{x\,dx}{\sin x+\cos x}$.

9 Find the following integrals:

(*i*) $\displaystyle\int \frac{dx}{x^2+x}$; (*ii*) $\displaystyle\int \frac{dx}{x^2-x}$; (*iii*) $\displaystyle\int \frac{dx}{x^3-x^2}$;

(*iv*) $\displaystyle\int \frac{dx}{x^3+x}$; (*v*) $\displaystyle\int \frac{e^x}{1+e^x}\,dx$; (*vi*) $\displaystyle\int \frac{dx}{1+e^x}$;

(*vii*) $\displaystyle\int \frac{x\,dx}{x^4+1}$; (*viii*) $\displaystyle\int (\ln x)^2\,dx$; (*ix*) $\displaystyle\int x^2 \tan^{-1} x\,dx$;

(*x*) $\displaystyle\int \frac{dx}{2+\cos x}$; (*xi*) $\displaystyle\int \frac{dx}{x\sqrt{(x^2-1)}}$; (*xii*) $\displaystyle\int \sqrt{\left(\frac{1-x}{1+x}\right)}\,dx$.

10 Evaluate:

(*i*) $\displaystyle\int_0^1 \frac{dx}{(x+1)(x^2+1)}$; (*ii*) $\displaystyle\int_0^1 \frac{x\,dx}{(x^2+1)^2}$; (*iii*) $\displaystyle\int_0^1 \frac{dx}{(x^2+1)^2}$;

(*iv*) $\displaystyle\int_0^1 \frac{dx}{(x^2+1)^{3/2}}$; (*v*) $\displaystyle\int_0^{\frac{1}{2}\pi} \sin^5\theta\cos^2\theta\,d\theta$;

(*vi*) $\displaystyle\int_0^{\frac{1}{2}\pi} \sin^2\theta\cos^6\theta\,d\theta$; (*vii*) $\displaystyle\int_1^e x^2 \ln x\,dx$;

(*viii*) $\displaystyle\int_0^1 \sqrt{(4-x^2)}\,dx$; (*ix*) $\displaystyle\int_0^{\frac{1}{2}\pi} \frac{d\theta}{2+\cos\theta}$;

(*x*) $\displaystyle\int_0^{\frac{1}{4}\pi} \frac{\cos\theta-\sin\theta}{\cos\theta+\sin\theta}\,d\theta$;

(*xi*) $\displaystyle\int_{-\frac{1}{4}\pi}^{\frac{1}{4}\pi} \frac{d\theta}{3\sin^2\theta+\cos^2\theta}$ (put $t=\tan\theta$);

(*xii*) $\displaystyle\int_2^3 \sqrt{\left(\frac{x-2}{4-x}\right)}\,dx$ (put $x = 2\sin^2\theta + 4\cos^2\theta$).

9.8 Differential equations

We have seen that the process of integration is the reverse of differentiation and so requires the discovery of a function having the given rate of change. Frequently, however, we do not know either the function or its rate of change, but merely how they are related.

When, for instance in 9.2, we were investigating the cooling of bathwater, we simply knew that after time t the excess temperature θ and its rate of decrease $-\dfrac{d\theta}{dt}$ were connected by the equation

$$\frac{d\theta}{dt} = -k\theta \tag{1}$$

and in the same section it was mentioned that the current x flowing in a particular electrical circuit obeyed the equation

$$\frac{d^2x}{dt^2} + 2\frac{dx}{dt} + 5x = 0. \tag{2}$$

Similarly the rate of growth of the mass m of an organism frequently depends on the equation

$$\frac{dm}{dt} = km, \tag{3}$$

and the rate of spread of a disease in a population, of which a fraction x is infected, is often governed by the equation

$$\frac{dx}{dt} = kx(1 - x). \tag{4}$$

Any such equation which connects the rates of change of variables with their values is called a *differential equation*. Equations (1), (3), and (4) are differential equations *of the first order* and equation (2) is *of the second order*, so called from the degree of their highest derivatives.

In particular, such a first-order equation as

$$\frac{dy}{dx} = x^2$$

immediately leads to

$$y = \tfrac{1}{3}x^3 + c.$$

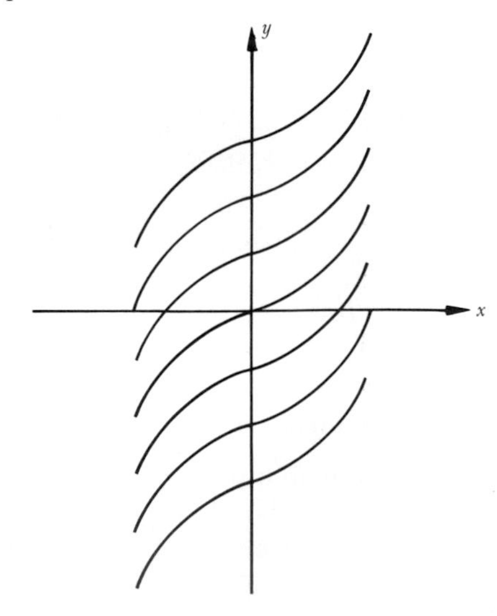

This is called its general solution; and by giving c a particular value we can obtain particular solutions, such as

$$y = \tfrac{1}{3}x^3 + 2 \quad \text{or} \quad y = \tfrac{1}{3}x^3 - 4.$$

Graphically, we see that the *general solution* is represented by a *family of curves*, and that each *particular solution* is an *individual member* of this family.

Two further examples will illustrate this relationship.

Example 1

If $x^2 + y^2 = a^2$,

form a differential equation independent of a.

As a varies, the equation

$$x^2 + y^2 = a^2$$

represents a family of circles, all with centre at the origin.

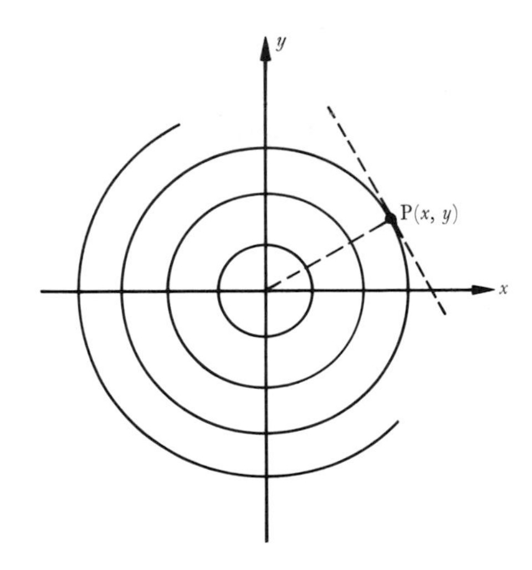

But $\quad x^2 + y^2 = a^2$

$$\Rightarrow \quad 2x + 2y\frac{dy}{dx} = 0$$

$$\Rightarrow \quad x + y\frac{dy}{dx} = 0.$$

This is the required differential equation, which can be written as

$$\frac{y}{x}\frac{dy}{dx} = -1;$$

and since the gradient of OP is y/x, this equation is simply expressing the fact that the tangent to each circle at every point is always perpendicular to the radius.

Example 2

A point P is moving along a straight line and after time t its displacement x is given by

$$x = A \sin 2t + B \cos 2t.$$

Form a differential equation which is independent of A and B.

As we here need to eliminate two arbitrary constants, we shall have to differentiate twice and form a second order differential equation.

Now $x = A \sin 2t + B \cos 2t$

$$\Rightarrow \quad \frac{dx}{dt} = 2A \cos 2t - 2B \sin 2t$$

$$\Rightarrow \quad \frac{d^2x}{dt^2} = -4A \sin 2t - 4B \cos 2t.$$

So $\dfrac{d^2x}{dt^2} = -4x.$

This, therefore, is the required equation, showing that for all values of A and B the acceleration is always $-4x$, and we see that

$$x = A \sin 2t + B \cos 2t$$

is a general solution of this equation.

More frequently, however, we start with the differential equation and need to find appropriate solutions. In this case, for instance, the equation

$$\frac{d^2x}{dt^2} = -4x$$

might have arisen from investigation of a buoy bobbing up and down in the sea, and our task would generally be to show that this leads to a solution of the form

$$x = A \sin 2t + B \cos 2t,$$

representing a periodic oscillation.

Exercise 9.8a

1 Verify that the following differential equations have the solutions indicated (where A, B, ε are arbitrary constants):

(*i*) $\dfrac{dx}{dt} = 3x;\ x = A\,e^{3t}.$

(*ii*) $x\dfrac{dy}{dx} = 2y;\ y = Ax^2.$

(*iii*) $y\dfrac{dy}{dx} = x;\ x^2 - y^2 = A.$

(*iv*) $\dfrac{dx}{dt} = x(1 - x);\ x = \dfrac{1}{1 + A\,e^{-t}}.$

(*v*) $\dfrac{d^2x}{dt^2} + 3\dfrac{dx}{dt} + 2x = 0;\ x = A\,e^{-t} + B\,e^{-2t}.$

(*vi*) $\dfrac{d^2x}{dt^2} + 2\dfrac{dx}{dt} + 5x = 1;\ x = A\,e^{-t}\sin(2t + \varepsilon) + \frac{1}{5}.$

2 Find differential equations which are satisfied at all points of all members of the following families of curves (where A, B, a, c, etc. define the individual members of the families):

(*i*) $xy = c$; (*ii*) $y = A\,e^x$; (*iii*) $y^2 = ax$;
(*iv*) $x^2 - y^2 = a^2$; (*v*) $y = Ax^2 + Bx$; (*vi*) $x = A\,e^t + B\,e^{-t}$.

We have now begun to see the relationship between a differential equation and its solutions. As with algebraic equations, the task of the mathematician is usually twofold:
(*i*) to express a given situation—whether mechanical, electrical, chemical, biological, economic, or whatever—in a suitable mathematical form (usually called a *model*), which very frequently leads to a differential equation; and,
(*ii*) to find the appropriate solution of such a differential equation. Very often this has to be done numerically, but we shall begin to discuss the problem by taking three special types of differential equation for which it is usually possible to find a general solution.

First order equations with separable variables

A typical first order equation is

$$\frac{dy}{dx} = f(x, y),$$

and if $f(x, y)$ is defined at every point of the x–y plane, it follows that at each point a corresponding gradient and direction are known, just as in a magnetic field plotted by a compass needle.

Example 3

$$\frac{dy}{dx} = \frac{x}{y}.$$

The values of the gradient $\frac{dy}{dx}$ can be indicated by a table:

	x →	−3	−2	−1	0	1	2	3
y	3	−1	$-\frac{2}{3}$	$-\frac{1}{3}$	0	$\frac{1}{3}$	$\frac{2}{3}$	1
	2	$-\frac{3}{2}$	−1	$-\frac{1}{2}$	0	$\frac{1}{2}$	1	$\frac{3}{2}$
	1	−3	−2	−1	0	1	2	3
	0							
	−1	3	2	1	0	−1	−2	−3
	−2	$\frac{3}{2}$	1	$\frac{1}{2}$	0	$-\frac{1}{2}$	−1	$-\frac{3}{2}$
	−3	1	$\frac{2}{3}$	$\frac{1}{3}$	0	$-\frac{1}{3}$	$-\frac{2}{3}$	−1

These can then be plotted at the different points of the plane and already we can begin to see the family of solutions, looking like the lines of force of a magnetic field.

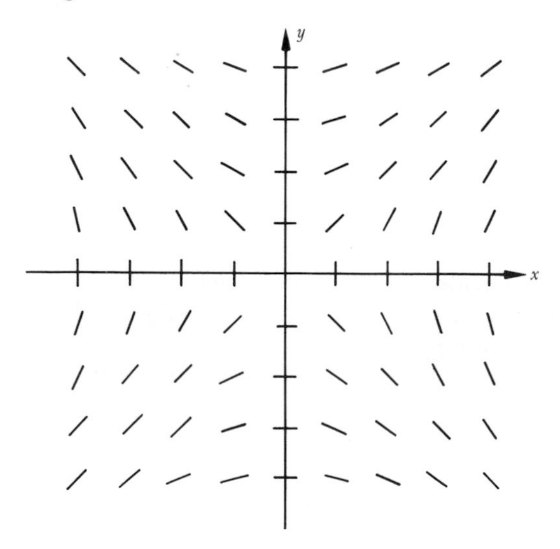

But we can also approach the problem quite differently. For the equation $\frac{dy}{dx} = \frac{x}{y}$ can be written as $y\frac{dy}{dx} = x$

$$\Rightarrow \quad \int y\frac{dy}{dx}\,dx = \int x\,dx$$

$$\Rightarrow \qquad \int y\,\mathrm{d}y = \int x\,\mathrm{d}x$$

$$\Rightarrow \qquad \tfrac{1}{2}y^2 = \tfrac{1}{2}x^2 + c$$

$$\Rightarrow \qquad y^2 = x^2 + 2c.$$

This, therefore, is the *general solution* of the given equation; and we can obtain different *particular solutions* by giving different values to the constant c.

Graphically, we see that each particular solution is a certain curve (in fact a rectangular hyperbola, each one having two branches), and the general solution represents a family of such hyperbolas. Finally, the sketch of these hyperbolas should be compared with our original 'compass-needle' figure.

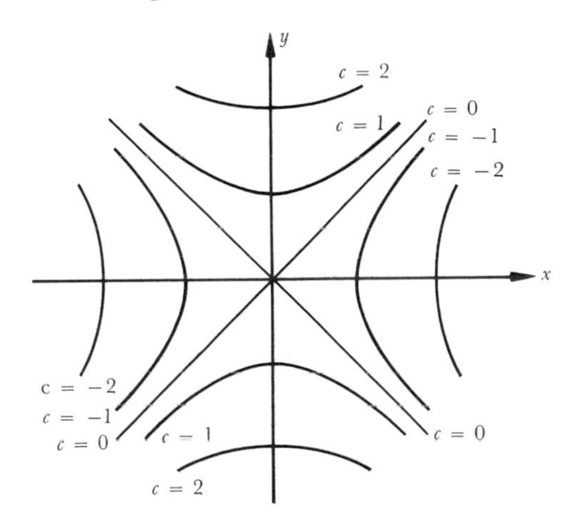

$$\begin{aligned}
c &= 0 &\Rightarrow\quad y^2 &= x^2\\
c &= 1 &\Rightarrow\quad y^2 &= x^2 + 2\\
c &= 2 &\Rightarrow\quad y^2 &= x^2 + 4\\
c &= -1 &\Rightarrow\quad y^2 &= x^2 - 2\\
c &= -2 &\Rightarrow\quad y^2 &= x^2 - 4, \text{ etc.}
\end{aligned}$$

In the above solution we could have proceeded a little more rapidly by saying that

$$\frac{\mathrm{d}y}{\mathrm{d}x} = \frac{x}{y} \tag{1}$$

can be written $\quad y\,\mathrm{d}y = x\,\mathrm{d}x \qquad$ (1a)

and then as $\quad \displaystyle\int y\,\mathrm{d}y = \int x\,\mathrm{d}x \qquad$ (2)

$$\Rightarrow \qquad \tfrac{1}{2}y^2 = \tfrac{1}{2}x^2 + c$$

$$\Rightarrow \qquad y^2 = x^2 + 2c.$$

Although the statement (1a) has not (as yet) any proper meaning, we have already justified the argument from (1) to (2), and (1a) clearly serves as a very useful intermediary. This is usually known as *separating the variables*: any first order equation which can similarly be expressed as

$$f(y)\,\mathrm{d}y = g(x)\,\mathrm{d}x$$

is said to have *separable variables*, and its general solution can be written as

$$\int f(y)\,\mathrm{d}y = \int g(x)\,\mathrm{d}x.$$

Example 4

Solve the differential equation

$$\frac{\mathrm{d}y}{\mathrm{d}x} = -\frac{y}{x}.$$

Here again, the variables are separable and we can write

$$\frac{\mathrm{d}y}{y} = -\frac{\mathrm{d}x}{x}$$

$$\Rightarrow \quad \ln y = -\ln x + c$$

$$\Rightarrow \quad \ln xy = c$$

$$\Rightarrow \quad xy = a \quad (\text{where } a = e^c).$$

The solution-curves are therefore:

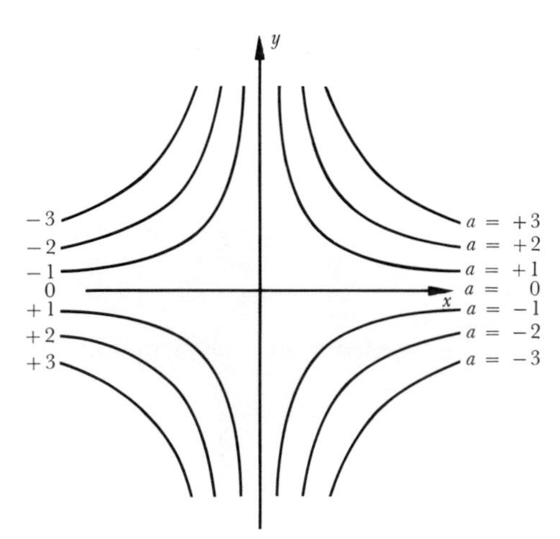

Lastly, we see that there is a relationship between the two examples we have chosen,

$$\frac{\mathrm{d}y}{\mathrm{d}x} = \frac{x}{y} \quad \text{and} \quad \frac{\mathrm{d}y}{\mathrm{d}x} = -\frac{y}{x}.$$

Since $x/y \times y/x = -1$, it is clear that the directions at any point of the solution-curves of the two equations must be perpendicular, and this is very clearly seen when the figures are superimposed:

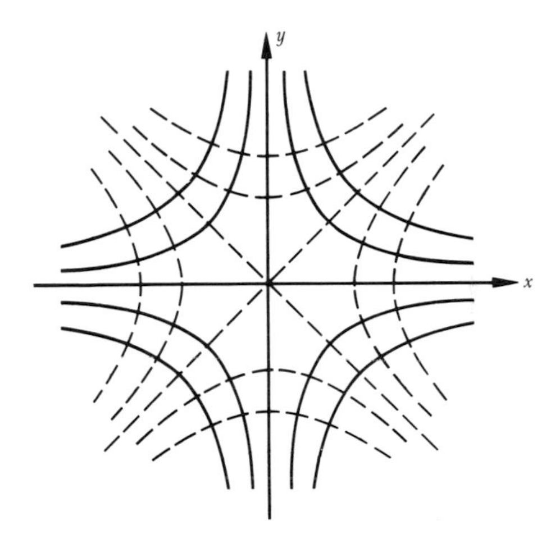

Such pairs of families of curves which are perpendicular everywhere they meet are called *orthogonal families* and are outstandingly important not only as *lines of forces* and their corresponding *equipotentials*, but throughout applied mathematics. There is indeed a particularly simple illustration of them on a map as the *contours* and the *lines of greatest slope* of a hillside, and this last diagram can be regarded as the map of a saddle-point with rainfall streaming from the hillsides in the North-East and South-West into the valleys in the North-West and South-East:

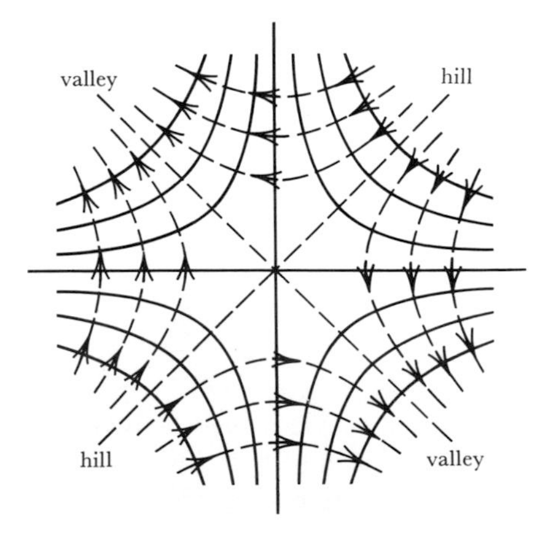

Example 5

A stone falls vertically from rest against an air resistance which is proportional to its velocity v, so that its acceleration is $g - kv$, where g and k are constants.

(*i*) Write down differential equations connecting v with t, the time; and v with x, the distance fallen.
(*ii*) Find expressions for v in terms of t and x in terms of t.
(*iii*) Find the terminal velocity, as $t \to \infty$.

(*i*) The downwards acceleration

$$= \frac{dv}{dt} = \frac{dx}{dt} \times \frac{dv}{dx} = v\frac{dv}{dx}.$$

So $$\frac{dv}{dt} = g - kv \tag{1}$$

and $$v\frac{dv}{dx} = g - kv. \tag{2}$$

(*ii*) From (1), we see that

$$\frac{dv}{g - kv} = dt \quad \Rightarrow \quad -\frac{1}{k}\ln(g - kv) = t + c.$$

But when $t = 0$, $v = 0$; so $c = -\frac{1}{k}\ln g$.

Hence $$-t = \frac{1}{k}\ln(g - kv) - \frac{1}{k}\ln g = \frac{1}{k}\ln\frac{g - kv}{g}$$

$$\Rightarrow \quad \ln\left(1 - \frac{kv}{g}\right) = -kt \quad \Rightarrow \quad 1 - \frac{kv}{g} = e^{-kt}$$

$$\Rightarrow \quad v = \frac{g}{k}(1 - e^{-kt}). \tag{3}$$

Hence $$\frac{dx}{dt} = \frac{g}{k}(1 - e^{-kt})$$

$$\Rightarrow \quad x = \frac{g}{k}\left(t + \frac{1}{k}e^{-kt}\right) + d.$$

But when $t = 0$, $x = 0$; so $d = -\frac{g}{k^2}$.

Hence $$x = \frac{gt}{k} + \frac{g}{k^2}(e^{-kt} - 1). \tag{4}$$

Finally, we see from (3) that as $t \to \infty$, $v \to g/k$, which is therefore called its terminal velocity. (And the same value can be found more simply from equation (1) if we suppose that the acceleration eventually tends to zero.)

Example 6

The rate at which a tree takes in its 'food' is proportional to the area of its rooting surface. The food is used
(*i*) to maintain the tree — this requires an amount proportional to its volume, and
(*ii*) for it to grow — this requires an amount proportional to the rate of increase of volume.

If the complete tree, including roots, retains a constant shape show that its height h satisfies a differential equation of the form

$$\frac{dh}{dt} + Ah = B$$

and that the tree will tend to a maximum height.

A tree is planted when it is 2 m high and its initial rate of growth in height is 0.3 m per year. Its final height is 100 m. How long does it take to reach a height of 50 m? (M.E.I.)

This example shows how a so-called 'mathematical model' can be set up to describe a given situation; whether or not the model is valid can then be tested from the accuracy of its predictions.

Suppose that after time t the volume of the tree is V and the area of its rooting surface is S. Then, since the tree retains a constant shape,

$$S = ah^2, \qquad V = bh^3 \quad \text{(where } a, b \text{ are constants)}.$$

Also the rate at which it takes in its food is kS.

$$kS = cV + d\frac{dV}{dt} \quad \text{(where } k, c, d \text{ are constants)}$$

$$\Rightarrow \quad kah^2 = cbh^3 + d \times 3bh^2\frac{dh}{dt}$$

$$\Rightarrow \quad ka = bch + 3bd\frac{dh}{dt}$$

$$\Rightarrow \quad \frac{dh}{dt} + Ah = B. \tag{1}$$

When $t = 0$, $h = 2$ and $\dfrac{dh}{dt} = 0.3$, so

$$0.3 + 2A = B. \tag{2}$$

Also, as the tree nears its final height, $h \to 100$ and $\dfrac{dh}{dt} \to 0$; so

$$100A = B. \tag{3}$$

From (2) and (3), $A = 0.3/98$ and $B = \frac{30}{98}$.

Hence $$\frac{dh}{dt} + \frac{0.3}{98}h = \frac{30}{98}$$

$$\Rightarrow \quad 98\frac{dh}{dt} = 0.3\ (100 - h)$$

$$\Rightarrow \quad \frac{98\ dh}{100 - h} = 0.3\ dt$$

$$\Rightarrow \quad -98 \ln (100 - h) = 0.3t + A.$$

But when $t = 0$, $h = 2$; so

$$-98 \ln 98 = A$$

$$\Rightarrow \quad 0.3t = 98 \ln \frac{98}{100 - h}. \tag{4}$$

Putting $h = 50$,

$$0.3t = 98 \ln \tfrac{98}{50}$$

$$\Rightarrow \quad t = \frac{98}{0.3} \ln 1.96$$

$$= \frac{98 \times 0.672\,9}{0.3} \approx 220.$$

So the tree would take just over 220 years to grow to a height of 50 m.

Moreover, from (4) we can obtain a more general formula for its height h after t years.

For $$\ln \frac{100 - h}{98} = -\frac{0.3t}{98}$$

$$\Rightarrow \quad \frac{100 - h}{98} = e^{-0.3t/98}$$

$$\Rightarrow \quad h = 100 - 98\ e^{-0.3t/98}$$

$$\Rightarrow \quad h = 100 - 98\ e^{-0.00306t}.$$

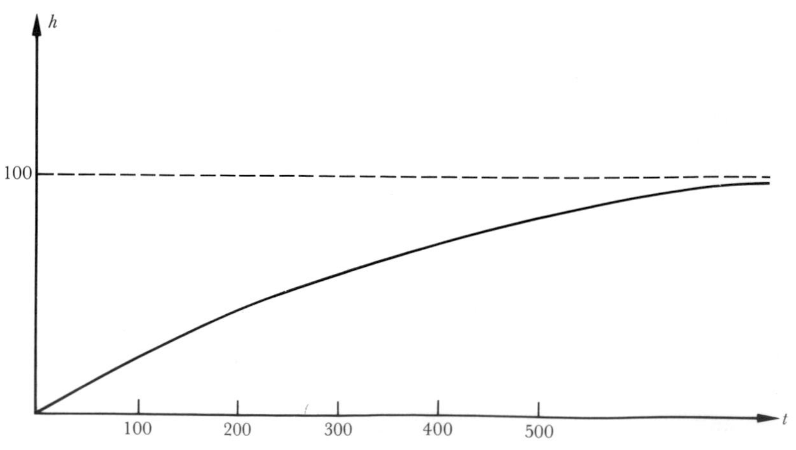

Exercise 9.8b

1 For each of the following equations,
(*a*) illustrate the equation by means of a compass-needle diagram;
(*b*) find and illustrate its general solution:

(*i*) $\dfrac{dy}{dx} = y$; (*ii*) $\dfrac{dy}{dx} = -xy$; (*iii*) $\dfrac{dy}{dx} = 2x$;

(*iv*) $\dfrac{dy}{dx} = -y^2$; (*v*) $\dfrac{dy}{dx} = \dfrac{y}{x}$; (*vi*) $\dfrac{dy}{dx} = -\dfrac{2x}{y}$.

2 For each of the following families of curves,
(*a*) eliminate the parameter c to form a differential equation which represents the family (i.e., is true at every point of every curve);
(*b*) illustrate by means of a compass-needle diagram;
(*c*) find the differential equation representing the orthogonal family;
(*d*) find and illustrate the general solution of this equation:

(*i*) $y = cx$; (*ii*) $y = x + c$; (*iii*) $y = cx^2$;

(*iv*) $y^2 = x + c$; (*v*) $y = c\,e^{-x}$; (*vi*) $y = \dfrac{c}{x^2}$.

3 (*i*) Solve the differential equation

$$\frac{dy}{dx} = \tan y \cot x,$$

given that, when $x = \frac{1}{2}\pi$, $y = \frac{1}{6}\pi$. (O.C.)

(*ii*) Find the solution of the differential equation

$$x\frac{dy}{dx} = \sin y$$

(for $x > 0$) which passes through the point $(2, \frac{1}{2}\pi)$. (S.M.P.)

4 The size S of a population at time t satisfies approximately the differential equation $\dfrac{dS}{dt} = kS$, where k is a constant. Integrate this equation to find S as a function of t.

The population numbered 32 000 in the year 1900 and had increased to 48 000 by 1970. Estimate what its size will be (correct to the nearest 1 000) in the year 2000. (S.M.P.)

5 If a ship's mooring rope is wrapped round a rough bollard, with coefficient of friction μ, and is on the point of slipping, it can be shown that its tension T an angle θ from its taut end is such that

$$\frac{dT}{d\theta} = -\mu T.$$

Deduce that

$$T = T_0\, e^{-\mu\theta},$$

where T_0 is the tension from the ship. If $\mu = \frac{1}{2}$, show that one complete turn will reduce this tension in the ratio $1:e^{\pi}$ (or approximately $1:23$).

6 If the rate of erosion of mass of a spherical pebble is proportional to its surface area and it is half-eroded after 100 years, how much longer will it last?

7 Show that the acceleration $\dfrac{dv}{dt}$ of a particle moving in a straight line can be written, in terms of its velocity v and its displacement x from a point of the line, in the form $v\dfrac{dv}{dx}$.

At a distance x km from the centre of the Earth the gravitational acceleration in km s^{-2} is given by the formula c/x^2 where $c = 4 \times 10^5$. If a lunar vehicle 10 000 km from the centre of the Earth is moving directly away from it at a speed of 10 km s^{-1}, at what distance will its speed be half that value? (S.M.P.)

8 A rumour is spreading through a large city at a rate which is proportional to the product of the numbers of those who have heard it and of those who have not heard it, so that if x is the fraction who have heard it after time t,

$$\frac{dx}{dt} = kx(1 - x).$$

If initially a fraction c of the population has heard the rumour, deduce that

$$x = \frac{c}{c + (1 - c)\, e^{-kt}}.$$

If 10% have heard the rumour at noon and another 10% by 3.00 p.m., find x as a function of t. What further proportion would you expect to have heard it by 6.00 p.m.?

9 The inhabitants of a country are in two racial groups which do not mix reproductively. Each group's population increases at a rate proportional to the population. At a certain stage 80% are of type A, type B will double in 50 years, but the whole population will take 100 years to double. How long will it take group A to double? How long will it take before the two groups are equal in numbers? (M.E.I.)

10 Heat is supplied to an object at a constant rate Wa joules s^{-1}; heat is lost to its surroundings at a rate of $bW\theta$ joules s^{-1}, where W is the number of joules needed to raise the temperature of the object by 1 °C, θ °C is the difference in temperature between the object and its surroundings, and a, b are constants. Obtain a differential equation to connect θ with the time t.

Use this mathematical model to solve the following problem. An electric kettle, open to a room temperature of 20 °C, reaches 70 °C in 3 min after being switched on, and reaches boiling-point in a further 3 min. Find the time taken to reach boiling-point from 20 °C if it is completely heat insulated.

What deficiencies has the mathematical model in representing the actual physical situation? (M.E.I.)

11 The food energy taken in by a human body goes partly to increase the mass and partly to fulfil the requirements of the body; these daily requirements are taken to be proportional to the mass M. The rate of increase of mass is proportional to the number of joules available for this. Write down a differential equation connecting M, the time t and the daily intake of joules $f(t)$.

A man's mass is 100 kg; if he took in no energy he would reduce his weight by 10% in 10 days. How long would it take him to reduce by this amount if, instead, he took in exactly half the number of joules needed to keep his mass constant at 100 kg. (M.E.I.)

12 A stone is thrown vertically upwards with velocity u, and air resistance is proportional to the square of its speed ($= kv^2$ per unit mass).

Find:

(*i*) the differential equation connecting its velocity v and the height x it has then reached;

(*ii*) x in terms of v;

(*iii*) its maximum height;

(*iv*) its terminal (or limiting) velocity.

13 A flask contains 10 dm^3 of a solution of water and a thoroughly dissolved chemical. Each minute, 2 dm^3 of water flow into the flask and 3 dm^3 of solution flow out of the flask. Write down an expression for the number of kilograms of chemical per cubic decimetre in the solution after t min. If there are x kg of chemical in the flask at this time, show that

$$\frac{dx}{dt} = \frac{-3x}{10 - t}.$$

Find the general solution of this equation. If there were 1 kg of chemical in the flask at time $t = 0$, find the mass of chemical in the flask after 5 min. (A.E.B.)

14 The rate of increase of thickness of ice on a pond is inversely proportional to the thickness of ice already present. It is known that, when the thickness of the ice is x cm and the temperature of the air above the ice is $-\alpha$ °C, the rate is $(\alpha/14\,400\,x)$ cm s^{-1}. Form an appropriate differential equation, and hence show that, if the air temperature is -10 °C, the time taken for the thickness to increase from 5 cm to 6 cm is a little more than 2 h. (C.)

Linear equations of first order

A particularly important class of differential equations consists of those which are *linear*, where y and its derivatives occur only in linear combinations even though their coefficients may be functions of x. So a linear equation of the first order can be written as

$$f(x)\frac{dy}{dx} + g(x)y = h(x). \tag{1}$$

Example 7

$$x^2\frac{dy}{dx} + 2xy = 1.$$

We quickly notice that this is a particularly convenient equation since it can be written as

$$\frac{d}{dx}(x^2y) = 1;$$

and, as the left-hand side is an exact derivative, the original equation is also called *exact*.

Proceeding with the solution, we see that

$$x^2y = x + A$$

$$\Rightarrow \quad y = \frac{1}{x} + \frac{A}{x^2}.$$

If we now return to the more general linear equation (1) we see that it can be written as

$$\frac{dy}{dx} + \frac{g(x)}{f(x)}y = \frac{h(x)}{f(x)}$$

$$\text{or} \quad \frac{dy}{dx} + Py = Q \tag{2}$$

where P and Q are functions of x.

This is usually taken as the standard form of the first-order linear equation. Unfortunately (unlike Example 7) it is not usually exact, so we now investigate whether there is any way in which it can be made into an exact equation.

If we multiply each side of (2) by another function, denoted by R, it becomes

$$R\frac{dy}{dx} + PRy = QR,$$

and we see that the left-hand side would be our exact derivative if only it could be written as $\dfrac{d}{dx}(Ry)$.

But $\dfrac{d}{dx}(Ry) = R\dfrac{dy}{dx} + \dfrac{dR}{dx}y.$

So $R\dfrac{dy}{dx} + PRy$ is an exact derivative provided that

$$\frac{dR}{dx} = PR$$

$$\Leftrightarrow \quad \frac{dR}{R} = P\,dx$$

$$\Leftrightarrow \quad \ln R = \int P\,dx$$

$$\Leftrightarrow \quad R = e^{\int P\,dx}.$$

As this function enables us to transform (2) into an exact equation, $e^{\int P\,dx}$ is usually known as an *integrating factor*. By this means the equation becomes

$$e^{\int P\,dx}\frac{dy}{dx} + P\,e^{\int P\,dx}y = Q\,e^{\int P\,dx}$$

$$\Rightarrow \quad \frac{d}{dx}(e^{\int P\,dx}y) = Q\,e^{\int P\,dx}$$

and the problem is reduced to one of ordinary integration.

Two further examples will make this clear:

Example 8

$\dfrac{dy}{dx} = x + y.$

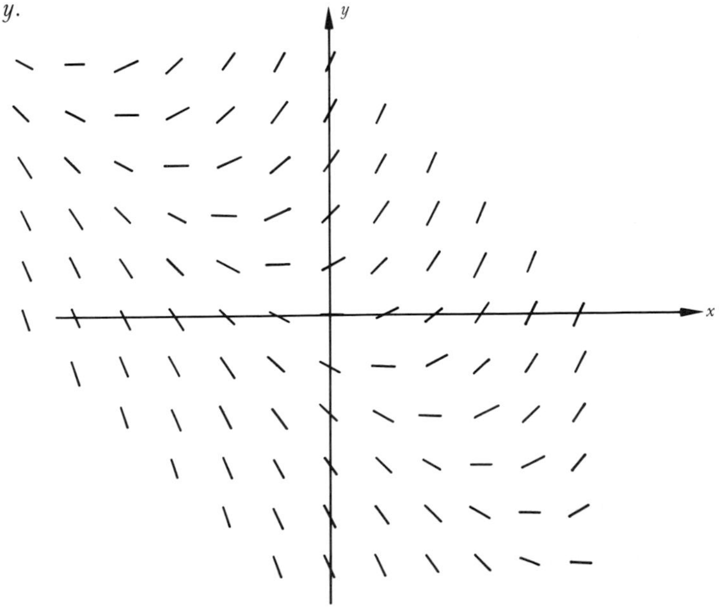

Unlike Example 7, this is not an exact equation. But it can be written as

$$\frac{dy}{dx} - y = x,$$

so that a possible integrating factor is

$$e^{\int P\,dx} = e^{\int -1\,dx} = e^{-x}.$$

Hence $\quad e^{-x}\dfrac{dy}{dx} - e^{-x}y = x\,e^{-x}$

$$\Rightarrow \qquad \frac{d}{dx}(e^{-x}y) = x\,e^{-x}$$

$$\Rightarrow \qquad e^{-x}y = \int x\,e^{-x}\,dx$$

$$= -x\,e^{-x} - e^{-x} + A$$

$$\Rightarrow \qquad y = A\,e^{x} - x - 1.$$

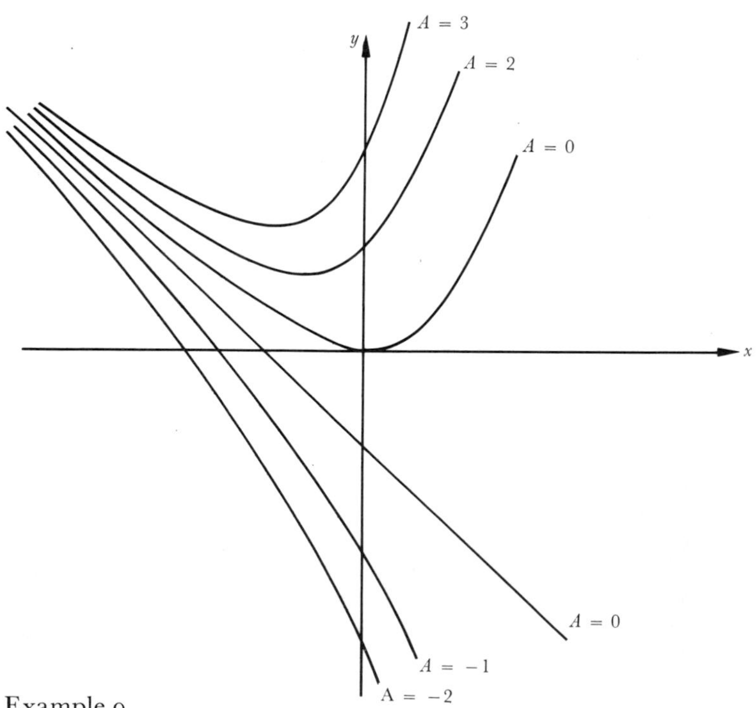

Example 9

$$\sin x\frac{dy}{dx} + 2y\cos x = 1.$$

In standard form, this becomes

$$\frac{dy}{dx} + (2\cot x)y = \operatorname{cosec} x$$

and the necessary integrating factor is

$$e^{\int 2 \cot x \, dx} = e^{2 \ln \sin x} = e^{\ln \sin^2 x} = \sin^2 x.$$

So the equation becomes

$$(\sin^2 x) \frac{dy}{dx} + (2 \sin x \cos x) y = \sin x$$

$$\Rightarrow \qquad \frac{d}{dx} (y \sin^2 x) = \sin x$$

$$\Rightarrow \qquad y \sin^2 x = -\cos x + A$$

$$\Rightarrow \qquad y = \frac{A - \cos x}{\sin^2 x}.$$

Exercise 9.8c

1 Find the general solutions of the following equations:

(*i*) $\dfrac{dy}{dx} = x - y;$ (*ii*) $x \dfrac{dy}{dx} = x + y;$

(*iii*) $x \dfrac{dy}{dx} - 2y = (x - 2) e^x;$ (*iv*) $\dfrac{dy}{dx} + y = x^2.$

2 Find the solutions of the following equations such that $y = 1$ when $x = 0$:

(*i*) $(1 - x^2) \dfrac{dy}{dx} - xy = x;$ (*ii*) $(1 + x) \dfrac{dy}{dx} = x(y + 1).$

3 The current i in a certain electric circuit satisfies the equation

$$\frac{di}{dt} + i = \sin t.$$

If it is known that $i = 0$ when $t = 0$, find i in terms of t. To what does this approximate as $t \to \infty$?

4 A radioactive substance P decays and changes (without loss of mass) into a substance Q, which itself similarly changes into a third substance R. R suffers no further change. The masses of P, Q, and R present at time t are given by p, q, and r respectively. The rates of change are such that

$$\frac{dp}{dt} = -2p \quad \text{and} \quad \frac{dr}{dt} = q. \tag{1}$$

Show that

$$\frac{dq}{dt} = 2p - q.$$

Initially (at time $t = 0$) there is 1 g of substance P and none of substance Q. Integrate equation (1) and hence show that q satisfies the differential equation

$$\frac{dq}{dt} + q = 2\,e^{-2t}. \qquad (2)$$

Show that (2) may be written in the form

$$\frac{d}{dt}(q\,e^{t}) = 2\,e^{-t},$$

and integrate to find q as a function of t. Hence prove that, at any subsequent time, there is never more than $\frac{1}{2}$ g of Q present. (S.M.P.)

5

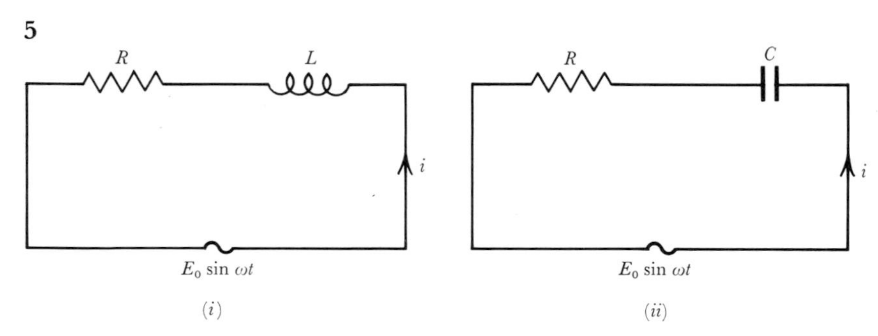

(*i*) (*ii*)

In each of the above circuits an alternating potential $E_0 \sin \omega t$ is applied and the current i satisfies the following equations (respectively), where R, L, C are constants, known as resistance, inductance, and capacitance.

$$(i)\ L\frac{di}{dt} + Ri = E_0 \sin \omega t; \qquad (ii)\ R\frac{di}{dt} + \frac{1}{C}i = \omega E_0 \cos \omega t.$$

In each case, find:

(*a*) the appropriate integrating factor;
(*b*) the general solution for i, as a function of t;
(*c*) the ultimate 'steady' alternating current (when $t \to \infty$ and any 'transient' terms have faded away).

[It may be assumed — or proved by repeated integration by parts — that

$$\int e^{kt} \sin \omega t\, dt = \frac{e^{kt}}{k^2 + \omega^2}(k \sin \omega t + \omega \cos \omega t)$$

and $$\int e^{kt} \cos \omega t\, dt = \frac{e^{kt}}{k^2 + \omega^2}(\omega \sin \omega t - k \cos \omega t)].$$

Linear equations with constant coefficients

A particularly important (and fortunately simple) linear differential equation is one in which the coefficients of y and its derivatives are all constants. We shall consider the second-order equation

$$a\frac{d^2y}{dx^2} + b\frac{dy}{dx} + cy = f(x) \qquad (1)$$

where a, b, c are constants.

The solution of such equations can be divided into two parts:

(a) finding a *general* solution of the *associated equation*:

$$a\frac{d^2y}{dx^2} + b\frac{dy}{dx} + cy = 0 \qquad (2)$$

in which $f(x)$ has been replaced by zero. This solution is known as the *complementary function* ($c.f.$).

(b) finding a *particular* solution of the original equation (1). This is called a *particular integral* ($p.i.$).

These processes will be clear when we take particular examples, but the general method can be summarized as follows:

(a) *The complementary function*

To find a general solution of the associated equation

$$a\frac{d^2y}{dx^2} + b\frac{dy}{dx} + cy = 0, \qquad (1)$$

we put $y = e^{mx}$ and observe that this is a solution provided that

$$am^2 e^{mx} + bm e^{mx} + c e^{mx} \equiv 0$$

$$\Leftrightarrow \qquad am^2 + bm + c = 0.$$

So if this equation† has roots α and β, it follows that $e^{\alpha x}$ and $e^{\beta x}$ are both solutions of equation (2).

Moreover,

if $\qquad u = A e^{\alpha x} + B e^{\beta x}$ (where A, B are any constants),

then $\qquad \frac{du}{dx} = A\alpha e^{\alpha x} + B\beta e^{\beta x}$

and $\qquad \frac{d^2u}{dx^2} = A\alpha^2 e^{\alpha x} + B\beta^2 e^{\beta x}$.

† The very important case when $am^2 + bm + c = 0$ does not have any real roots will be considered in 10.5. Meanwhile, we see that if it has identical roots α, α, then

$$u = A e^{\alpha x} + B e^{\alpha x} = (A + B) e^{\alpha x},$$

which effectively contains only one arbitrary constant $A + B$, rather than two independent constants A and B. But in this case it can quite easily be shown that $x e^{\alpha x}$ is another solution, so that

$$u = A e^{\alpha x} + Bx e^{\alpha x}$$

is also a solution, thus restoring two independent constants.

Hence $a\dfrac{d^2u}{dx^2} + b\dfrac{du}{dx} + cu$

$$= a(A\alpha^2 e^{\alpha x} + B\beta^2 e^{\beta x}) + b(A\alpha e^{\alpha x} + B\beta e^{\beta x}) + c(A e^{\alpha x} + B e^{\beta x})$$

$$= A(a\alpha^2 + b\alpha + c) e^{\alpha x} + B(a\beta^2 + b\beta + c) e^{\beta x}$$

$$= 0 + 0 = 0.$$

So $u = A e^{\alpha x} + B e^{\beta x}$ is also a solution of (2). This is therefore the complementary function.

(b) A particular integral

The discovery of a *particular* solution of the original equation is frequently a matter of trial and error, and we shall restrict ourselves to the simplest cases. But once we have found such a p.i. we can immediately combine it with the c.f. to provide a *general* solution of the original equation:

For suppose that $v(x)$ is a particular integral.

Then $a\dfrac{d^2v}{dx^2} + b\dfrac{dv}{dx} + cv = f(x).$

But the complementary function $u(x)$ was such that

$$a\frac{d^2u}{dx^2} + b\frac{du}{dx} + cu = 0.$$

Hence $a\dfrac{d^2}{dx^2}(u + v) + b\dfrac{d}{dx}(u + v) + c(u + v) = f(x)$

and we see that $u(x) + v(x)$ is a general solution of the original equation. So for such a linear equation,

general solution = complementary function + particular integral

Example 10

Find a general solution of the differential equation

$$\frac{d^2y}{dx^2} + \frac{dy}{dx} - 6y = x + 1.$$

(a) Complementary function

The associated equation is

$$\frac{d^2y}{dx^2} + \frac{dy}{dx} - 6y = 0$$

and $y = e^{mx}$ will be a solution provided that

$$\begin{aligned} m^2 e^{mx} + m\,e^{mx} - 6\,e^{mx} &\equiv 0 \\ \Leftrightarrow \qquad m^2 + m - 6 &= 0 \\ \Leftrightarrow \qquad (m-2)(m+3) &= 0 \\ \Leftrightarrow \qquad m &= 2 \quad \text{or} \quad -3. \end{aligned}$$

So the c.f. is

$$y = A\,e^{2x} + B\,e^{-3x}.$$

(*b*) *Particular integral*

If we seek for y as a function of x such that

$$\frac{d^2y}{dx^2} + \frac{dy}{dx} - 6y \equiv x + 1,$$

we naturally first look amongst simple linear functions of the form

$$y = ax + b.$$

Then $\dfrac{dy}{dx} = a$

and $\dfrac{d^2y}{dx^2} = 0.$

$$\begin{aligned} \text{So} \qquad \frac{d^2y}{dx^2} + \frac{dy}{dx} - 6y &\equiv x + 1 \\ \Leftrightarrow \qquad 0 + a - 6(ax + b) &\equiv x + 1 \\ \Leftrightarrow \quad a = -\tfrac{1}{6} \quad \text{and} \quad a - 6b = 1 \quad &\Rightarrow \quad b = -\tfrac{7}{36}. \end{aligned}$$

So a p.i. is $y = -\frac{1}{6}x - \frac{7}{36}$

and a general solution of the original equation is

$$y = A\,e^{2x} + B\,e^{-3x} - \tfrac{1}{6}x - \tfrac{7}{36}.$$

We can clearly also apply this method to similar first-order equations. For instance, the equation of Example 8 can be written as

Example 11

$$\frac{dy}{dx} - y = x$$

and we can immediately obtain the necessary complementary function and particular integral:

(*a*) *Complementary function*

Put $y = e^{mx}$ in the associated equation

$$\frac{dy}{dx} - y = 0.$$

Then $\quad m\,e^{mx} - e^{mx} \equiv 0$

$\Rightarrow \qquad m = 1$

so that the c.f. is $\quad y = A\,e^x.$

(b) Particular integral

First we try $y = ax + b$. This would be a p.i. provided that

$$a - (ax + b) \equiv x$$

$$\Rightarrow \quad -ax + (a - b) \equiv x$$

$$\Rightarrow \quad a = -1, \quad a - b = 0$$

$$\Rightarrow \quad a = b = -1.$$

So a p.i. is $\qquad y = -x - 1.$

Hence the general solution of the given equation is

$$y = A\,e^x - x - 1.$$

Example 12

If an alternating potential is applied to a certain electric circuit then (using suitable units) the current x after time t is such that

$$\frac{d^2x}{dt^2} + 4\frac{dx}{dt} + 3x = \sin t.$$

Find the general solution.

(a) Complementary function

The associated equation is

$$\frac{d^2x}{dt^2} + 4\frac{dx}{dt} + 3x = 0$$

and $x = e^{mt}$ is a solution provided that

$$m^2\,e^{mt} + 4m\,e^{mt} + 3\,e^{mt} \equiv 0$$

$$\Leftrightarrow \quad m^2 + 4m + 3 = 0$$

$$\Leftrightarrow \quad m = -1 \quad \text{or} \quad -3.$$

So the c.f. is $\qquad x = A\,e^{-t} + B\,e^{-3t}.$

(b) Particular integral

Perhaps we would first try to find a function of the type

$x = a \sin t$

which satisfies the original equation for all values of t.

But this would entail

$$-a \sin t + 4a \cos t + 3a \sin t \equiv \sin t$$
$$\Rightarrow \qquad 2a \sin t + 4a \cos t \equiv \sin t.$$

This identity is clearly impossible, so we now try

$$x = a \sin t + b \cos t.$$

In this case $\dfrac{dx}{dt} = a \cos t - b \sin t$

and $\dfrac{d^2x}{dt^2} = -a \sin t - b \cos t.$

So $(-a \sin t - b \cos t) + 4(a \cos t - b \sin t) + 3(a \sin t + b \cos t) \equiv \sin t$

$\Rightarrow \quad (2a - 4b) \sin t + (4a + 2b) \cos t \equiv \sin t$

$$\Rightarrow \quad \left.\begin{aligned} 2a - 4b &= 1 \\ 4a + 2b &= 0 \end{aligned}\right\} \Rightarrow \quad a = \tfrac{1}{10}, \quad b = -\tfrac{1}{5}$$

and we see that

$x = \frac{1}{10} \sin t - \frac{1}{5} \cos t$ is a p.i.

Hence a general solution is

$$x = A\,e^{-t} + B\,e^{-3t} + \tfrac{1}{10}(\sin t - 2 \cos t).$$

In a particular case, the constants A and B would usually be determined by the initial conditions, e.g., from one's knowledge of the current x at $t = 0$ and how it was changing. But it is noticeable that both terms of the complementary function $A\,e^{-t} + B\,e^{-2t}$ tend to zero as $t \to \infty$; and this happens very rapidly, so that in every case the steady-state current after the initial surge (or other transient effect) is

$$x \approx \tfrac{1}{10}(\sin t - 2 \cos t).$$

Exercise 9.8d

1 Find the general solutions of:

(*i*) $\dfrac{d^2y}{dx^2} - 4\dfrac{dy}{dx} + 3y = 0;$

(*ii*) $\dfrac{d^2y}{dx^2} - 4\dfrac{dy}{dx} - 5y = 0;$ (*iii*) $\dfrac{d^2y}{dx^2} - 4y = 0;$

(*iv*) $\dfrac{d^2y}{dx^2} - 4\dfrac{dy}{dx} = 0;$ (*v*) $\dfrac{d^2y}{dx^2} - 4\dfrac{dy}{dx} + 4y = 0.$

2 Verify that the given expressions are particular integrals of the corresponding equations, and hence find their general solutions:

(*i*) $\dfrac{d^2y}{dx^2} - 3\dfrac{dy}{dx} + 2y = 6;\quad y = 3.$

(*ii*) $\dfrac{d^2y}{dx^2} - 3\dfrac{dy}{dx} + 2y = x^2;\quad y = \frac{1}{4}(2x^2 + 6x + 7).$

(*iii*) $\dfrac{d^2x}{dt^2} + 3\dfrac{dx}{dt} + 2x = e^t;\quad x = \frac{1}{6}e^t.$

(*iv*) $\dfrac{d^2x}{dt^2} + 3\dfrac{dx}{dt} + 2x = \sin t;\quad x = \frac{1}{10}(\sin t - 3\cos t).$

3 For each of the following equations find the complementary function, a particular integral and the general solution:

(*i*) $\dfrac{d^2y}{dx^2} - 2\dfrac{dy}{dx} - 3y = x;$

(*ii*) $\dfrac{d^2y}{dx^2} + 2\dfrac{dy}{dx} - 3y = e^{-x};$

(*iii*) $\dfrac{d^2x}{dt^2} - 3\dfrac{dx}{dt} + 2x = \sin t;$

(*iv*) $\dfrac{d^2\theta}{dt^2} + 3\dfrac{d\theta}{dt} + 2\theta = \cos t.$

4 Find the solutions of the following equations which satisfy the given initial conditions:

(*i*) $\dfrac{d^2y}{dx^2} - 3\dfrac{dy}{dx} + 2y = 0;\quad y = 3$ and $\dfrac{dy}{dx} = 4$ when $x = 0$;

(*ii*) $\dfrac{d^2x}{dt^2} = 4x;\quad x = 0$ and $\dfrac{dx}{dt} = 4$ when $t = 0$;

(*iii*) $\dfrac{d^2\theta}{dt^2} + \dfrac{d\theta}{dt} - 2\theta = 4;\quad \theta = 0$ and $\dfrac{d\theta}{dt} = -1$ when $t = 0$.

5 The needle of a certain heavily damped instrument makes an angle θ with its zero position after time t, and it can be shown that

$$\frac{d^2\theta}{dt^2} + 5\frac{d\theta}{dt} + 6\theta = 0.$$

Find an expression for θ in terms of t if initially (i.e., when $t = 0$) the needle was at the zero position but moving at 2 rad s^{-1}. Hence find the maximum deflection of the needle and sketch the graph of θ against t.

6

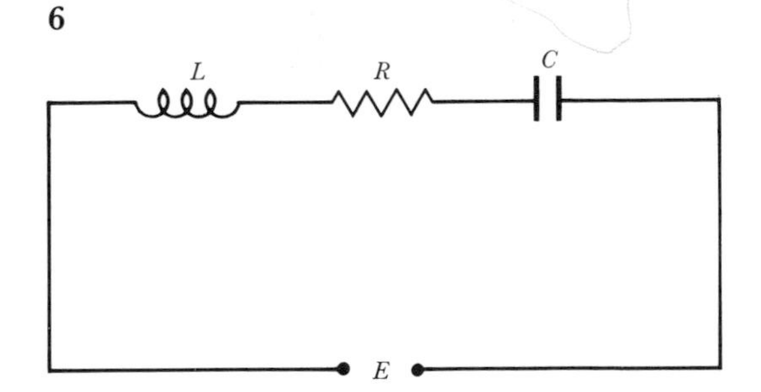

If a potential E is applied to a simple circuit with inductance L, resistance R and capacitance C, it can be shown that the charge q on the capacitor after time t must satisfy the equation

$$L\frac{d^2q}{dt^2} + R\frac{dq}{dt} + \frac{q}{C} = E.$$

Find the general solution of this equation for the circuit in which $L = 5 \times 10^{-3}$ (henry), $R = 6$ (ohm), $C = 10^{-3}$ (farad) when the applied potential is 100 (volts), and show that the charge on the capacitor very soon becomes $\frac{1}{10}$ C.

7 In Exercise 9.8c no 5 (*i*) and (*ii*), (*a*) find the corresponding complementary functions and appropriate particular integrals.

Hence proceed to parts (*b*) and (*c*).

9.9 Hyperbolic functions

Earlier in this chapter we introduced the two functions e^x and e^{-x} and we now proceed to consider their semi-difference and semi-sum. For reasons which will soon become clear, these are called $\sinh x$ and $\cosh x$, the *hyperbolic sine* and *hyperbolic cosine* (or, more familiarly, 'shine' and 'cosh').

So
$$\boxed{\begin{aligned}\sinh x &= \tfrac{1}{2}(e^x - e^{-x})\\ \cosh x &= \tfrac{1}{2}(e^x + e^{-x})\end{aligned}}$$

Firstly, we see that

$$\sinh(-x) = \tfrac{1}{2}(e^{-x} - e^x) = -\sinh x$$

and $\cosh(-x) = \tfrac{1}{2}(e^{-x} + e^x) = \cosh x;$

$\Rightarrow$ $\sinh x$ is an odd function

and $\cosh x$ is an even function.

This is also clear from considering the graphs of e^x and e^{-x}, and of

$\sinh x$ and $\cosh x$:

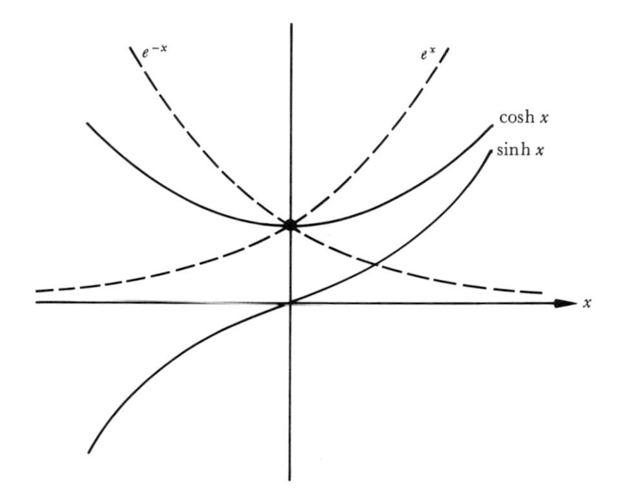

The first similarity with the trigonometric functions is noticed if we find the derivatives of $\sinh x$ and $\cosh x$.

$$\text{For}\quad \frac{\mathrm{d}}{\mathrm{d}x}(\sinh x) = \frac{\mathrm{d}}{\mathrm{d}x}\{\tfrac{1}{2}(e^x - e^{-x})\}$$

$$= \tfrac{1}{2}(e^x + e^{-x}) = \cosh x$$

$$\frac{\mathrm{d}}{\mathrm{d}x}(\cosh x) = \frac{\mathrm{d}}{\mathrm{d}x}\{\tfrac{1}{2}(e^x + e^{-x})\}$$

$$= \tfrac{1}{2}(e^x - e^{-x}) = \sinh x.$$

Further, we notice that

$$\cosh^2 x = \tfrac{1}{4}(e^x + e^{-x})^2 = \tfrac{1}{4}(e^{2x} + 2 + e^{-2x})$$

$$\sinh^2 x = \tfrac{1}{4}(e^x - e^{-x})^2 = \tfrac{1}{4}(e^{2x} - 2 + e^{-2x}).$$

$$\text{So}\quad \cosh^2 x - \sinh^2 x = 1$$

$$\text{and}\quad \cosh^2 x + \sinh^2 x = \tfrac{1}{2}(e^{2x} + e^{-2x})$$

$$= \cosh 2x.$$

$$\text{Also}\quad \sinh 2x = \tfrac{1}{2}(e^{2x} - e^{-2x})$$

$$= \tfrac{1}{2}\{(e^x)^2 - (e^{-x})^2\}$$

$$= \tfrac{1}{2}(e^x - e^{-x})(e^x + e^{-x})$$

$$= 2 \times \tfrac{1}{2}(e^x - e^{-x}) \times \tfrac{1}{2}(e^x + e^{-x})$$

$$\Rightarrow\quad \sinh 2x = 2\sinh x \cosh x.$$

Summarising these elementary properties, we see close similarities, as well as subtle differences, between the hyperbolic and the trigonometric functions, the only property of the latter without parallel being that of periodicity:

$$\frac{d}{dx}(\sinh x) = \cosh x \qquad \frac{d}{dx}(\cosh x) = \sinh x$$
$$\cosh^2 x - \sinh^2 x = 1$$
$$\sinh 2x = 2 \sinh x \cosh x$$
$$\cosh 2x = \cosh^2 x + \sinh^2 x$$

Lastly, and again by analogy with the trigonometric functions, we define

$$\tanh x = \frac{\sinh x}{\cosh x} \qquad \coth x = \frac{\cosh x}{\sinh x}$$
$$\operatorname{sech} x = \frac{1}{\cosh x} \qquad \operatorname{cosech} x = \frac{1}{\sinh x}.$$

Exercise 9.9a

1 Show that the hyperbola

$$\frac{x^2}{a^2} - \frac{y^2}{b^2} = 1$$

can be represented parametrically by the hyperbolic functions

$$x = a \cosh \theta, \qquad y = b \sinh \theta.$$

Find $\frac{dy}{dx}$ in terms of θ.

2 Show that:

(*i*) $\tanh x = \frac{e^{2x} + 1}{e^{2x} - 1}$;

(*ii*) $\tanh x$ is an odd function;

(*iii*) $\frac{d}{dx}(\tanh x) = \operatorname{sech}^2 x$;

and sketch the graph of $\tanh x$.

3 Differentiate with respect to x:

(*i*) $\sinh 2x$; (*ii*) $\cosh \frac{1}{2}x$; (*iii*) $x \cosh x$; (*iv*) $\ln \sinh x$;
(*v*) $\ln \cosh x$; (*vi*) $\operatorname{sech} x$; (*vii*) $\operatorname{cosech} x$; (*viii*) $\ln \tanh \frac{1}{2}x$.

4 Find the following integrals:

(*i*) $\int \sinh x \, dx$; (*ii*) $\int \cosh 2x \, dx$; (*iii*) $\int \operatorname{sech}^2 x \, dx$;

(*iv*) $\int \tanh x \, dx$; (*v*) $\int x \cosh x \, dx$; (*vi*) $\int x \operatorname{sech}^2 x \, dx$.

5 Calculate:
(*i*) the area beneath the curve $y = \cosh x$ between $x = -1$ and $x = +1$;
(*ii*) the volume formed when this area is rotated about the x-axis.

6 Use Maclaurin series to find the power series for:
(*i*) $\sinh x$; (*ii*) $\cosh x$.

7 A bead is at rest on a straight smooth horizontal wire which is then made to rotate with constant angular velocity ω about one of its ends. It can be shown that its distance r from the point of rotation will satisfy the equation

$$\frac{d^2 r}{dt^2} - \omega^2 r = 0.$$

Show:
(*i*) that the general solution of this equation is $r = A\,e^{\omega t} + B\,e^{-\omega t}$

and (*ii*) that if initially $r = a$, then $r = a \cosh \omega t$

(*iii*) that if the wire has rotated through an angle θ, then $\theta = \omega t$ and the polar equation of the bead's path is $r = a \cosh \theta$.

Inverse hyperbolic functions

We see from a graph that the function $\sinh x$ sets up a 1–1 mapping of the real numbers R on to R, so that its inverse function is defined throughout R and we can write:

$$y = \sinh x \quad \Leftrightarrow \quad x = \sinh^{-1} y.$$

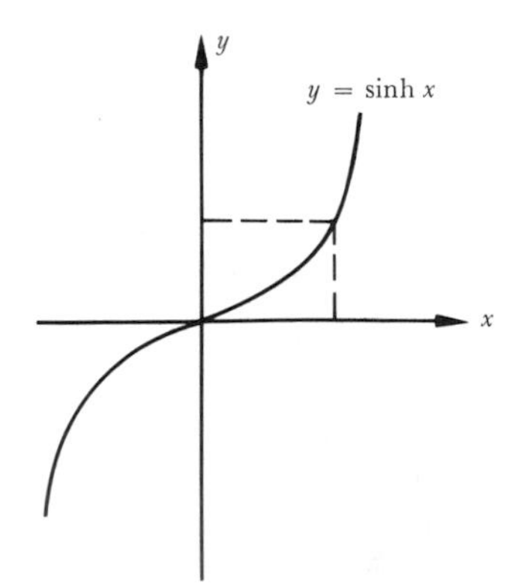

Now $\quad y = \sinh x = \frac{1}{2}(e^x - e^{-x})$

$\Rightarrow \quad e^{2x} - 2y\, e^x - 1 = 0$

$\Rightarrow \quad e^x = y + \sqrt{(y^2 + 1)} \quad (\text{since } e^x > 0)$

$\Rightarrow \quad x = \ln [y + \sqrt{(y^2 + 1)}].$

So $\quad \sinh^{-1} y = \ln [y + \sqrt{(y^2 + 1)}];$

and just as sinh is related to the exponential function, so $\sinh^{-1}$ is related to the logarithmic function.

The function $\cosh x$ is slightly more awkward, for it maps all $x \in R$ on to the range $y \geqslant 1$, and each of the numbers of the range (except $y = 1$) arises from two equal and opposite members of the domain.

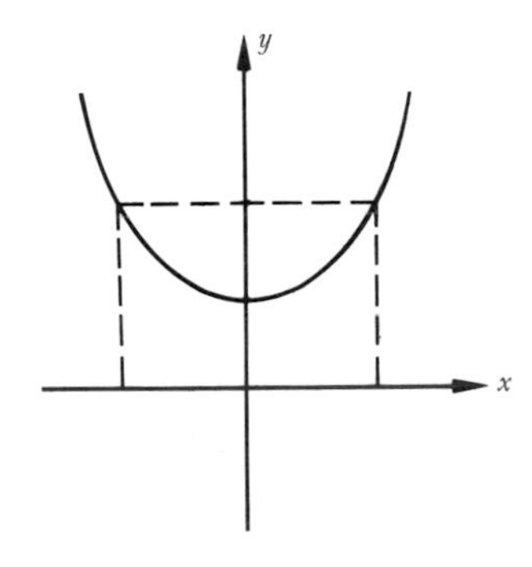

If, however, we restrict the domain to $x \geqslant 0$, we can obtain an inverse function, and write

$y = \cosh x \quad (x \geqslant 0) \quad \Rightarrow \quad x = \cosh^{-1} y \quad (y \geqslant 1).$

Now $\quad y = \cosh x = \frac{1}{2}(e^x + e^{-x})$

$\Rightarrow \quad e^{2x} - 2y\, e^x + 1 = 0$

$\Rightarrow \quad (e^x)^2 - 2y\, e^x + 1 = 0$

$\Rightarrow \quad e^x = y \pm \sqrt{(y^2 - 1)}$

$\Rightarrow \quad x = \ln [y \pm \sqrt{(y^2 - 1)}].$

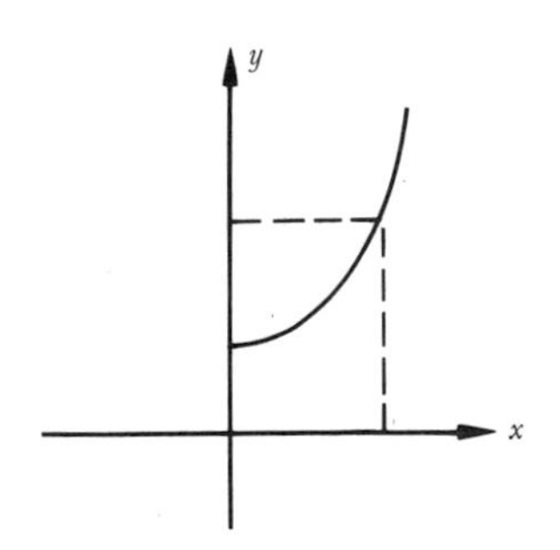

But $$y - \sqrt{(y^2 - 1)} = \frac{1}{y + \sqrt{(y^2 - 1)}}$$

$$\Rightarrow \quad \ln \{y - \sqrt{(y^2 - 1)}\} = \ln \frac{1}{y + \sqrt{(y^2 - 1)}}$$
$$= -\ln [y + \sqrt{(y^2 - 1)}].$$

Hence $x = \pm\ln [y + \sqrt{(y^2 - 1)}]$.

But $x \geqslant 0$, so $x = +\ln [y + \sqrt{(y^2 - 1)}]$

$$\Rightarrow \quad \cosh^{-1} y = \ln [y + \sqrt{(y^2 - 1)}].$$

We can now write these inverse hyperbolic functions as

$$\sinh^{-1} x = \ln [x + \sqrt{(x^2 + 1)}]$$
$$\cosh^{-1} x = \ln [x + \sqrt{(x^2 - 1)}]$$

Their main importance is in the calculation of two standard integrals:

(*i*) $\displaystyle\int \frac{dx}{\sqrt{(x^2 + 1)}}$; (*ii*) $\displaystyle\int \frac{dx}{\sqrt{(x^2 - 1)}}$.

(*i*) In $\displaystyle\int \frac{dx}{\sqrt{(x^2 + 1)}}$,

put $x = \sinh u \Rightarrow dx = \cosh u \, du$.

Then $$\int \frac{dx}{\sqrt{(x^2 + 1)}} = \int \frac{\cosh u \, du}{\sqrt{(\sinh^2 u + 1)}}$$
$$= \int \frac{\cosh u \, du}{\cosh u}$$
$$= \int du$$
$$= u + c$$
$$= \sinh^{-1} x + c.$$

(*ii*) In $\displaystyle\int \frac{dx}{\sqrt{(x^2 - 1)}}$,

put $x = \cosh u \Rightarrow dx = \sinh u \, du$.

Then $$\int \frac{dx}{\sqrt{(x^2 - 1)}} = \int \frac{\sinh u \, du}{\sqrt{(\cosh^2 u - 1)}}$$
$$= \int \frac{\sinh u \, du}{\sinh u}$$

$$= \int du$$

$$= u + c$$

$$= \cosh^{-1} x + c.$$

Hence

$$\int \frac{dx}{\sqrt{(x^2+1)}} = \sinh^{-1} x + c$$

$$\int \frac{dx}{\sqrt{(x^2-1)}} = \cosh^{-1} x + c$$

Exercise 9.9b

1 Express as logarithms:

(*i*) $\sinh^{-1} \frac{3}{4}$; (*ii*) $\cosh^{-1} 2$; (*iii*) $\tanh^{-1} x$.

2 Find the following integrals:

(*i*) $\int \frac{dx}{\sqrt{(x^2+4)}}$; (*ii*) $\int \frac{dx}{\sqrt{(x^2-9)}}$; (*iii*) $\int \frac{dx}{\sqrt{(4x^2+9)}}$;

(*iv*) $\int \frac{dx}{\sqrt{(9x^2-4)}}$; (*v*) $\int \frac{dx}{\sqrt{(x^2+2x)}}$; (*vi*) $\int \frac{dx}{\sqrt{(x^2+2x+2)}}$;

(*vii*) $\int \sinh^{-1} x \, dx$; (*viii*) $\int \cosh^{-1} x \, dx$;

(*ix*) $\int \sqrt{(x^2+1)} \, dx$; (*x*) $\int \sqrt{(x^2-4)} \, dx$.

3 Evaluate:

(*i*) $\int_0^2 \frac{dx}{\sqrt{(x^2+1)}}$; (*ii*) $\int_2^3 \frac{dx}{\sqrt{(x^2-4)}}$;

(*iii*) $\int_1^2 \sqrt{(x^2-1)} \, dx$; (*iv*) $\int_0^1 \sqrt{(1+4x^2)} \, dx$.

4 (*i*) Find $\int \frac{dx}{a^2 - x^2}$ by two methods: the use of partial fractions, and the substitution $x = \tanh u$ [see also no. 1 (*iii*)].

(*ii*) If a particle falls from rest through a medium whose resistance is proportional to the square of the velocity, it can be shown (in the usual notation) that

$$\frac{dv}{dt} = v\frac{dv}{dx} = g - kv^2.$$

Hence find v in terms of x; v in terms of t; and the limiting value of the velocity.

9.10 Lengths of curves and areas of surfaces

We have already used the method of integration to find both plane areas bounded by curves and also volumes bounded by surfaces of revolution, and we now complete our account by a short investigation into the lengths of such curves and the areas of such surfaces. A refined treatment of these topics is difficult as it is necessary to seek protection against the curve being too crinkly or the surface being too corrugated, so we shall restrict ourselves to an introductory look at smooth curves and at the surfaces formed by their revolution.

The length of a curve

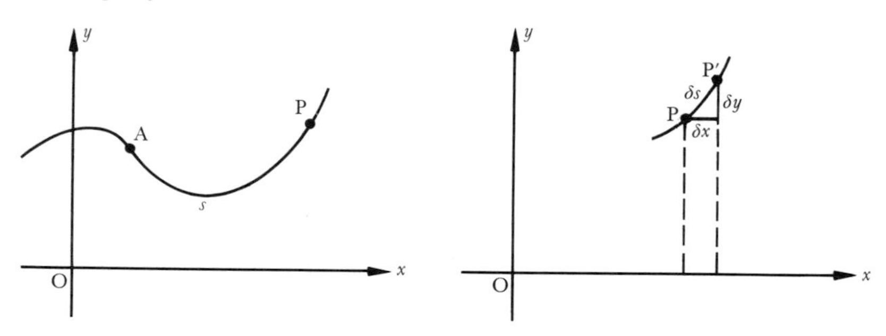

Suppose that a fixed point A is taken on the given curve and that the arc-length $\overset{\frown}{AP}$ from A to a point P (x, y) of the curve is s.

Let us further suppose that P′$(x + \delta x, y + \delta y)$ is a neighbouring point of the curve and that $\overset{\frown}{PP'} = \delta s$.

Then $(\delta s)^2 \approx (\delta x)^2 + (\delta y)^2$ (1)

$$\Rightarrow \quad \left(\frac{\delta s}{\delta x}\right)^2 \approx 1 + \left(\frac{\delta y}{\delta x}\right)^2.$$

So it appears (and, provided the curve is smooth, this can be fully justified) that:

$$\left(\frac{\mathrm{d}s}{\mathrm{d}x}\right)^2 = 1 + \left(\frac{\mathrm{d}y}{\mathrm{d}x}\right)^2$$

$$\Rightarrow \quad \frac{\mathrm{d}s}{\mathrm{d}x} = \sqrt{1 + \left(\frac{\mathrm{d}y}{\mathrm{d}x}\right)^2}$$

$$\Rightarrow \quad s = \int \sqrt{1 + \left(\frac{\mathrm{d}y}{\mathrm{d}x}\right)^2}\,\mathrm{d}x.$$

Alternatively, if x and y are known in terms of a parameter t,

$$\left(\frac{\delta s}{\delta t}\right)^2 \approx \left(\frac{\delta x}{\delta t}\right)^2 + \left(\frac{\delta y}{\delta t}\right)^2$$

$$\Rightarrow \quad \left(\frac{ds}{dt}\right)^2 = \left(\frac{dx}{dt}\right)^2 + \left(\frac{dy}{dt}\right)^2$$

$$\Rightarrow \quad s = \int \sqrt{\left(\frac{dx}{dt}\right)^2 + \left(\frac{dy}{dt}\right)^2}\, dt.$$

Hence
$$\boxed{s = \int \sqrt{1 + \left(\frac{dy}{dx}\right)^2}\, dx = \int \sqrt{\left(\frac{dx}{dt}\right)^2 + \left(\frac{dy}{dt}\right)^2}\, dt}$$

Example 1

Find an expression in terms of x for the length s of the curve $y = c \cosh x/c$, measured from its lowest point. (It can be shown that this curve, known as a *catenary*, is the curve in which a uniform flexible chain hangs when suspended between two points. The constant $c = T_0/w$, where T_0 is the tension at its lowest point and w is its weight per unit length.)

$$y = c \cosh \frac{x}{c}$$

$$\Rightarrow \quad \frac{dy}{dx} = \sinh \frac{x}{c}$$

$$\Rightarrow \quad s = \int_0^x \sqrt{1 + \left(\frac{dy}{dx}\right)^2}\, dx$$

$$= \int_0^x \sqrt{\left(1 + \sinh^2 \frac{x}{c}\right)}\, dx$$

$$\int_0^x \cosh \frac{x}{c}\, dx$$

$$= \left[c \sinh \frac{x}{c}\right]_0^x$$

$$\Rightarrow \quad s = c \sinh \frac{x}{c}.$$

(Hence, further, we see that

$$y^2 - s^2 = c^2 \cosh^2 \frac{x}{c} - c^2 \sinh^2 \frac{x}{c} = c^2$$

$$\Rightarrow \quad y^2 = s^2 + c^2.)$$

Example 2

A wheel of radius a is rolling in a straight line along a horizontal road. Find the total distance travelled by a point on its circumference between successive contacts with the road.

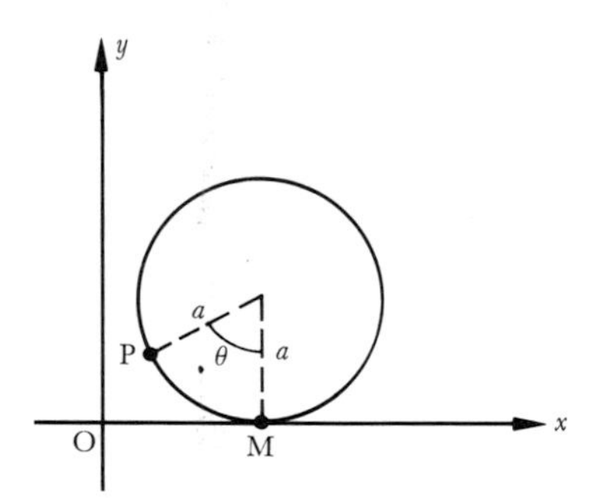

Suppose that the point P was initially in contact with the road at O, but that the wheel has since rolled through an angle θ. If M is its new point of contact, since the wheel has been rolling,

$$OM = \widehat{PM} = a\theta.$$

Hence $\quad x = a\theta - a \sin \theta$

and $\quad y = a - a \cos \theta.$

So the curve traced by P (known as a *cycloid*) can be expressed parametrically as

$$x = a(\theta - \sin \theta), \qquad y = a(1 - \cos \theta),$$

and we need to find the length of this curve between the points $\theta = 0$ and $\theta = 2\pi$.

Now $\dfrac{dx}{d\theta} = a(1 - \cos \theta), \qquad \dfrac{dy}{d\theta} = a \sin \theta.$

$$\text{So} \quad s = \int_0^{2\pi} \sqrt{\left(\frac{dx}{d\theta}\right)^2 + \left(\frac{dy}{d\theta}\right)^2}\, d\theta$$

$$= a \int_0^{2\pi} \sqrt{(1 - \cos \theta)^2 + \sin^2 \theta}\, d\theta$$

$$= a \int_0^{2\pi} \sqrt{(2 - 2\cos\theta)}\, d\theta$$

$$= a \int_0^{2\pi} \sqrt{(4 \sin^2 \tfrac{1}{2}\theta)}\, d\theta$$

$$= 2a \int_0^{2\pi} \sin \tfrac{1}{2}\theta \, d\theta$$

$$= 2a \left[-2 \cos \tfrac{1}{2}\theta\right]_0^{2\pi}$$

$$= 2a[-2(-1) + 2(1)]$$

$\Rightarrow \quad s = 8a.$

Exercise 9.10a

1 Sketch the following curves and find their lengths between the given points:

(*i*) $y = x^{3/2}$, from $(0, 0)$ to $(4, 8)$;
(*ii*) $y = \ln \sec x$, from $x = 0$ to $\frac{1}{4}\pi$;
(*iii*) $x = 2at^3$, $y = 3at^2$, from $t = 0$ to $t = 1$;
(*iv*) $x = at^2$, $y = 2at$, from $t = 0$ to $t = 1$.

2 Sketch the *astroid*

$x = a \cos^3 t, \quad y = a \sin^3 t$

and find its total length.

The area of a surface of revolution

We begin by investigating the simplest case, of a conical surface being formed by the revolution of a straight line. Now if a complete cone of base-radius r and slant-length l is cut along a generator, it can be unwrapped to form a circular sector of radius l and arc $2\pi r$.

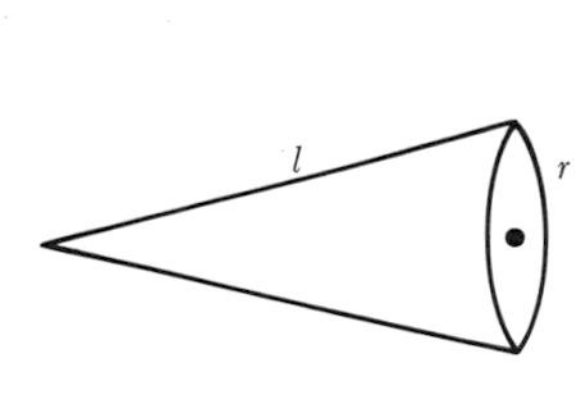

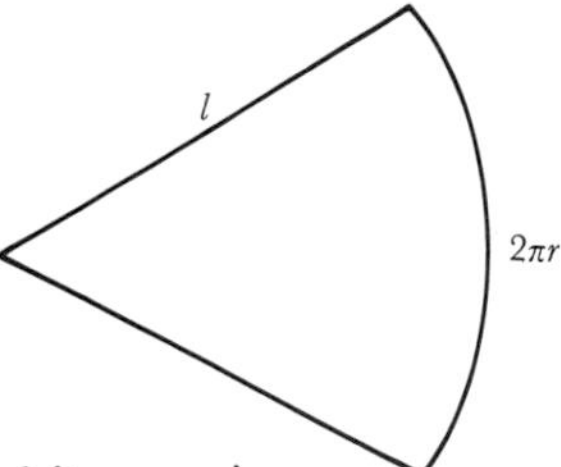

So the angle of the sector is $2\pi r/l$, and its area is

$$\tfrac{1}{2} l^2 \left(\frac{2\pi r}{l}\right) = \pi r l$$

$\Rightarrow$ **Area of curved surface of cone $= \pi r l$**

We now proceed to consider the section of a conical surface whose end-faces have radii R, r, and whose slant-length is l:

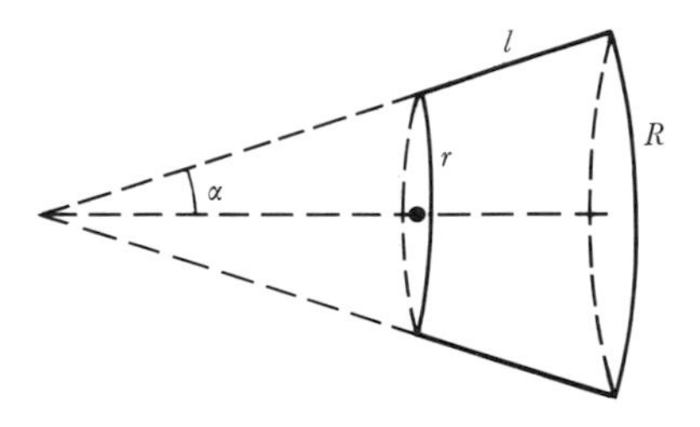

If the semi-vertical angle of the cone is α, then the area of its curved surface

$$= \pi R(R \operatorname{cosec} \alpha) - \pi r(r \operatorname{cosec} \alpha)$$
$$= \pi(R + r)(R - r) \operatorname{cosec} \alpha$$
$$= \pi(R + r)l$$
$$= 2\pi \frac{R + r}{2} l.$$

Hence

Area of curved surface of section = average circumference × slant length

We can now move to the case of a surface formed by revolving a curve about Ox:

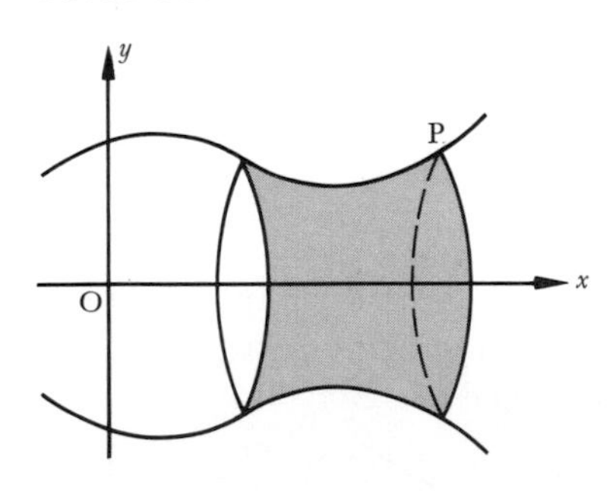

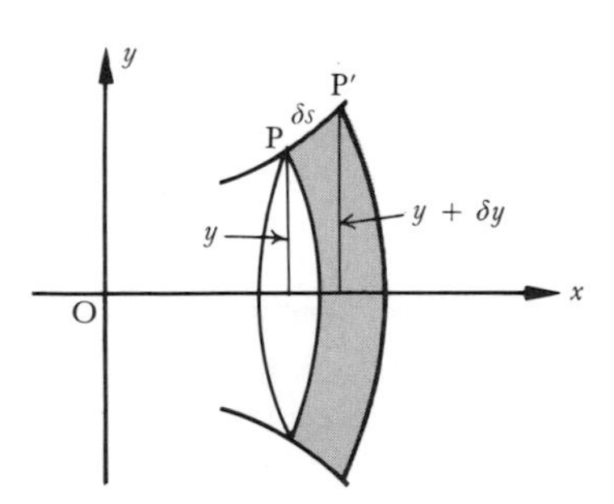

If S denotes the area of surface generated, we see from consideration of a small element that

$$\delta S \approx 2\pi(y + \tfrac{1}{2}\,\delta y)\,\delta s \approx 2\pi y\,\delta s$$

$$\Rightarrow \quad \frac{dS}{ds} = 2\pi y$$

$$\Rightarrow \quad \boxed{S = \int 2\pi y \, ds = \int 2\pi y \sqrt{1 + \left(\frac{dy}{dx}\right)^2}\, dx}$$

Example 3

Find the area of the cap cut off by the plane $x = a$ when the parabola $y^2 = 4ax$ is rotated about Ox.

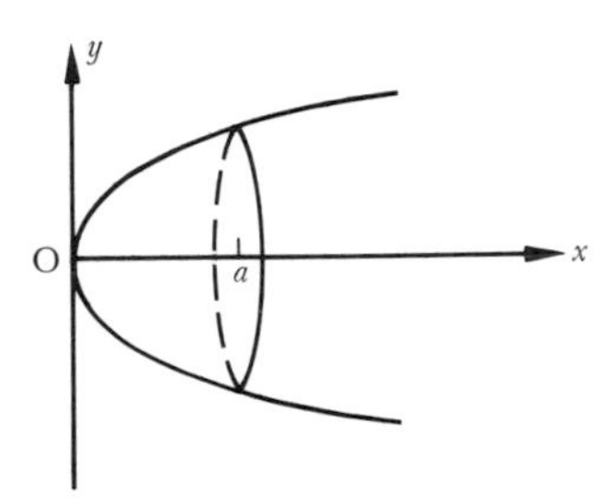

$$y^2 = 4ax$$

$$\Rightarrow \quad y = \pm 2\sqrt{(ax)}$$

$$\Rightarrow \quad \frac{\mathrm{d}y}{\mathrm{d}x} = \pm\sqrt{\frac{a}{x}}.$$

$$\text{Hence area} = \int 2\pi y \,\mathrm{d}s$$

$$= \int_0^a 2\pi y \sqrt{1 + \left(\frac{\mathrm{d}y}{\mathrm{d}x}\right)^2}\,\mathrm{d}x$$

$$= 2\pi \int_0^a \sqrt{(ax)}\sqrt{\left(1 + \frac{a}{x}\right)}\,\mathrm{d}x$$

$$= 4\pi\sqrt{a}\int_0^a \sqrt{(x + a)}\,\mathrm{d}x$$

$$= 4\pi\sqrt{a}\left[\tfrac{2}{3}(x + a)^{3/2}\right]_0^a$$

$$= 4\pi\sqrt{a}\left[\frac{2 \times 2\sqrt{2}}{3} - \frac{2}{3}\right]a^{3/2}$$

$$\Rightarrow \quad \text{Required area} = \frac{8(2\sqrt{2} - 1)}{3}\pi a^2.$$

Example 4

Investigate the surface area of the portion of a sphere between two parallel planes.

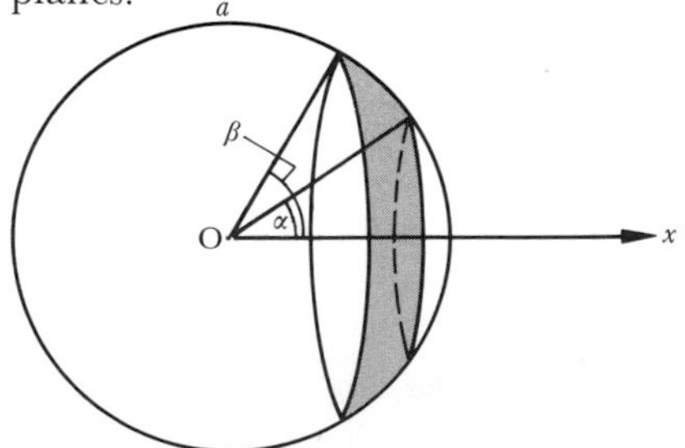

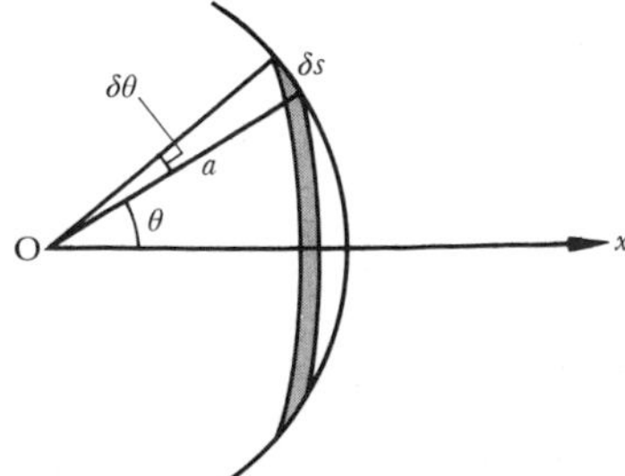

Suppose that we need to find the area between two circles whose declinations from Ox are α, β respectively.

$$\begin{aligned}\text{Required area} &= \int 2\pi y \, \mathrm{d}s \\ &= \int_{\alpha}^{\beta} 2\pi a \sin\theta \times a \, \mathrm{d}\theta \\ &= 2\pi a^2 \int_{\alpha}^{\beta} \sin\theta \, \mathrm{d}\theta \\ &= 2\pi a^2 [-\cos\theta]_{\alpha}^{\beta} \\ &= 2\pi a^2 (\cos\alpha - \cos\beta).\end{aligned}$$

Now if t is the thickness of this portion, $t = a(\cos\alpha - \cos\beta)$

$\Rightarrow$ Required area $= 2\pi at$,

and so is dependent only upon the distance between the parallel planes and not upon their position.

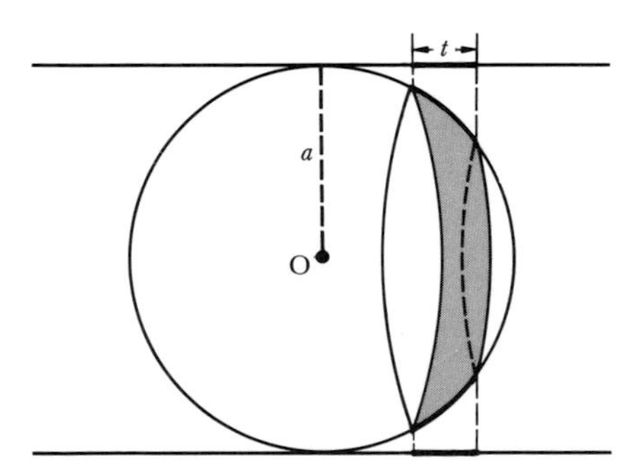

We can also consider the circumscribing cylinder and immediately see that the corresponding area on the cylinder is also $2\pi at$, so we obtain Archimedes' theorem† that the area of a 'zone' on a sphere is equal to the corresponding zone on the circumscribing cylinder (or, in other words, that if a spherical hard-boiled egg were put through an egg-slicer whose wires were all equidistant, each portion would have the same area of curved surface!).

Exercise 9.10b

1 Find the areas of the surfaces of revolution formed by rotating

(*i*) the catenary $y = c\cosh(x/c)$ about the x-axis, from $x = 0$ to $x = c$;

(*ii*) the parabola $y^2 = 4ax$ about the x-axis, from $x = 0$ to $x = a$;

(*iii*) the astroid $x = a\cos^3\theta$, $y = a\sin^3\theta$ about Ox;

† Archimedes of Syracuse (287–212 BC), who requested that this diagram be inscribed on the splendid tomb which was erected for him by the Romans.

(*iv*) the cycloid $x = a(\theta - \sin\theta)$, $y = a(1 - \cos\theta)$ about Ox, from $\theta = 0$ to 2π.

9.11 Areas and lengths in polar coordinates

If we are given the equation of a curve in Cartesian coordinates, we have seen how the inclination ψ of its tangent to Ox, the area A beneath the curve and its length s can all be derived from an examination of the diagrams

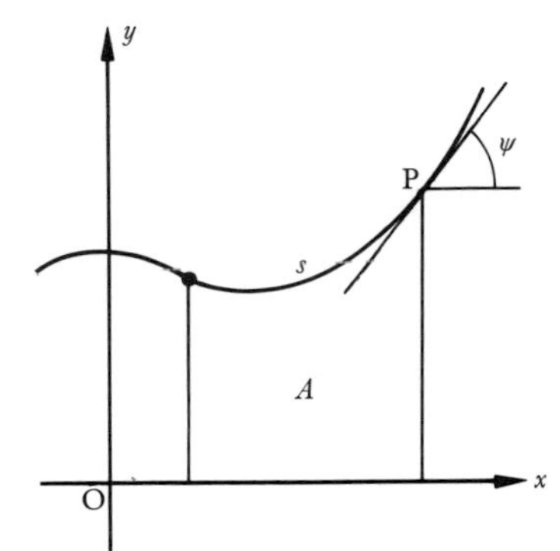

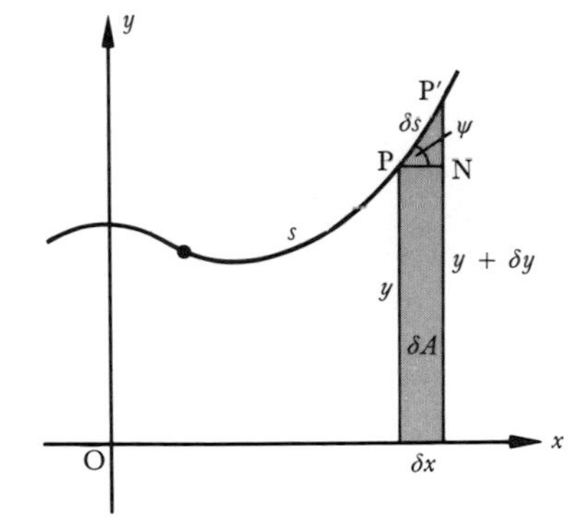

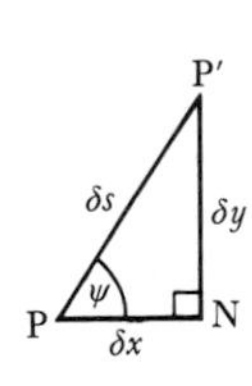

from which we immediately see that

$$\delta A \approx y\,\delta x; \qquad \tan\psi \approx \frac{\delta y}{\delta x}; \qquad (\delta s)^2 \approx (\delta x)^2 + (\delta y)^2$$

$$\Rightarrow \quad A = \int y\,\mathrm{d}x; \qquad \tan\psi = \frac{\mathrm{d}y}{\mathrm{d}x}; \qquad s = \int \sqrt{1 + \left(\frac{\mathrm{d}y}{\mathrm{d}x}\right)^2}\,\mathrm{d}x.$$

Now in polar coordinates, instead of the inclination ψ which the tangent makes with Ox, it is usual to consider its inclination ϕ to the radius (or, if preferred, its *angle of trail*):

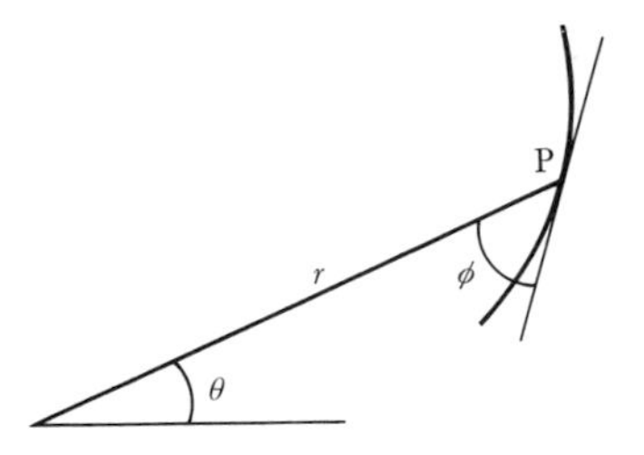

So, just as in Cartesian coordinates, we can find formulae for A, ϕ, and s by considering the figures formed as $\mathrm{P}(r, \theta)$ moves along the curve to its neighbouring position $\mathrm{P}'(r + \delta r, \theta + \delta\theta)$:

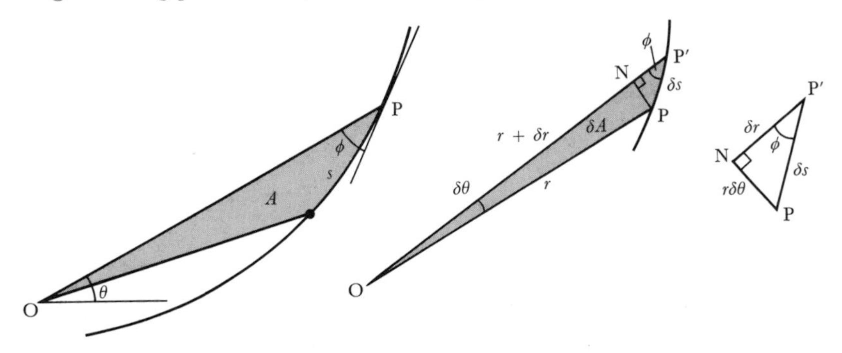

From the second figure, it is clear that

$$\delta A \approx \triangle \mathrm{OPP}' = \tfrac{1}{2}r(r + \delta r)\sin\delta\theta$$

$$\Rightarrow \quad \delta A \approx \tfrac{1}{2}r^2\,\delta\theta.$$

Furthermore, since PN has been drawn perpendicular to OP′ it follows that

$$ON = r\cos\delta\theta \approx r$$

and $\quad PN = r\sin\delta\theta \approx r\,\delta\theta.$

So in the elemental $\triangle \mathrm{P'PN}$,

$$PN = r\,\delta\theta, \qquad P'N \approx \delta r, \qquad \angle \mathrm{NP'P} \approx \phi.$$

Hence $\quad \tan\phi \approx \dfrac{r\,\delta\theta}{\delta r}$

and $\qquad (\delta s)^2 \approx (\delta r)^2 + (r\,\delta\theta)^2.$

Summarising these three, we see that

$$\delta A \approx \tfrac{1}{2}r^2\,\delta\theta; \qquad \tan\phi \approx r\frac{\delta\theta}{\delta r}; \qquad \delta s \approx \sqrt{(\delta r)^2 + (r\,\delta\theta)^2}$$

$$\Rightarrow \quad \boxed{A = \int \tfrac{1}{2}r^2\,\mathrm{d}\theta; \qquad \tan\phi = r\frac{\mathrm{d}\theta}{\mathrm{d}r}; \qquad s = \int\sqrt{\left(\frac{\mathrm{d}r}{\mathrm{d}\theta}\right)^2 + r^2}\,\mathrm{d}\theta}$$

Example 1

If $r = e^{\theta \cot \alpha}$ (where α is a constant), find A, ϕ, and s in terms of r (measuring A and s from the point where $\theta = 0$).

$$(i)\quad A = \int_0^\theta \tfrac{1}{2}r^2\, d\theta$$
$$= \int_0^\theta \tfrac{1}{2}\, e^{2\theta \cot \alpha}\, d\theta$$
$$= \left[\frac{e^{2\theta \cot \alpha}}{4 \cot \alpha}\right]_0^\theta$$
$$= \tfrac{1}{4} \tan \alpha \,(e^{2\theta \cot \alpha} - 1)$$
$$= \tfrac{1}{4}(r^2 - 1) \tan \alpha.$$

$$(ii)\quad \tan \phi = r\frac{d\theta}{dr}$$
$$= \frac{r}{dr/d\theta} = \frac{e^{\theta \cot \alpha}}{\cot \alpha\, e^{\theta \cot \alpha}}$$
$$\Rightarrow \quad \tan \phi = \tan \alpha$$
$$\Rightarrow \quad \phi = \alpha.$$

Hence ϕ, the angle of trail, is constant, so that the curve is known as an equiangular spiral:

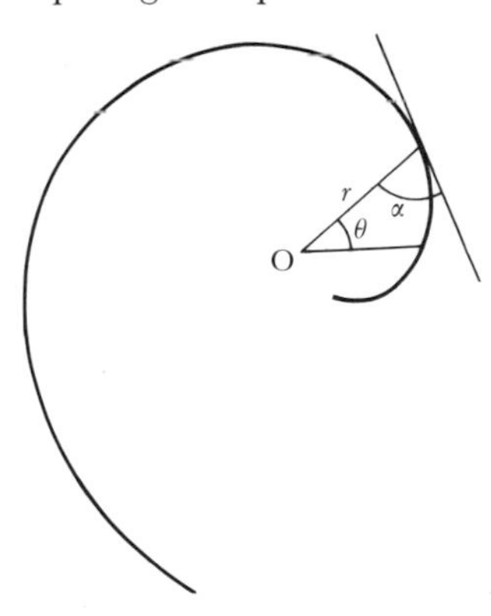

$$(iii)\quad s = \int_0^\theta \sqrt{\left[\left(\frac{dr}{d\theta}\right)^2 + r^2\right]}\, d\theta$$
$$= \int_0^\theta \sqrt{(\cot^2 \alpha\, e^{2\theta \cot \alpha} + e^{2\theta \cot \alpha})}\, d\theta$$
$$= \operatorname{cosec} \alpha \int_0^\theta e^{\theta \cot \alpha} d\theta$$
$$= \operatorname{cosec} \alpha \left[\frac{e^{\theta \cot \alpha}}{\cot \alpha}\right]_0^\theta = \sec \alpha\, (e^{\theta \cot \alpha} - 1).$$

Exercise 9.11

1 Sketch each of the following curves, finding also their total lengths, their total areas, and ϕ in terms of θ.
(*i*) $r = a \cos \theta$;
(*ii*) $r = a(1 + \cos \theta)$, a *cardioid*.

2 Sketch the *lemniscate* $r^2 = a^2 \sin 2\theta$. Find its total area and also ϕ in terms of θ.

3 Sketch the *trefoil* $r^2 = a^2 \sin 3\theta$ and find the area of each petal.

4 Sketch the *Archimedean spiral* $r = a\theta$. Find the area of the sector and the length of arc between the points where $\theta = 0$ and $\theta = 1$.

Miscellaneous problems 9

1 (*i*) Calculate $(1 + 1/n)^n$

for $n = 1, 2, 3, 5, 10, 100.$

Does it appear to tend to a limit as $n \to \infty$?

(*ii*) If $n = 1/h$, show that

$$\ln\left(1 + \frac{1}{n}\right)^n = \frac{\ln(1 + h)}{h} = \frac{\ln(1 + h) - \ln 1}{h}.$$

Letting $n \to \infty$ (and so $h \to 0$), show that

$$\ln\left(1 + \frac{1}{n}\right)^n \to 1, \text{ and hence that } \left(1 + \frac{1}{n}\right)^n \to e.$$

(*iii*) Prove similarly that, as $n \to \infty$,

$$\left(1 + \frac{x}{n}\right)^n \to e^x.$$

2 By the method of two staircases (4.2), show that

$$\int_1^{10^6} \frac{dx}{x} \text{ lies between } \sum_{r=2}^{10^6} \frac{1}{r} \text{ and } \sum_{r=1}^{10^6 - 1} \frac{1}{r}.$$

Hence show that the sum of the reciprocals of the first million integers lies between 13.81 and 14.82.

3 By considering the graph of $1/x$ or otherwise, show that

$$\ln n - \ln(n-1) > \frac{1}{n} \quad \text{and} \quad \ln n - \ln(n-1) < \frac{1}{2}\left(\frac{1}{n} + \frac{1}{n-1}\right),$$

where n is an integer greater than 1.

The function $f(n)$ is defined by

$$f(n) = 1 + \frac{1}{2} + \frac{1}{3} + \cdots + \frac{1}{n} - \ln n.$$

Show that $f(n)$ decreases as n increases and that $f(n) - \dfrac{1}{2n}$ increases as n increases. Deduce that $f(n)$ tends to a finite limit as n tends to infinity. (C.S.)

4 Show that

$$\int_1^n \ln x \, dx < \sum_{r=1}^{n} \ln r < \int_1^{n+1} \ln x \, dx$$

and deduce that the geometric mean of the first n positive integers is approximately n/e. (C.S.)

5 (*i*) Investigate the behaviour as $x \to \infty$ of the ratios

$$\frac{\ln x}{x}, \quad \frac{\ln x}{\sqrt{x}}, \quad \frac{\ln x}{\sqrt[1\,000]{x}}.$$

(*ii*) Show that if $t > 1$,

$$\frac{1}{t} < \frac{1}{\sqrt{t}}$$

and hence that

$$\ln x < 2(\sqrt{x} - 1)$$

and $$\frac{\ln x}{x} < 2\left(\frac{1}{\sqrt{x}} - \frac{1}{x}\right).$$

Deduce that as $x \to \infty$,

$$\frac{\ln x}{x} \to 0.$$

(*iii*) Show that

$$\frac{\ln x}{\sqrt[1\,000]{x}} = 1\,000\,\frac{\ln \sqrt[1\,000]{x}}{\sqrt[1\,000]{x}}$$

and use (*ii*) to deduce its limit as $x \to \infty$.

(*iv*) More generally, show that as $x \to \infty$, $\ln x \to \infty$ more slowly than any positive power (however small) of x.

6 (*i*) Investigate the behaviour as $x \to \infty$ of

$$\frac{e^x}{x}, \quad \frac{e^x}{x^2}, \quad \frac{e^x}{x^{1\,000}}.$$

(*ii*) Show that

$$\ln \frac{e^x}{x} = x\left(1 - \frac{\ln x}{x}\right)$$

and use the result of no. 5 (*ii*) to show that as $x \to \infty$,

$$\ln \frac{e^x}{x} \to \infty$$

and hence that $\dfrac{e^x}{x} \to \infty$.

(*iii*) Deduce that as $x \to \infty$,

$$\frac{e^x}{x^{1\,000}} \to \infty.$$

(*iv*) More generally, show that as $x \to +\infty$, $e^x \to +\infty$ faster than *any* positive power of x; and that $e^{-x} \to 0$ faster than *any* negative power of x.

7 If $\quad I_n(x) = \displaystyle\int_0^x t^n\, e^{-t}\, dt$

show that $I_n(x) = -x^n e^{-x} + nI_{n-1}(x)$.

If $I_n(x) \to I_n$ as $x \to +\infty$, use the result of no. **6** to show that

$$I_n = nI_{n-1}$$

and hence that, if $n \in Z^+$,

$$I_n = n!$$

8 (*i*) Sketch the graph of the function e^{-x^2} and illustrate the integral

$$I = \int_{-\infty}^{\infty} e^{-x^2}\, dx.$$

Can you calculate this integral?

(*ii*) Illustrate by a surface the function of two variables $f(x, y)$, where

$$f(x, y) = e^{-x^2 - y^2}.$$

(*iii*) Show that the volume under this surface can be represented in Cartesian coordinates as

$$\lim \sum \sum e^{-x^2 - y^2}\, \delta x\, \delta y = \int\!\!\int e^{-x^2 - y^2}\, dx\, dy$$

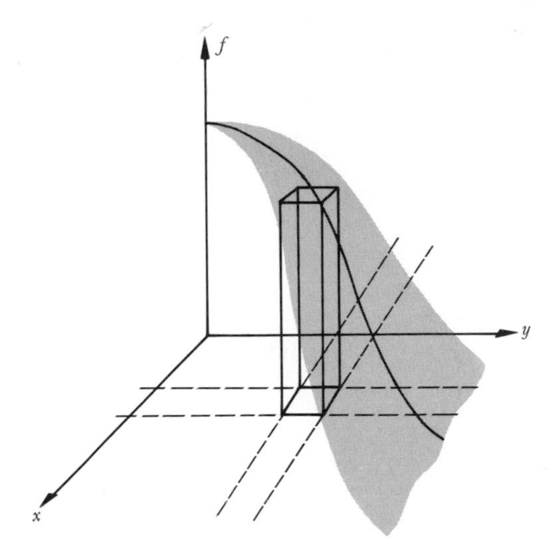

and in polar coordinates as

$$\sum\sum e^{-r^2}\,(\delta r)\,(r\,\delta\theta) = \int\int r\,e^{-r^2}\,\mathrm{d}r\,\mathrm{d}\theta;$$

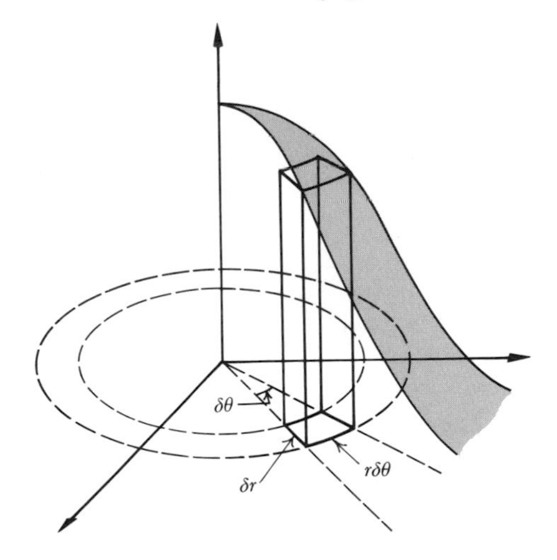

and hence that

$$\left(\int_{-\infty}^{\infty} e^{-x^2}\,\mathrm{d}x\right)\left(\int_{-\infty}^{\infty} e^{-y^2}\,\mathrm{d}y\right) = \int_{r=0}^{\infty} r\,e^{-r^2}\,\mathrm{d}r \int_{\theta=0}^{2\pi} \mathrm{d}\theta.$$

(*iv*) Finally show that $\int_0^\infty r\,e^{-r^2}\,\mathrm{d}r = \frac{1}{2}$ and hence calculate I.

9 Show that, if

$$f(x) = \sum_1^N a_n \sin nx, \qquad (1)$$

then $a_n = \dfrac{2}{\pi}\displaystyle\int_0^{\pi} f(x)\sin nx\,dx.$ (2)

Assuming that the function

$$f(x) = x(\pi - x), \quad (0 \leqslant x \leqslant \pi)$$

can be expressed as an infinite series†

$$\sum_1^{\infty} a_n \sin nx,$$

and that the coefficients are still given by the formula (2), show that in this case

$$a_{2m} = 0, \qquad a_{2m+1} = \frac{8}{\pi(2m+1)^3}$$

and hence sum the series

$$1 - \frac{1}{3^3} + \frac{1}{5^3} - \frac{1}{7^3} + \cdots. \qquad \text{(C.S.)}$$

10 In a population of N rats, n initially have a disease, the rest have not yet caught it. If a rat recovers from the disease it becomes immune. At a later time, t, the population is therefore composed of $I(t)$ rats ill with the disease, $D(t)$ dead rats, $R(t)$ rats who have already recovered and $S(t)$ who are still susceptible.

It is observed that at any stage

$$\frac{dR}{dt} = aI, \qquad \frac{dD}{dt} = bI, \qquad \frac{dS}{dt} = -cSI$$

where a, b, and c are constants, and $N - n > (a + b)/c$.

Prove that I increases until S falls to $(a + b)/c$, and then decreases. Prove also that the maximum value of I is

$$N - \frac{a+b}{c}\ln\frac{ec(N-n)}{a+b}. \qquad \text{(O.S.)}$$

11 Two tribes live on a desert island. Each tribe breeds only among itself and, in isolation, the population of each would increase at the same constant rate. But each tribe is eaten (by the other) at a rate equal to a constant, b, times the population of the other tribe. At a certain time the total population of both tribes together is N_0. Investigate how the total population varies with time subsequently. (O.S.)

12 JAMES: 'π is the most important constant in mathematics.'

JOHN: 'No, e is.'

Continue the discussion. (C.S.)

† A typical example of a *Fourier series*, so called after their discoverer, J. B. J. Fourier (1768–1830).

10

Complex numbers

'I met a man recently who told me that, so far from believing in the square root of minus one, he did not even believe in minus one. This is at any rate a consistent attitude. There are certainly many people who regard $\sqrt{2}$ as something perfectly obvious, but jib at $\sqrt{-1}$. This is because they think that they can visualise the former as something in physical space, but not the latter. Actually, $\sqrt{-1}$ is a much simpler concept.'

E. C. Titchmarsh, *Mathematics for the General Reader.*

10.1 Introduction

The problem of complex numbers arises immediately in attempts to solve such quadratic equations as $x^2 + 6x + 13 = 0 \Rightarrow x = -3 \pm \sqrt{(9 - 13)}$, which is seemingly impossible since there does not apparently exist a number whose square is -4. Understandably, many people give up at this point and say in desperation that this quadratic has no solution, meaning 'has no solution *among the reals*'. But a pupil who had learned to solve quadratic equations by factorisation might equally well say that the equation $x^2 - 2x - 4 = 0$ cannot be solved, on the grounds that he cannot factorise the left-hand side. In fact, he would find, a little later in his mathematical education, that $x = 1 + \sqrt{5}$ and $1 - \sqrt{5}$ are solutions of the latter equation, and probably feel quite happy about the existence of the numbers $1 + \sqrt{5}$ and $1 - \sqrt{5}$. When at the earlier stage he had given up in his attempt to solve $x^2 - 2x - 4 = 0$ because it would not factorise, what he really meant was that it had no solutions *among the rationals*, because he was attempting to find *rational* factors of $x^2 - 2x - 4$. In fact, of course, it does 'factorise', as $(x - 1 + \sqrt{5})(x - 1 - \sqrt{5})$, but only with the use of irrationals.

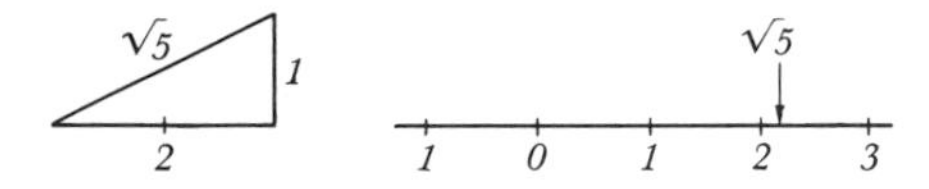

Now it is true, as Professor Titchmarsh implied in the quotation above, that $\sqrt{5}$ is a far more complicated concept than $\sqrt{-1}$, and we have already shown, in 1.5, that such numbers as $\sqrt{5}$ are irrational in the sense that they cannot be expressed as the ratio of integers. Yet most people are happier with $\sqrt{5}$ than with $\sqrt{-1}$. Why? Perhaps simply because $\sqrt{5}$ can be *seen*, with a place in the real number line? By contrast, $\sqrt{-1}$ and the roots of $x^2 + 6x + 13 = 0$ do not appear on the number line, and so it is tempting to suppose they do not exist.

Now the mathematician's attitude at this stage is to say 'It would be a great pity if we had for ever to suffer the irregularity of having some quadratic equations with two roots and others with none. How much better if *all* quadratics were to have two roots! We shall not let an equation like $x^2 + 6x + 13 = 0$ beat us! Let us *invent* a "number" $\sqrt{-1}$, which we shall call i, and suppose that it obeys all the usual laws of algebra (which will be stated in full in 14.1)'. Upon this bold hypothesis we should then say that the roots of $x^2 + 6x + 13 = 0$ were $-3 \pm \sqrt{-4} = -3 + 2\sqrt{-1} = -3 \pm 2i$. Notice that the word 'number' applied to $\sqrt{-1}$ is in inverted commas, because we must not think of it in the same terms as we have thought of numbers hitherto, as a length or a measure. Rather is it a number in the sense that it *behaves* like ordinary numbers, and for some time we may use i without being too scrupulous about its meaning. This is what happened historically: for centuries mathematicians 'played with i', treating it as an 'ordinary' number and using it to obtain extremely valuable results. For example,

$$\begin{aligned}
(a^2 + b^2)(c^2 + d^2) &= (a^2 - i^2b^2)(c^2 - i^2d^2) \\
&= (a + ib)(a - ib)(c + id)(c - id) \\
&= (a + ib)(c + id) \times (a - ib)(c - id) \\
&= (ac + i^2bd + ibc + iad)(ac + i^2bd - ibc - iad) \\
&= [ac - bd + i(bc + ad)][ac - bd - i(bc + ad)] \\
&= (ac - bd)^2 + (bc + ad)^2.
\end{aligned}$$

The final result is quite valid, even though the working involved the use of the 'invented' number i (the word 'imaginary' is a somewhat unfortunate choice). And in many other situations, mathematicians have found that perfectly respectable results can be derived — often more shortly — by the use of i, and this provides strong encouragement for *belief* in this extension of the number system resulting from such a bold hypothesis.

A word of warning is necessary about bold hypotheses. Sometimes mathematicians have to accept defeat. For example, the equation $ax = b$ always has a solution, *provided* $a \neq 0$. What a pity, exclaims the true mathematician, that this exception should have to be tolerated; that there should be just one number 0 which does not possess a reciprocal. Let us therefore invent a new number, ∞, such that $\infty \times 0 = 1 = 0 \times \infty$.

This time, unfortunately, the bold hypothesis must be rejected. For, starting with $0 \times 1 = 0 \times 2$ and multiplying both sides on the left by ∞, we have $\infty \times (0 \times 1) = \infty \times (0 \times 2)$

$\Rightarrow$ $(\infty \times 0) \times 1 = (\infty \times 0) \times 2$ (associativity of multiplication)

$\Rightarrow$ $1 \times 1 = 1 \times 2$ (since $\infty \times 0 = 1$, by definition)

$\Rightarrow$ $1 = 2$, a contradiction.

So we have run into trouble, and the assumption, that a reciprocal for 0 can be dreamed up, is untenable.

However, it can be shown that no such inconsistency arises in the case of the invention of i, and that everything proceeds happily. Nevertheless, a rigorous justification of the invention must be provided at some stage. Historically, this came early in the nineteenth century after i had already had a long and useful innings. For as early as the sixteenth century mathematicians were experimenting with the square root of minus one, and had observed that consistent and acceptable results could be obtained, though they were unable to justify these procedures. Later mathematicians such as Euler (1707–83), continuing the work of John Bernoulli (1667–1748) and de Moivre (1667–1754) developed complex numbers as powerful mathematical tools; whilst, at the end of the eighteenth century, Wessel and Argand gave their celebrated geometrical interpretation. But it was not till the beginning of the nineteenth century that Gauss (1777–1855) laid a sound logical foundation for complex numbers, and then proceeded to develop the theory of functions of a complex variable.

Now the invention of i to provide solutions for all quadratic equations may possibly give the impression that further inventions may be necessary to solve cubic equations, and possibly that even more weird and wonderful numbers might have to be dreamed up to deal with equations of higher degree; and what further miracles of man's genius would be needed to assign meanings to such expressions as $(2 - i)^i$ and $\sqrt[i]{[\sin (5 - 2i)^{2/3}]}$? Happily, such further inventions are not necessary; the extension of the number system by the single step of adjoining the square root of -1 enables all these situations to be covered! The roots of equations of any degree of complication, and numbers of any degree of 'complexity', can be expressed in the same simple form, $x + iy$, where x and y are real, just like the roots of the quadratic equation $x^2 + 6x + 13 = 0$ that we examined at the outset. It is this remarkable property of playing a unifying role in mathematics that endows i with such interest and importance.

Our work in the present chapter will therefore rest on the assumption of a number i satisfying $i^2 = -1$, and we define a 'complex number' to be the sum of two parts:

$$z = x + iy,$$

where x and y are real. x is called the 'real part' of the complex number, and the second term, iy, is the so-called 'imaginary part'. We also write

$$x = R(z), \qquad iy = I(z).$$

In the manipulation of complex numbers, we simply assume that the symbol i will behave exactly as an ordinary number, and that in any manipulations i^2 can be replaced by -1.

We observe immediately that the sum and difference of two complex numbers are also complex numbers; or, as we shall say in 14.2, the set of complex numbers $C = \{x + iy : x, y \in R\}$ is *closed* under addition and subtraction. For if $z_1 = x_1 + iy_1$ and $z_2 = x_2 + iy_2$,

$$\text{then} \quad z_1 + z_2 = x_1 + iy_1 + x_2 + iy_2$$
$$= (x_1 + x_2) + i(y_1 + y_2),$$

which is also a complex number; and similarly for subtraction. It is less obvious that the set is closed under multiplication and division.

$$\text{However} \quad z_1 z_2 = (x_1 + iy_1)(x_2 + iy_2) = (x_1x_2 - y_1y_2) + i(x_1y_2 + x_2y_1),$$

$$\text{while} \qquad \frac{z_1}{z_2} = \frac{x_1 + iy_1}{x_2 + iy_2} = \frac{(x_1 + iy_1)(x_2 - iy_2)}{(x_2 + iy_2)(x_2 - iy_2)}$$
$$= \frac{x_1x_2 + y_1y_2}{x_2^2 + y_2^2} + i\frac{x_2y_1 - x_1y_2}{x_2^2 + y_2^2},$$

both of which are in the form $x + iy$. Thus the repeated processes of addition, subtraction, multiplication and division by numbers other than zero will lead every time to another complex number, and the system is self-contained.

Moreover, we note that two complex numbers are equal if and only if their real and imaginary parts are separately equal:

$$\text{For} \quad z_1 = z_2 \;\Rightarrow\; x_1 - x_2 = i(y_2 - y_1)$$
$$\Rightarrow\; (x_1 - x_2)^2 = -(y_2 - y_1)^2$$
$$\Rightarrow\; (x_1 - x_2)^2 + (y_1 - y_2)^2 = 0$$
$$\Rightarrow\; x_1 = x_2 \quad \text{and} \quad y_1 = y_2$$

(since each of the perfect squares of real numbers must be positive or zero). The converse is trivial; and we have also the corollary that

$$z = 0 \;\Leftrightarrow\; x = 0 \quad \text{and} \quad y = 0.$$

Example

In each of the following cases express the given number in the form $x + iy$:

$$(i)\quad \frac{7-3i}{(2-4i)^2} = \frac{7-3i}{4-16i+16i^2} = \frac{7-3i}{-16i-12} = \frac{7-3i}{-4(3+4i)}$$

$$= \frac{(7-3i)(3-4i)}{-4(3+4i)(3-4i)}$$

(multiplying top and bottom by $3 - 4i$ in order to obtain a real denominator)

$$= \frac{21-37i+12i^2}{-4(9-16i^2)} = \frac{9-37i}{-100} = -0.09 + 0.37i.$$

$$(ii)\quad (1+i)^{-3} - (1-i)^{-3} = \frac{1}{1+3i+3i^2+i^3} - \frac{1}{1-3i+3i^2-i^3}$$

$$= \frac{1}{-2+2i} - \frac{1}{-2-2i}$$

$$= \frac{1}{2}\left(\frac{1}{1+i} - \frac{1}{1-i}\right)$$

$$= \frac{1}{2}\frac{1-i-1-i}{(1+i)(1-i)} = \frac{-2i}{4} = -\tfrac{1}{2}i.$$

$$(iii)\quad \frac{z}{z-1} = \frac{x+iy}{x+iy-1} = \frac{(x+iy)(x-1-iy)}{(x-1+iy)(x-1-iy)}$$

$$= \frac{x^2+y^2-x-iy}{(x-1)^2+y^2} = \frac{x^2+y^2-x}{(x-1)^2+y^2} + \frac{-y}{(x-1)^2+y^2}\,i;$$

or we might proceed:

$$\frac{z}{z-1} = 1 + \frac{1}{z-1} = 1 + \frac{x-1-iy}{(x-1+iy)(x-1-iy)}, \quad \text{etc.}$$

Exercise 10.1

1 Express the following in the form $x + iy$:

(i) $(-1-i) + (-2-5i)$;
(ii) $(4-3i) - (6+5i)$;
(iii) $5i(1-2i)$;
(iv) $(4+3i)(2-i)$;
(v) $(7-2i)(-1-3i)$;
(vi) $(1-2i)^2$, $(1-2i)^4$, $(1-2i)^8$;
(vii) $(2-3i)^2 - (4+i)^2$;
(viii) $(1+2i)^3$;
(ix) $(5-6i)(5+6i)$;
(x) $(7-i)(7+i)$;

(*xi*) $(x - i)^5 \quad (x \in R)$;
(*xii*) $(x + iy)^6 \quad (x, y \in R)$;
(*xiii*) Use the result of (*x*) to simplify $\dfrac{1}{7 - i}$ and $\dfrac{1}{7 + i}$.

2 Express in the form $x + iy$:

(*i*) $\dfrac{1}{2 + i}$; (*ii*) $\dfrac{2}{3 - i}$; (*iii*) $\dfrac{1}{(i - 2)(1 - 3i)}$;

(*iv*) $\dfrac{7 + 4i}{3 - 2i}$; (*v*) $\dfrac{i + 1}{i - 1}$; (*vi*) $(1 + 2i)^{-3}$

(*vii*) $\dfrac{(2 - i)^2(1 + i)}{1 - 3i}$; (*viii*) $\dfrac{1}{\cos\theta - i\sin\theta}$.

3 Find integers p, q such that $(3 + 7i)(p + iq)$ is purely imaginary.

4 If $z = x + iy$ $(x, y \in R)$, express in the form $X + iY$:

(*i*) z^3; (*ii*) z^{-2}; (*iii*) $z + \dfrac{1}{z}$.

5 Show that $(4 - 5i)(2 + 3i) = 23 + 2i$, and deduce factors of $23 - 2i$. Hence show that $(4^2 + 5^2)(2^2 + 3^2) = 23^2 + 2^2$.
Use a similar method to express $(4^2 + 7^2)(9^2 + 5^2)$ as the sum of two squares.

6 Solve the equation for z: $\dfrac{z}{3 + 4i} + \dfrac{z - 1}{5i} = \dfrac{5}{3 - 4i}$.

7 Find real values of x and y such that $(1 + 3i)x + (2 - 5i)y + 2i = 0$.

8 Evaluate $\sqrt{(1 + i)}$ and $\sqrt{(5 - 12i)}$ by letting the result be $x + iy$ in each case.

9 Verify that $x = 2$, $y = -1$ satisfy the simultaneous equations
$x^3 - 3xy^2 = 2, \qquad 3x^2y - y^3 = -11.$
Hence find the cube root of $2 - 11i$.

10 Find real values of x and y to satisfy
$$\frac{x}{1 - 2i} + \frac{y}{3 - 2i} = \frac{-5 + 6i}{1 + 8i}.$$

11 Solve the equations:
(*i*) $z^2 + 4z + 29 = 0$; (*ii*) $4z^2 - 12z + 25 = 0$;
(*iii*) $z^2 + 2iz + 1 = 0$; (*iv*) $z^2 + iz = 2$;
(*v*) $z^3 + 10z^2 + 37z + 42 = 0$; (*vi*) $(2 + i)z^2 - z + (2 - i) = 0$.

12 Find the quadratic equations whose roots are:
(*i*) $3 + \sqrt{5}, 3 - \sqrt{5}$; (*ii*) $-3 + i, -3 - i$;
(*iii*) $2 + \sqrt{3}i, 2 - \sqrt{3}i$; (*iv*) $-i, 1 + 2i$.,

13 Find the sum and the product of the roots z_1, z_2 of the equations
(*i*) $z^2 - 5z + 7 = 0$; (*ii*) $z^2 + z + 7 = 0$.
Obtain in each case the value of $z_1^2 + z_2^2$. What do you notice, and how do you account for it?

14 Sum the series:
(*i*) $1 + i + i^2 + i^3 + \cdots + i^n$;
(*ii*) $1 - 2i + 4i^2 - 8i^3 + \cdots + (-2i)^{n-1}$;
(*iii*) $1 + 2i + 3i^2 + 4i^3 + \cdots + ni^{n-1}$.

15 Express $\dfrac{x}{x^2 + 1}$ in partial fractions in the form $\dfrac{A}{x + i} + \dfrac{B}{x - i}$.

Repeat for the expression $\dfrac{1}{x^2 + 1}$.

10.2 The Argand diagram

We have seen that any complex number consists essentially of a pair of real numbers: $z = x + iy$, where x and y are real. In view of this aspect of a complex number as an ordered pair of real numbers, it is natural that we should choose to represent them by the points in a plane using a pair of coordinate axes Ox, Oy, and by assigning the complex number $x + iy$ to the point whose coordinates are (x, y). It is clear that this sets up a 1, 1 correspondence between the complex numbers and the points of the Cartesian plane. For example,

i	is represented by the point	$(0, 1)$,
$-2 + 3i$		$(-2, 3)$,
$-5i$		$(0, -5)$,
-4		$(-4, 0)$;

and conversely, the point $(1, -\frac{1}{2})$ represents the complex number $1 - \frac{1}{2}i$, and so on. This method of representing the complex numbers is referred to as the Argand diagram after the French mathematician J. R. Argand (1768–1822). Ox and Oy are called the 'real' and 'imaginary' axes, for those complex numbers of the form $x + 0i$, i.e., the real numbers, are represented by points of the x-axis, whose rôle is thus seen to be simply that of the real number line; while those complex numbers whose 'real parts' are zero appear in the Argand diagram on the y-axis, and are sometimes loosely referred to as 'purely imaginary', among them being the number i itself.

We have already emphasised that it is a mistake to try and think of i as a number in the sense of a 'magnitude', and it is only after the Argand diagram has been established that it becomes possible to acquire any sort of visual insight into the nature of i.

If i is regarded as a geometric operator, or an instruction to perform a certain geometrical transformation, then the equation $i^2 = -1$ may be translated to mean that the operation i, when performed twice in succession, is equivalent to the operation represented by multiplication by the number -1, which causes a *reversal of direction*. For if $P(x, y)$ represents $x + iy$, then the effect of multiplication by -1 is to replace $x + iy$ by $-x - iy$ (see p. 106 'negative'), and this *reverses* the vector ***OP***, rotating it through 180°. Since, therefore, i^2 causes a 180° rotation in the plane, it is a short step to interpret i as a 90° rotation, for then the equation $i^2 = -1$ states that two successive quarter-turns are equivalent to a half-turn. Whether the quarter-turn is clockwise or anti-clockwise is arbitrary; with the usual choice of Cartesian axes, it will be observed that, if U represents the 'unit' point, $1 + 0i$ (or the real number 1), then OU must be given an anti-clockwise quarter-turn to take up the position OI, where I represents the number i, so that with the usual system of axes the sense of the 90° rotation representing i is *anti-clockwise*.

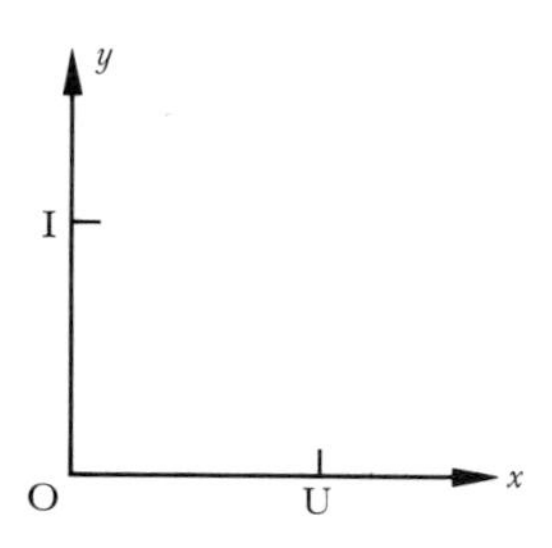

More generally, if
$$z = x + iy,$$
then
$$iz = i(x + iy) = -y + ix;$$

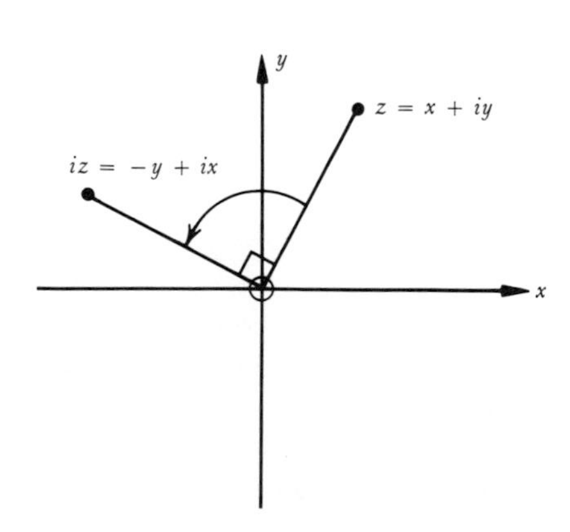

so again we see that multiplication of any complex number by i leads to a rotation of the vector through 90°. What is the effect of multiplication by $-i$?

Addition and subtraction in the Argand diagram

If $z_1 = x_1 + iy_1$ and $z_2 = x_2 + iy_2$.

Then $z_1 + z_2 = (x_1 + x_2) + i(y_1 + y_2)$.

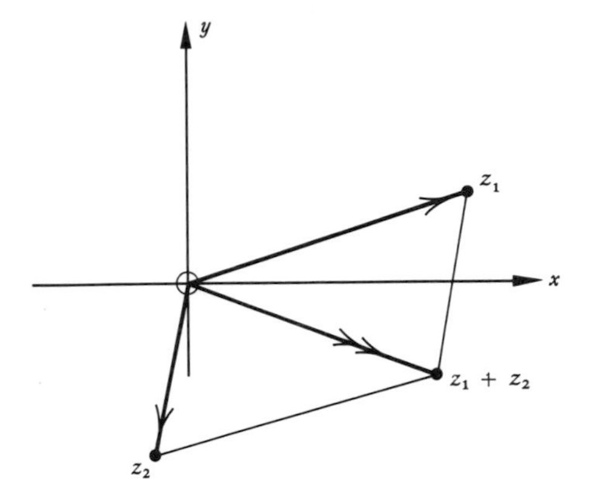

So complex numbers are added like displacements, or like position vectors, according to a parallelogram law. Indeed, the above equation might have been written as

$$\begin{pmatrix} x_1 \\ y_1 \end{pmatrix} + \begin{pmatrix} x_2 \\ y_2 \end{pmatrix} = \begin{pmatrix} x_1 + x_2 \\ y_1 + y_2 \end{pmatrix};$$

this means we can regard complex numbers in the Argand diagram to be represented as well by position vectors as by points. But it must be emphasised that a complex number z corresponds not only to $\boldsymbol{OZ}$ but to the whole class of equivalent vectors which are equal and parallel to $\boldsymbol{OZ}$ (or, as we shall say in Chapter 14, the additive group of complex numbers is *isomorphic* with that of the two-dimensional vectors).

Continued addition

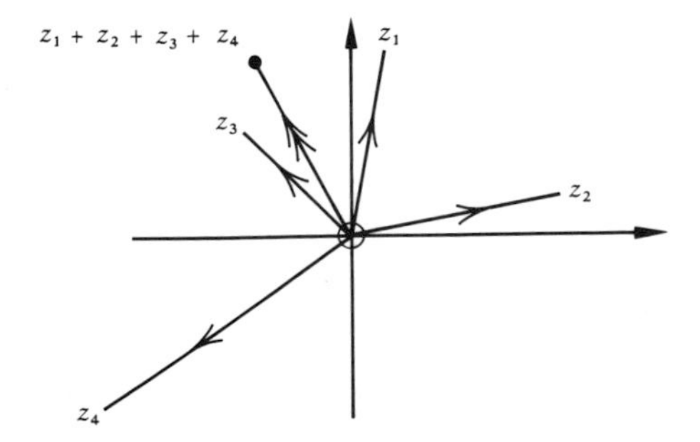

When several complex numbers are being added, there is no need for them to be combined laboriously in pairs using the parallelogram law repeatedly. Any number of vectors may be added by 'nose-to-tail' addition, the order in which they are added being immaterial in view of the associativity and commutativity of vector addition. Such an addition of complex numbers z_1, z_2, z_3, z_4 is illustrated by the vector polygons overleaf.

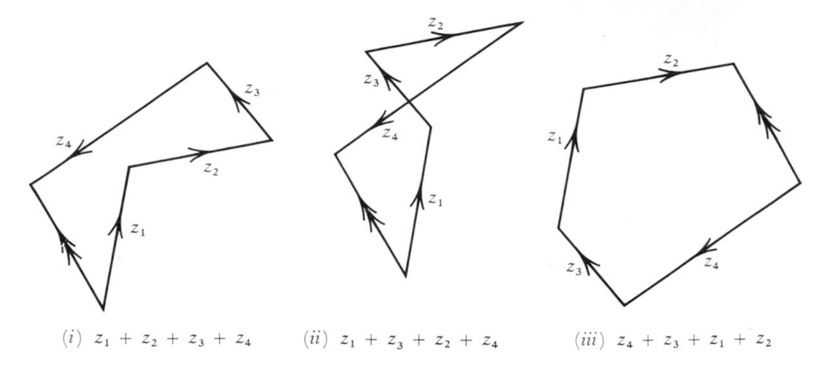

(*i*) $z_1 + z_2 + z_3 + z_4$ (*ii*) $z_1 + z_3 + z_2 + z_4$ (*iii*) $z_4 + z_3 + z_1 + z_2$

The negative of a complex number

If $z = x + iy$, then $-z = -x - iy$, so that $-z$ is obtained geometrically from z by means of a half-turn about the origin.

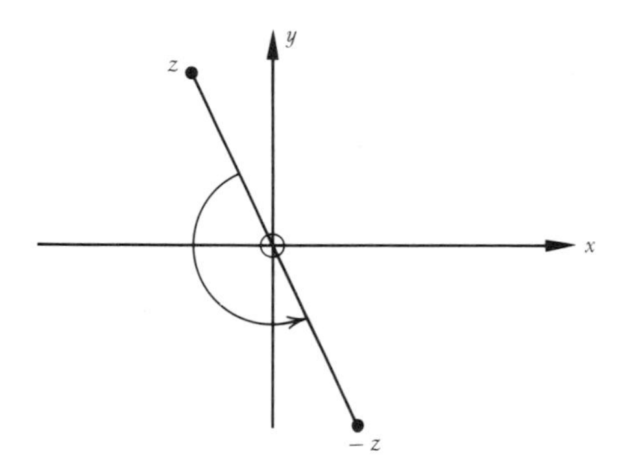

The subtraction of complex numbers

When we say, in the field of real numbers, that $7 - 4 = 3$, this is open to two interpretations:
(*i*) 3 is the number which, when added to 4, makes 7: $3 + 4 = 7$.

More generally, $a - b = c \Leftrightarrow b + c = a$;

(*ii*) 3 is the sum of 7 and -4: $7 + (-4) = 3$.

More generally, $a - b = a + (-b)$, or subtraction is equivalent to the *addition of the negative*, or *additive inverse* (see 14.1).

Similarly in the field of complex numbers, $z_1 - z_2$ has two interpretations, each of which can be seen in the Argand diagram:

(*i*) $z_1 - z_2$ is represented by that vector which, when added to z_2, gives z_1, and (*ii*) $z_1 - z_2$ is the sum of z_1 and $(-z_2)$, and when z_1 and $(-z_2)$ are added by the parallelogram law to give $z_1 - z_2$ it is immediately clear that the vector joining the origin to this point is equivalent to that joining z_2 to z_1.

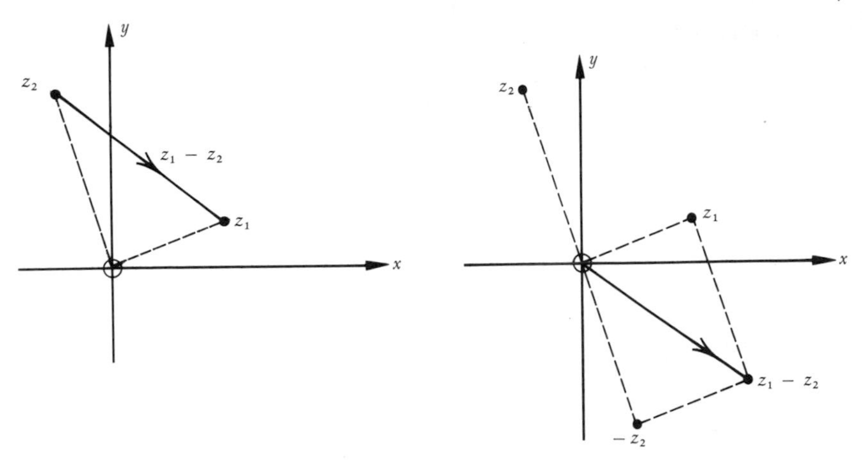

The first of the two geometrical interpretations is more useful in practice, and we may say generally that, if complex numbers $a, b, c, \ldots$ are represented by points $A, B, C, \ldots$, then

$a - b$	is represented by	$\boldsymbol{BA}$	
$b - a$		$\boldsymbol{AB}$	(note the 'reversal')
$q - p$		$\boldsymbol{PQ}$	

and so on.

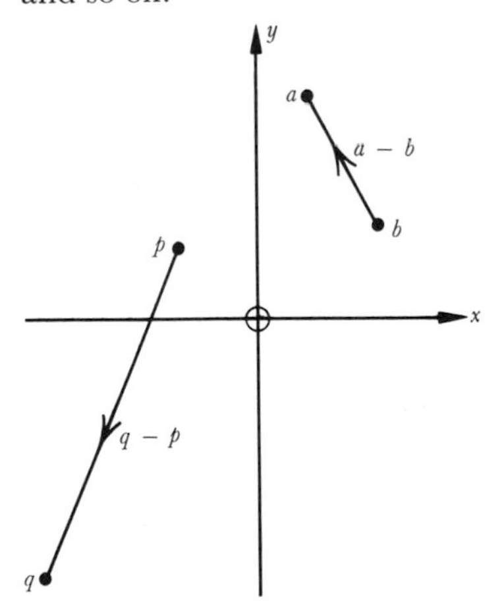

Multiplication by a real number

It is evident in general that if two complex numbers z_1, z_2 are so situated that the vectors $\boldsymbol{OZ}_1$ and $\boldsymbol{OZ}_2$ have the *same direction*, then one is a real multiple of the other. For example, if $z_1 = 3 + 5i$, and $z_2 = 9 + 15i$, then $\boldsymbol{OZ}_2 = 3\boldsymbol{OZ}_1$, these vectors having the same direction. Similarly, if $z_3 = -1\frac{1}{2} - 2\frac{1}{2}i$ then we have $\boldsymbol{OZ}_3 = -\frac{1}{2}\boldsymbol{OZ}_1$: these two vectors have the same direction, but the minus sign informs us of their opposite senses.

Exercise 10.2a

1 Plot the following points on the Argand diagram:
(*i*) $1 - i$; (*ii*) $-2 + 3i$; (*iii*) $-2 - 3i$; (*iv*) $4 - i$; (*v*) $-12 + 5i$; (*vi*) $\cos \frac{1}{3}\pi + i \sin \frac{1}{3}\pi$; (*vii*) $\cos \pi + i \sin \pi$; (*viii*) $-\frac{1}{2}\sqrt{3} + \frac{1}{2}i$.

2 If $z = 1 + i$, simplify $z^2, z^3, z^4, \ldots,$ and mark these points on the Argand diagram.

3 If $z_1 = 2 - i$, $z_2 = 1 + 3i$, mark on the Argand diagram the points

$$z_1, \quad z_2, \quad z_1 + z_2, \quad z_1 - z_2, \quad z_2 - z_1, \quad z_1 z_2, \quad \frac{z_1}{z_2}, \quad \frac{z_2}{z_1}, \quad z_1^2, \quad \text{etc.}$$

4 Let $z = -5 + 12i$. Plot the following points on the Argand diagram:

$$z, \quad iz, \quad z - 1, \quad 3z, \quad -3z, \quad \frac{1}{z}, \quad z^2.$$

Mark also the points z', $-z'$ and zz' where $z' = -5 - 12i$.

5 Show that the points a, b, c, d ($\in C$) form a parallelogram if and only if the sum of two of them is equal to the sum of the other two.

6 Give a geometrical interpretation of $a - b = +i(c - d)$ where $a, b, c, d \in C$.

7 If A, B represent complex numbers a, b, obtain the points which represent:
(*i*) $\frac{1}{2}(a + b)$; (*ii*) $\frac{1}{2}a + \frac{1}{2}b$; (*iii*) $a + \frac{1}{2}(b - a)$; (*iv*) $b + \frac{1}{2}(a - b)$.

8 ABCD is a rhombus with $AC = 2BD$. If $b = 3 + i$, $d = 1 - 3i$, find a and c.

9 The centre of a square is the point $-2 + i$, and one vertex is $1 + 3i$. Find the others.

10 ABCD is a square, with $a = 3 + i$, $b = 4 - 2i$. Find possible positions of C and D.

11 ABCD is a square with $a = -2 + 5i$, $c = 4 - 8i$. Find b and d.

12 If $a = 2 - i$, $b = 4 - 3i$, find two values of c so that triangle ABC is equilateral.

Division of a line in a given ratio

If P is the mid-point of AB, it is also the mid-point of the other diagonal of the parallelogram formed by OA and OB as adjacent sides, so that $p = \frac{1}{2}(a + b)$.

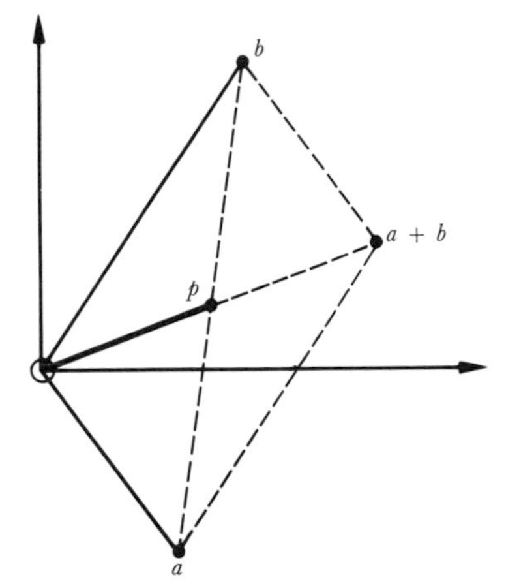

More generally, if P divides AB in the ratio $\beta:\alpha$, then since AP and PB have the same direction, and their lengths are in the ratio $\beta:\alpha$, it follows that

$$\alpha \boldsymbol{AP} = \beta \boldsymbol{PB}$$

so that $\alpha(p - a) = \beta(b - p)$.

Hence $(\alpha + \beta)p = \alpha a + \beta b \Rightarrow p = \dfrac{\alpha a + \beta b}{\alpha + \beta}$.

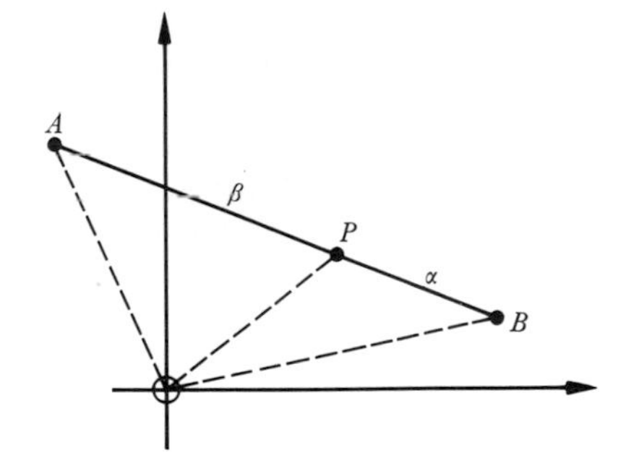

Note that (*a*) when $\alpha = \beta = 1$, we get $p = \frac{1}{2}(a + b)$ as the mid-point of AB; (*b*) when either α or β is negative, the point P will lie outside the line segment AB, in the extension of either AB or BA; (*c*) we must exclude the case $\alpha = -\beta$; (*d*) we choose to name the two parts of the ratio so that β and α are reckoned from A and B respectively. The reason for this is explained elsewhere (see 11.1). Briefly, P is the centre of mass of α at A and β at B respectively.

Centroid of a triangle

If G is the centroid of triangle ABC whose medians are AP, BQ, CR, we have $p = \frac{1}{2}(b + c)$. Now G divides each median in the ratio 1 : 2 so that

$PG : GA = 1:2 \Rightarrow g = \frac{2}{3}p + \frac{1}{3}a = \frac{2}{3} \times \frac{1}{2}(b + c) + \frac{1}{3}a = \frac{1}{3}(a + b + c).$

Equating real and imaginary parts would give us the coordinates of G as

$[\frac{1}{3}(x_1 + x_2 + x_3), \frac{1}{3}(y_1 + y_2 + y_3)],$

where A, B, and C are (x_1, y_1), (x_2, y_2), and (x_3, y_3).

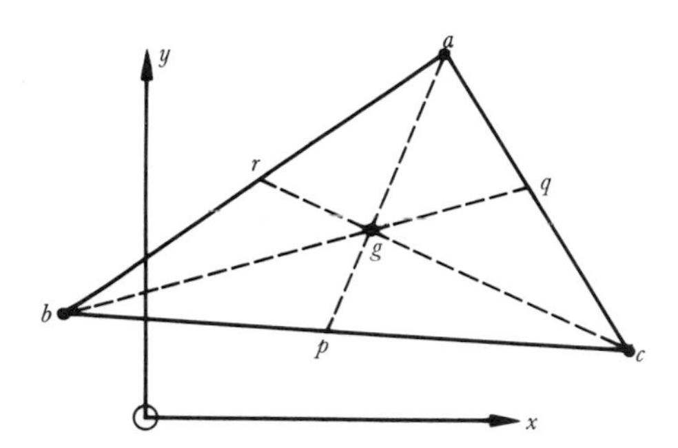

Exercise 10.2b

1 Find the centroid of the triangle ABC where $a = -2i$, $b = 4 + i$, $c = -10 - 5i$.

2 ABCD is any quadrilateral. Prove by using complex numbers that the lines joining the mid-points of AB and CD, of AC and BD, and of AD and BC all bisect each other.

3 ABCD is a parallelogram. P is the mid-point of BC, BD and AP intersect at F. Prove by using complex numbers that F trisects both BD and AP.

4 Given any three complex numbers a, b, c, show that there exist real numbers α, β, γ such that $a\alpha + b\beta + c\gamma = 0$. Show further that A, B, and C are collinear when $\alpha + \beta + \gamma = 0$.

5 The triangle ABC is equilateral. Show that

$c = \frac{1}{2}(a + b) \pm \frac{1}{2}\sqrt{3}(a - b)i$

and express c also in terms of ω where $\omega = \cos \frac{2}{3}\pi + i \sin \frac{2}{3}\pi$. Find also the complex number representing the centre of the equilateral triangle and show that

$a^2 + b^2 + c^2 - bc - ca - ab = 0$

is a necessary and sufficient condition for the triangle ABC to be equilateral (which may be used to prove 'Napoleon's theorem', see 10.2).

6 Given points A, B, C representing complex numbers a, b, c, show how to construct the positions of:

$a + b$, $3a + 2b$, $b - 2a$, $a - c$, $a + b + c$,
$a + b - c$, $a + 2b - 3c$.

Polar form, modulus, and argument

We have already seen that the addition of complex numbers has an immediate interpretation on the Argand diagram in terms of the addition of vectors. When it comes to multiplication, if $z = z_1 z_2$,

$$\text{then} \quad z = z_1 z_2 = (x_1 + iy_1)(x_2 + iy_2)$$
$$= (x_1 x_2 - y_1 y_2) + i(x_1 y_2 + x_2 y_1), \quad \text{so that}$$
$$\Rightarrow \quad \left.\begin{aligned} x &= x_1 x_2 - y_1 y_2 \\ y &= x_1 y_2 + x_2 y_1 \end{aligned}\right\}.$$

These formulae for the Cartesian coordinates of the point representing the product do not help us to locate it readily on the Argand diagram, and in order to provide a simple interpretation it is necessary to use polar co-ordinates.

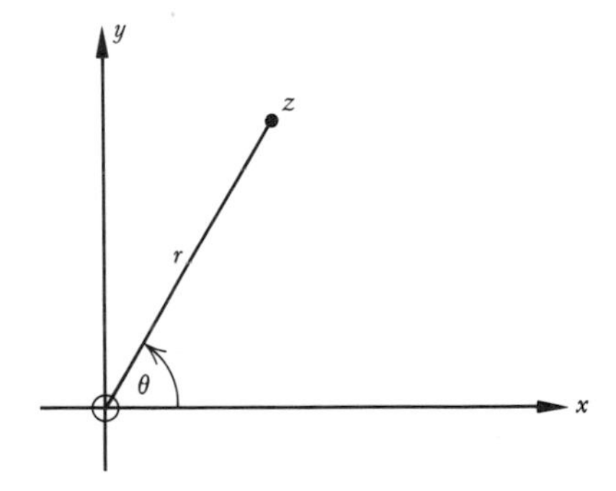

We have, for all non-zero values of $z(= x + iy)$,

$$x = r \cos \theta, \qquad y = r \sin \theta$$

where r (= the distance OZ) is called the *modulus* of z, written $|z|$,

and θ (= $\angle$XOZ) is called the *argument* (or *amplitude* in some older books), written arg z.

Hence $z = x + iy$ may now be expressed as $z = r(\cos \theta + i \sin \theta)$, which we shall usually abbreviate to $z = r \operatorname{cis} \theta$,

$$\text{where} \quad r = |z| = \sqrt{(x^2 + y^2)}, \quad \text{and} \quad \theta = \arg z = \tan^{-1} \frac{y}{x}.$$

(It must be emphasised that, although when working with polar coordinates it is possible for r to be negative, here r is essentially positive.)

So, for example, if $z = 2 - 5i$,

$$\text{then} \quad |z| = \sqrt{(2^2 + (-5)^2)} = \sqrt{29}$$

$$\arg z \text{ (in fourth quadrant)} = \tan^{-1}(-5/2) = 291^\circ\, 48'$$

$$\text{and} \quad 2 - 5i = \sqrt{29}(\cos 291^\circ\, 48' + i \sin 291^\circ\, 48')$$
$$= \sqrt{29} \operatorname{cis} 291^\circ\, 48'.$$

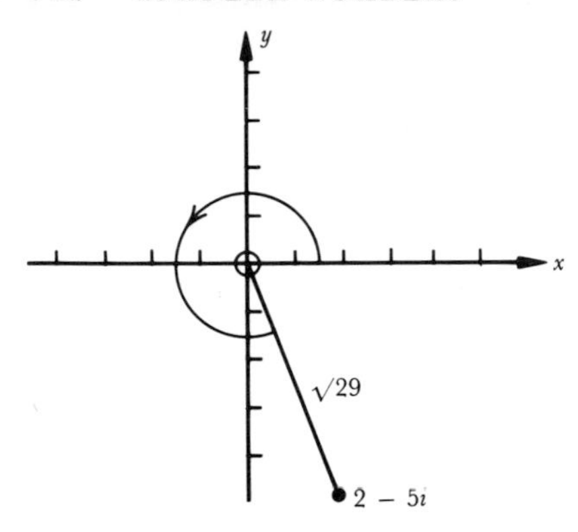

But there is clearly an infinity of possible values of arg z, such as $651^\circ\ 48'$, $1\ 011^\circ\ 48'$, $-68^\circ\ 12'$, $-428^\circ\ 12'$, etc. (all of which are equivalent *modulo* 360°); so of these we usually select the *principal value* θ, such that $-\pi < \theta \leqslant \pi$, the principal value of the argument in the example above being $-68^\circ\ 12'$.

Similarly, with the complex number $-2i$,

$|-2i| = 2, \quad$ and $\arg(-2i) = -\frac{1}{2}\pi.$

So $\quad -2i = 2(\cos(-\frac{1}{2}\pi) + i\sin(-\frac{1}{2}\pi))$

$\qquad\qquad = 2 \text{ cis } (-\frac{1}{2}\pi).$

Summarising,

$$z = x + iy = r(\cos\theta + i\sin\theta) = r \text{ cis } \theta$$

$$Rz = x = r\cos\theta \qquad r = |z| = \sqrt{(x^2 + y^2)} \quad \text{(modulus)}$$

$$Iz = iy = ir\sin\theta \qquad \theta = \arg z = \tan^{-1} y/x \quad \text{(argument)}$$

(and for its principal value, $-\pi < \theta \leqslant \pi$)

Example 1

Express in polar form:

$(i)\ -\sqrt{3} + i; \quad (ii)\ 2 - 3i; \quad (iii)\ \dfrac{4 + 3i}{12i - 5}; \quad (iv)\ \dfrac{(i - 2)^2(3i + 1)}{i - 3}.$

In all these examples, and in the following exercises, it is a great help to plot the complex numbers on the Argand diagram, as this enables the modulus and argument readily to be 'read off'.

$(i)\quad -\sqrt{3} + i = 2(-\frac{1}{2}\sqrt{3} + \frac{1}{2}i) = 2(\cos 150^\circ + i\sin 150^\circ);$

$(ii)\ |2 - 3i| = \sqrt{(2^2 + 3^2)} = \sqrt{13},$

$\quad \arg(2 - 3i) = \tan^{-1}(-\frac{3}{2})$ (4th quadrant) $= -56^\circ\ 19'.$

So $2 - 3i = \sqrt{13} \text{ cis } (-56° \, 19')$

(*iii*) $$\frac{4 + 3i}{12i - 5} = \frac{(4 + 3i)(-5 - 12i)}{(-5 + 12i)(-5 - 12i)} = \frac{16 - 63i}{169}$$

$$= \frac{65}{169}\left(\frac{16}{65} - \frac{63}{65}i\right) \quad (\text{for } 16^2 + 63^2 = 65^2)$$

$$= \frac{65}{169} \text{ cis } \alpha;$$

where $\alpha = \tan^{-1}(-\frac{63}{16})$ in the fourth quadrant.

(*iv*) $$\frac{(i - 2)^2(3i + 1)}{i - 3} = \frac{(3 - 4i)(1 + 3i)}{-3 + i} = \frac{(15 + 5i)(-3 - i)}{(-3 + i)(-3 - i)}$$

$$= \frac{-40 - 30i}{10} = -4 - 3i = 5 \text{ cis } 216° \, 52'.$$

Example 2

If $z = \text{cis } \theta$, express in polar form:

(*i*) $1 - z$; (*ii*) $\dfrac{2z}{1 - z^2}$; (*iii*) $\dfrac{z + 1}{z - 1}$.

(*i*) $$\begin{aligned} 1 - \cos\theta - i\sin\theta &= 2\sin^2 \tfrac{1}{2}\theta - 2i\sin\tfrac{1}{2}\theta\cos\tfrac{1}{2}\theta \\ &= 2\sin\tfrac{1}{2}\theta(\sin\tfrac{1}{2}\theta - i\cos\tfrac{1}{2}\theta) \\ &= 2\sin\tfrac{1}{2}\theta \text{ cis}(-\tfrac{1}{2}\pi + \tfrac{1}{2}\theta); \end{aligned}$$

(*ii*) $$\frac{2z}{1 - z^2} = \frac{2 \text{ cis } \theta}{1 - (\cos\theta + i\sin\theta)^2}$$

$$= \frac{2 \text{ cis } \theta}{1 - \cos^2\theta - 2i\sin\theta\cos\theta + \sin^2\theta}$$

$$= \frac{2 \text{ cis } \theta}{2\sin\theta(\sin\theta - i\cos\theta)}$$

$$= \frac{(\cos\theta + i\sin\theta)(\sin\theta + i\cos\theta)}{\sin\theta(\sin\theta - i\cos\theta)(\sin\theta + i\cos\theta)}$$

$$= \frac{i(\cos^2\theta + \sin^2\theta)}{\sin\theta(\sin^2\theta + \cos^2\theta)} = i \text{ cosec } \theta;$$

$$(iii)\quad \frac{z+1}{z-1} = \frac{\text{cis}\,\theta + 1}{\text{cis}\,\theta - 1} = \frac{\cos\theta + i\sin\theta + 1}{\cos\theta + i\sin\theta - 1}$$

$$= \frac{2\cos^2\tfrac{1}{2}\theta + 2i\sin\tfrac{1}{2}\theta\cos\tfrac{1}{2}\theta}{-2\sin^2\tfrac{1}{2}\theta + 2i\sin\tfrac{1}{2}\theta\cos\tfrac{1}{2}\theta}$$

$$= \frac{\cos\tfrac{1}{2}\theta(\cos\tfrac{1}{2}\theta + i\sin\tfrac{1}{2}\theta)}{-\sin\tfrac{1}{2}\theta(\sin\tfrac{1}{2}\theta - i\cos\tfrac{1}{2}\theta)}$$

$$= \frac{-\cot\tfrac{1}{2}\theta(\cos\tfrac{1}{2}\theta + i\sin\tfrac{1}{2}\theta)(\sin\tfrac{1}{2}\theta + i\cos\tfrac{1}{2}\theta)}{\sin^2\tfrac{1}{2}\theta + \cos^2\tfrac{1}{2}\theta}$$

$$= -i\cot\tfrac{1}{2}\theta.$$

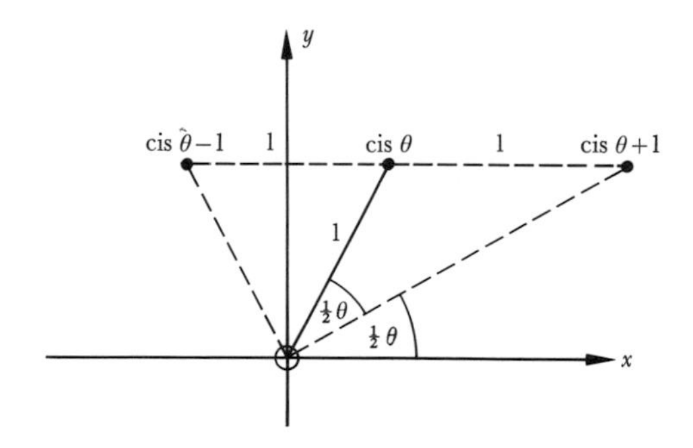

Note that the two questions above may be done more easily by using the device:

$$\begin{aligned}\text{cis}\,\theta + 1 &= \text{cis}\,\tfrac{1}{2}\theta(\text{cis}\,\tfrac{1}{2}\theta + \text{cis}(-\tfrac{1}{2}\theta)) = \text{cis}\,\tfrac{1}{2}\theta \times 2\cos\tfrac{1}{2}\theta\\ &= 2\cos\tfrac{1}{2}\theta\,\text{cis}\,\tfrac{1}{2}\theta\\ \text{cis}\,\theta - 1 &= \text{cis}\,\tfrac{1}{2}\theta(\text{cis}\,\tfrac{1}{2}\theta - \text{cis}(-\tfrac{1}{2}\theta)) = \text{cis}\,\tfrac{1}{2}\theta \times 2i\sin\tfrac{1}{2}\theta\\ &= 2i\sin\tfrac{1}{2}\theta\,\text{cis}\,\tfrac{1}{2}\theta = 2\sin\tfrac{1}{2}\theta\,\text{cis}(\tfrac{1}{2}\theta + \tfrac{1}{2}\pi).\end{aligned}$$

We shall return to these again on pp. 122, 137.

Inequalities

Since $z_1 + z_2$ is represented by a side of the triangle, it is immediately clear that

$$\boxed{|z_1 + z_2| \leqslant |z_1| + |z_2|}$$

(and the equality would only occur if the triangle collapsed into a straight line).

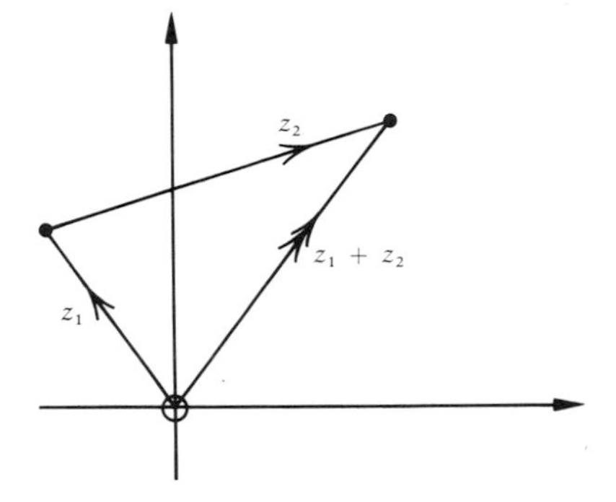

More generally, we see that

$$|z_1 + z_2 + \cdots + z_n| \leqslant |z_1| + |z_2| + \cdots + |z_n|,$$

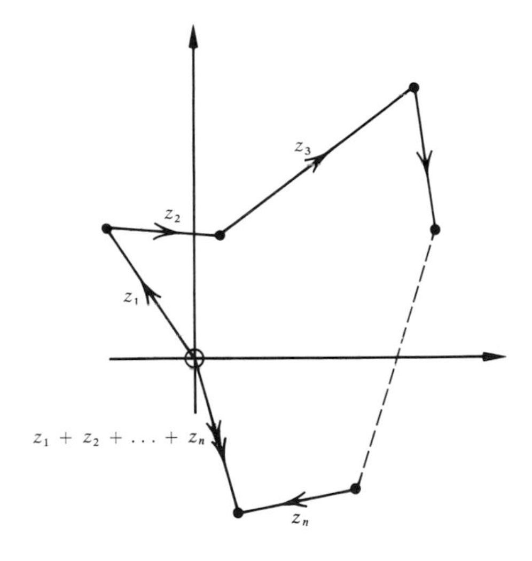

while the two further results

$$|z_1 - z_2| \geqslant |z_1| - |z_2| \quad \text{and} \quad |z_1 - z_2| \leqslant |z_1| + |z_2|$$

also have simple geometrical counterparts.

Exercise 10.2c

1 Express the following complex numbers in the polar form

$$r(\cos\theta + i\sin\theta),$$

giving the principal value of the argument, and illustrating on the Argand diagram:

(*i*) -1; (*ii*) $1 - i$; (*iii*) $1 + \sqrt{3}i$; (*iv*) $-i$; (*v*) $-12 - 5i$;
(*vi*) $-3 + 2i$; (*vii*) $\sin\alpha + i\cos\alpha$; (*viii*) $\sin\alpha - i\cos\alpha$.

2 Express in the form $x + iy$:

(*i*) $4 \text{ cis } \frac{\pi}{6}$; (*ii*) $5 \text{ cis } \pi$; (*iii*) $10 \text{ cis } \left(-\frac{\pi}{4}\right)$; (*iv*) $8 \text{ cis } 165°$;

(*v*) $2 \text{ cis } 800°$; (*vi*) $6 \text{ cis } \frac{7\pi}{4}$.

3 If $a = 2 + i$, $b = 1 + 3i$, mark in the Argand diagram the points representing:

(*i*) ab; (*ii*) $\frac{a}{b}$; (*iii*) $\frac{b}{a}$; (*iv*) a^2; (*v*) b^2; (*vi*) $a^2 + b^2$;

and find the modulus and argument in each case.

4 If $a = \text{cis } \frac{2}{3}\pi$ and $b = 2 \text{ cis}(-\frac{1}{4}\pi)$, evaluate:

(*i*) $a^2, a^3, \ldots$; (*ii*) $b^2, b^3, \ldots$; (*iii*) $ab, \frac{a}{b}$.

Give the principal value of the arguments and mark each point in the Argand diagram.

5 Simplify $\text{cis } \theta \times \text{cis } \phi$. Use the result to write down

(*i*) $\text{cis } \frac{\pi}{6} \times \text{cis } \frac{4\pi}{3}$; (*ii*) $(\text{cis } 80°)^3$.

6 Solve the equations, giving the roots in polar form:

(*i*) $z^2 - z + 1 = 0$; (*ii*) $z^2 + z + 1 = 0$;
(*iii*) $9z^2 + 12z + 5 = 0$; (*iv*) $z^2 - z + 1 - i = 0$.

7 Find quadratic equations whose roots are:

(*i*) $\text{cis } 120°, \text{cis } 240°$; (*ii*) $\text{cis } \frac{\pi}{5}, \text{cis } -\frac{\pi}{5}$;

(*iii*) $\cos \alpha + i \sin \alpha$, $\cos \alpha - i \sin \alpha$.

8 By using the identity $(a^2 + b^2)(c^2 + d^2) = (ac - bd)^2 + (bc + ad)^2$, prove that $|z_1||z_2| = |z_1 z_2|$.

9 Using only the definition $|z|^2 = x^2 + y^2$, prove that $|z_1 + z_2| \leqslant |z_1| + |z_2|$.

10 If $|z_1| = |z_2|$, prove that $(z_1 + z_2)/(z_1 - z_2)$ is purely imaginary. Interpret geometrically.

11 Prove that $|z_1 + z_2|^2 + |z_1 - z_2|^2 = 2|z_1|^2 + 2|z_2|^2$, and give a geometrical interpretation of this result.

12 Show that the equation $|z| = 1$ represents a circle of unit radius and centre O. Describe what the following equations represent:

(*i*) $|z - i| = 2$; (*ii*) $|z + 1| = |z - 2i|$;

(*iii*) $|z + 3i| < 2$; (*iv*) $|z| < 4$ and $|z - 2| > 1$;
(*v*) $|z - 3| + |z + 3| = 10$.

13 Illustrate the regions:
(*i*) $|z - i| > |z + i|$; (*ii*) $|z| > 3 > |z - 3|$.

14 Given that $|z - 2 - i| < 5$, show geometrically that

$8 < |z - 14 - 6i| < 18$.

15 Prove that $\left|\dfrac{1}{1 - z}\right| \leqslant \dfrac{1}{1 - |z|}$.

Multiplication of complex numbers

We admitted earlier that the rule for multiplication, when expressed in Cartesian form $x + iy$, is cumbersome and does not lend itself to simple geometrical interpretation. When the polar form is used, however, a much simpler result emerges:

For
$$\begin{aligned}&\operatorname{cis}\theta_1 \times \operatorname{cis}\theta_2\\ &= (\cos\theta_1 + i\sin\theta_1)(\cos\theta_2 + i\sin\theta_2)\\ &= (\cos\theta_1\cos\theta_2 - \sin\theta_1\sin\theta_2) + i(\sin\theta_1\cos\theta_2 + \cos\theta_1\sin\theta_2)\\ &= \cos(\theta_1 + \theta_2) + i\sin(\theta_1 + \theta_2)\\ &= \operatorname{cis}(\theta_1 + \theta_2).\end{aligned}$$

So
$$\boxed{\operatorname{cis}\theta_1\operatorname{cis}\theta_2 = \operatorname{cis}(\theta_1 + \theta_2)}$$

Hence if $z_1 = r_1 \operatorname{cis}\theta_1$ and $z_2 = r_2\operatorname{cis}\theta_2$,

it follows that

$$z_1z_2 = r_1\operatorname{cis}\theta_1 \times r_2\operatorname{cis}\theta_2 = r_1r_2\operatorname{cis}(\theta_1 + \theta_2).$$

So, in multiplying complex numbers, we simply multiply their moduli and add their arguments:

$$\boxed{|z_1z_2| = |z_1||z_2| \quad\text{and}\quad \arg(z_1z_2) = \arg z_1 + \arg z_2}$$

Evidently the rule may be extended for the product of more than two complex numbers:

$$\begin{aligned}z_1z_2z_3 &= r_1\operatorname{cis}\theta_1 \times r_2\operatorname{cis}\theta_2 \times r_3\operatorname{cis}\theta_3 = r_1r_2\operatorname{cis}(\theta_1 + \theta_2) \times r_3\operatorname{cis}\theta_3\\ &= r_1r_2r_3\operatorname{cis}(\theta_1 + \theta_2 + \theta_3);\end{aligned}$$

and more generally, $z_1z_2\ldots z_n = r_1r_2\ldots r_n\operatorname{cis}(\theta_1 + \theta_2 + \cdots + \theta_n)$.

So $|z_1z_2z_3\ldots z_n| = |z_1||z_2||z_3|\ldots|z_n|$
and $\arg(z_1z_2z_3\ldots z_n) = \arg z_1 + \arg z_2 + \arg z_3 + \cdots + \arg z_n$.

As a general rule, whenever one is concerned with the *multiplication* of complex numbers, it is usually advisable to have them in modulus-argument form. If one wished to evaluate the very simple product $(2 + i)(1 + 3i)$ $(= -1 + 7i)$, it would hardly be worthwhile changing into polars; but it is nevertheless instructive to do so as an illustration:

$$\begin{aligned} z_1 &= 2 + i = \sqrt{5} \text{ cis } 26^\circ\ 34'; \qquad z_2 = 1 + 3i = \sqrt{10} \text{ cis } 71^\circ\ 34' \\ \Rightarrow \quad z_1 z_2 &= \sqrt{50} \text{ cis}(26^\circ\ 34' + 71^\circ\ 34') = \sqrt{50} \text{ cis } 98^\circ\ 8' \\ &= 5\sqrt{2} \text{ cis } 98^\circ\ 8', \end{aligned}$$

which will be found to agree with $-1 + 7i$. However, if one wished to evaluate $(1 + 3i)^7$, it would be well worthwhile using the polar form to avoid the work with the binomial theorem. Comparing the two methods:

$$\begin{aligned} (a)\ (1 + 3i)^7 &= 1 + {}^7C_1 3i + {}^7C_2 (3i)^2 + {}^7C_3(3i)^3 + {}^7C_4(3i)^4 \\ &\quad + {}^7C_5(3i)^5 + {}^7C_6(3i)^6 + (3i)^7 \\ &= 1 + 21i + 189i^2 + 945i^3 + 2\,835i^4 + 5\,103i^5 \\ &\quad + 5\,103i^6 + 2\,187i^7 \\ &= 1 + 21i - 189 - 945i + 2\,835 + 5\,103i - 5\,103 \\ &\quad - 2\,187i \quad (\text{putting } i^2 = -1) \\ &= -2\,456 + 1\,992i; \end{aligned}$$

$$\begin{aligned} (b)\ (1 + 3i)^7 &= (\sqrt{10} \text{ cis } 71^\circ\ 34')^7 = (\sqrt{10})^7 \text{ cis}(71^\circ\ 34' + 71^\circ\ 34' \\ &\quad + \cdots + 71^\circ\ 34') \quad (\text{adding the argument 7 times}) \\ &= 1\,000\sqrt{10} \text{ cis } 500^\circ\ 58' = 1\,000\sqrt{10} \text{ cis } 140^\circ\ 58' \\ &\approx 3\,162(-0.776\,7 + 0.629\,7i) = -2\,457 + 1\,991i. \end{aligned}$$

The second answer is less accurate, due to the use of 4-figure tables; but its advantages are plain, and would of course increase if the exponent, 7, had been larger.

The unit circle

An important special case is that of numbers of *unit modulus* (i.e., of the form $z = 1 \text{ cis } \theta$) lying on the unit circle, $|z| = 1$. These are multiplied by simply adding their arguments, i.e., by performing a rotation about the origin on the Argand diagram. This will be clear since

$$z_1 z_2 = \text{cis } \theta_1 \text{ cis } \theta_2 = \text{cis}(\theta_1 + \theta_2),$$

so that in an Argand diagram $z_1 z_2$ may be obtained either by rotating about O from position z_1 through θ_2, or from z_2 through θ_1 (positive anti-clockwise). In particular, multiplication of any complex number by $i(= \text{cis } 90^\circ)$ is represented by a positive rotation through a right angle.

This leads us to consider the general interpretation of multiplication on the Argand diagram. Suppose we consider again the product $z_1 z_2$ where $z_1 = 2 + i$, $z_2 = 1 + 3i$. Then $z_1 z_2 = \sqrt{5} \text{ cis } \alpha \times \sqrt{10} \text{ cis } \beta$, where $\alpha = 26^\circ\ 34'$, $\beta = 71^\circ\ 34'$. If we start with z_1, we may multiply by z_2 in

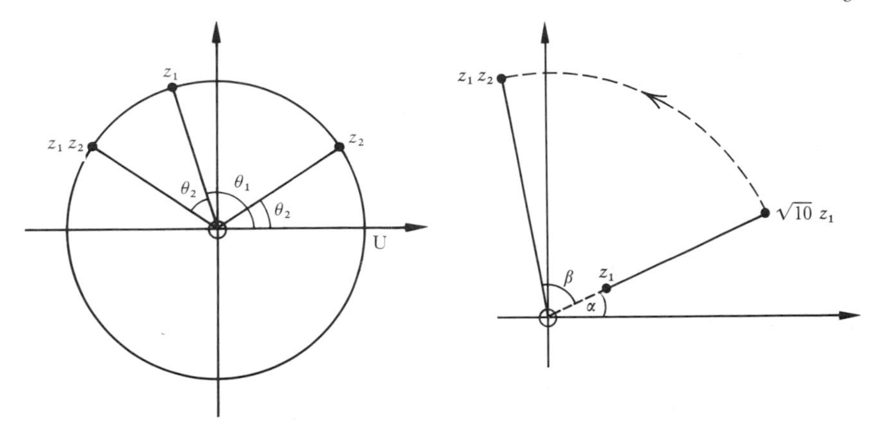

two stages: first by $\sqrt{10}$, then by cis β. Multiplication by $\sqrt{10}$ clearly corresponds to an enlargement from the origin by a factor $\sqrt{10}$ and multiplication by cis β causes the argument to be increased by β to the value $\alpha + \beta$, i.e., corresponds to a rotation through β, which brings us to the final position $z_1\sqrt{10}$ cis $\beta = z_1 z_2$. More generally, multiplication by r cis θ corresponds to enlargement from the origin by a factor r, coupled with a rotation, through θ about O. Evidently the two transformations — enlargement and rotation, both about O — commute, and the composite transformation is known as a *spiral similarity*.

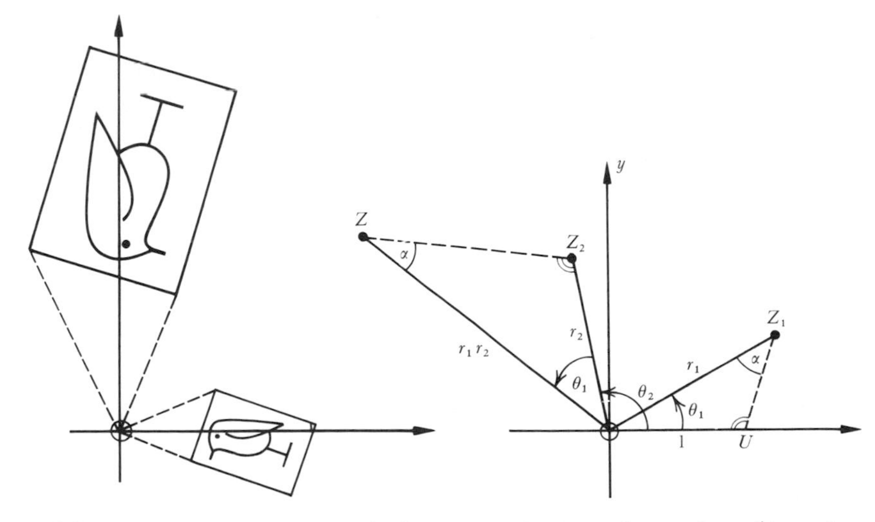

We now see how a geometrical construction may be used to determine the position of the point Z representing the product $z_1 z_2$ when the points Z_1, Z_2 representing z_1, z_2 are given. Letting U be the unit point, we have $OU = 1$, $OZ_1 = r_1$, $OZ_2 = r_2$, $OZ = r_1 r_2$ so that $OZ/OZ_2 = OZ_1/OU$. Since each of $\angle$s UOZ_1, Z_2OZ is equal to θ_1, it follows that $\triangle$s UOZ_1, Z_2OZ are similar, so that the position of Z may be constructed by making $\angle Z_2OZ = \angle UOZ_1$ and $\angle OZ_2Z_1 = \angle OUZ_1$.

Reciprocal of a complex number

We have already seen that the reciprocal of the complex number $x + iy$ is $(x - iy)/(x^2 + y^2)$, the result being obtained by the use of the conjugate $x - iy$. However, it is somewhat easier to consider the reciprocal of a complex number when it is expressed in polar form.

For $$\frac{1}{z} = \frac{1}{r \operatorname{cis} \theta} = \frac{\operatorname{cis}(-\theta)}{r \operatorname{cis} \theta \operatorname{cis}(-\theta)} = \frac{\operatorname{cis}(-\theta)}{r \operatorname{cis} 0} = \frac{1}{r} \operatorname{cis}(-\theta).$$

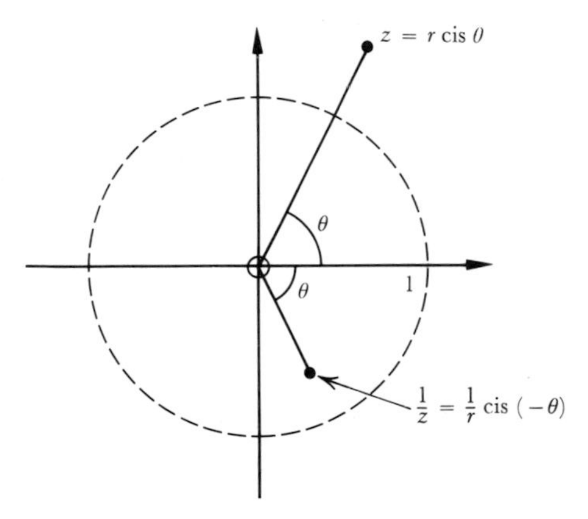

Note that cis θ and cis $(-\theta)$ are reciprocals, i.e., that

$(\cos \theta + i \sin \theta)(\cos \theta - i \sin \theta) = 1.$

To plot the point on the Argand diagram representing the reciprocal of a given complex number z, we simply make $\angle$UOZ$' = \theta$, measured clockwise [since $\arg(1/\bar{z}) = -\theta$], and $OZ' = 1/r$, where U is the unit point. Since $OZ/OU = OU/OZ'$, and $\angle$UOZ $= \angle$ZOU, triangles ZUO, UOZ$'$ are similar; so to perform the construction it is only necessary to make $\angle$UOZ$' = \angle$UOZ and $\angle$OUZ$' = \angle$OZU. Of course, Z represents the reciprocal of the complex number represented by Z$'$: reciprocals are interchangeable.

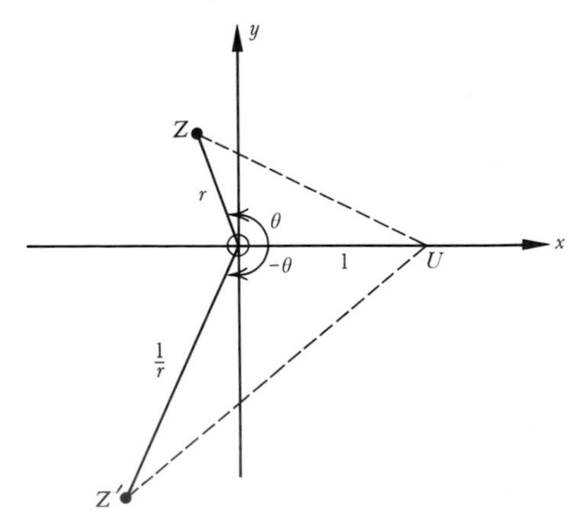

Division of complex numbers

If we require $\frac{z_1}{z_2}$, this may be written $z_1 \times \frac{1}{z_2}$, and we can now use our knowledge of reciprocals:

$$\frac{z_1}{z_2} = r_1 \operatorname{cis} \theta_1 \times \frac{1}{r_2} \operatorname{cis}(-\theta_2) = \left(r_1 \times \frac{1}{r_2}\right) \operatorname{cis}[\theta_1 + (-\theta_2)]$$

$$= \frac{r_1}{r_2} \operatorname{cis}(\theta_1 - \theta_2).$$

So the rule for division is to divide the moduli, and *subtract* the arguments — the reverse of the rule for multiplication. Or we may write:

$$\left|\frac{z_1}{z_2}\right| = \frac{|z_1|}{|z_2|}; \quad \arg \frac{z_1}{z_2} = \arg z_1 - \arg z_2.$$

This is what one would expect since division is the inverse process of multiplication. A useful exercise would be to evolve a construction for arriving at the point representing $\frac{z_1}{z_2}$ starting from given positions of z_1, z_2. This will in effect be a combination of the processes in the previous sections.

Example 3

Evaluate $\dfrac{(i-2)^2(3i+1)}{i-3}$ using polar form (compare p. 113)

$$i - 2 = -2 + i = \sqrt{5} \operatorname{cis} 153° \, 26'$$

$$\Rightarrow \quad (i-2)^2 = 5 \operatorname{cis} 306° \, 52'.$$

Also, $$3i + 1 = \sqrt{10} \operatorname{cis} 71° \, 34'.$$

So $$(i-2)^2(3i+1) = 5\sqrt{10} \operatorname{cis} 378° \, 26'.$$

Finally, $$i - 3 = -3 + i = \sqrt{10} \operatorname{cis} 161° \, 34'$$

$$\Rightarrow \quad \frac{(i-2)^2(3i+1)}{i-3} = \frac{5\sqrt{10} \operatorname{cis} 378° \, 26'}{\sqrt{10} \operatorname{cis} 161° \, 34'}$$

$$= 5 \operatorname{cis} 216° \, 52' = -4 - 3i.$$

Example 4

Simplify $\dfrac{1 - \operatorname{cis} n\theta}{1 - \operatorname{cis} \theta}$ and $\left(\dfrac{1 - \operatorname{cis} n\theta}{1 - \operatorname{cis} \theta}\right)^3$

$$\frac{1 - \operatorname{cis} n\theta}{1 - \operatorname{cis} \theta} = \frac{1 - \cos n\theta - i \sin n\theta}{1 - \cos \theta - i \sin \theta} = \frac{2 \sin^2 \frac{1}{2}n\theta - 2i \sin \frac{1}{2}n\theta \cos \frac{1}{2}n\theta}{2 \sin^2 \frac{1}{2}\theta - 2i \sin \frac{1}{2}\theta \cos \frac{1}{2}\theta}$$

$$= \frac{\sin \frac{1}{2}n\theta\,(\sin \frac{1}{2}n\theta - i \cos \frac{1}{2}n\theta)}{\sin \frac{1}{2}\theta\,(\sin \frac{1}{2}\theta - i \cos \frac{1}{2}\theta)}$$

$$= \frac{\sin \frac{1}{2}n\theta \sin \frac{1}{2}\theta \times -i\,(\cos \frac{1}{2}n\theta + i \sin \frac{1}{2}n\theta)}{\sin \frac{1}{2}\theta \sin \frac{1}{2}\theta \times -i\,(\cos \frac{1}{2}\theta + i \sin \frac{1}{2}\theta)}$$

$$= \frac{\sin \frac{1}{2}n\theta}{\sin \frac{1}{2}\theta} \times \frac{\operatorname{cis} \frac{1}{2}n\theta}{\operatorname{cis} \frac{1}{2}\theta} = \frac{\sin \frac{1}{2}n\theta}{\sin \frac{1}{2}\theta} \operatorname{cis} \tfrac{1}{2}(n - 1)\theta.$$

It follows that $\left(\dfrac{1 - \operatorname{cis} n\theta}{1 - \operatorname{cis} \theta}\right)^3 = \left(\dfrac{\sin \frac{1}{2}n\theta}{\sin \frac{1}{2}\theta}\right)^3 \operatorname{cis} \tfrac{3}{2}(n - 1)\theta,$

the argument being trebled when the number is cubed.

Alternatively,

$$\frac{1 - \operatorname{cis} n\theta}{1 - \operatorname{cis} \theta} = \frac{\operatorname{cis} \frac{1}{2}n\theta[\operatorname{cis} \frac{1}{2}n\theta - \operatorname{cis}(-\frac{1}{2}n\theta)]}{\operatorname{cis} \frac{1}{2}\theta[\operatorname{cis} \frac{1}{2}\theta - \operatorname{cis}(-\frac{1}{2}\theta)]}$$

$$= \frac{\operatorname{cis} \frac{1}{2}n\theta \times 2i \sin \frac{1}{2}n\theta}{\operatorname{cis} \frac{1}{2}\theta \times 2i \sin \frac{1}{2}\theta} = \frac{\sin \frac{1}{2}n\theta}{\sin \frac{1}{2}\theta} \times \operatorname{cis} \tfrac{1}{2}(n - 1)\theta.$$

{Writing $1 \pm \operatorname{cis} \theta$ in the form $\operatorname{cis} \frac{1}{2}\theta[\operatorname{cis}(-\frac{1}{2}\theta) \pm \operatorname{cis} \frac{1}{2}\theta]$ is often a very useful device.}

* Example 5

Show that if triangle ABC is equilateral with the sense of rotation from A to B to C being anti-clockwise, then $a + b\omega + c\omega^2 = 0$, where $\omega = \operatorname{cis} \frac{2}{3}\pi$; and conversely.

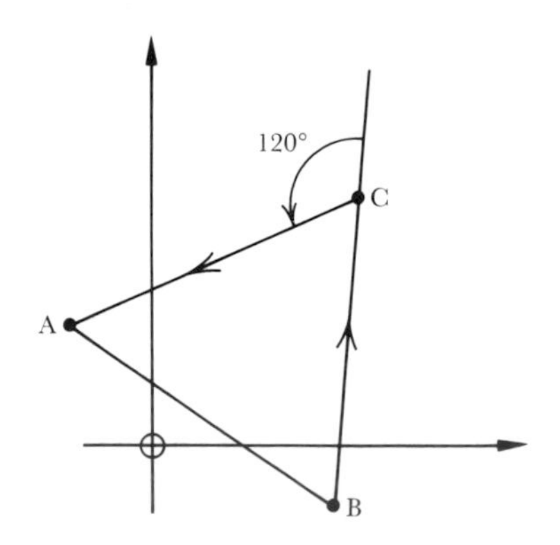

Since $\triangle$ABC is equilateral, ***CA*** is obtained from ***BC*** by a pure rotation through $120° = \frac{2}{3}\pi$ about C.

So $\quad a - c = (c - b) \operatorname{cis} \frac{2}{3}\pi,$

$\Rightarrow \quad a - c = \omega(c - b),$ with the notation above,

$\Rightarrow \quad a + \omega b - (1 + \omega)c = 0.$

Now $\quad \omega^3 = 1 \Rightarrow (\omega - 1)(\omega^2 + \omega + 1) = 0.$

But $\quad \omega \neq 1,$ so $\omega^2 + \omega + 1 = 0,$

(which may also be deduced from the fact that the three points $1, \omega, \omega^2$ form an equilateral triangle about the origin). Replacing $(1 + \omega)$ by $-\omega^2$ gives immediately the required result, that $a + \omega b + \omega^2 c = 0$.

Conversely, if it is known that $a + \omega b + \omega^2 c = 0$, then we may write

$$a + \omega b + (-1 - \omega)c = 0 \Rightarrow a - c = \omega(c - b),$$
$$\Rightarrow a - c = (c - b) \operatorname{cis} \tfrac{2}{3}\pi.$$

Hence ***CA*** is obtained from ***BC*** by a positive 120° rotation, and $\triangle$ABC is equilateral, in an anti-clockwise sense. The reader may also verify that the condition $a + \omega^2 b + \omega c = 0$ is necessary and sufficient for the triangle ABC to be equilateral in the opposite or clockwise sense.

*Example 6

Prove Napoleon's theorem, that if equilateral $\triangle$s BCP, CAQ, ABR are drawn on the sides of any triangle ABC, with P, Q, R outside the triangle, then the centres of these three triangles form another equilateral triangle.

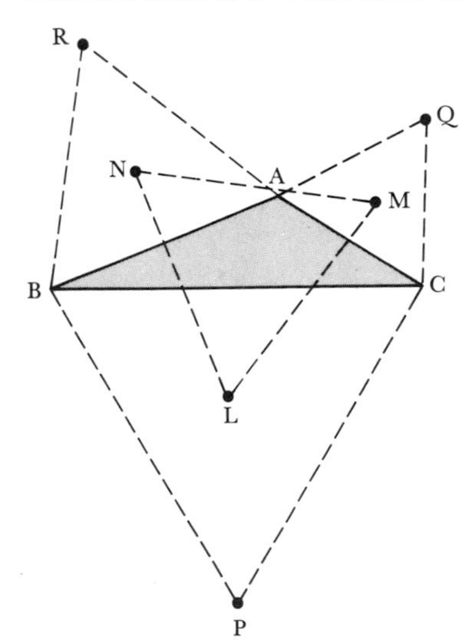

If $\triangle$PCB is equilateral with anti-clockwise sense,

$p + \omega c + \omega^2 b = 0.$

Similarly,

$$q + \omega a + \omega^2 c = 0,$$
$$r + \omega b + \omega^2 a = 0.$$

But if L, M, N are the centres of $\triangle$s PCB, QAC, RBA, respectively, then

$$3l = p + c + b,$$
$$3m = q + a + c,$$
$$3n = r + b + a.$$

So $3(l + \omega m + \omega^2 n)$

$$= (p + c + b) + \omega(q + a + c) + \omega^2(r + b + a)$$
$$= (p + \omega c + \omega^2 b) + \omega(q + \omega a + \omega^2 c) + \omega^2(r + \omega b + \omega^2 a)$$
$$= 0.$$

Hence $\triangle$LMN is equilateral.

Exercise 10.2d

1 Simplify:

(*i*) $4 \text{ cis } 150° \times 5 \text{ cis } 80°$;

(*ii*) $\dfrac{\cos\theta - i\sin\theta}{\cos\phi + i\sin\phi}$;

(*iii*) $\text{cis }\theta \times \text{cis}(-\theta)$;

(*iv*) $\left(\cos\dfrac{3\pi}{4} + i\sin\dfrac{3\pi}{4}\right)^2$;

(*v*) $(i - 1)(1 - \sqrt{3}i)$ by first expressing each in the form $r \text{ cis }\theta$;

(*vi*) $\left(\text{cis }\dfrac{7\pi}{10}\right)^5$;

(*vii*) $(1 + \cos\alpha + i\sin\alpha)(1 + \cos\alpha - i\sin\alpha)$;

(*viii*) $(1 - \cos\beta + i\sin\beta)^2$;

(*ix*) $\text{cis}(-\theta) \div \text{cis } 2\theta$;

(*x*) $\cos\theta + i\sin\theta + \dfrac{1}{\cos\theta - i\sin\theta}$;

(*xi*) $\text{cis }\alpha \times \text{cis }\beta \div (\text{cis }\gamma)^2$;

(*xii*) $\dfrac{1 - \cos\theta + i\sin\theta}{1 + \cos\theta + i\sin\theta}$.

2 Prove that $|z_1 z_2 z_3 \ldots z_n| = |z_1|\,|z_2|\,|z_3| \ldots |z_n|$

and that $\arg(z_1 z_2 z_3 \ldots z_n)$

$$= \arg z_1 + \arg z_2 + \arg z_3 + \cdots + \arg z_n.$$

3 If $z = \text{cis}\,\theta$, express in polar form:

$(i)\ 1 + z;\quad (ii)\ \dfrac{z - 1}{z + 1};\quad (iii)\ \dfrac{z}{z^2 + 1};\quad (iv)\ \dfrac{z + i}{z - i}.$

4 Given the position of a point z on the Argand diagram, show how to construct the positions of:

$3z,\quad -2z,\quad |z|,\quad iz,\quad -iz,\quad -2iz,\quad z^2,\quad iz^2,\quad z^2 + 2i,\quad (z - 1)^2.$

5 If $c = (1 - k)a + kb$, prove that when k is real, C lies on AB and divides AB in the ratio $k:(1 - k)$; while when k is complex, the triangle ABC has $AB:AC = 1:|k|$, and the angle of turn from $\boldsymbol{AB}$ to $\boldsymbol{AC} = \arg k$.

6 Show that the triangles ABC and A′B′C′ are directly similar if and only if

$$\frac{a - c}{b - c} = \frac{a' - c'}{b' - c'}$$

and that this may be expressed as $\sum a'(b - c) = 0$; or as

$$\begin{vmatrix} 1 & 1 & 1 \\ a & b & c \\ a' & b' & c' \end{vmatrix} = 0 \quad \text{(see 13.5)}.$$

7 Use the result in the first part of no. **6** above to deduce the condition for triangle ABC to be equilateral, in the form:

$a^2 + b^2 + c^2 - bc - ca - ab = 0$ (compare Exercise 10.2b, no. 5).

Furthermore, show that this is equivalent to

$(a + \omega b + \omega^2 c)(a + \omega^2 b + \omega c) = 0$ (compare Example 5).

8 Given a triangle ABC, points L, M, N are chosen so that Δ's BCL, CAM, ABN are all directly similar. Prove that ΔABC and ΔLMN have the same centroid (use the first part of no. **6**).

10.3 Complex conjugates

We have already noted that the roots of the quadratic equation $x^2 + 6x + 13 = 0$ are $-3 \pm 4i$, and that these complex numbers appear on the Argand diagram as mirror images in the real axis.
Complex numbers having this property are called *conjugate*.

More generally, if

$z = x + iy = r\,\text{cis}\,\theta,$

we define the complex conjugate of z, denoted z^*, as

$z^* = x - iy = r\,\text{cis}\,(-\theta),$

and on the Argand diagram z and z^* will be reflections in the real axis. We might also write $(x + iy)^* = x - iy$.

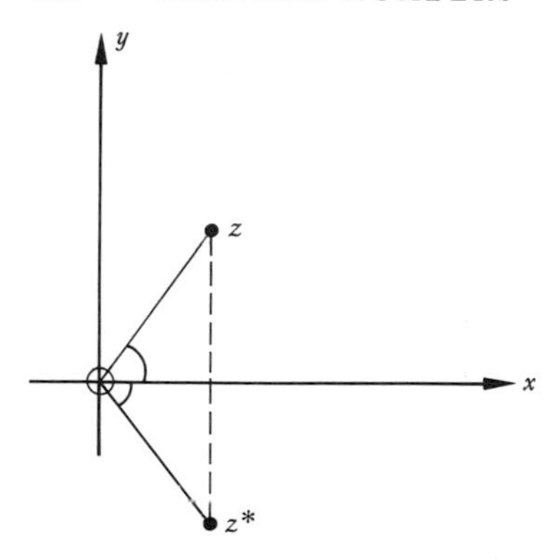

Note that:

(*a*) When $y = 0$ (i.e., the complex number is real) then z is its own conjugate: real numbers are *self-conjugate*.
(*b*) When $|z| = 1$, $z = \text{cis}\,\theta$, $z^* = \text{cis}\,(-\theta)$, and in this case conjugates are reciprocals.
(*c*) Both the sum and the product of a pair of conjugates are real numbers:

$$z + z^* = (x + iy) + (x - iy) = 2x = 2R(z)$$
$$zz^* = (x + iy)(x - iy) = x^2 + y^2 = |z|^2;$$

however, by contrast,

$$z - z^* = 2iy = 2I(z).$$

The reader can quickly verify that for any complex number z,

$$(z^*)^* = z \quad \text{and} \quad (-z)^* = -z^*;$$

also $(z_1 \pm z_2)^* = z_1^* \pm z_2^*$ and $(z_1 z_2)^* = z_1^* z_2^*$.

The result $zz^* = |z|^2$ has already proved useful when simplifying expressions, by applying it to make the denominator of a quotient real. For example,

$$\frac{1 + 3i}{2 + i} = \frac{(1 + 3i)(2 - i)}{(2 + i)(2 - i)} = \frac{5 + 5i}{5} = 1 + i.$$

As we observed at the beginning of this section, the roots of the quadratic equation $z^2 + 6z + 13 = 0$ are a pair of conjugate complex numbers. Clearly this is bound to happen when solving any quadratic equation

$az^2 + bz + c = 0$, if a, b, c are real and $b^2 < 4ac$.

For $$z = \frac{-b \pm \sqrt{(b^2 - 4ac)}}{2a} = -\frac{b}{2a} \pm \frac{\sqrt{(4ac - b^2)}}{2a}\,i,$$

which is a conjugate pair provided that the coefficients a, b, c are real. (To show the necessity for this last condition, we can consider the equation

$$z^2 + z(-1 + i) + (2 + i) = 0$$

$$\Rightarrow \quad (z - i)(z - 1 + 2i) = 0$$
$$\Rightarrow \quad z = i \quad \text{or} \quad 1 - 2i,$$

which are certainly *not* conjugate.)

Again, the equation

$z^3 - z^2 - z - 15 = 0$ (with real coefficients) has roots $z = 3, -1 + 2i, -1 - 2i$, i.e., one real root and a pair of conjugate complex roots.

Furthermore, the equation
$z^5 - 2z^4 + 4z^3 - 2z^2 + 8 = 0$ (with real coefficients)
may be written as
$(z + 1)(z^2 + 4)(z^2 - 2z + 2) = 0$
and so has roots -1, $\pm 2i$, $1 \pm i$, and again the complex roots occur in conjugate pairs.

We now prove that, for any polynomial equation with real coefficients, the complex roots occur in conjugate pairs.

Let the polynomial be

$$f(z) = a_n z^n + a_{n-1} z^{n-1} + \cdots + a_1 z + a_0$$

and suppose that z is a root, so that $f(z) = 0$. We therefore have to prove that z^* is also a root.

$$\begin{aligned} \text{Now} \quad f(z^*) &= a_n(z^*)^n + a_{n-1}(z^*)^{n-1} + \cdots + a_1 z^* + a_0 \\ &= a_n^*(z^*)^n + a_{n-1}^*(z^*)^{n-1} + \cdots + a_1^* z^* + a_0^* \end{aligned}$$

(since all the a_n are real and therefore self conjugate).

$$\text{But} \quad a_n^*(z^*)^n = a_n^*(z^n)^* = (a_n z^n)^*, \quad \text{etc.}$$

$$\begin{aligned} \text{Hence} \quad f(z^*) &= (a_n z^n)^* + (a_{n-1} z^{n-1})^* + \cdots + (a_1 z)^* + a_0^* \\ &= (a_n z^n + a_{n-1} z^{n-1} + \cdots + a_1 z + a_0)^* \\ &= \{f(z)\}^* \\ &= 0^* \\ &= 0. \end{aligned}$$

So whenever z is a root, so is z^*; and we see that the roots of such equations with real coefficients always occur in conjugate pairs (or as real numbers, which are self-conjugate).

Example

Solve the equation

$$z^3 - 3z^2 - 19z + 13 = 0,$$

given that one root is $2 + 3i$.

As the coefficients are real, another root is $2 - 3i$. So if the third root is α, the sum of the roots of the equation gives

$$\alpha + (2 + 3i) + (2 - 3i) = -(-3) = 3$$
$$\Rightarrow \quad \alpha = -1.$$

Hence the roots are -1, $2 \pm 3i$, each of which may be checked in the original equation.

Exercise 10.3

1 Prove that:

(*i*) $(z_1 + z_2)^* = z_1^* + z_2^*$;

(*ii*) $(z_1 - z_2)^* = z_1^* - z_2^*$;

(*iii*) $(z_1 z_2)^* = z_1^* z_2^*$;

(*iv*) $\left(\dfrac{z_1}{z_2}\right)^* = \dfrac{z_1^*}{z_2^*}$.

2 Extend the above results to:

(*i*) $(\sum z_r)^* = \sum z_r^*$; (*ii*) $(\prod z_r)^* = \prod z_r^*$.†

3 Prove that $z_1^* z_2 = (z_1 z_2^*)^*$, and verify by putting:

$z_1 = r_1 \text{ cis } \theta_1$; $z_2 = r_2 \text{ cis } \theta_2$.

4 If $z = r \text{ cis } \theta$, show that

$r = \sqrt{(zz^*)}$, $\theta = \frac{1}{2}(\arg z - \arg z^*)$

and also that $z^2 + (z^*)^2 = 2|z|^2 \cos 2\theta$.

Prove that $\arg z = \theta \Rightarrow z = z^* \text{ cis } 2\theta$.

Is the converse true?

5 Use conjugates instead of moduli and arguments to express:

(*i*) $|z - 1| = 3|z|$; (*ii*) $\arg(z - 1) - \arg(z + 1) = \frac{2}{3}\pi$.

6 If z_1 and z_2 are complex numbers such that $z_1 + z_2$ and $z_1 z_2$ are both real, prove that $z_1^* = z_2$.

7 Prove that the area of the triangle OAB is $\dfrac{1}{4i}(a^*b - ab^*)$, and that the area of the triangle ABC is

$$\frac{1}{4i}\begin{vmatrix} 1 & 1 & 1 \\ a & b & c \\ a^* & b^* & c^* \end{vmatrix}$$ (see also 13.5).

8 Given that $1 - i$ is a root of

$2z^3 - 7z^2 + 10z - 6 = 0$,

find the other roots.

† $\prod$ denoting *product*, just as $\sum$ denotes sum.

9 If $2 + i$ is a root of

$$2z^3 - 9z^2 + 14z - 5 = 0,$$

find the other roots.

10 Given that $z + n$ ($n \in Z^+$) is a factor of $z^3 + 6z^2 + 16z + 16$, find n and then factorise completely.

11 Solve the equation

$$z^4 - 3z^3 + 4z^2 - 3z + 1 = 0,$$

given that one root is $\frac{1}{2}(1 - \sqrt{3}i)$. Solve also by making the substitution $w = z + 1/z$.

12 Solve the equation

$$z^4 - 6z^2 + 25 = 0.$$

13 Solve $z^4 - 6z^3 + 23z^2 - 34z + 26 = 0$, given that one root is $1 + i$.

14 Show that the equation of the circle

$$(x - \alpha)^2 + (y - \beta)^2 = \rho^2$$

can be written as

$$zz^* - p^*z - pz^* + pp^* - \rho^2 = 0,$$

where $p = \alpha + i\beta$.

15 If O is the orthocentre of the triangle formed by a, b, c, prove that:

$$bc^* + b^*c = ca^* + c^*a = ab^* + a^*b.$$

16 Prove that the reflection of A in the line BC is the point A′ given by

$$a' = \frac{a^*(b - c) + b^*c - bc^*}{b^* - c^*}.$$

17 In Exercise **10.2*a*** no. **6**, we obtained the condition for two triangles to be *directly* similar. What is the condition for *opposite* similarity?

10.4 de Moivre's theorem†

Having established means of obtaining the sum, difference, product, and quotient of two complex numbers and their interpretation on the Argand diagram, we now investigate their squares, cubes, square roots, cube roots, and other powers. Initially we shall restrict ourselves to the *rational* powers of complex numbers of unit modulus, i.e., to $(\text{cis}\,\theta)^n$, where $n \in Q$.

Case 1 $n \in Z^+$

We know that $(\text{cis}\,\theta)^2 = \text{cis}\,\theta \times \text{cis}\,\theta = \text{cis}\,(\theta + \theta) = \text{cis}\,2\theta$

† After Abraham de Moivre (1667–1754), born in France but resident in England for most of his life.

$$(\text{cis}\,\theta)^3 = \text{cis}\,\theta \times (\text{cis}\,\theta)^2 = \text{cis}\,\theta \times \text{cis}\,2\theta$$
$$= \text{cis}\,3\theta, \quad \text{and so on.}$$

More generally, $\forall n \in Z^+$, $(\text{cis}\,\theta)^n = \text{cis}\,n\theta$, which can be proved formally by induction.

Case 2 $n \in Z^-$

Since $\text{cis}\,\theta \times \text{cis}\,(-\theta) = 1$,

$$(\text{cis}\,\theta)^{-1} = \text{cis}\,(-\theta).$$

Again,
$$(\text{cis}\,\theta)^{-2} = \frac{1}{(\text{cis}\,\theta)^2} = \frac{1}{\text{cis}\,2\theta} = \text{cis}\,(-2\theta)$$
$$(\text{cis}\,\theta)^{-3} = \frac{1}{(\text{cis}\,\theta)^3} = \frac{1}{\text{cis}\,3\theta} = \text{cis}\,(-3\theta), \quad \text{etc.}$$

So, more generally, $(\text{cis}\,\theta)^n = \text{cis}\,n\theta$, where $n = -1, -2, -3, \ldots$ (i.e., $\forall n \in Z^-$).

Case 3 $n = 0$

$(\text{cis}\,\theta)^0 = 1 = \text{cis}\,0.$

So again, $(\text{cis}\,\theta)^n = \text{cis}\,n\theta$ when $n = 0$.

Case 4 $n = p/q$, a rational fraction, with $p, q \in Z$

As $q \in Z$ $\left(\text{cis}\,\frac{p}{q}\,\theta\right)^q = \text{cis}\left(\frac{p\theta}{q} \times q\right)$ using the result above, p and q being integers

$$= \text{cis}\,p\theta = (\text{cis}\,\theta)^p.$$

So $\text{cis}\,\frac{p}{q}\theta$ is a qth root of $(\text{cis}\,\theta)^p$, i.e., a value of $(\text{cis}\,\theta)^{p/q}$.

Summarising these results, we have de Moivre's theorem, that

> if $n \in Z$, $(\text{cis}\,\theta)^n = \text{cis}\,n\theta$;
> and if $n \in Q$, $\text{cis}\,n\theta$ is one of the values of $(\text{cis}\,\theta)^n$

It is necessary to say 'one of the values of', for we shall see later that $(\text{cis}\,\theta)^{p/q}$ has in fact q distinct values. At the moment, we note that, since

$$(\text{cis}\,\tfrac{1}{4}\pi)^2 = \text{cis}\,\tfrac{1}{2}\pi \text{ and } (\text{cis}\,\tfrac{5}{4}\pi)^2 = \text{cis}\,\tfrac{5}{2}\pi = \text{cis}\,\tfrac{1}{2}\pi,$$

then, $(\text{cis}\,\tfrac{1}{2}\pi)^{1/2} = \text{either cis}\,\tfrac{1}{4}\pi \text{ or cis}\,\tfrac{5}{4}\pi$,

there being exactly *two* square roots of a complex number, which (not unexpectedly) are a pair of negatives. The fact that there are three cube roots may be more surprising. However, it is easily verified that cis 120°

and cis 240° are both cube roots of 1. Moreover, you may verify that $2 - i$, $-2 + i$, $1 + 2i$, and $-1 - 2i$ are all fourth roots of $-7 - 24i$.

De Moivre's theorem tells us that an effect of raising a complex number to the power n is to multiply its argument by n. This is a result of far-reaching importance, leading to powerful and elegant solutions of many problems, and we now proceed to consider some of its applications.

Example 1

Use de Moivre's theorem to find formulae for $\cos 4\theta$ and $\sin 4\theta$ in terms of $\cos\theta$ and $\sin\theta$.

(Here, and in subsequent examples, we shall use the abbreviation $c = \cos\theta$, $s = \sin\theta$).

$$\begin{aligned}\cos 4\theta + i\sin 4\theta &= (c + is)^4 = c^4 + 4c^3is + 6c^2i^2s^2 + 4ci^3s^3 + i^4s^4\\ &= (c^4 - 6c^2s^2 + s^4) + i(4c^3s - 4cs^3).\end{aligned}$$

Equating real and imaginary parts,

$$\begin{aligned}\cos 4\theta &= c^4 - 6c^2s^2 + s^4 = c^4 - 6c^2(1 - c^2) + (1 - c^2)^2\\ &= 8c^4 - 8c^2 + 1\\ \sin 4\theta &= 4c^3s - 4cs^3 = 4cs(c^2 - s^2).\end{aligned}$$

The method may be generalised to obtain $\cos n\theta$ and $\sin n\theta$ in terms of c and s, starting from $(c + is)^n = \cos n\theta + i\sin n\theta$. Using the binomial theorem and equating real and imaginary parts, we get

$$\begin{aligned}\cos n\theta &= c^n - {}^nC_2c^{n-2}s^2 + {}^nC_4c^{n-1}s^4 + \cdots\\ \sin n\theta &= nc^{n-1}s - {}^nC_3c^{n-3}s^3 + {}^nC_5c^{n-5}s^5 - \cdots.\end{aligned}$$

It follows that

$$\begin{aligned}\tan n\theta &= \frac{\sin n\theta}{\text{cis } n\theta} = \frac{nc^{n-1}s - {}^nC_3c^{n-3}s^3 + \cdots}{c^n - {}^nC_2c^{n-2}s^2 + \cdots}\\ &= \frac{nt - {}^nC_3t^3 + {}^nC_5t^5 - \cdots}{1 - {}^nC_2t^2 + {}^nC_4t^4 - \cdots},\end{aligned}$$

(dividing numerator and denominator by c^n, and using $t = \tan\theta$).

In particular, for example,

$$\tan 5\theta = \frac{5t - 10t^3 + t^5}{1 - 10t^2 + 5t^4}; \qquad \tan 4\theta = \frac{4t - 4t^3}{1 - 6t^2 + t^4}, \quad \text{etc.}$$

Example 2

Use de Moivre's theorem to investigate the cube roots of i.

$$i = \text{cis } \tfrac{1}{2}\pi; \qquad \sqrt[3]{i} = i^{1/3} = (\text{cis } \tfrac{1}{2}\pi)^{1/3},$$

so one of the values is

$$\text{cis}\,\tfrac{1}{6}\pi = \cos\tfrac{1}{6}\pi + i\sin\tfrac{1}{6}\pi = \tfrac{1}{2}(\sqrt{3} + i).$$

However, we may also regard i as having argument $5\pi/2$, $9\pi/2, \ldots$, and when de Moivre's theorem is applied we obtain $i^{1/3} = (\text{cis}\,5\pi/2)^{1/3} = \text{cis}\,5\pi/6$, and again $i^{1/3} = (\text{cis}\,9\pi/2)^{1/3} = \text{cis}\,3\pi/2$. The addition of further revolutions to the argument of i cannot produce any different values of $i^{1/3}$; for example, $(\text{cis}\,13\pi/2)^{1/3} = \text{cis}\,13\pi/6 = \text{cis}\,\pi/6$, which has already been obtained. Again, if we regard i as having argument $-3\pi/2$, this gives $[\text{cis}\,(-3\pi/2)]^{1/3} = \text{cis}\,(-\pi/2) = \text{cis}\,3\pi/2$, also already obtained. Thus the three cube roots of i will be seen to have arguments separated by multiples of $2\pi/3$, and so appear on the Argand diagram as the vertices of an equilateral triangle, at the points P_1, P_2, P_3.

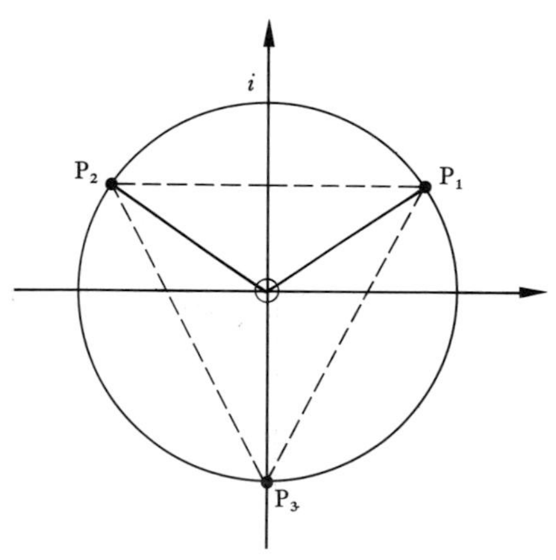

Example 3

Find the values of $(-1 + i)^{3/5}$.

Now $\quad -1 + i = \sqrt{2}\,\text{cis}\,135°$.

So, by de Moivre's theorem, one of the values of $(-1 + i)^{3/5}$ is

$(\sqrt{2}\,\text{cis}\,135°)^{3/5} = 2^{0.3}\,\text{cis}\,81°$.

However, any number of revolutions may be added to the argument of the original number, which may therefore be written $\sqrt{2}\,\text{cis}\,(135° + 360°\,n)$, where n is any integer.

$$\text{Thus} \quad (-1 + i)^{3/5} = \{\sqrt{2}\,\text{cis}\,(135° + 360°\,n)\}^{3/5}$$
$$= 2^{0.3}\,\text{cis}\,(81° + 216°\,n),$$

i.e. $2^{0.3}\,\text{cis}\,81°$, $2^{0.3}\,\text{cis}\,297°$, $2^{0.3}\,\text{cis}\,513°$, $2^{0.3}\,\text{cis}\,729°$, $2^{0.3}\,\text{cis}\,945°$, etc; so, reducing the arguments modulo 360°, we see that

$2^{0.3}\,\text{cis}\,81°$, $2^{0.3}\,\text{cis}\,297°$, $2^{0.3}\,\text{cis}\,153°$, $2^{0.3}\,\text{cis}\,9°$, $2^{0.3}\,\text{cis}\,225°$

are five distinct values.

These are shown in the figure, where the arrows show the route followed by adding 216° repeatedly to the argument. It is also clear that they form a regular pentagon, and that precisely five values exist.

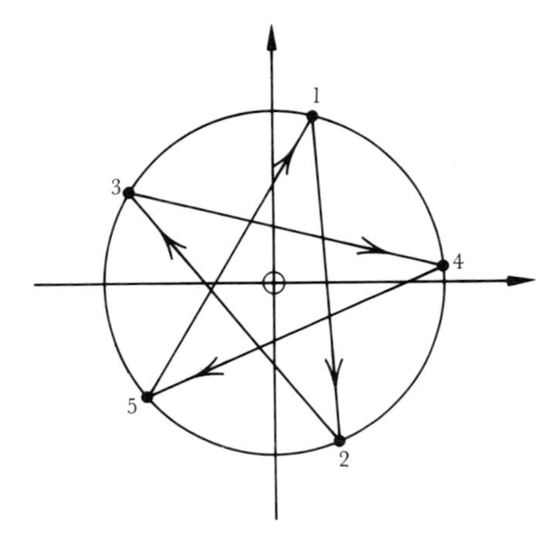

At this stage it is interesting to reflect on what has been achieved. When i was introduced to represent $\sqrt{-1}$, it might have been expected that further inventions would be required in order to represent $\sqrt{i}$, $\sqrt[3]{-1}$ and $\sqrt[4]{(1-i)}$, and that a whole range of 'hyper-complex' numbers would have to be developed. But de Moivre's theorem immediately shows that this will not be necessary, and that the simple invention of i as $\sqrt{-1}$ is entirely sufficient for dealing with all such complications.

Exercise 10.4a

1 Express in the form $r \text{ cis } \theta$ where the argument has its principal value:

(*i*) $(\text{cis } 72°)^{14}$; (*ii*) $(\text{cis } 100°)^{-5}$;
(*iii*) $(28 + 96i)^4$; (*iv*) $(\cos 80° - i \sin 80°)^{3.5}$;
(*v*) $(\sqrt{3} - i)^{7/4}$; (*vi*) $(1 + i)^{10}$ (two methods);
(*vii*) $(\sqrt{3} - i)^{20}$; (*viii*) $\sqrt[3]{(1 + i)}$;
(*ix*) $\sqrt{(1 - i\sqrt{3})}$; (*x*) $(\sin \theta - i \cos \theta)^5$;
(*xi*) $(i + \tan \theta)^n$; (*xii*) $\dfrac{(\cos 5\alpha + i \sin 5\alpha)^6 (\cos 3\alpha - i \sin 3\alpha)^7}{(\cos \alpha + i \sin \alpha)^4 (\cos 2\alpha - i \sin 2\alpha)^{-2}}$.

2 Find $\sqrt{i}$: (*i*) by letting its value be $x + iy$; (*ii*) by expressing in polar form. Find also values of $\sqrt[4]{i}$ and of $\sqrt[5]{(-i)}$.

3 Obtain $\sin 3\theta$ in terms of $\sin \theta$, and $\cos 3\theta$ in terms of $\cos \theta$ by using de Moivre's theorem. Hence show that if $\sin 3\theta + \cos 3\theta = 0$, then either $\tan \theta = 1$ or $\sin 2\theta = -\frac{1}{2}$.

4 Find the principal value of the arguments of $(\text{cis } 50°)^5$, $(\text{cis } 122°)^5$, and $[\text{cis } (-166°)]^5$. Can you find two other complex numbers which give the same result?

5 Find as many as possible values of $(1 + i)^{1/6}$.

6 (*i*) Solve $z^3 - 1 = 0$, expressing the three roots in polar form.
(*ii*) By writing -1 as cis π, obtain the roots of $z^3 + 1 = 0$.
(*iii*) Find all the values of $(\text{cis } 150°)^{1/3}$.

7 Find formulae for: (*i*) $\cos 6\theta$ in terms of $\cos \theta$; (*ii*) $\sin 7\theta/\sin \theta$ in terms of $\cos \theta$; (*iii*) $\tan 5\theta$ in terms of $\tan \theta$; (*iv*) $\sin 6\theta/\sin \theta$ in terms of $\cos \theta$.

8 (*i*) If $z + z^{-1} = 2 \cos \theta$, show that $z^n + z^{-n} = 2 \cos n\theta$, and find $z^n - z^{-n}$.
(*ii*) If $z + z^{-1} = -1$, find the values of $z^n + z^{-n}$ for $n = 2, 3, 4, \ldots$.

9 If $\cos \alpha + \cos \beta + \cos \gamma = 0$ and $\sin \alpha + \sin \beta + \sin \gamma = 0$, prove that α, β and γ are separated by $\frac{2}{3}\pi$. Generalise in the case of the equations

$$\sum_1^n \cos \alpha_r = 0, \qquad \sum_1^n \sin \alpha_r = 0.$$

10 Prove that $(1 + \cos \theta - i \sin \theta)^n = 2^n \cos^n \frac{1}{2}\theta\,(\cos \frac{1}{2}n\theta - i \sin \frac{1}{2}n\theta)$. Hence *write down* the value of $(1 + \cos \theta + i \sin \theta)^n$ and simplify the sum of these two expressions.

11 Using the expansion of $\tan 5\theta$ from no. **7** above, show that the roots of

$$t^5 + 5t^4 - 10t^3 - 10t^2 + 5t + 1 = 0 \quad \text{are} \quad \tan (4n - 1)\frac{\pi}{20} \quad (n \in Z),$$

and obtain five distinct roots.
Use the expansion of $\tan 6\theta$ to evaluate $\tan \pi/12$.

12 Use the expansion of $\tan 5\theta$ in terms of t ($= \tan \theta$) obtained in no. **7** to solve the equation

$$t^5 - 10t^4 - 10t^3 + 20t^2 + 5t - 2 = 0.$$

13 Solve the equation $\sin 5\theta = \sin^5 \theta$.

14 Using the expansion of $\cos 5\theta$, deduce that $\sin 18° = \frac{1}{4}(\sqrt{5} - 1)$.

15 Prove that

$$\sin 7\theta = 7c^6 s - 35c^4 s^3 + 21c^2 s^5 - s^7 \quad \text{where } c = \cos \theta,\ s = \sin \theta.$$

Hence solve the equation

$$7x^3 - 35x^2 + 21x - 1 = 0$$

giving three distinct roots.

The n*th roots of a complex number*

The above examples illustrate how, in general, the nth roots of a complex number may be found. Writing the number in the form $z = r \text{ cis } \theta$, we have $(r \text{ cis } \theta)^{1/n} = r^{1/n} \text{ cis } (\theta/n)$ as one of the roots. To obtain the other roots, we write $z = r \text{ cis } (\theta + 2\pi k)$ $(k \in Z)$. Then $z^{1/n} = r^{1/n} \text{ cis } (\theta + 2\pi k)/n$.

It is evident that n different values are obtained by k taking the values $0, 1, 2, 3, \ldots, n - 1$, and that this exhausts the distinct values: when k takes the value n (i.e. when $2\pi n$ is added to the argument of z) the argument of $z^{1/n}$ will have been augmented by 2π, so repeating the value corresponding to $k = 0$. So the nth roots lie in the Argand diagram on the circle of radius $r^{1/n}$ separated by a consecutive angular distance of $2\pi/n$, and form a regular n-gon.

In particular, the nth roots of *unity* are given by

$$(\text{cis } 2\pi k)^{1/n} = \text{cis } \frac{2\pi k}{n}, \quad (k = 0, 1, 2, \ldots, n - 1),$$

and these form a regular n-gon in the Argand diagram. If we let $\omega = \text{cis } \frac{2\pi}{n}$, then the n roots may be denoted $1, \omega, \omega^2, \omega^3, \ldots, \omega^{n-1}$, with $\omega^n = 1$. It is clear that $1 + \omega + \omega^2 + \omega^3 + \cdots + \omega^{n-1} = 0$; and furthermore, since $1, \omega, \omega^2, \omega^3, \ldots, \omega^{n-1}$ are the roots of $z^n - 1 = 0$,

$$z^n - 1 \equiv (z - 1)(z - \omega)(z - \omega^2)(z - \omega^3)\ldots(z - \omega^{n-1}).$$

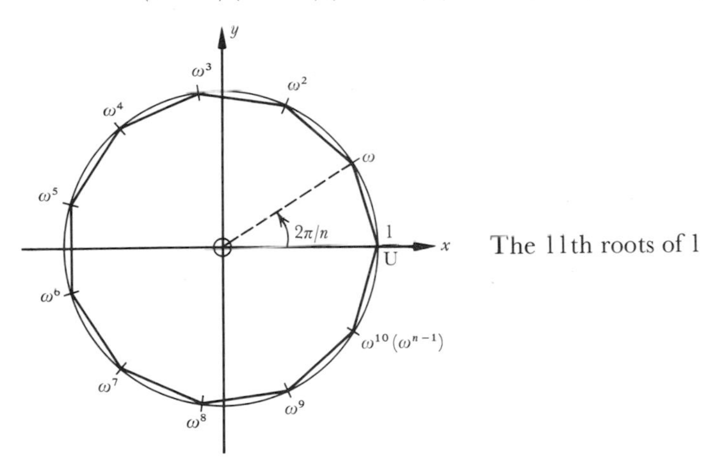

The 11th roots of 1

The particular case $n = 3$ is important, and we have 1, cis $\frac{2}{3}\pi$ and cis $(-\frac{2}{3}\pi)$ as the cube roots of unity. It has already been seen how the cube roots of unity can be applied to the solution of a geometrical problem such as Napoleon's theorem, and in Chapter 13 we shall also see how they may be used to evaluate a determinant. Meanwhile we provide some simple examples on the nth roots of a number:

Example 4

Solve the equation $z^6 + 1 = 0$

$z^6 = -1 = \text{cis}(2\pi k + \pi)$.

So, by de Moivre's theorem, $z = [\text{cis}(2\pi k + \pi)]^{1/6} = \text{cis } \frac{\pi(2k + 1)}{6}$.

Taking $k = 0, 1, 2, 3, 4, 5$ we obtain $z = \text{cis}\ \pi/6, \text{cis}\ \pi/2, \text{cis}\ 5\pi/6, \text{cis}\ 7\pi/6, \text{cis}\ 3\pi/2, \text{cis}\ 11\pi/6$

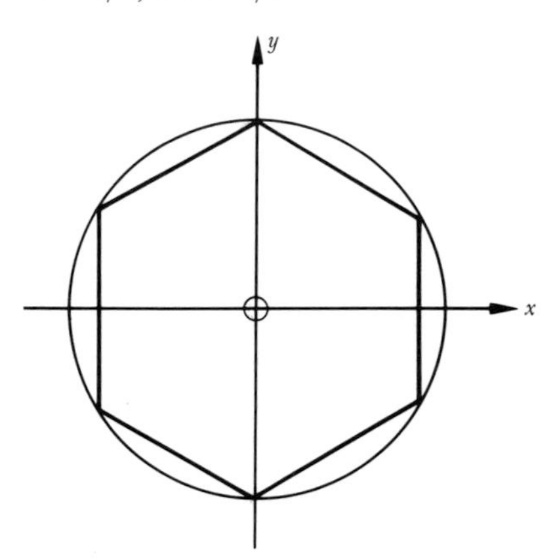

$$\Rightarrow \quad z = \pm i, \quad \pm\ \text{cis}\frac{\pi}{6}, \quad \text{or} \quad \pm\ \text{cis}\frac{5\pi}{6}.$$

Example 5

Solve the equation $(z - 1)^5 + (z + 1)^5 = 0$

$$\left(\frac{z-1}{z+1}\right)^5 = -1 = \text{cis}\,(2k+1)\pi.$$

So $\dfrac{z-1}{z+1} = \text{cis}\,\dfrac{2k+1}{5}\pi \quad (k = 0, 1, 2, 3, 4).$

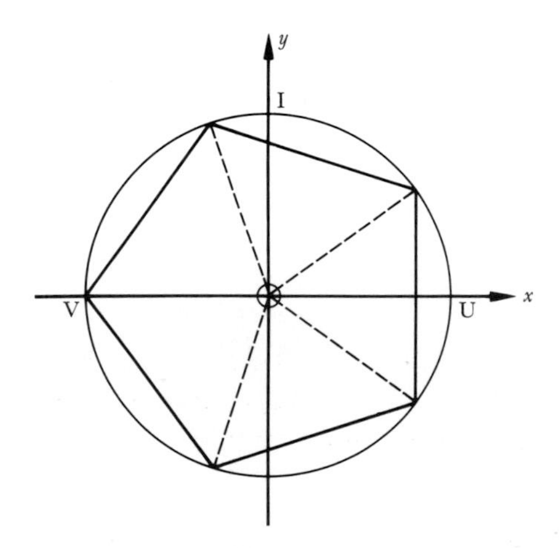

Writing $\alpha = \dfrac{2k+1}{5}\pi,$

$$z = \frac{1 + \text{cis}\ \alpha}{1 - \text{cis}\ \alpha}$$

$$= \frac{\operatorname{cis}\frac{1}{2}\alpha[\operatorname{cis}\frac{1}{2}\alpha + \operatorname{cis}(-\frac{1}{2}\alpha)]}{-\operatorname{cis}\frac{1}{2}\alpha[\operatorname{cis}\frac{1}{2}\alpha - \operatorname{cis}(-\frac{1}{2}\alpha)]}$$

$$= \frac{2\cos\frac{1}{2}\alpha}{-2i\sin\frac{1}{2}\alpha} = i\cot\tfrac{1}{2}\alpha = i\cot\frac{(2k+1)\pi}{10}.$$

So the roots are (putting $k = 0, 1, 2, 3, 4$):

$$z = i\cot\frac{\pi}{10},\quad i\cot\frac{3\pi}{10},\quad i\cot\frac{5\pi}{10},\quad i\cot\frac{7\pi}{10},\quad i\cot\frac{9\pi}{10};$$

$$\Rightarrow\quad z = i\cot\frac{\pi}{10},\quad i\cot\frac{3\pi}{10},\quad 0,\quad -i\cot\frac{3\pi}{10},\quad -i\cot\frac{\pi}{10};$$

$$\Rightarrow\quad z = 0,\quad \pm i\cot\frac{\pi}{10},\quad \pm i\cot\frac{3\pi}{10}.$$

(We expect an equation of degree 5 to have 5 roots. Note that there is one real root, together with two pairs of conjugate complex roots.)

*Example 6

Factorise $x^{2n} - 2x^n\cos n\alpha + 1$.

If $x^{2n} - 2x^n\cos n\alpha + 1 = 0$,

then $(x^n - \cos n\alpha)^2 + \sin^2 n\alpha = 0$

giving $x^n = \cos n\alpha \pm i\sin n\alpha = \operatorname{cis}[\pm(n\alpha + 2k\pi)]$

$$\Rightarrow\quad x = \operatorname{cis}\left[\pm\left(\alpha + \frac{2k\pi}{n}\right)\right].$$

For $k = 0, 1, 2, \ldots, n-1$, we obtain $2n$ roots, and these also occur in conjugate complex pairs. So the expression may now be expressed in factors:

$$x^{2n} - 2x^n\cos n\alpha + 1 = \prod_0^{n-1}\left[x - \operatorname{cis}\left(\alpha + \frac{2k\pi}{n}\right)\right]\left[x - \operatorname{cis}\left(-\alpha - \frac{2k\pi}{n}\right)\right]$$

$$= \prod_0^{n-1}\left[x^2 - 2x\cos\left(\alpha + \frac{2k\pi}{n}\right) + 1\right],$$

combining the conjugate pairs of factors. Thus the expression has been converted to the product of n *real* quadratic factors, and once again we have achieved a result in real algebra by the use of complex numbers.

Further examples on the use of de Moivre's theorem

Example 7

Express $\cos^4\theta$ and $\sin^5\theta$ in terms of the sines and cosines of multiple angles.

We note that $z = \text{cis}\,\theta \;\Rightarrow\; z^{-1} = \text{cis}\,(-\theta)$.

So $z + z^{-1} = 2\cos\theta$ and $z - z^{-1} = 2i\sin\theta$.

Also $z^n = \text{cis}\,n\theta$ and $z^{-n} = \text{cis}\,(-n\theta)$.

So $z^n + z^{-n} = 2\cos n\theta$; $z^n - z^{-n} = 2i\sin n\theta$.

Hence $\left(z + \dfrac{1}{z}\right)^4 = z^4 + 4z^2 + 6 + 4z^{-2} + z^{-4}$

$$\Rightarrow \quad (2\cos\theta)^4 = (z^4 + z^{-4}) + 4(z^2 + z^{-2}) + 6$$

$$= 2\cos 4\theta + 4 \times 2\cos 2\theta + 6$$

$$\Rightarrow \quad 16\cos^4\theta = 2\cos 4\theta + 8\cos 2\theta + 6$$

$$\Rightarrow \quad \cos^4\theta = \tfrac{1}{8}(\cos 4\theta + 4\cos 2\theta + 3).$$

Also $\left(z - \dfrac{1}{z}\right)^5 = (z^5 - z^{-5}) - 5(z^3 - z^{-3}) + 10(z - z^{-1})$

$$\Rightarrow \quad (2i\sin\theta)^5 = 2i\sin 5\theta - 5 \times 2i\sin 3\theta + 10 \times 2i\sin\theta$$

$$\Rightarrow \quad 32\sin^5\theta = 2\sin 5\theta - 10\sin 3\theta + 20\sin\theta \quad (\text{since } i^4 = 1)$$

$$\Rightarrow \quad \sin^5\theta = \tfrac{1}{16}(\sin 5\theta - 5\sin 3\theta + 10\sin\theta).$$

Example 8

Find the sum of the series

$$S = \sin\alpha + \sin(\alpha + \beta) + \sin(\alpha + 2\beta) + \cdots + \sin[\alpha + (n-1)\beta].$$

Letting
$$C = \cos\alpha + \cos(\alpha + \beta) + \cos(\alpha + 2\beta) + \cdots + \cos[\alpha + (n-1)\beta],$$

we have
$$C + iS = \text{cis}\,\alpha + \text{cis}\,(\alpha + \beta) + \text{cis}\,(\alpha + 2\beta) + \cdots + \text{cis}\,[\alpha + (n-1)\beta];$$

which may be written

$$C + iS = \text{cis}\,\alpha + \text{cis}\,\alpha\,\text{cis}\,\beta + \text{cis}\,\alpha\,\text{cis}\,2\beta + \cdots + \text{cis}\,\alpha\,\text{cis}\,(n-1)\beta$$

$$= \text{cis}\,\alpha[1 + \text{cis}\,\beta + (\text{cis}\,\beta)^2 + \cdots + (\text{cis}\,\beta)^{n-1}]$$

$$= \frac{\text{cis}\,\alpha[1 - (\text{cis}\,\beta)^n]}{1 - \text{cis}\,\beta}, \quad \text{being the sum of a g.p.}$$

$$= \frac{\text{cis}\,\alpha(1 - \text{cis}\,n\beta)}{1 - \text{cis}\,\beta}, \quad \text{using de Moivre's theorem.}$$

Using the result of p. 122, this gives

$$C + iS = \frac{\sin\frac{1}{2}n\beta}{\sin\frac{1}{2}\beta}\,\text{cis}\,[\alpha + \tfrac{1}{2}(n-1)\beta].$$

Finally, equating real and imaginary parts,

$$C = \frac{\sin\frac{1}{2}n\beta}{\sin\frac{1}{2}\beta}\cos[\alpha + \tfrac{1}{2}(n-1)\beta]; \qquad S = \frac{\sin\frac{1}{2}n\beta}{\sin\frac{1}{2}\beta}\sin[\alpha + \tfrac{1}{2}(n-1)\beta].$$

In the special case when $\alpha = \beta = \theta$, we get

$$\cos\theta + \cos 2\theta + \cdots + \cos n\theta = \frac{\sin\frac{1}{2}n\theta}{\sin\frac{1}{2}\theta}\cos\tfrac{1}{2}(n+1)\theta.$$

Exercise 10.4b

1 Show that $(1 + 3i)^4 = 28 - 96i$, and find the other fourth roots of $28 - 96i$.

2 Find:
(*i*) the cube roots of $\frac{1}{8}(\cos 60° + i\sin 60°)$;
(*ii*) the fifth roots of $4\sqrt{2}\,\text{cis}(-80°)$;
(*iii*) the eighth roots of $\cos 160° - i\sin 160°$;
(*iv*) the ninth roots of i;
(*v*) the four values of $(1 - \sqrt{3}\,i)^{3/4}$;
(*vi*) the sixth roots of $i - \sqrt{3}$;
(*vii*) $(-i)^{3/4}$;
(*viii*) four values of z to satisfy $z^4 + 1 = i$.

3 Solve the equations:
(*i*) $z^3 + 8 = 0$;
(*ii*) $z^6 + z^5 + z^4 + z^3 + z^2 + z + 1 = 0$;
(*iii*) $z^4 - z^3 + z^2 - z + 1 = 0$;
(*iv*) $z^5 - 1 = 0$;
(*v*) $z^6 + 22z^3 + 125 = 0$.

4 Solve the equations:
(*i*) $(2z - 1)^5 = (z - 2)^5$; (*ii*) $(z + i)^4 = 81z^4$;
(*iii*) $z^6 - 2z^3 + 2 = 0$ (*iv*) $z^9 + z^5 - z^4 - 1 = 0$

In Exercises **5** to **9**, take ω to be a complex cube root of 1.

5 Prove that:
(*i*) $(1 + \omega)(1 + \omega^2) = 1$; (*ii*) $(1 + \omega^2)^3 = -1$;
(*iii*) Simplify $(1 + \omega)(1 + 2\omega)(1 + 3\omega)(1 + 5\omega)$.

6 Prove by reference to the Argand diagram that $1 + 2\omega^2$ is purely imaginary.

7 Form the quadratic equations whose roots are:
(*i*) $3 + \omega,\ 3 + \omega^2$; (*ii*) $2\omega - \omega^2,\ 2\omega^2 - \omega$.

8 Prove that $1 + \omega^n + \omega^{2n}$ takes the values 3 or 0 according as n is or is not divisible by 3.

9 From the identity

$$a^2 + b^2 + c^2 - bc - ca - ab \equiv (a + b\omega + c\omega^2)(a + b\omega^2 + c\omega)$$
$(a, b, c \in C)$,

show that if $\sum a^2 = \sum bc$, then either $a - c = \omega(c - b)$ or else $a - c = \omega^2(c - b)$ and interpret in the Argand diagram.

10 Express the following in terms of multiple angles:
(*i*) $\cos^5 \theta$; (*ii*) $\sin^6 \theta$; (*iii*) $\sin^7 \theta$;
(*iv*) $\sin^p \theta \cos^q \theta$ in the cases where $(p, q) = (3, 4), (4, 3), (4, 6), (3, 7)$, etc.

11 Use the results of no. **10** to find the integrals

$$\int \sin^6 \theta \, d\theta \quad \text{and} \quad \int_0^{\frac{1}{2}\pi} \sin^4 \theta \cos^6 \theta \, d\theta.$$

12 Prove that

$$1 + \cos \alpha + \cos 2\alpha + \cdots + \cos (n - 1)\alpha = \frac{\sin \frac{1}{2}n\alpha \cos \frac{1}{2}(n - 1)\alpha}{\sin \frac{1}{2}\alpha}$$

and that $$\frac{\sin \alpha + \sin 2\alpha + \cdots + \sin n\alpha}{\cos \alpha + \cos 2\alpha + \cdots + \cos n\alpha} = \tan \tfrac{1}{2}(n + 1)\alpha.$$

Obtain formulae for $\cos \alpha - \cos 3\alpha + \cos 5\alpha - \cdots \cos (2n - 1)\alpha$
and $\sin \alpha - \sin 3\alpha + \sin 5\alpha - \cdots \sin (2n - 1)\alpha$.

13 Use the result of no. **12** to find a formula for the sum of the series $\sin \alpha + 2 \sin 2\alpha + 3 \sin 3\alpha + \cdots + n \sin n\alpha$.

14 Find the sum to infinity of the series
$S = \cos \theta \sin \theta + \cos^2 \theta \sin 2\theta + \cos^3 \theta \sin 3\theta + \cdots$
(consider the g.p. $\sum \cos^r \theta \operatorname{cis} r\theta$).

10.5 Euler's equation: cis $\theta = e^{i\theta}$

Let us now consider the function $y = \cos x + i \sin x$. Continuing to treat i like any other number, we have, by differentiation:

$$\frac{dy}{dx} = -\sin x + i \cos x = i(\cos x + i \sin x).$$

Hence $\dfrac{dy}{dx} = iy$

$$\Rightarrow \quad i\frac{dx}{dy} = \frac{1}{y}$$

$$\Rightarrow \quad ix = \ln y + c.$$

But when $x = 0$, $y = 1$. So $c = 0$.

$$\Rightarrow \quad ix = \ln y$$

$$\Rightarrow \quad y = e^{ix}.$$

So $\boxed{\operatorname{cis} x = e^{ix}}$

It may be objected that we have nowhere defined the meaning of a number such as e^z when z is complex; but the reader should not be deterred by such inhibitions! Indeed, the above paragraph may be regarded as providing, if not a definition, at least a reasonable exposition of the *meaning* of e^{ix}, making it consistent with the familiar processes of mathematics. For example, the rule for the multiplication of complex numbers: $\text{cis}\,\theta_1 \times \text{cis}\,\theta_2 = \text{cis}\,(\theta_1 + \theta_2)$ may now be re-stated: $e^{i\theta_1} \times e^{i\theta_2} = e^{i(\theta_1+\theta_2)}$, which is just what we should expect in a well-behaved system, 'imaginary' powers of e obeying the same rules as real powers. Indeed, the rule for addition of *arguments* (10.2) when multiplying complex numbers is now seen to be no more than a disguised version of the rule for addition of indices, or *logarithms*!

Again, de Moivre's theorem may now be rewritten in the perhaps more familiar form $(e^{ix})^n = e^{inx}$; and, as further support for the truth of the statement $e^{ix} = \text{cis}\,x$, it will be remembered that we obtained the following infinite series by using Maclaurin's expansion:

$$\cos x = 1 - \frac{x^2}{2!} + \frac{x^4}{4!} - \cdots; \qquad \sin x = x - \frac{x^3}{3!} + \frac{x^5}{5!} - \cdots.$$

Then
$$\text{cis}\,x = \cos x + i \sin x = \left(1 - \frac{x^2}{2!} + \frac{x^4}{4!} - \cdots\right) + i\left(x - \frac{x^3}{3!} + \frac{x^5}{5!} - \cdots\right)$$

(for all real x)

$$= 1 + ix - \frac{x^2}{2!} - \frac{ix^3}{3!} + \frac{x^4}{4!} + \cdots$$

$$= 1 + (ix) + \frac{(ix)^2}{2!} + \frac{(ix)^3}{3!} + \cdots = e^{ix},$$

being the infinite series for e^z, with z substituted by ix. This further encourages us to accept the validity of the expansion of the exponential function as an infinite series in cases when the exponent is imaginary.

The above is neither a complete nor a conclusive argument: a great deal of analysis, involving the use of the exponential series $e^z = 1 + z + z^2/2! + \cdots$, is needed to make the definition of e^{ix} watertight. Meanwhile, however, we shall proceed on the basis that the above broad outline makes sense.

The most remarkable result may be obtained by putting $x = \pi$:

$$e^{i\pi} = \text{cis}\,\pi = -1 + 0i = -1,$$

so we have derived the equation due to the great mathematician Euler:

$$\boxed{e^{i\pi} + 1 = 0,}$$

which unites in one spectacular relationship

the five most important numbers in mathematics: 0, 1, π, e and i.

Sense or nonsense? It must be true! *Certum est quia impossibile est:* Nothing so incredible could be an invention!

Example 1

Integrals such as $\int e^{ax} \cos bx \, dx$ and $\int e^{ax} \sin bx \, dx$ are usually found by two applications of integration by parts, a tedious procedure. We now look at a way of avoiding this tedium by using the exponential form of the complex number, working the method in a numerical case.

Let $C = \displaystyle\int e^{-3x} \cos 2x \, dx$ and $S = \displaystyle\int e^{-3x} \sin 2x \, dx$.

$$\text{Then} \quad C + iS = \int e^{-3x} (\cos 2x + i \sin 2x) \, dx = \int e^{-3x} e^{2ix} \, dx$$

$$= \int e^{(-3+2i)x} \, dx = \frac{e^{(-3+2i)x}}{-3+2i} = \frac{-3-2i}{13} e^{-3x} e^{2ix}$$

$$= \frac{-3-2i}{13} e^{-3x} (\cos 2x + i \sin 2x).$$

Separating the real and imaginary parts, we obtain C and S:

$$C = \frac{e^{-3x}}{13}(-3 \cos 2x + 2 \sin 2x) \quad \text{and} \quad S = \frac{e^{-3x}}{13}(-2 \cos 2x - 3 \sin 2x)$$

(arbitrary constants have been omitted). These perfectly real results have been achieved by the use, not only of complex numbers, but of exponential functions of complex numbers. Their truth should be verified by direct differentiation.

Exponential functions of the general complex number

If $z = x + iy$, then $e^z = e^{x+iy} = e^x e^{iy} = e^x \operatorname{cis} y$

$\Rightarrow \qquad |e^z| = e^x$ and $\arg e^z = y$.

Moreover, we may now say that $\ln (e^x \operatorname{cis} y) = x + iy$,

or $\ln (r \operatorname{cis} \theta) = \ln r + i\theta \; (+ 2\pi n i)$.

For example,

$$\ln(-3) = \ln(3 \operatorname{cis} \pi) = \ln 3 + i\pi.$$

Note that the logarithm of a (complex) number has an infinity of values corresponding to the infinity of arguments, e.g., $\ln(-3) = \ln 3 + 3\pi i$, or $\ln 3 - 5\pi i$, etc.

As a further application, suppose we want to find i^i. We may write $i = \text{cis}\,\frac{1}{2}\pi = e^{i\frac{1}{2}\pi}$.

So $i^i = (e^{\frac{1}{2}i\pi})^i = e^{-\frac{1}{2}\pi} \approx 0.207\ldots$, the value, surprisingly, being *real.*

Trigonometric functions as exponential functions; hyperbolic functions

Since $\cos x + i \sin x = e^{ix}$,

and $\cos x - i \sin x = e^{-ix}$,

we deduce, by adding and subtracting, that

$$\cos x = \frac{e^{ix} + e^{-ix}}{2}; \qquad \sin x = \frac{e^{ix} - e^{-ix}}{2i}$$

and discover the astonishing revelation that trigonometric functions, which started life as ratios of sides of a right-angled triangle, are really exponential functions in disguise. Moreover, these exponential versions of the trigonometric functions bear a striking resemblance to the definitions of the hyperbolic functions (see 9.9):

$$\cosh x = \frac{e^x + e^{-x}}{2}, \qquad \sinh x = \frac{e^x - e^{-x}}{2}.$$

Hence we may now state the following relations connecting the circular and hyperbolic functions, which the reader may verify by direct substitution:

$$\begin{array}{ll} \cosh ix = \cos x & \sinh ix = i \sin x \\ \cos ix = \cosh x & \sin ix = i \sinh x \end{array}$$

We have already, in 9.9, noted the close resemblance between trigonometric and hyperbolic formulae, e.g., for $\cos(a+b)$ and $\cosh(a+b)$, and we are now in a position to see how this arises.

$$\begin{aligned} \text{For} \quad \cosh(a+b) &= \cos i(a+b) \\ &= \cos ia \cos ib - \sin ia \sin ib \\ &= \cosh a \cosh b - (i \sinh a)(i \sinh b) \\ \Rightarrow \quad \cosh(a+b) &= \cosh a \cosh b + \sinh a \sinh b, \end{aligned}$$

compared with

$$\cos(a+b) = \cos a \cos b - \sin a \sin b.$$

More generally, we can summarise these resemblances in Osborn's rule, that the formulae correspond exactly provided that the sign is changed whenever there is a product of two sine or two sinh functions. Thus, again,

the formulae

$$\cos 2x = 1 - 2\sin^2 x \qquad \cos^2 x + \sin^2 x = 1$$
$$\cosh 2x = 1 + 2\sinh^2 x \qquad \cosh^2 x - \sinh^2 x = 1$$

$$\sin 2x = 2\sin x\cos x$$
$$\sinh 2x = 2\sinh x\cosh x$$

may all be quoted in support of Osborn's rule.

We may also see that our statements regarding $\sin\theta$ and $\cos\theta$ as exponential functions are consistent with the various formulae in 'real' trigonometry. For example, $\sin(\theta - \phi) = \sin\theta\cos\phi - \cos\theta\sin\phi$; and if we substitute the exponential versions in the R.H.S., we get

$$\frac{e^{i\theta} - e^{-i\theta}}{2i} \times \frac{e^{i\phi} + e^{-i\phi}}{2} - \frac{e^{i\theta} + e^{-i\theta}}{2} \times \frac{e^{i\phi} - e^{-i\phi}}{2i}$$

$$= \frac{1}{4i}(e^{i\theta+i\phi} - e^{-i\theta+i\phi} + e^{i\theta-i\phi} - e^{-i\theta-i\phi} - e^{i\theta+i\phi} - e^{-i\theta+i\phi}$$
$$+ e^{i\theta-i\phi} + e^{-i\theta-i\phi})$$

$$= \frac{1}{4i}(2e^{i(\theta-\phi)} - 2e^{-i(\theta-\phi)}) = \frac{e^{i(\theta-\phi)} - e^{-i(\theta-\phi)}}{2i},$$

which is the exponential version of $\sin(\theta - \phi)$. A similar verification may be achieved with any of the trigonometrical formulae.

With the above equipment, we may now even make sense of trigonometrical functions of complex numbers:

$$\sin z = \sin(x + iy) = \sin x\cos iy + \cos x\sin iy$$
$$= \sin x\cosh y + \cos x(i\sinh y)$$
$$= \sin x\cosh y + i\cos x\sinh y,$$

so that $\sin z$ is a complex number, and we have separated it into real and imaginary parts.

Example 2

Find $\cos^{-1} 2$.

Suppose $\cos^{-1} 2 = x + iy$.

Then $\cos(x + iy) = 2$

$\Rightarrow \quad \cos x\cosh y - i\sin x\sinh y = 2$

$\Rightarrow \quad \cos x\cosh y = 2 \qquad (i)$

and $\quad \sin x\sinh y = 0 \qquad (ii)$

From (ii), $\sin x = 0$ or $\sinh y = 0$.

Case 1

$\sinh y = 0 \Rightarrow y = 0 \Rightarrow \cosh y = 1.$

Substituting in (i), $\cos x = 2$.

But x is real, so this is impossible.

Case 2

$\sin x = 0 \quad \Rightarrow \quad x = n\pi\,(n \in Z) \quad \Rightarrow \quad \cos x = (-1)^n$.

Substituting in (i), $(-1)^n \cosh y = 2$.

But y is real, so $\cosh y \geqslant 1 \quad \Rightarrow \quad n$ is even $(= 2k)$, and $\cosh y = 2$.

Hence $x = 2k\pi$ and $y = \cosh^{-1} 2 = \pm\ln(2 + \sqrt{3})$
and finally $\cos^{-1} 2 = 2k\pi \pm i \ln(2 + \sqrt{3})$.

Example 3

For a circuit containing resistance R, inductance L and capacitance C, in series, the differential equation satisfied by Q, the quantity of electricity which has passed at time t, is $L\ddot{Q} + R\dot{Q} + Q/C = V_0 \cos \omega t$ (where $V_0 \cos \omega t$ is an alternating input E.M.F.). Suppose that for a particular circuit, the differential equation arising is

$$\ddot{Q} + 6\dot{Q} + 25Q = 60 \cos 5t,$$

and that we wish to find the current flowing at time t.

Seeking first the complementary function (see 9.8), we obtain the quadratic equation $k^2 + 6k + 25 = 0$, so that $k = -3 \pm 4i$.

Thus the c.f. is
$$\begin{aligned} Q &= A\,e^{(-3+4i)t} + B\,e^{(-3-4i)t} = e^{-3t}(A\,e^{4it} + B\,e^{-4it}) \\ &= e^{-3t}[A(\cos 4t + i \sin 4t) + B(\cos 4t - i \sin 4t)] \\ &= e^{-3t}(C \cos 4t + D \sin 4t), \end{aligned}$$

where $C = A + B$ and $D = i(A - B)$ are a fresh pair of arbitrary constants to replace the pair A, B.

The particular integral will be of the form $Q = L \cos 5t + M \sin 5t$. Direct substitution back into the differential equation enables us to find the values of L, M, namely $L = 0$, $M = 2$, so the complete solution of the differential equation is

$$Q = e^{-3t}(C \cos 4t + D \sin 4t) + 2 \sin 5t.$$

Hence the current $I = \dot{Q} = e^{-3t}(E \cos 4t + F \sin 4t) + 10 \cos 5t$, where E, F are yet another replacement pair of arbitrary constants. Their values could be found from two pieces of information, such as the initial charge and current. However, since the factor e^{-3t} diminishes rapidly with time, this term, arising as the complementary function, soon ceases to be of any consequence, and is known as the 'transient'. The term $10 \cos 5t$ gives the value of the permanent current resulting in the circuit from the application of the given alternating input.

Exercise 10.5

1 Express in the form $x + iy$:

(*i*) $e^{\frac{1}{2}i\pi}$; (*ii*) $e^{-i\pi}$; (*iii*) $e^{-1+\frac{1}{2}i\pi}$; (*iv*) $e^{i} - e^{-i}$; (*v*) $\sinh(2 - 5i)$; (*vi*) $\cos(3 + i)$; (*vii*) $\ln(-8)$; (*viii*) $\ln 3i$; (*ix*) $\ln(\sqrt{3} + i)$; (*x*) $\ln(-\sqrt{3} - i)$.

2 Prove that $e^{x-iy} = (\cosh x + \sinh x)(\cos y - \text{i} \sin y)$.

3 Using the exponential definitions of sin and cos, prove the formula $\sin(\theta + \phi) = \sin\theta\cos\phi + \cos\theta\sin\phi$ and other trigonometrical formula.

4 Use $\text{cis}\, 3x = e^{3ix}$ to integrate $e^{-2x}\cos 3x$ and $e^{-2x}\sin 3x$.

5 If $\sin(x + iy) = r\,\text{cis}\,\theta$, prove that $r = \frac{1}{2}(\cosh 2y - \cos 2x)$ and $\tan\theta = \cot x \tanh y$.

6 Find the real and imaginary parts and the moduli of:

(*i*) $\sin z$; (*ii*) $\cos z$; (*iii*) $\tan z$, where $z = x + iy$;

(*iv*) Express the real part of $\tan z$ as $\dfrac{\sin 2x}{\cos 2x + \cosh 2y}$.

7 Find z so that $\sin z = -3$.

8 If $\sinh(x + iy) = e^{\frac{1}{3}i\pi}$, find x and y.

9 Express $u = e^{x}\cos y$, $v = e^{x}\sin y$ as a single equation in $z(= x + iy)$ and $w(= u + iv)$. Repeat for $u = \cosh x \cos y$, $v = \sinh x \sin y$.

10 Given that $\tanh z = w$, express $\tanh 2x$ and $\tan 2y$ in terms of u and v, where $z = x + iy$ and $w = u + iv$.

11 Functions $c(x)$ and $s(x)$ are defined by the formulae

$c(x) = \dfrac{1}{2}(e^{ix} + e^{-ix})$ and $s(x) = \dfrac{1}{2i}(e^{ix} - e^{-ix})$, prove that

$[c(x)]^2 + [s(x)]^2 = 1$ and that $[c(x)]^2 - [s(x)]^2 = c(2x)$.

12 Give the general solution of the differential equations:

(*i*) $\dfrac{d^2y}{dx^2} - 8\dfrac{dy}{dx} + 25y = 0$; (*ii*) $\dfrac{d^2y}{dx^2} - 2\dfrac{dy}{dx} + 2y = 0$;

(*iii*) $\dfrac{d^2x}{dt^2} = 6 - 4x$; (*iv*) $\dfrac{d^2x}{dt^2} + 9x = \sin 4t$.

13 Obtain the solutions of the following differential equations subject to the stated boundary conditions (dots indicate differentiation with respect to time):

(*i*) $\ddot{x} - 6\dot{x} + 10x = 0$ given that $x = 0$ and $\dot{x} = 3$ at $t = 0$;

(*ii*) $\ddot{y} + 9y = 4$, where $y = 0$, $\dot{y} = 2$ at $t = 0$.

*10.6 Transformations in the Argand diagram

Suppose that z is a typical point of a configuration in the Argand diagram, and let $z' = z + a$, where a is a fixed complex number.

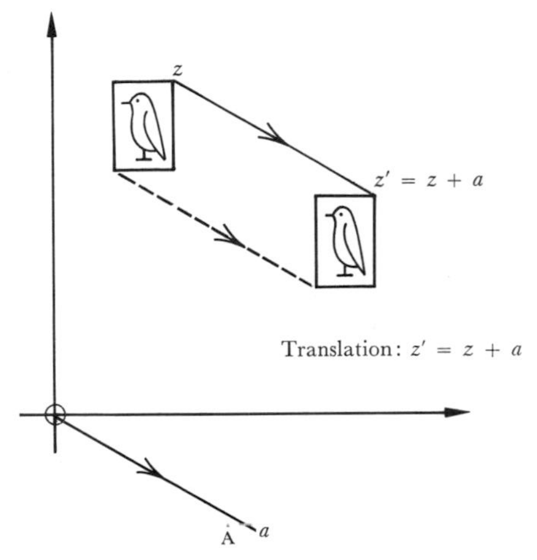

Translation: $z' = z + a$

Every point of the configuration undergoes a similar displacement specified by the (constant) vector $\boldsymbol{OA}$. The transformation $z' = z + a$ therefore represents a *translation*. For example, $z' = z + 2 - 3i$ represents a translation which carries the whole figure bodily 2 units in the positive x-direction and 3 units in the negative y-direction.

Consider next the transformation represented by $z' = \rho z$, where ρ is real and positive.

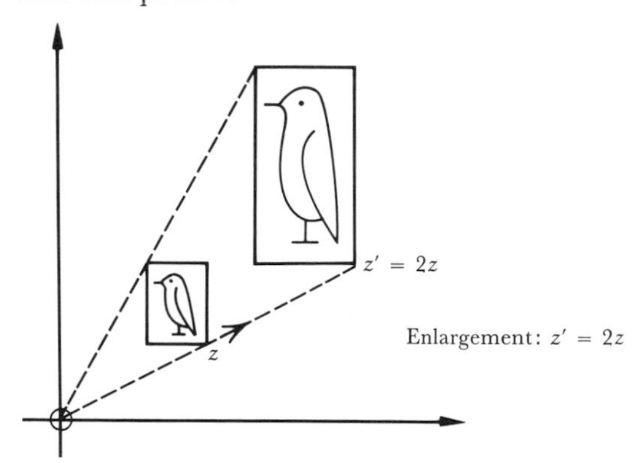

Enlargement: $z' = 2z$

Its effect is to multiply the modulus of z by a factor ρ, whilst its argument is unchanged. For example, if $\rho = 3$, the point $2 \text{ cis } 50°$ would map into the point $6 \text{ cis } 50°$, while the point $2 - 5i$ would map into $6 - 15i$. Thus the transformation is an enlargement of the original configuration from the origin in the ratio $\rho : 1$, the image of the configuration being *similar* to the original. The transformation is known as an *enlargement* (or *magnification*, *similarity*, *dilatation*, *homothety*) even when $\rho < 1$; and when $\rho = 1$, we obtain the *identity transformation*.

We have already seen in 10.2 that the transformation $z' = z \operatorname{cis} \phi$ has the effect of a pure rotation about the origin, since the modulus is unchanged, while the argument is increased by the angle ϕ.

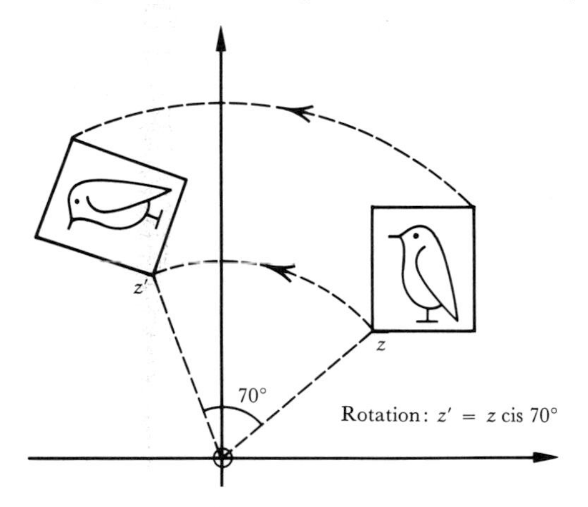

The final image will be congruent with the original. Note that when $\phi = \frac{1}{2}\pi$, we obtain $z' = iz$, and we have already remarked that multiplication by i may be interpreted geometrically as a positive quarter-turn. For rotation about a point *other than the origin*, say about the point A, remembering that vector $\boldsymbol{AZ}$ is represented by $z - a$, and $\boldsymbol{AZ'}$ by $z' - a$, we have

$$z' - a = (z - a) \operatorname{cis} \phi.$$

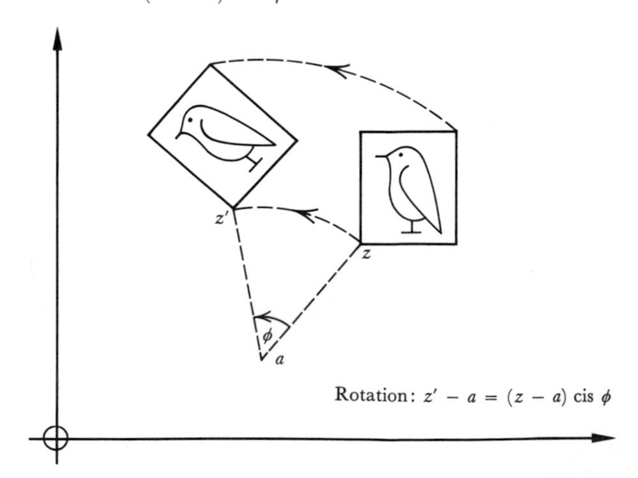

For example, if $z' = iz + 4$, we may write $z' - a = i(z - a)$

where $a = ia + 4$, so that $a = \dfrac{4}{1 - i} = 2(1 + i)$.

Thus $z' - 2 - 2i = i(z - 2 - 2i)$,

which represents a rotation through $90°$ about the point $2 + 2i$.

All the above transformations have been 'direct' in the sense that the clockwise or anti-clockwise sense of description of a figure is preserved in each case. A mirror reflection, however, is an 'opposite' transformation, the figure now being 'turned over'. The simplest case is reflection in the real axis, given by the equation $z' = z^*$. Reflection in the imaginary axis would be represented by $z' = -z^*$ and reflections in other lines will be given by equations of the form $z' = az^* + b$ $(a, b \in C)$. However, the latter equation does not always give a pure reflection; for in the simple case $z' = z^* + \alpha$, where α is real, there is a reflection in the real axis combined with a translation parallel to it. Such a transformation is known as a 'glide reflection'.

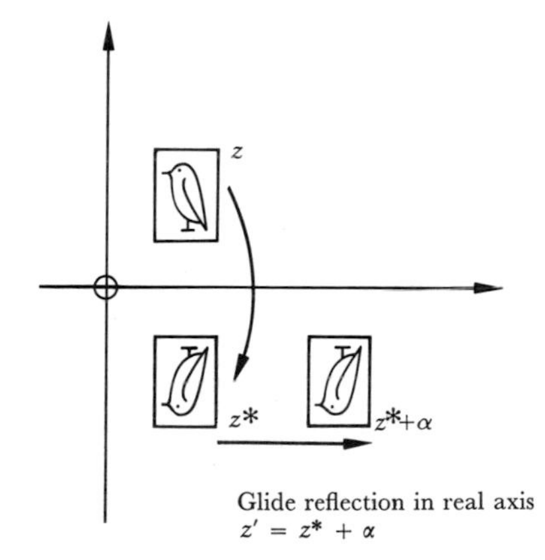

Glide reflection in real axis
$z' = z^* + \alpha$

Again, $z' = \rho z^*$, where ρ is real $(\neq 1)$, includes a similarity as well as a reflection.

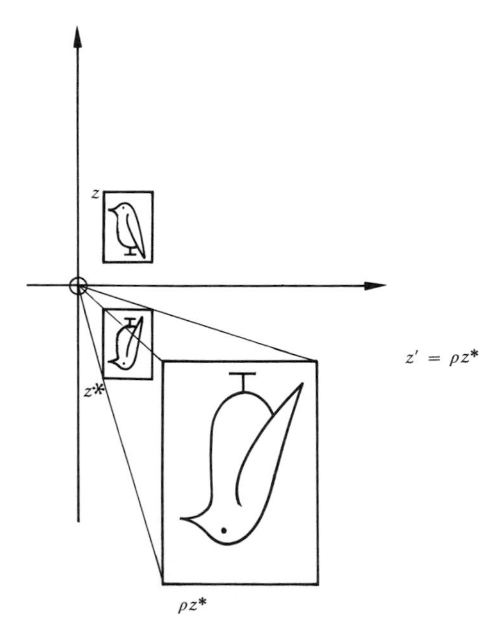

$z' = \rho z^*$

Composition of transformations

Argand diagram methods provide a simple means of finding the result of combining successive transformations in the plane. For example, a rotation $\frac{1}{2}\pi$ about O followed by a translation $+4$ is represented by $z' = iz + 4$, and we have already seen that this is equivalent to a single rotation about the point $2 + 2i$.

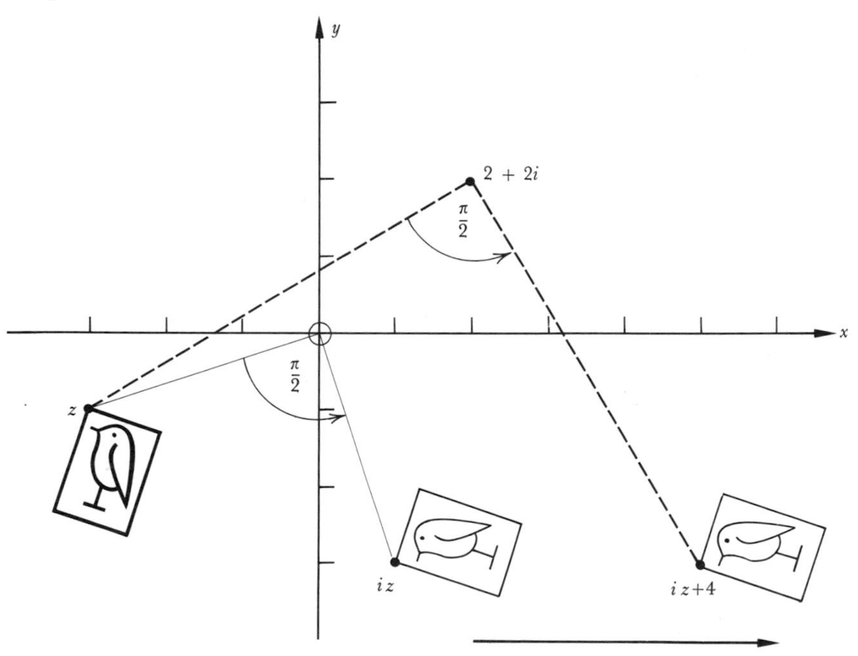

It should be noted that the two transformations do not commute. Fir if the translation were followed by the rotation, we would get $z' = i(z + 4)$, which is a rotation about $-2 + 2i$.

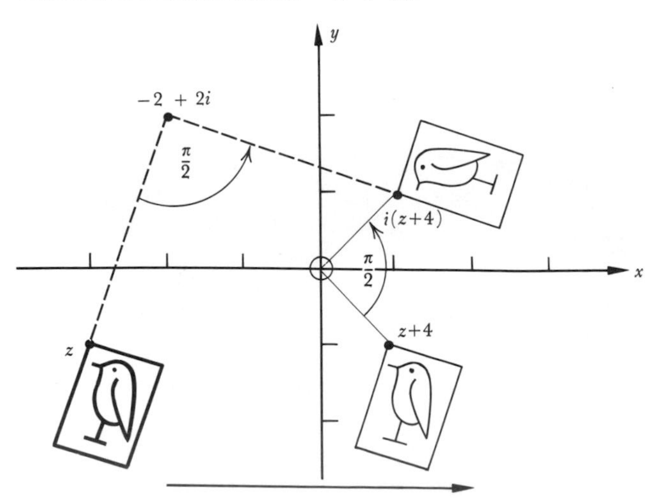

On the other hand, an enlargement $z' = \rho z$ in the origin, together with a rotation $z' = z \operatorname{cis} \phi$ (*also about the origin*) *do* commute, giving the composite transformation $z' = \rho \operatorname{cis} \phi \times z$; i.e., $z' = az$, where a is complex. This type of transformation we have already met, as a *spiral similarity*.

Next consider the enlargement $z' = 3z$ in the origin, together with the half-turn $z' - i = -(z - i)$ about i. If the enlargement is done first, we get

$$z' - i = -(3z - i) \quad \Rightarrow \quad z' = -3z + 2i.$$

But if they are commuted,

$$z' = 3[i + -(z - i)] \quad \Rightarrow \quad z' = -3z + 6i.$$

In the first case, we may write $z' = -3z + 2i$ in the form

$$z' - a = -3(z - a), \text{ where } 4a = 2i,$$

giving $a = \frac{1}{2}i$; i.e., $z' - \frac{1}{2}i = -3(z - \frac{1}{2}i)$, which is a spiral similarity about the point $\frac{1}{2}i$. You should find the centre of the spiral similarity in the second case when the half-turn is done first.

The above discussion opens up the whole study of plane transformations, and the interested reader may find a more thorough treatment of elementary transformation geometry of the plane in an article by F. J. Budden in the *Mathematical Gazette*, Feb. 1969, pp. 132–158. Here we must content ourselves with leaving further work to the exercises, though we shall continue the development of the subject when dealing with functions of a complex variable in 10.8, where we shall be concerned with some simple cases of non-linear functions.

Exercise 10.6

1 Interpret the following transformations, and find the invariant points where possible:

(*i*) $z' = z + 2 - 4i$; (*ii*) $z' = 2 - 4i - z$;
(*iii*) $z' = 5z - 2i$; (*iv*) $z' = 2iz$;
(*v*) $z' = (3 - 4i)z$; (*vi*) $z' = (5 \operatorname{cis} 100°)z$;
(*vii*) $z' + 2 = \frac{1}{2}(z + 2)$; (*viii*) $z' = z(1 - i)$;
(*ix*) $z' = iz - 1$.

2 Write down an equation of the form $z' = f(z)$ to represent the following:

(*i*) A translation $+3$ in the direction of the imaginary axis.
(*ii*) A spiral similarity with ratio 3:1 and angle of rotation $+135°$ about O.
(*iii*) A spiral similarity with ratio 3:1 and angle of rotation $+135°$ about i.
(*iv*) A rotation through $-\frac{1}{2}\pi$ about the point $1 - i$.
(*v*) A half-turn about $4 + i$.

(*vi*) An enlargement 1:4 in the point $2 + i$.
(*vii*) A magnification $\times 3$ in the origin followed by a translation -2 in the direction of the imaginary axis.
(*viii*) The same as in (*vii*) but in the reverse order.
(*ix*) An enlargement in the ratio 2:1 in the point i followed by a rotation through $-\frac{1}{4}\pi$ about the point 2.
(*x*) The same as in (*ix*) but in the reverse order.

3 Analyse the transformations:
(*i*) $z' = iz + 1 - i$; (*ii*) $z' = iz^* + 1 - i$.

4 By using complex numbers show that successive half-turns about a and b are equivalent to a translation $2\boldsymbol{AB}$.

5 Interpret the following transformations:
(*i*) $z' = z^*$; (*ii*) $z' = iz^*$; (*iii*) $z' = z^* \operatorname{cis} 2\alpha$;
(*iv*) $z' = iz^* - 2$; (*v*) $z' = i(z^* - 2)$.

*10.7 Loci in the Argand diagram

As a complex number is an ordered pair of reals, it may be used to represent any two-dimensional vector, such as a displacement, or a force, or an impedance operator in the theory of alternating currents. Such vectors are subject to change: a particle moves in a plane so that the complex number representing its position traces out a curve, the resistance or the frequency of a circuit is altered with the result that its impedance is altered, and the corresponding points on the Argand diagram move. We shall therefore be concerned with the *locus* of a point z in the Argand diagram, moving subject to certain conditions. The equations of such curves may in certain cases be expressible algebraically in terms of z. For example, a circle centre O and radius ρ may be written

$$|z| = \rho \quad \text{or} \quad zz^* = \rho^2,$$

while the half-line through O making angle θ with the real axis has equation $\arg z = \theta$. Similarly, the equation $\arg(z + 2i) = 300°$ represents the half-line terminating at the point $-2i$ and proceeding in the direction 300°.

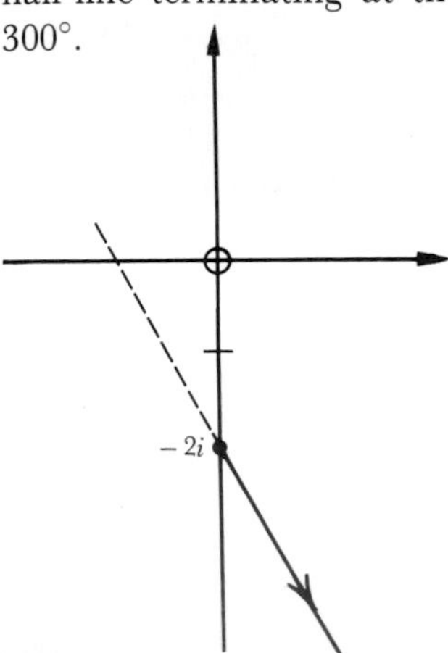

The equation of any circle, centre A and radius ρ, may be written

$$|z - a| = \rho \quad \Rightarrow \quad (z - a)(z - a)^* = \rho^2$$
$$\Rightarrow \quad zz^* - za^* - z^*a + (aa^* - \rho^2) = 0.$$

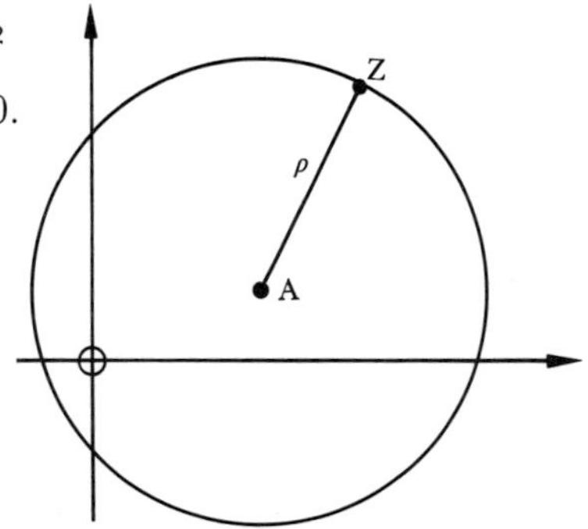

This can be rewritten in the usual Cartesian form,

$$x^2 + y^2 + 2gx + 2fy + c = 0$$

by replacing z and z^* by $x + iy$ and $x - iy$.

Circle equations may alternatively be written in the form

$$\left|\frac{z - a}{z - b}\right| = \lambda \quad (\lambda \in R^+),$$

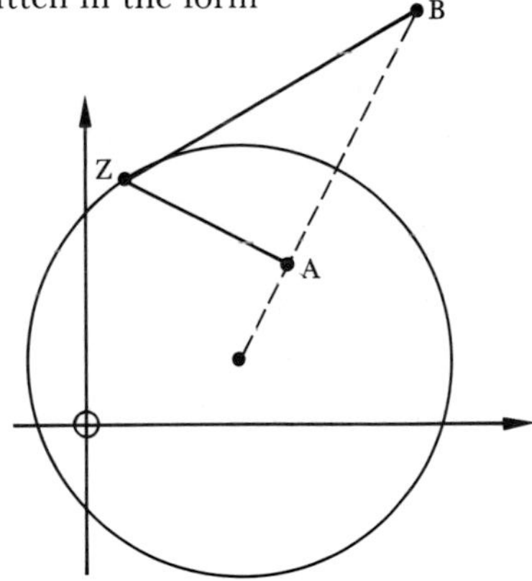

such an equation expressing the geometrical property that the point Z moves so that the ratio of its distances from two fixed points A and B is λ, the locus under such conditions being called the 'circle of Apollonius'. Likewise, the equation

$$\arg\left(\frac{z - a}{z - b}\right) = \phi \ (\in R)$$

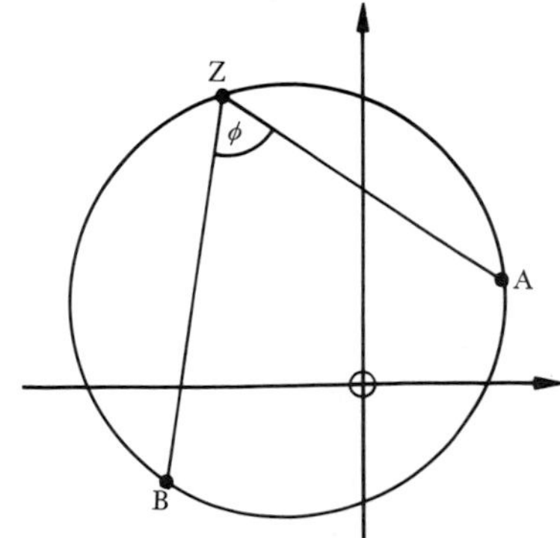

represents a circular arc terminated by the points A, B; for $\arg \dfrac{z-a}{z-b}$ is the angle of turn from $\boldsymbol{BZ}$ to $\boldsymbol{AZ}$, and so the equation expresses the property that this angle is constant. In our figure we show the cases where $\phi = 50°, 130°, -130°$, and $-50°$.

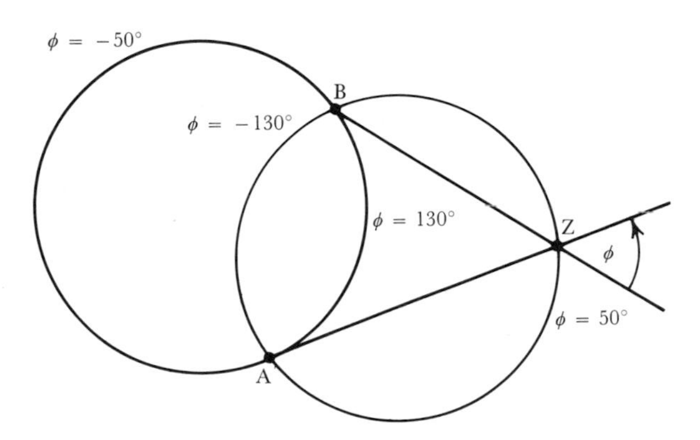

We may invent other examples of loci based upon our geometrical knowledge; e.g., the locus of Z so that $AZ + BZ =$ constant is an ellipse, and this may be expressed in complex numbers as $|z - a| + |z - b| = \lambda$ ($\lambda \in R^+$). Again, the inequality $|z| < 1$ represents the whole of the interior region of the unit circle; while $1 < |z + i| \leqslant 3$ represents the region lying between the circles centred at the point $-i$ with radii 1 and 3, including the circumference of the latter circle.

Exercise 10.7

1 Describe the loci when z moves in the Argand diagram subject to the following conditions:

(*i*) $|z| = 2$; (*ii*) $|z - 1| = |z + 1|$;

(*iii*) $|z - i| = |z + 1|$; (*iv*) $z = z^*$;

(*v*) $\arg(z - i) = -\frac{1}{3}\pi$; (*vi*) $|z| + |z - 1 + i| = 2$;

(*vii*) $\arg \dfrac{z-1}{z+1} = \frac{1}{2}\pi$; (*viii*) $|z| = 3|z - 1|$;

(*ix*) $\arg(z - 1) - \arg(z + i) = \pi$; (*x*) $|z - 1| + |z + 1| = 4$.

2 Shade the regions of the Argand diagram for which z is subject to the following restraints:

(*i*) $|z| < 3$; (*ii*) $|z + 2| > 2$; (*iii*) $|z + 4i| \geqslant 3$;

(*iv*) $|z - 3 + i| > 2$; (*v*) $|z - 1| + |z + 1| \leqslant 4$; (*vi*) $\left|\dfrac{z+2i}{z-2i}\right| \geqslant 1$.

3 From the equation $|z + 16| = 4|z + 1|$, deduce that $|z| = 4$, and interpret geometrically.

4 Show that $|z + 2i| = |2iz - 1|$ represents a circle, and find its centre and radius.

5 Express $|z/(z - i)| = 4$ using z and z^* instead of modulus signs. Repeat for $|(z - a)/(z - b)| = \lambda$.

6 If z moves along the semicircle connecting the points i, 1 and $-i$, describe the loci of the following:
(*i*) $3z$; (*ii*) $-2z$; (*iii*) iz; (*iv*) $2 + iz$;
(*v*) $2z^*$; (*vi*) z^2; (*vii*) $z^2 - 1$; (*viii*) $z^{1/2}$.

7 If z moves along the real axis from -1 to $+1$, find the locus of $(1 - iz)/(z - i)$.

8 If $(z - i)/(z - 1)$ is purely imaginary, show that the locus of z is a circle centre $\frac{1}{2}(1 + i)$, radius $2\sqrt{2}$.

9 If z describes the semicircle from the point 1 through the point i to the point -1, and thence the part of the real axis from -1 to $+1$, describe the locus of:
(*i*) $z + i$; (*ii*) $2z$; (*iii*) iz; (*iv*) z^2; (*v*) $1/z$.

10 If z describes the square 0, 1, $1 + i$, i, 0, describe the locus of:
(*i*) z^2; (*ii*) e^z; (*iii*) $\cosh z$.

*10.8 Functions of a complex variable

When x is a real variable, the function $f(x)$ is usually illustrated by a graph. Alternatively, however, it can be represented as a mapping from a set of values of x (its domain) onto a set of values of $f(x)$ (its range) by an arrow diagram.

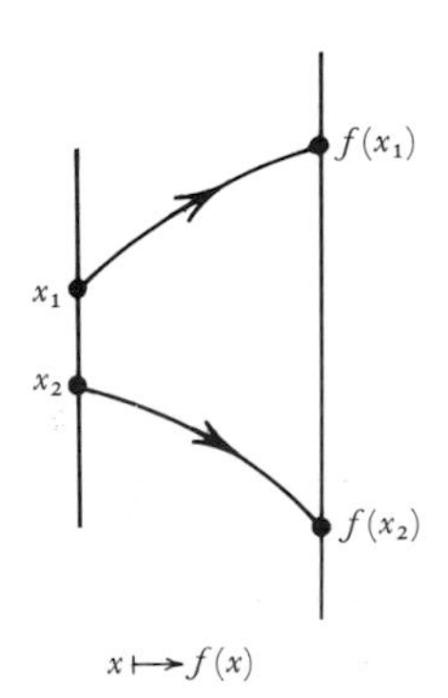

For instance,

$f: x \mapsto 2x + 1$

can be represented by

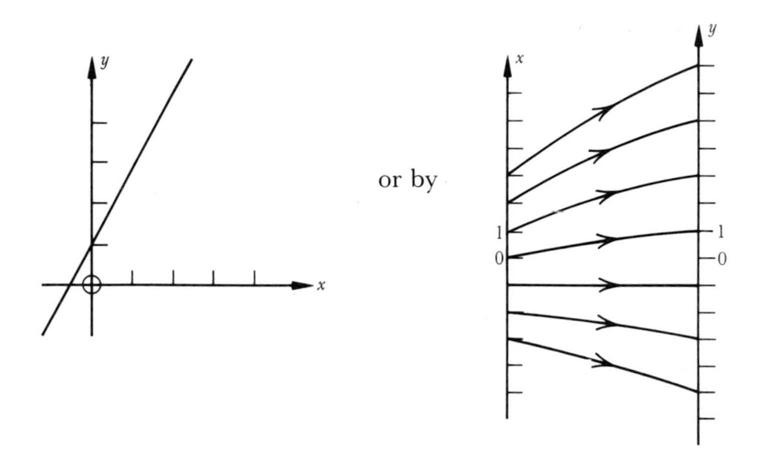

If z is a complex variable, the function $f(z)$ can no longer be shown in the first of these ways, as a graph, and has to be shown as a mapping, this time from a two-dimensional set of points (its domain) in one Argand diagram onto a set of points in another Argand diagram (its range).

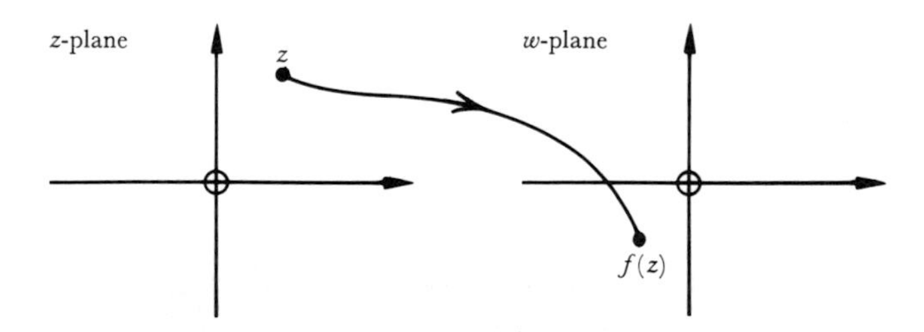

So, just as we can speak of $y = f(x)$ as a mapping $f: x \mapsto y$ of points from an x-line into a y-line, so we can also speak of $w = f(z)$ as a mapping $f: z \mapsto w$ of points from a z-plane into a w-plane.

For example, the function $w = 2z + 3 - i$ maps points as follows:

	O	U	I	J	V	W	A	B	K
z	0	1	i	$-i$	-1	$1+i$	2	-2	$-1\frac{1}{2}+\frac{1}{2}i$
w	$3-i$	$5-i$	$3+i$	$3-3i$	$1-i$	$5+i$	$7-i$	$-1-i$	0
	O′	U′	I′	J′	V′	W′	A′	B′	K′ = O

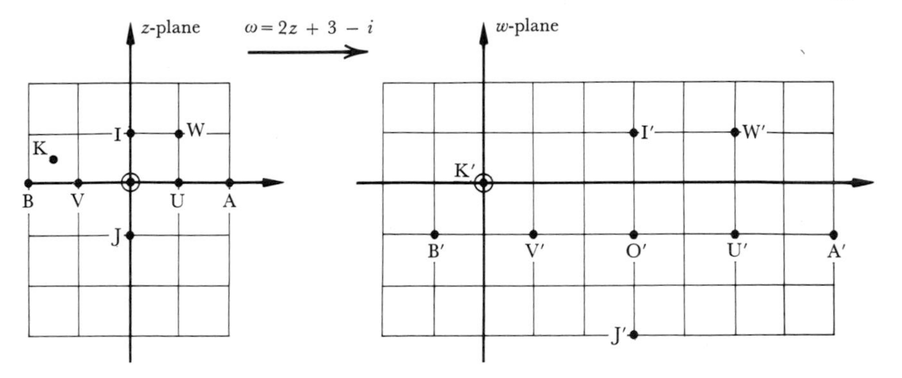

But it is rather simpler in this case to think of the z-plane and w-plane as being superimposed, and to regard $w = 2z + 3 - i$ as carrying the point z to a new position, first by the *enlargement* in the origin $z' = 2z$, followed by the *translation* $w = z' + 3 - i$. If z belongs to a configuration, then the whole of the configuration is carried by the transformation to the new position.

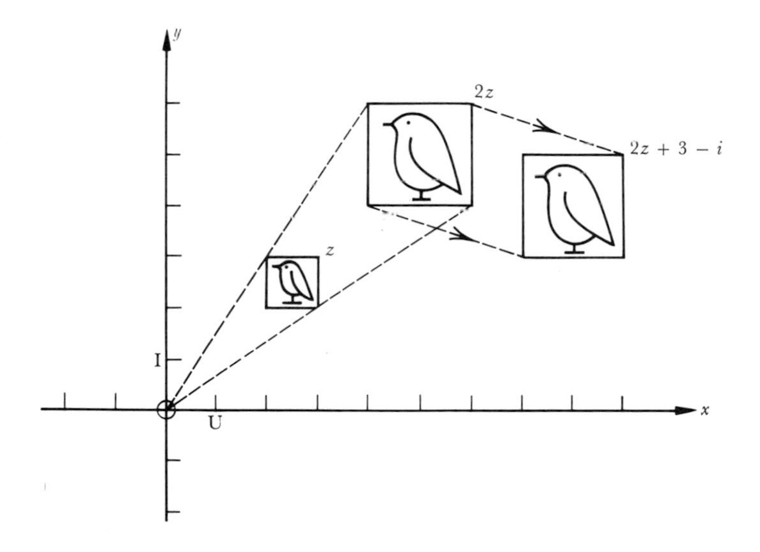

Such mappings are extremely important when the theory of complex numbers is applied to physical problems, ranging from the investigation of streamlines of fluids to that of lines of electric force and equipotentials, and to the study of the behaviour of oscillating circuits. Within mathematics itself, functions of a complex variable are equally important: exploration of what meaning can be given to such ideas as continuity, differentiation and integration has led to one of its most fertile areas. But this is a journey which cannot be undertaken here, and we confine ourselves to investigating a few examples.

The general linear function

It is clear that the general linear transformation

$$w = az + b \quad (a, b \in C)$$

or $\quad w = (\rho \text{ cis } \phi)z + b \quad (\text{where } a = \rho \text{ cis } \phi)$

may be analysed into basic elementary steps by splitting the formula as follows:

$$\left.\begin{array}{ll} z_1 = \rho z & \text{an enlargement in O} \\ z_2 = z_1 \text{ cis } \phi = az & \text{a rotation about O} \end{array}\right\} \text{spiral similarity}$$

$$w = z_2 + b = az + b \quad \text{a translation.}$$

Thus the general linear transformation may be composed of enlargements, rotations, and translations, and so must preserve straight lines. When $\rho = 1$, it will preserve distance (and so be an *isometric* transformation); when $\rho \neq 1$ the image will be *similar* to the original figure. The example $w = 2z + 3 - i$ was such a transformation, though the element of rotation was absent because the value of a $(= 2)$ was real (so that $\phi = 0$).

The reciprocal function, $w = 1/z$

The function $1/z$ is immediately seen to be quite different in that it does *not* preserve straight lines. We may see this easily by considering the three collinear points

U $(1 + 0i)$, S $(1 + i)$, T $(1 - i)$,

which have as images:

U′ $(1 + 0i)$, S′ $\frac{1}{2}(1 - i)$, T $\frac{1}{2}(1 + i)$;

and these do not lie on a straight line.

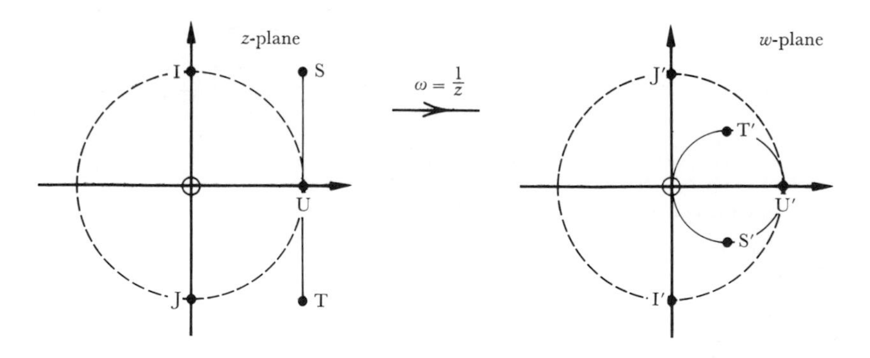

The transformation represented by $w = 1/z$ may be investigated geometrically, but this requires a knowledge of *inversion* and we content ourselves with a simple case, using analytical methods.

Example 1

If $w = \dfrac{1}{z}$, where $w = u + iv$ and $z = x + iy$,

then $u + iv = \dfrac{1}{x + iy} = \dfrac{x - iy}{x^2 + y^2}$,

so that $u = \dfrac{x}{x^2 + y^2}$ and $v = -\dfrac{y}{x^2 + y^2}$.

Consider a line $u = \text{constant} = \lambda$ in the w-plane.

Since $u = \dfrac{x}{x^2 + y^2} = \lambda$, we have

$x^2 + y^2 - \dfrac{1}{\lambda}x = 0$, and this is a circle in the z-plane.

Similarly the line $y = \mu$ in the z-plane will lead to the circle locus whose equation is

$u^2 + v^2 - \dfrac{1}{\mu}v = 0$ in the w-plane.

Example 2

If z traces out the circle $|z| = 1$ anti-clockwise from U back to U, find the locus of $w = \dfrac{z - 1}{z + 1}$.

This could be done by writing $w = 1 - \dfrac{2}{z + 1}$, and breaking this down into

$z_1 = z + 1$, a translation;

$z_2 = \dfrac{1}{z_1} = \dfrac{1}{z + 1}$, the reciprocal transformation;

$z_3 = -2z_2 = \dfrac{-2}{z + 1}$, a half-turn and enlargement;

$w = z_3 + 1 = 1 - \dfrac{2}{z + 1} = \dfrac{z - 1}{z + 1}$, a translation;

and then finding the geometrical effect of each of these on the unit circle.

Or we may take $z = \text{cis}\,\theta$, so that

$$w = \frac{\text{cis}\,\theta - 1}{\text{cis}\,\theta + 1} = \frac{\text{cis}\,\frac{1}{2}\theta - \text{cis}\,(-\frac{1}{2}\theta)}{\text{cis}\,\frac{1}{2}\theta + \text{cis}\,(-\frac{1}{2}\theta)} = \frac{2i\sin\frac{1}{2}\theta}{2\cos\frac{1}{2}\theta} = i\tan\tfrac{1}{2}\theta.$$

Thus, as θ goes from 0 to π, w traces out the imaginary axis from O to infinity; whilst as θ goes from π to 2π, w traces the other half of the imaginary axis from $-\infty$ up to the origin.

There is, however, a simple geometric solution. For if U, V are the points 1, -1 (our usual notation), then $z - 1$ and $z + 1$ are represented by the vectors ***UZ***, ***VZ***. The fact that these are *perpendicular* ($\angle$UZV being the angle in a semi-circle), means that their quotient w must be 'purely imaginary', or a real multiple of i. Hence w lies on the imaginary axis.

Alternatively we may rearrange the formula as $z = \dfrac{1 + w}{1 - w}$. Since $|z| = 1$, it follows that $|1 + w| = |1 - w|$, so that w is equidistant from the points -1 and $+1$, and therefore lies on the imaginary axis.

Example 3

If $w = z^2$, find the locus of z when w traces the lines $u =$ constant, $v =$ constant.

$$u + iv = (x + iy)^2 = x^2 - y^2 + 2ixy;$$

so $\quad u = x^2 - y^2, \qquad v = 2xy.$

Hence $\quad u =$ constant gives a family of rectangular hyperbolas,

$$x^2 - y^2 = \lambda \ (\in R),$$

$v =$ constant gives another family of rectangular hyperbolas,

$$xy = \mu \ (\in R).$$

All mutual intersections are perpendicular and the families are said to be orthogonal.

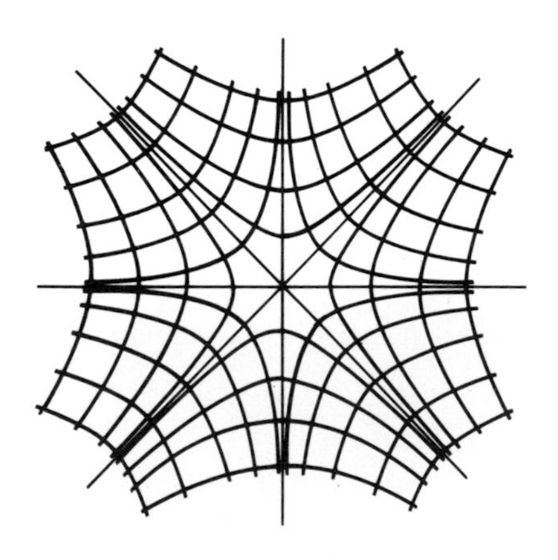

Example 4

Express $\quad u = e^{-x} \sin y, \quad v = e^{-x} \cos y \quad$ as a relation between w and z.

$$\begin{aligned} w = u + iv &= e^{-x}(\sin y + i \cos y) = e^{-x} i(\cos y - i \sin y) \\ &= i\, e^{-x} e^{-iy} = i\, e^{-z}. \end{aligned}$$

Polynomial functions: the fundamental theorem of algebra

Earlier in this chapter the reader may have had the uneasy feeling that there may just possibly exist a certain polynomial function of the form

$$f(z) \equiv a_0 z^n + a_1 z^{n-1} + \cdots + a_{n-1} z + a_n$$

with the property that there is no value of z whatsoever which makes $f(z) = 0$. Certainly there are quadratic equations which have no solutions in the field of *real* numbers, and it might reasonably have been supposed by analogy that there are also polynomial equations which have no solutions even in the much wider field of complex numbers. The reader should not be ashamed of harbouring such doubts, for the same question troubled many mathematicians for ages. It was finally resolved, and doubts dispelled, by the so-called *fundamental theorem of algebra*, which states that for *every* polynomial $f(z)$ there does indeed exist a value of z ($\in C$), which satisfies the equation $f(z) = 0$. In other words, every polynomial equation has at least one root in the field of complex numbers, and this is still true even when the coefficients, $a_0, a_1, \ldots, a_n$ are themselves complex. So the theorem, which is an example of what mathematicians call an *existence theorem*, is really a very powerful result; and, once proved, it is a simple matter to prove further that a polynomial of degree n has n roots in complex algebra (repeated roots being counted according to their multiplicity). But the proof of the fundamental theorem of algebra is difficult, involving sophisticated analytical concepts. For readers who wish to pursue the matter, a traditional proof may be found in Hardy, G. H., *A course in pure mathematics* (Appendix II) or in Dickson, L. E., *First course in the theory of equations* (Appendix). But more modern methods of algebraic topology enable the proof to be considerably shortened.

Exercise 10.8

1 If $w = (z - 1)/(z + 1)$ (as on p. 159), show that when w traces out the lines u = constant and v = constant, the locus of z is two systems of circles. Find also the locus of w as z traces the upper half of the circle diameter $-1, +1$.

2 If $w = (z - i)/(z + i)$ and lies below the real axis, show that w lies outside the unit circle $|w| = 1$. How will w move as z travels along the real axis from $-\infty$ to $+\infty$?

3 If $w = (z - 6i)/(z + 8)$, show that the locus of z is a circle when w traces the imaginary axis, but a line when w traces the real axis. Find the equations of the z-loci.

4 Examine the effect of the transformation $w = (z + 2)/(z + i)$ on the system of concentric circles having centres at the origin in the z-plane. What is the image of the exterior of the circle $|z| = 1$?

5 If $w = (z - a)/(z - b)$ verify that w is obtained by making triangle 0, 1, w similar to triangle z, b, a.

6 If $w = (z + 1)^2$ and z traces the unit circle, show that the locus of w has polar equation $r = 2(1 + \cos\theta)$ (a *cardioid*).

7 Show that the lines x = constant and y = constant in the z-plane are transformed by $w = \cosh z$ into sets of confocal ellipses and hyperbolas in the w-plane.

8 What is the effect of $w = e^z$ on the system of lines x = constant, y = constant in the z-plane?

9 If $w = 4/(z + 1)^2$ and $|z| = 1$, show that w describes the parabola whose polar equation is $r = \sec^2 \frac{1}{2}\theta$. Show that the interior of the circle maps into the exterior of the parabola.

10 Find the invariant points of the transformation $w = \dfrac{z + 5}{2 + i(z - 1)}$.

11 If $(1 - z)(1 - w) = 1$ and $\arg z = \frac{1}{4}\pi$, find the locus of w.

12 What is the effect of $w = \cos z$ on the lines x = constant, y = constant in the z-plane?

Miscellaneous problems 10

1 Solve the equation $z^4 + 10z^2 + 169 = 0$ by first solving for z^2, and then for z in the form $a + ib$ where a and b are real. Hence or otherwise, express $z^4 + 10z^2 + 169$ as the product of two real quadratic factors.

2 Show that for real x, $\dfrac{1}{1 + x^2} = \dfrac{1}{2i}\left(\dfrac{1}{x - i} - \dfrac{1}{x + i}\right)$.

Deduce that

$$\frac{d^{n-1}}{dx^{n-1}}\left(\frac{1}{1 + x^2}\right) = (-1)^{n-1}(n - 1)!\frac{1}{2i}\left(\frac{1}{(x - i)^n} - \frac{1}{(x + i)^n}\right).$$

Prove that, if $x - i = r(\cos\theta - i\sin\theta)$,

then $(x - i)^{-n} = r^{-n}(\cos n\theta + i\sin n\theta)$,

and obtain a similar expression for $(x + i)^{-n}$.

Hence obtain a formula for

$\dfrac{d^{n-1}}{dx^{n-1}}\left(\dfrac{1}{1 + x^2}\right)$ in terms of r and θ,

and deduce that for $0 < x < \frac{1}{2}\pi$,

$$\frac{d^n}{dx^n}(\tan^{-1} x) = (-1)^{n-1}\frac{(n - 1)!}{(1 + x^2)^{n/2}}\sin(n\cot^{-1} x).$$ (M.E.I.)

3 What geometrical transformation is given by

$z' - c = (z - c)(\cos\theta + i\sin\theta)$, where c is a complex constant?

If the transformation $z \mapsto z'$ corresponds to a turn of 120° about the point (1, 0), and the transformation $z' \mapsto z''$ corresponds to a turn of 60° about the point (−3, 0), show that $z'' = -z + 2\sqrt{3}i$, and hence express in geometrical terms the single transformation equivalent to the two transformations in the order given. (S.M.P.)

4 (*i*) Given that one root of the equation $z^4 - 4z^3 + 12z^2 + 4z - 13 = 0$ is $2 - 3i$, find the other three roots.
(*ii*) Given $z = 2(\cos\pi/6 + i\sin\pi/6)$ illustrate on the Argand diagram the points representing the complex numbers z, z^*, $(z^*)^2$, $(z^*)^2/|z|$.
S is the set of points in the first quadrant which lie on the circumference of the circle $|z| = 2$. Find the image of S under the mapping $z \mapsto (z^*)^2/|z|$. (M.E.I.)

5 The vertices of a triangle in the Argand diagram are given by the complex numbers a, b, c. Say whether the following are true or false, with reasons:
(*i*) The centroid is at the point $\frac{1}{3}(a + b + c)$.
(*ii*) The area of the triangle is $\frac{1}{2}[(b - c)^2 + (c - a)^2 + (a - b)^2]$.
(*iii*) The triangle is equilateral if and only if
$(a - b)(a - c) + (b - c)^2 = 0$. (O.S.)

6 Let $f(z) = \frac{1}{2}(z + z^{-1})$ where z is a non-zero complex number.
(*i*) If z describes a circle $|z| = r$ where r is a positive real constant, show that if $r \neq 1$, $f(z)$ describes an ellipse in the complex number plane. What happens if $r = 1$?
(*ii*) If z describes the half-line $\arg z = \theta$, where θ is constant, prove that, in general (i.e., if certain exceptional values of θ are excluded) $f(z)$ describes one branch of a hyperbola.
(*iii*) What are the exceptional cases in (*ii*)? Describe what happens for these exceptional cases. (O.S.)

7 If $|z - 1| = 1$, prove that $\arg(z - 1) = 2\arg z = \frac{2}{3}\arg(z^2 - z)$, and give a geometrical interpretation of these results. (O.S.)

8 Show that the roots of $(z + i)^5 = (z - i)^5$ are $\cot k\pi/5$ ($k = 1, 2, 3, 4$), and hence evaluate $\cot^2\pi/5 + \cot^2 2\pi/5$, and $\cot\pi/5 \cot 2\pi/5$. Generalise the result and thence obtain

$$\sum_1^n \cot^2\frac{k\pi}{2n + 1} = \frac{1}{3}n(2n - 1).$$

9 Prove that the roots of

$$(z + 1)^n + (z - 1)^n = 0 \quad \text{are} \quad i\cot\frac{(2k + 1)\pi}{2n} \quad (k = 0, 1, 2, \ldots, n - 1).$$

Deduce that

$$\sum_{0}^{n-1} \cot \frac{(2k+1)\pi}{2n} = 0 \quad \text{and} \quad \sum_{0}^{n-1} \cot^2 \frac{(2k+1)\pi}{2n} = n^2 - n.$$

10 Show that

$$(1+x)^{2n} + (1-x)^{2n} = 2\prod_{1}^{n}\left(x^2 + \tan^2 \frac{(2k-1)\pi}{4n}\right),$$

and prove that

$$\sum_{1}^{n} \sec^2 \frac{(2k-1)\pi}{4n} = 2n^2.$$

11 Prove that $e^{2k\pi i} = 1$ for all integral values of k, and that $e^{\frac{1}{2}\pi i} = i$. Deduce that a value of i^i is $e^{-\frac{1}{2}\pi}$, and using the result $e^{2k\pi i} = 1$, find a formula for all possible values of i^i. (C.S.)

12 If $\cos z = w$, prove that

$$\frac{u^2}{\cos^2 x} - \frac{v^2}{\sin^2 x} = 1 \quad \text{and} \quad \frac{u^2}{\cosh^2 y} + \frac{v^2}{\sinh^2 y} = 1$$

and discuss interpretations in terms of the Argand diagram.

13 For a rotation through α and about O, we have

$$z' = z \operatorname{cis} \alpha \quad \text{or} \quad x' + iy' = (x+iy)(\cos\alpha + i\sin\alpha).$$

Equating real and imaginary parts gives:

$$\left.\begin{aligned} x' &= x\cos\alpha - y\sin\alpha \\ y' &= x\sin\alpha + y\cos\alpha \end{aligned}\right\}, \text{ so we get the rotation matrix } \begin{pmatrix} \cos\alpha & -\sin\alpha \\ \sin\alpha & \cos\alpha \end{pmatrix}.$$

Show that reflection in the line $y = x\tan\alpha$ can be represented

$$z' = z^* \operatorname{cis} 2\alpha.$$

By equating real and imaginary parts, find the matrix for this reflection.

11

Further vectors and mechanics

11.1 Points of subdivision

Suppose that two masses are attached to the ends of a light rod AB, 2 kg at A and 3 kg at B.

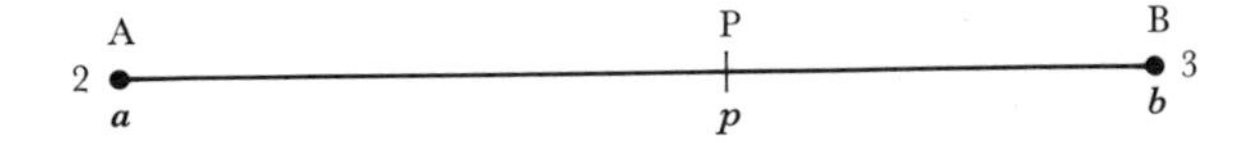

It is found from elementary experiments that such a rod will balance at the point P,

where $2\boldsymbol{AP} = 3\boldsymbol{PB}$,

so we call this the 'point of balance'.

If we now denote the position vectors (from an origin O) of A, B, P by $\boldsymbol{a}$, $\boldsymbol{b}$, $\boldsymbol{p}$,

$$\text{then} \quad 2(\boldsymbol{p} - \boldsymbol{a}) = 3(\boldsymbol{b} - \boldsymbol{p})$$

$$\Rightarrow \quad 5\boldsymbol{p} = 2\boldsymbol{a} + 3\boldsymbol{b}$$

$$\Rightarrow \quad \boldsymbol{p} = \frac{2\boldsymbol{a} + 3\boldsymbol{b}}{5}.$$

More generally, let us consider two masses, α at A and β at B.

Then their 'point of balance' would be at P, where

$$\alpha \boldsymbol{AP} = \beta \boldsymbol{PB}.$$

So $\quad \alpha(\boldsymbol{p} - \boldsymbol{a}) = \beta(\boldsymbol{b} - \boldsymbol{p})$

$\Rightarrow \quad (\alpha + \beta)\boldsymbol{p} = \alpha\boldsymbol{a} + \beta\boldsymbol{b}$

$\Rightarrow \quad \boldsymbol{p} = \dfrac{\alpha\boldsymbol{a} + \beta\boldsymbol{b}}{\alpha + \beta}.$

It is clear that only the ratio $\alpha:\beta$ is important. For if α and β were both multiplied by the same number κ,

$$\kappa\alpha\boldsymbol{AP} = \kappa\beta\boldsymbol{PB} \quad \Rightarrow \quad \alpha\boldsymbol{AP} = \beta\boldsymbol{PB}$$

and $\quad \boldsymbol{p} = \dfrac{\kappa\alpha\boldsymbol{a} + \kappa\beta\boldsymbol{b}}{\kappa\alpha + \kappa\beta} = \dfrac{\alpha\boldsymbol{a} + \beta\boldsymbol{b}}{\alpha + \beta},$

so that the position of P is unchanged.

When, however, the ratio changes, P will move along the line AB; and if either of the numbers α, β were allowed to be negative, P would fall outside the segment AB and be either beyond A or beyond B.

Example 1

(*i*) $\alpha = -2, \beta = +3.$

So $\quad -2\boldsymbol{AP} = 3\boldsymbol{PB} \quad$ and P is beyond B.

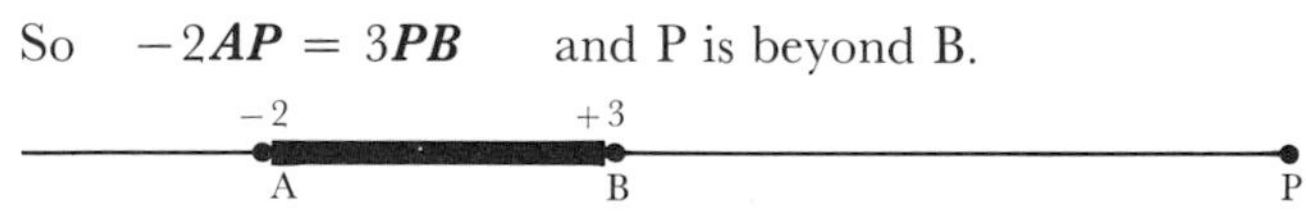

Furthermore, $\quad \boldsymbol{p} = \dfrac{-2\boldsymbol{a} + 3\boldsymbol{b}}{-2 + 3} = -2\boldsymbol{a} + 3\boldsymbol{b}.$

(*ii*) $\alpha = +2, \beta = -1.$

So $\quad 2\boldsymbol{AP} = -\boldsymbol{PB} \quad$ and P is beyond A.

Furthermore, $\quad \boldsymbol{p} = \dfrac{2\boldsymbol{a} - \boldsymbol{b}}{2 - 1} = 2\boldsymbol{a} - \boldsymbol{b}.$

The only exceptional case is when $\alpha = -\beta$.

Then $\quad \alpha\boldsymbol{AP} = -\alpha\boldsymbol{PB} \quad \Rightarrow \quad \boldsymbol{AP} = -\boldsymbol{PB},$

which is impossible when A, B are distinct

(and $\quad \boldsymbol{p} = \dfrac{\alpha\boldsymbol{a} - \alpha\boldsymbol{b}}{\alpha - \alpha} = \dfrac{\alpha\boldsymbol{a} - \alpha\boldsymbol{b}}{0}, \quad$ which is meaningless).

Exercise 11.1a

1 If A, B and P have position vectors $\boldsymbol{a}$, $\boldsymbol{b}$ and $\boldsymbol{p}$, find $\boldsymbol{p}$ in terms of $\boldsymbol{a}$ and $\boldsymbol{b}$ and illustrate by a sketch when:

(*i*) $\boldsymbol{AP} = \boldsymbol{PB}$; (*ii*) $3\boldsymbol{AP} = \boldsymbol{PB}$; (*iii*) $\boldsymbol{AP} = 3\boldsymbol{PB}$;
(*iv*) $3\boldsymbol{AP} = -\boldsymbol{PB}$; (*v*) $\boldsymbol{AP} = -3\boldsymbol{PB}$.

2 If A, B have position vectors $\boldsymbol{a}$, $\boldsymbol{b}$, illustrate the points whose position vectors are:

(*i*) $\frac{4}{5}\boldsymbol{a} + \frac{1}{5}\boldsymbol{b}$; (*ii*) $\frac{3}{5}\boldsymbol{a} + \frac{2}{5}\boldsymbol{b}$; (*iii*) $\frac{2}{5}\boldsymbol{a} + \frac{3}{5}\boldsymbol{b}$;
(*iv*) $\frac{1}{5}\boldsymbol{a} + \frac{4}{5}\boldsymbol{b}$; (*v*) $2\boldsymbol{a} - \boldsymbol{b}$; (*vi*) $3\boldsymbol{a} - 2\boldsymbol{b}$;
(*vii*) $4\boldsymbol{a} - 3\boldsymbol{b}$; (*viii*) $2\boldsymbol{b} - \boldsymbol{a}$; (*ix*) $3\boldsymbol{b} - 2\boldsymbol{a}$;
(*x*) $4\boldsymbol{b} - 3\boldsymbol{a}$.

3 The points A and B have position vectors $6\boldsymbol{i} - 5\boldsymbol{j}$ and $4\boldsymbol{i} - 3\boldsymbol{j}$ respectively. Show that the mid-point M of AB is collinear with the two points X, Y whose position vectors are $2\boldsymbol{i} - 6\boldsymbol{j}$ and $11\boldsymbol{i}$ respectively. (O.C.)

Points of subdivision

If $\alpha\boldsymbol{AP} = \beta\boldsymbol{PB}$,

then $\alpha AP = \beta PB$

and $\dfrac{AP}{PB} = \dfrac{\beta}{\alpha}$;

so P is said to *divide* AB in the ratio $\beta:\alpha$.

Example 2

(*i*) $\alpha = 2, \beta = 3$.

So $2\boldsymbol{AP} = 3\boldsymbol{PB}$

$\Rightarrow \dfrac{AP}{PB} = \frac{3}{2}$

and P divides AB in the ratio $3:2$.

(*ii*) $\alpha = -2, \beta = 3$.

So $2\boldsymbol{AP} = -3\boldsymbol{PB}$

$\Rightarrow \dfrac{AP}{PB} = -\frac{3}{2}$

and P divides AB in the ratio $3:-2$.

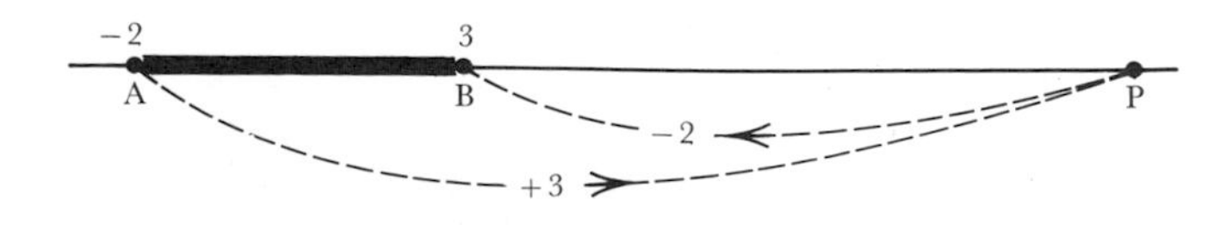

(*iii*) $\alpha = 2, \beta = -1$.

So $-2\boldsymbol{AP} = \boldsymbol{PB}$

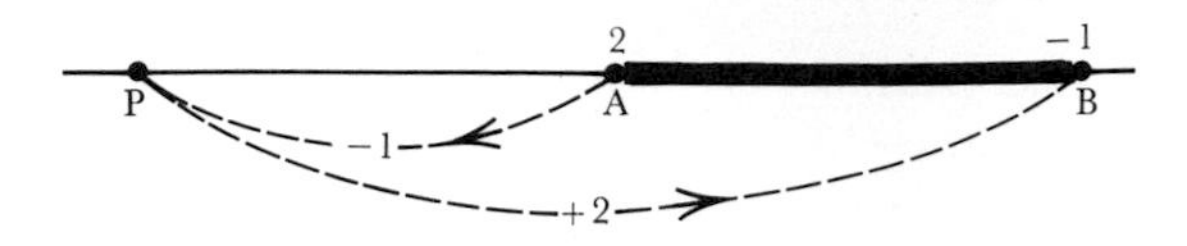

$\Rightarrow \dfrac{AP}{PB} = -\frac{1}{2}$

and P divides AB in the ratio $-1:2$.

Summarising, we see that the following statements are all equivalent:

> P is the centre of mass of α at A and β at B
>
> $\alpha\boldsymbol{AP} = \beta\boldsymbol{PB}$
>
> $\boldsymbol{p} = \dfrac{\alpha\boldsymbol{a} + \beta\boldsymbol{b}}{\alpha + \beta}$
>
> P divides AB in the ratio $\beta:\alpha$

Example 3

Find the position vectors of the points P, Q which divide AB in the ratio $1:2$ and $-5:2$ respectively.

(*i*) P divides AB in the ratio $1:2$

$\Rightarrow$ P is the 'point of balance' of 2 at A and 1 at B

$$\Rightarrow \quad \boldsymbol{p} = \frac{2\boldsymbol{a} + 1\boldsymbol{b}}{2 + 1} = \tfrac{2}{3}\boldsymbol{a} + \tfrac{1}{3}\boldsymbol{b}.$$

(*ii*) Q divides AB in the ratio $5:-2$

$\Rightarrow$ Q is the 'point of balance' of -2 at A and 5 at B

$$\Rightarrow \quad \boldsymbol{q} = \frac{-2\boldsymbol{a} + 5\boldsymbol{b}}{-2 + 5} = -\tfrac{2}{3}\boldsymbol{a} + \tfrac{5}{3}\boldsymbol{b}.$$

Exercise 11.1b

1 The vertices of $\triangle$ABC have position vectors $\boldsymbol{a}$, $\boldsymbol{b}$, $\boldsymbol{c}$. Find, in terms of $\boldsymbol{a}$, $\boldsymbol{b}$, $\boldsymbol{c}$, the position vectors of

(*i*) D. E. F. the mid-points of BC, CA, AB respectively;

(*ii*) the point X which divides the line AD in the ratio $2:1$, and the corresponding points Y, Z dividing BE, CF in the same ratio.

Hence show that the medians AD, BE, CF are concurrent (at a point called the *centroid* of $\triangle$ABC).

2 The vertices A, B, C of $\triangle$ABC have position vectors $\boldsymbol{a}$, $\boldsymbol{b}$, $\boldsymbol{c}$. Find, in terms of $\boldsymbol{a}$, $\boldsymbol{b}$ and $\boldsymbol{c}$, the position vectors of

(*i*) P, which divides BC in the ratio $3:2$;

(*ii*) Q, which divides CA in the ratio $1:3$;

(*iii*) R, which divides AB in the ratio $2:1$;

(*iv*) X, which divides AP in the ratio 5:1;
(*v*) Y, which divides BQ in the ratio 2:1;
(*vi*) Z, which divides CR in the ratio 1:1.
Hence show that AP, BQ, CR are concurrent.

3 In no. **2** find also the position vectors of
(*i*) U, which divides BC in the ratio −3:2;
(*ii*) V, which divides CA in the ratio −1:3;
(*iii*) W, which divides AB in the ratio −2:1.
Hence show that:
U also divides QR in the ratio −3:4;
V also divides RP in the ratio −5:3;
and W also divides PQ in the ratio −4:5.

4 If the points A, B, C have position vectors $\boldsymbol{a}$, $\boldsymbol{b}$, $\boldsymbol{c}$ from an origin O, show that the equation

$$\boldsymbol{r} = t\boldsymbol{a} + (1 - t)\boldsymbol{b},$$

where t is a parameter, represents the straight line AB, and find the equation of BC.

Find the equation of the straight line joining L the mid-point of OA to M the mid-point of BC. Find also the position vector of the point in which the line LM meets the straight line joining the mid-point of OB to the mid-point of AC. (L.)

5 OABC is a square of side $2a$. $\boldsymbol{i}, \boldsymbol{j}$ are unit vectors along OA, OC. The mid-point of AB is L; the mid-point of BC is M; OL, AM meet at P; BP meets OA at N. Show that the segment OP can be measured by the vector $\lambda(2a\boldsymbol{i} + a\boldsymbol{j})$ and also by the vector $2a\boldsymbol{i} + \mu(2a\boldsymbol{j} - a\boldsymbol{i})$. Hence determine λ and μ. Prove that $ON = \frac{2}{3}OA$. (O.C.)

11.2 Centres of mass

In the last section, when masses α and β were placed at two points whose position vectors are $\boldsymbol{a}$ and $\boldsymbol{b}$, we referred to

$$\boldsymbol{p} = \frac{\alpha\boldsymbol{a} + \beta\boldsymbol{b}}{\alpha + \beta}$$

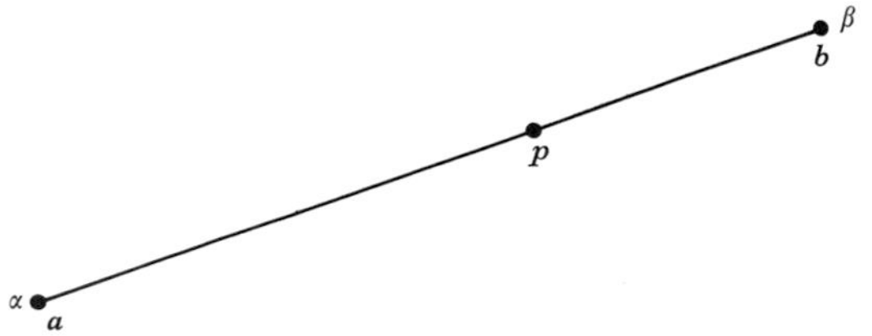

as their 'point of balance'.

Let us now suppose that we have a number of masses:

$m_1, m_2, m_3, m_4 \ldots$

at $\boldsymbol{r}_1, \boldsymbol{r}_2, \boldsymbol{r}_3, \boldsymbol{r}_4 \ldots$ respectively.

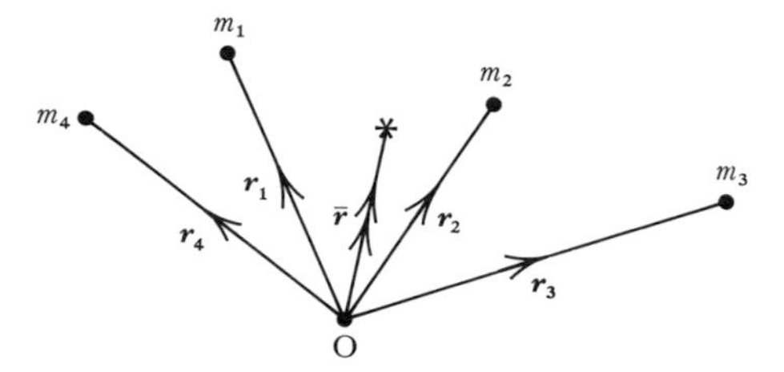

Let G be the point which has position vector $\bar{\boldsymbol{r}}$, where

$$\bar{\boldsymbol{r}} = \frac{m_1\boldsymbol{r}_1 + m_2\boldsymbol{r}_2 + m_3\boldsymbol{r}_3 + m_4\boldsymbol{r}_4 + \cdots}{m_1 + m_2 + m_3 + m_4 + \cdots}.$$

Now G is a definite point, but there is a serious risk that its position will depend on our choice of origin, O.

To discover whether or not this is so, we must take another origin O′ and see whether such a calculation based on O′ leads to a different point G′. Let us suppose that from the new origin O′, O has position vector $\boldsymbol{c}$, and $m_1, m_2, \ldots$ have position vectors $\boldsymbol{r}'_1, \boldsymbol{r}'_2, \ldots$.

Then $\quad \boldsymbol{r}'_1 = \boldsymbol{c} + \boldsymbol{r}_1$, etc.

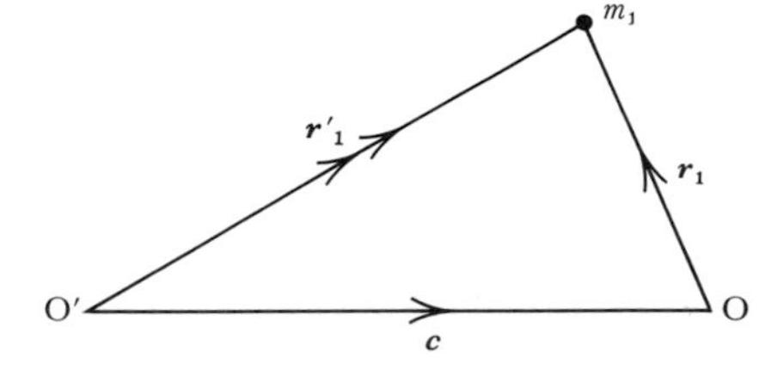

So the position vector of G′, calculated from O′, is

$$\begin{aligned}\bar{\boldsymbol{r}}' &= \frac{m_1\boldsymbol{r}'_1 + m_2\boldsymbol{r}'_2 + m_3\boldsymbol{r}'_3 + \cdots}{m_1 + m_2 + m_3 + \cdots}\\ &= \frac{m_1(\boldsymbol{c} + \boldsymbol{r}_1) + m_2(\boldsymbol{c} + \boldsymbol{r}_2) + m_3(\boldsymbol{c} + \boldsymbol{r}_3) + \cdots}{m_1 + m_2 + m_3 + \cdots}\\ &= \frac{(m_1 + m_2 + m_3 + \cdots)\boldsymbol{c} + (m_1\boldsymbol{r}_1 + m_2\boldsymbol{r}_2 + m_3\boldsymbol{r}_3 + \cdots)}{m_1 + m_2 + m_3 + \cdots}\\ &= \boldsymbol{c} + \frac{m_1\boldsymbol{r}_1 + m_2\boldsymbol{r}_2 + m_3\boldsymbol{r}_3 + \cdots}{m_1 + m_2 + m_3 + \cdots}.\end{aligned}$$

So $\quad \bar{\boldsymbol{r}}' = \boldsymbol{c} + \bar{\boldsymbol{r}}$.

Hence $\quad \boldsymbol{O'G'} = \boldsymbol{O'O} + \boldsymbol{OG} = \boldsymbol{O'G}$

$\Rightarrow \qquad$ G and G′ are identical.

So if masses $\quad m_1, m_2, m_3 \ldots$

are at points $\quad \boldsymbol{r}_1, \boldsymbol{r}_2, \boldsymbol{r}_3 \ldots,$

the point $$\boxed{\bar{\boldsymbol{r}} = \frac{m_1\boldsymbol{r}_1 + m_2\boldsymbol{r}_2 + m_3\boldsymbol{r}_3 + \cdots}{m_1 + m_2 + m_3 + \cdots}}$$

is *not* dependent on our choice of origin, but only on the given system of masses, and is therefore called their *centre of mass*. In the particular case when the masses are all equal this becomes

$$\bar{\boldsymbol{r}} = \frac{1}{n}(\boldsymbol{r}_1 + \boldsymbol{r}_2 + \cdots + \boldsymbol{r}_n),$$

which is also known as the *centroid* of the given points. So the centroid of $\boldsymbol{r}_1$ and $\boldsymbol{r}_2$ is at $\frac{1}{2}(\boldsymbol{r}_1 + \boldsymbol{r}_2)$, and of $\boldsymbol{r}_1, \boldsymbol{r}_2, \boldsymbol{r}_3$ is at $\frac{1}{3}(\boldsymbol{r}_1 + \boldsymbol{r}_2 + \boldsymbol{r}_3)$.

Finally, in terms of components, the centre of mass is

$$\begin{aligned}\bar{x}\boldsymbol{i} + \bar{y}\boldsymbol{j} + \bar{z}\boldsymbol{k} &= \frac{m_1(x_1\boldsymbol{i} + y_1\boldsymbol{j} + z_1\boldsymbol{k}) + m_2(x_2\boldsymbol{i} + y_2\boldsymbol{j} + z_2\boldsymbol{k}) + \cdots}{m_1 + m_2 + \cdots} \\ &= \frac{m_1x_1 + m_2x_2 + \cdots}{m_1 + m_2}\boldsymbol{i} + \frac{m_1y_1 + m_2y_2 + \cdots}{m_1 + m_2 + \cdots}\boldsymbol{j} \\ &\quad + \frac{m_1z_1 + m_2z_2 + \cdots}{m_1 + m_2 + \cdots}\boldsymbol{k}.\end{aligned}$$

So the centre of mass has coordinates

$$\boxed{\begin{aligned}\bar{x} &= \frac{m_1x_1 + m_2x_2 + \cdots}{m_1 + m_2 + \cdots}, \\ \bar{y} &= \frac{m_1y_1 + m_2y_2 + \cdots}{m_1 + m_2 + \cdots}, \quad \text{or} \quad \left(\frac{\sum mx}{\sum m}, \frac{\sum my}{\sum m}, \frac{\sum mz}{\sum m}\right). \\ \bar{z} &= \frac{m_1z_1 + m_2z_2 + \cdots}{m_1 + m_2 + \cdots}.\end{aligned}}$$

Example 1

A light rectangular frame of length 4 m, breadth 3 m and height 2 m has masses 1 kg, 2 kg, 3 kg attached at corners as shown in the figure. Where is their centre of mass?

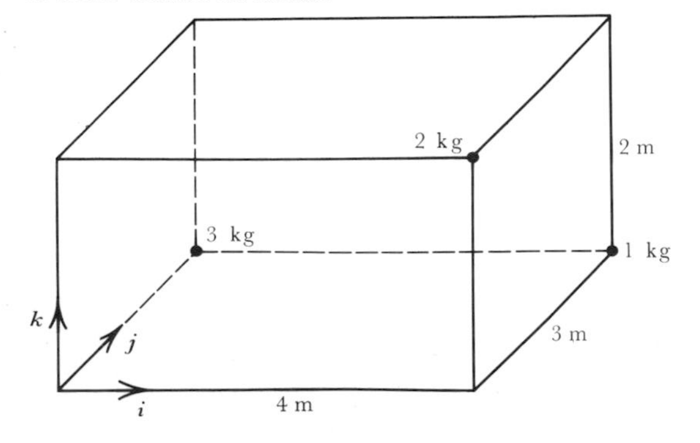

Taking unit vectors $\boldsymbol{i}, \boldsymbol{j}, \boldsymbol{k}$ as shown, the three masses are:

1 kg at the point $4\boldsymbol{i} + 3\boldsymbol{j}$
2 kg at the point $4\boldsymbol{i} + 2\boldsymbol{k}$
3 kg at the point $3\boldsymbol{j}$.

So their centre of mass is at

$$\bar{\boldsymbol{r}} = \frac{1 \times (4\boldsymbol{i} + 3\boldsymbol{j}) + 2 \times (4\boldsymbol{i} + 2\boldsymbol{k}) + 3 \times 3\boldsymbol{j}}{1 + 2 + 3}$$

$$= \frac{12\boldsymbol{i} + 12\boldsymbol{j} + 4\boldsymbol{k}}{6} = 2\boldsymbol{i} + 2\boldsymbol{j} + \tfrac{2}{3}\boldsymbol{k}.$$

Alternatively, taking axes Ox, Oy, Oz in the same directions as $\boldsymbol{i}, \boldsymbol{j}, \boldsymbol{k}$, the calculation could be set out:

m	x	y	z	mx	my	mz
1	4	3	0	4	3	0
2	4	0	2	8	0	4
3	0	3	0	0	9	0
$\sum m = 6$				$\sum mx = 12$	$\sum my = 12$	$\sum mz = 4$

So the centre of mass has coordinates

$$\bar{x} = \frac{\sum mx}{\sum m} = 2; \qquad \bar{y} = \frac{\sum my}{\sum m} = 2; \qquad \bar{z} = \frac{\sum mz}{\sum m} = \tfrac{2}{3}$$

and is therefore at the point $(2, 2, \tfrac{2}{3})$.

Example 2

If a square of side a is removed from the corner of a uniform square lamina of side $2a$, where is the centre of mass of the remaining piece?

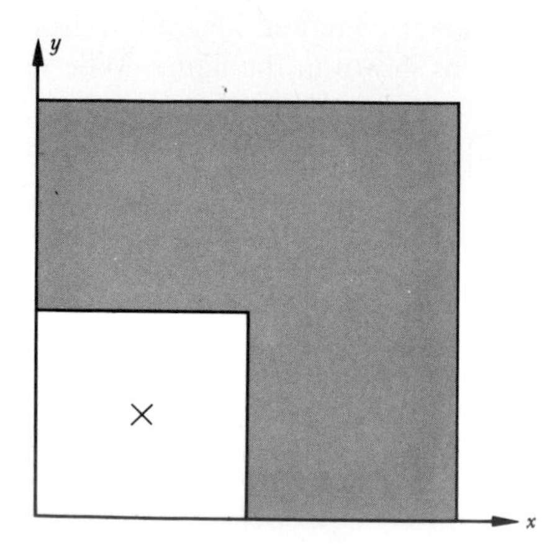

We can regard the remaining piece as being composed of a large square of mass 4 together with a smaller square of negative mass -1. So, using axes Ox, Oy as shown, we have:

	m	x	y	mx	my
Large square	4	a	a	$4a$	$4a$
Small square	-1	$\frac{1}{2}a$	$\frac{1}{2}a$	$-\frac{1}{2}a$	$-\frac{1}{2}a$
Final lamina	$\sum m = 3$			$\sum mx = \frac{7}{2}a$	$\sum my = \frac{7}{2}a$

So $\bar{x} = \dfrac{\sum mx}{\sum m} = \dfrac{7a}{6}, \quad \bar{y} = \dfrac{\sum my}{\sum m} = \dfrac{7a}{6}$

and the centre of mass lies at $(7a/6,\ 7a/6)$.

Exercise 11.2a

1 If the vertices of a tetrahedron ABCD have position vectors $\boldsymbol{a}, \boldsymbol{b}, \boldsymbol{c}, \boldsymbol{d}$, find the position vectors of

(*i*) the centroid P of $\triangle$BCD;

(*ii*) the point dividing AP in the ratio 3:1. What do you deduce about this and the three other similar lines?

(*iii*) the mid-point of the line joining the mid-points of AB and CD.

Hence find seven lines associated with the tetrahedron which are all concurrent. At what point do they meet?

2 If G, G′ are the centroids of $\triangle$ABC, $\triangle$A′B′C′, prove that

$\boldsymbol{AA'} + \boldsymbol{BB'} + \boldsymbol{CC'} = 3\boldsymbol{GG'}$.

3 Find the centres of mass of

(*i*) 1 kg at $\boldsymbol{a}$, 2 kg at $\boldsymbol{b}$, 3 kg at $\boldsymbol{c}$;

(*ii*) 1 g at (0, 0, 0), 4 g at (1, 4, 2), 3 g at (3, -1, -2), 2 g at (-4, 1, 6);

(*iii*) 1 kg at $\boldsymbol{i} + 2\boldsymbol{j}$, 4 kg at $\boldsymbol{j} - 2\boldsymbol{k}$, 5 kg at $-\boldsymbol{i} + 3\boldsymbol{j} + \boldsymbol{k}$.

4 If a cube of side 1 m is removed from the corner of a uniform solid cube of side 2 m, how far has its centre of mass been displaced?

5 A mass of 2 kg is placed at the point (4, 1). Find the masses which should be placed at (-1, 3) and (-2, -1) in order that the centroid of the three masses should be at the origin. (S.M.P.)

Continuous distributions of mass

So far, we have limited our investigations to systems consisting of a finite number of discrete masses. When there is a *continuous* distribution of mass, the method is very similar, except that we shall usually begin by dividing the given body into a series of elements.

Example 2

Uniform triangular lamina

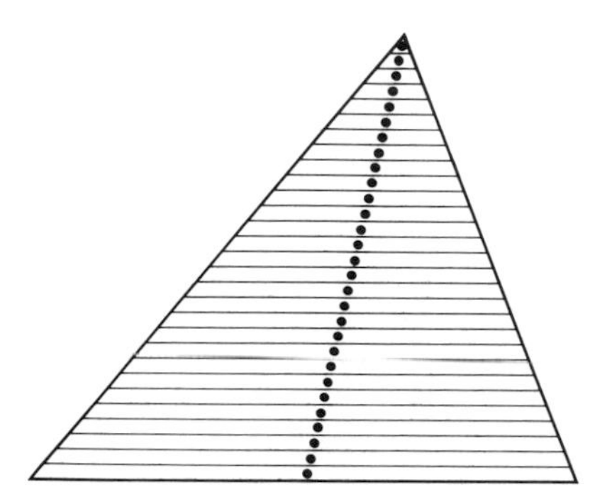

We first divide the lamina into a set of elemental strips parallel to one of its sides. Now the centre of mass of each strip is at its mid-point, and these are all collinear on a median. Hence the centre of mass of the lamina lies on this median, and similarly on the other two medians. It is therefore situated at the *centroid* of the given triangle.

Example 3

Find the centre of mass of:
(*i*) A uniform lamina (or plate) in the shape of the area between $y = x^2$, $y = 0$ and $x = 1$.
(*ii*) A uniform solid formed by rotating this area about the x-axis.

(*i*) *Uniform lamina*

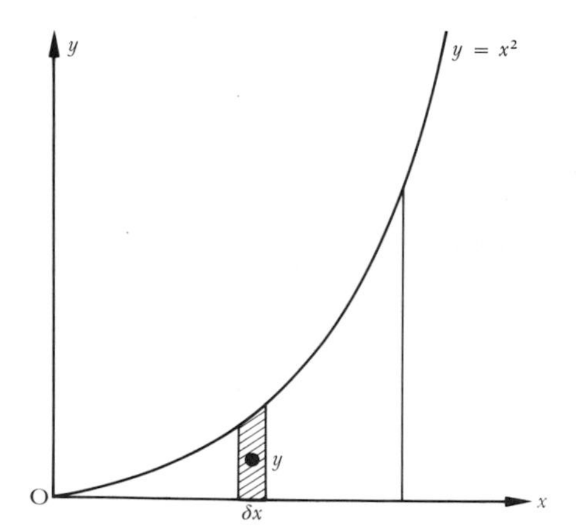

Suppose that the lamina has surface density (i.e., mass per unit area) σ, and that it is split into narrow strips of width δx and height y.

Then the strip shown has mass $m = \sigma y \, \delta x$ which can be regarded as concentrated at the point $(x, \frac{1}{2}y)$.

So $\quad \overline{x} = \dfrac{\sum mx}{\sum m} = \dfrac{\sum (\sigma y\ \delta x)\,x}{\sum (\sigma y\ \delta x)} = \dfrac{\sum xy\ \delta x}{\sum y\ \delta x}$

$\Rightarrow \quad \overline{x} = \dfrac{\int_0^1 xy\ \mathrm{d}x}{\int_0^1 y\ \mathrm{d}x} = \dfrac{\int_0^1 x^3\ \mathrm{d}x}{\int_0^1 x^2\ \mathrm{d}x} = \dfrac{\frac{1}{4}}{\frac{1}{3}} = \frac{3}{4}$

and $\quad \overline{y} = \dfrac{\sum my}{\sum m} = \dfrac{\sum (\sigma y\ \delta x)\frac{1}{2}y}{\sum (\sigma y\ \delta x)} = \dfrac{\frac{1}{2}\sum y^2\ \delta x}{\sum y\ \delta x}$

$\Rightarrow \quad \overline{y} = \dfrac{\frac{1}{2}\int_0^1 y^2\ \mathrm{d}x}{\int_0^1 y\ \mathrm{d}x} = \dfrac{\frac{1}{2}\int_0^1 x^4\ \mathrm{d}x}{\int_0^1 x^2\ \mathrm{d}x} = \dfrac{\frac{1}{10}}{\frac{1}{3}} = \frac{3}{10}.$

So the centre of mass is at $(\frac{3}{4}, \frac{3}{10})$.

(ii) Uniform solid

Suppose that the solid has volume density ρ and that it is sliced into thin discs of thickness δx and radius y.

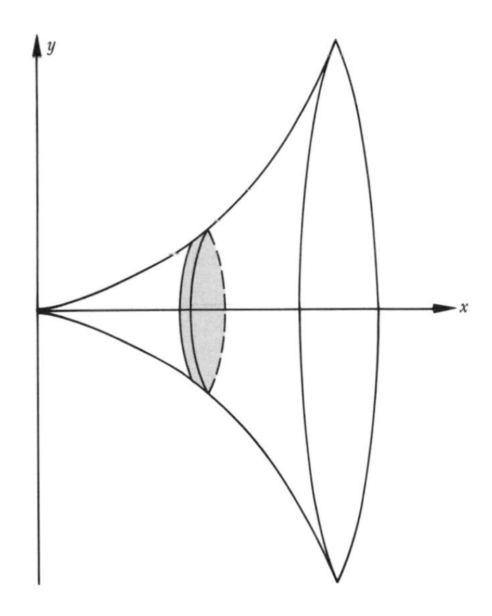

Then the mass of the disc shown $= m = \rho(\pi y^2\ \delta x) = \rho\pi y^2\ \delta x$.
Now if the centre of mass is at $(\overline{x}, \overline{y})$, we see by symmetry that $\overline{y} = 0$.

Also $\quad \overline{x} = \dfrac{\sum mx}{\sum m} = \dfrac{\sum (\rho\pi y^2\ \delta x)\,x}{\sum \rho\pi y^2\ \delta x} = \dfrac{\sum xy^2\ \delta x}{\sum y^2\ \delta x}$

$\Rightarrow \quad \overline{x} = \dfrac{\int_0^1 xy^2\ \mathrm{d}x}{\int_0^1 y^2\ \mathrm{d}x} = \dfrac{\int_0^1 x^5\ \mathrm{d}x}{\int_0^1 x^4\ \mathrm{d}x} = \dfrac{\frac{1}{6}}{\frac{1}{5}} = \frac{5}{6}.$

So the centre of mass is at $(\frac{5}{6}, 0)$.

Example 4

Find the centre of mass of:
(*i*) a uniform semi-circular wire of radius a;
(*ii*) a uniform semi-circular plate of radius a.

(*i*) *Semi-circular wire*

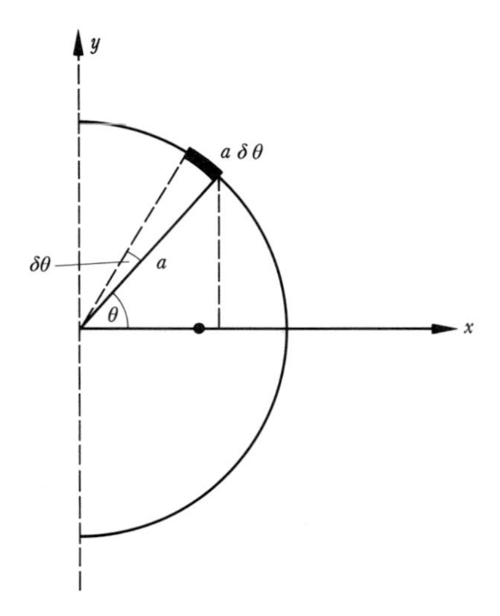

Take an axis Ox along the line of symmetry and consider a small section of wire which subtends angle $\delta\theta$ at the centre. The length of this section is $a\,\delta\theta$, so if its line density is ρ it will have mass $m = \rho a\,\delta\theta$.

Also the length of the wire is πa, so its total mass will be $\rho\pi a$.

By symmetry, $\bar{y} = 0$

$$\text{and}\quad \bar{x} = \frac{\sum mx}{\sum m} = \frac{\sum (\rho a\,\delta\theta)a\cos\theta}{\rho\pi a} = \frac{a}{\pi}\sum \cos\theta\,\delta\theta$$

$$\Rightarrow \quad \bar{x} = \frac{a}{\pi}\int_{-\pi/2}^{+\pi/2} \cos\theta\,\mathrm{d}\theta = \frac{a}{\pi}[\sin\theta]_{-\pi/2}^{+\pi/2} = \frac{2a}{\pi}.$$

So the centre of mass of the wire is at $(2a/\pi,\ 0)$, i.e., approximately at $(0.64a,\ 0)$.

(*ii*) *Semi-circular plate*

Method 1

Suppose that the plate is divided into strips of length $2y$ and width δx and that its surface density is σ.

Then the mass of strip $= \sigma(2y\,\delta x)$ and the total mass of the plate $= \frac{1}{2}\pi a^2\sigma$.

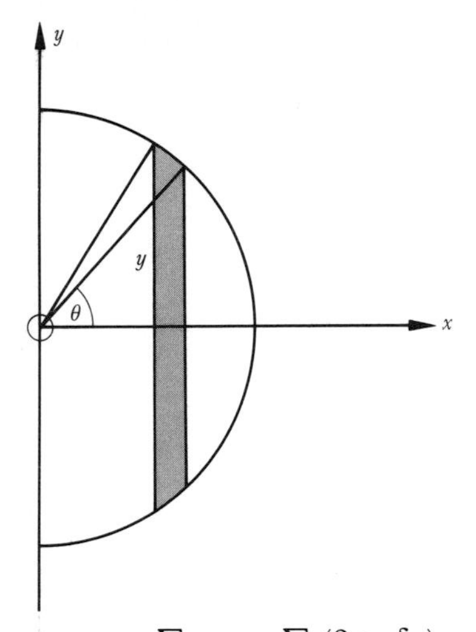

$$\text{So} \quad \bar{x} = \frac{\sum mx}{\sum m} = \frac{\sum (2\sigma y\ \delta x)x}{\frac{1}{2}\pi a^2 \sigma} = \frac{4}{\pi a^2} \sum xy\ \delta x$$

$$\Rightarrow \quad \bar{x} = \frac{4}{\pi a^2} \int_0^a xy\ \mathrm{d}x = \frac{4}{\pi a^2} \int_0^a x\sqrt{(a^2 - x^2)}\ \mathrm{d}x$$

$$= \frac{4}{\pi a^2} \left[-\tfrac{1}{3}(a^2 - x^2)^{3/2} \right]_0^a$$

$$= \frac{4}{\pi a^2} \left[0 - \left(-\tfrac{1}{3}a^3 \right) \right] = \frac{4a}{3\pi}.$$

Also, by symmetry, $\bar{y} = 0$.

So the centre of mass is at $(4a/3\pi, 0)$; or, approximately, $(0.42a, 0)$.

Method 2

This last result can be obtained more easily if we make use of the previous result for a semi-circular wire. For the semi-circular plate can be dissected into a large number of sectors and each of these sectors is approximately a triangle with its centre of mass at a distance $2a/3$ from O.

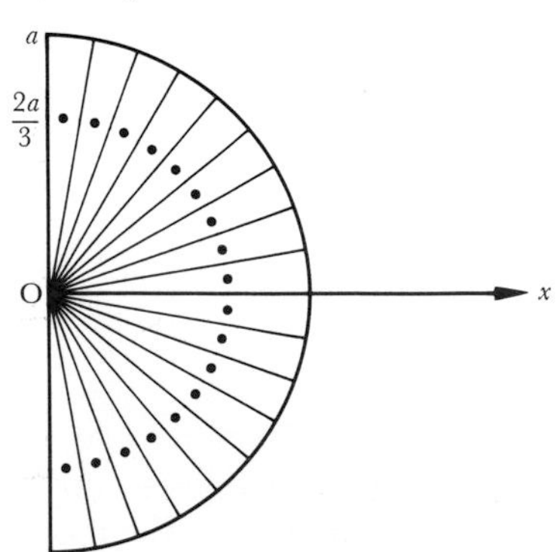

In this way we see that the uniform plate of radius a has the same centre of mass as a uniform wire of radius $2a/3$.

Hence $\bar{x} = \dfrac{2(2a/3)}{\pi} = \dfrac{4a}{3\pi}$.

Exercise 11.2b

1 Find the coordinates of the centres of mass of uniform laminae bounded by:
(*i*) $y = x^3$, $y = 0$, $x = 2$;
(*ii*) $y = x^2$, $y = 0$, $x = 1$, $x = 2$;
(*iii*) $y^2 = 4x$, $x = 9$;
(*iv*) $y = \dfrac{ax}{h}$, $y = 0$, $x = h$.

2 Find the coordinates of the centres of mass of the uniform solids formed by rotating the areas of no. **1** through $360°$ about the x-axis.

3 By rotating a quadrant of the curve $y = \sqrt{(a^2 - x^2)}$ about the x-axis, find the position of the centre of mass of a uniform solid hemisphere.

4 By slicing a uniform tetrahedron parallel to one of its faces, show that its centre of mass lies on the line joining the centroid of this face to the opposite vertex. Hence show that the centre of mass
(*i*) of a uniform tetrahedron lies at the centroid of its four vertices;
(*ii*) of a uniform cone (or pyramid) is at one-quarter of its height above the base (cf. **2** (*iv*));
(*iii*) of a uniform hemisphere shell is half-way from its centre to the middle-point of the shell (use the result of no. **3** and part (*ii*)).

5 A sphere of radius a is cut into two portions by a plane which is distant $\frac{1}{2}a$ from the centre of the sphere. Show by integration that the volume of the smaller of the two portions is $5\pi a^3/24$.

Calculate the distance of the centre of gravity of this smaller portion from the centre of the sphere. (O.C.)

6 Use Simpson's rule to find the approximate value of the x-coordinate of the centroid of the area bounded by the x-axis, the line $x = 4$, and a curve joining the points

x	0	1	2	3	4
y	0	1.05	2.21	3.50	4.92.

(O.C.)

11.3 Scalar products

So far we have confined ourselves to the addition and subtraction of vectors and to their multiplication by a scalar. We now move to two other operations which can be performed on a pair of vectors, and obtain their *scalar product* and *vector product*. These are both exceedingly important throughout applied mathematics and particularly for describing (respectively) the work done by a force and the moment of a force. In this section we shall concentrate upon the scalar product.

If $\boldsymbol{a}$ and $\boldsymbol{b}$ are two vectors with magnitudes a and b whose directions are separated by an angle θ, we define their *scalar product* as

$$\boldsymbol{a}.\boldsymbol{b} = ab \cos \theta.$$

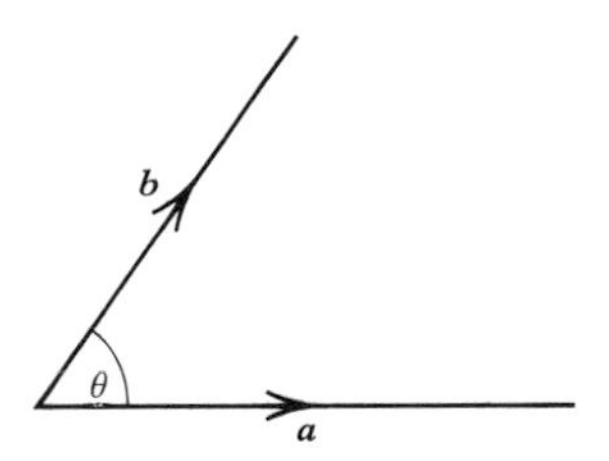

As $\cos(-\theta) = \cos \theta$ we need not be concerned about the direction in which to measure θ, and it is also apparent that

$$\begin{aligned}\boldsymbol{a}.\boldsymbol{b} &= a(b \cos \theta) = a \times \text{projection of } \boldsymbol{b} \text{ on } \boldsymbol{a}\\ &= b(a \cos \theta) = b \times \text{projection of } \boldsymbol{a} \text{ on } \boldsymbol{b}.\end{aligned}$$

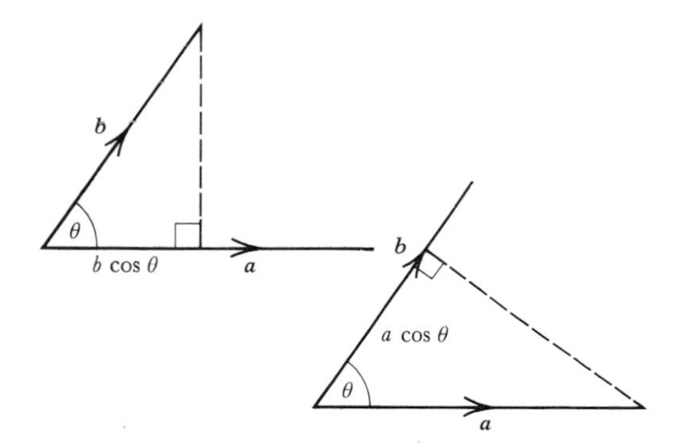

As an illustration, if $\boldsymbol{F}$ is a force and $\boldsymbol{c}$ is a displacement of its point of application, the scalar product $\boldsymbol{F}.\boldsymbol{c}$ is called the *work done by the force* in this displacement.

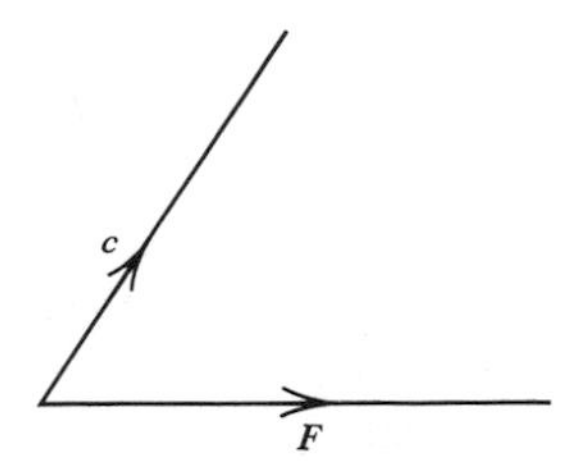

So the work can be regarded as

(projection of force $\boldsymbol{F}$ in direction of displacement $\boldsymbol{c}$)
$\times$ (magnitude of displacement),

or (projection of displacement $\boldsymbol{c}$ in direction of force $\boldsymbol{F}$)
$\times$ (magnitude of force);

and if $\boldsymbol{F}$ and $\boldsymbol{c}$ are perpendicular, then $\boldsymbol{F}.\boldsymbol{c} = 0$, so that no work is done when a force is displaced perpendicular to its line of action.

A number of results follow immediately from the definition of the scalar product:

(*i*) $\boldsymbol{a}.\boldsymbol{a} = aa\cos 0 = a^2$

$\boldsymbol{a}.\boldsymbol{a}$ is usually abbreviated as $\boldsymbol{a}^2$. So $\boldsymbol{a}^2 = a^2$.

(*ii*) $\boldsymbol{a}.\boldsymbol{b} = \boldsymbol{b}.\boldsymbol{a}$

(i.e., scalar product is *commutative*).

(*iii*) If $\boldsymbol{a}$ and $\boldsymbol{b}$ are non-zero vectors,

$$\boldsymbol{a}.\boldsymbol{b} = 0 \quad \Leftrightarrow \quad ab\cos\theta = 0 \quad \Leftrightarrow \quad \theta = \frac{\pi}{2}.$$

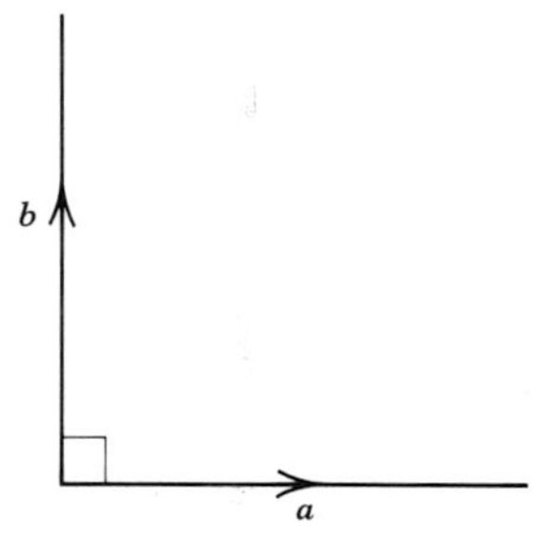

So $\boldsymbol{a}.\boldsymbol{b} = 0 \quad \Leftrightarrow \quad \boldsymbol{a}$ and $\boldsymbol{b}$ are perpendicular.

(*iv*) $\boldsymbol{a}.(\lambda\boldsymbol{b}) = a(\lambda b)\cos\theta = \lambda ab\cos\theta$

$\lambda(\boldsymbol{a}.\boldsymbol{b}) = \lambda(ab\cos\theta) = \lambda ab\cos\theta.$

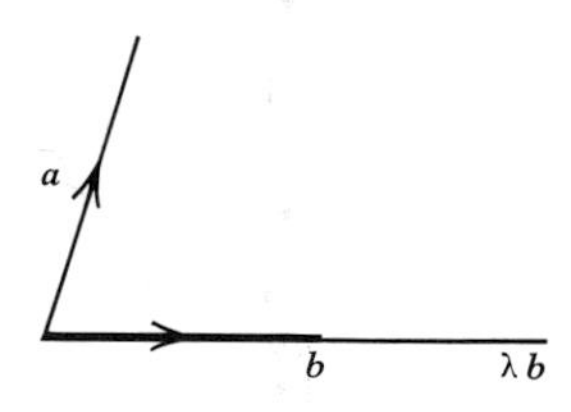

So $\boldsymbol{a}.(\lambda\boldsymbol{b}) = \lambda(\boldsymbol{a}.\boldsymbol{b})$

i.e., the scalar product is *distributive with respect to multiplication by a scalar.*

(This last result, like some of its predecessors and also the next, at first appears trivial. But it is worthwhile pausing to appreciate its importance, and to be grateful that when we are confronted with $\boldsymbol{a}.(2\boldsymbol{b})$, $(2\boldsymbol{a}).\boldsymbol{b}$ and $2(\boldsymbol{a}.\boldsymbol{b})$, we know without hesitation that they are all equal.)

(*v*) $(\boldsymbol{a} + \boldsymbol{b}).\boldsymbol{c} = \boldsymbol{a}.\boldsymbol{c} + \boldsymbol{b}.\boldsymbol{c}$,

i.e., scalar product is *distributive with respect to addition.*

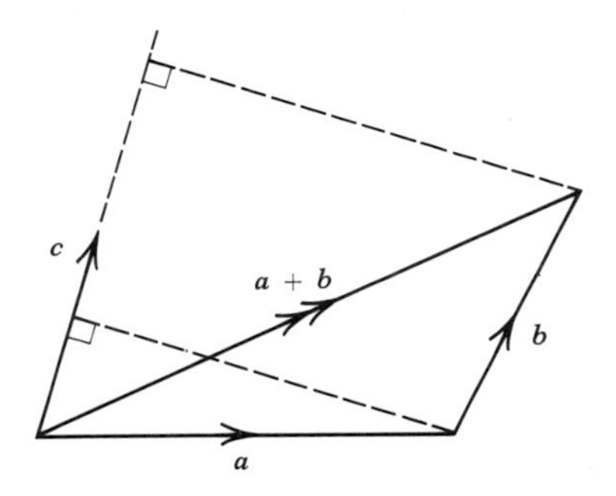

This result, too, is clear from a figure.

$$\begin{aligned}\text{For}\quad (\boldsymbol{a} + \boldsymbol{b}).\boldsymbol{c} &= (\text{projection of } \boldsymbol{a}+\boldsymbol{b} \text{ on } \boldsymbol{c}) \times c \\ &= (\text{projection of } \boldsymbol{a} \text{ on } \boldsymbol{c} + \text{projection of } \boldsymbol{b} \text{ on } \boldsymbol{c}) \times c \\ &= (\text{projection of } \boldsymbol{a} \text{ on } \boldsymbol{c}) \times c \\ &\quad + (\text{projection of } \boldsymbol{b} \text{ on } \boldsymbol{c}) \times c \\ &= \boldsymbol{a}.\boldsymbol{c} + \boldsymbol{b}.\boldsymbol{c}.\end{aligned}$$

So $\qquad (\boldsymbol{a} + \boldsymbol{b}).\boldsymbol{c} = \boldsymbol{a}.\boldsymbol{c} + \boldsymbol{b}.\boldsymbol{c}$

and, similarly, $\boldsymbol{a}.(\boldsymbol{b} + \boldsymbol{c}) = \boldsymbol{a}.\boldsymbol{b} + \boldsymbol{a}.\boldsymbol{c}$.

This result is particularly important in mechanics. For if $\boldsymbol{P}$ and $\boldsymbol{Q}$ are two forces and $\boldsymbol{c}$ is a certain displacement, then

$$(\boldsymbol{P} + \boldsymbol{Q}).\boldsymbol{c} = \boldsymbol{P}.\boldsymbol{c} + \boldsymbol{Q}.\boldsymbol{c}.$$

So, for any displacement, the work done by the resultant of two forces is the sum of the amounts of work done by the forces separately.

Example 1

If two pairs of opposite edges of a tetrahedron are perpendicular, so is the third.

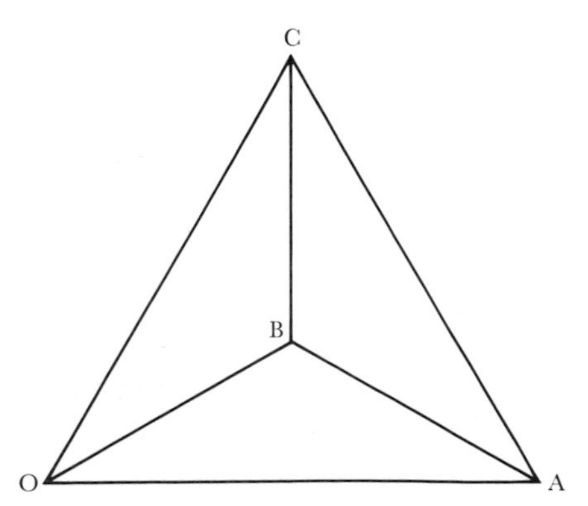

Take one of the vertices as the origin O, and let the position vectors of the others be $\boldsymbol{a}$, $\boldsymbol{b}$, $\boldsymbol{c}$.

Suppose that $OB \perp CA$ and $OC \perp AB$.

Then $\boldsymbol{b}.(\boldsymbol{a} - \boldsymbol{c}) = 0$ and $\boldsymbol{c}.(\boldsymbol{b} - \boldsymbol{a}) = 0$

$\Rightarrow$ $\boldsymbol{b}.\boldsymbol{a} - \boldsymbol{b}.\boldsymbol{c} = 0$ and $\boldsymbol{c}.\boldsymbol{b} - \boldsymbol{c}.\boldsymbol{a} = 0$

$\Rightarrow$ $\boldsymbol{a}.\boldsymbol{b} = \boldsymbol{b}.\boldsymbol{c}$ and $\boldsymbol{b}.\boldsymbol{c} = \boldsymbol{a}.\boldsymbol{c}$

$\Rightarrow$ $\boldsymbol{a}.\boldsymbol{b} = \boldsymbol{a}.\boldsymbol{c}$

$\Rightarrow$ $\boldsymbol{a}.(\boldsymbol{b} - \boldsymbol{c}) = 0$

$\Rightarrow$ $OA \perp BC$.

Finally, we see that in the particular case when the four points are coplanar, each is the *orthocentre* of the triangle formed by the other four.

Exercise 11.3a

1 Letting $\boldsymbol{a}$ and $\boldsymbol{b}$ be the position vectors of two points A and B and expanding the scalar product $(\boldsymbol{a} - \boldsymbol{b})^2$, prove the cosine rule for $\triangle$OAB.

2 Taking O as the centre of a circle, points with position vectors $\boldsymbol{a}$, $-\boldsymbol{a}$ at opposite ends of a diameter, and $\boldsymbol{r}$ anywhere else on the circle, prove that the angle in a semi-circle must be a right angle.

3 Let O be the middle point of the base BC of $\triangle$ABC and let A, B, C have position vectors $\boldsymbol{a}$, $\boldsymbol{b}$, $-\boldsymbol{b}$. By expanding $(\boldsymbol{a} - \boldsymbol{b})^2 + (\boldsymbol{a} + \boldsymbol{b})^2$, prove Apollonius' theorem, that

$$AB^2 + AC^2 = 2(AO^2 + BO^2).$$

4 AB, CD are two skew (non-coplanar) lines which have their mid-points at P, Q respectively. Prove that

$$AC^2 + AD^2 + BC^2 + BD^2 = AB^2 + CD^2 + 4PQ^2.$$

5 A point P lies in the plane of a triangle ABC, and G is the centroid this triangle. Prove that

$$PA^2 + PB^2 + PC^2 = 3PG^2 + \tfrac{1}{3}(BC^2 + CA^2 + AB^2).$$

What is the least value of $PA^2 + PB^2 + PC^2$ as P varies in the plane? What is the locus of P when P varies so that

$PA^2 + PB^2 + PC^2$ is constant? (O.C.)

Scalar product in component form

Suppose that $\boldsymbol{a}$ and $\boldsymbol{b}$ are expressed in terms of unit vectors $\boldsymbol{i}, \boldsymbol{j}$ and $\boldsymbol{k}$:

$$\boldsymbol{a} = a_1\boldsymbol{i} + a_2\boldsymbol{j} + a_3\boldsymbol{k}$$

and $$\boldsymbol{b} = b_1\boldsymbol{i} + b_2\boldsymbol{j} + b_3\boldsymbol{k}.$$

We can readily find $\boldsymbol{a}.\boldsymbol{b}$ in terms of these components.

Firstly we note that $\boldsymbol{i}, \boldsymbol{j}, \boldsymbol{k}$ are unit vectors, so $\boldsymbol{i}.\boldsymbol{i} = \boldsymbol{j}.\boldsymbol{j} = \boldsymbol{k}.\boldsymbol{k} = 1$; and $\boldsymbol{i}, \boldsymbol{j}, \boldsymbol{k}$ are perpendicular, so $\boldsymbol{i}.\boldsymbol{j} = \boldsymbol{i}.\boldsymbol{k} = 0$ etc.

Now $\quad \boldsymbol{a}.\boldsymbol{b} = (a_1\boldsymbol{i} + a_2\boldsymbol{j} + a_3\boldsymbol{k}).(b_1\boldsymbol{i} + b_2\boldsymbol{j} + b_3\boldsymbol{k})$

$$\Rightarrow \quad \boxed{\boldsymbol{a}.\boldsymbol{b} = a_1b_1 + a_2b_2 + a_3b_3}$$

In particular, $\quad \boldsymbol{a}^2 = \boldsymbol{a}.\boldsymbol{a} = a_1^2 + a_2^2 + a_3^2$

and $\quad \boldsymbol{b}^2 = \boldsymbol{b}.\boldsymbol{b} = b_1^2 + b_2^2 + b_3^2.$

Also, if $\boldsymbol{a} = l\boldsymbol{i} + m\boldsymbol{j} + n\boldsymbol{k}$ is a unit vector which makes angles θ, ϕ, ψ with the three axes, then

$$\cos\theta = \boldsymbol{a}.\boldsymbol{i} = l$$
$$\cos\phi = \boldsymbol{a}.\boldsymbol{j} = m$$
$$\cos\psi = \boldsymbol{a}.\boldsymbol{k} = n,$$

so that l, m, n are known as the *direction cosines* of $\boldsymbol{a}$.

These results are frequently of great use in finding the angle between two directions in space.

Example 2

If three corners of A, B, C of a crystal are known to be at points (3, 2, 1), (4, 0, 2), (5, 2, 4) respectively, we may wish to find the angle θ between $\boldsymbol{AB}$ and $\boldsymbol{AC}$.

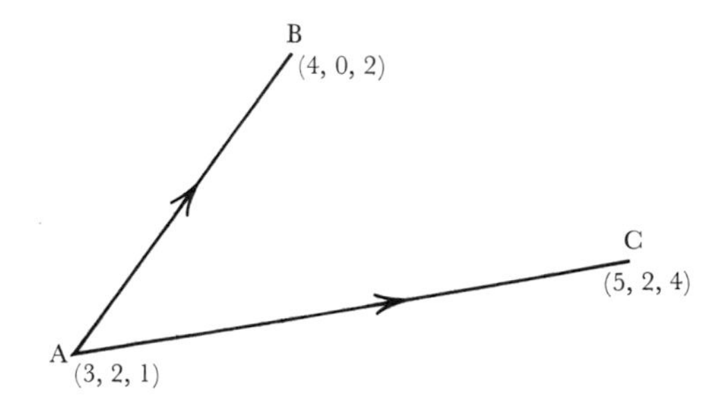

Taking unit vectors $\boldsymbol{i}, \boldsymbol{j}, \boldsymbol{k}$ in the direction of the three axes,

$\quad \boldsymbol{a} = \boldsymbol{AB} = \boldsymbol{i} - 2\boldsymbol{j} + \boldsymbol{k}$

and $\quad \boldsymbol{b} = \boldsymbol{AC} = 2\boldsymbol{i} + 3\boldsymbol{k}.$

Then $\quad a^2 = 1^2 + 2^2 + 1^2 = 6 \quad \Rightarrow \quad a = \sqrt{6}$

and $\quad b^2 = 2^2 + 0^2 + 3^2 = 13 \quad \Rightarrow \quad b = \sqrt{13}.$

Also $\quad \boldsymbol{a}.\boldsymbol{b} = 1\times 2 + -2\times 0 + 1\times 3 = 5.$

Now $\quad \boldsymbol{a}.\boldsymbol{b} = ab\cos\theta.$

So $\quad 5 = \sqrt{6} \times \sqrt{13} \cos \theta$

$$\Rightarrow \quad \cos \theta = \frac{5}{\sqrt{6}\sqrt{13}} = \frac{5}{\sqrt{78}} = \frac{5}{8.832} = 0.5661$$

$$\Rightarrow \quad \theta = 55^\circ\ 31'.$$

So $\boldsymbol{AB}$ and $\boldsymbol{AC}$ are inclined at 55° 31′.

Exercise 11.3b

1 Find the scalar products of the following pairs of vectors and hence the angle between them:

(*i*) $\boldsymbol{i} + \boldsymbol{j}$ and $\boldsymbol{i} + 2\boldsymbol{j}$;

(*ii*) $\boldsymbol{i} + \boldsymbol{j} + \boldsymbol{k}$ and $\boldsymbol{i} + \boldsymbol{j} - \boldsymbol{k}$;

(*iii*) $-\boldsymbol{i} + 2\boldsymbol{j} - \boldsymbol{k}$ and $\boldsymbol{i} - \boldsymbol{j} + \boldsymbol{k}$.

2

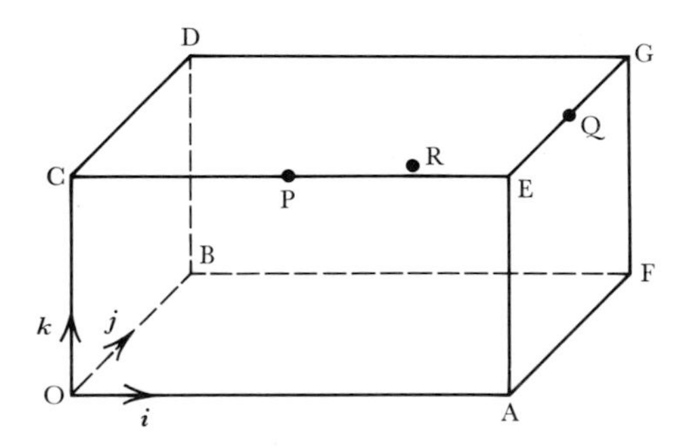

In the above rectangular box, $OA = 4$, $OB = 3$, $OC = 2$. Furthermore, P is the mid-point of CE; Q divides EG in the ratio 2:1; and R is the centre of rectangle BFGD.

For each of the following pairs of lines, find vectors which represent the two lines (in terms of $\boldsymbol{i}, \boldsymbol{j}, \boldsymbol{k}$) and hence the angle between the lines

(*i*) $\boldsymbol{OP}$ and $\boldsymbol{OD}$;

(*ii*) $\boldsymbol{AQ}$ and $\boldsymbol{AB}$;

(*iii*) $\boldsymbol{ER}$ and $\boldsymbol{RC}$;

(*iv*) $\boldsymbol{PR}$ and $\boldsymbol{QF}$.

3 Find whether the line through (1, 0, 2) and (4, 3, −1) meets the line through (0, 3, −1) and (5, −2, 4). Calculate the (acute) angle between their directions. (S.M.P.)

4 The rectangular cartesian coordinates of two points A and B are (1, 2) and (−1, 4) respectively. Show that the column vector $\begin{pmatrix} t \\ t \end{pmatrix}$ is always normal to AB and hence, or otherwise, find the coordinates of the foot of the perpendicular from the origin to AB. (C.)

5 Lines l, m have respectively the parametric equations

$$\begin{pmatrix} x \\ y \\ z \end{pmatrix} = t\begin{pmatrix} 2 \\ 1 \\ -1 \end{pmatrix} + \begin{pmatrix} 0 \\ 1 \\ 3 \end{pmatrix}, \qquad \begin{pmatrix} x \\ y \\ z \end{pmatrix} = u\begin{pmatrix} -2 \\ 1 \\ 1 \end{pmatrix} + \begin{pmatrix} 1 \\ 1 \\ -1 \end{pmatrix}.$$

A is the point on l with parameter t_1, B the point on m with parameter u_1. Write down an expression for the vector $\boldsymbol{AB}$.

Given that the line AB is perpendicular to both l and m, find the values of t_1 and u_1, and show that the length of AB is $7/\sqrt{5}$ units. (S.M.P.)

*11.4 Coordinate geometry of three dimensions

The equations of a line

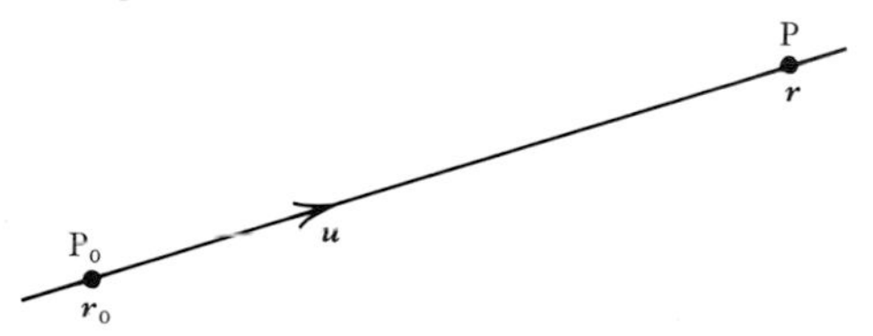

Suppose that P_0 is a point whose position vector is $\boldsymbol{r}_0$ and that a straight line passes through P_0 in the direction of the vector $\boldsymbol{u}$. Then if P, with position vector $\boldsymbol{r}$, lies on this line, it follows that

$\boldsymbol{r} = \boldsymbol{r}_0 + \lambda\boldsymbol{u}$ (where λ is a scalar).

Suppose now that the components of

$\boldsymbol{r}$ are (x, y, z), $\boldsymbol{r}_0$ are (x_0, y_0, z_0), and $\boldsymbol{u}$ are (l, m, n).

Then $x = x_0 + \lambda l$, $y = y_0 + \lambda m$, $z = z_0 + \lambda n$,

$$\Rightarrow \quad \boxed{\frac{x - x_0}{l} = \frac{y - y_0}{m} = \frac{z - z_0}{n},}$$

which are called the equations of the line.

Example 1

Find the vector and Cartesian equations of the line joining the points $(1, 2, 3)$ and $(2, 3, 5)$.

In this case, $\boldsymbol{r}_0 = \boldsymbol{i} + 2\boldsymbol{j} + 3\boldsymbol{k}$

$\boldsymbol{u} = \boldsymbol{i} + \boldsymbol{j} + 2\boldsymbol{k}.$

So $\boldsymbol{r} = (\boldsymbol{i} + 2\boldsymbol{j} + 3\boldsymbol{k}) + \lambda(\boldsymbol{i} + \boldsymbol{j} + 2\boldsymbol{k})$

$\Rightarrow \boldsymbol{r} = (1 + \lambda)\boldsymbol{i} + (2 + \lambda)\boldsymbol{j} + (3 + 2\lambda)\boldsymbol{k},$

and in Cartesian form:

$$x = 1 + \lambda, \qquad y = 2 + \lambda, \qquad z = 3 + 2\lambda$$

$$\Rightarrow \quad \frac{x-1}{1} = \frac{y-2}{1} = \frac{z-3}{2}.$$

Example 2

Find the angle between the lines

$$\frac{x}{1} = \frac{y-1}{2} = \frac{z-2}{3} \quad \text{and} \quad \frac{x}{2} = \frac{y+1}{3} = \frac{z+2}{4}.$$

These are in the directions of the vectors

$\boldsymbol{i} + 2\boldsymbol{j} + 3\boldsymbol{k}$ and $2\boldsymbol{i} + 3\boldsymbol{j} + 4\boldsymbol{k}$.

Now, if the angle between these vectors is θ, then

$$(\boldsymbol{i} + 2\boldsymbol{j} + 3\boldsymbol{k}).(2\boldsymbol{i} + 3\boldsymbol{j} + 4\boldsymbol{k}) = \sqrt{14} \times \sqrt{29} \cos\theta$$

$$\Rightarrow \qquad 20 = \sqrt{14} \times \sqrt{29} \cos\theta$$

$$\Rightarrow \qquad \cos\theta = \frac{20}{\sqrt{14} \times \sqrt{29}} = 0.969\,9$$

$$\Rightarrow \qquad \theta = 14^\circ\, 6'.$$

The equation of a plane

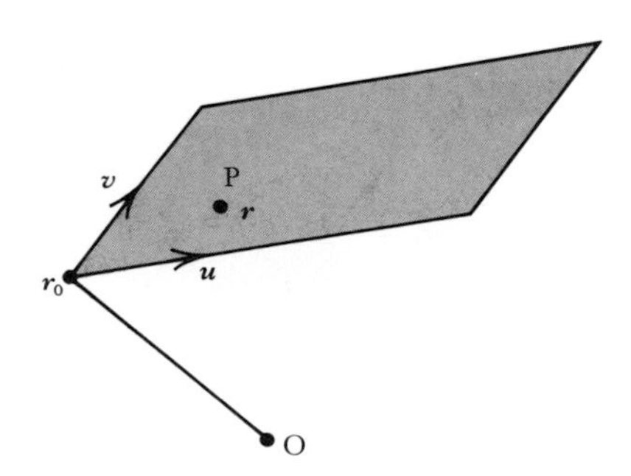

Suppose that P (with position vector $\boldsymbol{r}$) lies in a plane which passes through a fixed point P_0 (whose position vector is $\boldsymbol{r}_0$), and that $\boldsymbol{u}$, $\boldsymbol{v}$ are two non-parallel vectors in the plane.

Then $\boldsymbol{r} = \boldsymbol{r}_0 + \lambda\boldsymbol{u} + \mu\boldsymbol{v}$ (where λ, μ are scalars), which is therefore the vector equation of the plane.

Suppose further that the normal to the plane is in the direction $\boldsymbol{n}$.

Then $\boldsymbol{n}.(\boldsymbol{r} - \boldsymbol{r}_0) = 0$

$$\Rightarrow \qquad \boldsymbol{n}.\boldsymbol{r} = \boldsymbol{n}.\boldsymbol{r}_0.$$

If we now suppose that

$\boldsymbol{n} = a\boldsymbol{i} + b\boldsymbol{j} + c\boldsymbol{k}$, it follows that

$(a\boldsymbol{i} + b\boldsymbol{j} + c\boldsymbol{k}).(x\boldsymbol{i} + y\boldsymbol{j} + z\boldsymbol{k}) = \boldsymbol{n}.\boldsymbol{r}_0$;

and if we let $\boldsymbol{n}.\boldsymbol{r}_0 = -d$, we see that

$$ax + by + cz + d = 0,$$

which is the Cartesian equation of the plane.

Example 3

Find the Cartesian equation of the plane through the point (1, 2, 3) which is perpendicular to the line

$$\frac{x-1}{3} = \frac{y-2}{2} = \frac{z-3}{1}.$$

The direction of this line is given by the vector $3\boldsymbol{i} + 2\boldsymbol{j} + \boldsymbol{k}$, and the vector from (1, 2, 3) to (x, y, z) is

$(x - 1)\boldsymbol{i} + (y - 2)\boldsymbol{j} + (z - 3)\boldsymbol{k}$.

As these must be perpendicular, it follows that

$$3(x - 1) + 2(y - 2) + (z - 3) = 0$$
$$\Rightarrow \quad 3x + 2y + z - 10 = 0.$$

Exercise 11.4a

1 Find, in both vector and Cartesian forms, the equations of:
(*i*) the line joining (0, 1, 2) and (2, 1, 0);
(*ii*) the line joining (0, 1, 2) and (−2, −1, 0).

2 Find the equations of:
(*i*) the plane which passes through (1, 2, 3) and is perpendicular to the line $(x - 1)/2 = y + 1 = z/3$;
(*ii*) the plane which passes through (0, 1, 2) and is perpendicular to $\boldsymbol{i} + 2\boldsymbol{j} + 3\boldsymbol{k}$;
(*iii*) the line through (1, 1, 2) which is perpendicular to $2x + y + z = 4$.

3 Find the angles between the lines
(*i*) $\boldsymbol{r} = (1 + \lambda)\boldsymbol{i} + (1 - \lambda)\boldsymbol{j} + (2 + \lambda)\boldsymbol{k}$
and $\boldsymbol{r} = (1 - \mu)\boldsymbol{i} + (1 - 2\mu)\boldsymbol{j} + (1 + \mu)\boldsymbol{k}$;
(*ii*) $\dfrac{x-1}{2} = \dfrac{y-2}{1} = \dfrac{z}{3}$ and $\dfrac{x}{3} = \dfrac{y+1}{2} = \dfrac{z+2}{1}$.

Angles between planes and lines

We have already seen how to calculate the angle between two lines in space. We now proceed to calculate the angle between two planes; and, finally, the angle between a line and a plane.

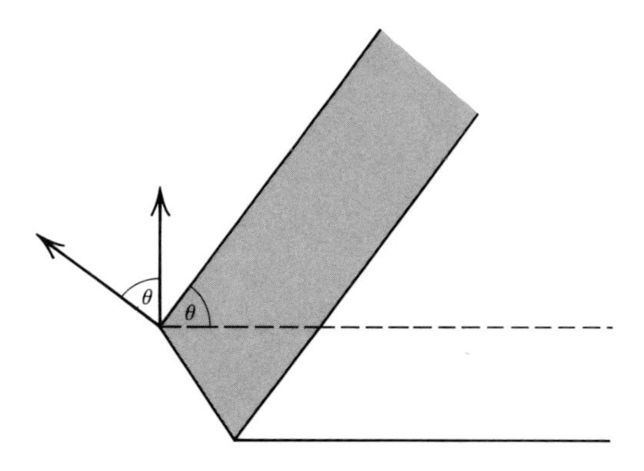

The angle between two planes can be defined as the angle between two lines in the planes both of which are perpendicular to the line of intersection of the planes. Hence it must also be the angle between the normals to the two planes.

Example 4

Find the angle between the planes

$3x + 2y + z = 0$ and $x + 2y + 4 = 0$.

The directions of the normals to these planes are given by

$3\boldsymbol{i} + 2\boldsymbol{j} + \boldsymbol{k}$ and $\boldsymbol{i} + 2\boldsymbol{j}$,

and the angle between these two vectors is θ, where

$$\cos\theta = \frac{(3\boldsymbol{i} + 2\boldsymbol{j} + \boldsymbol{k}).(\boldsymbol{i} + 2\boldsymbol{j})}{\sqrt{14} \times \sqrt{5}}$$

$$= \frac{7}{\sqrt{70}} = \sqrt{0.7} = 0.836\,7$$

$$\Rightarrow \quad \theta = 33°\ 12'.$$

The angle between a line and a plane is defined as the angle between the line (l) and its projection (p) on the plane (π).

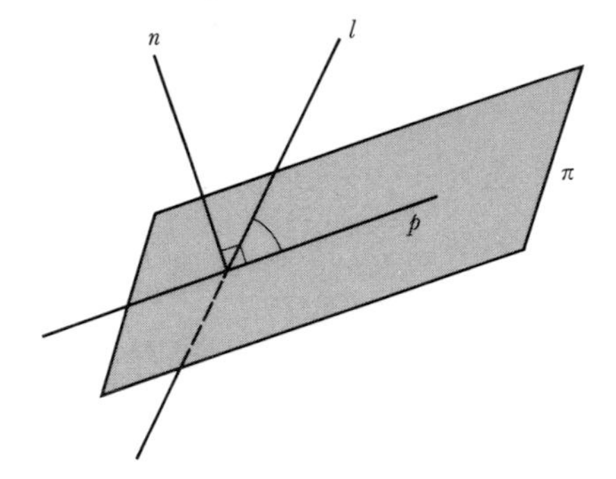

Hence it is conveniently calculated by finding the angle between the line and the normal n to the plane, and then taking its complement.

Example 4

Find the angle between the line

$$\frac{x-1}{2} = \frac{y-2}{3} = \frac{z-3}{4}$$

and the plane

$$4x + 3y + 2z + 1 = 0.$$

Now the line is in the direction of the vector $2\boldsymbol{i} + 3\boldsymbol{j} + 4\boldsymbol{k}$ and the normal to the plane is in the direction $4\boldsymbol{i} + 3\boldsymbol{j} + 2\boldsymbol{k}$.

So the angle θ between these directions is given by

$$\cos\theta = \frac{(2\boldsymbol{i} + 3\boldsymbol{j} + 4\boldsymbol{k}).(4\boldsymbol{i} + 3\boldsymbol{j} + 2\boldsymbol{k})}{\sqrt{29} \times \sqrt{29}}$$

$$= \tfrac{25}{29} = 0.862\,0$$

$$\Rightarrow \quad \theta = 30°\ 27'.$$

Hence the required angle is

$$90° - 30°\ 27' = 59°\ 33'.$$

The distance from a point to a plane

Find the perpendicular distance from the point (x_1, y_1, z_1) to the plane $ax + by + cz + d = 0$.

Let
$$\boldsymbol{r}_1 = x_1\boldsymbol{i} + y_1\boldsymbol{j} + z_1\boldsymbol{k}$$
$$\boldsymbol{r} = x\boldsymbol{i} + y\boldsymbol{j} + z\boldsymbol{k}$$
$$\boldsymbol{n} = a\boldsymbol{i} + b\boldsymbol{j} + c\boldsymbol{k}.$$

Then $\boldsymbol{n}.\boldsymbol{r} = ax + by + cz = -d =$ constant, so that the projection of $\boldsymbol{r}$ in the direction of $\boldsymbol{n}$ is a constant.

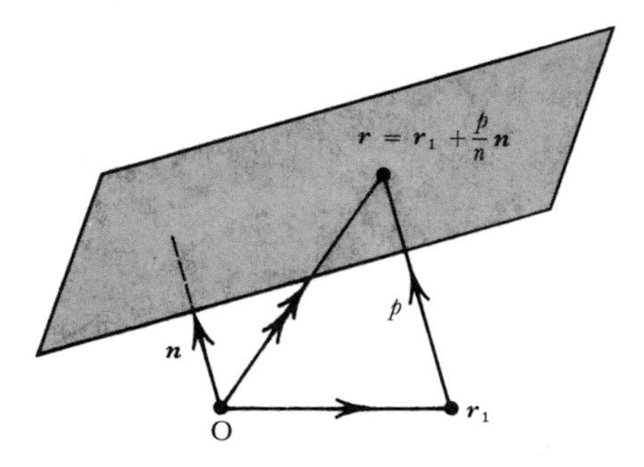

Hence $\boldsymbol{n}$ is normal to the plane. But the unit vector in this direction is $\boldsymbol{n}/n$; so if the required perpendicular distance is p, it follows that the foot of the perpendicular from P_1 to the plane is

$$\boldsymbol{r}_1 \pm p\frac{\boldsymbol{n}}{n} \quad \text{(the sign depending on the direction of } \boldsymbol{n}\text{).}$$

But this point lies in the plane.

$$\text{So} \quad \boldsymbol{n}.\left(\boldsymbol{r}_1 \pm p\frac{\boldsymbol{n}}{n}\right) + d = 0$$

$$\Rightarrow \quad \boldsymbol{n}.\boldsymbol{r}_1 \pm pn + d = 0$$

$$\Rightarrow \quad |p| = \frac{|\boldsymbol{n}.\boldsymbol{r}_1 + d|}{n}.$$

Now $\boldsymbol{n}.\boldsymbol{r}_1 = ax_1 + by_1 + cz_1$

and $n = \sqrt{(a^2 + b^2 + c^2)}$.

$$\text{So} \quad \boxed{|p| = \frac{|ax_1 + by_1 + cz_1 + d|}{\sqrt{(a^2 + b^2 + c^2)}}}$$

In exactly similar fashion in a plane, the perpendicular distance from a point $P_1(x_1, y_1)$ to the line $ax + by + c = 0$ is given by

$$\boxed{\frac{|ax_1 + by_1 + c|}{\sqrt{(a^2 + b^2)}}}$$

So, for example, the distances of the point $(1, 2)$ from the line

$3x + 4y + 1 = 0$ is

$$\frac{|3 \times 1 + 4 \times 2 + 1|}{\sqrt{(3^2 + 4^2)}} = \tfrac{12}{5} = 2.4.$$

Exercise 11.4b

1 Find pairs of vectors which are normal to the following planes, and hence the angles between them:
(*i*) $x + 2y + 3z = 6$ and $x - y - z = 4$;
(*ii*) $x - y - 3 = 0$ and $y + z - 1 = 0$.

2 Find the angles between the following planes and lines:

(*i*) $3x + 2y + z = 1$ and $\dfrac{x-1}{1} = \dfrac{y-2}{2} = \dfrac{z-3}{3}$;

(*ii*) $x + 2y - z = 3$ and $\dfrac{x-2}{1} = \dfrac{y-3}{-2} = \dfrac{z-1}{4}$.

3 (*i*) Calculate the perpendicular distances from (2, 1) to the lines (in the plane $z = 0$)

$3x + 4y + 5 = 0$ and $5x - 12y + 15 = 0$.

(*ii*) Find the locus of a point P (x, y) which is equidistant from these two lines. Describe this locus.

4 (*i*) Calculate the perpendicular distances from (1, 2, 3) to the planes

$x + 2y + 2z + 7 = 0$ and $2x + y + 2z - 1 = 0$.

(*ii*) Find the locus of a point P (x, y, z) which is equidistant from these two planes. Describe this locus.

5 A cube ABCDA′B′C′D′ has base ABCD and vertical sides AA′ etc. The mid-points of the edges AB, AD, AA′ are respectively P, Q, R. The cube is cut along the plane PQR.
(*i*) Find the angle the edge AA′ makes with the plane PQR.
(*ii*) Find the angle the plane PQR makes with each face of the cube.
(*iii*) Determine the ratio of the volume of the tetrahedron APQR to that of the remainder of the cube.

If the cube is also cut along the plane AB′C, draw a diagram of the triangle PQR and mark in the line of this second cut. (S.M.P.)

6 Find the perpendicular distance of the point (3, 0, 1) from the line whose Cartesian equation is

$$\frac{x-1}{3} = \frac{y+2}{4} = \frac{z}{12}.$$ (M.E.I.)

(First find the general point P of this line and then the condition for the vector from (3, 0, 1) to P to be perpendicular.)

7 Find the perpendicular (i.e., shortest) distance between the two skew lines:
$\boldsymbol{r} = (\boldsymbol{i} - 3\boldsymbol{j} + 3\boldsymbol{k}) + \lambda(-2\boldsymbol{i} + \boldsymbol{j} - 2\boldsymbol{k})$
$\boldsymbol{r} = (3\boldsymbol{i} - \boldsymbol{j} + 2\boldsymbol{k}) + \mu(2\boldsymbol{i} - 3\boldsymbol{j} - 2\boldsymbol{k})$.

8 Find the equation of the plane through A = (2, 2, −1) perpendicular to OA. Which point on this plane is closest to the point (2, −1, 2)? (S.M.P.)

*11.5 Vector products

If $\boldsymbol{a}$ and $\boldsymbol{b}$ are two vectors, we have seen that we can associate with them a scalar quantity, known as their scalar product. We now ask ourselves whether there is also a convenient *vector product* which we could usefully consider.

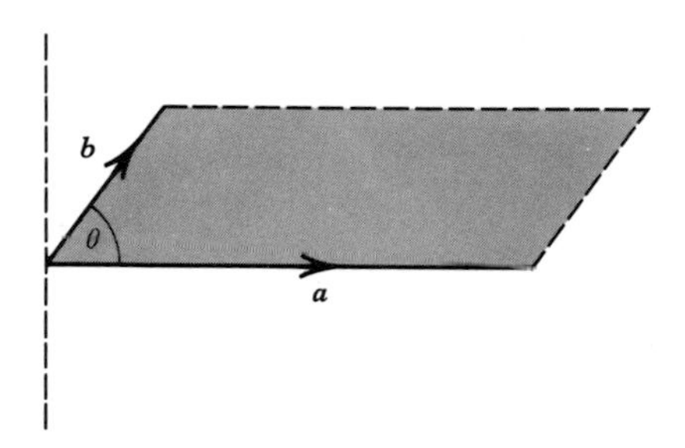

It is clear that the two vectors $\boldsymbol{a}$ and $\boldsymbol{b}$ fix a particular parallelogram, whose area is easily seen to be $ab \sin \theta$. Furthermore, the most natural direction to associate with this parallelogram would be the one which is perpendicular, or normal, to its plane. As for the particular *sense* (up or down this line), we can easily resolve our dilemma by following the movement of a right-handed screw which is turning from $\boldsymbol{a}$ to $\boldsymbol{b}$.

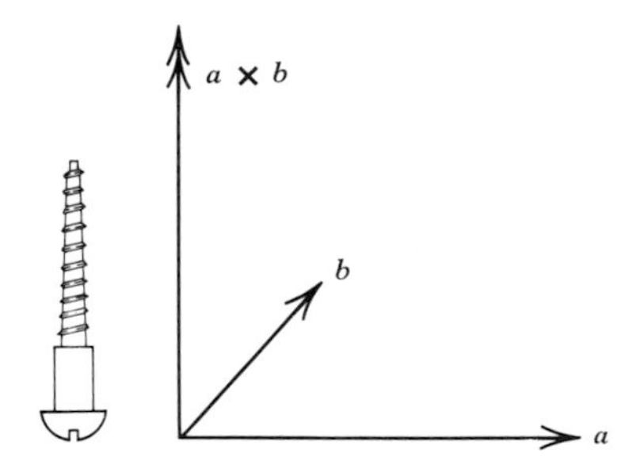

We therefore define the *vector product*, which we shall write† $\boldsymbol{a} \times \boldsymbol{b}$, as the vector

whose magnitude is $ab \sin \theta$;

whose direction is perpendicular to $\boldsymbol{a}$ and $\boldsymbol{b}$;

and whose sense is determined by a right-hand screw turning from $\boldsymbol{a}$ to $\boldsymbol{b}$.

The most common use of vector products occurs in mechanics, when a force $\boldsymbol{F}$ acts through a point whose position vector is $\boldsymbol{r}$ and whose moment about O is defined as $\boldsymbol{r} \times \boldsymbol{F}$:

In this figure it is seen that the magnitude of this moment is $rF \sin \theta = Fp$, and so is equal to the product of the force and the length of the perpendicular arm from O. The moment itself is a vector of this magnitude which acts perpendicular to the plane of $\boldsymbol{r}$ and $\boldsymbol{F}$. In our example, therefore, by

† Also denoted by $\boldsymbol{a} \wedge \boldsymbol{b}$.

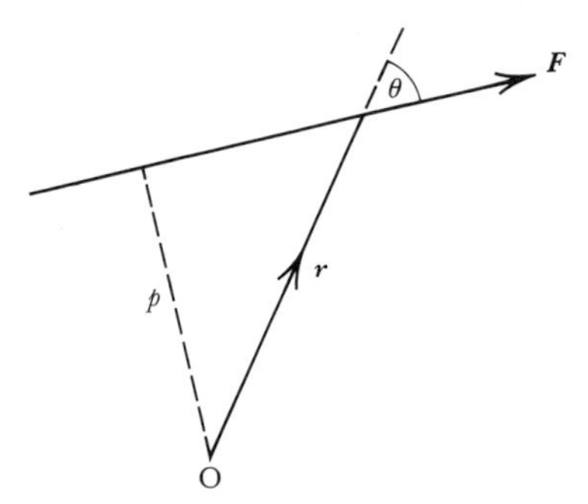

the right-hand screw rule, this moment acts *into* the paper and represents a turning effect which is in the plane of the paper and clockwise.

As with the scalar product, the properties of the vector product follow from its definition. Some are expected, like

$$\boldsymbol{a} \times (\lambda \boldsymbol{b}) = (\lambda \boldsymbol{a}) \times \boldsymbol{b} = \lambda(\boldsymbol{a} \times \boldsymbol{b})$$

and can easily be demonstrated in a figure. Others are more surprising,

like $\boldsymbol{a} \times \boldsymbol{a} = \boldsymbol{0}$ (since $aa \sin 0 = 0$)

and $\boldsymbol{a} \times \boldsymbol{b} = -\boldsymbol{b} \times \boldsymbol{a}$

(because a right-hand screw turning from $\boldsymbol{a}$ to $\boldsymbol{b}$ moves *against* a right-hand screw turning from $\boldsymbol{b}$ to $\boldsymbol{a}$). So vector products are *not* commutative.

Lastly, however, vector products (like scalar products) *are* distributive with respect to addition:

$$\boldsymbol{a} \times (\boldsymbol{b} + \boldsymbol{c}) = \boldsymbol{a} \times \boldsymbol{b} + \boldsymbol{a} \times \boldsymbol{c}.$$

We shall not establish this until the end of this section, though in passing we remark on its importance if $\boldsymbol{P}$ and $\boldsymbol{Q}$ are two forces.

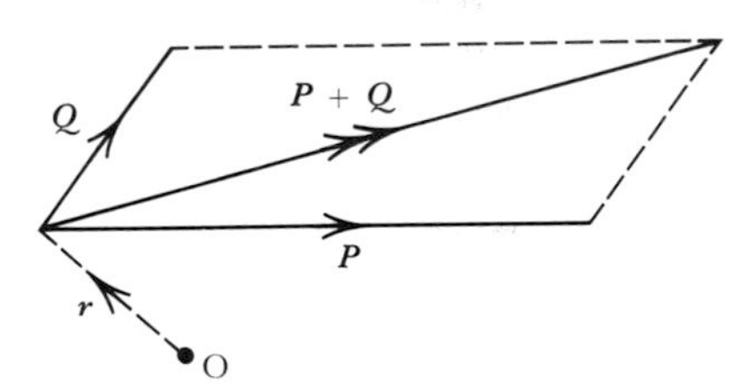

For the result means that

$$\boldsymbol{r} \times (\boldsymbol{P} + \boldsymbol{Q}) = \boldsymbol{r} \times \boldsymbol{P} + \boldsymbol{r} \times \boldsymbol{Q}.$$

i.e. that the sum of the moments of two forces is equal to the moment of their resultant.

Vector product in component form

If $\boldsymbol{a} = a_1\boldsymbol{i} + a_2\boldsymbol{j} + a_3\boldsymbol{k}$

and $\boldsymbol{b} = b_1\boldsymbol{i} + b_2\boldsymbol{j} + b_3\boldsymbol{k}$,

we can easily find the vector product $\boldsymbol{a} \times \boldsymbol{b}$.

For, firstly, we notice that

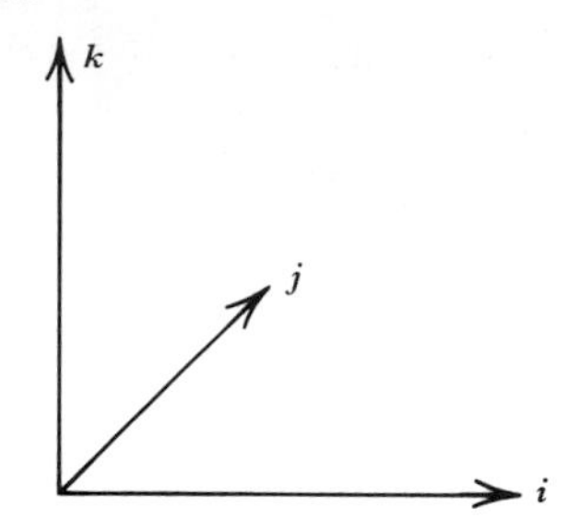

$$\boldsymbol{i} \times \boldsymbol{i} = \boldsymbol{j} \times \boldsymbol{j} = \boldsymbol{k} \times \boldsymbol{k} = \boldsymbol{0}$$

and that
$$\boldsymbol{i} \times \boldsymbol{j} = \boldsymbol{k} = -\boldsymbol{j} \times \boldsymbol{i}$$
$$\boldsymbol{j} \times \boldsymbol{k} = \boldsymbol{i} = -\boldsymbol{k} \times \boldsymbol{j}$$
$$\boldsymbol{k} \times \boldsymbol{i} = \boldsymbol{j} = -\boldsymbol{i} \times \boldsymbol{k}.$$

So
$$\boldsymbol{a} \times \boldsymbol{b} = (a_1\boldsymbol{i} + a_2\boldsymbol{j} + a_3\boldsymbol{k}) \times (b_1\boldsymbol{i} + b_2\boldsymbol{j} + b_3\boldsymbol{k})$$
$$= (a_2b_3 - a_3b_2)\boldsymbol{i} + (a_3b_1 - a_1b_3)\boldsymbol{j} + (a_1b_2 - a_2b_1)\boldsymbol{k}.$$

Example 1

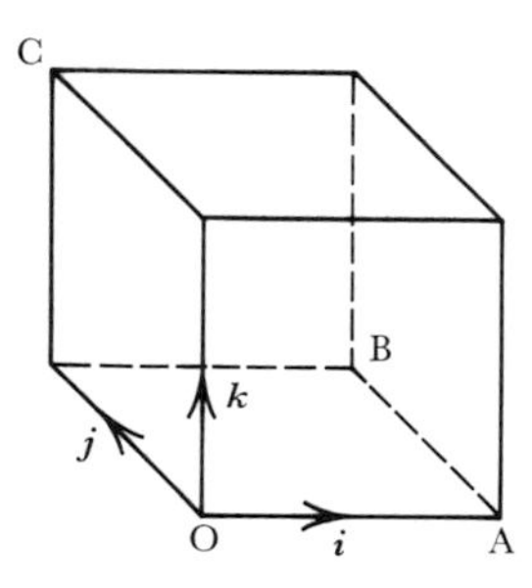

The position vectors of three points A, B, C are $\boldsymbol{a} = \boldsymbol{i}$, $\boldsymbol{b} = \boldsymbol{i} + \boldsymbol{j}$, $\boldsymbol{c} = \boldsymbol{j} + \boldsymbol{k}$. Find the angles between:
(*i*) the planes OBC and OAC;
(*ii*) the line OB and the plane ABC.

(*i*) Two vectors in plane OBC are $\boldsymbol{b}$ and $\boldsymbol{c}$, so the normal to the plane is along

$$\boldsymbol{b} \times \boldsymbol{c} = (\boldsymbol{i} + \boldsymbol{j}) \times (\boldsymbol{j} + \boldsymbol{k})$$
$$= \boldsymbol{i} \times \boldsymbol{j} + \boldsymbol{i} \times \boldsymbol{k} + \boldsymbol{j} \times \boldsymbol{j} + \boldsymbol{j} \times \boldsymbol{k}$$
$$= \boldsymbol{k} - \boldsymbol{j} + \boldsymbol{0} + \boldsymbol{i} = \boldsymbol{i} - \boldsymbol{j} + \boldsymbol{k};$$

and similarly the normal to OAC is along

$$\boldsymbol{a} \times \boldsymbol{c} = \boldsymbol{i} \times (\boldsymbol{j} + \boldsymbol{k}) = \boldsymbol{k} - \boldsymbol{j} = -\boldsymbol{j} + \boldsymbol{k}.$$

Now the angle θ between two planes is the angle between their normals, so

$$\cos\theta = \frac{(\boldsymbol{i} - \boldsymbol{j} + \boldsymbol{k}).(-\boldsymbol{j} + \boldsymbol{k})}{\sqrt{3} \times \sqrt{2}} = \frac{2}{\sqrt{6}} = \sqrt{\tfrac{2}{3}}$$

$$\Rightarrow \quad \theta = 35^\circ\ 16'.$$

(*ii*) Similarly, two vectors in plane ABC

are $\quad \boldsymbol{AB} = \boldsymbol{b} - \boldsymbol{a} = \boldsymbol{j}$

and $\quad \boldsymbol{AC} = \boldsymbol{c} - \boldsymbol{a} = -\boldsymbol{i} + \boldsymbol{j} + \boldsymbol{k}.$

So the normal to the plane is along

$$\boldsymbol{j} \times (-\boldsymbol{i} + \boldsymbol{j} + \boldsymbol{k}) = \boldsymbol{i} + \boldsymbol{k}.$$

Now $\quad \boldsymbol{OB} = \boldsymbol{b} = \boldsymbol{i} + \boldsymbol{j}.$

Hence the angle between $\boldsymbol{OB}$ and this normal is ϕ, where

$$\cos \phi = \frac{(\boldsymbol{i} + \boldsymbol{j}).(\boldsymbol{i} + \boldsymbol{k})}{\sqrt{2} \times \sqrt{2}} = \tfrac{1}{2}$$

$$\Rightarrow \quad \phi = 60°.$$

So the angle between $\boldsymbol{OB}$ and the plane itself is $90° - 60° = 30°$.

Scalar and vector triple products

We now proceed to ask if there is any way in which we can speak of the product of *three* vectors, $\boldsymbol{a}$, $\boldsymbol{b}$, $\boldsymbol{c}$.

Firstly, it is clear that the expression $\boldsymbol{a}.(\boldsymbol{b}.\boldsymbol{c})$ is meaningless, since $\boldsymbol{b}.\boldsymbol{c}$ is not a vector but a scalar and so cannot form a scalar product with $\boldsymbol{a}$. By contrast, however, $\boldsymbol{a}.(\boldsymbol{b} \times \boldsymbol{c})$ does exist and, as it is a scalar quantity, is called a *scalar triple product*. Furthermore, $\boldsymbol{a} \times (\boldsymbol{b} \times \boldsymbol{c})$ also exists and as it is a vector quantity is called a *vector triple product*.

Though these quantities are extremely important in applied mathematics, we cannot investigate them further and simply note the connection between scalar triple products and volumes:

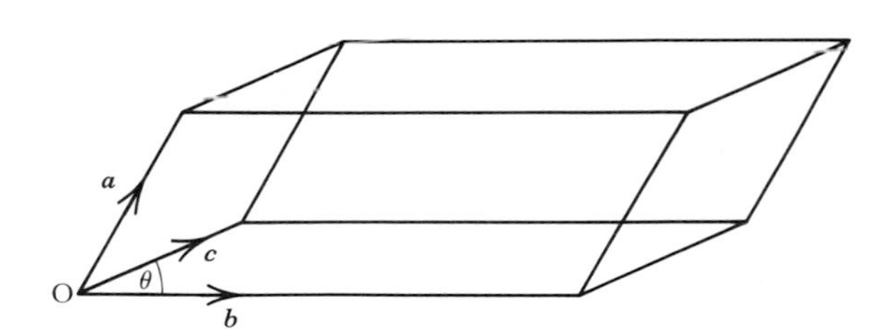

Suppose that a parallelepiped is formed from the three vectors $\boldsymbol{a}$, $\boldsymbol{b}$, $\boldsymbol{c}$, as shown in the figure.

Then $\boldsymbol{b} \times \boldsymbol{c}$ is a vector of magnitude $bc \sin \theta$, which is equal to the area A of the base parallelogram, in a direction normal to the base.

So $\boldsymbol{a}.(\boldsymbol{b} \times \boldsymbol{c})$ is of magnitude $\quad$ (projection of $\boldsymbol{a} \perp \boldsymbol{b}$ and $\boldsymbol{c}$) $\times A$

= perpendicular height $\times$ area of base

= volume of parallelepiped.

Hence $\quad$ volume of parallelepiped $= \boldsymbol{a}.(\boldsymbol{b} \times \boldsymbol{c})$

and it clearly follows that:

volume of parallelepiped is zero $\Leftrightarrow$ $\boldsymbol{a}$, $\boldsymbol{b}$, $\boldsymbol{c}$ are coplanar $\Leftrightarrow$ $\boldsymbol{a}.(\boldsymbol{b} \times \boldsymbol{c}) = 0$.

Furthermore, in components:

$$\boldsymbol{a} = a_1\boldsymbol{i} + a_2\boldsymbol{j} + a_3\boldsymbol{k}$$

$$\boldsymbol{b} \times \boldsymbol{c} = (b_2c_3 - b_3c_2)\boldsymbol{i} + (b_3c_1 - b_1c_3)\boldsymbol{j} + (b_1c_2 - b_2c_1)\boldsymbol{k}$$

$$\Rightarrow \quad \boldsymbol{a}.(\boldsymbol{b} \times \boldsymbol{c}) = a_1(b_2c_3 - b_3c_2) + a_2(b_3c_1 - b_1c_3) + a_3(b_1c_2 - b_2c_1).$$

So $\boldsymbol{a}$, $\boldsymbol{b}$, $\boldsymbol{c}$ are coplanar

$$\Leftrightarrow \quad a_1(b_2c_3 - b_3c_2) + a_2(b_3c_1 - b_1c_3) + a_3(b_1c_2 - b_2c_1) = 0$$

which is a result to which we shall return in chapter 13.

Exercise 11.5

1 If $\boldsymbol{a} = \boldsymbol{i} - \boldsymbol{j}$, $\boldsymbol{b} = \boldsymbol{i} + \boldsymbol{j} + \boldsymbol{k}$, $\boldsymbol{c} = \boldsymbol{k}$, calculate $\boldsymbol{b} \times \boldsymbol{c}$, $\boldsymbol{c} \times \boldsymbol{a}$, $\boldsymbol{a} \times \boldsymbol{b}$.

2 Denoting corresponding points in no. **1** by capital letters, find:
(*i*) vectors which are normal to the planes OBC, OCA, ABC;
(*ii*) the angle between the planes OBC and OCA;
(*iii*) the angles between these planes and the plane ABC;
(*iv*) the angles between these planes and the line AB.

3 In no. **1** find whether:
(*i*) $\boldsymbol{a}.(\boldsymbol{b} \times \boldsymbol{c}) = \boldsymbol{b}.(\boldsymbol{c} \times \boldsymbol{a}) = \boldsymbol{c}.(\boldsymbol{a} \times \boldsymbol{b})$;
(*ii*) $\boldsymbol{a} \times (\boldsymbol{b} \times \boldsymbol{c}) = (\boldsymbol{a} \times \boldsymbol{b}) \times \boldsymbol{c}$;
(*iii*) $\boldsymbol{a} \times (\boldsymbol{b} \times \boldsymbol{c}) = (\boldsymbol{a}.\boldsymbol{c})\boldsymbol{b} - (\boldsymbol{a}.\boldsymbol{b})\boldsymbol{c}$.
Test these results with other vectors $\boldsymbol{a}$, $\boldsymbol{b}$, $\boldsymbol{c}$.

4 Let the three sides of a triangle ABC have vectors $\boldsymbol{a}$, $\boldsymbol{b}$, $\boldsymbol{c}$ so that $\boldsymbol{a} + \boldsymbol{b} + \boldsymbol{c} = \mathbf{0}$.
Hence prove
(*i*) $\boldsymbol{b} \times \boldsymbol{c} = \boldsymbol{c} \times \boldsymbol{a} = \boldsymbol{a} \times \boldsymbol{b}$;
(*ii*) the sine formula for $\triangle$ABC.

* *The distributive law for vector products*†

$$\boldsymbol{a} \times (\boldsymbol{b} + \boldsymbol{c}) = \boldsymbol{a} \times \boldsymbol{b} + \boldsymbol{a} \times \boldsymbol{c}.$$

Firstly, we project $\boldsymbol{b}$ and $\boldsymbol{c}$ on to the plane perpendicular to $\boldsymbol{a}$, and let their projections be $\boldsymbol{b}_1$ and $\boldsymbol{c}_1$.

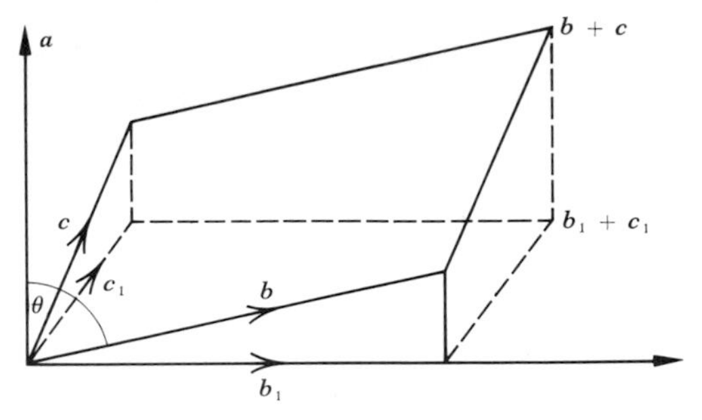

† Or, more strictly, that the operation of vector multiplication distributes over that of vector addition.

If the angle between $\boldsymbol{a}$ and $\boldsymbol{b}$ is θ, then $\boldsymbol{a} \times \boldsymbol{b}$ has magnitude $ab \sin \theta$.

But $\boldsymbol{a} \times \boldsymbol{b}_1$ has magnitude ab_1

$$= ab \cos(\tfrac{1}{2}\pi - \theta)$$
$$= ab \sin \theta.$$

Also, as $\boldsymbol{a}$, $\boldsymbol{b}$ and $\boldsymbol{b}_1$ are coplanar, $\boldsymbol{a} \times \boldsymbol{b}$ and $\boldsymbol{a} \times \boldsymbol{b}_1$ both lie along the perpendicular to this plane.

$$\left.\begin{array}{ll} \text{So} & \boldsymbol{a} \times \boldsymbol{b} = \boldsymbol{a} \times \boldsymbol{b}_1; \\ \text{similarly} & \boldsymbol{a} \times \boldsymbol{c} = \boldsymbol{a} \times \boldsymbol{c}_1, \\ \text{and} & \boldsymbol{a} \times (\boldsymbol{b} + \boldsymbol{c}) = \boldsymbol{a} \times (\boldsymbol{b}_1 + \boldsymbol{c}_1). \end{array}\right\} \quad (1)$$

We now concentrate on the plane perpendicular to $\boldsymbol{a}$, of which we can take a bird's eye view:

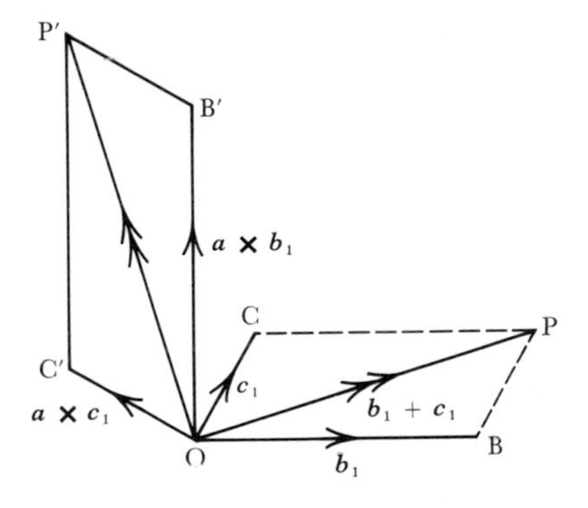

Now $\boldsymbol{a} \times \boldsymbol{b}_1$ has magnitude ab_1
and $\boldsymbol{a} \times \boldsymbol{c}_1$ has magnitude ac_1.

So the vectors

$\boldsymbol{a} \times \boldsymbol{b}_1$, $\boldsymbol{a} \times \boldsymbol{c}_1$, $\boldsymbol{a} \times (\boldsymbol{b}_1 + \boldsymbol{c}_1)$

are formed by first multiplying

$\boldsymbol{b}_1$, $\boldsymbol{c}_1$, $\boldsymbol{b}_1 + \boldsymbol{c}_1$

by a and then rotating through a right angle.

But it is clear that

$\boldsymbol{a} \times (\boldsymbol{b}_1 + \boldsymbol{c}_1)$

is the diagonal of the parallelogram formed by

$\boldsymbol{a} \times \boldsymbol{b}_1$ and $\boldsymbol{a} \times \boldsymbol{c}_1$.

So $\boldsymbol{a} \times (\boldsymbol{b}_1 + \boldsymbol{c}_1) = \boldsymbol{a} \times \boldsymbol{b}_1 + \boldsymbol{a} \times \boldsymbol{c}_1$,

from which it follows by (1) that

$\boldsymbol{a} \times (\boldsymbol{b} + \boldsymbol{c}) = \boldsymbol{a} \times \boldsymbol{b} + \boldsymbol{a} \times \boldsymbol{c}$.

*11.6 Plane kinematics

If a point P is moving in a plane, then we have already seen that (relative to a fixed point O) its position vector $\boldsymbol{r}$, velocity $\boldsymbol{v}$ and acceleration $\boldsymbol{a}$ can all be expressed in component form:

$$\boldsymbol{r} = x\boldsymbol{i} + y\boldsymbol{j},$$
$$\boldsymbol{v} = \dot{\boldsymbol{r}} = \dot{x}\boldsymbol{i} + \dot{y}\boldsymbol{j},$$
$$\boldsymbol{a} = \ddot{\boldsymbol{r}} = \ddot{x}\boldsymbol{i} + \ddot{y}\boldsymbol{j}.$$

These, of course, are based on Cartesian axes Ox, Oy, and the question now arises whether we can possibly obtain similar results using polar coordinates.

Let us start by supposing that at time t the point P has (relative to a fixed point O and a fixed base-line), position vector $\boldsymbol{r}$ and polar coordinates (r, θ), where r, θ are functions of t:

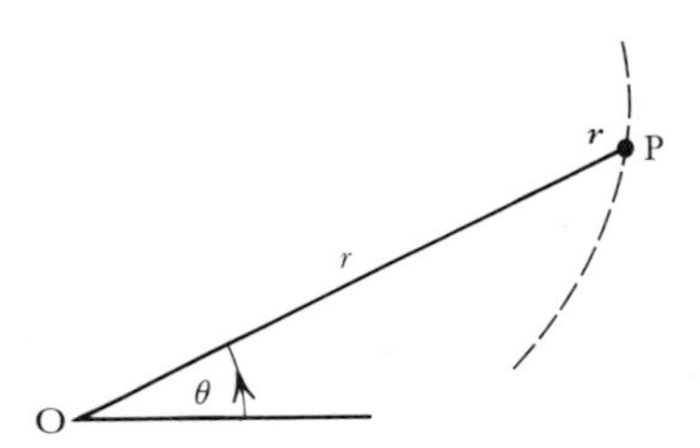

We now let $\boldsymbol{p}$ and $\boldsymbol{q}$ be unit vectors along and perpendicular to $\boldsymbol{OP}$; so that as P moves, $\boldsymbol{p}$ and $\boldsymbol{q}$ change direction, rotating with angular velocity $\dot{\theta}$.

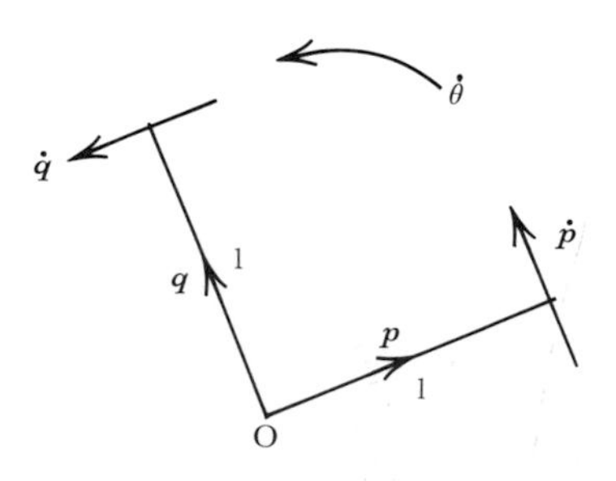

We immediately see that

$$\dot{\boldsymbol{p}} = \dot{\theta}\boldsymbol{q} \quad \text{and} \quad \dot{\boldsymbol{q}} = -\dot{\theta}\boldsymbol{p}.$$

But $\qquad \boldsymbol{r} = r\boldsymbol{p}.$

So $\qquad \dot{\boldsymbol{r}} = \dot{r}\boldsymbol{p} + r\dot{\boldsymbol{p}} = \dot{r}\boldsymbol{p} + r\dot{\theta}\boldsymbol{q}.$

Furthermore,
$$\begin{aligned}\ddot{\boldsymbol{r}} &= \ddot{r}\boldsymbol{p} + \dot{r}\dot{\boldsymbol{p}} + \dot{r}\dot{\theta}\boldsymbol{q} + r\ddot{\theta}\boldsymbol{q} + r\dot{\theta}\dot{\boldsymbol{q}} \\ &= \ddot{r}\boldsymbol{p} + \dot{r}\dot{\theta}\boldsymbol{q} + \dot{r}\dot{\theta}\boldsymbol{q} + r\ddot{\theta}\boldsymbol{q} - r\dot{\theta}^2\boldsymbol{p} \\ &= (\ddot{r} - r\dot{\theta}^2)\boldsymbol{p} + (r\ddot{\theta} + 2\dot{r}\dot{\theta})\boldsymbol{q}.\end{aligned}$$

Summarising,

$$\boldsymbol{r} = r\boldsymbol{p}$$
$$\dot{\boldsymbol{r}} = \dot{r}\boldsymbol{p} + r\dot{\theta}\boldsymbol{q}$$
$$\ddot{\boldsymbol{r}} = (\ddot{r} - r\dot{\theta}^2)\boldsymbol{p} + (r\ddot{\theta} + 2\dot{r}\dot{\theta})\boldsymbol{q}$$

Hence the *radial* and *transverse* components of the velocity and acceleration of P can be denoted:

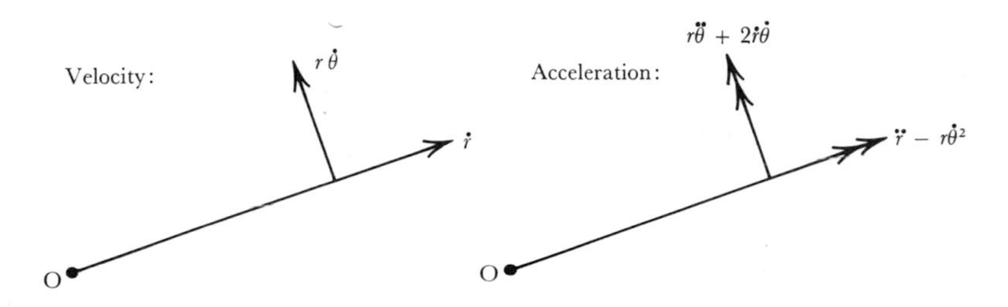

In the particular case when P is moving in a circle of radius a,

$r = a \Rightarrow \dot{r} = \ddot{r} = 0.$

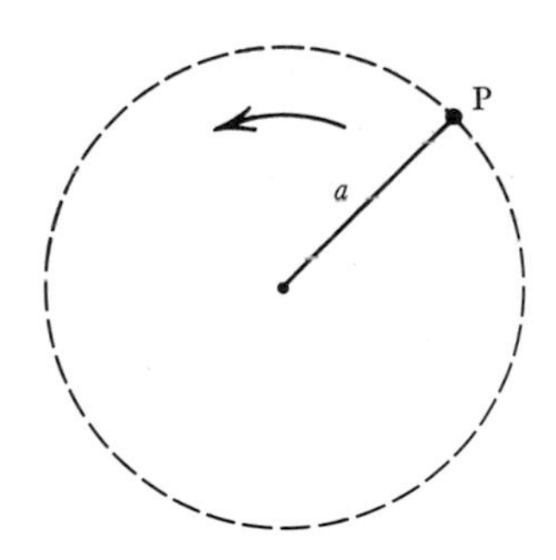

Hence the components of velocity and acceleration become

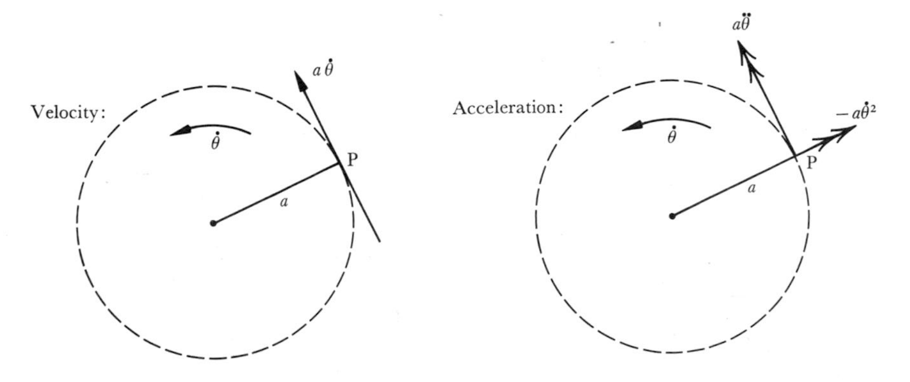

Furthermore, if P is moving with *constant* angular velocity $\dot{\theta}$ it follows that $\ddot{\theta} = 0$, so that the acceleration of P is again shown to be $a\dot{\theta}^2$ directed towards O.

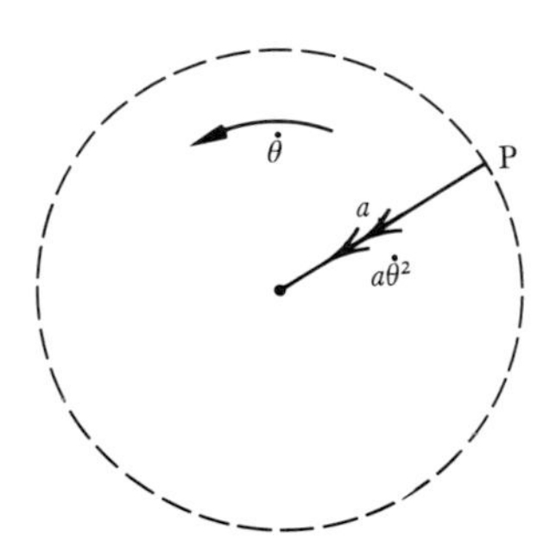

Example 3 (see also 8.1)

In 1611 John Kepler announced that each planet moves so that a line from the Sun 'sweeps out' area at a constant rate (i.e., so that equal areas are swept in equal times). Deduce that the planet's acceleration must be along this line.

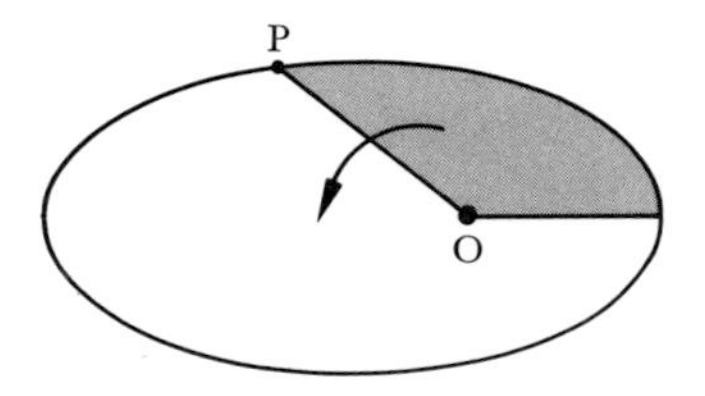

If we suppose that in time δt the planet moves from P to P′ and the additional area swept out is δA, then with the usual notation,

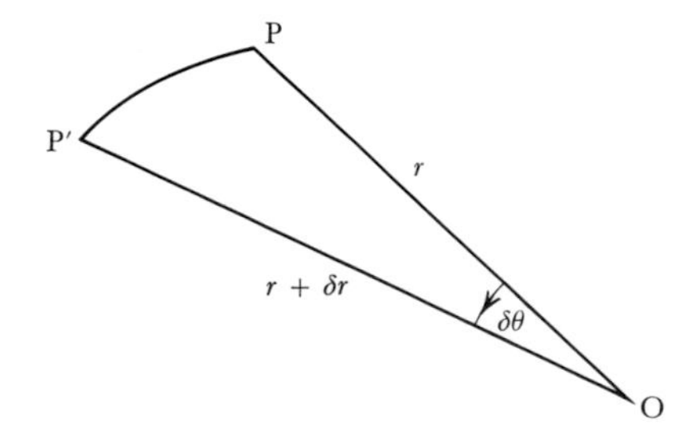

$$\delta A \approx \tfrac{1}{2}r^2\,\delta\theta$$

$$\Rightarrow\quad \frac{\delta A}{\delta t} \approx \tfrac{1}{2}r^2\,\frac{\delta\theta}{\delta t}$$

$$\Rightarrow\quad \frac{\mathrm{d}A}{\mathrm{d}t} = \tfrac{1}{2}r^2\dot{\theta}.$$

So if $\mathrm{d}A/\mathrm{d}t$ is constant, it follows that

$$\frac{\mathrm{d}}{\mathrm{d}t}(r^2\dot{\theta}) = 0$$

$$\Rightarrow\quad r^2\ddot{\theta} + 2r\dot{r}\dot{\theta} = 0$$

$$\Rightarrow\quad r\ddot{\theta} + 2\dot{r}\dot{\theta} = 0.$$

So the transverse component of the acceleration is zero, and the acceleration must be entirely radial. It was this discovery, although not expressed in this notation, that halted men's search for *transverse* forces *propelling* the planets, and caused Newton to propose forces *towards* the Sun *holding* them in orbit.)

Exercise 11.6

1 The polar coordinates r, θ of a particle after time t are given by

$r = e^{2t}, \qquad \theta = 2t.$

(*i*) Sketch its path, and state its polar equation.
(*ii*) Find the radial and transverse components of velocity and acceleration after time t.
(*iii*) Show that its direction of motion is always inclined at 45° to the radius and that its radial acceleration is zero.

2

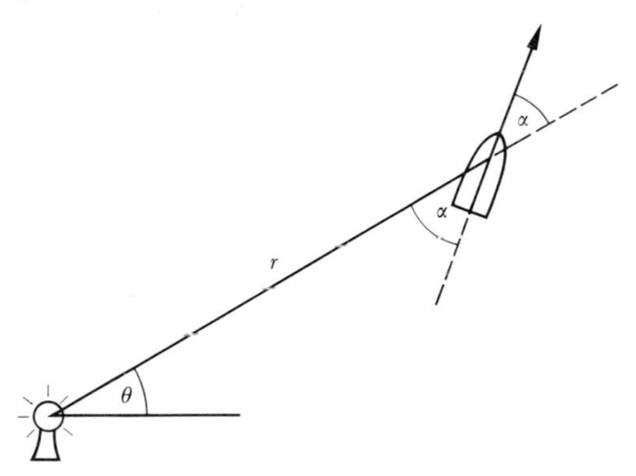

A ship sails away from a lighthouse keeping the light on its port quarter at a constant angle α.

Show that: (*i*) $\dfrac{r\dot{\theta}}{\dot{r}} = \tan\alpha$; (*ii*) $\dfrac{d\theta}{dr} = \dfrac{\tan\alpha}{r}$;

(*iii*) if $r = a$ when $\theta = 0$, then $r = a\, e^{\theta \cot\alpha}$. Sketch this path.

3

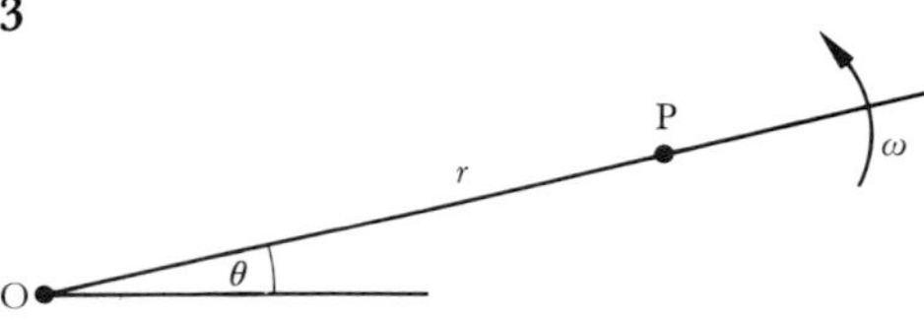

A bead is placed on a straight smooth horizontal wire which is rotating about its end O with constant angular velocity ω, so that $\theta = \omega t$. As there is no *radial* force on the bead, its *radial* acceleration must be zero. Hence show that

(*i*) $\ddot{r} - \omega^2 r = 0,$ and $r = A\,e^{\omega t} + B\,e^{-\omega t}$;

(*ii*) if the bead is initially at rest at a distance a from O (i.e., $t = 0, r = a$ and $\dot{r} = 0$), then $r = a \cosh \omega t$;

(*iii*) $\dot{r} = \omega\sqrt{(r^2 - a^2)}$.

4 Describe the curve given by

$$\boldsymbol{r} = \begin{pmatrix} 5\cos(t^2) \\ 5\sin(t^2) \end{pmatrix}, \text{ where } t \text{ denotes time.}$$

Find $\boldsymbol{a}\,(= \ddot{\boldsymbol{r}})$ and deduce the radial and transverse components of the acceleration.

A boy slides down a banister such that his path is given by

$$\boldsymbol{r} = \begin{pmatrix} 5\cos(t^2) \\ 5\sin(t^2) \\ kt^2 \end{pmatrix},$$

where the units are metres and k is a constant. Indicate on a diagram the shape of the banister and the magnitudes of the boy's acceleration components in the directions (*a*) towards the axis of the curve, (*b*) vertically downwards, (*c*) in a direction perpendicular to those in (*a*) and (*b*). (*Note*. The unit vectors for $\boldsymbol{r}$ are horizontal south, horizontal west, and vertically downwards.) (S.M.P.)

5 A bead is threaded on a smooth wire in the shape of a cycloid $x = a(\theta - \sin\theta), y = a(1 + \cos\theta)$, which is fixed in a vertical plane with the positive y-axis as the upward vertical. The bead is released from the position given by $\theta = 0$. Prove that, in the subsequent motion

(*i*) $d\theta/dt$ is constant;

(*ii*) the acceleration vector has constant magnitude. (O.C.)

11.7 Impulse and momentum

Suppose that a particle of mass m is pulled in a horizontal line by a constant force of magnitude F, and that in an interval of time t its speed changes from v_0 to v_1 with an acceleration of magnitude a.

Since the acceleration is uniform,

$$v_1 - v_0 = at.$$

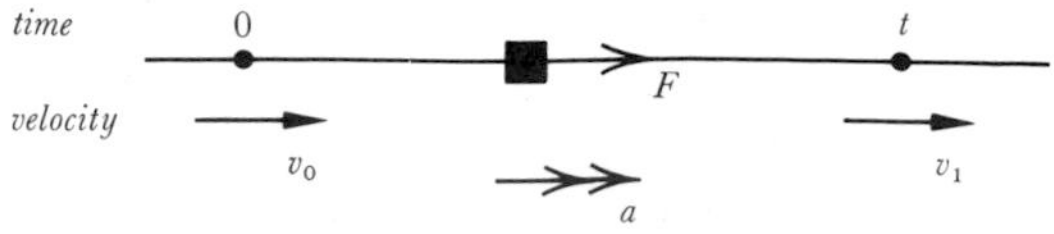

time 0 t

F

velocity

v_0 v_1

a

Also, by Newton's second law,

$F = ma.$

Hence $Ft = mv_1 - mv_0$.

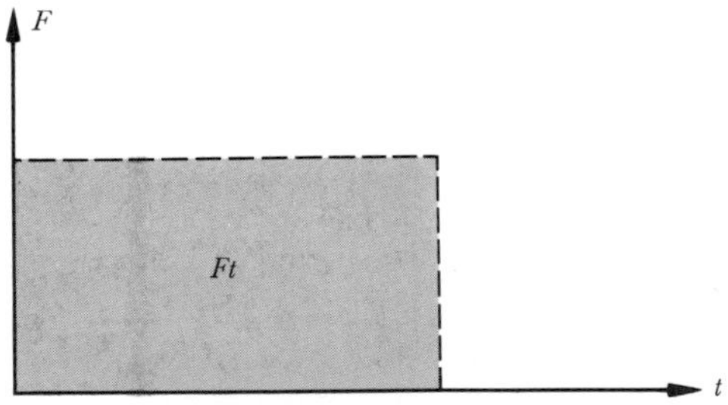

The expression Ft is called the *impulse of* F *during the interval* t; and if, further, the product of mass and velocity is called the *momentum*, we see that $mv_1 - mv_0$ is the increase in momentum.

So **Impulse of force during interval = Change of momentum**

In SI, the unit of impulse Ft is 1 N s (and, since 1 N = 1 kg m s^{-2}, is the same as 1 kg m s^{-1}, which is also the unit of momentum, mv).

Example 1

A stone of mass 2 kg is thrown vertically downwards from a high cliff with velocity 4 m s^{-1}. Find the impulse of its weight in the first two seconds of its fall and verify that this is equal to its increase in momentum.

Taking $g = 9.8$ m s^{-2}, the stone's weight is 19.6 N and the impulse of this weight is $19.6 \times 2 = 39.2$ N s.

Also, its velocity after 2 s $= 4 + 9.8 \times 2 = 23.6$ m s^{-1}.

So initial momentum $= 2 \times 4 = 8$ N s,

final momentum $= 2 \times 23.6 = 47.2$ N s,

and increase of momentum $= 39.2$ N s.

Impulse of a variable force

It frequently happens when a sudden blow is struck — for instance when a ball is hit by a cricket bat — that a very large force acts for a very small interval of time. In such cases it is often either impossible or unnecessary to investigate either the magnitude of the force or the duration of its action, and attention is confined to the impulse of the force and the consequent change of momentum. Moreover, the force will usually vary with time: the thrust of the bat on ball increases very rapidly from zero (when contact is first made at time t_0) to a maximum value, and then decreases just as rapidly to zero (when the ball finally leaves the bat at time t_1).

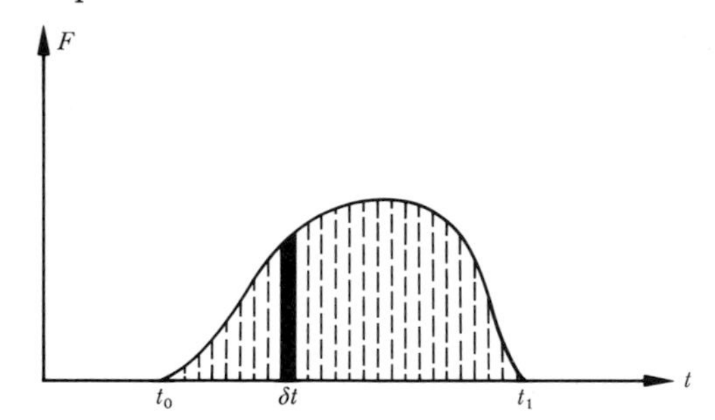

In such cases the total impulse during the interval will be the limit of the sum of small impulses $F\,\delta t$.

So $\quad$ impulse $= \lim \sum F\,\delta t = \int_{t_0}^{t_1} F\,\mathrm{d}t.$

Moreover, at any instant $\quad F = m\dfrac{\mathrm{d}v}{\mathrm{d}t}$

$$\Rightarrow \quad \int_{t_0}^{t_1} F\,\mathrm{d}t = \int_{t_0}^{t_1} m\frac{\mathrm{d}v}{\mathrm{d}t}\,\mathrm{d}t = [mv]_{v_0}^{v_1}.$$

So, again,

Impulse of force during interval = Change of momentum

Exercise 11.7a

1 Calculate the momentum (in N s) of:
(*i*) a car of mass 1 tonne travelling at 30 m s^{-1};
(*ii*) a lorry of mass 10 tonne travelling at 10 m s^{-1};
(*iii*) a pellet of mass 5 g moving at 150 m s^{-1};
(*iv*) an electron of mass 9×10^{-31} kg moving at 3×10^7 m s^{-1}.

2 Calculate (in SI units) the impulse of a force:
(*i*) 100 N acting for 3 s;
(*ii*) $6t$ N, from $t = 0$ to $t = 2$;
(*iii*) $4t^3$ N, from $t = 1$ to $t = 3$;
(*iv*) $2 \sin t$ N, from $t = 0$ to $t = \frac{1}{2}\pi$.

3 A car of mass 1 tonne moves from rest under the action of a tractive force and a resistance whose difference (the *effective force*) has a constant value of 400 N. Find:
(*i*) the impulse of the effective force in the first 5 s;
(*ii*) the resulting momentum;
(*iii*) the resulting velocity.

4 Repeat no. **3** if the effective force diminishes steadily from 400 N to zero.

5 Repeat no. **3** if the effective force after time t is $16(5 - t)^2$ N.

6 A lorry of mass 10 tonne is travelling at 30 m s^{-1} when its brakes are applied for a period of 2 s. If the braking force has the constant value 20 000 N, find:
(*i*) its impulse;
(*ii*) the final momentum;
(*iii*) the final velocity.

7 Repeat no. **6** if the braking force increases steadily to 20 000 N in the 2 s interval.

8 Repeat no. **6** if the braking force increases steadily from zero to 20 000 N in one second and then decreases steadily to zero after another second.

9 A ball of mass 1 kg strikes the cushion of a billiard table perpendicularly at a speed of 3 m s^{-1} and rebounds at 1 m s^{-1}. What is its change of momentum? If the total time of impact was 0.02 s and the force of the cushion is supposed constant, find the value of this force.

10 Repeat no. **9** on the different supposition that the force of the cushion increases steadily from zero and then decreases steadily to zero (all within 0.02 s). What is the maximum force?

11 A jet of water of radius 1 cm and speed 4 m s^{-1} plays directly on to a steel plate which completely destroys its momentum. Find:
(*i*) the loss of momentum of the water which strikes the plate in 1 s;
(*ii*) the force of the water on the plate.

12 A machine gun fires bullets of mass 30 g with muzzle velocity 500 m s^{-1} and at a rate of 200 per minute. What average force must the gunner exert to prevent recoil?

13 At a waterfall, a stream (which has no vertical velocity at the top of the waterfall) falls 5 m vertically on to a horizontal rock without rebounding. Given that 150 dm^3 of water pass over the waterfall each second, calculate the force exerted on the rock. (S.M.P.)

14 A horizontal jet of water delivering 5 kg s^{-1} at a speed of 10 m s^{-1} hits a vertical wall. Assuming that the water does not rebound, calculate the force on the wall. If allowance were made for rebound, would this force be increased or decreased? Give a reason for your answer. (C.)

15 A train of mass 1 000 tonne started from rest and the effective horizontal force was recorded at intervals of ten seconds:

t/s	0	10	20	30	40	50	60
$F/10^5$ N	1.62	1.45	1.36	1.24	1.07	0.91	0.73

Find the total impulse of this force and hence the final momentum and velocity of the train. (The use of Simpson's rule is recommended.)

Impulse and momentum as vectors

So far we have confined ourselves to the study of a body moving in a straight line. We can now consider the more general case of a body of mass m moving under the action of a variable force $\boldsymbol{F}$.

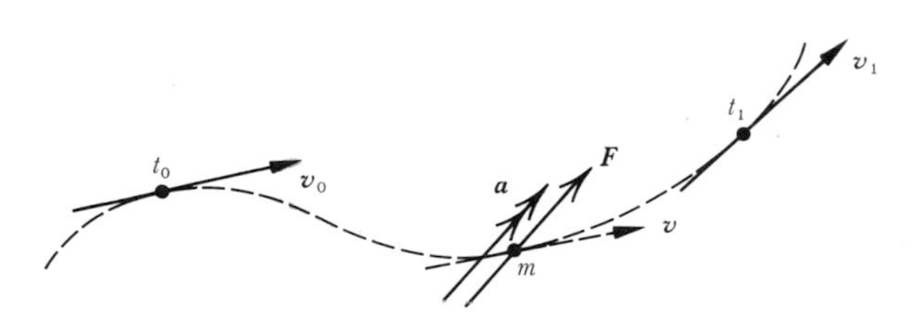

Suppose that after time t the particle has velocity $\boldsymbol{v}$ and acceleration $\boldsymbol{a}$; and that when

$$t = t_0, \qquad \boldsymbol{v} = \boldsymbol{v}_0;$$
$$t = t_1, \qquad \boldsymbol{v} = \boldsymbol{v}_1.$$

By Newton's second law,

$$\boldsymbol{F} = m\boldsymbol{a} = m\frac{\mathrm{d}\boldsymbol{v}}{\mathrm{d}t};$$

so that we naturally write

$$\int_{t_0}^{t_1} \boldsymbol{F}\,\mathrm{d}t = \int_{t_0}^{t_1} m\frac{\mathrm{d}\boldsymbol{v}}{\mathrm{d}t}\,\mathrm{d}t$$
$$= [m\boldsymbol{v}]_{t=t_0}^{t=t_1} = m\boldsymbol{v}_1 - m\boldsymbol{v}_0.$$

This is the first time that we have dared to speak of the integral of a vector quantity, but success with the differentiation of vectors is a strong incentive for such a step; and in any case, the expression

$$\int_{t_0}^{t_1} \boldsymbol{F}\,\mathrm{d}t$$

can be regarded as shorthand for the vector whose components are

$$\int_{t_0}^{t_1} X\,\mathrm{d}t, \qquad \int_{t_0}^{t_1} Y\,\mathrm{d}t, \qquad \int_{t_0}^{t_1} Z\,\mathrm{d}t$$

(where X, Y, Z are the components of $\boldsymbol{F}$).

We therefore see that the impulse of a force during an interval of time, like the momentum $m\boldsymbol{v}$, is essentially a vector quantity; and, again, that

Impulse of force during interval = Change of momentum

The 2 kg mass of Example 1 is now thrown horizontally with a speed of 4 m s^{-1}. Investigate the impulse of its weight and the change of its momentum in the first two seconds of its flight.

Taking unit vectors $\boldsymbol{i}$ horizontally and $\boldsymbol{j}$ vertically downwards, we see that

$$\text{weight} = 2 \times 9.8\boldsymbol{j} = 19.6\boldsymbol{j},$$
$$\Rightarrow \quad \text{impulse} = 19.6\boldsymbol{j} \times 2 = 39.2\boldsymbol{j}.$$

Also $\quad$ initial velocity $= 4\boldsymbol{i}$

and $\quad$ final velocity $= 4\boldsymbol{i} + 9.8 \times 2\boldsymbol{j}$
$\qquad = 4\boldsymbol{i} + 19.6\boldsymbol{j}$.

So $\quad$ initial momentum $= 4\boldsymbol{i} \times 2 = 8\boldsymbol{i}$

and $\quad$ final momentum $= (4\boldsymbol{i} + 19.6\boldsymbol{j}) \times 2$
$\qquad = 8\boldsymbol{i} + 39.2\boldsymbol{j}$.

Hence change in momentum $= 39.2\boldsymbol{j}$ and we confirm that this is equal to the impulse of the weight. This can also be summarised diagrammatically:

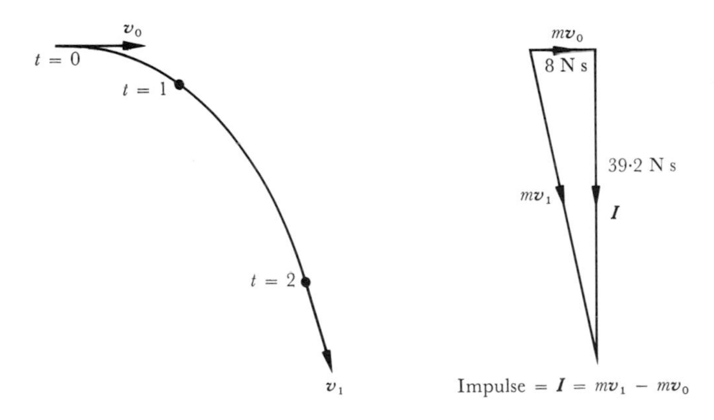

Example 3

Water flowing in a horizontal pipe of radius 2 cm at a speed of 3 m s^{-1} first goes round a sharp right-angled bend and then issues as a jet perpendicular to a fixed metal plate which destroys its momentum. Calculate:
(*i*) its thrust on the bend;
(*ii*) its thrust on the plate.

In 1 s the stretch of water coming out of the pipe has length 3 m and volume

$$\pi \times (0.02)^2 \times 3 = 12\pi \times 10^{-4} \text{ m}^3.$$

Now 1 m^3 of water has mass 10^3 kg. So the mass of this water is 1.2π kg, and its momentum is

$$1.2\pi \times 3 = 3.6\pi \text{ N s}.$$

Now after going round the bend its momentum changes direction through a right angle, and such a change in momentum must be caused by an impulse $\boldsymbol{I}$ such that

$$\boldsymbol{I} = m\boldsymbol{v}_1 - m\boldsymbol{v}_0.$$

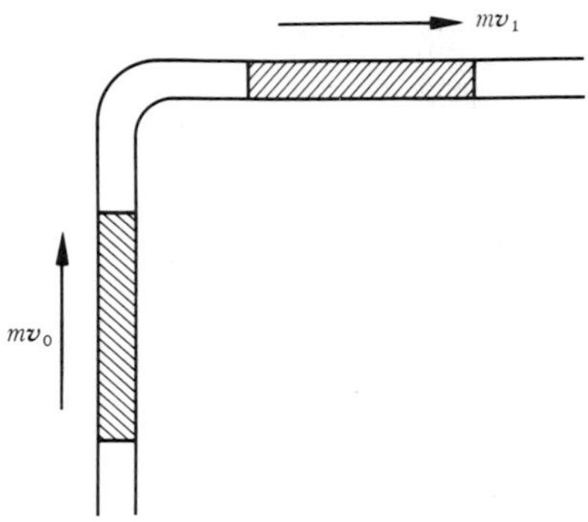

Hence the impulse $\boldsymbol{I}$ acts into the bend, at 45° to the two sections of pipe, and has magnitude

$$3.6\pi \times \sqrt{2} \approx 16 \text{ N s}.$$

But this is the impulse, over an interval of 1 s, of a force of 16 N.

The thrust of the water on the pipe, therefore, is of magnitude 16 N outwards from the bend at 135° with both sections.

When the jet strikes the metal plate, the momentum 3.6π N s is reduced to zero and this therefore necessitates an impulse of 3.6π N s which is caused by a thrust of 3.6π N acting for 1 second. So the thrust of the plate on the jet, and hence of the jet on the plate, has magnitude $3.6\pi \approx 11.4$ N.

Exercise 11.7b

1 Find (in SI) the impulse of:
(*i*) a force $6\boldsymbol{i} + 2\boldsymbol{j} + \boldsymbol{k}$, acting for 2 s;
(*ii*) a force $-\boldsymbol{i} - 3\boldsymbol{j} + 2\boldsymbol{k}$, acting for 3 s;
(*iii*) a force $t\boldsymbol{i} + (1 - t)\boldsymbol{j} + 3t^2\boldsymbol{k}$, from $t = 0$ to 2;
(*iv*) a force $\sin t\boldsymbol{i} + t\boldsymbol{j}$, from $t = 0$ to $\frac{1}{2}\pi$.

2 Find the final velocity of a mass of 2 kg which has initial velocity $4\boldsymbol{i} + 3\boldsymbol{j} + 2\boldsymbol{k}$ and is acted upon (on separate occasions) by the forces described in no. 1.

3 A body of mass 4 kg has initial velocity $4\boldsymbol{i} + 2\boldsymbol{j} - \boldsymbol{k}$. What constant force must act upon it so as to produce:
(*i*) velocity $3\boldsymbol{i} + 2\boldsymbol{j} + 4\boldsymbol{k}$ after 2 s;
(*ii*) velocity $\boldsymbol{i} + \boldsymbol{k}$ after 5 s.

4 A sudden downpour of r cm of rain, of density ρ kg m^{-3}, falls in one hour on an area of A km^2 and p per cent of this water runs straight off the ground into a river b m wide. What is the likely rise in level of the water in the river if this flows at v m s^{-1}.

A containing bank turns the river-flow through a right angle. What is the extra force on this bank due to the downpour? (M.E.I.)

Conservation of momentum

So far we have limited ourselves to the consideration of an impulse upon a single body and its consequent change of momentum. But it is when a number of bodies are connected or interacting that the concept of momentum becomes most valuable.

Suppose that A and B are two particles and that the only forces acting upon them are their mutual reactions $\boldsymbol{R}$ and $-\boldsymbol{R}$ which can (as when two billiard balls clash) be changing both in magnitude and direction.

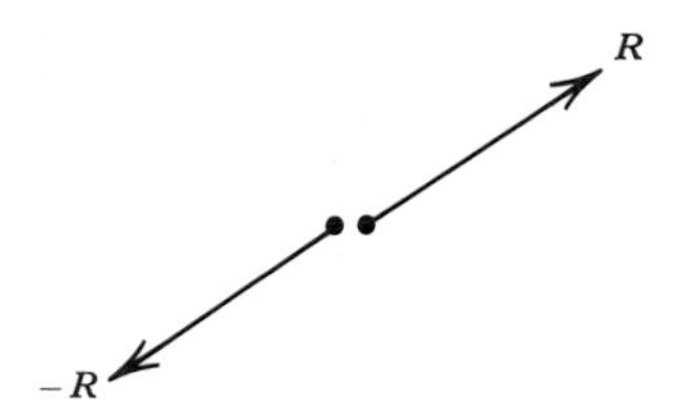

Over any interval of time,

$$\int_{t_0}^{t_1} \boldsymbol{R}\,\mathrm{d}t = \text{change in momentum of A},$$

$$\int_{t_0}^{t_1} -\boldsymbol{R}\,\mathrm{d}t = \text{change in momentum of B}.$$

$$\text{But} \quad \int_{t_0}^{t_1} \boldsymbol{R}\,\mathrm{d}t + \int_{t_0}^{t_1} -\boldsymbol{R}\,\mathrm{d}t = \int_{t_0}^{t_1} \boldsymbol{0}\,\mathrm{d}t = \boldsymbol{0}.$$

So the total change in momentum is zero, and we have verified the *principle of conservation of momentum* for a system which has no external forces acting upon it.

Example 4

A mass of 2 kg moving with velocity $\boldsymbol{i} - 2\boldsymbol{j} + 3\boldsymbol{k}$ coalesces with a mass of 8 kg which is moving with velocity $-\boldsymbol{i} + 3\boldsymbol{j} + 6\boldsymbol{k}$. Find the velocity of the combined mass of 10 kg.

If we let this final velocity be $\boldsymbol{v}$, we can apply the principle of conservation of energy and obtain:

$$\begin{aligned} 10\boldsymbol{v} &= 2(\boldsymbol{i} - 2\boldsymbol{j} + 3\boldsymbol{k}) + 8(-\boldsymbol{i} + 3\boldsymbol{j} + 6\boldsymbol{k}) \\ &= -6\boldsymbol{i} + 20\boldsymbol{j} + 54\boldsymbol{k} \\ \Rightarrow \quad \boldsymbol{v} &= -0.6\boldsymbol{i} + 2\boldsymbol{j} + 5.4\boldsymbol{k}. \end{aligned}$$

Example 5

A shell of mass m is fired with velocity $\boldsymbol{v}$ at an angle α to the horizontal. If the gun has mass M and recoils horizontally on smooth rails, find its speed of recoil, V.

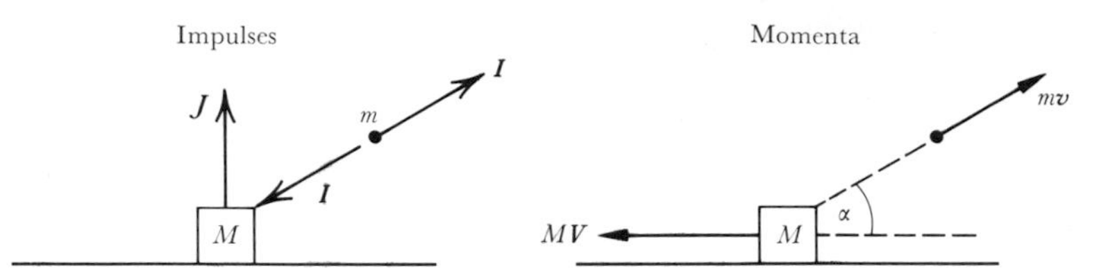

The forces acting for the short interval of the firing are:
(*i*) the weights $M\boldsymbol{g}$ and $m\boldsymbol{g}$, whose impulses over a very short interval are negligible;
(*ii*) the very large explosive force between gun and shell which has equal and opposite impulses $\boldsymbol{I}$ and $-\boldsymbol{I}$ on the gun and the shell;
(*iii*) the normal reaction between the rails and the gun which exerts a vertical impulse $\boldsymbol{J}$ on the gun.

So the only external impulse is $\boldsymbol{J}$, which has no horizontal component. Hence the horizontal component of momentum must be conserved.

$$\Rightarrow \quad mv \cos \alpha - MV = 0$$

$$\Rightarrow \quad V = \frac{mv \cos \alpha}{M}.$$

Exercise 11.7c

1 Find the velocity of the combined mass formed by the coalescing of:
(*i*) 1 kg moving at $4\boldsymbol{i} + 3\boldsymbol{j} - \boldsymbol{k}$,
and 2 kg moving at $\boldsymbol{i} + 3\boldsymbol{j} + 2\boldsymbol{k}$;

(*ii*) 3 kg moving at $\boldsymbol{i} + \boldsymbol{j} + \boldsymbol{k}$,
2 kg moving at $2\boldsymbol{i} + 3\boldsymbol{j} - 3\boldsymbol{k}$,
1 kg moving at $3\boldsymbol{i} + 2\boldsymbol{j} - 3\boldsymbol{k}$.

2 A mass of 2 kg moving with velocity $\begin{pmatrix} 3 \\ 5 \end{pmatrix}$ m s^{-1} collides and combines with a mass of 3 kg moving with velocity $\begin{pmatrix} 1 \\ -2 \end{pmatrix}$ m s^{-1}.

Calculate the velocity of the single mass so formed.
Find also the impulse which each mass has received. (S.M.P.)

3 A rocket of mass 10 kg is moving with a velocity whose three components (in m s^{-1}) are (200, 160, 4) and then separates into two parts. If the rear portion has mass 8 kg and velocity (100, 100, 0), find the new velocity of its front portion.

4 A bullet of mass 0.05 kg moving horizontally with velocity 321 m s^{-1} strikes a stationary block of mass 16 kg which is free to slide without rotation on a smooth horizontal plane. The bullet becomes embedded in the block after 0.01 s. Calculate, in newtons, the resistance, assumed uniform, of the block to penetration by the bullet.

5 A rocket of mass 40 tonnes ($= 4 \times 10^4$ kg) is mounted rigidly on a trolley of mass 10 tonne which is free to move horizontally. The rocket ejects mass 1 tonne horizontally in a burst lasting 5 seconds, giving the ejected matter a speed which may be taken as 2 km s^{-1} relative to the ground. Calculate the velocity of the rocket just after the ejection (*i*) neglecting resistances, and (*ii*) assuming resistance of 4×10^4 N.

If the same operation were carried out with the rocket mounted vertically and free to lift off its mounting show that, for a lift-off to occur, the duration of the burst (assumed uniform) must not exceed a certain critical time, and calculate this time. (The question of resistance does not arise in this calculation. Explain why.) (Take $g = 10$ m s^{-2}.) (M.E.I.)

6 Two stones, each of mass $5m$, are moving across a sheet of smooth ice at equal speeds of $10v$ in opposite directions on parallel paths, so that no collision is involved. A frog of mass m, travelling on one of the stones, leaps across to the other one, and in so doing deflects the stone he leaves through 30° and changes its speed to $8v$.

Find, by drawing and measurement or by calculation, (*i*) through what angle the other stone is deflected and its subsequent speed; (*ii*) the (vector) impulse the frog exerts on the stone on which he lands. (S.M.P.)

Newton's law of impact

Suppose that two bodies of mass m_1, m_2 collide and that the components of their velocities along the line of impact (i.e., along the common normal at the point of contact)

before the impact are u_1, u_2; (m_1 $\bullet\!\longrightarrow u_1$, m_2 $\bullet\!\longrightarrow u_2$)

and after the impact are v_1, v_2. (m_1 $\bullet\!\longrightarrow v_1$, m_2 $\bullet\!\longrightarrow v_2$)

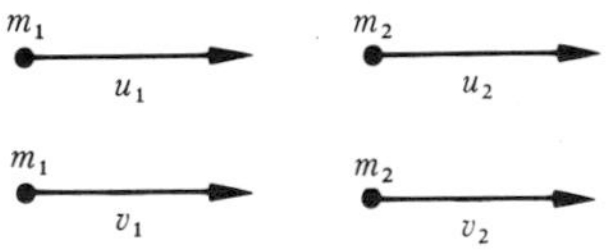

Then Newton's experimental law of impact states that

$$v_1 - v_2 = -e(u_1 - u_2),$$

where e is a constant for the pair of bodies known as their *resilience*, or *coefficient of restitution*.

Alternatively, this can be expressed as

$$v_2 - v_1 = e(u_1 - u_2),$$

that their relative velocity of separation is proportional to their relative velocity of approach.

If the second body is at rest, for instance when it is a fixed hard floor on to which the first body is dropped, it follows that $v_1 = -eu_1$; and it is quickly seen that e can range from 0 (in the case of a piece of putty dropped on to the floor) to nearly 1 (for a ball with a very good bounce, which is said to be almost *perfectly elastic*).

So $0 \leqslant e \leqslant 1$.

Example 6

A ball is dropped from a height of 5 m on to a hard horizontal pavement. If $e = \frac{1}{2}$, calculate:

(*i*) the height to which it returns after its first bounce;

(*ii*) the total distance travelled after repeated bounces;

(*iii*) the total time taken.

(Take $g = 10 \text{ m s}^{-2}$.)

(*i*) After falling from a height of 5 m the speed of the ball is

$$\sqrt{(2 \times 10 \times 5)} = 10 \text{ m s}^{-1};$$

and since $e = \frac{1}{2}$, it rebounds with a speed of $\frac{1}{2} \times 10 = 5 \text{ m s}^{-1}$.

Hence it will rebound to a height h, where

$$5^2 = 2 \times 10 \times h \quad \Rightarrow \quad h = 1.25 \text{ m};$$

(*ii*) Just as the height of the first rebound is $\frac{1}{4} \times 5$ m, so, similarly, the height of its second rebound is

$$\tfrac{1}{4} \times (\tfrac{1}{4} \times 5) = (\tfrac{1}{4})^2 \times 5 \text{ m}$$

and the height of its third rebound is

$$\tfrac{1}{4} \times (\tfrac{1}{4})^2 \times 5 = (\tfrac{1}{4})^3 \times 5 \text{ m, etc.}$$

Hence its total distance travelled (up and down) in making an infinite number of bounces

$$= 5 + 2\{\tfrac{1}{4} \times 5 + (\tfrac{1}{4})^2 \times 5 + (\tfrac{1}{4})^3 \times 5 + \cdots\}$$
$$= 5 + 2 \times \tfrac{1}{4} \times 5\{1 + \tfrac{1}{4} + (\tfrac{1}{4})^2 + \cdots\}$$
$$= 5 + \tfrac{5}{2}\frac{1}{1 - \frac{1}{4}} = 5 + \tfrac{5}{2} \times \tfrac{4}{3} = 8\tfrac{1}{3} \text{ m.}$$

(*iii*) The time taken to drop from 5 m $= 1$ s;

the time taken between first and second bounces $= 2 \times \frac{5}{10} = 1$ s;

the time taken between second and third bounces $= 2 \times \dfrac{\frac{5}{2}}{10} = \frac{1}{2}$ s; etc.

$$\text{Hence total time taken} = 1 + (1 + \tfrac{1}{2} + \tfrac{1}{4} + \cdots)$$
$$= 1 + 2 = 3 \text{ s.}$$

So the ball comes to rest (after an infinite number of bounces) after 3 s and travelling a total distance $8\frac{1}{3}$ m.

Example 7

A ball of mass 2 kg is moving with velocity 3 m s^{-1} and collides directly with another of mass 1 kg which is moving in the opposite direction with velocity 5 m s^{-1}. If $e = \frac{3}{4}$, find their velocities after impact.

Before impact: 2 →3   5← 1

After impact: 2 → v_1 1 → v_2

Let the final velocities be v_1, v_2 respectively. Then as there is no *external* force acting on the system, momentum is conserved:

$$2v_1 + v_2 = 2 \times 3 + 1 \times -5$$

$$\Rightarrow \quad 2v_1 + v_2 = 1.$$

Also, by Newton's law of impact,

$$v_2 - v_1 = \tfrac{3}{4} \times (5 + 3)$$

$$\Rightarrow \quad v_2 - v_1 = 6.$$

So $\left.\begin{aligned} 2v_1 + v_2 &= 1 \\ -v_1 + v_2 &= 6 \end{aligned}\right\}$ and

$$\Rightarrow \quad v_1 = -\tfrac{5}{3}, \qquad v_2 = \tfrac{13}{3}.$$

Hence the final velocities of the two balls are $-1\tfrac{2}{3}$ m s^{-1} and $4\tfrac{1}{3}$ m s^{-1}.

Exercise 11.7d

1 A ball of mass 4 kg moving at 3 m s^{-1} catches up with another ball of mass 2 kg moving at 1 m s^{-1} in the same straight line. If $e = \tfrac{3}{4}$, find their velocities after impact.

2 A ball of mass 5 kg moving at 4 m s^{-1} catches up with another ball of mass 2 kg moving at 3 m s^{-1} in the same straight line. If the second ball has velocity 4 m s^{-1} after impact find the coefficient of restitution and the final velocity of the first ball.

3 A ball of mass 4 kg moving with speed 2 m s^{-1} meets a ball of mass 2 kg moving in the opposite direction with speed 3 m s^{-1}. If $e = \tfrac{1}{2}$, find the final velocities of the two balls.

4 A ball of mass 3 kg moving at 5 m s^{-1} meets another ball of mass 6 kg moving in the opposite direction at 3 m s^{-1}. If the impact brings the second ball to rest, find the final velocity of the first ball and the coefficient of restitution.

5 A smooth sphere A of mass 1 kg moves with speed 5 m s^{-1} on a smooth horizontal plane directly towards a vertical wall. Before it hits the wall it is struck from behind by a smooth sphere B with the same radius as A but of mass 2 kg moving towards the wall along the same line as A with speed 10 m s^{-1}. Given that the coefficient of restitution between the spheres is $\tfrac{1}{2}$, find their new velocities.

The sphere A later strikes the wall and rebounds, the coefficient of restitution being $\frac{1}{4}$. It then collides again with B which has been following it towards the wall. Show that, after this collision, both A and B are again moving towards the wall, the speed of A being three times that of B. (M.E.I.)

6 Three identical imperfectly elastic smooth spheres A, B, C are at rest in a straight line, but not in contact, on a smooth horizontal plane in the order named. A is given a velocity towards B. Prove that there will be at least three collisions during the subsequent motion.

If the coefficient of restitution between each pair of spheres is $\frac{1}{2}$, determine whether there is a fourth collision. (L.)

7 The centres of three equal spheres A, B, C, at rest on a smooth horizontal table, lie on a straight line. The coefficient of restitution between A and B is e and between B and C is e'. The sphere A is projected to strike B directly and B then strikes C. Show that there are no further collisions if $e' < (3e - 1)/(1 + e)$. (W.)

8 A 'Supaball' when dropped on a hard floor displays a coefficient of restitution, e, of very nearly unity, and e may be assumed to be independent of the velocity of approach. When the ball is dropped vertically from a height of 2 m on to a hard floor it comes finally to rest 26.0 s after being released. What is e? What is the total distance that the ball has travelled? [g = 9.8 m s^{-2}]. (C.S.)

11.8 Work, energy, and power

Movement in a straight line

Suppose that a mass m is pulled along a smooth horizontal table by a constant force of magnitude F and that after moving a distance s its velocity has increased from v_0 to v_1.

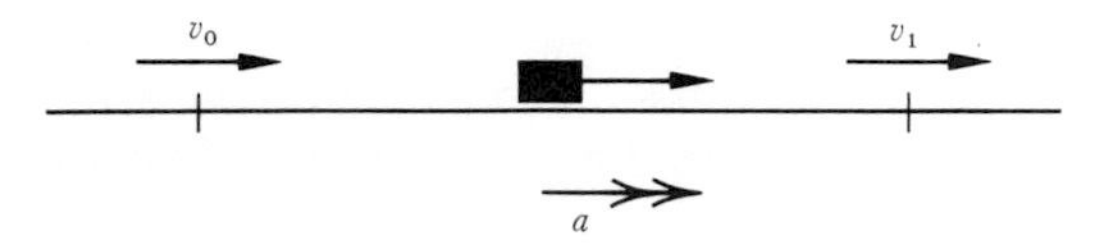

If the magnitude of the constant acceleration of m is a,

then $v_1^2 = v_0^2 + 2as$.

But, by Newton's second law,

$$F = ma$$

$$\Rightarrow \quad Fs = mas = \tfrac{1}{2}mv_1^2 - \tfrac{1}{2}mv_0^2.$$

The expression Fs is called the *work done by F in the displacement s*; and if, further, the expression $\frac{1}{2}mv^2$ is called the *kinetic energy* of a mass m moving with velocity v, we see that $\frac{1}{2}mv_1^2 - \frac{1}{2}mv_0^2$ is its increase in kinetic energy.

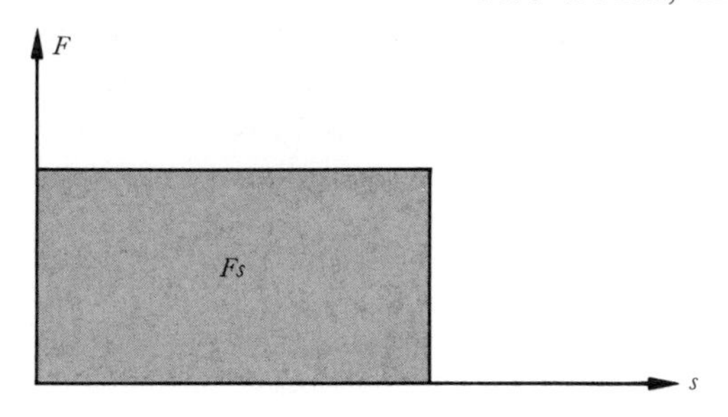

So

Work done by the force in the displacement = Increase of kinetic energy

In SI, the unit of work (and therefore of energy) will be 1 N m, which is called 1 joule (1 J).

Example 1

A car of mass 800 kg on a horizontal road accelerates from a speed of 10 m s^{-1} under the action of a tractive force of 500 N and against a resistance of 100 N. Find its speed when it has travelled a distance of 20 m.

Work done by tractive force $= 500 \times 20 = 10\,000$ J.

Work done by resistance $= -100 \times 20 = -2\,000$ J.

Hence total work $= 8\,000$ J.

If the final speed $= v$, the gain in kinetic energy

$$= \tfrac{1}{2} \times 800v^2 - \tfrac{1}{2} \times 800 \times 10^2$$
$$= 400(v^2 - 100).$$

So $8\,000 = 400(v^2 - 100)$

$\Rightarrow \quad v^2 - 100 = 20$

$\Rightarrow \quad v^2 = 120$

$\Rightarrow \quad v = 10.95.$

So final speed is 10.95 m s^{-1}.

Exercise 11.8a

1 A man pushes a chest of mass 20 kg through a distance of 6 m across a horizontal floor by means of a horizontal force of 50 N.

Find:

(*i*) the work done by this force;
(*ii*) the final velocity of the chest if the floor is smooth;
(*iii*) the final velocity if the coefficient of friction is 0.1.

2 Ignoring all resistances, what mean propulsive force is required to accelerate an aircraft of mass 10 tonne from rest to a take-off speed of 40 m s^{-1} on a runway of length 1 km? What force would be required if the take-off speed had to be doubled in the same length of runway?

3 A bullet of mass 20 g is travelling at a speed of 400 m s^{-1}. If it penetrates a block of wood to a depth of 5 cm, find the average resistance.

4 A stone of mass 2 kg is skidding across a sheet of ice at a speed of 10 m s^{-1}. If it comes to rest in 80 m, find:
(*i*) the work done by friction;
(*ii*) the coefficient of friction.

5 A lorry of mass 8 tonne needs to brake from a speed of 10 m s^{-1} in a distance of 80 m.

Find:
(*i*) the work done by the braking force;
(*ii*) the average braking force;
(*iii*) the average braking force required if this distance were to be halved;
(*iv*) the average braking force required if the initial speed were to be doubled.

6 Express a vehicle's braking distance d in terms of its speed v and maximum braking force F. Supposing that F is constant, sketch a graph showing d as a function of v.

7 A mass of 2 kg moving at 3 m s^{-1} catches up with a mass of 3 kg moving at 2 m s^{-1} in the same direction. Supposing that the bodies are perfectly inelastic ($e = 0$), find:
(*i*) their velocities after impact;
(*ii*) their total loss of kinetic energy.

8 Repeat no. **7** if the bodies are elastic, with $e = \frac{1}{2}$.

9 Repeat no. **7** if the bodies are perfectly elastic, with $e = 1$.

10 A nail of mass 10 g is driven 1 cm into a wall by a hammer of mass 100 g which strikes it with a speed of 2 m s^{-1}. Calculate the average resistance of the wall on the assumption:
(*i*) that there is no resilience between hammer and nail ($e = 0$);
(*ii*) that there is perfect resilience ($e = 1$).

Variable forces

When F is a variable force, the total work done in a displacement s will be the limit of the sum of small amounts of work $F \times \delta s$ and is represented by the area under the F, s graph:

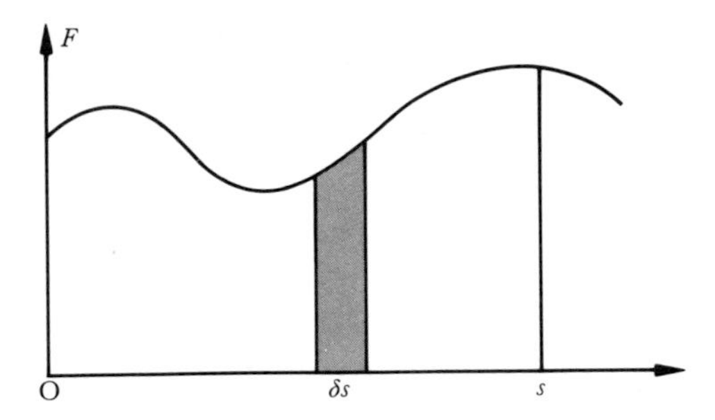

$$\text{Work} = \lim \sum F\,\delta s = \int_0^s F\,ds.$$

Elastic forces: Hooke's law

One of the commonest instances of a variable force is provided by a wire spring or an elastic string. It was first shown by Robert Hooke (1635–1703) that if a spring is extended from its natural length but remains within its elastic limits, the tension T is proportional to the extension x:

$$T = kx,$$

where k is a constant of the particular spring, known as its *stiffness*.†

(If, in the case of a spring, x is negative, then so is T: the spring is under compression and is exerting a negative tension, or *thrust*.)

When such a spring is extended, we see that T and x are in opposite directions, so the work done *by the tension* $T = -kx$ during this extension is

$$\int_0^x -kx\,dx = -[\tfrac{1}{2}kx^2]_0^x = -\tfrac{1}{2}kx^2,$$

whose magnitude can also be found from the force, distance graph:

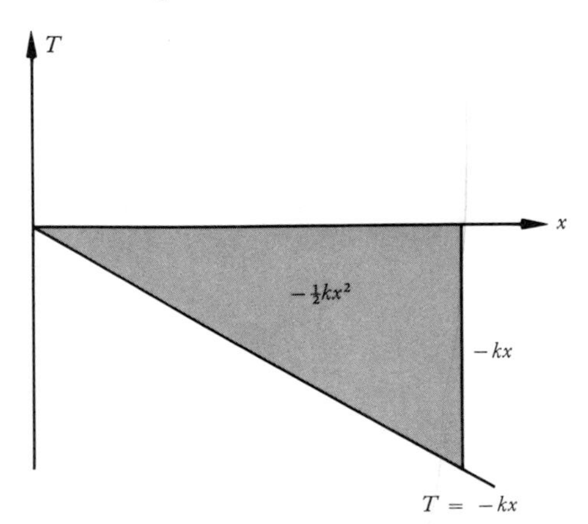

† If the natural length of the spring (or string) is l, then k is sometimes written as λ/l, where λ is called its *modulus of elasticity*. Hence $T = \lambda x/l$, and the corresponding elastic energy is $\lambda x^2/2l$.

Potential energy

Finally, we note that if our loaded elastic spring with extension x were to return to its natural length, the tension would clearly do a positive amount of work $\frac{1}{2}kx^2$. This, therefore, is called the *elastic potential energy* of the spring, being its stored capacity for doing work.

Similarly, if a mass m is at a height h above a given base-level, its weight mg would do an amount of work mgh if it were to return to its original level. This, therefore, is called its *gravitational potential energy* (referred to the given base-level) when at height h.

We therefore see that, for any displacement, the loss in potential energy (p.e.) is equal to the work done by the weight and so is equal to the gain in kinetic energy (k.e.):

loss in p.e. = gain in k.e.

Example 2

An elastic spring of unstretched length 1 m hangs from the ceiling of a room whose height is 3 m. When a 1 kg mass is attached, the spring extends to a length 2 m, and it is then pulled down further, so that the mass touches the floor. If it is then released, find its velocity at heights 1 m, 2 m above the floor, and calculate the kinetic energy, gravitational potential energy, and elastic potential energy in all three positions.

Let the stiffness of the spring be k. Then, taking $g = 10 \text{ m s}^{-2}$, the weight of the mass will be 10 N

and $\quad k \times 1 = 10 \quad \Rightarrow \quad k = 10.$

If we measure depth x from the unstretched position, we can now investigate:

(*i*) *The displacement from* $x = 2$ *to* $x = 1$.

Work done by weight $= -10 \times 1 = -10$ J.

Also tension in spring $= 10x$.

$$\text{So work done by spring} = \int_2^1 -10x \, \mathrm{d}x = [5x^2]_1^2 = 15 \text{ J.}$$

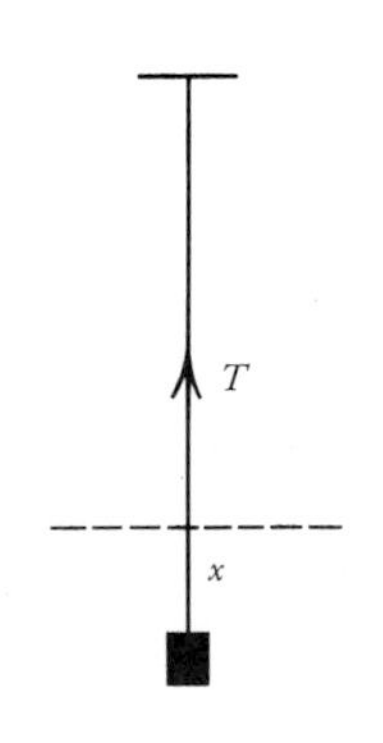

Furthermore, if velocity is v when $x = 1$, gain in k.e. $= \frac{1}{2} \times 1 \times v^2$.

$$\text{So} \quad -10 + 15 = \tfrac{1}{2}v^2$$
$$\Rightarrow \quad v^2 = 10$$
$$\Rightarrow \quad v = \sqrt{10} = 3.2 \text{ m s}^{-1}.$$

(*ii*) *The displacement from* $x = 2$ *to* $x = 0$

Work done by weight $= -10 \times 2 = -20$ J.

$$\text{Work done by spring} = \int_2^0 -10x \, dx$$
$$= [5x^2]_0^2 = 20 \text{ J}.$$

So if velocity is V when $x = 1$, gain in k.e. $= \frac{1}{2} \times 1 \times V^2 = \frac{1}{2}V^2$.

So $\quad -20 + 20 = \frac{1}{2}V^2$

$\Rightarrow \quad V = 0.$

Finally, if we measure gravitational p.e. from floor level and elastic p.e. from the unstretched position (so that elastic p.e. $= \frac{1}{2} \times kx^2 = 5x^2$), we can express the three forms of energy in the following table:

Depth from ceiling/m	*Gravitational p.e.*/J	*Elastic p.e.*/J	*k.e.*/J
1 ($x = 0$)	20	0	0
2 ($x = 1$)	10	5	5
3 ($x = 2$)	0	20	0

So we note that at each level the sum of the three forms of energy is 20 J.

Hence, in both our examples, the total energy of each system remains constant. More generally, this is true in any system, *provided that the work done by its forces to achieve any particular position is independent of the route by which this position is achieved.*

The general result is then known as the *principle of conservation of energy.*

It is, however, necessary to issue a warning that not all systems are as well-behaved (or '*conservative*') as these. If, for instance, we consider a mass being moved across a rough table from a fixed starting point to a particular position, it is clear that the work done by friction is entirely dependent on the route taken. So friction cannot have a potential and is a *non-conservative* force, so that it is impossible to have a principle of conservation of mechanical energy in this case. We shall meet many similar instances, such as the impact of particles which are not perfectly elastic, when the total mechanical energy of the system is seen to diminish, though this is accompanied by the production of other forms of energy, such as heat or sound which conserve the overall energy of the system.

Exercise 11.8b

1 If a spring has stiffness 200 N m^{-1}, find:

(*i*) the tension needed to extend it by 4 cm;
(*ii*) the force required to compress it by 5 mm;
(*iii*) its extension when pulled out by a force of 40 N;
(*iv*) its compression when pushed in by a force of 10 N.

2 In each case of no. **1**, find the elastic energy stored in the spring.

3 An elastic string is extended 2 mm by a tension of 4 N. Find:
(*i*) its stiffness;
(*ii*) its elastic energy.

4 A spring is compressed a distance of 1 cm by a force of 500 N. Find:
(*i*) its stiffness;
(*ii*) its elastic energy.

5 An elastic string has natural length 2 m and stiffness 50 N m^{-1}. Find:
(*i*) how far it stretches when a mass 20 kg hangs from it;
(*ii*) its elastic energy in this position;
(*iii*) the work done by its weight if the 20 kg was attached to the end of the string when unstretched and gently lowered into its equilibrium position. (Take $g \approx 10$ m s^{-2}.)

6 The particle of mass 20 kg in no. **5** is attached to the string and dropped vertically from its point of suspension. When the string has become taut and is extended by a further amount x, find:
(*i*) the loss in p.e. of the weight;
(*ii*) the elastic energy of the string;
(*iii*) the k.e. of the particle.
Hence find the maximum extension of the string. (Take $g \approx 10$ m s^{-2}.)

7 A nylon climbing rope of length 40 m is stretched by a force F. The relation between the force F and the corresponding extension x is given in the following table:

x/m	0	1	2	3	4	5	6	7	8	9	10
F/N	0	1 600	2 400	3 000	3 600	4 100	4 600	5 200	5 700	6 400	7 900

Use Simpson's rule to estimate the energy stored in the rope when it is extended by 10 m.

A climber of mass 64 kg is climbing on a vertical rock-face h m above the point to which he is securely attached by the rope of length 40 m described above. The climber slips and falls freely, being brought to rest for the first time by the rope when it has been extended by 10 m.

Calculate the value of h to the nearest metre.

Determine also the velocity of the climber at the moment the rope becomes taut, and hence, or otherwise, calculate the average deceleration of the climber from that moment until he comes to rest. (Take $g = 9.8$ m s^{-2}.)

(J.M.B.)

8 If the radius of the Earth is R, the gravitational attraction on a body of mass 1 kg at a distance x from the Earth's centre is

$$\frac{gR^2}{x^2} \quad \text{if } x \geqslant R; \qquad \frac{gx}{R} \quad \text{if } x \leqslant R.$$

Hence find the work done by this force (or *weight*):
(*i*) if the body came to the Earth's surface from a very great distance;
(*ii*) if the body could move from the Earth's surface in to its centre.

9 From no. **8** find:
(*i*) the minimum speed with which a body from outer space would approach the Earth;
(*ii*) the speed with which a body would reach the centre of the Earth if it could be dropped from its surface down a vertical shaft. (Take $g \approx 10\,\text{m}\,\text{s}^{-2}$, $R = 6.4 \times 10^6$ m.)

Work as a scalar product

So far, we have restricted ourselves to the work done by a force when its displacement is in its own direction, but we are now able to examine the more general case.

If a force $\boldsymbol{F}$ undergoes a displacement $\boldsymbol{s}$, the work done is defined as the scalar product $\boldsymbol{F}.\boldsymbol{s}$. We therefore see that work is a scalar of magnitude $Fs \cos \theta$ (where θ is the angle between $\boldsymbol{F}$ and $\boldsymbol{s}$) and so can be regarded as

$F \times$ component of $\boldsymbol{s}$ along $\boldsymbol{F}$

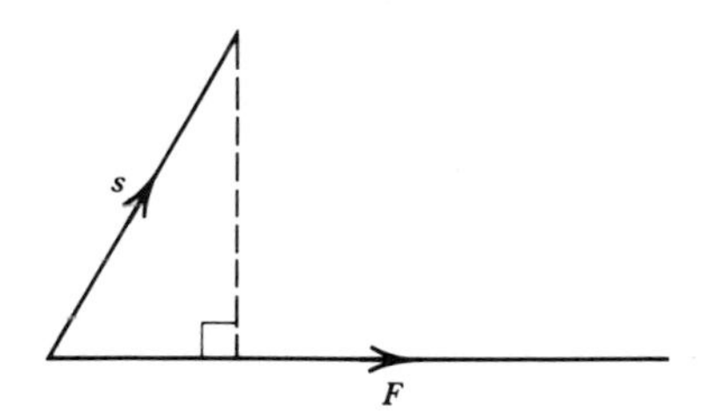

or $s \times$ component of $\boldsymbol{F}$ along $\boldsymbol{s}$.

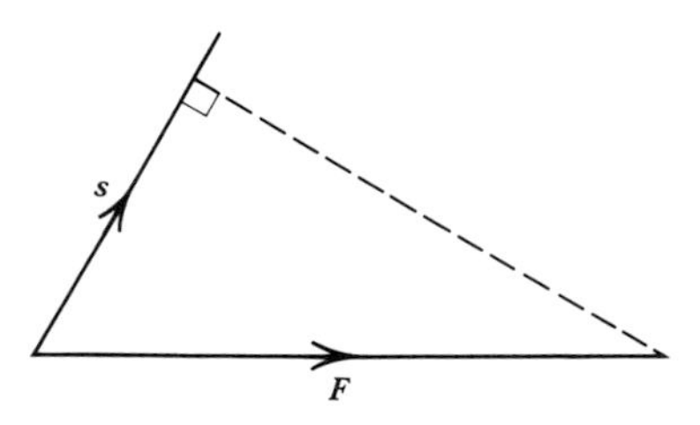

In particular:
(*i*) when $\theta = 0$, work $= +Fs$;

s F

(*ii*) when $\theta = \pi$, work $= -Fs$;

(*iii*) when $\theta = \frac{1}{2}\pi$, work $= 0$.

Hence the work done in any displacement perpendicular to the force is always zero. In particular, the work done by a normal reaction at a smooth surface is always zero.

Moreover, we have already seen (in 9.5) that the operation of scalar multiplication is distributive with respect to addition.

Hence $\boldsymbol{F}_1 . \boldsymbol{s} + \boldsymbol{F}_2 . \boldsymbol{s} = (\boldsymbol{F}_1 + \boldsymbol{F}_2) . \boldsymbol{s}$ and the sum of the amounts of work done by two forces is equal to the work done by their resultant.

Similarly, $\boldsymbol{F} . \boldsymbol{s}_1 + \boldsymbol{F} . \boldsymbol{s}_2 = \boldsymbol{F} . (\boldsymbol{s}_1 + \boldsymbol{s}_2)$ and the sum of the amounts of work done by a force in two displacements is equal to the work done by the force in the resultant displacement.

Furthermore, if $\boldsymbol{F}$ is a variable force, the work done in a small displacement $\delta \boldsymbol{s}$ will be δW, where

$$\delta W = \boldsymbol{F} . \delta \boldsymbol{s}.$$

Hence the total work can be written as

$$W = \lim \sum \boldsymbol{F} . \delta \boldsymbol{s} = \int \boldsymbol{F} . \mathrm{d}\boldsymbol{s}.$$

Alternatively, if the velocity and acceleration of the particle are $\boldsymbol{v}$, $\boldsymbol{a}$ respectively, we can write

$$\frac{\delta W}{\delta t} = \boldsymbol{F} . \frac{\delta \boldsymbol{s}}{\delta t} \approx \boldsymbol{F} . \boldsymbol{v}$$

$$\Rightarrow \quad \frac{\mathrm{d}W}{\mathrm{d}t} = \boldsymbol{F} . \boldsymbol{v}.$$

But, by Newton's second law,

$$\boldsymbol{F} = m\boldsymbol{a} = m \frac{\mathrm{d}\boldsymbol{v}}{\mathrm{d}t}$$

$$\Rightarrow \quad \frac{\mathrm{d}W}{\mathrm{d}t} = m \frac{\mathrm{d}\boldsymbol{v}}{\mathrm{d}t} . \boldsymbol{v} = \frac{\mathrm{d}}{\mathrm{d}t} (\tfrac{1}{2} m v^2)$$

$$\Rightarrow \quad W = [\tfrac{1}{2} m v^2]$$

$$\Rightarrow \quad \boxed{\int \boldsymbol{F} . \mathrm{d}\boldsymbol{s} = [\tfrac{1}{2} m v^2]}$$

and again we see that the work done by a force is equal to the increase in kinetic energy.

Example 3

A particle of mass 6 kg slides from rest down a rough plane which is inclined at 30° to the horizontal. If the resistance to motion is 12 N throughout, find the work done by the various forces when the mass has moved a distance 8 m and hence find its velocity at this point. (Take $g \approx 10$ m s^{-2}.)

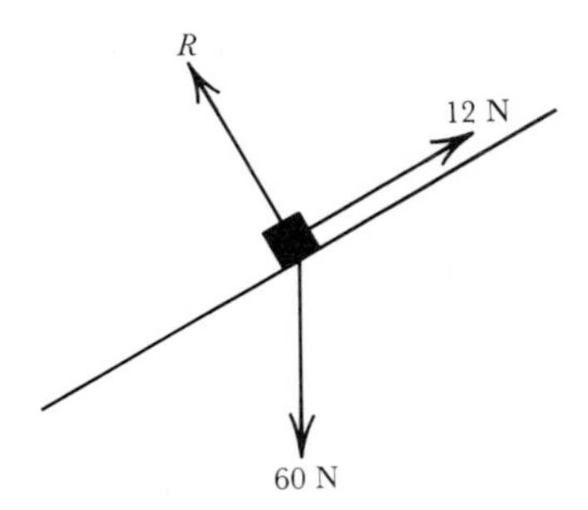

The three forces on the particle are:
(*i*) its weight $6 \times 10 = 60$ N;
(*ii*) the normal reaction R;
(*iii*) the resistance 12 N;
and the amounts of work done by these in the displacement 8 m are:
(*i*) $60 \times 8 \times \cos 60° = 240$ J;
(*ii*) zero (since force is perpendicular to displacement);
(*iii*) $-12 \times 8 = -96$ J.

Hence total work = 144 J.

But final k.e. $= \frac{1}{2} \times 6 \times v^2 = 3v^2$.

So $144 = 3v^2$

$\Rightarrow \quad v^2 = 48 \quad \Rightarrow \quad v = 6.9$ m s^{-1}.

Exercise 11.8c

1 Find the work done (in SI) when a force $\boldsymbol{i} + 3\boldsymbol{j} + \boldsymbol{k}$ acts on a particle and displaces it by an amount:
(*i*) $-\boldsymbol{i} + 2\boldsymbol{j} + 3\boldsymbol{k}$; (*ii*) $3\boldsymbol{i} - 2\boldsymbol{j} - \boldsymbol{k}$; (*iii*) $\boldsymbol{i} - \boldsymbol{k}$.

2 A car of mass 800 kg is travelling at 30 m s^{-1} with its engine switched off when it comes to a small hill. Find:
(*i*) its k.e. at the foot of the hill;
(*ii*) the maximum height of hill which the car can surmount (ignoring all resistances).

3 A boy slides down a smooth chute at a swimming pool. If its top is 6 m above the water and its bottom is 1 m above the water, find his speed:
(*i*) on leaving the chute;
(*ii*) on entering the water.

4 A box of mass 10 kg is attached to the end of a rope of length 5 m and hangs freely. If it is then given a horizontal impulse of 100 N s find:
(*i*) its initial speed;
(*ii*) its initial kinetic energy;
(*iii*) the height to which it will rise.

5 A marble of mass 20 g rests at the top of a smooth circular tube which has diameter 1 m and is in a vertical plane. If the marble is just displaced from rest, find its speed:
(*i*) when travelling vertically at the end of the horizontal diameter;
(*ii*) when passing through the lowest point of the tube;
(*iii*) when 25 cm above the lowest point.
Hence find the reactions of the tube on the marble at these three instants.

6 A boy of mass 40 kg swings on a rope of length 2 m which is just taut and attached to a point at the same height. When the rope is vertical, find:
(*i*) the kinetic energy of the boy;
(*ii*) the speed of the boy;
(*iii*) the tension in the rope.

7 A commando swings on a rope fixed at its upper end A. The rope, initially inclined at 60° to the downward vertical, remains taut and may be considered as inextensible and of negligible mass. The commando can be idealised as a particle at the end of the rope. If:
(*i*) his mass is 70 kg;
(*ii*) the rope is 10 m long;
(*iii*) he lets go when the rope is vertical;
(*iv*) A is 12 m above the ground; find his landing speed.
(Neglect air resistance, and take $g = 10 \text{ m s}^{-2}$.)

Which of (*i*), (*ii*), (*iii*), (*iv*) can be varied, one at a time, without affecting this speed? (S.M.P.)

8 A particle of mass m is slightly displaced from the top of a smooth hemispherical dome of radius a. If its speed is v when the radius to the particle makes an angle θ with the vertical, find:
(*i*) an equation connecting v with a, g and θ;
(*ii*) the normal reaction N in terms of a, g and θ;
(*iii*) the value of θ for which $N = 0$ (i.e., at which the particle leaves the surface).

Power

We know from everyday experience that when a certain task can be accomplished by either of two forces, it frequently happens that one of them can do the necessary work more quickly. Such a force is then said to have 'greater power'.

More precisely, we define the *power* P of a force as its *rate of doing work*: $P = \mathrm{d}W/\mathrm{d}t$.

But we have already seen that when a force $\boldsymbol{F}$ is moving its point of application with velocity $\boldsymbol{v}$,

$$\frac{dW}{dt} = \boldsymbol{F}.\boldsymbol{v}.$$

Hence $$\boxed{P = \frac{dW}{dt} = \boldsymbol{F}.\boldsymbol{v}}$$

So the unit of power in SI is 1 joule per second and is called 1 *watt* (W):

$$\boxed{1\ \text{W} = 1\ \text{J s}^{-1}}$$

Example 4

In British units the standard unit of power was 1 horse power (h.p.), being the rate of working required to lift a mass of 550 pounds through 1 foot in 1 second.

Taking 1 lb = 0.454 kg, 1 ft = 0.305 m, g = 9.81 m s^{-1}, find 1 h.p. in kilowatts.

In 1 s, a force working at a rate of 1 h.p. will lift 550 lb a height of 1 foot, i.e., 550 × 0.454 kg a height of 0.305 m.

So work done in 1 s

$= (550 \times 0.454 \times 9.81) \times 0.305$ J

$= 746$ J.

Hence 1 h.p. = 746 J s^{-1} = 746 W

$\Rightarrow$ 1 h.p. = 0.746 kW.

Example 5

Find the power of a pump which raises water from a lake at a rate of 5 m^3 s^{-1} and delivers it 6 m higher at a speed of 8 m s^{-1}.

In 1 s, the mass of water moved = 5×10^3 kg.

So weight of water moved = $5 \times 10^3 \times 9.81$ N and the work done in raising it through 6 m

$= 5 \times 10^3 \times 9.81 \times 6 = 2.94 \times 10^5$ J.

But this water is also given an amount of kinetic energy $= \frac{1}{2} \times (5 \times 10^3) \times 8^2 = 1.6 \times 10^5$ J.

So total work done in 1 s = 4.54×10^5 J and the power of the pump

$= 4.54 \times 10^5$ W

$= 454$ kW.

Exercise 11.8d

1 A pump delivers water at the rate of 900 kg min^{-1} from a reservoir to a nozzle 8 m above the surface of the reservoir. The water emerges from the nozzle with a speed of 12 m s^{-1}. Calculate the power, in kilowatts, of the pump. (C.)

2 A pump raises 1 000 kg of water each minute from a depth of 20 m and delivers it at a speed of 8 m s^{-1}. Assuming that the efficiency of the pump is 40%, calculate its power. (O.C.)

3 A railway train has mass 1.4×10^6 kg and the maximum power of the engine is 1.8×10^6 W. Calculate the maximum acceleration, in m s^{-2}, when the train is travelling on the level at a speed of 12 m s^{-1}, the total resistance being 9.6×10^4 N. (C.)

4 A car of power 50 kW has mass 800 kg, which is equally distributed over the four wheels. The coefficient of friction between the two driving wheels and the road is 0.3. If the road is level, and resistances to motion can be neglected, find the maximum acceleration of the car (in m s^{-2}):
(*i*) at 20 km h^{-1};
(*ii*) at 10 km h^{-1}. (O.C.)

5 An escalator takes passengers from a lower level to a higher level through a vertical height of 20 m. On average two passengers get on the escalator per second and two leave at the top per second. If the average mass of a passenger is taken to be 75 kg and the machinery has 60% efficiency, calculate, in watts, the average power output of the motor which drives the escalator. (Take g to be 9.8 m s^{-2}.) (C.)

6 A car of mass 1 000 kg is moving on a level road at a steady speed of 100 km h^{-1} with its engine working at 60 kW. Calculate in newtons the resistance to motion, which may be assumed to be constant.

The engine is now disconnected, the brakes applied, and the car comes to rest in 100 m. Find the additional retarding force arising from the application of the brakes.

If the engine is still disconnected, find the distance the car would run up a hill of inclination $\sin^{-1}\frac{1}{10}$ before coming to rest, starting at 100 km h^{-1}, when the same resistance and braking force are operating. (C.)

7 Sand pours over the lip of a chute and falls vertically onto a conveyor belt which is moving horizontally at a steady speed of 0.5 m s^{-1}. If the supply of sand is steady at the rate of 50 kg s^{-1}, what horizontal momentum is given to the sand each second? Find the force that is required to give the sand this momentum.

If 10 W is the power required to operate the conveyor belt without a load, find whether a motor of maximum power 25 W can deal with sand at the rate of supply of 50 kg s^{-1}.

If the chute were modified so that the sand was delivered with a horizontal component of velocity of 0.5 m s^{-1} in the direction of motion of the belt, find the maximum speed at which the belt could move the load, assuming that the power required for the belt itself is unaltered and the sand is delivered at the same rate as before. (M.E.I.)

8 A VTOL aircraft hovers by taking in air through a large intake and ejecting it at high velocity downwards. If the relative velocity of ejection is v m s^{-1} and the mass of the aircraft is M kg, find expressions for:
(*i*) the mass of air ejected per second for hovering;
(*ii*) the power required to hover.
Later, when the aircraft is flying as a normal aeroplane, the engines work at half this power and the air resistance varies as the square of the speed. The maximum steady speed attainable in level flights is V m s^{-1}. Show that at a speed of $\frac{3}{4}$ V m s^{-1} the engines would supply enough power for the aircraft to climb in normal flight at a rate of $37v/256$ m s^{-1}. (M.E.I.)

9 A car of mass m started along a horizontal road by an engine which works at a constant rate P. Ignoring all resistances to motion and letting the distance travelled, velocity and acceleration after time t be x, v, a respectively, find equations which connect:
(*i*) a, v; (*ii*) v, t; (*iii*) v, x; (*iv*) x, t.

10 An engine is fitted to a vehicle of mass m kg. It is known from tests that it will give constant tractive force F N up to a road-speed of U m s^{-1}, and at higher speeds will give constant power FU watt. Show that at speed v m s^{-1} $(v \geqslant U)$ the tractive force is FU/v N. Form the appropriate equations of motion for the vehicle if the resistance to motion at all speeds is given by kv^2 N, k being constant.
Show that there is a maximum attainable speed V m s^{-1} and find an expression for V in terms of k, U and F. Given that $m = 500$, $F = 2\ 500$, $U = 25$ and $V = 50$, determine the value of k. Calculate the acceleration at each of the speeds 5, 15, 25, 35 m s^{-1} and hence estimate the time taken to reach 40 m s^{-1} from rest, giving your result to the nearest second. Explain your method. (M.E.I.)

11.9 Oscillations

In 5.10 we spent a little time studying simple oscillations like alternating currents, together with their amplitudes, periods and frequencies. We can now take this investigation further, and shall begin by considering three other ways in which such oscillations can arise.

Example 1

If a spring of stiffness k hangs vertically from a fixed point, with its other end attached to a mass m, investigate its equation of motion at a point when it is pulled down a further distance x.

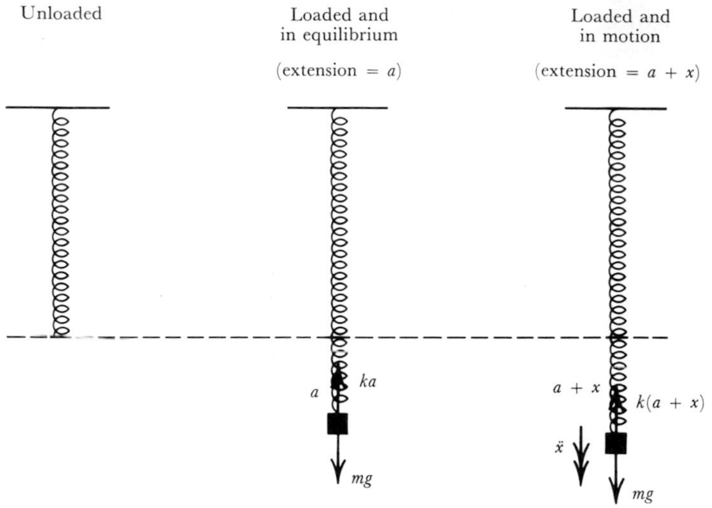

Unloaded

Loaded and in equilibrium (extension $= a$)

Loaded and in motion (extension $= a + x$)

a ka mg

$a + x$ $k(a + x)$ $\ddot{x}$ mg

By Hooke's law, when extension $= a$, the tension $= ka$. So, since the mass is then in equilibrium, $ka = mg$.

Also, when extension $= a + x$, tension $= k(a + x)$.

So the general equation of motion is

$$mg - k(a + x) = m\ddot{x}.$$

But $\quad ka = mg.$

So $\quad -kx = m\ddot{x}$

$$\Rightarrow \qquad \ddot{x} = -\frac{k}{m}x.$$

Hence the acceleration of the mass is proportional to its extra displacement and, of course, in the opposite direction: the tension in the spring acts as a *restoring force*, attempting to re-establish equilibrium.

Example 2

Investigate the equation of motion of a cylindrical canister buoy, bobbing up and down at sea.

We shall suppose that the buoy has constant cross-sectional area A and height h, and that in equilibrium it is submerged to a depth d. If the density of the buoy is ρ and of the sea is σ, we can investigate its motion when it is submerged by a further amount x.

In each case, both of equilibrium and of submersion by a further amount x, there are just two forces acting:

(i) the *weight* of the buoy.

Now $\quad$ its volume $= Ah$

so $\quad$ its mass $= Ah\rho$

and $\quad$ its weight $= Ah\rho g$;

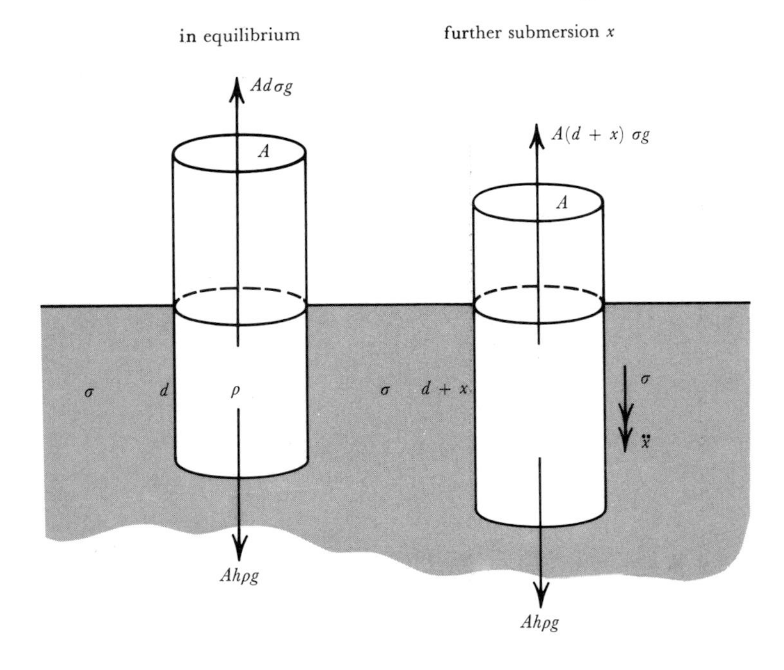

(*ii*) the *buoyancy*, or *upthrust* of water on the buoy, which Archimedes showed to be equal to the weight of the water displaced.

Now in the equilibrium position:

$$\begin{aligned}
&\text{volume of water displaced} = Ad \\
\Rightarrow\quad &\text{mass of water displaced} = Ad\sigma \\
\Rightarrow\quad &\text{weight of water displaced} = Ad\sigma g \\
\Rightarrow\quad &\text{buoyancy} = Ad\sigma g;
\end{aligned}$$

and after submersion by a further amount x, the buoyancy is $A(d + x)\sigma g$.

At the equilibrium position, we see that

$$\begin{aligned}
&Ad\sigma g = Ah\rho g \\
\Rightarrow\quad &d\sigma = h\rho;
\end{aligned}$$

and at the more general position, using Newton's second law,

$$\begin{aligned}
&Ah\rho g - A(d + x)\sigma g = Ah\rho\ddot{x} \\
\Rightarrow\quad &-Ax\sigma g = Ah\rho\ddot{x} \\
\Rightarrow\quad &\ddot{x} = -\frac{\sigma g}{\rho h}x.
\end{aligned}$$

So, again, we see that the acceleration of the buoy is proportional to its displacement from the equilibrium position and in the opposite direction.

As our final example, we now investigate an oscillation which, unlike the previous cases, is not along a straight line.

Example 3 (Simple pendulum)

A mass m is suspended from a fixed point by means of a light inextensible string of length l and swings to and fro in a vertical plane.

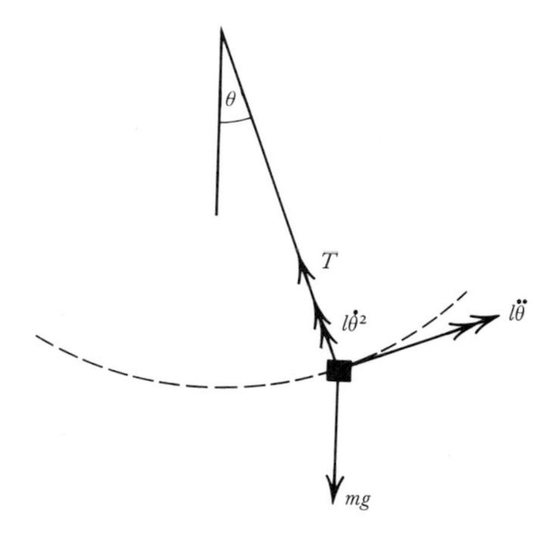

When the string makes an angle θ with the vertical, the acceleration of the mass has two components:

$l\ddot{\theta}$ perpendicular to the string,

and $l\dot{\theta}^2$ along the string.

Furthermore, there are just two forces acting:

(*i*) the weight mg,

and (*ii*) the tension T in the string.

Hence, by Newton's second law and resolving perpendicular to the string:

$$-mg \sin \theta = ml\ddot{\theta}$$

$$\Rightarrow \qquad \ddot{\theta} = -\frac{g}{l} \sin \theta.$$

In this case, therefore, $\ddot{\theta}$ is proportional not to θ, but to $\sin \theta$. However, if the string is swinging through a small angle, we know that $\sin \theta \approx \theta$.

Hence $\ddot{\theta} \approx -\frac{g}{l}\theta.$

Simple harmonic motion

In all three examples the final equation of motion was of the form

$$\ddot{x} = -kx \quad (k > 0),$$

which is evidently of considerable importance. Any motion which can be modelled by such an equation is called *simple harmonic motion* (*or s.h.m.*).

As $k > 0$, we can let $k = \omega^2$ (so that ω is another constant) and obtain the equation:

$$\ddot{x} = -\omega^2 x$$

This proves to be a more convenient form of the equation and is therefore adopted as standard.

We can now proceed to find its solutions by either of two methods:

Method 1

$$\ddot{x} = -\omega^2 x$$

$$\Rightarrow \quad 2\dot{x}\ddot{x} = -2\omega^2 x\dot{x}.$$

If we now integrate with respect to t, it follows that

$$\dot{x}^2 = -\omega^2 x^2 + A.$$

Supposing that when $\dot{x} = 0$ the value of x is a (i.e., that a is the *amplitude* of the oscillation), it follows that

$$0^2 = -\omega^2 a^2 + A$$

$$\Rightarrow \quad A = \omega^2 a^2.$$

So $\dot{x}^2 = \omega^2(a^2 - x^2)$.

Hence $\dfrac{dx}{dt} = \pm\omega\sqrt{(a^2 - x^2)}$

$$\Rightarrow \quad \pm\frac{dx}{\sqrt{(a^2 - x^2)}} = \omega\, dt$$

$$\Rightarrow \quad \pm\sin^{-1}\frac{x}{a} = \omega t + \varepsilon \quad \text{(where } \varepsilon \text{ is a constant angle)}$$

$$\Rightarrow \quad \frac{x}{a} = \pm\sin(\omega t + \varepsilon)$$

$$\Rightarrow \quad x = \pm a\sin(\omega t + \varepsilon).$$

Now $\quad -\sin(\omega t + \varepsilon) = \sin(\omega t + \varepsilon + \pi)$,

so $\quad x = a\sin(\omega t + \varepsilon) \quad$ or $\quad a\sin(\omega t + \varepsilon + \pi)$.

But ε is itself an arbitrary constant, so the second of these general solutions is the same as the first.

Hence $x = a\sin(\omega t + \varepsilon)$ and any simple harmonic motion can be described by an ordinary sine oscillation. The constant ε is called the *angle of phase*, or simply the *phase*.

Moreover, $\sin(\omega t + \varepsilon)$ takes the same value when ωt is increased by 2π, i.e., when t is increased by $2\pi/\omega$.

So the *period* of the s.h.m. is $2\pi/\omega$ (and its *frequency* is $\omega/2\pi$).

Summarising, we see that the s.h.m. has the properties:

$$\ddot{x} = -\omega^2 x$$
$$\dot{x}^2 = \omega^2(a^2 - x^2)$$
$$x = a\sin(\omega t + \varepsilon)$$

amplitude $= a$; period $= \dfrac{2\pi}{\omega}$; phase angle $= \varepsilon$

Method 2

Alternatively, $\ddot{x} = -\omega^2 x$ can be written

$$\frac{d^2x}{dt^2} + \omega^2 x = 0.$$

Now $x = A\,e^{mt}$ is a solution, provided that

$$m^2 + \omega^2 = 0 \quad \Rightarrow \quad m = \pm i\omega.$$

Hence the complementary function is

$$\begin{aligned} x &= A\,e^{i\omega t} + B\,e^{-i\omega t} \\ &= A(\cos \omega t + i\sin \omega t) + B(\cos \omega t - i\sin \omega t) \\ &= C\sin \omega t + D\cos \omega t, \end{aligned}$$

where C, D (like A, B) are arbitrary constants.

If we now choose a, ε so that

$$a\cos\varepsilon = C$$
$$a\sin\varepsilon = D,$$

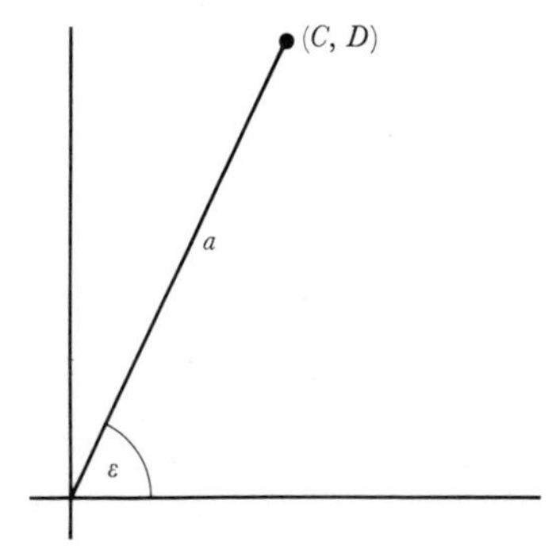

it follows that

$$x = a(\sin\omega t\cos\varepsilon + \cos\omega t\sin\varepsilon)$$
$$\Rightarrow \quad x = a\sin(\omega t + \varepsilon).$$

Hence $\dot{x} = a\omega\cos(\omega t + \varepsilon)$

and it follows immediately that

$$\dot{x}^2 = \omega^2(a^2 - x^2).$$

In particular, we can now calculate the periods of the oscillations in the above examples:

Spring of stiffness k supporting particle of mass m:

$$\omega = \sqrt{\frac{k}{m}} \quad \Rightarrow \quad \text{period} = 2\pi\sqrt{\frac{m}{k}}.$$

Buoy of density ρ in sea of density σ:

$$\omega = \sqrt{\frac{\sigma g}{\rho h}} \quad \Rightarrow \quad \text{period} = 2\pi\sqrt{\frac{\rho h}{\sigma g}}.$$

Simple pendulum of length l:

$$\omega = \sqrt{\frac{g}{l}} \quad \Rightarrow \quad \text{period} = 2\pi\sqrt{\frac{l}{g}}.$$

Example 4

A particle is describing simple harmonic motion in a straight line. When its distance from O, the centre of its path, is 3 m, its velocity is 16 m s^{-1} towards O and its acceleration is 48 m s^{-2} towards O. Find:
(*i*) the period of the motion;
(*ii*) the amplitude of the motion;
(*iii*) the time taken by the particle to reach O;
(*iv*) the velocity of the particle as it passes through O.

With the usual notation, $\ddot{x} = -\omega^2 x$.

But when $x = 3$, $\ddot{x} = -48$.

So $-48 = -3\omega^2 \quad \Rightarrow \quad \omega = 4$.

Hence period $= \dfrac{2\pi}{\omega} = \dfrac{\pi}{2} \approx 1.57$ s.

(*ii*) Furthermore, $\dot{x}^2 = \omega^2(a^2 - x^2) = 16(a^2 - x^2)$.

But when $x = 3$, $\dot{x} = -16$.

So $\quad 256 = 16(a^2 - 9)$

$\Rightarrow \quad a^2 - 9 = 16 \quad \Rightarrow \quad a = 5$.

So amplitude $= 5$ m.

(*iii*) Moreover, measuring t from the instant when $x = 0$, we know that $x = a \sin \omega t = 5 \sin 4t$.

So letting $x = 3$, we see that

$3 = 5 \sin 4t \quad \Rightarrow \quad 4t = \sin^{-1} 0.6 = 0.6434 \; (= 36° \; 52')$

$\Rightarrow \quad t = 0.16$ s.

So time taken $= 0.16$ s.

(*iv*) Finally, $\dot{x}^2 = 16(25 - x^2)$

and putting $x = 0$

we obtain $\dot{x}^2 = 16 \times 25$

$\Rightarrow \qquad \dot{x} = 4 \times 5 = 20.$

So velocity of particle as it passes through O is 20 m s^{-1}.

Example 5

A particle of mass m hangs at rest from the end of a vertical string which is attached to a fixed point and stretched by an amount e. If it is then given a further extension a and then released, find the period of its subsequent oscillations if $\quad (i)\ a < e; \quad (ii)\ a > e.$

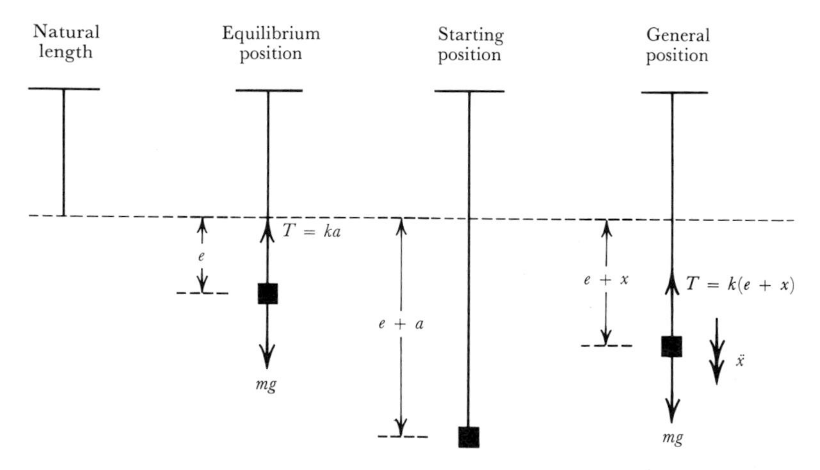

Let the stiffness of the string be k. Then, by considering the position of equilibrium,

$$mg = ke \quad \Rightarrow \quad k = \frac{mg}{e}.$$

At the general position, applying Newton's second law:

$$mg - k(e + x) = m\ddot{x}$$

$$\Rightarrow \qquad -kx = m\ddot{x}$$

$$\Rightarrow \qquad \ddot{x} = -\frac{k}{m}x = -\frac{g}{e}x.$$

So the particle executes simple harmonic motion with $\omega^2 = g/e$, provided only that the string remains taut. It is therefore now necessary to distinguish two cases:

$(i)\ a < e.$

In this case, the string remains taut, so the period of the oscillation is

$$\frac{2\pi}{\omega} = \frac{2\pi}{\sqrt{g/e}} = 2\pi\sqrt{\frac{e}{g}}.$$

(*ii*) $a > e$.

In this case the string becomes slack as when $x = -e$ the particle is still travelling upwards. So the full cycle consists of three parts:

(*a*) *Particle below equilibrium position* $(x > 0)$:

$$\text{time taken} = \frac{\pi}{\omega} = \pi\sqrt{\frac{e}{g}}.$$

(*b*) *Particle above equilibrium position but string still taut* $(-e < x < 0)$:

as the motion is simple harmonic, and if we measure time from the instant of passing through the equilibrium position,

$$x = -a \sin \omega t = -a \sin \sqrt{\left(\frac{g}{e}\right)}\, t.$$

Putting $x = -e$,

$$-e = -a \sin \sqrt{\left(\frac{g}{e}\right)}\, t \quad \Rightarrow \quad t = \sqrt{\left(\frac{e}{g}\right)} \sin^{-1} \frac{e}{a}.$$

So time taken $= 2\sqrt{\dfrac{e}{g}} \sin^{-1} \dfrac{e}{a}.$

(*c*) *String slack:*

The string becomes slack when $x = -e$.

But $\dot{x}^2 = \omega^2(a^2 - x^2)$.

So at this point $\dot{x} = \sqrt{\dfrac{g(a^2 - e^2)}{e}}.$

Hence time taken with string slack

$$= \frac{2\dot{x}}{g} = 2\sqrt{\frac{a^2 - e^2}{ge}}.$$

So total time taken

$$= \pi\sqrt{\frac{e}{g}} + 2\sqrt{\frac{e}{g}} \sin^{-1} \frac{e}{a} + 2\sqrt{\frac{a^2 - e^2}{ge}}$$

$$= \sqrt{\frac{e}{g}}\left\{\pi + 2 \sin^{-1} \frac{e}{a} + 2\sqrt{\frac{a^2 - e^2}{e^2}}\right\}.$$

Exercise 11.9

1 A spring has natural length 4 m and stiffness 20 N m^{-1}. One end is attached to a fixed point O of a smooth horizontal plane and the other end to a particle of mass 5 kg which is extended by a further distance of 3 m

and then released to oscillate in a straight line. After time t the extension of the spring is x. Taking $g = 10 \text{ m s}^{-2}$, find:

(*i*) the equation of motion of the particle;

(*ii*) the period of its oscillation;

(*iii*) x in terms of t;

(*iv*) the maximum speed of the particle.

2 A vertical light spring has two equal masses fixed to it at its ends, as shown. When the lower mass lies on a table, the spring is compressed a distance d by the weight of the upper mass. If the upper mass is now pushed down a further distance $2.5d$ and then released, examine whether the lower mass will at some time leave the table.

(You may assume that if the lower mass were fixed, the upper mass would oscillate symmetrically about its equilibrium position.)

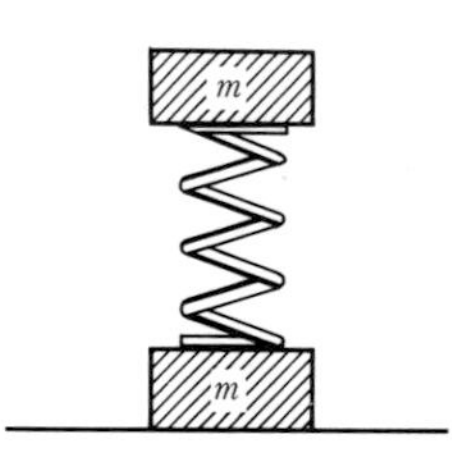

3 A boat is tossing (moving vertically up and down) in s.h.m. so that at one extreme position the weight of a passenger appears to be halved.

(*i*) What is the ratio of apparent weight to actual weight at the other extreme?

(*ii*) If the total range of motion is 10 m, what is the period of a complete oscillation?

(*iii*) What is the ratio of the volumes of water displaced by the boat at the two extremes?

(*iv*) During what fraction of the time is the apparent weight within 10 per cent of the real weight? (M.E.I.)

4 A particle of mass 2 kg is hung from the end of a vertical string and stretches it by 10 cm. Find the period of the oscillation if it is then pulled down a further distance of:

(*i*) 5 cm; (*ii*) 15 cm;

and then released. (Take $g \approx 10 \text{ m s}^{-2}$.)

5 One end O of a light elastic string OA, of natural length $4a$ and modulus $3mg$, is attached to a fixed point of a smooth horizontal table. A particle of mass m is attached to the other end A. The particle is pulled along the table until it is at a distance $5a$ from O and is then released with the string straight. Show that the particle first reaches O after a time

$$(\pi + 8)\sqrt{\frac{a}{3g}}.$$

(O.C.)

6 A man stands on a horizontal rough plank which performs a horizontal simple harmonic motion of period T and amplitude a. The coefficient of friction between the man and the plank is μ. Show that he will not slip on the plank if

$$\mu g T^2 \geqslant 4\pi^2 a. \qquad \text{(O.C.)}$$

7 Two particles of equal mass on a smooth horizontal table are connected by an elastic string of natural length a and modulus of elasticity equal to the weight of one particle. If the particles are held at rest at a distance $3a$ apart and are then released simultaneously, find the time which elapses before they collide.

8 A particle P, of mass m, is fastened to one end of a light string, of natural length a and modulus of elasticity λ, whose other end is fixed at O. A particle Q, also of mass m, is fastened to one end of a light inextensible string of length a whose other end is fastened to P. Initially the system is at rest with OPQ in a vertical straight line and OP of length a, so that P is at the middle point of OQ. The system is now released. Prove, by considerations of energy or otherwise, that the length of OP is never less than a.

The length of OP at time t is $a + x$. Prove that,

$$\text{if} \quad n^2 = \frac{\lambda}{2ma}, \quad \text{then} \quad \frac{d^2x}{dt^2} + n^2x = g.$$

Show that

$$x = \frac{g}{n^2}(1 - \cos nt) \text{ and deduce the tension at time } t \text{ in the string PQ.}$$

(C.)

9 A particle of mass m hangs at rest from a light inextensible string of length l. The particle is suddenly struck a small blow of impulse mv, perpendicular to the string. Find the amplitude of the subsequent motion and the time taken for the particle to rise to a height equal to one-half of its greatest height, measured from the initial position. (M.E.I.)

10 A particle of mass m is attached to the mid-point of a stretched light string whose tension is T and whose ends are fixed to two points distance l apart on a smooth horizontal plane. Show that if the particle is moved a distance x perpendicular to the string, the restoring force is $2T \sin \theta$, where θ is the angle which the parts of the string make with their original line.

Hence find the equation of motion when θ is small, and the period of oscillation.

*11.10 General statics: moments, couples, and equilibrium

So far we have limited our study to that of a single particle. But we are often concerned with the behaviour of collections of particles, both at rest and in motion. These might range from the particles of a highly compressible gas injected into a turbine to those of water, nearly incompressible, in waves at sea. But the simplest such collection, with which we shall be principally concerned, is one in which the distances between every pair of particles remains constant, and this we know as a *rigid body*. First, however, we must develop the idea of *moment of a force*.

Moments of forces

When we considered a particular force, of known magnitude and direction, acting on a *single* particle, there was never any doubt about its line of action: it necessarily had to pass through the particle. But if we now consider a collection of particles, say a chair, it is a common experience that a particular force can have a very different effect if its line of action is shifted: the reaction of a man on a chair might in one position produce stability, whilst if exactly the same force acts at a different position the chair might topple. In the case of a force acting on a single particle it was unnecessary to consider the *turning effect* of the force, because it had none. But when we consider a collection of particles, we must clearly study the turning effects of the individual forces involved.

To take a simple example, let us suppose that a beam of length 2 m is freely hinged at one of its ends O so that the beam is in a horizontal position, and let us further suppose that a force of magnitude 1 000 N acts vertically upwards at its other end.

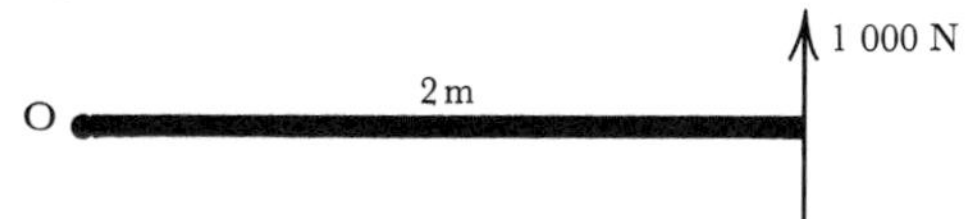

Now it is clear that the force of 1 000 N exerts a turning effect about O, and we know that this effect can be altered

(*a*) by increasing, or decreasing, the magnitude of the force;
(*b*) by altering its direction;
(*c*) by altering the position of its line of action, perhaps by moving it nearer to O.

The reader will already be familiar with a measure of such a turning effect called *the moment about O of the given force*, which is influenced by such changes. This is usually defined as the product of the magnitude of the force and its perpendicular distance from O:

moment about O *of* $\boldsymbol{F} = pF$.

Let us now consider what will happen if we radically alter the direction of the force. For instance suppose that instead of acting vertically, the force remains perpendicular to the rod but acts in a horizontal direction:

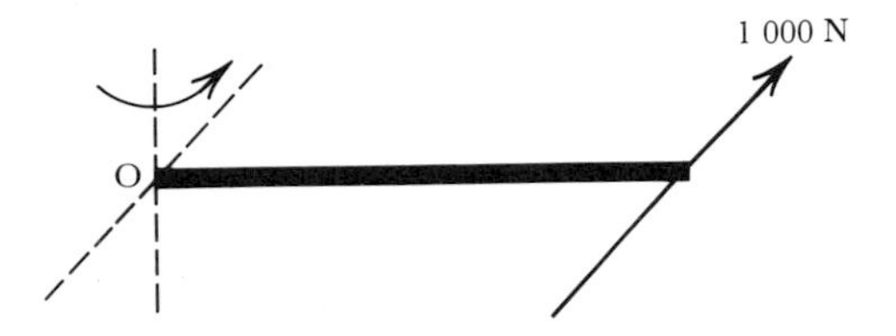

It is immediately apparent that such a change alters the turning effect from being in a vertical plane to being in a horizontal plane, as shown by the arrows in the two figures.

We therefore see that a turning effect is defined only when we know not only its magnitude and point of application but also the plane in which it is acting. A bowler, for instance, may give equal turning effects to three successive balls, one with top spin, the next as an off-break and the last as a leg-break: their distinguishing features are the planes in which they are made to spin. Now each of these planes is perfectly defined by the direction of its normal so we see that for the turning effect to be fully defined we need to know both a magnitude and also a *direction*, that of the normal to the plane in which it acts. It is, therefore, no surprise that the moment of a force should emerge not as a scalar quantity, but as a *vector*.

Moment as a vector product

A force and its line of action are completely defined by:

(*i*) a vector $\boldsymbol{F}$ representing the force; and

(*ii*) the position vector $\boldsymbol{r}$ of *any* point on its line of action.

Let us now consider the vector product $\boldsymbol{r} \times \boldsymbol{F}$.

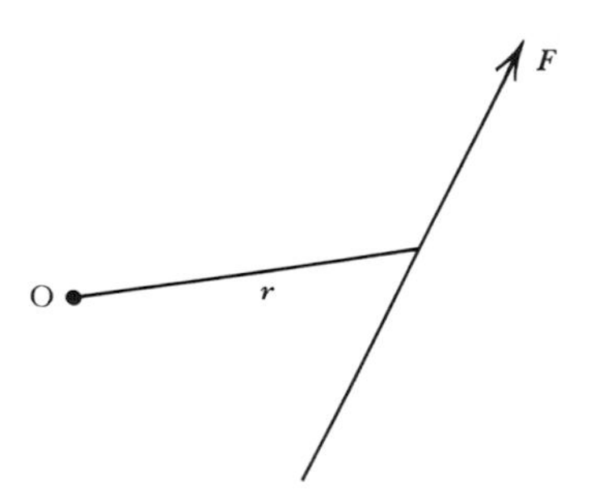

First, we notice that the vector $\boldsymbol{r} \times \boldsymbol{F}$ does *not* depend on the particular point which was chosen on its line of action. For suppose that, instead of the point P (with position vector $\boldsymbol{r}$), we had chosen another point P′ (with position vector $\boldsymbol{r}'$):

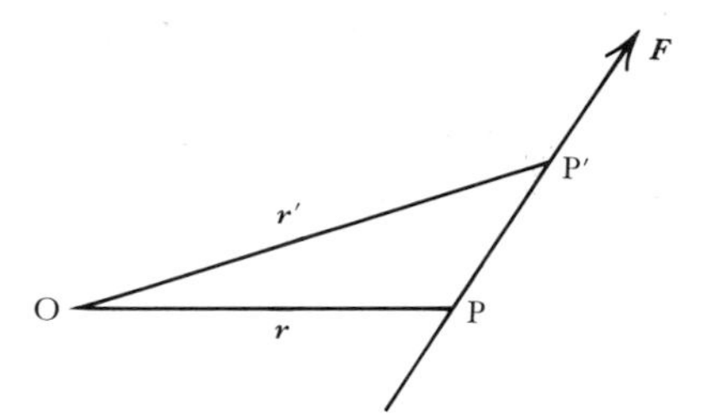

Now $\boldsymbol{PP'}$ is in the same direction as $\boldsymbol{F}$.

So $\qquad \boldsymbol{PP'} = \lambda \boldsymbol{F}, \quad$ where λ is a scalar

$\Rightarrow \qquad \boldsymbol{r'} - \boldsymbol{r} = \lambda \boldsymbol{F}$

$\Rightarrow \quad (\boldsymbol{r'} - \boldsymbol{r}) \times \boldsymbol{F} = \lambda \boldsymbol{F} \times \boldsymbol{F} = 0$

$\Rightarrow \qquad \boldsymbol{r'} \times \boldsymbol{F} = \boldsymbol{r} \times \boldsymbol{F}.$

Hence the vector product $\boldsymbol{r} \times \boldsymbol{F}$ is independent of the particular point chosen on $\boldsymbol{F}$ and depends only on $\boldsymbol{F}$ and its line of action.

Furthermore, if the angle between $\boldsymbol{r}$ and $\boldsymbol{F}$ is θ, we see that the magnitude of $\boldsymbol{r} \times \boldsymbol{F}$ is

$rF \sin \theta = (r \sin \theta)F = pF.$

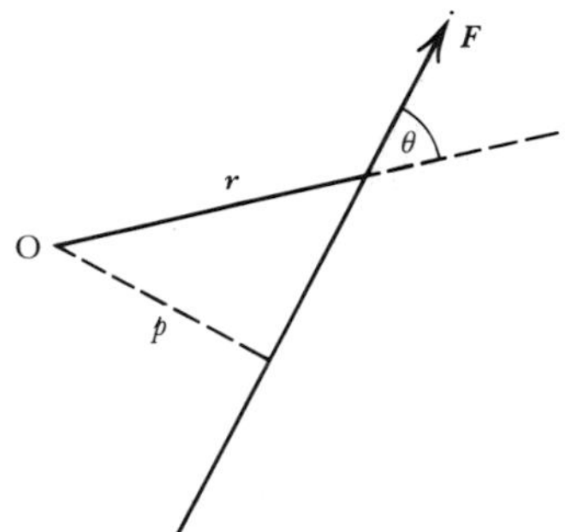

Lastly, the direction of $\boldsymbol{r} \times \boldsymbol{F}$ is perpendicular to $\boldsymbol{r}$ and to $\boldsymbol{F}$ and so is perpendicular to the plane containing the line of action and O.

In summary, it is evident that $\boldsymbol{r} \times \boldsymbol{F}$ depends only upon the vector $\boldsymbol{F}$ and its line of action, that its magnitude is the product of F and its perpendicular distance from O, and that its direction corresponds to a turning effect in the plane containing $\boldsymbol{F}$ and O. We therefore now *re-define* the moment about O of the force $\boldsymbol{F}$ as the vector $\boldsymbol{r} \times \boldsymbol{F}$:

moment about O of the force $\boldsymbol{F} = \boldsymbol{r} \times \boldsymbol{F}$

In particular, we notice that if $\boldsymbol{F}$ goes through O, its moment about O is

$\boldsymbol{0} \times \boldsymbol{F} = \boldsymbol{0}.$

Also $\quad \boldsymbol{r} \times (-\boldsymbol{F}) = -(\boldsymbol{r} \times \boldsymbol{F}),$

so that the moments about O of two equal and opposite forces are themselves equal and opposite.

More generally, an extremely valuable result follows directly from the fact that vector products are distributive with respect to addition. For suppose that $\boldsymbol{P}$, $\boldsymbol{Q}$ are two forces which meet in a point and that their resultant is $\boldsymbol{R}$. Let the position vector of their point of intersection be $\boldsymbol{r}$.

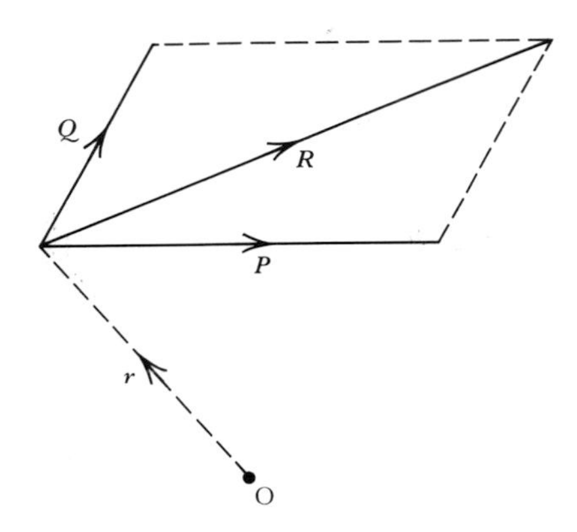

Then $\boldsymbol{r} \times \boldsymbol{P} + \boldsymbol{r} \times \boldsymbol{Q} = \boldsymbol{r} \times (\boldsymbol{P} + \boldsymbol{Q}) = \boldsymbol{r} \times \boldsymbol{R}$.

So

> the sum of the moments about any point of two concurrent forces is equal to the moment of their resultant.

Equivalent systems of forces

It sometimes happens that two apparently different systems of forces have:
(*i*) their vector sums equal;
(*ii*) their moments equal about every chosen point.

In such a case we say that the two systems are *equivalent*. In particular, the resultant of two concurrent forces has the same moment about any point as the sum of the moments of its constituent forces. So the resultant of two concurrent forces is equivalent to the forces themselves.

It can now easily be shown that two sets of forces are equivalent if:
(*i*) their vector sums are equal, and
(*ii*) their moments are equal about any *one* particular point.

For let us take this point as origin and let the two sets be

$$\boldsymbol{F}_i \text{ at } \boldsymbol{r}_i \quad (i = 1, 2, \ldots, n)$$

and $\quad \boldsymbol{F}'_i$ at $\boldsymbol{r}'_i \quad (i = 1, 2, \ldots, n')$.

Then $\sum \boldsymbol{F}_i = \sum \boldsymbol{F}'_i$ and $\sum \boldsymbol{r}_i \times \boldsymbol{F}_i = \sum \boldsymbol{r}'_i \times \boldsymbol{F}'_i$.

So the moment of the first set of forces about another point $\boldsymbol{c}$

$$= \sum (\boldsymbol{r}_i - \boldsymbol{c}) \times \boldsymbol{F}_i$$

$$= \sum \boldsymbol{r}_i \times \boldsymbol{F}_i - \boldsymbol{c} \times \sum \boldsymbol{F}_i$$

$$= \sum \boldsymbol{r}'_i \times \boldsymbol{F}'_i - \boldsymbol{c} \times \sum \boldsymbol{F}'_i$$

$$= \sum (\boldsymbol{r}'_i - \boldsymbol{c}) \times \boldsymbol{F}'_i$$

= moment of the second set of forces about $\boldsymbol{c}$.

Parallel forces

The question now arises whether we can find such an equivalent force, or resultant, when the two forces are not concurrent but parallel.

Let us consider two parallel forces $k_1\boldsymbol{F}$ and $k_2\boldsymbol{F}$ acting through points $\boldsymbol{r}_1$ and $\boldsymbol{r}_2$.

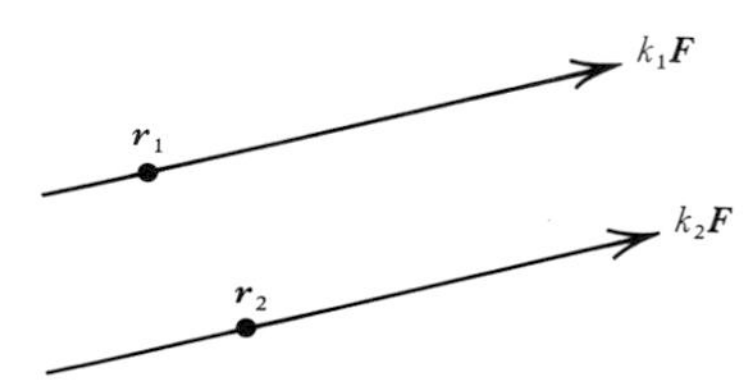

Now if a single force is equivalent to the original pair, it must be

$$k_1\boldsymbol{F} + k_2\boldsymbol{F} = (k_1 + k_2)\boldsymbol{F}.$$

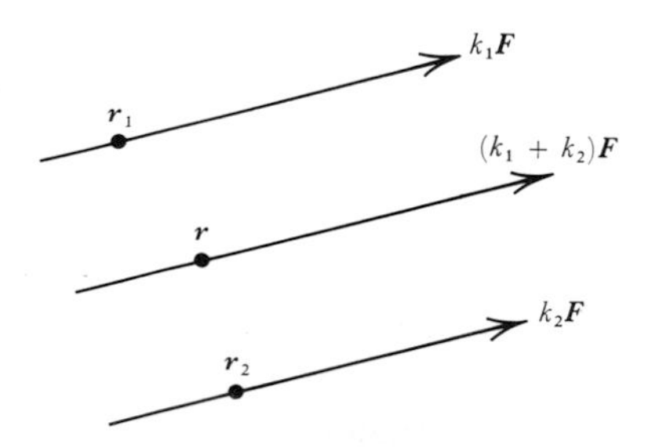

We now take a point $\boldsymbol{r}$ on the line of action of this force. Since it is equivalent to the original forces, we know that

$$\boldsymbol{r} \times (k_1 + k_2)\boldsymbol{F} = \boldsymbol{r}_1 \times k_1\boldsymbol{F} + \boldsymbol{r}_2 \times k_2\boldsymbol{F}$$

$$\Leftrightarrow \quad (k_1 + k_2)\boldsymbol{r} \times \boldsymbol{F} = (k_1\boldsymbol{r}_1 + k_2\boldsymbol{r}_2) \times \boldsymbol{F}.$$

In particular, we see that this equation is true if

$$(k_1 + k_2)\boldsymbol{r} = k_1\boldsymbol{r}_1 + k_2\boldsymbol{r}_2$$

$$\Leftrightarrow \quad \boldsymbol{r} = \frac{k_1\boldsymbol{r}_1 + k_2\boldsymbol{r}_2}{k_1 + k_2},$$

which is the point dividing the line joining $\boldsymbol{r}_1$ and $\boldsymbol{r}_2$ in the ratio $k_2 : k_1$.

As O could have been taken at any point whatsoever, it is clear that the original forces $k_1\boldsymbol{F}$ and $k_2\boldsymbol{F}$ are equivalent to a force $(k_1 + k_2)\boldsymbol{F}$ acting along a line which separates their lines of action in the ratio $k_2 : k_1$. So this force is again known as their resultant.

Centre of gravity

We now suppose that a system of particles consists of masses $m_1, m_2, m_3, \ldots$ at points $\boldsymbol{r}_1, \boldsymbol{r}_2, \boldsymbol{r}_3, \ldots$.

Then these masses have weights $m_1\boldsymbol{g}, m_2\boldsymbol{g}, m_3\boldsymbol{g} \ldots$ acting at points $\boldsymbol{r}_1, \boldsymbol{r}_2, \boldsymbol{r}_3 \ldots$.

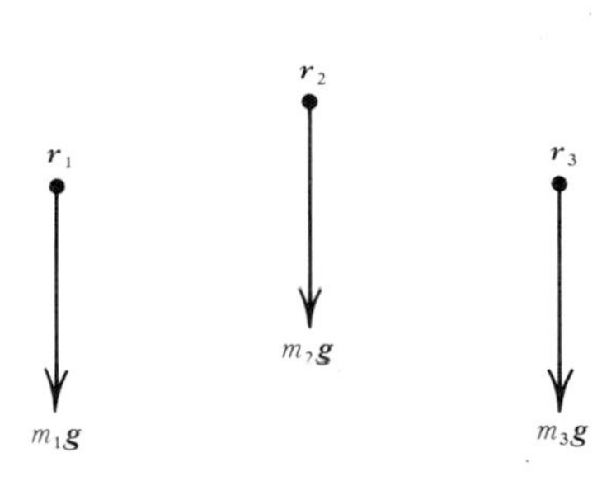

Hence, by a simple extension of the above result, we see that they are equivalent to a weight of $(m_1 + m_2 + m_3 + \cdots)\boldsymbol{g}$ acting at the point

$$\frac{m_1\boldsymbol{r}_1 + m_2\boldsymbol{r}_2 + m_3\boldsymbol{r}_3 + \cdots}{m_1 + m_2 + m_3 + \cdots},$$

which we have already met as the centre of mass.

Hence the weights of the separate particles are equivalent to their total weight acting through the centre of mass, which is therefore also called the centre of gravity of the system.

Couples

There is, however, one exceptional case of two parallel forces, when $k_1 = -k_2$ and the forces are equal and opposite with different lines of action. Such a system is called a *couple*.

We now consider a couple consisting of two forces, $\boldsymbol{F}$ at $\boldsymbol{r}_1$ and $-\boldsymbol{F}$ at $\boldsymbol{r}_2$.

If a point P is taken with position vector $\boldsymbol{p}$, the moment of the couple about P is

$$\begin{aligned} &(\boldsymbol{r}_1 - \boldsymbol{p}) \times \boldsymbol{F} + (\boldsymbol{r}_2 - \boldsymbol{p}) \times -\boldsymbol{F} \\ &= \boldsymbol{r}_1 \times \boldsymbol{F} - \boldsymbol{p} \times \boldsymbol{F} - \boldsymbol{r}_2 \times \boldsymbol{F} + \boldsymbol{p} \times \boldsymbol{F} \\ &= (\boldsymbol{r}_1 - \boldsymbol{r}_2) \times \boldsymbol{F}. \end{aligned}$$

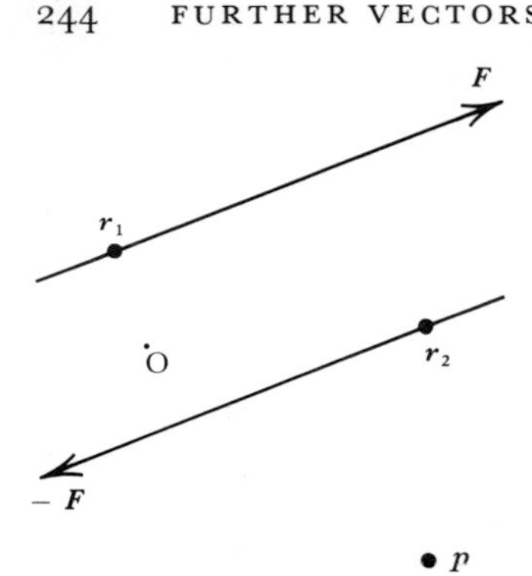

As this is independent of the position of P, it is simply called the *moment of the couple*. Note that the moment of a single force about a point on its line of action is zero, so a couple cannot possibly be equivalent to a single force and is therefore said to be *irreducible*.

Example 1

Coplanar forces $\boldsymbol{F}_1$, $\boldsymbol{F}_2$, $\boldsymbol{F}_3$ act through points whose position vectors are $\boldsymbol{r}_1$, $\boldsymbol{r}_2$, $\boldsymbol{r}_3$ respectively and

$$\boldsymbol{F}_1 = 4\boldsymbol{i} + 3\boldsymbol{j}, \qquad \boldsymbol{r}_1 = 3\boldsymbol{i} + 2\boldsymbol{j},$$
$$\boldsymbol{F}_2 = 5\boldsymbol{i} - 8\boldsymbol{j}, \qquad \boldsymbol{r}_2 = -\boldsymbol{i} + \boldsymbol{j},$$
$$\boldsymbol{F}_3 = -2\boldsymbol{i} + 6\boldsymbol{j}, \qquad \boldsymbol{r}_3 = 2\boldsymbol{i} - \boldsymbol{j}.$$

(*i*) Show that these forces reduce to a single force and give its magnitude and direction.
(*ii*) Find a vector equation or the Cartesian equation of the line of action of the resultant.
(*iii*) Find the single force through the point $\boldsymbol{r} = \boldsymbol{i} + \boldsymbol{j}$ and the couple which are together equivalent to this system. (W.)

(*i*) If these forces reduce to a single force, it must be

$$\boldsymbol{F}_1 + \boldsymbol{F}_2 + \boldsymbol{F}_3 = 7\boldsymbol{i} + \boldsymbol{j},$$

and so have magnitude $\sqrt{50}$ and be in the direction making $\tan^{-1}\frac{1}{7} \approx 8^\circ\, 8'$ with Ox.

(*ii*) If the line of action of its resultant passes through $\boldsymbol{r} = x\boldsymbol{i} + y\boldsymbol{j}$,

$$\text{then} \quad \boldsymbol{r} \times (\boldsymbol{F}_1 + \boldsymbol{F}_2 + \boldsymbol{F}_3) = \boldsymbol{r}_1 \times \boldsymbol{F}_1 + \boldsymbol{r}_2 \times \boldsymbol{F}_2 + \boldsymbol{r}_3 \times \boldsymbol{F}_3$$
$$\Rightarrow \quad (x\boldsymbol{i} + y\boldsymbol{j}) \times (7\boldsymbol{i} + \boldsymbol{j}) = (3\boldsymbol{i} + 2\boldsymbol{j}) \times (4\boldsymbol{i} + 3\boldsymbol{j})$$
$$+ (-\boldsymbol{i} + \boldsymbol{j}) \times (5\boldsymbol{i} - 8\boldsymbol{j})$$
$$+ (2\boldsymbol{i} - \boldsymbol{j}) \times (-2\boldsymbol{i} + 6\boldsymbol{j})$$
$$\Rightarrow \quad (x - 7y)\boldsymbol{k} = (1 + 3 + 10)\boldsymbol{k} = 14\boldsymbol{k}$$
$$\Rightarrow \quad x - 7y = 14.$$

(*iii*) Let the required force be $\boldsymbol{F}$ and the required couple be $\boldsymbol{G}$.

Again, $\boldsymbol{F} = \boldsymbol{F}_1 + \boldsymbol{F}_2 + \boldsymbol{F}_3 = 7\boldsymbol{i} + \boldsymbol{j}$.

Furthermore,

$$\boldsymbol{r} \times (\boldsymbol{F}_1 + \boldsymbol{F}_2 + \boldsymbol{F}_3) + \boldsymbol{G} = \boldsymbol{r}_1 \times \boldsymbol{F}_1 + \boldsymbol{r}_2 \times \boldsymbol{F}_2 + \boldsymbol{r}_3 \times \boldsymbol{F}_3$$

$$\Rightarrow \quad (\boldsymbol{i} + \boldsymbol{j}) \times (7\boldsymbol{i} + \boldsymbol{j}) + \boldsymbol{G} = 14\boldsymbol{k}$$

$$\Rightarrow \quad -6\boldsymbol{k} + \boldsymbol{G} = 14\boldsymbol{k}$$

$$\Rightarrow \quad \boldsymbol{G} = 20\boldsymbol{k}.$$

So required force is $7\boldsymbol{i} + \boldsymbol{j}$ and required couple is $20\boldsymbol{k}$.

Exercise 11.10a

1 Find the moments:

	About the point,	*Of the force,*	*Acting at*
(*i*)	$\boldsymbol{0}$	$\boldsymbol{i} + \boldsymbol{j} - \boldsymbol{k}$	$\boldsymbol{i} - \boldsymbol{k}$
(*ii*)	$\boldsymbol{0}$	$\boldsymbol{i} + \boldsymbol{j} - 2\boldsymbol{k}$	$\boldsymbol{i} + 2\boldsymbol{j} + \boldsymbol{k}$
(*iii*)	$\boldsymbol{i} + \boldsymbol{j}$	$\boldsymbol{i} + 2\boldsymbol{j} - \boldsymbol{k}$	$\boldsymbol{i} + \boldsymbol{j} - \boldsymbol{k}$
(*iv*)	$\boldsymbol{i} + 2\boldsymbol{j} - 3\boldsymbol{k}$	$3\boldsymbol{i} + 2\boldsymbol{j} - 3\boldsymbol{k}$	$3\boldsymbol{j} + \boldsymbol{k}$
(*v*)	$\boldsymbol{i} - \boldsymbol{j}$	$-2\boldsymbol{i} + 3\boldsymbol{j} + \boldsymbol{k}$	$\boldsymbol{j} + 2\boldsymbol{k}$
(*vi*)	$2\boldsymbol{i} + \boldsymbol{j} - \boldsymbol{k}$	$-\boldsymbol{i} + 4\boldsymbol{j} - \boldsymbol{k}$	$-\boldsymbol{i} + \boldsymbol{j} + 2\boldsymbol{k}$

2 Determine in each of the following cases whether the two sets of forces are equivalent:

(*i*) $\boldsymbol{i}$ acting at $\boldsymbol{j}$ and $\boldsymbol{j}$ acting at $\boldsymbol{i}$, $\boldsymbol{i} + \boldsymbol{j}$ acting at $2\boldsymbol{j}$;
(*ii*) $-\boldsymbol{i}$ acting at $\boldsymbol{j}$ and $\boldsymbol{j}$ acting at $\boldsymbol{i}$, $-\boldsymbol{i} + \boldsymbol{j}$ acting at $2\boldsymbol{j}$;
(*iii*) $2\boldsymbol{j}$ acting at O and $\boldsymbol{j}$ acting at $3\boldsymbol{i}$, $3\boldsymbol{j}$ acting at $\boldsymbol{i}$;
(*iv*) $4\boldsymbol{i}$ acting at O and $-\boldsymbol{i}$ acting at $-3\boldsymbol{j}$, $3\boldsymbol{i}$ acting at $-4\boldsymbol{j}$;
(*v*) the couple formed by $-\boldsymbol{i}$ at $\boldsymbol{j}$ and $\boldsymbol{i}$ at $2\boldsymbol{j}$, and that formed by $-\boldsymbol{j}$ at $2\boldsymbol{i}$ and $\boldsymbol{j}$ at $\boldsymbol{i}$;
(*vi*) the couple formed by $-\boldsymbol{i}$ at $\boldsymbol{j}$ and $\boldsymbol{i}$ at $-\boldsymbol{j}$, and that formed by $-\boldsymbol{j}$ at $\boldsymbol{i}$ and $\boldsymbol{j}$ at $\boldsymbol{k}$.

3 Show that each of the following systems of coplanar forces is equivalent either to a single force or a single couple (which may be zero). In each case find the equivalent force (including its line of action) or the equivalent couple:

(*i*) $\boldsymbol{i} + \boldsymbol{j}$ at $\boldsymbol{i} - \boldsymbol{j}$, $3\boldsymbol{i} - \boldsymbol{j}$ at $2\boldsymbol{i}$, $2\boldsymbol{i} + 3\boldsymbol{j}$ at $2\boldsymbol{j}$;
(*ii*) $\boldsymbol{i} + \boldsymbol{j}$ at $\boldsymbol{i} - \boldsymbol{j}$, $-3\boldsymbol{i} + \boldsymbol{j}$ at $\boldsymbol{i} + \boldsymbol{j}$, $2\boldsymbol{i} - 2\boldsymbol{j}$ at $-\boldsymbol{i} + 2\boldsymbol{j}$;
(*iii*) $2\boldsymbol{j}$ at $\boldsymbol{i} - \boldsymbol{j}$, $-\boldsymbol{i}$ at $\boldsymbol{j}$, $\boldsymbol{i}$ at $3\boldsymbol{j}$, $-2\boldsymbol{j}$ at O;
(*iv*) $2\boldsymbol{i} + 3\boldsymbol{j}$ at $\boldsymbol{i}$, $-3\boldsymbol{i} + 4\boldsymbol{j}$ at $\boldsymbol{j}$, $4\boldsymbol{i} + 5\boldsymbol{j}$ at $\boldsymbol{i} + \boldsymbol{j}$, $-\boldsymbol{i} + 3\boldsymbol{j}$ at $-\boldsymbol{i} - 2\boldsymbol{j}$.

4 ABCD is a square of side a. The anti-clockwise moment of a set of forces in the plane of the square is $10aP$ about A, $-20aP$ about B and $10aP$ about C. Taking the x-axis along AB and the y-axis along AD, determine the resultant of the set of forces and the equation of its line of action.

Determine the couple and the single force at the centre of the square which would be equivalent to this resultant.

5 (*i*) Show that three forces represented in magnitude and direction by $k\boldsymbol{AB}$, $k\boldsymbol{BC}$, $k\boldsymbol{CA}$ and acting along the sides AB, BC, CA respectively of a triangle are equivalent to a couple of magnitude $2k$ times the area of the triangle.
(*ii*) Use the result (*i*) to show that four forces represented in magnitude and direction by $k\boldsymbol{AB}$, $k\boldsymbol{BC}$, $k\boldsymbol{CD}$, $k\boldsymbol{DA}$ and acting along the sides AB, BC, CD, DA respectively of a plane quadrilateral are equivalent to a couple of magnitude $2k$ times the area of the quadrilateral. (M.E.I.)

Statics of a system : equilibrium

Any system of particles (which may, or may not, comprise a rigid body) is said to be *in equilibrium* if, under the forces acting, it can remain at rest.

Let us consider a system in equilibrium which consists of just three particles of masses m_1, m_2, m_3 whose position vectors are $\boldsymbol{r}_1$, $\boldsymbol{r}_2$, $\boldsymbol{r}_3$.

The forces acting on these particles can be of two kinds:
(*i*) *external forces*, which we call $\boldsymbol{F}_1$, $\boldsymbol{F}_2$, $\boldsymbol{F}_3$, acting on the particles m_1, m_2, m_3 respectively; and
(*ii*) *internal forces*, between the pairs of particles. We let the force exerted upon m_1 by m_2 be $\boldsymbol{F}_{12}$, and the force exerted upon m_2 by m_1 be $\boldsymbol{F}_{21}$, etc. So the complete set of forces can be represented:

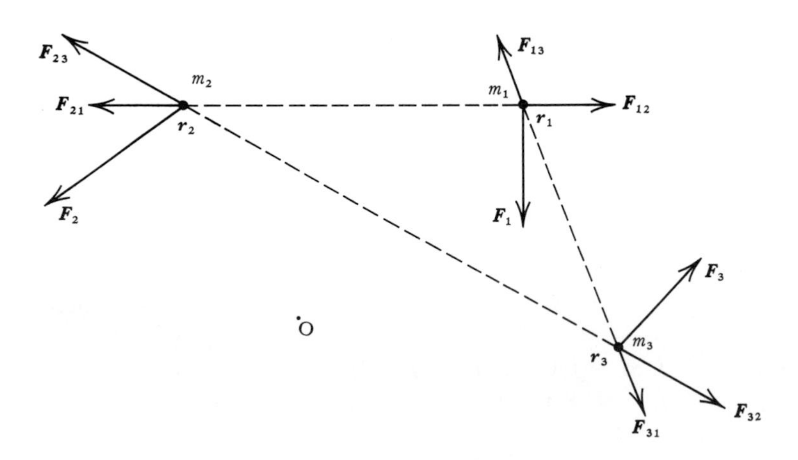

Now, by Newton's third law, every pair of internal forces must be equal and opposite.

So $\boldsymbol{F}_{12} + \boldsymbol{F}_{21} = \boldsymbol{0}$.

Moreover, since these forces have the same line of action, their moments about O must be equal and opposite, and

$$\boldsymbol{r}_1 \times \boldsymbol{F}_{12} + \boldsymbol{r}_2 \times \boldsymbol{F}_{21} = \boldsymbol{0}.$$

If we now look at the equilibrium of the particles separately, we see that

$$\begin{aligned}
&\boldsymbol{F}_1 \qquad\qquad + \boldsymbol{F}_{13} + \boldsymbol{F}_{12} = \boldsymbol{0} \\
&\boldsymbol{F}_2 + \boldsymbol{F}_{23} \qquad\qquad + \boldsymbol{F}_{21} = \boldsymbol{0} \\
&\boldsymbol{F}_3 + \boldsymbol{F}_{32} + \boldsymbol{F}_{31} \qquad\qquad = \boldsymbol{0}.
\end{aligned}$$

We now add these equations and remember that

$$\boldsymbol{F}_{23} + \boldsymbol{F}_{32} = \boldsymbol{0}, \quad \text{etc.,}$$

so that

$$\boldsymbol{F}_1 + \boldsymbol{F}_2 + \boldsymbol{F}_3 = \boldsymbol{0}.$$

Hence if a system is in equilibrium the sum of the external forces must be zero, just as when they act upon a single particle.

Moreover, we can also take the vector products of each of the above expressions with $\boldsymbol{r}_1$, $\boldsymbol{r}_2$, $\boldsymbol{r}_3$ respectively, and obtain

$$\begin{aligned}
&\boldsymbol{r}_1 \times \boldsymbol{F}_1 \qquad\qquad\qquad + \boldsymbol{r}_1 \times \boldsymbol{F}_{13} + \boldsymbol{r}_1 \times \boldsymbol{F}_{12} = \boldsymbol{0} \\
&\boldsymbol{r}_2 \times \boldsymbol{F}_2 + \boldsymbol{r}_2 \times \boldsymbol{F}_{23} \qquad\qquad\qquad + \boldsymbol{r}_2 \times \boldsymbol{F}_{21} = \boldsymbol{0} \\
&\boldsymbol{r}_3 \times \boldsymbol{F}_3 + \boldsymbol{r}_3 \times \boldsymbol{F}_{32} + \boldsymbol{r}_3 \times \boldsymbol{F}_{31} \qquad\qquad\qquad = \boldsymbol{0}.
\end{aligned}$$

We now add these three equations and remember that

$$\begin{aligned}
&\boldsymbol{r}_1 \times \boldsymbol{F}_{12} + \boldsymbol{r}_2 \times \boldsymbol{F}_{21} = \boldsymbol{0} \\
\text{(and } &\boldsymbol{r}_1 \times \boldsymbol{F}_{13} + \boldsymbol{r}_3 \times \boldsymbol{F}_{31} = \boldsymbol{0}, \\
&\boldsymbol{r}_2 \times \boldsymbol{F}_{23} + \boldsymbol{r}_3 \times \boldsymbol{F}_{32} = \boldsymbol{0}).
\end{aligned}$$

Hence $\boldsymbol{r}_1 \times \boldsymbol{F}_1 + \boldsymbol{r}_2 \times \boldsymbol{F}_2 + \boldsymbol{r}_3 \times \boldsymbol{F}_3 = \boldsymbol{0}.$

So the sum of the moments about O of the external forces must be zero.

Now from the start of this section, O has been any fixed point which it is wished to use as an origin. It therefore follows that the sum of the moments of the external forces about *any* point must be zero.

For simplicity, we limited ourselves to the consideration of just three particles. But the whole of the above analysis can immediately be extended to cover any number, finite or infinite. Summarising, we can see that:

> For any system of particles to be in equilibrium:
> (*i*) the sum of the external forces must be zero; and
> (*ii*) the sum of the moments of the external forces about any fixed point must also be zero.

Example 2

A uniform beam of length 10 m and weight 2 000 N has a man of weight 600 N standing on one end. If the beam is supported at two points 2 m from each end, find the reactions at these supports.

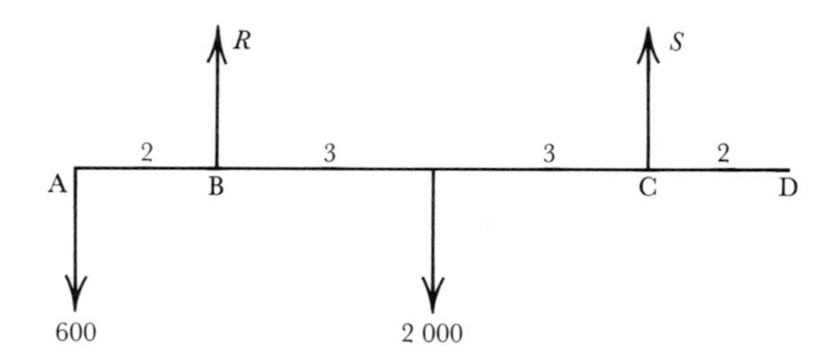

Letting the two reactions (in newtons) be R, S and taking moments about C,

$$6R = 8 \times 600 + 3 \times 2\,000 = 10\,800$$

$$\Rightarrow \quad R = 1\,800.$$

Also, taking moments about B,

$$2 \times 600 + 6S = 3 \times 2\,000$$

$$\Rightarrow \quad 6S = 4\,800$$

$$\Rightarrow \quad S = 800.$$

So the two reactions are 1 800 N and 800 N (and we note, as we would expect, that $R + S = 600 + 2\,000$).

Example 3

A lightweight ladder of length 8 m is inclined at 70° to the horizontal, resting upon a rough horizontal path and a smooth vertical wall. If a man climbs this ladder and it begins to slip when he is three-quarters of the way up, find the coefficient of friction at its lower end.

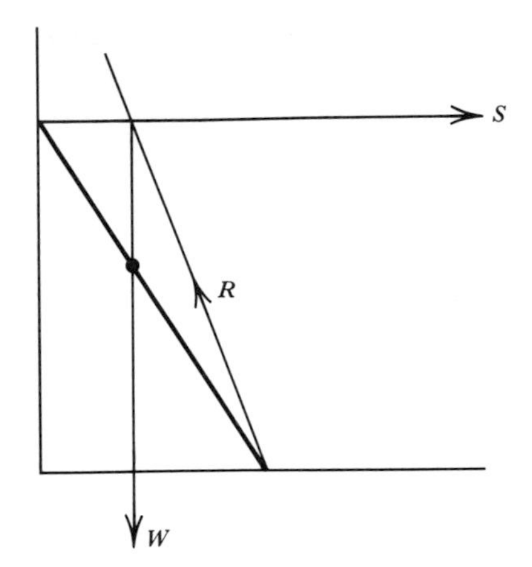

If we neglect the weight of the ladder itself, there are just three external forces acting on the system of the man and ladder:

(*i*) $\boldsymbol{W}$, the weight of the man;

(*ii*) $\boldsymbol{S}$, the normal reaction at the smooth wall;

(*iii*) $\boldsymbol{R}$, the reaction at the ground which (on account of a frictional component) can act at an inclination to the normal.

Now when the system is in equilibrium these three forces must have zero movement about every point. But $\boldsymbol{S}$ and $\boldsymbol{W}$ clearly have zero moment about their point of intersection, so $\boldsymbol{R}$ must also pass through this point.

Hence as the man gradually climbs the ladder, the reaction $\boldsymbol{R}$ makes an increasing angle with the vertical; until, finally, when the man is three-quarters of the way up, this angle becomes λ, the angle of friction, and the ladder starts to slip.

If we now look more closely at this limiting position, we see that (in the notation of the figure)

$$h = 8 \sin 70^\circ$$

$$\text{and} \quad x = 6 \cos 70^\circ$$

$$\Rightarrow \quad \tan \lambda = \frac{x}{h} = \frac{6 \cos 70^\circ}{8 \sin 70^\circ}$$

$$= \tfrac{3}{4} \cot 70^\circ$$

$$= \tfrac{3}{4} \times 0.36$$

$$= 0.27.$$

So $\mu = \tan \lambda = 0.27$.

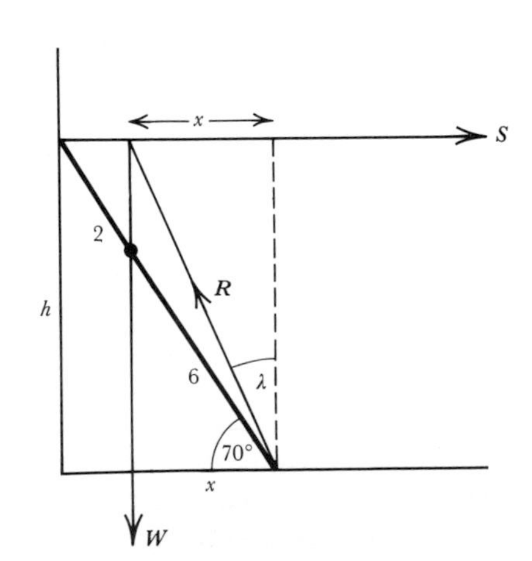

Alternatively, we can let the components of $\boldsymbol{R}$ be F and N, and then proceed as follows:

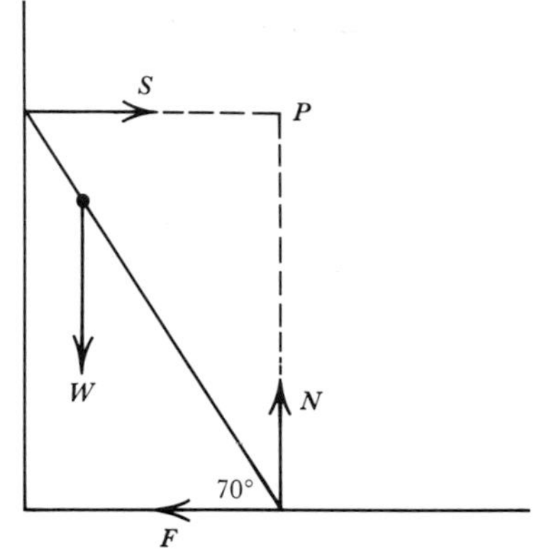

Resolving vertically, $N = W$.

Taking moments about the point P,

$$F \times 8 \sin 70° = W \times 6 \cos 70°$$

$$\Rightarrow \quad F = \frac{3W}{4} \cot 70°$$

$$\Rightarrow \quad \frac{F}{N} = \tfrac{3}{4} \cot 70° = 0.27.$$

Hence, if slipping takes place at this point, $\mu = 0.27$.

Exercise 11.10b

1 A beam AB, of length 22 m and mass 150 kg has its centre of gravity at C, 11.5 m from A. It is supported in a horizontal position by two trestles at D and E, where $AD = 7.5$ m and $AE = 15$ m. Find the pressure on each trestle. (Taking $g \approx 10 \text{ m s}^{-2}$.)

Find the mass of the heaviest man who can sit at either end of the beam without tilting it. (O.C.)

2

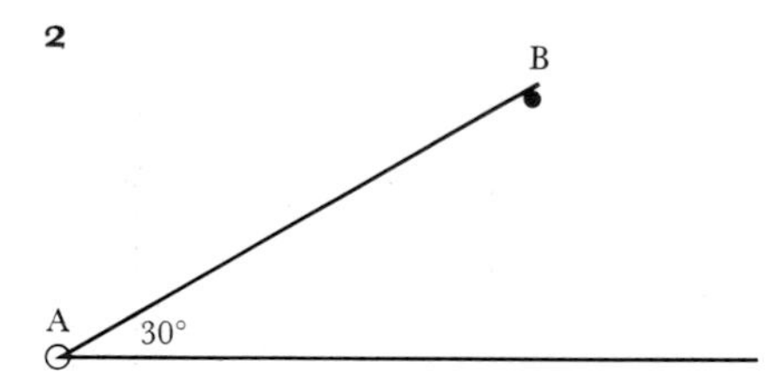

A uniform rod of length 1 m and mass 50 g is freely hinged at A and has its other end resting on a smooth peg B. The rod in this position makes an angle of 30° with the horizontal.

Calculate:

(*i*) the reaction at B;

(*ii*) the horizontal component of the reaction at A.

(Take $g \approx 9.81 \text{ m s}^{-2}$.) (C.)

3 A uniform beam ABCD, where $AB = a$, $BC = 3a$ and $CD = a$, is of weight $4W$ and rests horizontally in equilibrium on supports at B and C. Loads of weight xW, yW are hung from A, D respectively. Calculate, in terms of x, y and W, the load borne by each support. If the sum of the loads borne by both supports cannot exceed $20W$, indicate by shading on the plane of the coordinate axes Oxy the region within which the point (x, y) must lie. Deduce that the greatest value of y is 14. (O.C.)

4 A rod AB, hinged at A, is 4 m long and of mass 5 kg, with its centre of mass 3 m from A. A rope is attached to B, passes over a smooth pulley 3 m above A and supports a mass M hanging freely. A mass m hangs from B and keeps AB in a horizontal position.
(*i*) When $m = 3$, find M; and,
(*ii*) when $M = 20$, find m.

5 A non-uniform rod AB, of length a and weight W, rests making an angle β with the horizontal with its lower end B on a rough horizontal table. The centre of mass of the rod is at G where $AG = \frac{1}{3}a$. Equilibrium is maintained by a horizontal string attached to A. Show that the coefficient of friction between the rod and the table cannot be less than $\frac{2}{3}\cot\beta$. Find the tension in the string. (O.C.)

6 A uniform beam AB rests in limiting equilibrium with A against a rough vertical wall and with B on rough horizontal ground. The vertical plane through the beam is perpendicular to the wall. The coefficient of friction at A is $\frac{1}{3}$ and the coefficient of friction at B is $\frac{1}{2}$. Calculate, correct to the nearest degree, the inclination of the beam to the horizontal. (O.C.)

7 A uniform ladder is leaning against a vertical wall on horizontal ground. The coefficients of friction at the wall and at the ground are equal, and the ladder is on the point of slipping. Show that the total reaction at the wall must be at right angles to the total reaction at the ground.

Draw a figure with the ladder inclined at 60° to the ground; by means of a circle with the ladder as diameter, mark on the figure the point where the two reactions intersect. (O.C.)

8 A circular cylinder of weight W is held with its axis horizontal on a rough plane inclined at 30° to the horizontal by a cord wrapped round it, one end of the cord being fixed to the cylinder, the other leading away tangentially, at right angles to the axis of the cylinder. Determine the tension of the cord and the least possible coefficient of friction between the cylinder and the plane
(*i*) if the cord leads away horizontally;
(*ii*) if it leads away vertically upwards;
(*iii*) if it leads away in such a direction as to make the tension a minimum.

9 A uniform hemispherical shell rests with its curved surface against a rough vertical wall and a rough horizontal floor. If the equilibrium is limiting at both points of contact when the plane of the base of the shell makes an angle θ with the horizontal, prove that

$$\sin\theta = \frac{2\mu(1 + \mu)}{1 + \mu^2},$$

where μ is the coefficient of friction at both points of contact.
Prove that such a position is impossible if $\mu > \sqrt{2} - 1$.
(The centre of mass of the shell bisects the radius of symmetry.) (O.C.)

10 The mass per unit length of a rod AB of length l varies linearly from m at the end A to $2m$ at the end B. Prove by means of the integral calculus that the total mass of the rod is $3\ ml/2$ and find the position of the centre of gravity.

An inextensible string of length a is attached to a point P of the rod at a distance a $(< l)$ from the end A. The other end of the string is fixed to the point C on a rough vertical wall and the end A of the rod rests against this wall so that A is below C and AC $= a$. Find the tension in the string and the frictional force at the point A, and show that this vanishes if $l = 18a/5$.

(M.E.I.)

11 A uniform rod AB is suspended from a fixed point O of a rough vertical wall by a light inextensible string of length l which is fastened to B. The end A of the rod rests against the wall vertically below O with A below the level of B. If the rod is in equilibrium (not necessarily limiting) with OA $= d$ and angle AOB $= \theta$, find the tangent of the angle of inclination to the horizontal of the resultant reaction on the rod at A.

If the angle of friction at A is λ and equilibrium is limiting, with A about to slide downwards, show that

$$\cos(\theta - \lambda) = \frac{2d}{l}\cos\lambda.$$

12 O, A, B, C are the four corners of a square lamina OABC of side a. Forces act in the plane of the lamina as follows: $6P$ along $\boldsymbol{AO}$, P along $\boldsymbol{OC}$, $2P$ along $\boldsymbol{BC}$ and $7P$ along $\boldsymbol{BA}$.

(*i*) Find the magnitude of the resultant of this system of forces, and the equation of its line of action referred to OA, OC as x- and y-axes.

(*ii*) It is desired to maintain equilibrium by applying a force and a couple at one corner of the square. Find which corner should be selected so that the moment of the couple is as small as possible, and state the moment of this least couple.

(*iii*) Find the couple which must be applied to the lamina so that the resultant of the four original forces and the couple acts in a line passing through the centre of the square.

(M.E.I.)

*11.11 General dynamics and rotation about a fixed axis

Dynamics of a system of particles

Suppose that a system of masses is in motion and that a typical mass m_i at point $\boldsymbol{r}_i$ is under the action of an external force $\boldsymbol{F}_i$ and an internal force $\boldsymbol{I}_i$.

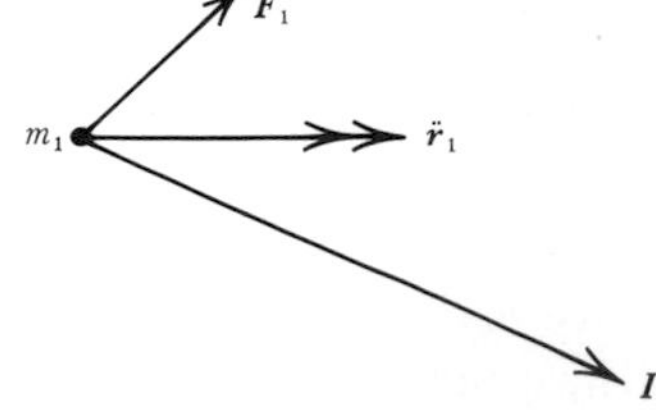

Then, by Newton's second law,

$$\boldsymbol{F}_i + \boldsymbol{I}_i = m_i\ddot{\boldsymbol{r}}_i.$$

Adding these equations over all the particles of the system,

$$\sum \boldsymbol{F}_i + \sum \boldsymbol{I}_i = \sum m_i \ddot{\boldsymbol{r}}_i. \tag{1}$$

But, applying Newton's third law,

$$\sum \boldsymbol{I}_i = \boldsymbol{0}$$

and if the total mass of the system is M $(= \sum m_i)$ and the centre of mass is at $\bar{\boldsymbol{r}}$, then

$$M\bar{\boldsymbol{r}} = \sum m_i \boldsymbol{r}_i$$

$$\Rightarrow \quad M\ddot{\bar{\boldsymbol{r}}} = \sum m_i \ddot{\boldsymbol{r}}_i.$$

Hence equation (1) becomes

$$\sum \boldsymbol{F}_i = M\ddot{\bar{\boldsymbol{r}}},$$

so that the centre of mass moves as though all the mass and all the external forces were concentrated upon it.

When, for example, a diving champion makes a spectacular dive from a high board, his movement may be a complicated mixture of twists and turns and somersaults, and his body is certainly far from rigid. Nevertheless, the only external forces are the weights of the various particles that make up his body, and we see from the above result that his centre of mass moves as though all these weights and all his mass were concentrated there: in other words, his centre of mass (which might itself be varying relative to his body as he curls up, 'jack-knifes', or stretches out) is bound to move in a perfect parabola.

Exercise 11.11a

1 A girl standing on perfectly smooth ice loses her balance. What is the path of her centre of mass? If a friend helped to save her as she fell, what would be the path of their joint centre of mass?

2 Find the position vector of the centre of mass of particles of masses 4, 3, 2, 3 units at rest at the points

$$\boldsymbol{i} + \boldsymbol{j}, \quad 2\boldsymbol{i} - \boldsymbol{j}, \quad 2\boldsymbol{i} + \boldsymbol{j}, \quad 2\boldsymbol{i} + 3\boldsymbol{j}$$

respectively. If each mass is acted upon by a force directed towards the origin and proportional to its distance from the origin, find the direction of the initial acceleration of the centre of mass. (L.)

Rotation about a fixed axis: moments of inertia

It is beyond our present scope to investigate further the general motion of a system of particles, and we must restrict ourselves to the special case of a rigid body which is rotating about a fixed axis:

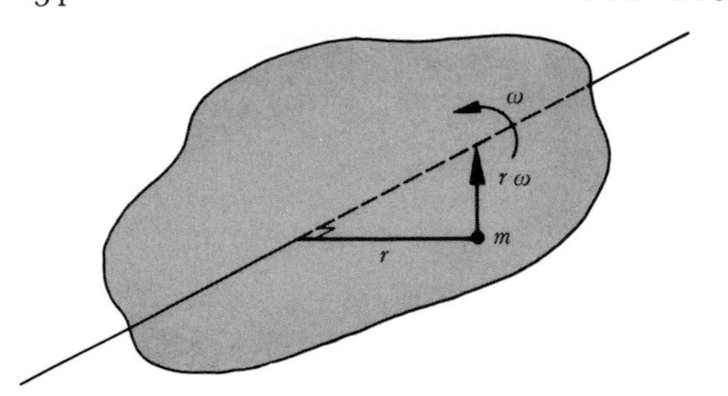

Suppose that a rigid body is rotating with angular velocity ω about a fixed axis and that a typical particle of mass m is at a perpendicular distance r from this axis.

Then the speed of m is $r\omega$ and its kinetic energy is $\frac{1}{2}mr^2\omega^2$.

So the total kinetic energy of the body is $\sum \frac{1}{2}mr^2\omega^2 = \frac{1}{2}(\sum mr^2)\omega^2$.

Now the expression $\sum mr^2$ clearly depends only on the masses m and the way in which they are distributed in the body relative to the given axis. It is usually denoted by I, called the *moment of inertia of the body about the given axis*.

So $\boxed{\text{k.e.} = \frac{1}{2}I\omega^2, \quad \text{where } I = \sum mr^2}$

Comparing the two expressions

k.e. $= \frac{1}{2}mv^2$ and k.e. $= \frac{1}{2}I\omega^2$,

it might be expected that I will, for rotational motion, play a part very similar to that taken by mass for linear motion; and just as mass can be regarded as a measure of a particle's inertia, or reluctance to accelerate when under the action of a given force, so the moment of inertia of a body about a given axis is a measure of its reluctance to accelerate rotationally when under the action of a given moment.

Our first task, however, is to calculate the moments of inertia of a number of different rigid bodies.

Circular hoop rotating about a perpendicular axis through its centre

Suppose that a circular hoop of mass M and radius a is rotating about a perpendicular axis through its centre.

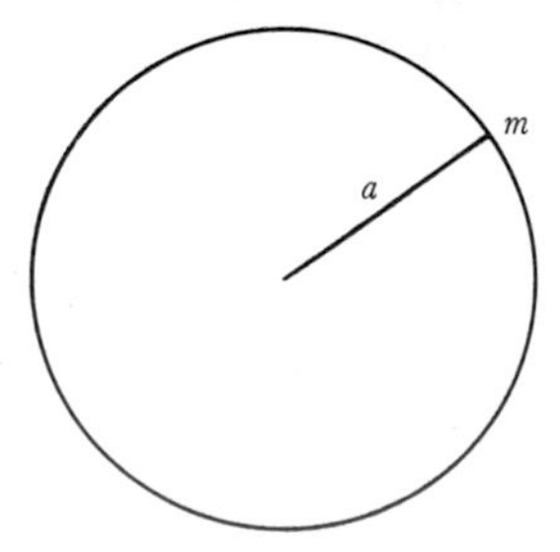

If a typical particle has mass m,

$I = \sum mr^2.$

But, for every particle, $r = a$;

so $I = \sum ma^2 = (\sum m)a^2$

$\Rightarrow \quad I = Ma^2.$

Uniform circular disc, about a perpendicular axis through its centre

Suppose again that the body has mass M and radius a, and also that its surface density is σ.

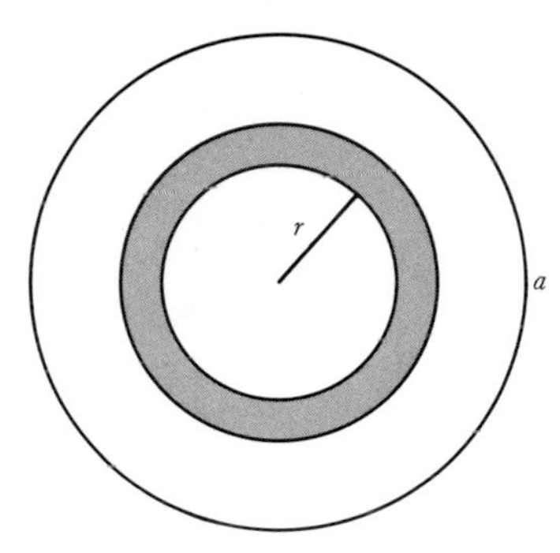

We begin by dividing the disc into a series of concentric hoops, and suppose that a typical hoop has radius r and thickness δr.

Then area of hoop $= 2\pi r\ \delta r$

and its mass $= 2\pi\sigma r\ \delta r.$

Hence m.i. of hoop $= (2\pi\sigma r\ \delta r)r^2$

$= 2\pi\sigma r^3\ \delta r$

and m.i. of disc $= \lim \sum 2\pi\sigma r^3\ \delta r$

$= \int_0^a 2\pi\sigma r^3\ \mathrm{d}r$

$= \frac{1}{2}\pi\sigma a^4.$

But $M = \sigma \times \pi a^2 = \pi\sigma a^2.$

So m.i. $= \frac{1}{2}Ma^2.$

Uniform rod

Suppose that a uniform rod has mass M and length $2a$. Its moment of inertia clearly depends on the chosen axis, and we shall consider two cases, both of them perpendicular to the rod and through its end and mid-point respectively.

(*a*) *About perpendicular axis through an end*

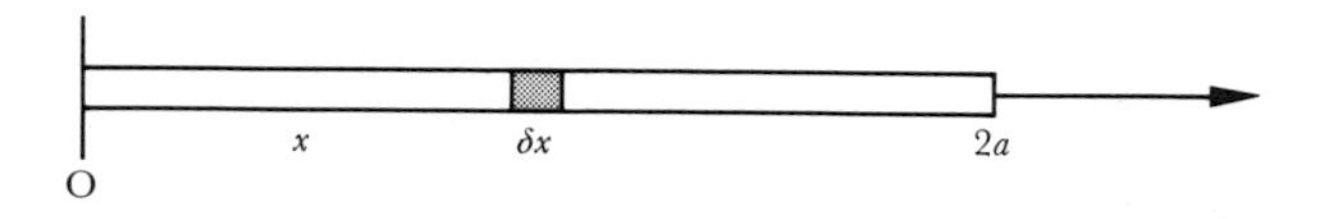

Let linear density of rod be ρ.

Then $\quad M = 2\rho a.$

Taking short elements of length δx and mass $\rho\, \delta x$, we see that

$$\begin{aligned}\text{m.i.} &= \lim \sum \rho\, \delta x \times x^2 \\ &= \int_0^{2a} \rho x^2\, \mathrm{d}x \\ &= \left[\tfrac{1}{3}\rho x^3\right]_0^{2a} = \frac{8\rho a^3}{3}.\end{aligned}$$

But $\quad M = 2\rho a.$

So $\quad \text{m.i.} = \dfrac{4Ma^2}{3}.$

(*b*) *About perpendicular axis through its mid-point*

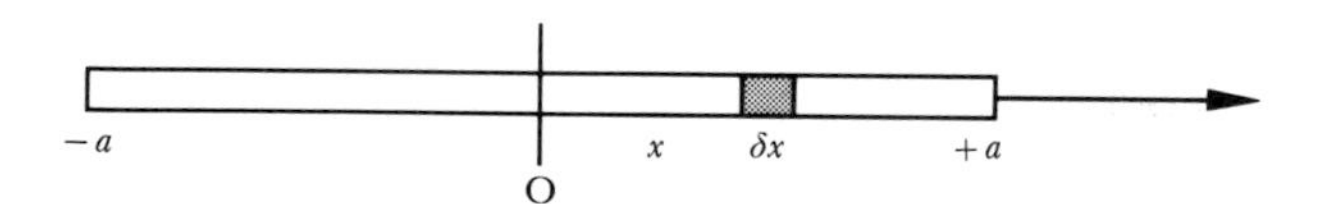

In a similar way,

$$\begin{aligned}\text{m.i.} &= \lim \sum \rho\, \delta x \times x^2 \\ &= \int_{-a}^{a} \rho x^2\, \mathrm{d}x = \frac{2\rho a^3}{3}.\end{aligned}$$

So $\quad \text{m.i.} = \tfrac{1}{3}Ma^2.$

Radius of gyration

Since $\quad I = \sum mr^2,$

we can also express it as $M\kappa^2$, where κ is a length, called the *radius of gyration* of the body about this axis.

Hence $\quad \boxed{I = M\kappa^2}$

and we see that

$$\kappa^2 = \frac{\sum mr^2}{\sum m}$$

is precisely the variance of the distances r of masses m from the given axis, so that κ is their standard deviation (or 'root mean square').

Thus, for a uniform circular disc about a perpendicular axis through its centre,

$$M\kappa^2 = \tfrac{1}{2}Ma^2$$

$$\Rightarrow \quad \kappa = \frac{a}{\sqrt{2}} \approx 0.71a,$$

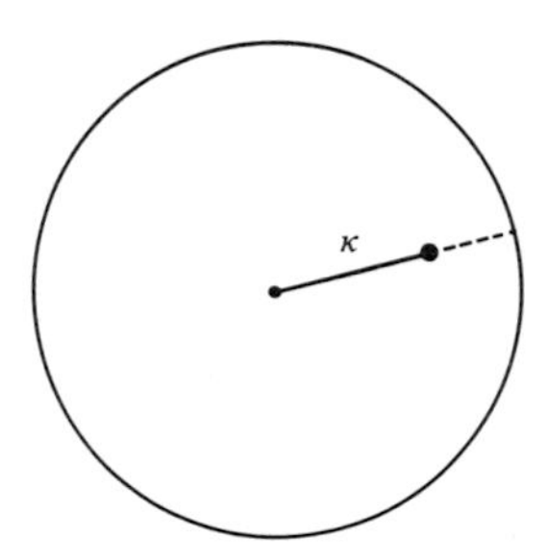

and for a uniform rod of length $2a$ about a perpendicular axis through its centre,

$$M\kappa^2 = \tfrac{1}{3}Ma^2$$

$$\Rightarrow \quad \kappa = \frac{a}{\sqrt{3}} \approx 0.58a.$$

Two general theorems

In the calculation of moments of inertia it is sometimes convenient to make use of a general theorem, and here we shall establish two: one for finding the moment of inertia about an axis in terms of the moment of inertia about a parallel axis through the centre of mass, and the other which is applicable only to a lamina.

Parallel axis theorem

Suppose that the moment of inertia of a body about a certain axis is I_O and about a parallel axis through the centre of mass is I_G.

We shall suppose that a typical mass m of the body lies at a point P and that a section is taken through P perpendicular to the axes (which in our figure are supposed to be perpendicular to the page) meets them in points O and G.

Taking an origin at O, we let

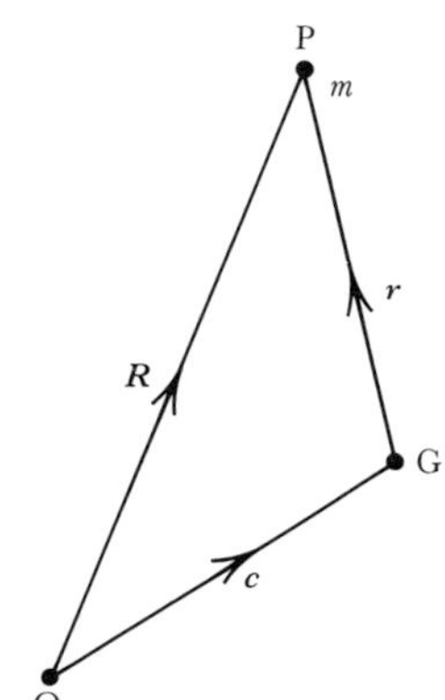

$\boldsymbol{OG} = \boldsymbol{c}, \qquad \boldsymbol{GP} = \boldsymbol{r}, \qquad \boldsymbol{OP} = \boldsymbol{R}.$

Then $I_O = \sum mR^2 = \sum m\boldsymbol{R}^2$

$$= \sum m(\boldsymbol{r} + \boldsymbol{c})^2$$
$$= \sum m\boldsymbol{r}^2 + 2\sum m\boldsymbol{r}.\boldsymbol{c} + \sum m\boldsymbol{c}^2$$
$$= \sum m\boldsymbol{r}^2 + 2\left(\sum m\boldsymbol{r}\right).\boldsymbol{c} + \left(\sum m\right)\boldsymbol{c}^2.$$

We now let the total mass of the body, $\sum m = M$.

Also, since the centre of mass lies on the axis through G,

$\sum m\boldsymbol{r} = \boldsymbol{0}.$

So $I_O = \sum m\boldsymbol{r}^2 + 2\boldsymbol{c}.\boldsymbol{0} + \left(\sum m\right)\boldsymbol{c}^2$

$$\Rightarrow \boxed{I_O = I_G + Mc^2}$$

So, for example, we could have found the moment of inertia of a uniform rod about its end from that about its centre, simply by writing

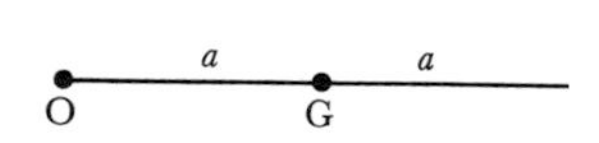

$$I_O = I_G + Ma^2$$
$$= \tfrac{1}{3}Ma^2 + Ma^2$$
$$= \tfrac{4}{3}Ma^2.$$

And the moment of inertia of a uniform hoop of mass M and radius a about a perpendicular axis through a point of its circumference

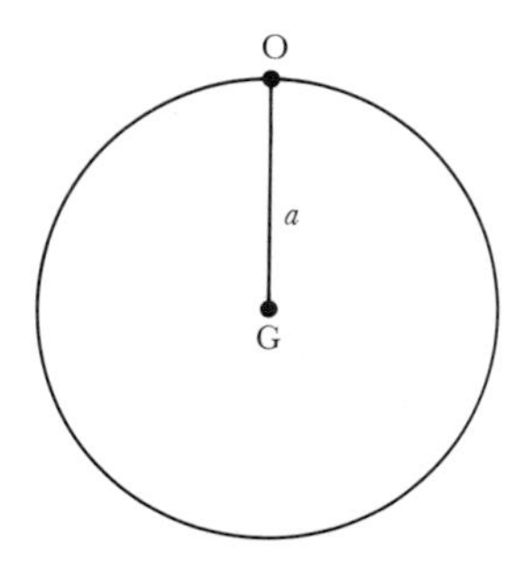

$$= I_G + Ma^2$$
$$= Ma^2 + Ma^2 = 2Ma^2.$$

'Perpendicular axes' theorem for a lamina

Suppose that perpendicular axes Ox, Oy, Oz are taken with Ox, Oy in the plane of a lamina and Oz perpendicular to its plane. Denoting moments of inertia about these axes by I_x, I_y, I_z,

$$\begin{aligned}
I_O &= \sum mr^2 \\
&= \sum m(y^2 + x^2) \\
&= \sum my^2 + \sum mx^2 \\
&= I_x + I_y.
\end{aligned}$$

So $\boxed{I_z = I_x + I_y}$

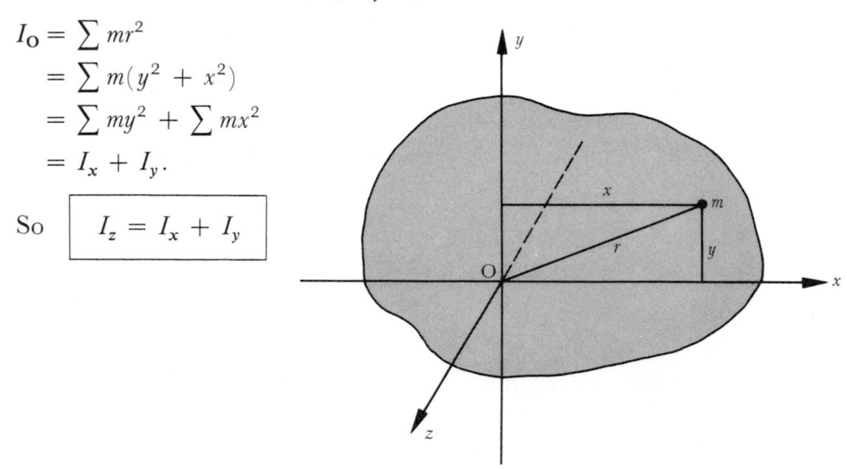

Hence, for example, we can immediately find the moment of inertia I of a uniform disc about a diameter simply by observing that, since $I_x = I_y$,

$$\begin{aligned}
I + I &= \tfrac{1}{2}Ma^2 \\
\Rightarrow \quad I &= \tfrac{1}{4}Ma^2.
\end{aligned}$$

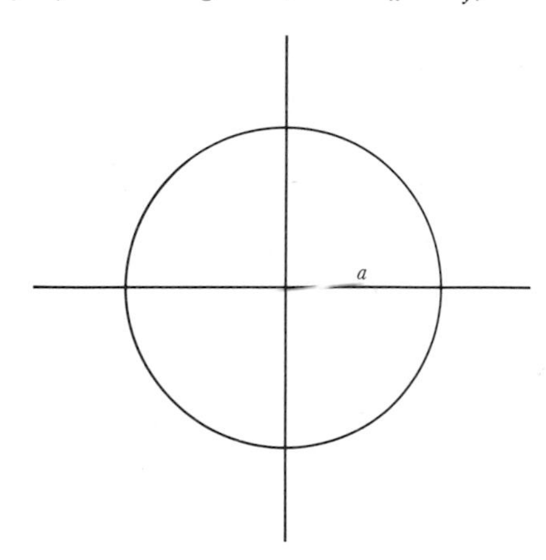

Similarly, we know that a uniform rectangular lamina of mass M and sides $2a$, $2b$ can be dissected into strips parallel to its sides, so that its moments of inertia about parallel axes through its centre are

$\tfrac{1}{3}Ma^2$ and $\tfrac{1}{3}Mb^2$.

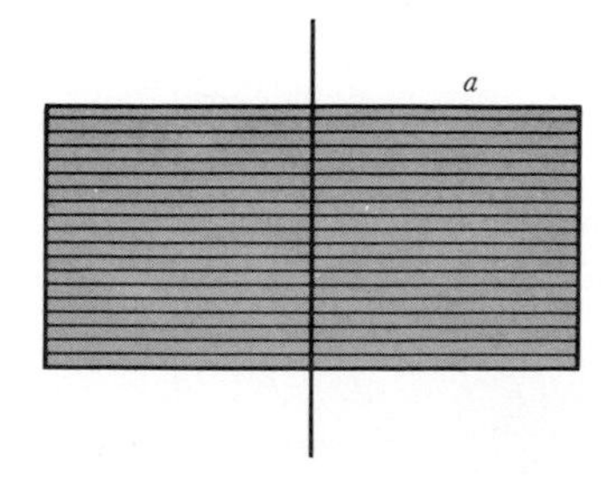

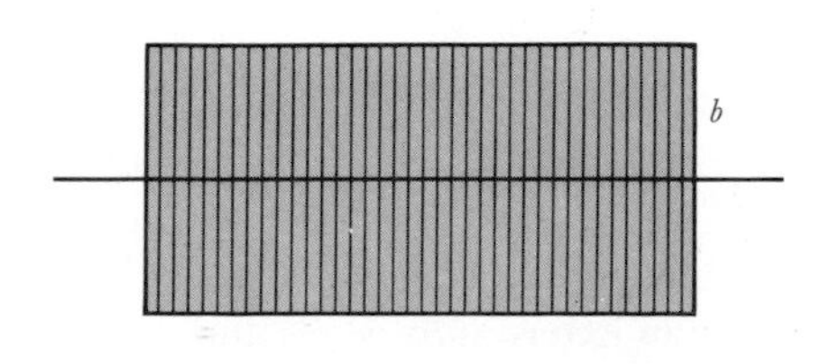

Hence, using the perpendicular axes theorem, we see that its moment of inertia about a perpendicular axis through its centre is

$$\tfrac{1}{3}Ma^2 + \tfrac{1}{3}Mb^2$$
$$= \tfrac{1}{3}M(a^2 + b^2).$$

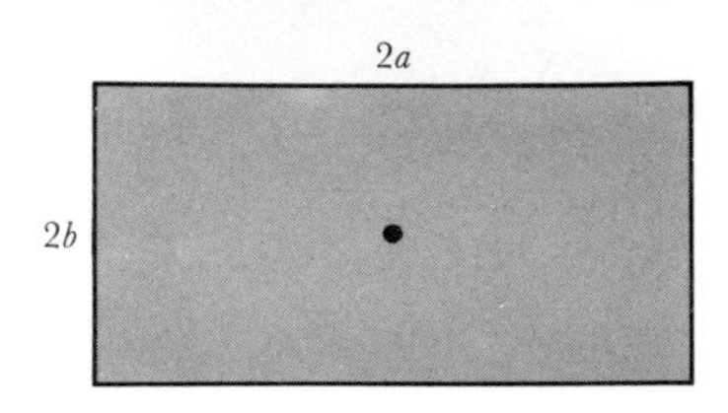

Exercise 11.11b

Calculate the moments of inertia of:

1 A uniform rectangular lamina of mass M and sides $2a$, $2b$ about a perpendicular axis through a corner.

2 A uniform circular hoop of mass M and radius a about:
(*i*) a diameter;
(*ii*) a tangent;
(*iii*) a perpendicular axis through a point on its circumference.

3 A uniform circular lamina of mass M and radius a about:
(*i*) a tangent;
(*ii*) a perpendicular axis through a point on its circumference.

4 A uniform sphere of mass M and radius a about:
(*i*) a diameter;
(*ii*) a tangent.

5 A uniform spherical shell of mass M and radius a about:
(*i*) a diameter;
(*ii*) a parallel line through a point of the shell.

(Hint: $\sum m(y^2 + z^2) = \sum m(z^2 + x^2) = \sum m(x^2 + y^2)$ and $x^2 + y^2 + z^2 = a^2$.)

6 A uniform solid cone of mass M, height h and base-radius a, about:
(*i*) its axis of symmetry;
(*ii*) a perpendicular axis through its vertex;
(*iii*) a diameter of its base.

Dynamics of rotating bodies

Suppose that a body is rotating with angular velocity ω about a fixed axis, that the moment of the external forces about this axis has magnitude L and the moment of inertia of the body about this axis is I.

Further, suppose that on a mass m at a distance r from the axis there are acting transversely:
(*i*) an external force P; and
(*ii*) an internal force Q.

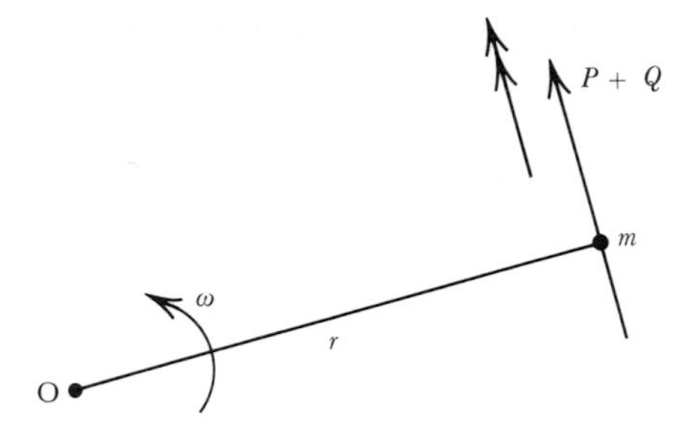

Then $\quad P + Q = mr\dot{\omega}$

$\Rightarrow \quad r(P + Q) = mr^2\dot{\omega}$

$\Rightarrow \quad \sum rP + \sum rQ = (\sum mr^2)\dot{\omega} = I\dot{\omega}.$

But since the forces Q are internal, it follows from Newton's third law that $\sum rQ = 0$.

Hence $\sum rP = I\dot{\omega}$.

But $\sum rP = L$.

So $\boxed{L = I\dot{\omega}}$

Now for a small rotation $\delta\theta$ the work done by P is $Pr\,\delta\theta$. So in a finite displacement, the work done by P is $\int Pr\,d\theta$; and the sum of these amounts of work for all the external forces is

$$\sum \int Pr\,d\theta = \int (\sum Pr)\,d\theta = \int L\,d\theta.$$

But $\quad L = I\dot{\omega} = I\dfrac{d\theta}{dt}\dfrac{d\omega}{d\theta} = I\omega\dfrac{d\omega}{d\theta}.$

So $\quad \displaystyle\int L\,d\theta = \int I\omega\frac{d\omega}{d\theta}\,d\theta = \int I\omega\,d\omega$

$$\Rightarrow \quad \boxed{\int L\,d\theta = [\tfrac{1}{2}I\omega^2]}$$

and again we see that the total work done by the external forces is equal to the gain in kinetic energy.

Example 1

A uniform flywheel of mass 200 kg and radius 0.5 m is initially at rest and is made to rotate about its fixed axis by a rope which is wrapped round the flywheel for three complete turns and then pulled with a steady force of 200 N. Find:

(*i*) the angular acceleration of the flywheel;

(*ii*) its final angular velocity.

(*i*) As the flywheel is uniform,

$$\text{m.i.} = \tfrac{1}{2} \times 200 \times (0.5)^2 = 25 \text{ kg m}^2.$$

But the moment about its axis of the tension in the rope

$$= 200 \text{ N} \times 0.5 \text{ m} = 100 \text{ Nm}.$$

So the angular acceleration $\dot{\omega}$ is given by

$$100 = 25\dot{\omega} \quad \Rightarrow \quad \dot{\omega} = 4.$$

Hence the angular acceleration is 4 rad s^{-2}.

(*ii*) As this angular acceleration is constant throughout three complete revolutions ($= 6\pi$ rad), the final angular velocity ω is such that

$$\omega^2 = 0^2 + 2 \times 4 \times 6\pi = 48\pi$$

$$\Rightarrow \quad \omega = 12.3.$$

Hence the final angular velocity is 12.3 rad s^{-1}.

Alternatively, the total work done by the tension

$$= 100 \text{ Nm} \times 6\pi \text{ rad} = 600\pi \text{ J}$$

$$\text{and final k.e.} \quad = \tfrac{1}{2} \times 25 \times \omega^2 \text{ J}.$$

$$\text{So} \quad \tfrac{1}{2} \times 25\omega^2 = 600\pi$$

$$\Rightarrow \quad \omega^2 = 48\pi$$

$$\Rightarrow \quad \omega = 12.3.$$

Example 2

A flywheel whose moment of inertia is 1.01 kg m^2 is rotating at approximately 1 200 rev min^{-1}, the value oscillating between 1.0 per cent above and 1.0 per cent below this figure. Calculate in joules the difference between the maximum and minimum values of the kinetic energy. If the change from minimum to maximum occurs in one-hundredth of a second calculate the average power input in kilowatts during this period.

Power to the flywheel is cut off at 1 200 rev min and the wheel is brought to rest by a constant couple of 80 N m: calculate how many revolutions the wheel makes in coming to rest. (M.E.I.)

$$\text{Maximum angular velocity} = 1\,212 \text{ rev min}^{-1}$$

$$= \frac{1\,212 \times 2\pi}{60} \text{ rad s}^{-1}$$

$$= 40.4\pi \text{ rad s}^{-1}.$$

Minimum angular velocity $= 1\ 188$ rev min^{-1}
$= 39.6\pi$ rad s^{-1}.

So difference of kinetic energy

$= \frac{1}{2} \times 1.01 \times \{(40.4\pi)^2 - (39.6\pi)^2\}$
$= \frac{1}{2} \times 1.01 \times 80\pi \times 0.8\pi$
$= 0.505 \times 64\pi^2 \approx 319$ J.

Hence average power input in 0.01 s

$$= \frac{319}{0.01} = 31\ 900 \text{ W} = 31.9 \text{ kW}.$$

When power is cut,

k.e. $= \frac{1}{2} \times 1.01 \times (40\pi)^2 = 808\pi^2$ J.

Now if the angle through which the flywheel turns before coming to rest is α, the work done by the braking couple is -80α N m.

But work done = gain in k.e.

So $-80\alpha = -808\pi^2$

$\Rightarrow \alpha = 10.1\,\pi^2$ rad ≈ 15.9 rev.

Exercise 11.11c

1 Calculate the angular momentum and the kinetic energy of:
(*i*) a uniform hoop of mass 8 kg and radius $\frac{1}{2}$ m which is spinning in its own plane about its centre with angular velocity 6 rad s^{-1};
(*ii*) a uniform disc of mass 6 kg and radius 20 cm which is spinning in its own plane about its centre with angular velocity 4 rad s^{-1};
(*iii*) the same disc rotating at the same speed, but about a diameter;
(*iv*) a uniform bar of mass 3 kg and length 4 m rotating about one of its ends with angular velocity 6 rad s^{-1};
(*v*) a hollow cylinder of mass 100 kg and radius 40 cm, rotating about its axis at 100 revolutions/min;
(*vi*) the earth, assuming it to be a uniform sphere of mass 6×10^{24} kg and radius 6.4×10^6 m.

2 A uniform flywheel has mass m and radius a, and is rotating freely about its axis with angular velocity w. If a constant braking force F is then applied tangentially to its surface, find:
(*i*) its angular deceleration;
(*ii*) the time it takes to stop;
(*iii*) the angle through which it turns.

3 A bicycle is held with one of its wheels rotating freely with angular velocity 200 rev/min. The wheel has mass 2 kg and radius 25 cm, and when the brakes are lightly applied it is brought to rest in three revolutions. Assuming that the mass of the wheel and the constant braking force are both concentrated at its circumference, find the magnitude of this force. For how long would it need to act; and if $\mu = 0.1$, what is the total contact force between the brakes and the rim?

4 A gyroscope consists of a uniform disc of mass 2 kg and diameter 10 cm mounted on a light axle of diameter 4 mm. A string is wrapped round the axle 20 times and pulled sharply with a constant force. If this pull takes 2 s to unwrap the string, find:
(*i*) the angular acceleration of the disc;
(*ii*) its final angular velocity;
(*iii*) the tension in the string.

5 A rope 2 m long is wound round the axle of a flywheel and pulled with a constant force of 400 N. When the string is unwound the flywheel is rotating at 200 rev/min. Find its moment of inertia.

6 A uniform rod of length $2a$ and mass M is freely pivoted at one end and is initially in a vertical position above its pivot. If it then topples over, find its loss of potential energy by the time it is (*i*) horizontal, (*ii*) vertically below the pivot; and so find its angular velocity at these two positions. Finally, by considering the moments of its weight, find the corresponding angular accelerations.

7 A uniform cylindrical flywheel of mass 1 000 kg and radius 0.4 m is being accelerated by an electric motor which is working at a constant rate of 10 kW. Ignoring all resistances, find the angular velocity of the flywheel after one minute.

8 A heavy uniform rod of mass m and length l is suspended from one end and oscillates freely about its position of equilibrium. Find:
(*i*) the moment of its weight about the point of suspension when the rod makes an angle θ with the downward vertical;
(*ii*) the equation of motion when the rod is in this position;
(*iii*) an approximate equation of motion when θ is small;
(*iv*) the period of small oscillations.

9 A light thread is wrapped round a uniform cylinder of mass m and radius a which is free to rotate about a horizontal axis. A mass M is tied to the end of this string and allowed to fall from rest in a vertical line. After it has fallen a distance x, find:
(*i*) its speed v;
(*ii*) its acceleration a.

10 A uniform circular disc, of mass m and radius a, is free to rotate about a frictionless horizontal axis through its centre and perpendicular to its plane. One end of a light string is attached to a point on the circumference of the disc and part of the string is wound on the circumference. The other end of the string carries a particle A of mass $2m$ hanging freely. The system is released from rest and a restoring couple of moment $mga\theta$ acts on the disc when θ is the angular displacement of the disc. Write down the equation of motion of the disc about its centre and the equation of motion of A. Hence show that the tension in the vertical part of the string is $\frac{2}{5}(1 + 2\theta)mg$. Show also that

$$5a\frac{\mathrm{d}^2\theta}{\mathrm{d}t^2} = 4g - 2g\theta.$$

Deduce that the motion of A is simple harmonic and find the period of this motion.

(You may assume that part of the string always remains wound on the disc during the motion.) (O.C.)

11.12 Dimensions

It will have been noticed that this introduction to mechanics has been based entirely upon the three fundamental concepts, or *dimensions*, of mass, length and time. If we now use the symbol [ ] to indicate the dimensions of a quantity, we can write

$$[\text{mass}] = \mathbf{M}, \qquad [\text{length}] = \mathbf{L}, \qquad [\text{time}] = \mathbf{T}.$$

Furthermore, we can proceed to find the dimensions of other quantities in terms of **M**, **L**, **T**, such as

$$[\text{area}] = [\text{length} \times \text{length}] = \mathbf{L}^2,$$

$$[\text{volume}] = [\text{area} \times \text{length}] = \mathbf{L}^3,$$

$$[\text{density}] = \left[\frac{\text{mass}}{\text{volume}}\right] = \frac{\mathbf{M}}{\mathbf{L}^3} = \mathbf{ML}^{-3},$$

$$[\text{velocity}] = \left[\frac{\text{length}}{\text{time}}\right] = \frac{\mathbf{L}}{\mathbf{T}} = \mathbf{LT}^{-1},$$

$$[\text{acceleration}] = \left[\frac{\text{velocity}}{\text{time}}\right] = \frac{\mathbf{LT}^{-1}}{\mathbf{T}} = \mathbf{LT}^{-2},$$

$$[\text{force}] = [\text{mass} \times \text{acceleration}] = \mathbf{MLT}^{-2},$$

$$[\text{moment of a force}] = \mathbf{MLT}^{-2} \times \mathbf{L} = \mathbf{ML}^2\mathbf{T}^{-2},$$

$$[\text{impulse}] = [\text{force} \times \text{time}] = \mathbf{MLT}^{-2} \times \mathbf{T} = \mathbf{MLT}^{-1},$$

$$[\text{momentum}] = [\text{mass} \times \text{velocity}] = \mathbf{MLT}^{-1},$$

$$[\text{work}] = [\text{force} \times \text{distance}] = \mathbf{ML}^2\mathbf{T}^{-2},$$

$$[\text{kinetic energy}] = [\text{mass} \times (\text{velocity})^2] = \mathbf{ML}^2\mathbf{T}^{-2},$$

$$[\text{power}] = \left[\frac{\text{work}}{\text{time}}\right] = \mathbf{ML}^2\mathbf{T}^{-3}.$$

It will be seen that the dimensions of impulse are identical (or *consistent*) with those of momentum, and that the dimensions of work are consistent with those of energy. This notion of consistency can be used to check, and sometimes to derive, equations in mechanics.

Example 1

A piano wire has mass m, length l, and tension F, and a student knows that its period of vibration is

$$\text{either} \quad 2\pi\sqrt{\left(\frac{ml}{F}\right)} \quad \text{or} \quad 2\pi\sqrt{\left(\frac{F}{ml}\right)}.$$

Which is more likely to be correct?

Since $[m] = \mathbf{M}$, $[l] = \mathbf{L}$, $[F] = \mathbf{MLT}^{-2}$, it follows that the dimensions of these two possibilities are

$$\sqrt{\left(\frac{\mathbf{ML}}{\mathbf{MLT}^{-2}}\right)} \quad \text{or} \quad \sqrt{\left(\frac{\mathbf{MLT}^{-2}}{\mathbf{ML}}\right)}$$

$$= \sqrt{\mathbf{T}^2} = \mathbf{T} \qquad\qquad = \sqrt{\mathbf{T}^{-2}} = \mathbf{T}^{-1}.$$

Hence it is clear that the second possibility for a period of dimension $\mathbf{T}$ would not be consistent, so that the likelier (and in fact correct) formula is $2\pi\sqrt{(ml/F)}$.

Example 2

A simple pendulum consists of a bob of mass m attached to a string of length l and the pendulum is swinging at a point where the acceleration due to gravity is g. Investigate, by means of dimensions, the way in which the period T depends upon m, l, and g.

Firstly, we *suppose* that the period T depends jointly upon m, l, and g in such a way that:

$T \propto m^{\alpha}l^{\beta}g^{\gamma}$, where α, β, γ are constants.

$$\text{But} \quad [m^{\alpha}l^{\beta}g^{\gamma}] = \mathbf{M}^{\alpha}\mathbf{L}^{\beta}(\mathbf{LT}^{-2})^{\gamma}$$
$$= \mathbf{M}^{\alpha}\mathbf{L}^{\beta+\gamma}\mathbf{T}^{-2\gamma}$$

$$\text{and} \qquad [T] = \mathbf{T}.$$

So for dimensional consistency,

$$\mathbf{M}^{\alpha}\mathbf{L}^{\beta+\gamma}\mathbf{T}^{-2\gamma} = \mathbf{T}$$

$$\Rightarrow \qquad \alpha = 0$$
$$-2\gamma = 1 \quad \Rightarrow \quad \gamma = -\tfrac{1}{2}.$$
$$\beta + \gamma = 0 \quad \Rightarrow \quad \beta = \tfrac{1}{2}$$

$$\text{Hence} \quad T \propto m^0 l^{\frac{1}{2}} g^{-\frac{1}{2}}$$

$$\Rightarrow \qquad T \propto \sqrt{\left(\frac{l}{g}\right)}.$$

It will be recognised that although this does not constitute a *proof*, it is nevertheless a highly convenient approach to such an investigation, which is confirmed by the results of p. 233 that $T = 2\pi\sqrt{(l/g)}$.

Exercise 11.12

1 State the dimensions, in terms of **M**, **L**, **T**, of the following quantities:

(*i*) pressure (force per unit area);
(*ii*) line density;
(*iii*) surface density;
(*iv*) frequency;
(*v*) angular velocity;
(*vi*) angular acceleration;
(*vii*) angle;
(*viii*) moment of inertia;
(*ix*) rate of loss of mass (e.g., of a rocket);
(*x*) constant of gravitation (see 8.11);
(*xi*) stiffness of a spring (sce 11.8);
(*xii*) surface tension (energy per unit area).

2 Use the method of dimensions to predict how:

(*i*) The tension T in a string depends upon the mass m of a particle which is being whirled round on its end in a circle of radius r and with speed v.
(*ii*) The height h reached by a stone will depend upon its mass m, the energy E with which it is projected vertically and the acceleration g due to gravity.
(*iii*) The speed v of sound in a gas will depend upon its pressure p, its density ρ and the acceleration g due to gravity.
(*iv*) The frequency f of oscillation of a light spring depends upon its stiffness k, the mass m which it is supporting and the acceleration g due to gravity.
(*v*) The velocity v of waves in a deep liquid depends upon its density ρ, their wavelength λ and the acceleration g due to gravity.

3 Use the method of dimensions to find an expression for the thrust of moving air on a fixed obstacle, assuming it to be the form $T = kv^{\alpha}\rho^{\beta}A^{\gamma}$ where v is the air velocity, ρ its density, A the frontal area which the obstacle presents to it, and k is a non-dimensional constant. Explain briefly the nature of a dynamical argument which leads us when using absolute units of force to expect the experimental value of k always to be less than unity.

Under certain conditions the compressibility c of the air can affect the thrust appreciably, c being defined as fractional change in volume produced per unit change in pressure. Find the dimensions of c and show that we can now obtain a dimensionally correct formula for T by replacing k by any polynomial in a quantity z defined as $v\sqrt{(c\rho)}$. (M.E.I.)

Miscellaneous problems 11

1 Let O be the centre of the circumcircle of $\triangle$ABC whose vertices have position vectors $\boldsymbol{a}$, $\boldsymbol{b}$ and $\boldsymbol{c}$. Show that the point H, with position vector $\boldsymbol{h} = \boldsymbol{a} + \boldsymbol{b} + \boldsymbol{c}$, is the orthocentre of $\triangle$ABC (i.e. AH $\perp$ BC, BH $\perp$ CA, CH $\perp$ AB) and hence show that the centroid G divides OH in the ratio 1:2.

2 The sides BC, CA, AB of $\triangle$ABC are divided in the same ratio by P, Q, R respectively. Investigate the centroid of $\triangle$PQR.

3 O is a point in the plane of a triangle ABC and AO, BO, CO meet BC, CA, AB respectively in points P, Q, R. Using bold letters to denote position vectors from O, prove

(*i*) that it is possible to find scalars α, β, γ such that

$$\alpha\boldsymbol{a} + \beta\boldsymbol{b} + \gamma\boldsymbol{c} = \boldsymbol{0};$$

(*ii*) that $\boldsymbol{p}$ can be expressed as

$$-\frac{\alpha}{\beta + \gamma}\boldsymbol{a} \equiv \frac{\beta\boldsymbol{b} + \gamma\boldsymbol{c}}{\beta + \gamma};$$

(*iii*) that P divides BC in the ratio γ/β;

(*iv*) *Ceva's theorem*, that

$$\frac{BP}{PC} \times \frac{CQ}{QA} \times \frac{AR}{RB} = 1.$$

Finally, verify Ceva's theorem in the three cases when O is the centroid, the orthocentre, the circumcentre, of $\triangle$ABC.

4 If a line cuts the sides BC, CA, AB (extended if necessary) of a triangle ABC prove *Menelaus' theorem*, that

$$\frac{BP}{PC} \times \frac{CQ}{QA} \times \frac{AR}{RB} = -1.$$

5 ABC, A′B′C′ are two skew lines and $AB:BC = A'B':B'C'$. Prove that the mid-points of AA′, BB′, CC′ are collinear.

6 The vertices of a tetrahedron are A, B, C, D and the points P, Q, R, S divide the segments AB, BC, CD, DA in the ratios $p:1$, $q:1$, $r:1$, $s:1$ where p, q, r, s are all positive. Show that P, Q, R, S are coplanar if $pqrs = 1$.

Show that, if also $p = r$ and $q = s$, then PQRS is a parallelogram. (C.S.)

7 The base of a solid hemisphere of radius a is firmly attached to one plane end, also of radius a, of a solid cylinder of length l, thus forming one solid of revolution. The material is the same and uniform throughout the composite body. Find the position of its centre of gravity.

If the body can rest in equilibrium with any point of the surface of the hemi-spherical portion resting on a horizontal plane, show that $a = l\sqrt{2}$. (M.E.I.)

8 Prove that if the sum of the squares of two opposite edges of a tetrahedron is equal to the sum of the squares of another pair of opposite edges, then the remaining pair of opposite edges are perpendicular.

9 OA, OB, OC are three lines through the point O and the angles BOC, COA, AOB are α, β, γ respectively. Calculate $\cos^2 \theta$, where θ is the angle between the line OA and the plane OBC. (C.S.)

10 A ball of mass m is thrown into the air with initial velocity $\boldsymbol{u}$, and during its flight it experiences a resistance from the air given by the vector $-mc\boldsymbol{v}$, where $\boldsymbol{v}$ is the velocity of the ball at that instant. Write down a differential equation for the velocity as a function of the time, and verify that this equation and the initial conditions are both satisfied if

$$\boldsymbol{v} = \frac{1 - e^{-ct}}{c}\boldsymbol{g} + e^{-ct}\boldsymbol{u},$$

where $\boldsymbol{g}$ is the vector acceleration due to gravity.

Find an expression for $\boldsymbol{r}$, the displacement of the ball from its initial position at time t; and show that, if c is small, then this is approximately equal to

$$\boldsymbol{r}^* - c(\tfrac{1}{6}t^3\boldsymbol{g} + \tfrac{1}{2}t^2\boldsymbol{u}),$$

where $\boldsymbol{r}^*$ is the position that it would occupy in the absence of air resistance. (M.E.I.)

11 A particle of unit mass is attracted to a point O with a force $-9\boldsymbol{r}$, where $\boldsymbol{r}$ is the position vector of the particle relative to O. Initially the particle is at a point with position vector $16\boldsymbol{i}$ and moving with velocity $24\boldsymbol{j}$, where $\boldsymbol{i}$ and $\boldsymbol{j}$ are fixed unit vectors in perpendicular directions. Verify that the subsequent motion is described by the equation

$$\boldsymbol{r} = 16\boldsymbol{i}\cos 3t + 8\boldsymbol{j}\sin 3t,$$

and interpret this equation geometrically.

If there were also a resistance to motion of magnitude ten times the speed, obtain a differential equation for the motion and show that it has a solution in the form

$$\boldsymbol{r} = \boldsymbol{A}\,e^{-t} + \boldsymbol{B}\,e^{-9t}.$$

With the same initial conditions as before, find the values of $\boldsymbol{A}$ and $\boldsymbol{B}$, and investigate the nature of the motion for $t > 0$. (S.M.P.)

12 A particle of mass m is projected with velocity V at an angle α to the horizontal. The air resistance is mk times the speed of the particle, where k is a constant, and opposes the direction of motion at all times. Prove that the equations of motion of the particle can be put in the form

$$\ddot{x} + k\dot{x} = 0$$

$$\ddot{y} + k\dot{y} + g = 0$$

where x and y are respectively the horizontal and vertical distances from the point of projection.

Show that

$$x = \frac{V}{k}\cos\alpha(1 - e^{-kt})$$

and find a similar expression for y.

Sketch the path of the projectile. (M.E.I.)

13 A steady uniform stream of air of density ρ and speed u strikes at right angles a plane surface of area A and proceeds, after the contact, with its effective speed still in the original direction and equal to ku (where $k > 0$). Derive an expression for the thrust of the air on the area, stating carefully the principles on which your derivation is based.

A householder has a garage with rectangular doors each 1.20 m wide by 2.25 m high, turning about vertical hinges at their edges. On a windy day he attempts to keep one door open in such a position that the wind is blowing at right angles to the door. To do this he puts a brick on the ground at the outer corner of the door, hoping that the friction of the brick on the ground will be sufficient for the task. Given the data listed below, calculate the greatest wind-speed for which the brick will be effective: state any additional assumptions you find it necessary to make in the course of your work.

Mass of brick = 3.0 kg; coefficient of friction = 0.4; density of air = 1.10 kg m^{-3}; $k = 0.6$; $g = 10$ m s^{-2}. (M.E.I.)

14 A shell of mass M is at rest in space, when it bursts into two fragments, the energy released being E. Show that the relative speed of the fragments after separation cannot be less than $2\sqrt{(2E/M)}$.

Explain how your conclusion is affected if the shell is moving initially with speed U. (C.S.)

15 A space craft of mass 10^4 kg is going round the moon in a circular orbit at a height of 10^6 m. Find the time of the orbit, taking g on the moon as 1.60 m s^{-2} and the diameter of the moon as 4.14×10^6 m.

To escape from the moon the kinetic energy of the body has to be doubled. What is the magnitude and direction of the least impulse which can be applied to the body to enable it to escape?

If this impulse is spread over a period of one minute, what is the apparent weight of a 100 kg man inside the space craft during this time? (M.E.I.)

16 A particle of mass m is attached to the middle point of a light elastic string of natural length a and modulus mg. The ends of the string are attached to two fixed points A and B, A being at a distance $2a$ vertically above B. Prove that the particle can rest in equilibrium at a depth $5a/4$ below A and find the period of oscillation if it is displaced slightly in the vertical direction.

17 A mountaineer falls over a cliff. He is attached to a rope which, providentially, stretches so that he just touches the ground at the foot of the cliff. Find the height of the cliff and the time taken for the mountaineer to reach the ground (in terms of his mass, the length of the unstretched rope and its elastic modulus). (C.S.)

18 Three particles A, B and C each of mass m lie at rest in a straight line on a smooth horizontal table, joined by equal taut strings AB, BC. A horizontal impulse P, is given to B in a direction perpendicular to the line of the strings. Describe the nature of the motion up to the instant when A and C meet, and calculate the loss of energy of the system if these masses do not separate after colliding. (M.E.I.)

19 A pump working effectively at P kW delivers water through a nozzle of area A cm^2 at a speed of v m s^{-1}, raising the water h m in the process. Obtain a formula for P in terms of A, v, h and g; assume that 1 m^3 of water has a mass of 10^3 kg.

The jet is directed horizontally at right angles to a vertical blade of a water wheel which is initially at rest. The wheel has a moment of inertia I about its axis which is horizontal and the point of contact of the water is c m from the axle; the water falls vertically after striking the blade.

What is the initial angular acceleration of the wheel? At what rate is energy being given to the wheel initially? (M.E.I.)

20 A tank containing a liquid of density ρ has a hole in the bottom. The speed u at which the liquid escapes (averaged over the cross-section of the hole) is given by $u = \lambda \rho V g$, where V is the volume of liquid in the tank and λ is a constant. Find the dimensions of λ in terms of mass, length and time.

Given that the hole has cross-sectional area A, write down a differential equation for V as a function of t, in terms of the various constants. (S.M.P.)

21 The velocity of propagation of waves of wavelength l on the surface of a liquid, is proportional to

$$\sqrt{\frac{\sigma}{\rho l}},$$

where σ is the surface tension and ρ the density. Deduce the physical dimensions of σ.

Lord Rayleigh showed that the period, T, of vibration of a small liquid drop, when given a slight distortion, depended only on σ, ρ and the radius r of its spherical equilibrium shape. Assuming that

$$T = C\sigma^{\alpha}\rho^{\beta}r^{\delta},$$

where C, α, β and γ are dimensionless constants, find α, β and γ. (C.S.)

12

Probability distributions and statistics

12.1 Probability distributions and generators

Take two coins, spin them as a pair ten times and record your results: two heads (HH), two tails (TT), or one of each (HT). When the author performed this experiment he obtained the sequence

HT, HT, TT, HT, HH, HH, HT, HT, HH, HT,

which can conveniently be classified according to the frequency f_r with which the number of heads obtained is x_r:

No. of heads (x_r)	Frequency (f_r)
$x_1 = 0$	$f_1 = 1$
$x_2 = 1$	$f_2 = 6$
$x_3 = 2$	$f_3 = 3$
	$n = 10$

In chapter 7 we called this a frequency distribution, and saw that its mean m is given by:

$$m = \frac{1}{n}\sum f_r x_r = \tfrac{1}{10}(1 \times 0 + 6 \times 1 + 3 \times 2) = 1.2$$

and its variance s^2 by

$$\begin{aligned} s^2 &= \frac{1}{n}\sum f_r(x_r - m)^2 \\ &= \frac{1}{n}\sum f_r x_r^2 - m^2 \end{aligned}$$

$= \frac{1}{10}(1 \times 0^2 + 6 \times 1^2 + 3 \times 2^2) - 1.2^2$

$= 1.8 - 1.44$

$= 0.36.$

If instead of recording frequencies, we were to tabulate the *relative* frequencies $f_r/n = g_r$ we obtain:

No. of heads (x_r)	Relative frequency (g_r)
$x_1 = 0$	$g_1 = \frac{1}{10}$
$x_2 = 1$	$g_2 = \frac{6}{10}$
$x_3 = 2$	$g_3 = \frac{3}{10}$

Furthermore, we can see that

$$m = \sum g_r x_r$$

and $$s^2 = \sum g_r(x_r - m)^2 = \sum g_r x_r^2 - m^2.$$

So far we have made no assumptions about the coins and have not remarked upon any expectations which we might have had. Our task has been simply to record a set of events and to describe them as conveniently as possible. If, however, we now make assumptions about the coins and the way in which they are spun, we can use the probability theory of chapter 7 to predict our expectations. Suppose, for instance, we assume that the coins are both completely unbiased, and they are spun with absolute fairness; then the probabilities p_r of obtaining x_r heads are clearly:

No. of heads (x_r)	Probability (p_r)
$x_1 = 0$	$p_1 = \frac{1}{4}$
$x_2 = 1$	$p_2 = \frac{1}{2}$
$x_3 = 2$	$p_3 = \frac{1}{4}$

This set of probabilities is referred to as a *probability distribution*, and the set of assumptions (which must be expressed in precise probability terms) from which it arises is called the *probability model.*

If different assumptions were made, we would have a different probability model, and so a different probability distribution. For instance if one of the coins was double-headed, and the other was fair and unbiased, then the probability distribution would be as follows:

x_r	p_r
$x_1 = 0$	$p_1 = 0$
$x_2 = 1$	$p_2 = \frac{1}{2}$
$x_3 = 2$	$p_3 = \frac{1}{2}$

One of the major tasks of statistics is to compare an observed frequency distribution with the probability distribution which arises from a particular probability model, so equivalent measures to the mean and variance of a frequency distribution are required for a probability distribution. To maintain a clear distinction between the two distributions, such probability measures are denoted by the Greek letters μ, σ^2.

Now $m = \sum g_r x_r$ and $s^2 = \sum g_r(x_r - m)^2$,

so $\mu = \sum p_r x_r$ and $\sigma^2 = \sum p_r(x_r - \mu)^2$.

As in statistics, so with probability distributions, μ is called the *mean*, or *expected value* of x, σ^2 its *variance* and σ its *standard deviation*.

So for the probability distribution with two unbiased coins,

$$\mu = \tfrac{1}{4} \times 0 + \tfrac{1}{2} \times 1 + \tfrac{1}{4} \times 2 = 1$$

$$\sigma^2 = \tfrac{1}{4} \times 1^2 + \tfrac{1}{2} \times 0^2 + \tfrac{1}{4} \times 1^2 = 0.5$$

(whereas in our trial, $m = 1.2$ and $s^2 = 0.36$).

[It should be pointed out at this stage that in the calculation of μ and σ, just as for m and s, the values of x_r are not restricted to integers: see, for instance, Exercise 12.1a, no. **8**.]

In summary:

Statistics	*Probability*
Relative frequency $g_r = \dfrac{f_r}{n}$	Probability p_r
Mean $m = \sum_r \dfrac{f_r}{n} x_r$	Mean (expected value) $\mu = \sum_r p_r x_r$
Variance $s^2 = \begin{cases} \sum_r \dfrac{f_r}{n}(x_r - m)^2 \\ \sum_r \dfrac{f_r}{n} x_r^2 - m^2 \end{cases}$	Variance $\sigma^2 = \begin{cases} \sum_r p_r(x_r - \mu)^2 \\ \sum_r p_r x_r^2 - \mu^2 \end{cases}$

Exercise 12.1a

1 Calculate the expected value and variance of the number of heads showing when three unbiased coins are tossed together. Conduct a series of 20 such trials and calculate the mean and variance for the frequency distribution that you obtain. Compare your theoretical and experimental results.

2 Calculate:
(*i*) the variance of the score obtained on throwing a single die;
(*ii*) the variance of the total score obtained on throwing two dice;
(*iii*) the variance of the average score on two dice.
How are your answers to (*i*) and (*ii*) and to (*ii*) and (*iii*) related? (Assume that the dice are 'fair'.)

3 A computer is made to produce randomly the numbers 0, 1, 2, . . ., 9. What is the expected value and variance of numbers so produced?

4 On the basis of past evidence, it is estimated that the probabilities of a certain type of plant having 4, 5, 6, 7, 8, 9 leaves are 0.13, 0.21, 0.38, 0.16, 0.09, 0.03 respectively. Find, correct to 2 decimal places, the expected value and variance of the number of leaves on such a plant.

5 Two unbiased dice are rolled and the greater score (or either if they are the same) is recorded. State the set of possible scores and the probabilities associated with these scores. Find the expected value (i.e., theoretical mean) of the recorded scores. (S.M.P.)

6 A boy spins a coin until he obtains a head, but impatiently gives up if he does not succeed in five attempts. What are the expected value and variance of the number of times he spins the coin?

7 What are the mean and variance of the number of different factors (other than 1 and the number itself) of an integer chosen at random in the range 1 to 30?

8 A box contains 100 tokens which differ in mass, but are otherwise identical. 20 of the tokens have a mass of 4.8 g each, 35 a mass of 5.2 g, 25 a mass of 5.7 g, 15 a mass of 6.5 g, and the remaining 5 have a mass of 8.0 g each. Calculate (to 2 decimal places) the expected value and variance of the mass of a token chosen at random from the box.

9 An examination question consists of two parts, A and B, and the probability of a pupil getting part A correct is $\frac{2}{3}$. If he gets A correct, the probability of getting B correct is $\frac{3}{4}$; otherwise it is $\frac{1}{6}$. There are three marks for a correct solution to part A, two marks for part B, and a bonus mark if both parts are correct. Calculate the expected value and variance of the pupil's total mark for the question.

10 In a game where the gambler rolls two dice, on a £1 stake the casino pays out £10 for a double six, and £3 for a score of seven (the stake money being returned as well), and for any other score the gambler loses his money. What is the casino's expected profit (to the nearest penny) on a £100 stake?

11 Two small piles of cards contain respectively the 1, 2, 3, 4, 5, 6 of diamonds and the 7, 8, 9, 10 of diamonds. Verify that these sets of numbers have expectations $3\frac{1}{2}$ and $8\frac{1}{2}$ respectively, and that the variances are $2\frac{11}{12}$ and $1\frac{1}{4}$ respectively.

One card is drawn at random from each of the piles and the product of the numbers so formed is calculated. Find the expected value and variance of this product.

Find also the probability that the product from any one draw will exceed its expectation. (C.)

12 A motorist drives into town and has the choice of two car parks. It takes him 20 minutes to drive from his home to car park A, which is never full, and then it takes him 15 minutes to walk to the office. If he decides to try to get to car park B it takes him 25 minutes to get there from his home. The probability that car park B is not full is p, and in this case he can walk to the office in 5 minutes. If car park B is full he drives back to car park A, which takes 5 minutes. If he always drives first to car park B find the probability distribution of the time T, in minutes, taken to get to work. Find the expected value of T, showing that its value is less than 35 if p exceeds $\frac{2}{3}$. Find also the variance of T.

If however, the motorist tosses a coin in order to decide whether to go straight to car park A or to try car park B first, find, in this case, the probability distribution of T, and the expected value of T. (C.)

Probability generators

Suppose that we want to find the mean and variance of the geometric probability distribution (see 7.4) of the number of throws required of a 'fair' die until a six is obtained.

The probability distribution is given by:

No. of throws	Probability
$x_1 = 1$	$p_1 = \frac{1}{6}$
$x_2 = 2$	$p_2 = \frac{1}{6} \times \frac{5}{6} = \frac{5}{36}$
$x_3 = 3$	$p_3 = \frac{1}{6} \times (\frac{5}{6})^2 = \frac{25}{216}$
. . . . . .	. . . . . .
$x_r = r$	$p_r = \frac{1}{6} \times (\frac{5}{6})^{r-1}$

Hence $\mu = \frac{1}{6} \times 1 + \frac{5}{36} \times 2 + \frac{25}{316} \times 3 + \cdots$

and $\sigma^2 = (\frac{1}{6} \times 1^2 + \frac{5}{36} \times 2^2 + \frac{25}{316} \times 3^2 + \cdots) - \mu^2$.

Direct calculation of such infinite sums will be lengthy, to say the least, and even where the number of outcomes is not infinite, calculation of μ and σ^2 can be very tedious. But, particularly when the variables x_r are integers, calculations can be eased considerably by using the *probability generator* (or as it is sometimes known, the *probability generating function*):

Let the outcomes which have associated random variables 0, 1, 2, . . . have probabilities denoted by $p_0, p_1, p_2, \ldots$. Then the probability generator $G(t)$ is defined by

$$G(t) \equiv \sum p_r t^r \equiv p_0 + p_1 t + p_2 t^2 + \cdots + p_r t^r + \cdots.$$

In the probability generator, t is a dummy variable: it has no significance in itself, and any other letter would do just as well. We merely use the powers of t to pick out equivalent probabilities — the coefficient of t^n in $G(t)$ giving the probability of obtaining the value n for the random variable. It should also be noted that $G(t)$ may be a finite or infinite polynomial in t.

It is possible also to define a probability generator when the random variables are negative or non-integral (see Exercise 12.1b, no. **13**). In such a case, of course, the generator will not be a polynomial, but nevertheless the properties of $G(t)$ which we will shortly derive will still hold. In practice, however, such probability generators are not often used.

Clearly the probability generator provides us with a means of displaying a complete probability distribution in a single expression. For instance, the distribution for the number of heads obtained on throwing two unbiased coins is summarised by the generator:

$$G(t) = \tfrac{1}{4} + \tfrac{1}{2}t + \tfrac{1}{4}t^2.$$

It is not immediately obvious how a probability generator can be used to evaluate μ and σ^2, but we shall see shortly that they are calculated from the first and second derivatives of $G(t)$. Meanwhile, to aid differentiation, the generator will need to be written in as simple a form as possible, so let us simplify the generator of our geometric distribution:

$$\begin{aligned} G(t) &= \tfrac{1}{6}t + \tfrac{1}{6} \times \tfrac{5}{6}t^2 + \tfrac{1}{6} \times (\tfrac{5}{6})^2 t^3 + \cdots + \tfrac{1}{6} \times (\tfrac{5}{6})^{r-1} t^r + \cdots \\ &= \tfrac{1}{6}t\left[1 + \frac{5t}{6} + \left(\frac{5t}{6}\right)^2 + \cdots + \left(\frac{5t}{6}\right)^{r-1} + \cdots\right]. \end{aligned}$$

Now the contents of the square brackets can be recognised as a geometric progression whose sum is

$$\lim_{n\to\infty}\left[\frac{1-(5t/6)^n}{1-5t/6}\right],$$

and since t has no significance in itself, we can choose it such that $|5t/6| < 1$,

so that $\left(\dfrac{5t}{6}\right)^n \to 0$ as $n \to \infty$.

Hence $G(t) = \tfrac{1}{6}t \times \dfrac{1}{1 - 5t/6} = \dfrac{t}{6 - 5t}$.

Derivation of μ, σ^2 from the probability generator

When the random variables $x_1, x_2, \ldots$ are $0, 1, 2, \ldots$ the formulae for expected value and variance take the form:

$$\mu = \sum rp_r \qquad \sigma^2 = \sum r^2 p_r - \mu^2.$$

Now $G(t) = \sum p_r t^r$

$$\begin{aligned} &\Rightarrow \quad G'(t) = \sum rp_r t^{r-1} && \Rightarrow \quad G'(1) = \sum rp_r \\ &\Rightarrow \quad G''(t) = \sum r(r-1)p_r t^{r-2} && \Rightarrow \quad G''(1) = \sum r(r-1)p_r \\ & && \Rightarrow \quad G''(1) + G'(1) = \sum r^2 p_r. \end{aligned}$$

Hence $\mu = \sum rp_r = G'(1)$

and $\sigma^2 = \sum r^2 p_r - \mu^2 = G''(1) + G'(1) - [G'(1)]^2$.

Summarising:

$$\boxed{\begin{aligned} \mu &= G'(1) \\ \sigma^2 &= G''(1) + G'(1) - [G'(1)]^2 \end{aligned}}$$

Returning now to our example,

$$G(t) = \frac{t}{6 - 5t}$$

$$\Rightarrow \quad G'(t) = \frac{6 - 5t + 5t}{(6 - 5t)^2}$$

$$= \frac{6}{(6 - 5t)^2} \quad \Rightarrow \quad G'(1) = 6$$

$$\Rightarrow \quad G''(t) = \frac{60}{(6 - 5t)^3} \quad \Rightarrow \quad G''(1) = 60$$

$$\Rightarrow \quad \mu = 6 \quad \text{and} \quad \sigma^2 = 60 + 6 - 6^2$$

$$= 30.$$

Hence the expected number of throws required to obtain a six is 6, and the standard deviation of the number of throws is $\sqrt{30}$.

Binomial distribution

We saw in 7.4 that if the chance of a particular event occurring in a single trial is p, then in n such independent trials, the probabilities of $0, 1, 2, 3, \ldots, n$ occurrences are given by the terms of the binomial expansion

$(q + p)^n$, where $q = 1 - p$.

These probabilities, therefore, are also the coefficients of successive powers of t in the expansion of

$$\boxed{G(t) = (q + pt)^n}$$

which is therefore the probability generator of the binomial distribution.

$$\text{Now} \quad G(t) = (q + pt)^n$$

$$\Rightarrow \quad G'(t) = np(q + pt)^{n-1}$$

$$\Rightarrow \quad \mu = G'(1) = np(q + p)^{n-1} = np, \quad \text{since } q + p = 1.$$

$$\text{Furthermore} \quad G''(t) = n(n - 1)p^2(q + pt)^{n-2}$$

$$\Rightarrow \quad G''(1) = n(n - 1)p^2$$

$$\Rightarrow \quad \sigma^2 = G''(1) + G'(1) - [G'(1)]^2$$

$$= n^2p^2 - np^2 + np - n^2p^2$$

$$= np(1 - p)$$

$$= npq.$$

$$\text{Hence} \quad \boxed{\mu = np \qquad \sigma = \sqrt{npq}}$$

The result that, for a binomial distribution,

$$G(t) = (q + pt)^n = (q + pt)(q + pt)\ldots(q + pt),$$

where $q + pt$ is the generator of a single trial, illustrates the fact that for a probability distribution arising from a series of *independent* (and not necessarily identical) trials, the overall probability generator is the product of the generators of each of the separate trials.

Example

In a certain large city, it is known that $\frac{1}{3}$ of the voters support the radical party. In a sample of 12 voters, what is the expected value and standard deviation of the number of radicals?

We have therefore a binomial probability situation with $n = 12$, $p = \frac{1}{3}$ and $q = \frac{2}{3}$.

So $\mu = np = 12 \times \frac{1}{3} = 4$

and $\sigma = \sqrt{npq} = \sqrt{(12 \times \frac{1}{3} \times \frac{2}{3})} = \sqrt{\frac{8}{3}} \approx 1.63.$

Exercise 12.1b

1 Calculate the probability generator for the number of tosses of an unbiased coin needed to obtain the first head, and hence evaluate the mean and variance of the number of tosses.

2 Calculate the expected value and standard deviation of the number of heads obtained when an unbiased coin is tossed (*i*) 4, (*ii*) 36, (*iii*) 100 times.

3 In the next general election suppose that 36% of the electors intend to vote Liberal. Find, in terms of n, the mean and standard deviation of the percentage of those who intend to vote Liberal in samples of size n.

(S.M.P.)

4 Six players take it in turns to cut a pack of cards (excluding jokers), and each time the cards are replaced before the next player cuts. Calculate the expected value and standard deviation of the total numbers of spades turned up when each player has cut.

5 The probability of a pupil arriving at school late on any given day is $\frac{1}{10}$. What is the probability of his being punctual for a whole week (i.e., 5 school days)? Calculate the mean and variance of the number of days he will be late in a school term consisting of 14 weeks (i.e., 70 days). Also calculate the expected number of completely punctual weeks in the term.

6 Seedlings are planted in 10 rows of six each. The probability of a seedling dying before it flowers is $\frac{1}{8}$. Calculate the mean and variance of the number of rows in which all the seedlings flower.

7 Eight per cent of eggs sold by a certain grocer are brown.
(*i*) Calculate the mean and variance of the number of brown eggs obtained in a carton of 12.

(*ii*) By finding the probability of a carton containing no brown eggs, and using a suitable probability generator, find how many cartons a housewife can expect to buy before she finds her first brown egg.

8 Prove that for a binomial probability distribution arising from n trials, the maximum possible value of the variance is $n/4$.

9 An electronics firm packs the resistors that it produces in boxes of 800. On average one component in 100 is faulty. Calculate the expected number of defective resistors in a box, and the variance of this number. How could you quickly obtain a good approximation to the variance in this case? What would the percentage error (of the actual value) be if this approximation were used?

10 A box contains a large number of screws. The screws are very similar in appearance, but are in fact of 3 different types, A, B and C, which are present in equal numbers. For a given job only screws of type A are suitable. If 4 screws are chosen at random, find the probability that:

(*i*) exactly two are suitable;

(*ii*) at least two are suitable.

If 20 screws are chosen at random, find the expected value and variance of the number of suitable screws. (C.)

11 In a machine game of chance, when a lever is pulled, one of the numbers 1, 2, 3 appears in a window. The lever is pulled 5 times and the total score is recorded. If the probabilities associated with the numbers 1, 2, 3 are $\frac{1}{6}, \frac{1}{3}, \frac{1}{2}$ respectively, write down an expression for the probability generator $G(t)$ for the possible total scores. Evaluate $G(-1)$ and, hence or otherwise, find the probability that the total score is even. (S.M.P.)

12 By using the fact that the probability generator for the number of throws of a die required to obtain a six is $t/(6 - 5t)$, or otherwise, obtain the probability generator for the number of throws required to obtain two sixes (not necessarily consecutively). Hence calculate the expected value and variance of the number of throws required to obtain two sixes.

13 A multiple-choice examination paper consists of 25 questions, each with 5 possible answers, only one of which is correct. A student gains 4 marks for a correct answer, and loses one mark for an incorrect answer.

(*i*) An entirely ignorant student has to guess the answer for each question. Find the probability generator for the number of marks he obtains on the paper, and hence calculate his expected total mark, and the standard deviation of that mark. (Hint: first find the probability generator for the marks on a single question.)

(*ii*) A rather more intelligent student has twice as much chance of choosing the correct answers as he has of choosing any one of the wrong answers; find the expected value and standard deviation of his total score on the paper.

14 A petrol company gives away one medal with every purchase of petrol. Each medal is equally likely to be any one of a set of n medals. At a stage when a customer already has s different medals, show that the expected number of purchases to get the first medal different from those already held is $n/(n - s)$. Deduce that the expected number of purchases for a complete set is

$$n\left(1 + \frac{1}{2} + \frac{1}{3} + \cdots + \frac{1}{n}\right). \qquad \text{(C.)}$$

12.2 Continuous probability distributions

Statement 1: The probability of throwing a 5 with a fair die is $\frac{1}{6}$.

Statement 2: The probability of a voter, chosen at random in London, being a Liberal is $\frac{1}{6}$.

Statement 3: The probability of the new machine turning out a component of length 241 mm is $\frac{1}{6}$.

Superficially the three statements above seem very similar. They all make predictions about what may occur if we conduct certain trials, and we presume that there is some foundation, theoretical or experimental, for each. But let us examine the statements separately, and in a little more detail.

Statement 1 is presumably based on theoretical considerations (themselves supported by past experience) and uses the notion of 'equal likelihood'. It tells us that if we conduct a large number of trials, about $\frac{1}{6}$ of the outcomes will be 5's; and that, in general, the larger the number of trials, the closer we get to this fraction.

Statement 2 can only be based on statistical evidence, and tells us that in a large, randomly chosen group of voters, we could expect about $\frac{1}{6}$ to be Liberals.

Both of these statements are concerned with *discrete* (i.e., definite and distinct) outcomes: when a die is thrown there are only six distinct results possible, and similarly a voter has political views which can be expressed in certain distinct choices (and we would regard 'don't know' as a category of choice).

Statement 3 however, is concerned with length, a *continuous* quantity, and presumably tells us that if, for instance, we look at 150 components produced by the machine, about 25 will be 241 mm in length. But if we take these 150 components, and measure each by a micrometer to 0.001 mm, then we would certainly not expect 25 of them to be exactly 241 mm in length, and the statement in its present form is meaningless. Perhaps it is more likely to mean that if measurements are taken *to the nearest millimetre*, then the probability of a length of 241 mm is $\frac{1}{6}$?

So although we can never be absolutely certain about the exact length of any object, and must say that

$$p(\text{length} = 241 \text{ mm}) = 0,$$

nevertheless our original statement can meaningfully be written as:

$$p(240.5 \leqslant \text{length} \leqslant 241.5) = \tfrac{1}{6}.$$

Indeed for any continuous quantity, we cannot talk (as we did for discrete quantities) about the probability of it attaining a specific value. We can only meaningfully talk about the probability of length, time, etc., lying *within a certain range of values.*

Cumulative probability

Even though, for a continuous variable like length, we cannot use $p(\text{length} = 241 \text{ mm})$, it would clearly be valuable if we could associate some form of probability with any given length. This is done by use of the notion of cumulative probability.

The *cumulative probability* of length 241 is defined to be the probability that the length is less than (or equal to) 241, and is written

$$\Phi(241) = p(\text{length} \leqslant 241).$$

More generally, the *cumulative probability function*† for any variable x is defined by

$$\Phi(X) = p(x \leqslant X),$$

so that $0 \leqslant \Phi(X) \leqslant 1$ for all X.

(It should be noted that we can also meaningfully talk about the cumulative probability of a discrete variable; and for instance if we throw a single die, $\Phi(5) = \frac{5}{6}$. But clearly this notion is less useful, and is less frequently used.)

Example 1

In a race, the times of the competitors are measured by stop-watch, to the nearest $\frac{1}{10}$ of a second. What is the cumulative probability function for error in measured time?

Let the error be denoted by x, which is uniformly distributed over the interval -0.05 to 0.05.

Therefore $\Phi(-0.05) = 0$ and $\Phi(X) = 0$ for any $X \leqslant -0.05$

$\Phi(0.05) = 1$ and $\Phi(X) = 1$ for any $X \geqslant 0.05$

and $\Phi(X)$ increases steadily in value from 0 to 1 over the interval

† Also called a *distribution function.*

$-0.05 \leqslant X \leqslant 0.05.$

So $\Phi(X) = 10(X + 0.05), \quad \text{if } -0.05 \leqslant X \leqslant 0.05.$

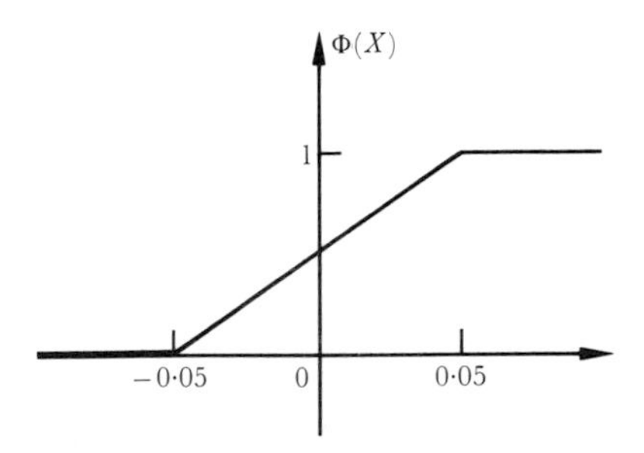

Example 2

Particles are emitted uniformly from a radioactive source in the corner of a room. A screen (of theoretically infinite length) is positioned one metre away from the source, so that all particles must strike the screen. What is the cumulative probability function of the distance of the point of impact from the bottom of the screen?

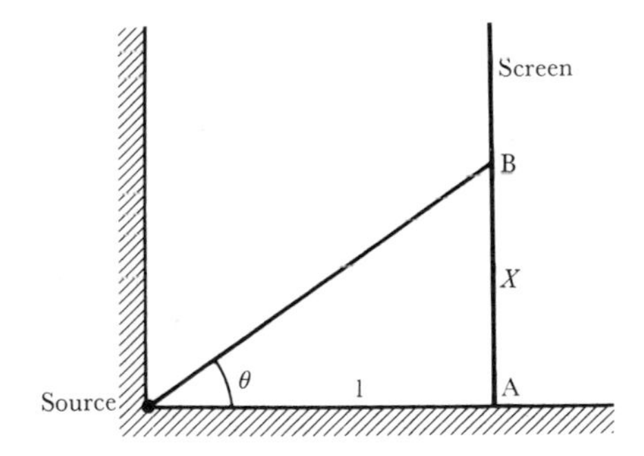

Since the particles are emitted uniformly over the complete arc of $\frac{1}{2}\pi$ radians,

$$\Phi(X) = p \text{ (particle striking AB)}$$

$$= \text{fraction of total angle } \tfrac{1}{2}\pi \text{ subtended by AB}$$

$$= \frac{\theta}{\frac{1}{2}\pi} = \frac{2\theta}{\pi}.$$

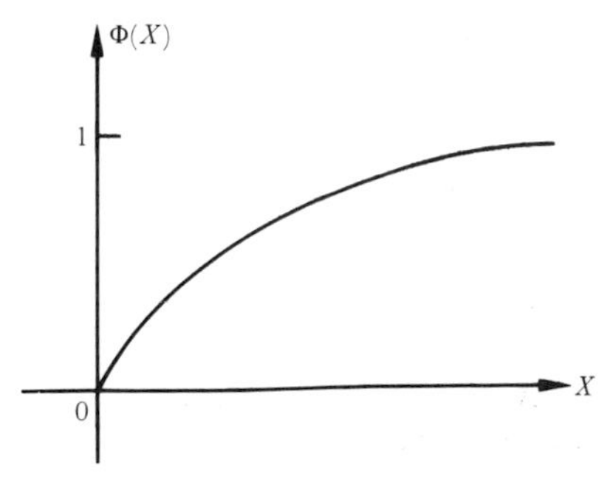

But since $\tan\theta = \dfrac{X}{1} = X$

then $\theta = \tan^{-1} X.$

Hence $\Phi(X) = \dfrac{2}{\pi}\tan^{-1} X$

in the interval $X \geqslant 0.$

Probability density function

The cumulative probability function, however, does have the disadvantage of often being cumbersome to use in practice. From its graph, for instance, we cannot easily picture the overall probability distribution, and we normally use instead the *probability density function* (p.d.f.), derived as follows:

By definition of $\Phi(X)$, we know that

$$p(a \leqslant x \leqslant b) = \Phi(b) - \Phi(a).$$

$$\text{Hence} \quad p(X \leqslant x \leqslant X + \delta X) = \Phi(X + \delta X) - \Phi(X)$$
$$= \frac{\Phi(X + \delta X) - \Phi(X)}{\delta X} \delta X.$$

Now as $\delta X \to 0$, we know that $\dfrac{\Phi(X + \delta X) - \Phi(X)}{\delta X} \to \Phi'(X)$.

$$\text{Hence} \quad p(X \leqslant x \leqslant X + \delta X) \approx \Phi'(X)\, \delta X.$$

Clearly the function $\Phi'(x)$ is very useful, since it enables us to find the probability of x lying in any small interval; and since the probability is equal to this function multiplied by the width of the interval, the function is referred to as the probability density function, and is rewritten $\phi(x)$.

Since $\phi(x) = \Phi'(x)$, it follows that

$$\Phi(X) = \int_L^X \phi(x)\, dx, \quad \text{where } L \text{ is the least possible value of } x.$$

Furthermore, the probability that x lies in the interval $[a, b]$ is clearly seen to be

$$\int_a^b \phi(x)\, dx$$

and is therefore represented by the area shaded on the diagram.

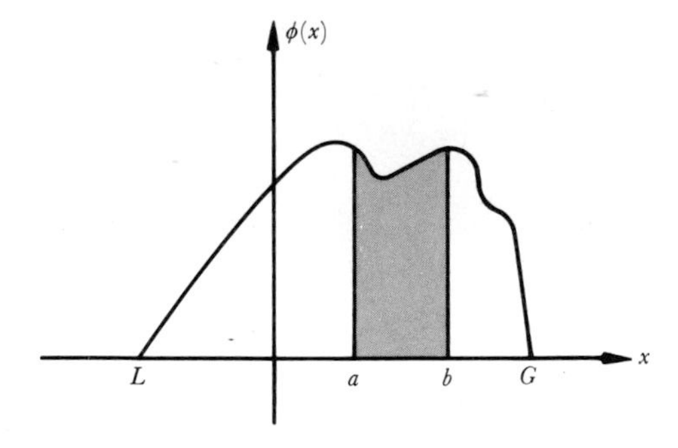

It therefore follows, since $\Phi(x)$ is an increasing function and $\phi(x)$ is its derivative, that $\phi(x) \geqslant 0$; and, since the total probability must be equal to 1, that

$$\int_L^G \phi(x)\,\mathrm{d}x = 1,$$

where L and G are the least and greatest possible values of x.

Mechanical analogies

There are several analogies to probability density, and perhaps the two most obvious are velocity and mass density:

(*i*) Probability is calculated from probability density in exactly the same way as distance is calculated from velocity. For if t denotes time and $v(t)$ is the velocity at time t, then $\int_a^b v(t)\,\mathrm{d}t$ is the distance travelled between times a and b, just as $\int_a^b \phi(x)\,\mathrm{d}x$ is the probability of x lying between a and b; and these integrals are represented as shaded areas under their respective graphs:

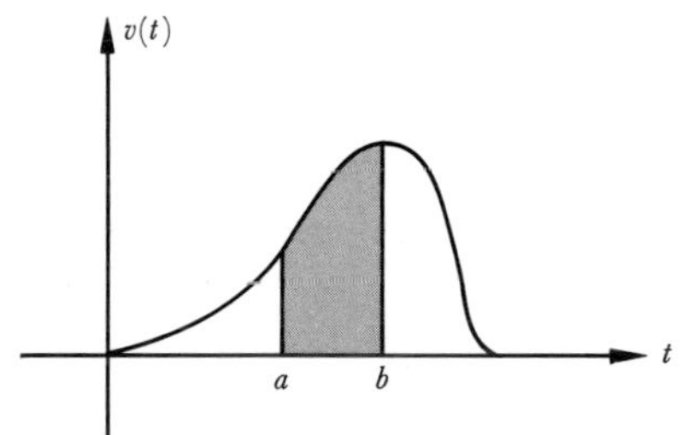

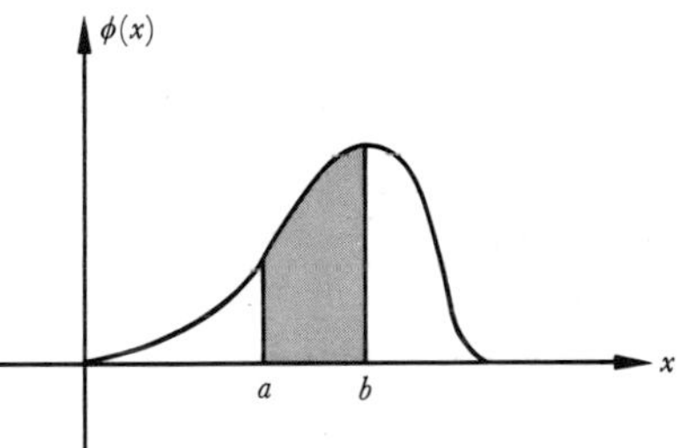

(*ii*) Similarly, just as density measures the rate of build-up of mass, so probability density measures the rate of build-up of probability. Suppose we have a metal rod of non-uniform material which has mass 1 kg, length 2 m and constant cross-section (so that we can regard the density as mass per unit length). Then if the density is graphed as

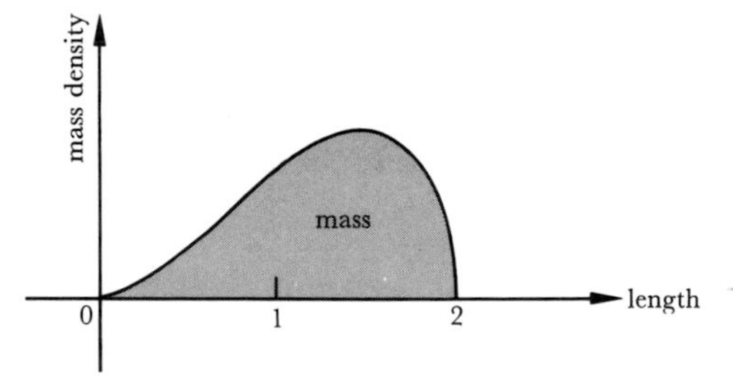

we see that most of the mass is concentrated to the right of the rod's mid-point, just as in the equivalent probability distribution:

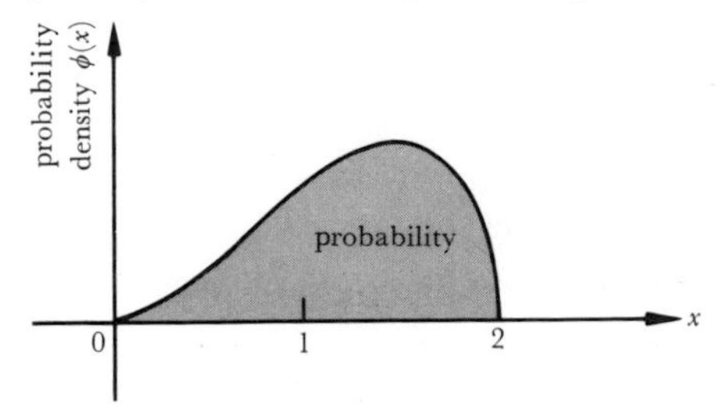

Returning to Example 1,

$$\Phi(x) = 10(x + 0.05) \quad \text{for } -0.05 \leqslant x \leqslant 0.05$$

then
$$\phi(x) = \Phi'(x) = \begin{cases} 10 & \text{for } -0.05 \leqslant x \leqslant 0.05 \\ 0 & \text{elsewhere.} \end{cases}$$

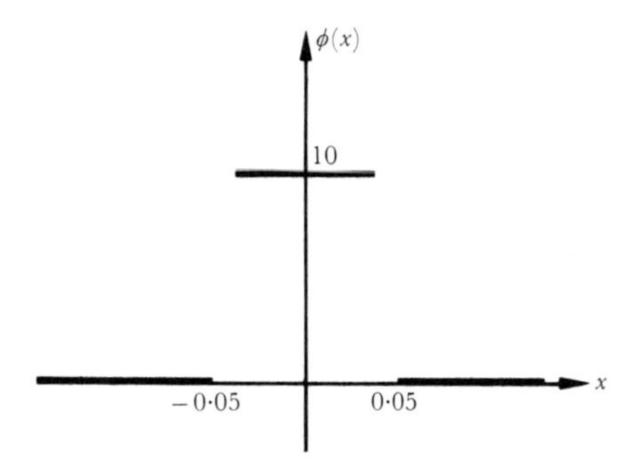

This continuous probability distribution is therefore analogous to the discrete rectangular distribution discussed in chapter 7.

Similarly, in Example 2,

$$\Phi(x) = \frac{2}{\pi}\tan^{-1} x \qquad \text{for } x \geqslant 0.$$

So
$$\phi(x) = \Phi'(x) = \begin{cases} \dfrac{2}{\pi(1 + x^2)} & \text{for } x \geqslant 0 \\ 0 & \text{for } x < 0. \end{cases}$$

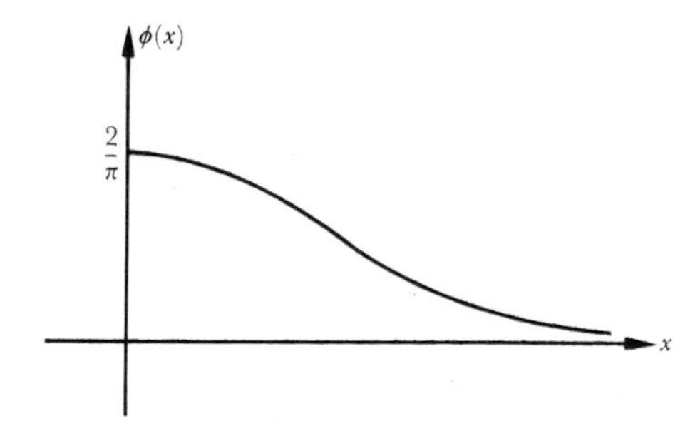

Example 3

For a circular disc of radius 4 cm, what is the probability density function of the distance (r) from the centre of a point chosen at random on the disc? Use the p.d.f. to calculate the probability that a point lies between 2 and 3 cm from the centre of the disc.

In this case it is easier first of all to calculate $\Phi(r)$.

Now $$\begin{aligned}\Phi(r) &= p(\text{distance from centre} \leqslant r)\\ &= \text{fraction of total area whose distance} \leqslant r\\ &= \frac{\pi r^2}{\pi \times 4^2}\\ &= \frac{r^2}{16}\end{aligned}$$

$$\Rightarrow \qquad \phi(r) = \Phi'(r) = \frac{r}{8}$$

and $$\int_0^4 \phi(r)\,\mathrm{d}r = \int_0^4 \frac{r}{8}\,\mathrm{d}r = \left[\frac{r^2}{16}\right]_0^4 = 1,$$ as required.

Now, using $\phi(r)$,

$$p(2 \leqslant r \leqslant 3) = \int_2^3 \frac{r}{8}\,\mathrm{d}r = \left[\frac{r^2}{16}\right]_2^3 = \tfrac{5}{16}.$$

Exercise 12.2a

1 A p.d.f. is given by

$$\phi(x) = \begin{cases} \dfrac{x+1}{4} & \text{for } 0 \leqslant x \leqslant a \\ 0 & \text{for } x < 0 \text{ and } x > a. \end{cases}$$

By using the property of the p.d.f. that $\int \phi(x)\,\mathrm{d}x = 1$, find the value of a. Obtain an expression for the cumulative probability function.

Calculate: (*i*) $p(x < \frac{1}{2})$; (*ii*) $p(x > 1)$.

2 Verify that

$$\phi(x) = \begin{cases} \dfrac{3}{x^4} & \text{for } x \geqslant 1 \\ 0 & \text{for } x < 1. \end{cases} \qquad \text{is a p.d.f.}$$

Find the value of X such that $p(x < X) = 0.5$.

Calculate: (*i*) $p(x > 10)$; (*ii*) $p(2 < x < 4)$.

3 A p.d.f. is given by

$$\phi(x) = \begin{cases} \frac{1}{6}x & \text{for } 0 \leqslant x \leqslant 3 \\ \frac{1}{2}(4 - x) & \text{for } 3 \leqslant x \leqslant 4 \\ 0 & \text{for } x < 0 \text{ and } x > 4. \end{cases}$$

Sketch the graph of $\phi(x)$.

Calculate:

(*i*) the probability that x occurs in the interval $[1, 2]$;

(*ii*) the probability that $x > 2$.

Obtain the cumulative probability function, and hence, or otherwise, find the median of the distribution.

4 The p.d.f. of a distribution is given by $\phi(x) = k(1 - x^2)$ for $-1 \leqslant x \leqslant 1$, and $\phi(x) = 0$ elsewhere.

Find the value of k, and hence calculate:

(*i*) $p(-\frac{1}{2} \leqslant x \leqslant \frac{1}{2})$; (*ii*) $p(x > \frac{1}{3})$.

Obtain an expression for the cumulative probability function of the distribution.

5 The p.d.f. of a distribution is given by $\phi(x) = a \sin \pi x$ for $0 \leqslant x \leqslant 1$, and $\phi(x) = 0$ elsewhere.

Find the value of a, and obtain an expression for the cumulative probability function.

Calculate: (*i*) $p(x < \frac{1}{3})$; (*ii*) $p(\frac{1}{2} < x < \frac{2}{3})$.

6 The probability that a transistor in a radio lasts less than t hours is $1 - e^{-t/2000}$. Find the p.d.f. for the lifetime of a transistor.

(*i*) What is the probability that a transistor lasts more than 4 000 hours?

(*ii*) What is the probability that a transistor ceases to function after 2 000 hours of use but before 3 000 hours.

(*iii*) If a radio contains 8 transistors, what is the probability that none of them fails before 1 000 hours of use?

Parameters of a continuous distribution

In a discrete probability distribution, the expected value and variance were defined by

$$\mu = \sum_r x_r p_r \quad \text{and} \quad \sigma^2 = \sum_r (x_r - \mu)^2 p_r = \sum_r x_r^2 p_r - \mu^2.$$

In a continuous distribution, the probability corresponding to p_r is given by $\phi(x)\,\delta x$.

So $\mu = \lim \sum x\phi(x)\,\delta x$ and $\sigma^2 = \lim \sum (x - \mu)^2 \phi(x)\,\delta x = \lim \sum x^2\phi(x)\,\delta x - \mu^2$

$$\Rightarrow \quad \mu = \int_L^G x\phi(x)\,dx \quad \text{and} \quad \sigma^2 = \int_L^G (x - \mu)^2\phi(x)\,dx = \int_L^G x^2\phi(x)\,dx - \mu^2.$$

Example 4

The p.d.f. of the lifetime (t), in minutes, of atoms of a radioactive element is given by $\phi(t) = 3\,e^{-3t}$. Find the expected lifetime of an atom, its variance, and the half-life of the element (the time it takes for half of the material to decay).

The expected lifetime, μ, is given by

$$\mu = \int_0^\infty 3t\, e^{-3t}\, dt.$$

Integrating by parts,

$$\begin{aligned}\mu &= \left[-t\, e^{-3t}\right]_0^\infty + \int_0^\infty e^{-3t}\, dt\\ &= 0 + \left[-\frac{e^{-3t}}{3}\right]_0^\infty\\ &= \tfrac{1}{3} \text{ min,}\end{aligned}$$

$$\begin{aligned}\text{and}\quad \sigma^2 &= \int_0^\infty 3t^2\, e^{-3t}\, dt - \tfrac{1}{9}\\ &= \left[-t^2\, e^{-3t}\right]_0^\infty + \int_0^\infty 2t\, e^{-3t}\, dt - \tfrac{1}{9}\\ &= 0 + \tfrac{2}{3} \times \tfrac{1}{3} - \tfrac{1}{9}\\ &= \tfrac{1}{9} \text{ min}^2.\end{aligned}$$

Now if T is the half-life, then

$$\begin{aligned} & \int_0^T 3\, e^{-3t}\, dt = \tfrac{1}{2}\\ \Rightarrow\quad & \left[-e^{-3t}\right]_0^T = \tfrac{1}{2}\\ \Rightarrow\quad & 1 - e^{-3T} = \tfrac{1}{2}\\ \Rightarrow\quad & e^{-3T} = \tfrac{1}{2}\\ \Rightarrow\quad & e^{3T} = 2\\ \Rightarrow\quad & 3T = \ln 2\\ \Rightarrow\quad & T = \tfrac{1}{3} \ln 2 \approx 0.23 \text{ min.}\end{aligned}$$

Exercise 12.2b

1–5 Calculate the mean and variance of each of the probability distributions defined in Exercise 12.2a nos. **1–5**.

6 A random variable x has cumulative probability function

$$\Phi(x) = \begin{cases} 0 & (x \leqslant a)\\ \dfrac{x-a}{b-a} & (a < x < b).\\ 1 & (x \geqslant b)\end{cases}$$

Find the p.d.f. $\phi(x)$, and sketch the graph of $\phi(x)$. Obtain the mean and variance of x. (C.)

7 The probability density function, $\phi(x)$, of a random variable x, is given by

$$\phi(x) = \alpha x(4 - x) \quad \text{if } 0 < x < 4.$$
$$= 0 \text{ otherwise.}$$

Find the value of α, and hence find the mean and variance of x. What is the probability that x lies between 0 and 1? (A.E.B.)

8 A random variable x has the cumulative probability function

$$\Phi(x) = \begin{cases} 0 & (x < 0) \\ kx^3 & (0 \leqslant x \leqslant 2) \\ 1 & (x > 2) \end{cases}$$

where k is a constant. Find the mean, median, and variance of x. (M.E.I.)

9 A probability distribution has the probability density function

$$\phi(x) = \begin{cases} 0 & x < 2 \\ k\, e^{-\lambda x} & x \geqslant 2, \quad k \text{ constant.} \end{cases}$$

Find, in terms of λ, the mean, median and standard deviation of the distribution. (M.E.I.)

10 The probability that a light bulb lasts longer than t hours is $e^{-t/\mu}$. Find the probability density function for the lifetime of a bulb.

Show that the mean lifetime is μ.

If the mean lifetime is 1 500 hours, how unlikely is it that the bulb will last more than 3 000 hours?

If the manufacturer wants to ensure that less than one in a thousand bulbs fail before 5 hours, what is the lowest mean lifetime he can allow his bulbs to have? (S.M.P.)

11 A random variable x has the probability distribution

$$\phi(x)\,\mathrm{d}x = C\,\mathrm{d}x \quad (-2a \leqslant x \leqslant -a,\ a \leqslant x \leqslant 2a)$$
$$\phi(x)\,\mathrm{d}x = 0 \quad \text{elsewhere.}$$

Sketch the distribution. Obtain the value of the constant C, and the standard deviation, σ, of the distribution.

Find the probabilities of obtaining values of x

(*i*) within one standard deviation of the mean;

(*ii*) within two standard deviations of the mean.

Find the value of k such that $p(|x| < k\sigma) = 0.95$. (M.E.I.)

12 A mathematical model for the fraction x of the sky covered with cloud $(0 < x < 1)$ assigns to this a p.d.f.

$$\phi(x) = \frac{k}{\sqrt{(x - x^2)}}.$$

Calculate:

(*i*) the value of k;
(*ii*) the expected fraction covered by cloud;
(*iii*) the probability that not more than $\frac{1}{4}$ of the sky is covered.

[*Hint*: Your integrations may be made easier by using the substitution $x = \sin^2 \theta$. You may assume that this substitution is valid, even though the function to be integrated may be discontinuous at the ends of the interval of integration.] (S.M.P.)

13 The probability density function of a distribution is given by

$$\phi(x) = \frac{e^{-x}x^{\lambda-1}}{(\lambda-1)!} \quad (x \geqslant 0, \lambda \text{ integer} > 0).$$

Find the expected value and variance of x.

Sketch $\phi(x)$ when $\lambda = 2$. (M.E.I.)

Histograms

In section 12.1 we saw how the relative frequency g_r corresponded to the probability p_r. Is it possible similarly therefore to find some statistical equivalent to the probability density $\phi(x)$? We shall see that such a measure arises from the use of what is known as a *histogram* to display data.

Consider the data shown below which gives the time, in seconds, between succeeding vehicles passing a given point on a main road, during a traffic survey. (Measurements were taken to the nearest second, so that a measurement registered as one second could actually be anywhere in the range 0.5 to 1.5 seconds. All the time intervals in the table are given in this way.)

Time interval	No. of vehicles (f_r)	Rel. frequency (g_r)
0.5–1.5	11	0.11
1.5–2.5	19	0.19
2.5–3.5	23	0.23
3.5–4.5	14	0.14
4.5–5.5	10	0.10
5.5–10.5	15	0.15
10.5–20.5	8	0.08
	100	1.00

These results can be displayed in a relative frequency diagram:

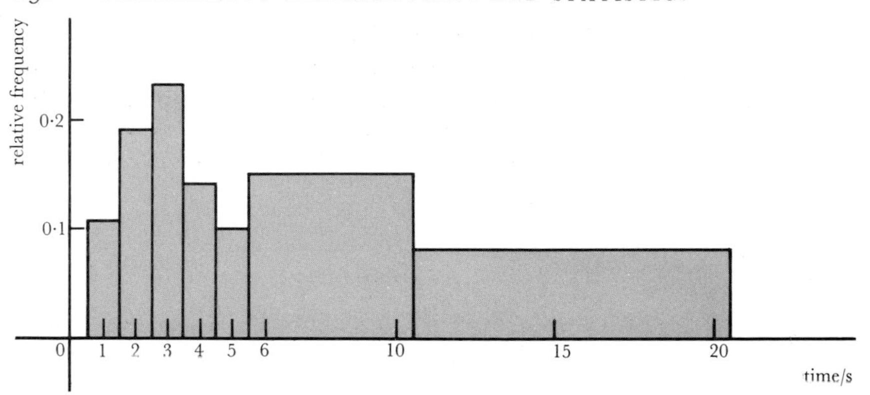

But it is immediately clear that this diagram gives a very misleading impression, and in order to correct this we require a diagram in which *areas* of the successive rectangles represent the corresponding relative frequencies. We therefore divide each relative frequency by the width of the corresponding interval. The resulting quantity is sometimes known as the *relative frequency density*, and the corresponding figure (whose total area is equal to 1) is a *histogram*:

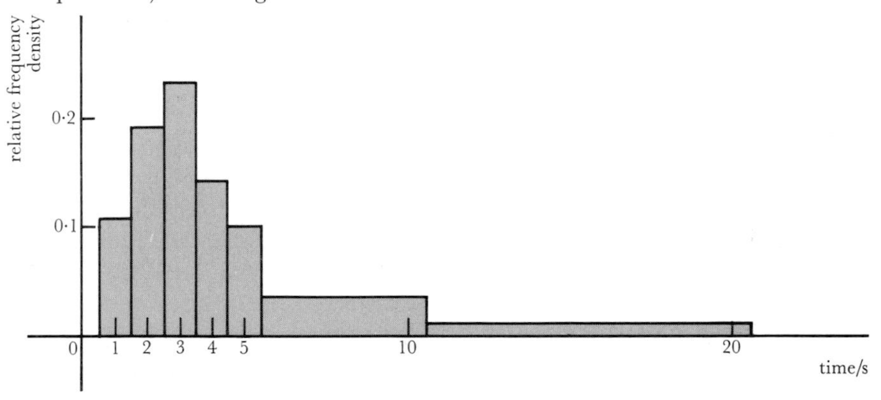

As a further extension of the traffic survey, 1 000 times between successive vehicles were measured to a greater accuracy, this time correct to 0.1 s, and the resulting histogram was:

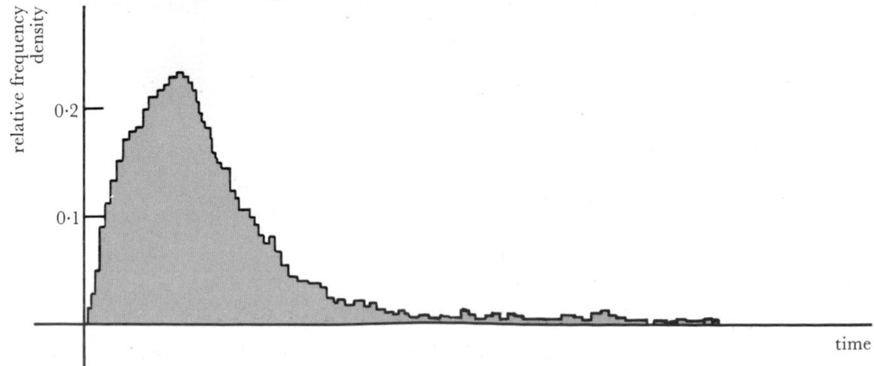

The general form of this histogram is obviously similar to that of its predecessor; but, because the time intervals are now much smaller, the outline of the histogram is correspondingly smoother. It is not difficult to see that as the time intervals become still smaller, the outline of the histogram will tend to a curve, which is therefore the statistical equivalent of the graph of the probability density function.

Exercise 12.2c

1 Conduct a traffic survey similar to that described in the text on a local main road, and record your results on a histogram.

2 The lifetimes (in completed weeks) of 100 mice of a certain species are given below.

Lifetime/weeks	0–9	10–39	40–49	50–59	60–69	70–79	80–99
No. of mice	8	3	7	18	31	22	11

Represent the data on a histogram.

3 Each of a group of 50 schoolboys attempts to throw a javelin for the first time. The distances they achieve (to the nearest metre) are given below.

Distance/m	0–19	20–29	30–34	35–39	40–49
No. of throws	5	11	18	13	3

Represent the data on a histogram.

4 120 people set out on a sponsored walk to raise money for charity. The distances they covered are given below.

Distance/km	less than 10	10–15	15–20	20–30	30–50
No. of walkers	12	15	25	41	27

Represent the data on a histogram.

5 From a sample of 100 marriages taking place at a registry office, the ages of the husbands are recorded below.

Age	16–20	21–25	26–30	31–40	41–60
No. of husbands	10	38	35	13	4

Represent the data on a histogram.

12.3 The Normal distribution

In the last section we derived methods of using probability models to describe continuous data. We now look in some detail at the most important (though by no means the only) continuous probability distribution. We shall first of all see, by looking at sets of statistical data, how it most frequently arises in practice. The data given in the following three examples is in each case continuous and each set of data is illustrated by its corresponding histogram.

1. Masses of five-penny pieces

100 five-penny pieces were weighed and their masses were classified into intervals of 0.05 g. The resulting frequency table, with corresponding relative frequencies and relative frequency densities was found to be as follows:

Mass/g	Frequency	Rel. freq.	r.f.d.
5.45–5.50	1	0.01	0.2
5.50–5.55	3	0.03	0.6
5.55–5.60	4	0.04	0.8
5.60–5.65	16	0.16	3.2
5.65–5.70	37	0.37	7.4
5.70–5.75	27	0.27	5.4
5.75–5.80	10	0.10	2.0
5.80–5.85	1	0.01	0.2
5.85–5.90	1	0.01	0.2
Total	100	1.00	

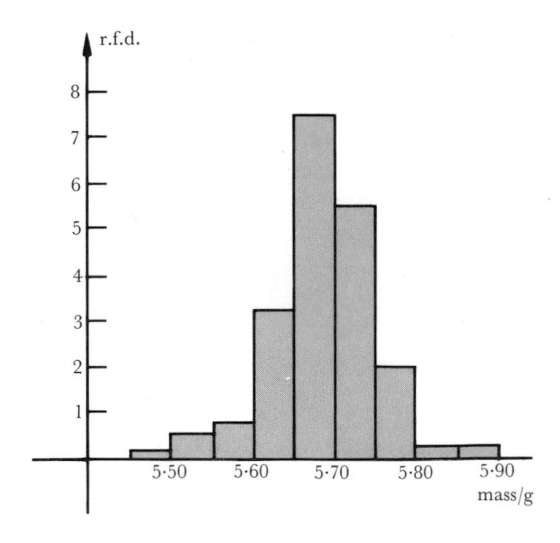

2. Lifetimes of electric light bulbs

The lifetimes of 200 electric light bulbs were classified into intervals of 200 hours, as follows:

Lifetime/h	Frequency	Rel. freq.	r.f.d. × 10^3
800–1 000	2	0.010	0.05
1 000–1 200	7	0.035	0.175
1 200–1 400	17	0.085	0.425
1 400–1 600	43	0.215	1.075
1 600–1 800	78	0.390	1.950
1 800–2 000	34	0.170	0.850
2 000–2 200	15	0.075	0.375
2 200–2 400	4	0.020	0.100
Total	200	1.000	

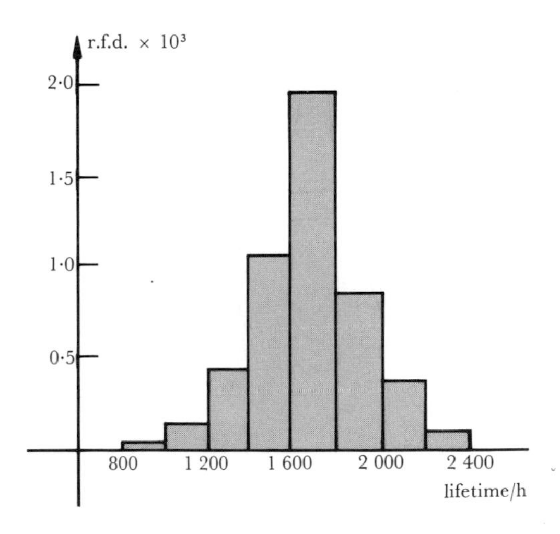

3. Heights of adult males

The heights of 1 000 adult males were measured and classified into intervals of 2 cm:

Height/cm	Frequency	Height/cm	Frequency	Height/cm	Frequency
150–152	2	166–168	101	182–184	23
152–154	3	168–170	116	184–186	16
154–156	5	170–172	125	186–188	7
156–158	9	172–174	112	188–190	2
158–160	16	174–176	102	190–192	2
160–162	29	176–178	93	192–194	1
162–164	55	178–180	61	194–196	0
164–166	79	180–182	40	196–198	1

From this table it is again easy to calculate the corresponding relative frequencies and relative frequency densities, which can be plotted as follows:

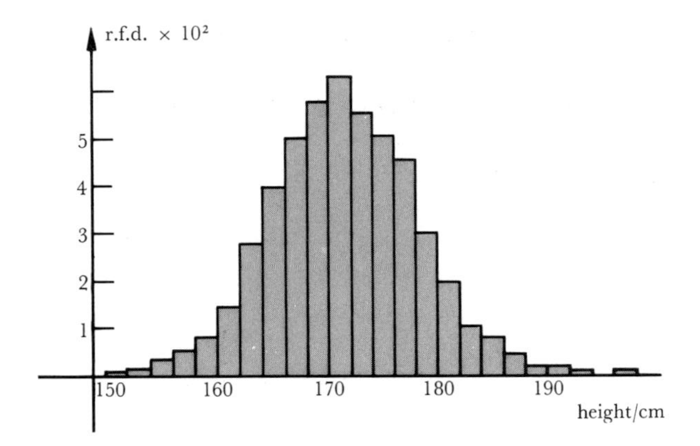

The three histograms appear to have the same general shape; they are approximately symmetrical, with most of their area concentrated in a central region, but tailing off fairly rapidly at each end. In the third example, where the total frequency and the number of class intervals are both greater than in the other two examples, the symmetry is more marked and the outline of the histogram could be approximated by a 'bell-shaped' curve. The histogram is reproduced below with such a curve superimposed, and similar (though less closely-fitting) curves are also superimposed on the other two histograms.

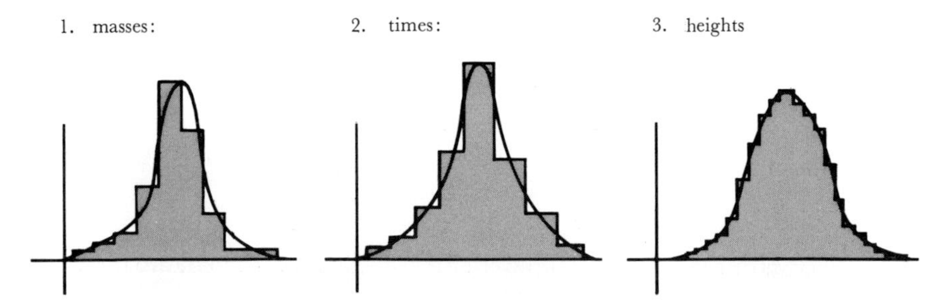

Though all have the same general bell-shape, these three curves do differ in certain ways: their axes of symmetry are in different positions, and one is 'short and wide' whilst another is 'tall and thin'. Such differences, of course, follow from the differences in position and spread (however they may be measured — though usually by m and s) of the original sets of statistical data. In fact these three curves which we have superimposed on the histograms were carefully chosen and are defined by a single function involving two parameters, their differences being due solely to differences of these parameters. This function is known as the *Normal probability function*, and satisfies all the conditions for a probability density function.

(The term *Gaussian function* is sometimes also used, after one of its earliest and most famous investigators, and Gauss himself called it the *error function*, from the way in which it originally arose.)

Continuous data which has a Normal p.d.f. is said to be Normally distributed. It must be strongly emphasised, however, that in many cases data are only *approximately* described by the Normal function. In our second set of data, for example, the lifetimes of electric light-bulbs, though having an approximately bell-shaped distribution are not *exactly* Normal in their distribution. Even so, this should not be regarded in any way as 'abnormal', for although the Normal distribution is extremely important, it is far from being universal. Indeed, perfect 'normality' is highly exceptional!

The equation of the Normal function will be derived later in the section, when further properties of the Normal distribution have been discussed. But for the present we will simply quote its probability density function as:

$$\phi(x) = \frac{1}{\sqrt{(2\pi)}\sigma} e^{-(x-\mu)^2/2\sigma^2}.$$

The parameters μ and σ can be shown, as might be suspected, to be the mean and standard deviation of the distribution.

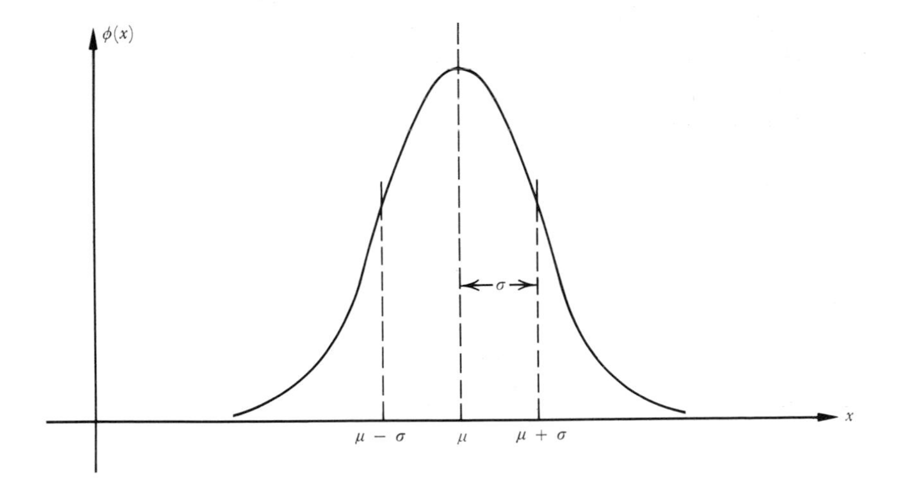

Properties of Normal probability curve

1 The curve is symmetrical about the line $x = \mu$.

2 The curve is asymptotic to the x-axis, i.e., $\phi(x) \to 0$ as $x \to \pm\infty$.

3 The total area beneath the curve equal to 1. This is, of course, a necessary condition for $\phi(x)$ to be a p.d.f., and also accounts for the value of the constant $1/\sqrt{(2\pi)}\sigma$ in the equation for $\phi(x)$.

4 Just over 68% of the area beneath the curve lies between $\mu - \sigma$ and

$\mu + \sigma$, so that approximately $\frac{2}{3}$ of the data lies within one standard deviation of the mean.

Similarly $95\frac{1}{2}\%$ of the area lies between $\mu - 2\sigma$ and $\mu + 2\sigma$,

and $99\frac{3}{4}\%$ lies between $\mu - 3\sigma$ and $\mu + 3\sigma$.

The standard Normal function

We now develop a single method of dealing with all Normal probability distributions, irrespective of the values of their parameters μ and σ. This is done by standardising the variable, that is transforming it so that its new mean and standard deviation become respectively 0 and 1.

Example 1

The marks in a mathematics examination are found to be approximately Normally distributed with mean 56 and standard deviation 18. Standardise the marks and find the standardised equivalent to a mark of 70.

(Examination marks, of course, are discrete, rather than continuous, variables; nevertheless, as we shall see in later examples, a continuous probability curve can give us a good approximation to a discrete statistical distribution.)

Let us denote the examination marks by the variable x. We first reduce the mean to zero by subtracting 56 from each mark. So x becomes $x - 56$.

As the mean of our new variable, $x - 56$, is now at the origin, and standard deviation measures spread from the mean, we reduce the standard deviation to 1 by dividing each new mark by 18. So $x - 56$ becomes $(x - 56)/18$. Therefore $(x - 56)/18$ is now a standardised Normal variable.

So a mark of 70 becomes $\dfrac{70 - 56}{18} = 0.78.$

If we now look at our Normal probability curve, in geometrical terms the standardisation process can be regarded as two successive transformations:

(*i*) a translation with vector $\begin{pmatrix} -\mu \\ 0 \end{pmatrix}$;

(*ii*) a two-way stretch with factor $1/\sigma$ parallel to the x-axis and factor σ parallel to the y-axis (so that the total area under the curve remains constant).

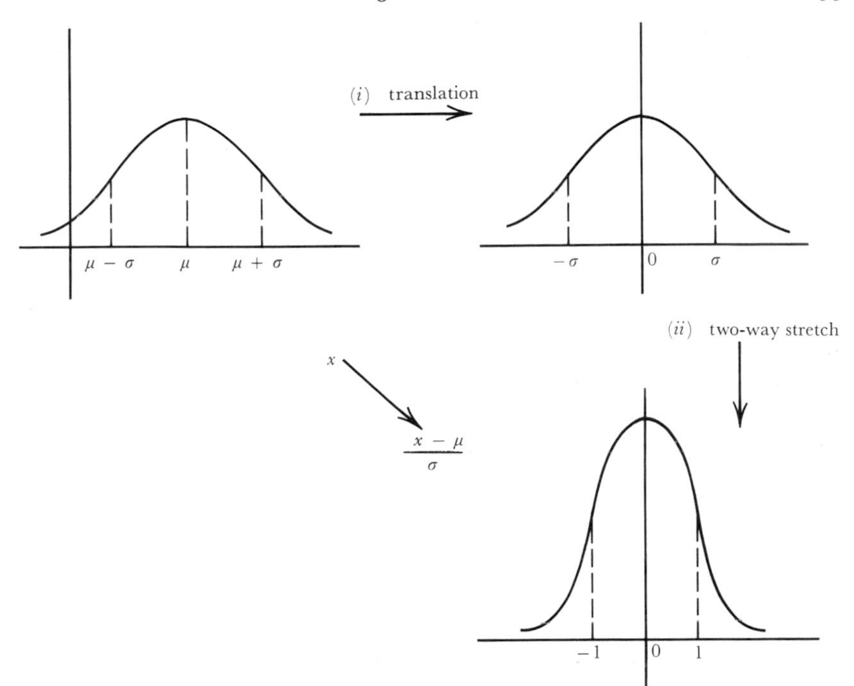

Hence the general standardised variable now becomes $(x - \mu)/\sigma$.

Writing the standardised variable $(x - \mu)/\sigma$ as t, the standard Normal function is therefore given by:

$$\phi(t) = \frac{1}{\sqrt{2\pi}} e^{-t^2/2}.$$

Since the p.d.f., $\phi(t)$, is not integrable, in practice we use the equivalent cumulative probability function $\Phi(t)$.

In geometrical terms, $\Phi(t)$ represents the shaded area on the diagram:

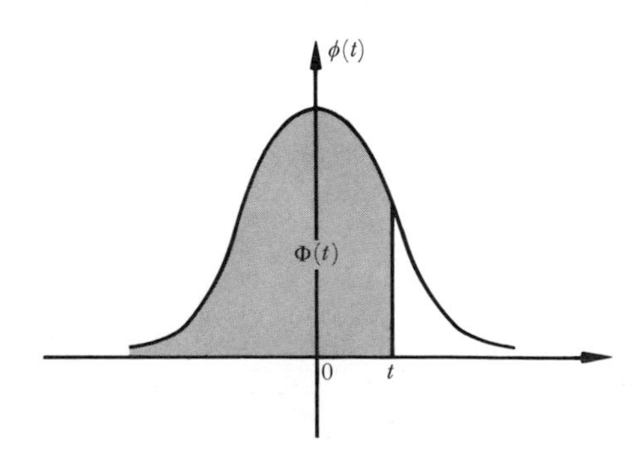

Values of $\Phi(t)$ are tabulated below:

t	$\Phi(t)$	t	$\Phi(t)$	t	$\Phi(t)$	t	$\Phi(t)$
0.0	0.500 0	1.0	0.841 3	2.0	0.977 2	3.0	0.998 65
0.1	0.539 8	1.1	0.864 3	2.1	0.982 1	3.1	0.999 03
0.2	0.579 3	1.2	0.884 9	2.2	0.986 1	3.2	0.999 31
0.3	0.617 9	1.3	0.903 2	2.3	0.989 3	3.3	0.999 52
0.4	0.655 4	1.4	0.919 2	2.4	0.991 8	3.4	0.999 66
0.5	0.691 5	1.5	0.933 2	2.5	0.993 8	3.5	0.999 77
0.6	0.725 7	1.6	0.945 2	2.6	0.995 3	3.6	0.999 84
0.7	0.758 0	1.7	0.955 4	2.7	0.996 5	3.7	0.999 89
0.8	0.788 1	1.8	0.964 1	2.8	0.997 4	3.8	0.999 93
0.9	0.815 9	1.9	0.971 3	2.9	0.998 1	3.9	0.999 95
1.0	0.841 3	2.0	0.977 2	3.0	0.998 6	4.0	0.999 97

Two points should be noted about the use of such a table of cumulative probabilities:

(i) *Intermediate values*

More accurate statistical tables are generally available, but otherwise linear interpolation must be used to evaluate $\Phi(t)$ for values of t not given in the table. However, the values of $\Phi(t)$ so obtained should be regarded as accurate only to 3 decimal places for values of $t \leqslant 3$, e.g.,

$$\begin{aligned}\Phi(1.37) &\approx \Phi(1.3) + \tfrac{7}{10}[\Phi(1.4) - \Phi(1.3)] \\ &\approx 0.903\,2 + \tfrac{7}{10} \times 0.016\,0 \\ &\approx 0.903\,2 + 0.011\,2 \\ &\approx 0.914, \text{ to 3 d.p.}\end{aligned}$$

(ii) *Negative values*

No negative values of t are given in the table since $\Phi(-t)$ can be calculated very easily from $\Phi(t)$.

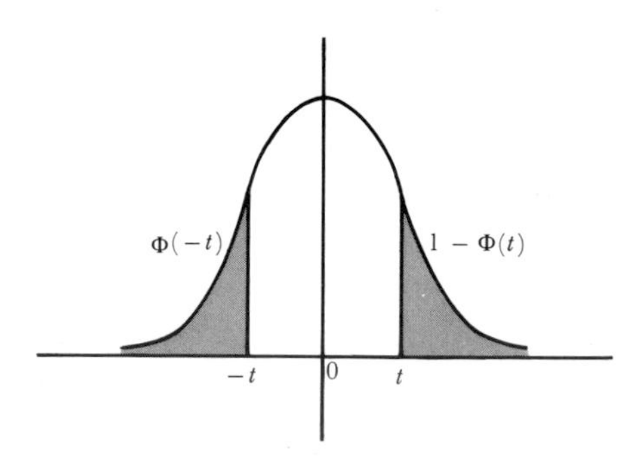

By symmetry, the two shaded areas on the diagram are equal. The areas represent $\Phi(-t)$ and $1 - \Phi(t)$, so it follows that:

$$\Phi(-t) = 1 - \Phi(t).$$

So, for instance,
$$\begin{aligned}\Phi(-0.6) &= 1 - \Phi(0.6)\\ &= 1 - 0.725\ 7\\ &= 0.274\ 3.\end{aligned}$$

The following two examples illustrate practical applications of the Normal distribution:

Example 2

Assuming that intelligence quotients are Normally distributed with mean 100 and standard deviation 15, calculate:
(*i*) the probability that a person chosen at random has an I.Q. between 88 and 118;
(*ii*) the percentage of the population with I.Q. greater than 130.
To standardise the variable in this case, we subtract 100 and divide by 15. Therefore:

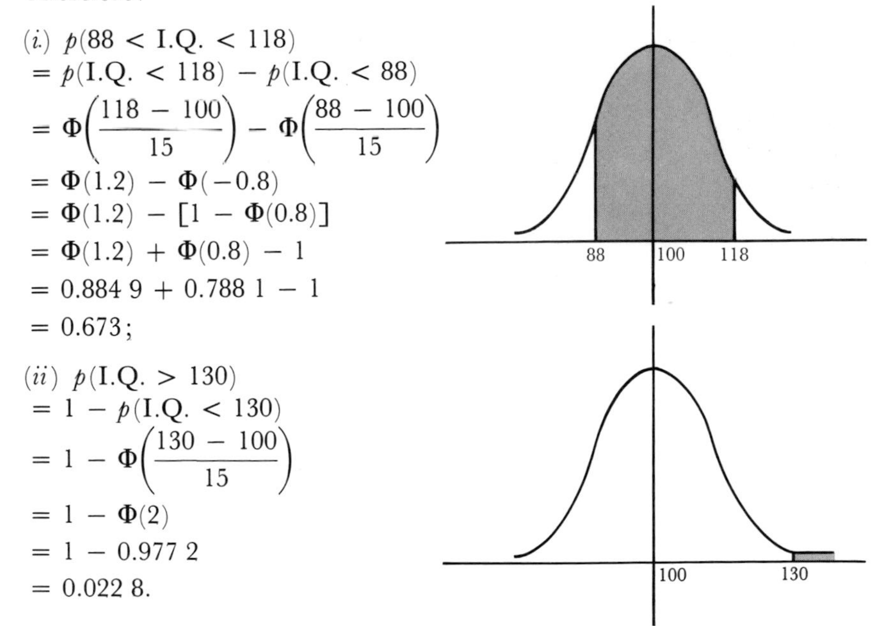

(*i*) $p(88 < \text{I.Q.} < 118)$
$$\begin{aligned}&= p(\text{I.Q.} < 118) - p(\text{I.Q.} < 88)\\ &= \Phi\left(\frac{118 - 100}{15}\right) - \Phi\left(\frac{88 - 100}{15}\right)\\ &= \Phi(1.2) - \Phi(-0.8)\\ &= \Phi(1.2) - [1 - \Phi(0.8)]\\ &= \Phi(1.2) + \Phi(0.8) - 1\\ &= 0.884\ 9 + 0.788\ 1 - 1\\ &= 0.673;\end{aligned}$$

(*ii*) $p(\text{I.Q.} > 130)$
$$\begin{aligned}&= 1 - p(\text{I.Q.} < 130)\\ &= 1 - \Phi\left(\frac{130 - 100}{15}\right)\\ &= 1 - \Phi(2)\\ &= 1 - 0.977\ 2\\ &= 0.022\ 8.\end{aligned}$$

So approximately 2.3% of the population have I.Q. greater than 130.

Example 3

Machine components have lengths which are Normally distributed with mean 57.2 mm. It is found that 13% of the components have lengths greater then 57.4 mm.

(*i*) What is the standard deviation of the component lengths?
(*ii*) What length will be exceeded by 20% of the components?

(*i*) We will denote the standard deviation by σ.

Now $p(\text{length} < 57.4) = 1 - p(\text{length} > 57.4)$

$$\Rightarrow \quad \Phi\left(\frac{57.4 - 57.2}{\sigma}\right) = 1 - 0.13$$

$$\Rightarrow \quad \Phi\left(\frac{0.2}{\sigma}\right) = 0.87.$$

But from the Normal probability table, $\Phi(1.13) = 0.87$.

So $\dfrac{0.2}{\sigma} = 1.13$

$$\Rightarrow \quad \sigma = \frac{0.2}{1.13} = 0.18 \text{ mm (correct to 2 d.p.)}.$$

(*ii*) Let the required length be denoted by X

then $p(\text{length} > X) = 0.2$

$\Rightarrow \quad p(\text{length} < X) = 0.8$

$$\Rightarrow \quad \Phi\left(\frac{X - 57.2}{0.18}\right) = 0.8.$$

But (correct to 2 d.p.) $\Phi(0.84) = 0.8$

$$\Rightarrow \quad \frac{X - 57.2}{0.18} = 0.84$$

$$\Rightarrow \quad X - 57.2 = 0.84 \times 0.18 = 0.15$$

$$\Rightarrow \quad X = 57.35 \text{ mm}.$$

Exercise 12.3a

1 The mean weight of 100 sixth-formers at a certain school is 63 kg and the standard deviation is 7 kg. Assuming that the weights are approximately normally distributed, estimate how many sixth-formers weigh: (*i*) more than 70 kg; (*ii*) less than 73 kg; (*iii*) less than 60 kg; (*iv*) between 61 and 65 kg.

2 The times taken by a large group of students to complete a project are approximately normally distributed, with mean 30 hours and standard deviation 4 hours. Estimate the percentage of students who:
(*i*) spend more than 36 hours on the project;
(*ii*) finish the project in less than 25 hours;
(*iii*) finish the project in less than 20 hours.

3 A factory produces ball-bearings whose masses are normally distributed with mean 45.32 g and standard deviation 0.17 g. What is the probability

that a ball-bearing chosen at random will have a mass: (*i*) more than 45.40 g; (*ii*) more than 45.50 g; (*iii*) less than 45.00 g; (*iv*) between 45.20 and 45.45 g?

4 In a certain book the frequency function for the number of words per page may be taken as approximately Normal with mean 800 and standard deviation 50. If I choose three pages at random, what is the probability that none of them has between 830 and 845 words? (S.M.P.)

5 The daily delivery of mail at a biological research station follows a time pattern conforming to the normal distribution, with a mean time of arrival at 8.40 a.m. and with a standard deviation of 20 minutes.

Estimate the number of occasions during the 250 working days in the year when the mail arrives

(*i*) before the main gates open, at 8.00 a.m.;
(*ii*) after the arrival of the office staff, at 8.20 a.m.;
(*iii*) during the Director's daily meeting with heads of research sections (9.00 a.m.–9.20 a.m.). (A.E.B.)

6 The heights in centimetres of a sample of 700 six-month old babies, attending a post-natal clinic, are given in the following frequency table:

Height/cm	62	63	64	65	66	67	68	69	70	71	72	Total
Frequency	25	35	52	84	120	135	101	61	40	33	14	700

Calculate the mean and standard deviation of the heights.

Assuming that the heights of all such babies are normally distributed about this mean with this standard deviation, estimate:

(*i*) the percentage of all six-month old babies of height 70 cm or more;
(*ii*) the height that will be exceeded by 60% of all babies of this age.

7 The table shows the frequency f with which x α-particles were radiated from a source in a given time, the values of x being integers only.

x	0	1	2	3	4	5	6	7	8	9	10
f	30	37	71	126	130	108	67	5	3	2	1

Calculate the mean value of x and its standard deviation. Assuming that the sample fits a normal distribution, find the value of x that will be exceeded on 70% of occasions. (M.E.I.)

8 If x is Normally distributed with mean μ and standard deviation σ, calculate the value of λ such that:

(*i*) $p(|x - \mu| < \lambda\sigma) = 0.5$;
(*ii*) $p(|x - \mu| < \lambda\sigma) = 0.9$.

9 Packets of soap-powder are filled in such a way that the masses of their contents are normally distributed with mean 520 g and standard deviation 15 g. The cartons are nominally of 500 g.

(*i*) Calculate (correct to 1 d.p.) the percentage of packets that will be 'under-weight'.
(*ii*) Calculate the percentage of packets containing more than 5% above their nominal contents.
(*iii*) If the mean mass of soap-powder can be altered, without affecting the standard deviation, what should the new mean be to ensure that only 2% of packets are under-weight?

10 Hens' eggs have mean mass 60 g with standard deviation 15 g, and the distribution may be taken as Normal. Eggs of mass less than 45 g are classified as 'small'. The remainder are divided into 'standard' and 'large', and it is desired that these should occur with equal frequency. Suggest the mass at which the division should be made (correct to the nearest gram). (S.M.P.)

11 In a cross-country race the times taken by competitors to complete the course are approximately normally distributed with mean 47 minutes and standard deviation 9 minutes. The race organisers wish to split the competitors, on the basis of their times, into 5 equal categories, very slow, slow, average, fast and very fast, so that there will be an equal number of competitors (as far as possible) in each category.
(*i*) What is the least time in which a competitor must complete the race to be recorded as 'very fast'?
(*ii*) Between what times will a competitor be recorded as 'average'?
(*iii*) Between what times will a competitor be regarded as slow? (Give answers in minutes, correct to 1 d.p.)

12 A machine produces components to any required length specification with a standard deviation of 1.40 mm. At a certain setting it produces to a mean length of 102.30 mm. Assuming the distribution of lengths to be Normal, calculate:
(*i*) what percentage would be rejected as less than 100 mm long;
(*ii*) to what value, to the nearest 0.01 mm, the mean should be adjusted if this rejection rate is to be 1%;
(*iii*) whether at the new setting more than 1% of components would exceed 107 mm in length. (M.E.I.)

13 Machined components are accepted if they pass through a gauge of 1.040 cm and do not pass through a gauge of 0.960 cm. It was found that over a period of production the percentages of components rejected by the larger and smaller gauges were 3.5 and 1.5 respectively. Assuming that the distribution of the dimension tested is normal, find the mean and the standard deviation. (O.C.)

14 A manufacturer hopes, for the sake of his reputation, to produce an article of such quality that not more than 5% of the articles may be expected to last for less than 6 months; and for reasonable economy in manufacturing costs, that not more than 15% may be expected to last for more than 20 months. He can control quality to give (assuming normal distribution) a pre-determined average μ and standard deviation σ for the durability of his product. At what values of μ and σ should he aim? (C.)

15 Skulls may be classified into 3 types according to an index based on their length–breadth ratio:

type A, under 75;
type B, between 75 and 80;
type C, over 80.

A large number of skulls are examined and it is found that 58% are of type A, 38% of type B and 4% of type C. Assuming that the length–breadth ratio is normally distributed, determine the mean and standard deviation of the length–breadth ratio of the skulls. (C.)

16 A man leaves home at 8.00 a.m. every morning in order to arrive at work at 9.00 a.m. He finds that over a long period he is late once in forty times. He then tries leaving home at 7.55 a.m. and finds that over a similar period he is late once in one hundred times. Assuming that the time of his journey has a normal distribution, before what time should he leave home in order not to be late more than once in two hundred times? (S.M.P.)

17 In a particular school 1 460 pupils were present on a particular day. By 8.40 a.m. 80 pupils had already arrived, and at 9.00 a.m. 12 pupils had not arrived but were on their way to school. By assuming that the frequency function of arrival times approximates to Normal form, use tables to estimate:
(*i*) the time by which half of those eventually present had arrived;
(*ii*) the standard deviation of the times of arrival.

If registration occurred at 8.55 a.m. how many would not have arrived by then?

If each school entrance permitted a maximum of 30 pupils per minute to enter, find the minimum number of entrances required to cope with the 'peak' minute of arrival. (S.M.P.)

18 The length of a certain mass-produced item is a normal variable with mean 24.3 mm and standard deviation 0.4 mm. Items are rejected if their lengths are below 23.7 mm or above 25.2 mm. Calculate the proportion of items that are rejected, and the proportion of rejects that are too long. The machine making these items is adjusted with the intention of reducing the proportion rejected; the standard deviation of length cannot be altered, but the mean can. State what value of the mean leads to the minimum proportion of rejects, and give the value of this proportion. (C.)

The limiting binomial distribution

The Normal probability distribution has so far arisen as a suitable model to describe certain statistical data, drawn from a wide variety of sources. We now see how it also arises purely from considerations of probability, as the limiting case (for large values of n) of the binomial distribution.

The probability diagrams below illustrate the binomial distributions with $p = \frac{1}{2}$ and $\frac{1}{5}$ for values of $n = 4$, 10 and 20.

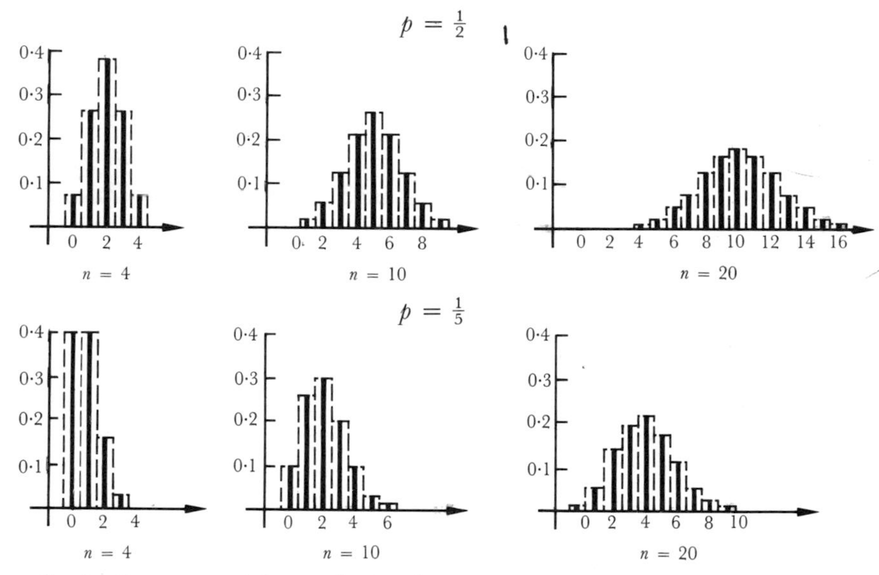

In the first case, with $p = \frac{1}{2}$, the distinctive normal bell-shape is immediately apparent, even for relatively small values of n. And for $n = 20$ the outline of the probability diagram approximates very closely to the Normal curve. With a symmetrical binomial distribution perhaps this is not so surprising, but with an apparently asymmetrical distribution, such as we have with $p = \frac{1}{5}$, the result is much more striking. Indeed, for small values of n (e.g., $n = 4$), the probability diagram bears no resemblance at all to the symmetrical Normal shape; but when $n = 10$ we see that the diagram is much less skew, and when $n = 20$ we once again have a diagram whose outline is very similar to the characteristic Normal shape. It is not difficult to envisage that as n becomes larger and larger, and the horizontal and vertical scales are suitably adjusted (maintaining a total area of 1), the diagram tends closer and closer to the Normal curve. The two diagrams with $n = 20$ are reproduced below, with Normal curves superimposed.

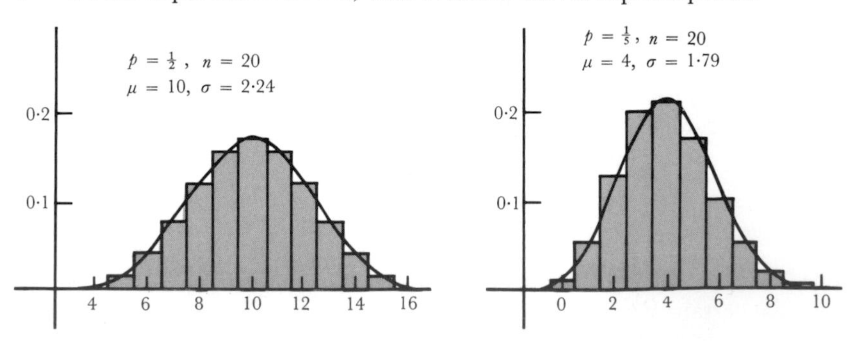

We know, from section 12.1, that for a binomial distribution

$$\mu = np \quad \text{and} \quad \sigma = \sqrt{npq},$$

which therefore gives us the values of μ and σ indicated on the diagram.

* *Derivation of the Normal probability function*

We can now derive the equation, stated earlier in this section, of the Normal probability function, by regarding it as the limit of the binomial distribution.

The mathematics involved in this derivation is not particularly easy, and it should not be regarded as essential to follow the working through in detail.

We have seen that as $n \to \infty$ the profile of the binomial distribution tends:

(*i*) to move to the right (since $\mu = np$);

(*ii*) to spread (since $\sigma = \sqrt{(npq)}$);

and (*iii*) to become flatter.

Now the first of these can be controlled (or *standardised*) by letting

$$u = r - np,$$

and the second by letting

$$t = \frac{u}{\sqrt{(npq)}} = \frac{r - np}{\sqrt{(npq)}}.$$

But this second transformation is equivalent to diminishing all widths by a factor $1/\sqrt{npq}$; so if the area beneath the profile is to remain 1, each of the ordinates must be stretched by a corresponding factor $\sqrt{(npq)}$ which (hopefully) will also counteract the flattening.

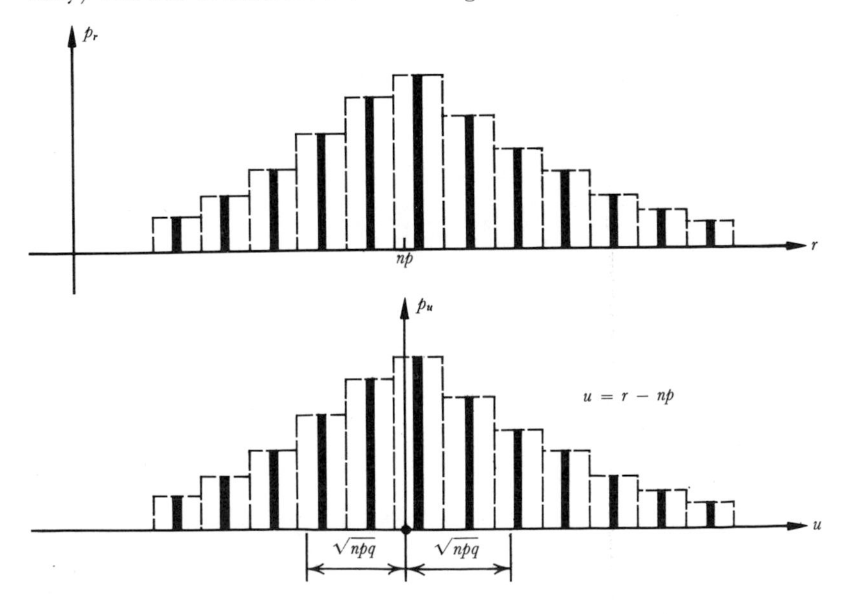

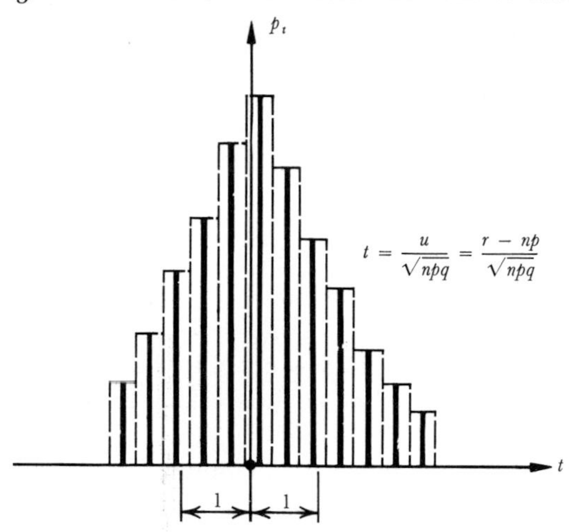

By this means we can bring each distribution into a standard position with $\mu = 0$ and $\sigma = 1$, and so can more easily investigate the alterations of its shape (rather than of its position and spread) as $n \to \infty$.

If we now consider the two neighbouring ordinates p_r and p_{r+1} of the original distribution, we see that each is stretched by a factor $\sqrt{(npq)}$, whilst the distance between them is reduced from 1 to $1/\sqrt{(npq)}$. Furthermore, we suppose that as these ordinates draw closer together, the distribution tends to one with probability density function $\phi(t)$:

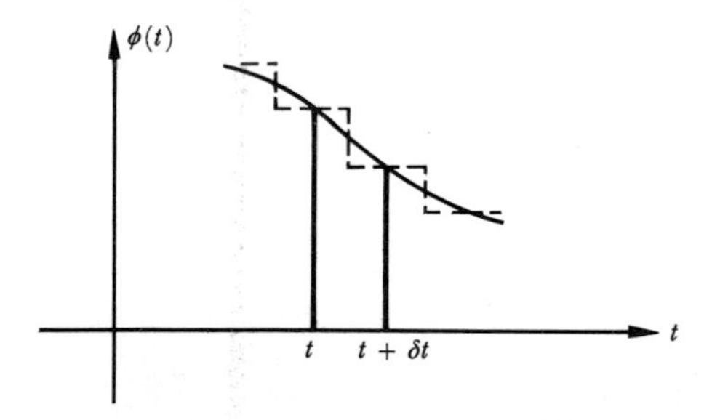

If the ordinates p_r and p_{r+1} are transformed into $\phi(t)$ and $\phi(t + \delta t)$, it follows that

$$\phi(t) = \sqrt{(npq)}\, p_r \quad \text{and} \quad \delta t = \frac{1}{\sqrt{(npq)}}.$$

$$\phi(t + \delta t) = \sqrt{(npq)}\, p_{r+1}.$$

Hence $\delta\phi = \phi(t + \delta t) - \phi(t)$

$$\approx \sqrt{(npq)}\,(p_{r+1} - p_r)$$

$$\Rightarrow \qquad \frac{\delta\phi}{\phi} \approx \frac{p_{r+1} - p_r}{p_r} = \frac{p_{r+1}}{p_r} - 1.$$

But $p_r = \binom{n}{r} p^r q^{n-r}$ and $p_{r+1} = \binom{n}{r+1} p^{r+1} q^{n-r-1}$.

So $$\frac{\delta\phi}{\phi} \approx \frac{\binom{n}{r+1}p^{r+1}q^{n-r-1}}{\binom{n}{r}p^r q^{n-r}} - 1$$

$$\approx \frac{(n-r)p}{(r+1)q} - 1$$

$$\approx \frac{np - rp - rq - q}{(r+1)q} = \frac{np - r - q}{(r+1)q}.$$

But $r = np + \sqrt{(npq)}t$ and $\delta t = \dfrac{1}{\sqrt{(npq)}}$.

So $$\frac{1}{\phi}\frac{\delta\phi}{\delta t} \approx \frac{[-\sqrt{(npq)}t - q]\sqrt{(npq)}}{[np + \sqrt{(npq)}t + 1]q}$$

$$\approx \frac{-npt - \sqrt{(npq)}}{np + \sqrt{(npq)}t + 1}$$

$$\approx \frac{-t - \sqrt{(q/np)}}{1 + \sqrt{(q/np)}t + 1/np}.$$

Letting $n \to \infty$ (and so $\delta t \to 0$),

$$\frac{1}{\phi}\frac{\mathrm{d}\phi}{\mathrm{d}t} = -t$$

$$\Rightarrow \quad \ln \phi(t) = -\tfrac{1}{2}t^2 + \ln c$$

$$\Rightarrow \quad \phi(t) = c\, e^{-\frac{1}{2}t^2}.$$

Hence $$\int_{-\infty}^{\infty} \phi(t)\, \mathrm{d}t = c \int_{-\infty}^{\infty} e^{-\frac{1}{2}t^2}\, \mathrm{d}t.$$

But $\int_{-\infty}^{\infty} \phi(t)\, \mathrm{d}t = 1$, since $\phi(t)$ is a p.d.f., and we have also seen (Miscellaneous problems, chapter 9) that $\int_{-\infty}^{\infty} e^{-x^2}\, \mathrm{d}x = \sqrt{\pi}$, which implies that $\int_{-\infty}^{\infty} e^{-\frac{1}{2}t^2}\, \mathrm{d}t = \sqrt{(2\pi)}$.

Hence $$c = \frac{1}{\sqrt{(2\pi)}}$$

and the limiting curve is

$$\boxed{\phi(t) = \frac{1}{\sqrt{(2\pi)}} e^{-\frac{1}{2}t^2}}$$

Finally, we can reverse the process of standardisation by the transformation

$$t = \frac{x - \mu}{\sigma}$$

and thereby obtain the probability density function of the more general Normal distribution as

$$\phi(x) = \frac{1}{\sqrt{(2\pi)}\,\sigma} e^{-(x-\mu)^2/2\sigma^2}$$

We are now able to put the limiting property of the Normal distribution to practical use, as illustrated in the following example.

Example 4

Forty per cent of university mathematics students are female. Assuming that these young ladies are scattered randomly throughout the country, what is the probability that a mathematics department of 100 students contains at least 50 women?

The probability model that describes this situation is a binomial one, with $n = 100$, $p = 0.4$, $q = 0.6$.

The required probability is therefore given by:

$$\begin{aligned} p\,(\text{at least 50 women}) &= p_{50} + p_{51} + \cdots + p_{100} \\ &= \binom{100}{50}(0.4)^{50}(0.6)^{50} \\ &\quad + \binom{100}{51}(0.4)^{51}(0.6)^{49} + \cdots + \binom{100}{100}(0.4)^{100}. \end{aligned}$$

The evaluation of such a probability would be, to say the least, tedious. But to ease the calculation, we can now use the Normal approximation to this binomial distribution, with

$$\mu = np = 100 \times 0.4 = 40$$

and $\sigma = \sqrt{npq} = \sqrt{(100 \times 0.4 \times 0.6)} = 4.9.$

Now if we denote the number of female students by f, then we require $p(f \geqslant 50) = 1 - p(f \leqslant 49)$.

Using the Normal distribution therefore it would seem reasonable to say that

$$p(f \leqslant 49) = \Phi\left(\frac{49 - 40}{4.9}\right),$$

but this is not quite true and the discrepancy is illustrated in the diagram opposite.

The shaded areas represent, on the left $p(f \leqslant 49)$ and on the right $\Phi[(49 - 40)/(4.9)]$. The difference between them is fairly evident, and arises from the fact that we are using a continuous curve to approximate to a discrete probability distribution. What we need to do is to shift the boundary of the shaded area on our right-hand diagram half a unit to the right. We are then considering $\Phi[(49.5 - 40)/(4.9)]$, which gives a better approximation to $p(f \leqslant 49)$. Such a modification is known as a *continuity correction*. (For very large values of n the correction is insignificant, and so may be disregarded.)

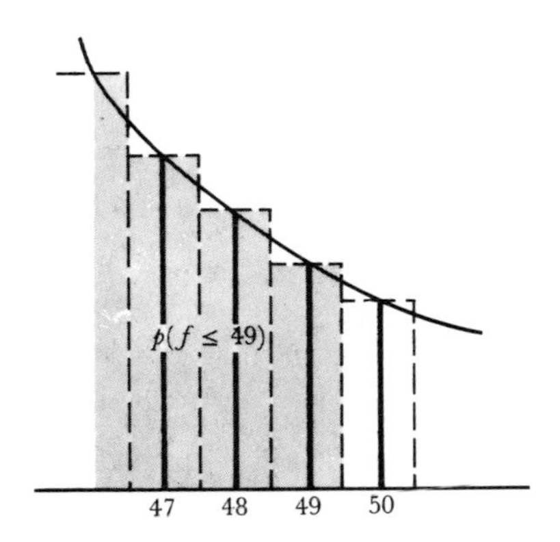

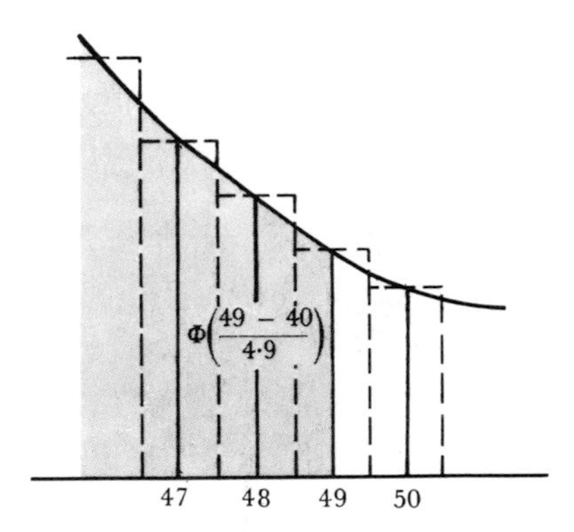

$$\begin{aligned} \text{Hence} \quad p(f \leqslant 49) &= \Phi\left(\frac{9.5}{4.9}\right) \\ &= \Phi(1.94) \\ &= 0.974 \\ \Rightarrow \quad p(f \geqslant 50) &= 1 - 0.974 = 0.026. \end{aligned}$$

So the probability of a department of 100 containing at least 50 women is 0.026.

For a general binomial distribution, where n is large, and the probability of 'success' $= p$.

$$p\text{ (at most } N \text{ successes)} = \Phi\left(\frac{N + \frac{1}{2} - np}{\sqrt{npq}}\right).$$

Exercise 12.3b

1 If an unbiased coin is tossed 100 times, what is the probability that
(*i*) there will be more than 60 heads?
(*ii*) there will be at least 45 and at most 55 heads?
(*iii*) there will be fewer than 43 heads?

2 If an unbiased coin is tossed 1 000 times, what is the probability that
(*i*) there will be more than 600 heads?
(*ii*) there will be at least 450 and at most 550 heads?
(*iii*) there will be fewer than 520 heads?

3 If a 'fair' die is thrown 300 times, what is the probability that
(*i*) there will be more than 60 sixes?
(*ii*) there will be fewer than 45 sixes?

4 In an examination which consists of 100 questions, a student has a probability of 0.6 of getting each question correct. The student fails the examination if he obtains a mark less than 55, and obtains a distinction for a mark, of 68 or more. Calculate:

(*i*) the probability that he fails the examination;
(*ii*) the probability that he obtains a distinction.

5 If it is known that 30% of people are short-sighted, what is the probability that more than 33% of a sample, chosen at random, will be short-sighted if:

(*i*) the sample consists of 100 people;
(*ii*) the sample consists of 1 000 people;
(*iii*) the sample consists of 10 000 people.

6 It is known that 72% of TV viewers watch a particularly popular programme. What is the probability that in a sample of 500 viewers, chosen at random,

(*i*) more than 350 watch the programme?
(*ii*) more than 375 watch the programme?
(*iii*) fewer than 340 watch the programme?

7 A playing card is drawn at random from a full pack, containing no jokers, its suit is recorded and it is then returned to the pack; the pack is then shuffled and the procedure repeated. In 60 such draws calculate, using a Normal approximation, the probability that:

(*i*) no more than 12 hearts appear;
(*ii*) more than 20 spades are obtained.

8 A confectionary firm produces three types of toffees, liquorice, nut and plain, and mixes them together in the ratio 1:2:5 before packing them into boxes. If there are 80 toffees to a box, what percentage of boxes will contain:

(*i*) more than 25 nut toffees?
(*ii*) less than 58 plain toffees?
(*iii*) more nut and liquorice than plain toffees?

9 In a certain manufacturing process a 10% rate of defectives is regarded as just tolerable. It is decided to accept a day's batch if in a random sample of size n taken from it the proportion of defectives does not exceed 12%. Calculate the value of n if there is a 0.05 probability of rejecting a batch which is in fact producing 10% of defectives.

The percentage of defectives rises to 14%. The procedure described above is used with a sample of size 600; calculate, to 2 decimal places, the probability that as a result of the test the batch is accepted. (M.E.I.)

The central limit theorem

A computer may be used to produce the binary digits 0 and 1 in a random fashion. Then if x denotes the value of the digit so obtained,

$$p(x = 0) = p(x = 1) = \tfrac{1}{2}.$$

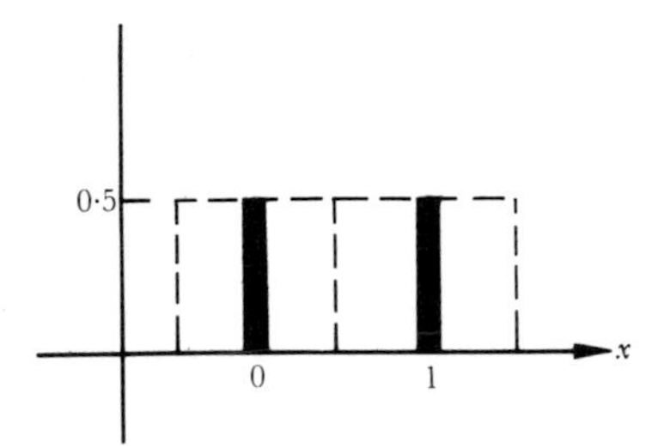

For this distribution,

$$\mu_x = \tfrac{1}{2} \times 0 + \tfrac{1}{2} \times 1 = 0.5$$

and $\quad \sigma_x{}^2 = \tfrac{1}{2}(1 - \tfrac{1}{2})^2 + \tfrac{1}{2}(0 - \tfrac{1}{2})^2 = 0.25$

$\Rightarrow \quad \sigma_x = 0.5.$

Now suppose that instead of looking at the digits individually, we group them together in sets of, say, 10. And for each group we will consider a new variable, y, the sum of the digits.

y can therefore take values from 0 to 10, and since each digit can have only one of two values, y has a binomial probability distribution with $n = 10, p = \tfrac{1}{2}, q = \tfrac{1}{2}$.

Hence $\quad \mu_y = np = 5$

and $\quad \sigma_y = \sqrt{npq} = \sqrt{2.5}.$

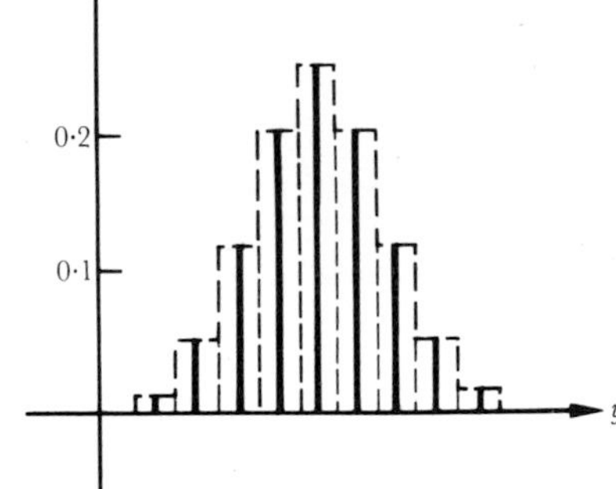

y represents the sum of 10 digits. We now define a third variable, z, to be the average of each group of digits, so that $z = \tfrac{1}{10}y$.

Since z is a constant fraction of y, its probability distribution will have the same shape as that of y, shown in the diagram, but the horizontal scale will be reduced by a factor 10.

So $\quad \mu_z = \tfrac{1}{10}\mu_y = 0.5$

and $\quad \sigma_z = \tfrac{1}{10}\sigma_y = \tfrac{1}{10}\sqrt{2.5} = \dfrac{0.5}{\sqrt{10}}.$

Hence $\quad \mu_z = \mu_x \quad$ and $\quad \sigma_z = \dfrac{\sigma_x}{\sqrt{10}}.$

Therefore if we consider the averages of groups of 10 digits, the expected value is the same as that of the digits taken individually, but the standard deviation of the averages (or 'sample means') is a fraction $1/\sqrt{10}$ (i.e., about $\frac{1}{3}$) of the standard deviation of the individual digits.

Moreover, despite the fact that the original distribution was rectangular, the distribution of sample means appears to be approaching the Normal distribution. This represents a special case of one of the most important and remarkable theorems in statistics, known as *the central limit theorem*, that if samples of size n are drawn at random from *any* background population then, as $n \to \infty$, the distribution of the sample means tends to the Normal distribution. (Although we generally refer to the sample *means* in the central limit theorem, the theorem clearly also holds for the sample *sums*.)

Having already discovered the limiting property of the binomial distribution it is perhaps hardly surprising that the theorem should hold for our example of the binary digits. But suppose we consider the sample means of sets of 10 random digits in the range 0–9; the frequency diagram below illustrates the sample means, suitably grouped, of 50 samples taken from a table of random digits.

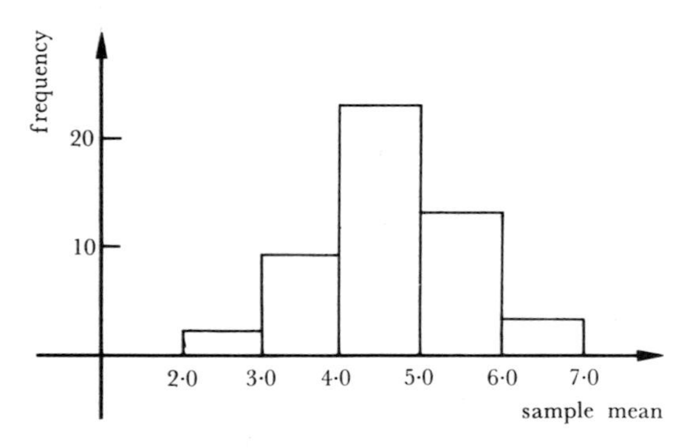

The sample means in this case do not have a distribution in any way binomial; nevertheless, as predicted by the central limit theorem, the Normal shape of the distribution is again apparent.

So, in yet another way, is emphasised the importance of the Normal distribution.

Exercise 12.3c

1 Using a table of random digits (0–9), obtain the sample means of 100 sets of ten digits. Display these sample means on a frequency diagram similar to that shown in the text.

2 What would have been the values of the expected value and standard deviation of the sample mean of the binary digits, described in the text, if the sample had been: (*i*) of size 25; (*ii*) of size 100?

3 A computer is now programmed to produce randomly the digits 0, 2. Calculate the values of μ and σ for this simple rectangular distribution, and hence obtain the expected value and standard deviation of the mean of a sample of size 10.

4 The digits 1, 2, 3, 4 are produced randomly. By applying the results obtained for binary digits, calculate the expected value and standard deviation of the mean of a sample of 100 such numbers. Hence, using the central limit theorem, estimate the probability that the mean of such a sample exceeds 2.7.

12.4 The Poisson distribution

In 12.3 we discovered that, for large values of n, the Normal distribution gave us a good approximation to the binomial distribution, and is in fact the limit of the binomial distribution as $n \to \infty$.

Consider, however, the binomial probabilities with $n = 1\,000$ and $p = 0.001$ and their Normal approximations.

Probability	Binomial	Normal
p_0	0.367 7	0.241 7
p_1	0.368 1	0.383 1
p_2	0.184 0	0.241 7
p_3	0.061 3	0.060 5
p_4	0.015 3	0.005 9
p_5	0.003 0	0.000 2
p_6	0.000 5	0.000 0

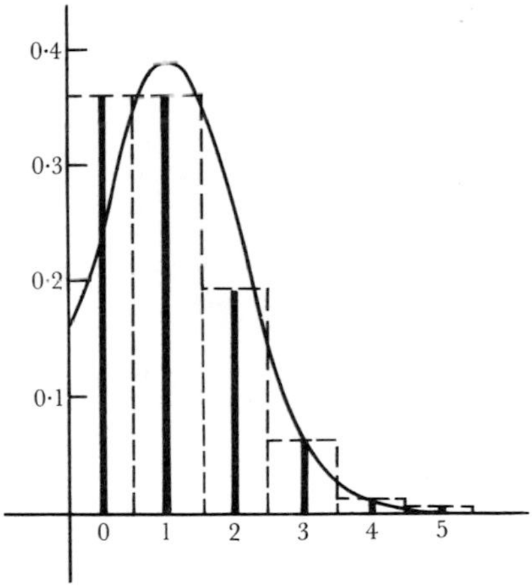

0·4
0·3
0·2
0·1
0 1 2 3 4 5

Here, despite the fact that n is very large, the Normal distribution seems to give us a poor approximation. This apparent contradiction of our previous result arises from the fact that in section 12.3 we assumed that as n became very large, so did the mean, $\mu(= np)$, with the effect that the binomial distribution became approximately symmetrical about this mean. For the distribution illustrated above, however, this is not so; for, even though n is large,

$$\mu = np = 1\,000 \times 0.001 = 1,$$

a relatively small value, and the distribution cannot be symmetrical about μ.

In fact the limiting Normal property of the binomial distribution holds only if

$$\mu = np \to \infty \quad \text{as } n \to \infty$$

(an assumption which was made in the last section). In many cases, of course, this is so, but if np is constant as $n \to \infty$, then the limit of the binomial is quite different, as we shall now show.

We know that for the general binomial distribution,

$$\begin{aligned}
p_r &= \binom{n}{r} p^r(1-p)^{n-r} \\
&= \frac{n(n-1)\ldots(n-r+1)}{r!} p^r(1-p)^{n-r} \\
&= \frac{1\left(1-\frac{1}{n}\right)\cdots\left(1-\frac{r-1}{n}\right)n^r}{r!} p^r(1-p)^{n-r} \\
&= \left(1-\frac{1}{n}\right)\cdots\left(1-\frac{r-1}{n}\right)\frac{(np)^r}{r!}(1-p)^{n-r} \\
&= \left(1-\frac{1}{n}\right)\cdots\left(1-\frac{r-1}{n}\right)\frac{\mu^r}{r!}\left(1-\frac{\mu}{n}\right)^{n-r}.
\end{aligned}$$

Now since r is finite, as $n \to \infty$

$$\left(1-\frac{1}{n}\right)\cdots\left(1-\frac{r-1}{n}\right) \to 1$$

and
$$\begin{aligned}
\left(1-\frac{\mu}{n}\right)^{n-r} &= 1-(n-r)\frac{\mu}{n}+\frac{(n-r)(n-r-1)}{2!}\frac{\mu^2}{n^2}-\cdots \\
&\qquad \text{(a finite series)} \\
&= 1-\frac{n-r}{n}\mu+\frac{(n-r)(n-r-1)}{n^2}\frac{\mu^2}{2!}-\cdots \\
&\to 1-\mu+\frac{\mu^2}{2!}-\frac{\mu^3}{3!}+\cdots \qquad \text{(an infinite series)} \\
&= e^{-\mu}.
\end{aligned}$$

Therefore, in this limiting case,

$$\boxed{p_r = \frac{\mu^r}{r!}e^{-\mu}}$$

and the distribution so defined is known as the *Poisson probability distribution.*

The distribution can also be obtained by considering its probability generator:

The probability generator of the general binomial distribution is given by

$$G(t) = (q + pt)^n$$
$$= (1 + p(t-1))^n \quad \text{since } p + q = 1$$
$$= \left(1 + \frac{\mu(t-1)}{n}\right)^n \quad \text{since } \mu = np.$$

Now we know (Miscellaneous problems, chapter 9) that if x is finite,

$$\left(1 + \frac{x}{n}\right)^n \rightarrow e^x \quad \text{as } n \rightarrow \infty.$$

So $G(t) \rightarrow e^{\mu(t-1)}$ as $n \rightarrow \infty$.

Now $$e^{\mu(t-1)} = e^{\mu t} \times e^{-\mu}$$
$$= \left(1 + \mu t + \frac{\mu^2 t^2}{2!} + \cdots + \frac{\mu^r t^r}{r!} + \cdots\right) e^{-\mu}$$
$$= e^{-\mu} + \mu e^{-\mu} t + \cdots + \frac{\mu^r}{r!} e^{-\mu} t^r + \cdots;$$

so the coefficient of t^r gives $p_r = \dfrac{\mu^r}{r!} e^{-\mu}$.

Returning to the original binomial distribution, if we now use the Poisson distribution with $\mu = 1$, we obtain probabilities which are very good approximations to the binomial, and these equivalent binomial and Poisson probabilities are set out in the table below.

Probability	Binomial	Poisson
p_0	0.367 7	0.367 9
p_1	0.368 1	0.367 9
p_2	0.184 0	0.183 9
p_3	0.061 3	0.061 3
p_4	0.015 3	0.015 3
p_5	0.003 0	0.003 1
p_6	0.000 5	0.000 5

So we now have, under different circumstances, another limit to the binomial distribution, but with the important difference that this limiting distribution, like the binomial itself, but unlike the Normal distribution, is discrete.

The probability diagrams for the Poisson distribution with $\mu = 0.5$, 2 and 5 respectively, are illustrated overleaf.

It can be seen from these diagrams that, as μ becomes larger, the Poisson distribution becomes more symmetrical in shape and, like the binomial, approximates to the Normal distribution.

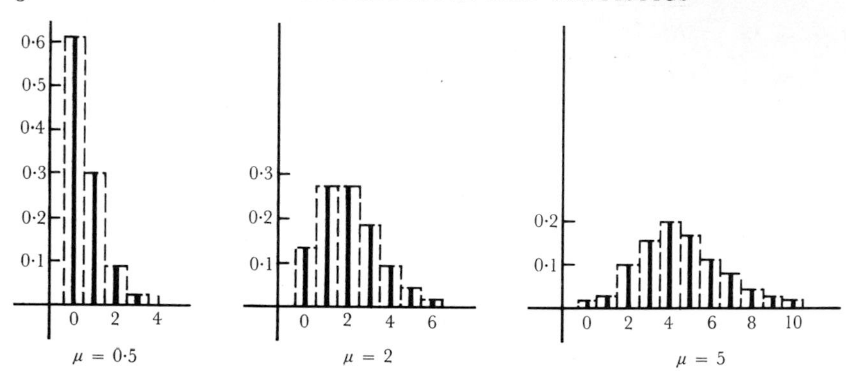

Calculation of Poisson probabilities

In the actual computation of probabilities, the Poisson distribution has one big advantage over the Normal and binomial distributions; it involves the use of only one parameter, μ, as opposed to two in each of the other cases (μ, σ for the Normal and n, p for the binomial). And though the formula for p_r may seem at first glance rather daunting, in practice the calculation of probabilities can be carried out very conveniently:

Suppose $\mu = 3$,

then

$$p_0 = e^{-3} = 0.049\ 8$$

$$p_1 = 3\,e^{-3} = 3p_0 = 0.149\ 4$$

$$p_2 = \frac{3^2\,e^{-3}}{2!} = \tfrac{3}{2}p_1 = 0.224\ 1$$

$$p_3 = \frac{3^3\,e^{-3}}{3!} = \tfrac{3}{3}p_2 = 0.224\ 1$$

$$p_4 = \frac{3^4\,e^{-3}}{4!} = \tfrac{3}{4}p_3 = 0.168\ 1$$

and $$p_{r+1} = \frac{3}{r+1}p_r \quad \left(\text{so in general} \quad p_{r+1} = \frac{\mu}{r+1}p_r\right).$$

Mean and variance of Poisson distribution

For the binomial distribution,

$$\mu = np$$

and $$\sigma^2 = npq = np(1-p).$$

Now we can regard the Poisson distribution as the limit of the binomial distribution as $n \to \infty$ and $np = \mu$ (so that $p \to 0$).

Hence $$\text{mean} = \mu$$

and $$\text{variance} = \lim_{n\to\infty} np(1-p) = \mu.$$

Applications of the Poisson distribution

We have derived the Poisson distribution as a limiting case of the binomial distribution. But it arises more fundamentally in its own right, whenever there is a continuous variable — say time or distance — and discrete events occur *randomly* and *independently*. In such a case, the number of events in a given interval has a Poisson distribution. (By imagining the continuous variable split into a very large number of equal small intervals, the reader can verify for himself that these conditions are consistent with those of a limiting binomial distribution.) Examples of such situations, in which the Poisson distribution is applicable, are provided by

flaws occurring in lengths of steel piping,

people arriving in a shop queue in a certain time interval,

and radio-active particles being emitted from a source in some time interval.

Consider the first of these examples:

Example 1

It is found experimentally that there is an average of one flaw in every 2 metres of a certain type of steel piping. If the piping is cut into 1-metre lengths what is the probability of 0, 1, 2 flaws in such a length? Find the equivalent probabilities if the pipe is cut into 4-metre lengths.

(*i*) For a 1-metre length of pipe, the expected number of flaws, $\mu = \frac{1}{2}$.

Hence $p_0 = e^{-\frac{1}{2}} = 0.606\ 5$.

Then using the relationship we have recently obtained between successive Poisson probabilities,

$$p_1 = \tfrac{1}{2}p_0 = 0.303\ 2$$

and $$p_2 = \frac{\frac{1}{2}p_1}{2} = 0.075\ 8.$$

(*ii*) For 4-metre lengths of pipe, $\mu = 2$.

Hence $p_0 = e^{-2} = 0.135\ 3$

$\Rightarrow$ $p_1 = 2p_0 = 0.270\ 6$

$\Rightarrow$ $p_2 = \dfrac{2p_1}{2} = 0.270\ 6.$

It was quite clear that Example 1 should be tackled by means of the Poisson distribution. By comparison, in the following example we can use either a binomial or a Poisson probability model, and the equivalent results for the two models are compared. In example 3 we fit a Poisson model to experimental data.

Example 2

A firm of wholesale fruit distributors has found that on average one apple in 50 is bruised on arrival from the growers. If the apples arrive in cartons of 100, calculate the probabilities of a carton having 0, 1, 2, 3, more than 3 bruised apples.

Using a binomial model, $n = 100$ and $p = \frac{1}{50}$.

Using a Poisson model, $\mu = 100 \times \frac{1}{50} = 2$.

The calculations for the two models are carried out simultaneously:

binomial		Poisson	
$p_0 = \left(\frac{49}{50}\right)^{100}$	$= 0.131\,6$	$p_0 = e^{-2}$	$= 0.135\,3$
$p_1 = \binom{100}{1}\left(\frac{49}{50}\right)^{99}\frac{1}{50}$	$= 0.270\,6$	$p_1 = 2p_0$	$= 0.270\,6$
$p_2 = \binom{100}{2}\left(\frac{49}{50}\right)^{98}\left(\frac{1}{50}\right)^2$	$= 0.272\,7$	$p_2 = p_1$	$= 0.270\,6$
$p_3 = \binom{100}{3}\left(\frac{49}{50}\right)^{97}\left(\frac{1}{50}\right)^3$	$= 0.181\,8$	$p_3 = \frac{2}{3}p_2$	$= 0.180\,4$
$p(>3)$	$= 0.143\,3$	$p(>3)$	$= 0.143\,1$

The results clearly are very similar but, since its calculations are so much easier, we would usually adopt the Poisson model.

Example 3

The following classical data (without which any account of Poisson probability would be incomplete!) gives the number of deaths by horse-kick of Prussian cavalrymen in ten corps during the years 1875–1894.

No. of deaths per corps per year	0	1	2	3	4
Frequency of occurrence	109	65	22	3	1

We will fit a Poisson probability model to the data, and calculate the equivalent theoretical frequencies.

The mean number of deaths per year is given by

$$m = \frac{109 \times 0 + 65 \times 1 + 22 \times 2 + 3 \times 3 + 1 \times 4}{200}$$

$$= \frac{122}{200}$$

$$= 0.61.$$

We will use this mean value as our parameter μ in the Poisson distribution. To obtain the theoretical (or expected) frequencies, we multiply the probabilities by the total frequency, 200, and round off to the nearest whole number.

No. of deaths	Probability	Expected frequency	Observed frequency
0	$e^{-0.61} = 0.543\,4$	109	109
1	$0.61\, p_0 = 0.331\,5$	66	65
2	$\frac{0.61}{2} p_1 = 0.101\,1$	20	22
3	$\frac{0.61}{3} p_2 = 0.020\,6$	4	3
4	$\frac{0.61}{4} p_3 = 0.003\,1$	1	1

******Alternative derivation of Poisson distribution*

To complete our account of the Poisson distribution, we now derive it directly from our basic assumptions.

Let us suppose that a series of events, such as the flashes of a Geiger counter, occur independently; and yet, over a long interval, at a uniform rate, so that in an interval of unit length the expected number of flashes is λ. Then in a very short interval δt, the probability

of a single flash is $\lambda \delta t$,

of two or more flashes is 0,

and of no flashes is $1 - \lambda \delta t$.

We now denote the probability of having r flashes in the first t units of time by $p_r(t)$. Then, if we consider a further small interval δt, it is clear that

$p(r$ flashes in time $t + \delta t)$

$= p(r$ flashes in time t *and* no flashes in interval $\delta t)$

$+ p(r - 1$ flashes in time t *and* 1 flash in interval $\delta t)$

$$\Rightarrow p_r(t + \delta t) = p_r(t)(1 - \lambda\, \delta t) + p_{r-1}(t)\lambda\, \delta t$$

$$\Rightarrow \frac{p_r(t + \delta t) - p_r(t)}{\delta t} + \lambda p_r(t) = \lambda p_{r-1}(t).$$

Letting $\delta t \to 0$, we obtain

$$p_r'(t) + \lambda p_r(t) = \lambda p_{r-1}(t)$$

$$\Rightarrow \quad \frac{\mathrm{d}}{\mathrm{d}t}[e^{\lambda t} p_r(t)] = \lambda\, e^{\lambda t} p_{r-1}(t)$$

$$\Rightarrow \quad [e^{\lambda t} p_r(t)]_0^t = \lambda \int_0^t e^{\lambda t} p_{r-1}(t)\, \mathrm{d}t.$$

But $p_r(0) = 0 \quad \text{if } r \neq 0,$

so $$e^{\lambda t} p_r(t) = \lambda \int_0^t e^{\lambda t} p_{r-1}(t)\, \mathrm{d}t. \tag{1}$$

Furthermore,

p (0 flashes in time $t + \delta t$)

$=$ p (0 flashes in time t *and* 0 flashes in interval δt)

$$\Rightarrow \quad p_0(t + \delta t) = p_0(t)(1 - \lambda\, \delta t)$$

$$\Rightarrow \quad \frac{p_0(t + \delta t) - p_0(t)}{\delta t} + \lambda p_0(t) = 0.$$

Letting $\delta t \to 0$,

$$p_0'(t) + \lambda p_0(t) = 0$$

$$\Rightarrow \quad \frac{\mathrm{d}}{\mathrm{d}t}[e^{\lambda t} p_0(t)] = 0$$

$$\Rightarrow \quad e^{\lambda t} p_0(t) = p_0(0).$$

But we certainly start with 0 flashes, so $p_0(0) = 1$.

Hence $$e^{\lambda t} p_0(t) = 1. \tag{2}$$

If we now put $P_r(t) = e^{\lambda t} p_r(t)$, we can rewrite equations (1) and (2) as

$$P_r(t) = \lambda \int_0^t P_{r-1}(t)\, \mathrm{d}t$$

and $P_0(t) = 1$.

So
$$P_1(t) = \lambda \int_0^t \mathrm{d}t = \lambda t \qquad \Rightarrow \quad p_1(t) = \lambda t\, e^{-\lambda t};$$

$$P_2(t) = \lambda \int_0^t \lambda t\, \mathrm{d}t = \frac{\lambda^2 t^2}{2} \qquad \Rightarrow \quad p_2(t) = \frac{\lambda^2 t^2}{2} e^{-\lambda t};$$

$$P_3(t) = \lambda \int_0^t \frac{\lambda^2 t^2}{2}\, \mathrm{d}t = \frac{\lambda^3 t^3}{6} \qquad \Rightarrow \quad p_3(t) = \frac{\lambda^3 t^3}{6} e^{-\lambda t};$$

and, more generally,

$$P_r(t) = \frac{\lambda^r t^r}{r!} \qquad \Rightarrow \quad p_r(t) = \frac{\lambda^r t^r}{r!} e^{-\lambda t}.$$

Finally, regarding t as fixed, we can write $p_r(t)$ as p_r; and since the expected number of flashes in this interval is λt, we write $\mu = \lambda t$ and obtain

$$p_r = \frac{\mu^r}{r!}e^{-\mu}.$$

Exercise 12.4

1 It is known that 0.1% of all people react adversely to a certain type of drug. What is the probability that out of a sample of 1 000 people:
(*i*) none will react to the drug;
(*ii*) just one person will react;
(*iii*) more than two people will react?

2 A hiker on a walking holiday in a country area estimates that in walking 150 kilometres he has passed a total of 60 public houses. Assuming that these public houses are distributed fairly randomly about the countryside, what is the probability of his being able to quench his thirst:
(*i*) within the next kilometre;
(*ii*) within the next five kilometres?

3 During a certain period of the day, an average of four people enter a supermarket each minute. Calculate the probability that:
(*i*) no-one enters during a particular minute;
(*ii*) three people enter during a particular minute;
(*iii*) five people enter during a particular minute;
(*iv*) no-one enters during a two-minute period.

4 Over a period of time, the number of serious road accidents in a city averages 1.8 per day. What is the probability of there being:
(*i*) no accidents on a given day;
(*ii*) more than two accidents on a given day;
(*iii*) fewer than five accidents in a three-day period?

5 The average proportion of bad eggs in an egg packing station is one in 2 000. The eggs are packed in boxes containing six eggs each.
(*i*) Evaluate to two significant figures the probability that a box contains one bad egg.
(*ii*) A housewife complains if she obtains two or more boxes, with one bad egg each, per 100 boxes. What is the probability that she complains?
(M.E.I.)

6 A manufacturer of electrical components knows that he can expect 0–2% of his products to be faulty. What is the maximum number of components he can pack into a box, if he wishes to ensure that at least 90% of the boxes contain no defectives?

7 During the Second World War 535 flying bombs fell on South London. The distribution of the numbers of hits in 576 areas, each of 0.25 km^2 is given in the table. Show that the aim was effectively random within the

South London area by calculating the expected frequencies for a Poisson distribution with the same average number of hits.

No. of hits	0	1	2	3	4	5	6 or more
Frequency	229	211	93	35	7	1	0

(M.E.I.)

8 The incidence of plumbing repairs in 78 council houses over a period of ten years is given in the table. Do the data support the view that there are good and bad tenants or the view that there are lucky and unlucky tenants?

No. of repairs	0	1	2	3	4	5	6	7	8	9	10
No. of houses	3	13	16	16	10	9	3	5	1	1	1

(Use a Poisson probability model and compare theoretical and observed frequencies.) (M.E.I.)

9 At a stage in the mass production of lamp holders, random samples, each of 40 articles, are examined and the number of defective articles recorded. The numbers of defective articles in each of 200 samples are shown in the following frequency table:

Defectives	0	1	2	3	4	5	6	7
No. of samples	29	56	42	42	23	7	0	1

Find the mean number of defectives per sample and the variance. Give reasons for thinking that the distribution approximates to a Poisson distribution.

Show that on this assumption there is a probability of about 5% of a sample containing more than four defectives and a probability of less than 1% of a sample containing more than six defectives. (O.C.)

10 Spot blemishes occur randomly along a steel wire. The number counted in consecutive centimetre lengths had the following distribution:

No. of spots	0	1	2	3	4	5	6	7
Frequency	102	150	112	56	21	6	2	1

If the count had been over consecutive 2 cm lengths, estimate the frequency of occurrence of 3 spots per 2 cm. (You may assume that the frequencies are close to a Poisson distribution.) (S.M.P.)

11 The following are the numbers of breakdowns of a machine in 40 12-hour periods:

0 4 0 1 1 0 0 2 2 0 1 1 0 1 1 0 1 3 2 0 2 1 2 2 1 1 0 0 2 0 0 1 0 2 1 1 3 1 0 2.

Calculate the mean number of breakdowns per 12-hour period and, using this value for μ in a Poisson probability model, calculate the probabilities of 0, 1, 2, 3, 4 breakdowns. Hence estimate the number of breakdown-free days in a year (taken to consist of 300 12-hour working days). Also estimate the number of days in the year when there will be more than 2 breakdowns.

12 Using probability generators, show that if x, y have independent Poisson distributions with means μ_x, μ_y then $x + y$ has a Poisson distribution with mean $\mu_x + \mu_y$.

13 Instruments A and B are set to record radiation from separate and independent sources. The number of particles recorded by the instruments in one second have Poisson distributions with means 2 and 4 respectively. Calculate the probabilities that:
(*i*) at least one particle is recorded on one of the two instruments in a second;
(*ii*) a total of 4 or more particles is recorded on the two instruments in a second. (O.C.)

14 A factory produces high-quality china dishes. Flaws occur in the dishes themselves with a Poisson probability distribution, mean μ, and flaws occur independently in the glazing, again with a Poisson probability distribution with mean μ. A dish is immediately rejected if it has a flaw of either kind. What is the probability that a dish is rejected? Hence find the probability that a rejected dish only had flaws in the glazing (expressing your answer in as simple a form as possible.)

15 In a simplified probability model of the service in a barber's shop it is supposed that all haircuts take exactly six minutes and that a fresh batch of customers arrives at six-minute intervals. The number of customers in a batch is described by a Poisson probability function, the mean number being 3. Any customer who cannot be served instantly goes away and has his hair cut elsewhere. The shop is open for 40 hours a week. Calculate the theoretical frequencies with which batches of 0, 1, 2, 3, 4, 5 and more than 5 customers will arrive per week.

The proprietor reckons that it costs him £25 a week to staff and maintain each chair in his shop, and he charges 25 p for each haircut. Calculate his expected weekly profit if he has (*i*) 3, (*ii*) 4, (*iii*) 5 chairs. (S.M.P.)

16 Show that when an event occurs randomly in time (i.e., the number of events occurring in a fixed interval of time has a Poisson distribution—with mean λ) then the time interval between successive occurrences has the exponential distribution:

$$\phi(x) = \lambda e^{-\lambda x} \quad (x > 0).$$

The mean number of calls per minute made on a certain telephone exchange is two. Find:
(*i*) the probability that in any given one minute interval not more than one call will be made;

(*ii*) the probability that in any given half-minute interval exactly two calls will be made;
(*iii*) the probability that there will be no call in the half-minute immediately following a call. (M.E.I.)

*12.5 Samples

In practical statistics we rarely consider individual readings or measurements in isolation; the normal procedure is to use a *sample*, or set of readings, and from these readings to obtain a *statistic*, or representative value. The most common, and generally most useful, statistic is the *mean*. Intuitively we expect it to give us a good approximation to the expected value (or mean), μ, of the background population, and we also feel that somehow the mean is a more reliable estimate of μ than is a single reading. We will now give some substance to these intuitive notions by extending our ideas of expectation to cover the expected value and variance of the mean of a sample. But before examining the mean itself, we must first establish some basic results in expectation algebra.

Expectation algebra

Suppose that x is a random variable,† and takes values

$$x_1, x_2, x_3, \ldots$$

with probabilities $p_1, p_2, p_3, \ldots.$

We now (as in chapter 7) use the notation $E[x]$ for the *expected value*, or *expectation* of x, so that

$$E[x] = \sum x_i p_i = \mu.$$

Furthermore, we denote the variance of x by $V[x]$ so that

$$\begin{aligned} V[x] &= \sum (x_i - \mu)^2 p_i = \sum x_i^2 p_i - \mu^2 \\ &= E[(x-\mu)^2] = E[x^2] - \mu^2. \end{aligned}$$

Hence
$$\boxed{\begin{aligned} E[x] &= \mu \\ V[x] &= E[(x-\mu)^2] = E[x^2] - \mu^2 \end{aligned}}$$
and

We now let a be constant, so that ax takes values $ax_1, ax_2, \ldots$ with probabilities $p_1, p_2, \ldots.$

Hence
$$\begin{aligned} E[ax] &= \sum ax_i p_i \\ &= a \sum x_i p_i \\ &= a\,E[x]. \end{aligned}$$

† It can be argued that what is generally referred to as a random variable is neither random nor variable. At this stage, however, it is undesirable to become too involved in such discussions, and since the term is in widespread use we shall continue to employ it.

$$\begin{aligned} \text{Similarly} \quad V[ax] &= \sum (ax_i - a\mu)^2 p_i \\ &= \sum a^2 (x_i - \mu)^2 p_i \\ &= a^2 \sum (x_i - \mu)^2 p_i \\ &= a^2 V[x]. \end{aligned}$$

Summarising,

$$\boxed{\begin{aligned} E[ax] &= aE[x] \\ V[ax] &= a^2 V[x] \end{aligned}}$$

The same results can be shown to hold for a continuous variable, using the probability density function, and integration instead of summation.

Example 1

If measurements are made in centimetres, the expected value and variance of length of leaf of a certain type of rhododendron are 5.5 cm and 2.4 cm^2 respectively. What will the expected value and variancc be if measurements are made in millimetres?

Let x denote the leaf length in centimetres,

then $\quad E[x] = 5.5 \quad$ and $\quad V[x] = 2.4.$

The length in millimetres is given by $10x$, so the expected length in millimetres is given by $E[10x]$

$$\begin{aligned} \text{and} \qquad E[10x] &= 10E[x] \\ &= 55. \end{aligned}$$

$$\begin{aligned} \text{Similarly} \quad V[10x] &= 10^2 V[x] \\ &= 240. \end{aligned}$$

These results of course are hardly surprising, and simply fortify the above expectation results with common sense.

Two random variables

Let x, y be two random variables, each of them discrete. (An equivalent analysis is also possible for continuous variables, but it involves the use of more advanced mathematics, with which the reader is probably not yet acquainted; it can be found in most specialised probability texts.)

Let $\quad x$ take possible values $x_1, x_2, \ldots, x_i, \ldots$
and $\quad y$ take possible values $y_1, y_2, \ldots, y_j, \ldots$

(a finite or infinite number in each case.)

We define the *joint probability distribution* of x and y by $\quad p(x = x_i, y = y_j)$, which can be written $p(x_i, y_j)$ or more simply p_{ij}.

Then $p(x_i) = \sum_j p_{ij}$, summing over all possible values of j,

and $p(y_j) = \sum_i p_{ij}$, summing over all possible values of i.

This situation can now be summarised in the table:

$x \backslash y$	y_1	y_2	$\dots$	y_j	$\dots$	
x_1	p_{11}	p_{12}		p_{1j}		$p(x_1)$
x_2	p_{21}	p_{22}		p_{2j}		$p(x_2)$
$\cdot$						
$\cdot$						
$\cdot$						
x_i	p_{i1}	p_{i2}		p_{ij}		$p(x_i)$
$\cdot$						
$\cdot$						
$\cdot$						
	$p(y_1)$	$p(y_2)$	$\dots$	$p(y_j)$	$\dots$	1

Using this notation, let us now investigate the expected value and variance of $x + y$, the sum of the two random variables.

Considering the values of $x + y$ which arise from all the combinations of values of x and y,

$$
\begin{aligned}
E[x + y] &= \sum_i \sum_j (x_i + y_j)p_{ij} \\
&= \sum_i \sum_j x_i p_{ij} + \sum_i \sum_j y_j p_{ij} \\
&= \sum_i \Big(x_i \big(\sum_j p_{ij}\big)\Big) + \sum_j \Big(y_j \big(\sum_i p_{ij}\big)\Big) \\
&= \sum_i x_i p(x_i) + \sum_j y_j p(y_j) \\
&= E[x] + E[y].
\end{aligned}
$$

Also $E[xy] = \sum_i \sum_j p_{ij} x_i y_j$.

But x, y are independent $\Rightarrow p_{ij} = p(x_i)\,p(y_j)$

$$
\begin{aligned}
\Rightarrow E[xy] &= \sum_i \sum_j p(x_i)\,p(y_j)\,x_i y_j \\
&= \sum_i p(x_i)\,x_i \sum_j p(y_j)\,y_j \\
&= E[x]\,E[y].
\end{aligned}
$$

$$\begin{aligned}\text{Moreover, } V[x+y] &= E[(x+y)^2] - (E[x+y])^2 \\ &= E[x^2 + 2xy + y^2] - (E[x] + E[y])^2 \\ &= \{E[x^2] - (E[x])^2\} + \{E[y^2] - (E[y])^2\} \\ &\quad + 2\{E[xy] - E[x]E[y]\} \\ &= V[x] + V[y] + 2Cov[x,y],\end{aligned}$$

where $Cov[x,y] = E[xy] - E[x]E[y]$ and is known as the *covariance* of x and y, which is a measure of their interdependence.

In particular, we have seen that x, y are independent

$\Rightarrow E[xy] = E[x]E[y]$.

$\Rightarrow Cov[x,y] = 0$

$\Rightarrow V[x + y] = V[x] + V[y]$.

Summarising: in general	$E[x + y] = E[x] + E[y]$
	$V[x + y] = V[x] + V[y] + 2\,Cov\,[x, y]$
but if x, y are independent	$V[x + y] = V[x] + V[y]$

Similarly, $E[x - y] = E[x] - E[y]$

$$V[x - y] = V[x] + V[y] - 2\,Cov\,[x, y];$$

and note that for independent x, y,

$$V[x - y] = V[x] + V[y];$$

so that the difference of two independent variables has exactly the same variance as their sum.

Example 2

A normal 'fair' 6-faced die is thrown together with a 'fair' tetrahedral (4-faced) die. What is the expected value and variance of the total score registered on the two dice? (For the tetrahedral die, the score registered is that of the face on which it lands.)

Let the score on the 6-faced die be denoted by x;

$$\text{then } E[x] = \tfrac{1}{6}(1 + 2 + 3 + 4 + 5 + 6) = 3.5,$$

and $$\begin{aligned} V[x] &= \tfrac{1}{6}(1^2 + 2^2 + 3^2 + 4^2 + 5^2 + 6^2) - 3.5^2 \\ &= \tfrac{91}{6} - 3.5^2 \\ &= 15.17 - 12.25 \\ &= 2.92. \end{aligned}$$

Now let the score on the 4-faced die be denoted by y

then $$\begin{aligned} E[y] &= \tfrac{1}{4}(1 + 2 + 3 + 4) \\ &= 2.5, \end{aligned}$$

and $$\begin{aligned} V[y] &= \tfrac{1}{4}(1^2 + 2^2 + 3^2 + 4^2) - 2.5^2 \\ &= \tfrac{30}{4} - 2.5^2 \\ &= 7.5 - 6.25 \\ &= 1.25. \end{aligned}$$

Now the total score is given by $x + y$

and $$\begin{aligned} E[x + y] &= E[x] + E[y] \\ &= 3.5 + 2.5 \\ &= 6 \end{aligned}$$

and if we assume that the two scores are independent, then

$$\begin{aligned} V[x + y] &= V[x] + V[y] \\ &= 2.92 + 1.25 \\ &= 4.17. \end{aligned}$$

So the expected value and variance of the total score are respectively 6 and 4.17.

Example 3

A firm which produces garden tools manufactures the blades and shafts (including handle) of its spades separately. The blade lengths (up to the point of attachment to the shaft) are approximately Normal, with mean 30 cm and standard deviation 1 cm, and the lengths of the shafts are also Normally distributed with mean 78 cm and standard deviation 5 cm. What percentage of spades, when assembled, have lengths exceeding 110 cm?

We will assume (without proof) that the sum of two Normally distributed variables is itself Normal. We will denote the length of the blade by x, and the length of the shaft by y. Therefore $x + y$ denotes the total length of the spade.

Hence $E[x + y] = E[x] + E[y] = 30 + 78 = 108$

and $V[x + y] = V[x] + V[y] = 5^2 + 1^2 = 26.$

So the mean total length $= 108$, and its standard deviation $= \sqrt{26} = 5.1$.

Now since we have assumed that $x + y$ is Normally distributed,

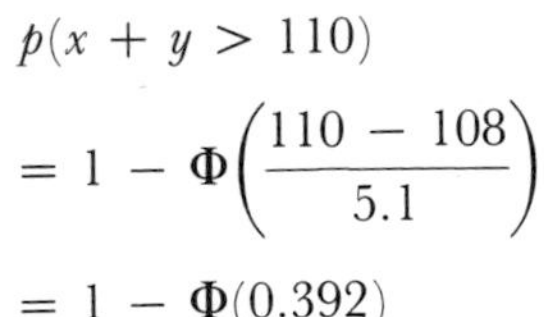

$$p(x + y > 110)$$

$$= 1 - \Phi\left(\frac{110 - 108}{5.1}\right)$$

$$= 1 - \Phi(0.392)$$

$$= 1 - 0.65 \quad \text{(to 2 d.p.)}$$

$$= 0.35.$$

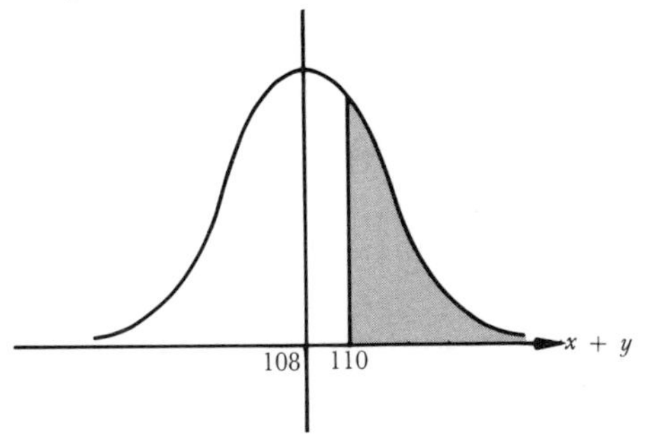

Therefore 35% of the spades are longer than 110 cm.

Exercise 12.5a

1 Prove that, if a is a constant,
(*i*) $E[x + a] = E[x] + a$;
(*ii*) $V[x + a] = V[x]$.

2 If x and y are random variables, and a, b are constants, obtain expressions, in terms of $E[x]$, $E[y]$, $V[x]$ and $V[y]$ for $E[ax + by]$ and $V[ax + by]$:
(*i*) when x, y are independent;
(*ii*) when x, y are not independent.

3 Extend no. **2** to three mutually independent random variables x, y, z and constants a, b, c to obtain expressions for:
(*i*) $E[ax + by + cz]$;
(*ii*) $V[ax + by + cz]$.

4 Find the expected total score obtained when three standard, 'fair' dice are thrown together. What is the variance of this score?

5 Two standard, 'fair' dice, one blue, the other red, are thrown together, and a total score is obtained by doubling the number shown on the blue die and adding to it the number on the red die. What are the expected value and variance of this score? Compare your results with those obtained in question **4**.

6 Two standard 'fair' dice are thrown, and their scores recorded. What are the expected value, and standard deviation of the difference between the two scores?

7 Alan and Bob play a game in which Alan uses a 6-faced die and Bob a 4-faced die (as described in the text in example 2). To make the game 'fair', they decide to double all Alan's scores and treble Bob's. Would you expect this to lead to a fair game? If not, who has the advantage? Evaluate the variance of each (increased) score on the single throw of a die.

8 Prove that if x, y are independent, $E[xy] = E[x] \times E[y]$. Hence calculate the expected product of the scores obtained on two standard unbiased dice.

9 If x and y are independent random single digit numbers other than zero, calculate the variance of
(*i*) x; (*ii*) $x + y$; (*iii*) $2x$; (*iv*) $2x - y$. (S.M.P.)

10 A man travels from his London office to his home by a tube journey from station A to station B. His walking times to A and from B add up to 5 minutes with negligible variation, the variable factors in the journey being as follows, measured in minutes:

	Mean time	Standard deviation
(*i*) waiting for a train	8	2.6
(*ii*) train journey	47	1.8

Assuming that these two factors are independent and normally distributed, find the mean time and standard deviation of his whole journey.

Estimate the probability of the *whole* journey taking
(*i*) less than 52 minutes;
(*ii*) more than 65 minutes;
(*iii*) between 57 and 62 minutes. (M.E.I.)

11 Three independent components are placed in an electrical circuit in such a way that, as soon as the first fails, the second (reserve) component is automatically switched in, and then when the second component itself fails, the third is automatically switched in. The circuit functions as long as one of the components works. If the lifetimes of the components are Normally distributed with mean 2 000 hours and standard deviation 300 hours, what is the probability that:
(*i*) the circuit will function for more than 6 500 hours;
(*ii*) the circuit will break down within 5 000 hours?

12 If $x_1, x_2, \ldots, x_n$ are independently distributed variables with means $\mu_1, \mu_2, \ldots, \mu_n$ and standard deviations $\sigma_1, \sigma_2, \ldots, \sigma_n$ respectively, write down the mean and standard deviation of $y = \lambda_1 x_1 + \lambda_2 x_2 + \cdots + \lambda_n x_n$ where $\lambda_1, \lambda_2, \ldots, \lambda_n$ are constants.

In the manufacture of certain metal tubes there are three distinct processes. The mean times for each process and the standard deviations of these times are given in the following table:

	Mean time/s	Standard deviation/s
Process 1	60	6
Process 2	10	1
Process 3	40	5

During the course of its manufacture each tube is subjected to processes 1 and 3 once each, and to process 2 five times. Assuming that the times spent at each stage are distributed independently, calculate the mean and standard deviation of the total time taken in the manufacture of a tube.

Assuming Normal variation, estimate:

(*i*) the percentage of tubes which would take less than 130 seconds to manufacture;

(*ii*) the percentage of tubes which would take more than 160 seconds to manufacture. (M.E.I.)

13 In a factory, cylindrical pins are made to fit into circular holes in blocks of metal. The diameter of a pin is a Normal variable with mean 9.80 mm and standard deviation 0.10 mm. The diameter of a hole is a Normal variable with mean 10.00 mm and standard deviation 0.12 mm. A pin and hole 'fit' if the diameter of the hole exceeds that of the pin by not more than 0.30 mm. Find the probability of getting a 'fit'. (Assume difference in diameters to be Normally distributed.) (C.)

14 Two independent random variables x_1 and x_2 have zero mean and the same variance, σ^2. Given that $y_1 = a_1x_1 + a_2x_2$ and $y_2 = b_1x_1 + b_2x_2$, where a_1, a_2, b_1, b_2 are constants, show:

(*i*) that $V[y_1] = (a_1^2 + a_2^2)\sigma^2$;

(*ii*) that the condition for $Cov\ [y_1, y_2] = 0$ is $a_1b_1 + a_2b_2 = 0$.

Obtain an expression for $V[z]$ where $z = \lambda_1y_1 + \lambda_2y_2$ and λ_1, λ_2 are constants.

Mean of n independent variables

We will consider a sample of n random variables $x_1, x_2, \ldots, x_n$, chosen independently from some background population. (Note that we are using the 'names' $x_1, x_2, \ldots, x_n$ in a somewhat different sense from that used previously, when they stood for the different values a single variable could take; convenience outweighs any possible ambiguity). Then each variable individually will have the same expected value μ and variance σ^2 as the background population. Let m be the mean of the sample.

$$\text{Then}\quad E[m] = E\left[\frac{1}{n}\sum_1^n x_i\right]$$

$$= \frac{1}{n}E[\sum x_i]$$

$$= \frac{1}{n}\sum_1^n E[x_i]$$

$$= \frac{1}{n}n\mu$$

$$= \mu$$

and $$V[m] = \sum V\left[\frac{x_i}{n}\right] = \frac{1}{n^2}\sum_1^n V[x_i] = \frac{1}{n^2}n\sigma^2 = \frac{\sigma^2}{n}.$$

So
$$\boxed{E[m] = \mu \quad \text{and} \quad V[m] = \frac{\sigma^2}{n}}$$

and the standard deviation of the mean (also known as the *standard error* of the mean) is given by $\sigma/\sqrt{n}$.

Hence we are absolutely justified in using m as an estimate of μ; and since the variance of the mean is less than that of individual readings, it follows, as expected, that there is less variation in the values of m than in the individual values of x_i.

Small and large samples

Although we have been able to specify how the expected value and variance of any sample mean may be obtained, for a small sample there is no immediate way of stating its probability distribution; each must be analysed individually.

For large samples however, we can use the central limit theorem, which tells us that any mean of a large sample has an approximate Normal distribution, *irrespective of the population from which it is drawn.*

Samples from a Normal distribution

Since the central limit theorem tells us that the mean of a large sample is Normally distributed, it seems reasonable to assume (as can be proved) that the mean of *any sample* (large or small) drawn from a population which is itself Normally distributed, will be exactly Normal with expected value μ and standard deviation $\sigma/\sqrt{n}$.

Example 4

A firm produces jars of jam in such a way that the net mass of jam per jar is known to be Normally distributed with mean 340 g and standard deviation 12 g. The jars are packed in cartons of 16. In what percentage of the cartons will the average mass of jam per jar be less than 336 g?

For a sample of 16 jars, the mean mass m will be Normally distributed with $\mu = 340$ and $\sigma = 12/\sqrt{16} = 3$.

$$\text{Therefore}\quad p(m < 336) = \Phi\left(\frac{336 - 340}{3}\right)$$
$$= \Phi(-1.33)$$
$$= 1 - \Phi(1.33)$$
$$= 1 - 0.91 \quad \text{(to 2 d.p.)}$$
$$= 0.09.$$

So the average mass of jam per jar will be less than 336 g in 9% of cartons.

Example 5

A laboratory obtains a certain chemical solution from the manufacturers in 100-cm^3 containers. The manufacturers state that on average the mass of impurity in each container is 3.8 mg, and its standard deviation is 0.9 mg. For a certain experiment the laboratory wants to ensure that there is at most a 5% chance that the mass of impurity in each 100 cm^3 of solution it uses exceeds 4.0 mg. How many containers of solution need to be mixed together to ensure this?

Suppose that we mix n containers of chemical. We will assume that n is large, so that the mean mass of impurity in the sample so obtained is approximately Normally distributed, with

$$\mu = 3.8 \text{ mg} \quad \text{and} \quad \sigma = 0.9/\sqrt{n} \text{ mg}.$$

$$\text{We require}\quad p(m > 4.0) \leqslant 0.05$$
$$\Rightarrow \quad p(m \leqslant 4.0) \geqslant 0.95$$
$$\Rightarrow \quad \Phi\left(\frac{4.0 - 3.8}{0.9/\sqrt{n}}\right) > \Phi(1.67)$$
$$\Rightarrow \quad \frac{0.2\sqrt{n}}{0.9} > 1.67$$
$$\Rightarrow \quad \sqrt{n} > 1.67 \times 4.5 = 7.52$$
$$\Rightarrow \quad n > 56.7.$$

So at least 57 containers of the solution need to be mixed together.

Exercise 12.5b

1 Assuming that the heights of male college students are Normally distributed with mean 173 cm and standard deviation 8 cm, what is the probability that the mean height of a sample of 40 students will
(*i*) exceed 175 cm;
(*ii*) lie between 172 and 174 cm?

2 The marks obtained in a mathematics examination are approximately Normally distributed with mean 54 and standard deviation 13. What is

the probability that the average mark for a group of 10 students
(*i*) exceeds 60;
(*ii*) is less than 50?

3 A certain industrial process takes a worker an average of 24 minutes to complete, and the standard deviation of the completion time is 6 minutes. If the process is repeated 100 times, use the central limit theorem to find approximately the probability that the mean process time exceeds 25 minutes.

4 Each schoolboy in a class of 36 spins an unbiased coin 100 times. What is the probability that the average number of heads obtained per boy
(*i*) is less than 49;
(*ii*) is greater than 52?
(A continuity correction is not necessary in this case.)

5 Capsules of a certain drug have a nominal net mass of 10 g. They are packed in batches of 100 capsules. A machine fills the capsules with the drug in such a way that the masses are normally distributed with mean net mass 10.06 g and standard deviation 0.2 g. Find the probability that:
(*i*) a batch does not contain any 'underweight' capsules;
(*ii*) the mean net mass of the capsules in a batch does not fall below the nominal net mass.

6 A coal merchant sells his coal in bags marked '50 kg'. In fact he claims that the average mass is 50 kg, with a standard deviation of 1 kg. A suspicious trade inspector has 60 of the bags weighed, and finds that their mean mass is 49.6 kg. What is the probability that such a result, or a more extreme one, could have arisen if the coal merchant's claim is true? Do you think that the inspector's suspicions were justified?

7 Electric light bulbs are produced with a mean life-time of 1 600 hours and a standard deviation of 250 hours. Calculate the minimum sample size necessary to ensure that there is at most a 1% chance of the average life-time of such a sample being less than 1 550 hours.

(Assume that the sample will be large enough for the mean to have an approximately Normal distribution.)

8 How large a sample would you take in order to estimate the mean of a population, so that the probability is 0.95 that the sample mean will not differ from the true mean of the population by more than 0.2 of the standard deviation of the population?

Unbiased estimates

If T is some statistical estimate calculated from data, and θ a probability parameter, then if $E[T] = \theta$, T is said to be an *unbiased estimate* of θ.

It should be noted, however, that an unbiased estimate is not necessarily a particularly good estimate — its variance, for instance, may be quite large. It is simply the best we have.

It therefore follows, since $E[m] = \mu$, that the sample mean m is an unbiased estimate of μ, the mean, or expected value, of the parent population.

Since m is an unbiased estimate of μ, it would seem to follow that s^2 should be an unbiased estimate of σ^2; after all, both are referred to as variances, and as we saw in 12.1, their forms are very similar. To pursue this further, we note that s^2 is itself a random variable, depending only on the sample from which it is calculated, and we can therefore consider its expected value.

Now $s^2 = \dfrac{1}{n}\sum (x_i - m)^2,$

and in this form it is a little difficult to manage, since we cannot say anything directly about $E[(x_i - m)^2]$. However, we do know that

$$E[(x_i - \mu)^2] = \sigma^2,$$

Also, recalling that x_i has mean m and variance s^2, it follows that $x_i - \mu$ has mean $m - \mu$ and variance s^2.

$$\text{So} \quad s^2 = \frac{1}{n}\sum (x_i - \mu)^2 - (m - \mu)^2$$

$$\begin{aligned}
\Rightarrow \qquad E[s^2] &= E\left[\frac{1}{n}\sum (x_i - \mu)^2\right] - E[(m - \mu)^2] \\
&= \sigma^2 - V[m] \\
&= \sigma^2 - \frac{\sigma^2}{n} \\
&= \frac{n-1}{n}\sigma^2.
\end{aligned}$$

So s^2 is *not* an unbiased estimate of σ^2, and in fact s^2 will tend to give a low value for σ^2 (as can readily be seen by considering the zero value of s^2 that would arise if the sample had only one member).

$$\text{Therefore} \quad E\left[\frac{n}{n-1}s^2\right] = \frac{n}{n-1} \times \frac{n-1}{n}\sigma^2 = \sigma^2,$$

so that an 'unbiased' estimate of σ is

$$\hat{\sigma} = \sqrt{\left(\frac{n}{n-1}\right)}\, s.$$

Exercise 12.5c

1 The lengths, in millimetres, of eight full-grown fish of a certain species were found to be: 56, 49, 68, 58, 63, 60, 55, 59. Obtain unbiased estimates for the mean and variance of the length of the species.

2 A sample of 15 leaves was obtained from a certain plant, and the areas of these leaves, in square centimetres, are given below:

4.60, 5.85, 5.13, 3.94, 5.61, 4.08, 4.71, 4.29, 6.01, 5.46, 5.80, 6.33, 4.43, 4.90, 5.09.

Obtain unbiased estimates of the mean and variance of leaf area (correct to 2 decimal places).

3 In a sample of bars of chocolate produced in a particular factory, the mean mass was found to be 101.3 g and the standard deviation 1.6 g. Obtain an unbiased estimate (correct to 2 decimal places) of the standard deviation of all bars of chocolate produced in the factory if the sample size was:
(*i*) 5; (*ii*) 20; (*iii*) 100.

4 t is an unbiased estimator for a parameter θ, and t_i, $i = 1, 2, \ldots, n$ are values of t obtained from n experiments.
(*i*) Prove that

$$\frac{1}{n}\sum_{i=1}^{n} t_i \text{ is an unbiased estimate of } \theta.$$

(*ii*) Find the relationship between the numbers λ_i, $(i = 1, \ldots n)$ if $\Sigma\, \lambda_i t_i$ is an unbiased estimate of θ.

5 A random variable x has expectation μ and variance σ^2. Find the variance of

$$\bar{x} = \frac{1}{n}\sum_{i=1}^{n} x_i$$

and the expected value of $s^2 = \dfrac{1}{n-1}\sum (x_i - \bar{x})^2$

where $x_1, x_2, \ldots, x_n$ are n independent observations from the distribution of x.

A further set of m independent observations $y_1, y_2, \ldots, y_m$ is taken from the same distribution and

$$t^2 = \frac{1}{m-1}\sum_{j=1}^{m} (y_j - \bar{y})^2.$$

If $u^2 = \lambda s^2 + (1 - \lambda)t^2$, $(0 < \lambda < 1)$, show that s^2, t^2 and u^2 have the same expectation and find λ such that $V[u^2]$ is a minimum.

$$\left(\text{Take} \quad V[s^2] = \frac{2\sigma^4}{n-1}.\right)$$ (M.E.I.)

*12.6 Significance and confidence

In a television advertisement, out of a sample of eight housewives interviewed leaving a supermarket, six were found to prefer the new washing powder, 'Squelch'. Is this convincing evidence that more housewives use Squelch than all other powders?

This question involves firstly practical and experimental considerations, and secondly mathematical considerations. The first thing we must know is how the eight housewives were chosen. Were they chosen randomly? (For the present we must be content with an intuitive notion of randomness.) Or did the interviewer particularly look out for shopping bags containing packets of Squelch? Such questions must be asked before any meaningful mathematical deductions can be made. However since we are primarily concerned with the mathematics of such situations, we will assume that these practical questions have been settled, and for the purposes of this example we will assume that the eight housewives were chosen randomly.

We shall not (and indeed we cannot) answer the question about preference for Squelch directly. Instead we invert the question and ask:

'Assuming that exactly 50% of housewives prefer Squelch to other powders, what is the probability that this result (six out of eight choices for Squelch), or a more extreme one, will occur by chance?'

(The phrase 'this result, *or a more extreme one*' though it seems to complicate the problem, does in fact provide us with a standard method for dealing with such problems and the necessity of its inclusion will become increasingly obvious.)

If this probability is large, then there is no reason to doubt the assumption, but if the probability is small then there is considerable evidence against our assumption; we become sceptical about the assumption that housewives are indifferent, and increasingly certain that they do indeed prefer Squelch to other powders.

The phrase 'it is not true that more housewives prefer Squelch to any other powder' could be interpreted mathematically in a number of ways. We will consider the most extreme case for which the statement holds, when exactly half prefer Squelch, so that the statement is only just true. (Apart from any other considerations of course, this will simplify the mathematics.) We therefore have a binomial situation, with 'success' being regarded as finding a housewife who prefers Squelch.

So $n = 8$, $p = \frac{1}{2}$, and the probability distribution is:

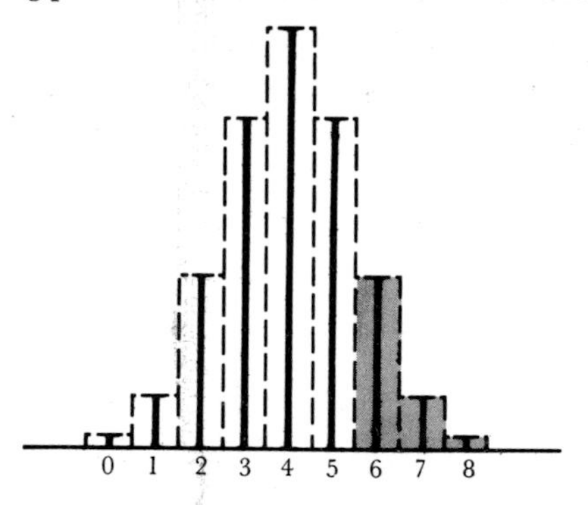

$$
\begin{aligned}
p(\geqslant 6 \text{ 'successes'}) &= \binom{8}{6} \times \left(\frac{1}{2}\right)^8 + \binom{8}{7} \times \left(\frac{1}{2}\right)^8 + \binom{8}{8} \times \left(\frac{1}{2}\right)^8 \\
&= 28 \times \frac{1}{2^8} + 8 \times \frac{1}{2^8} + 1 \times \frac{1}{2^8} \\
&= \frac{37}{256} \\
&= 0.145 \approx \tfrac{1}{7}.
\end{aligned}
$$

Therefore there is a probability of $\frac{1}{7}$ that six out of a sample of eight housewives would choose Squelch even when there is no overall preference for the powder. So we should regard the evidence of the advertisement as not particularly convincing.

This analysis is one example of a very simple *significance test.* The probability $\frac{1}{7}$ is a measure of the significance of the experimental result on the assumption of no overall preference. More generally, significance testing requires the use of two basic concepts: *null hypothesis* and *level of significance.*

Null hypothesis

This is the basic hypothesis upon which we build a probability model. It should be a precise statement, which lends itself to mathematical formulation. In our example the null hypothesis was that exactly half of housewives prefer Squelch. The term 'null' arises primarily from contexts such as medical research, where new treatments or drugs are tested for effectiveness, the null hypothesis being that such a treatment has no effect; and generally speaking a null hypothesis takes the form of just such a negative statement.

Levels of significance

Using the null hypothesis, we evaluate the probability of obtaining such an extreme value as our sample statistic: the smaller this probability, the greater is the evidence against the null hypothesis. We normally use three standard levels of significance: 5%, 1%, and 0.1%, corresponding to probabilities of 0.05, 0.01, and 0.001 respectively. If, for instance, we obtain a probability of 0.03, we say that this result is significant at the 5% level (but not at the 1% level), and we can therefore reject the null hypothesis at the 5% level. Which significance level is used generally depends on the nature of the experiment or research.

Errors

There are two distinct types of error which may arise from significance testing:

1 A sample may not provide significant evidence against the null hypothesis, and so the hypothesis may be accepted when in fact it is incorrect.

2 If we reject the null hypothesis at, say, the 5% level, then in 5 out of 100 trials the hypothesis could be correct, and the extreme probability will have occurred purely by chance. The incidence of this type of error can of course be reduced by using a higher level of significance, for example the 1% level.

But further detailed analysis of such errors is beyond the scope of this text, and for the present we must just accept that such unavoidable errors may arise.

The following two examples are typical significance tests.

Example 1

Before the introduction of a new drug, the success rate in treatment of patients for a certain disease was 43%. However, of 88 patients treated with the drug, 51 recovered. On the basis of this evidence, is the drug significantly effective in treating the disease?

Null hypothesis: That the drug has no effect, i.e., the probability of a patient recovering is still 0.43.

We therefore have a binomial situation, with $n = 88, p = 0.43$.

Since n is large we can approximate by a Normal distribution with

$$\mu = 88 \times 0.43 = 37.8$$

and $$\sigma = \sqrt{(88 \times 0.43 \times 0.57)} = 4.64.$$

Therefore
$$\begin{aligned} p(\text{51 or more recoveries}) &= 1 - p(\leqslant 50 \text{ recoveries}) \\ &= 1 - \Phi\left(\frac{50.5 - 37.8}{4.64}\right) \\ &\qquad \text{(using a continuity correction)} \\ &= 1 - \Phi(2.74) \\ &= 1 - 0.997 \\ &= 0.003. \end{aligned}$$

As $0.003 < 0.01$, this result is significant at the 1% level. So the probability, on the basis of the null hypothesis, of obtaining the observed result is very remote, and we can say with considerable confidence that the drug is effective.

Example 2

The machine described in example 1 of section 12.5, filling jars with jam such that the mass of jam is Normally distributed with $\mu = 340$ g and $\sigma = 12$ g, breaks down. After it is repaired, it is found that the average mass of jam in the first 100 jars it fills is 342 g. Has the performance of the machine significantly altered?

Null hypothesis: The machine is unchanged, i.e., the means of samples, size 100, are Normally distributed, with mean 340 and standard deviation $12/\sqrt{100} = 1.2$.

We need to find the probability of a result of 342, or one more extreme, occurring by chance. In this case, however, a more extreme result could reasonably be one less than 338, since we are not really concerned whether the mass of jam is increased or decreased, but simply whether it has changed, and the relevant probability is indicated by the shaded areas in the diagram.

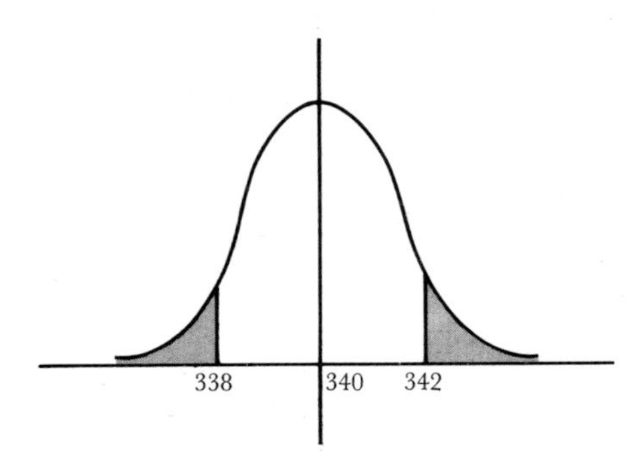

For fairly obvious reasons, therefore, this is referred to as a *two-tailed* (or two-sided) test, as opposed to the previous example which was one-tailed.

$$\text{Now} \quad p(338 \leqslant \text{mass} \leqslant 342) = 2\left[1 - \Phi\left(\frac{342 - 340}{1.2}\right)\right]$$
$$= 2[1 - \Phi(1.67)]$$
$$= 2[1 - 0.952]$$
$$= 0.096.$$

This is not significant at the 5% level, and so we can say that this evidence does not indicate any significant alteration in the machine's performance.

Exercise 12.6a

1 A coin is tossed eight times, and a head is obtained each time. Test, at the 1% level, the hypothesis that the coin is unbiased. (Use a two-tailed test.)

2 An amateur magician claims that he can read people's minds. To prove this he tells a person to think of one of the colours blue, yellow or red, and he then predicts what that colour is. In ten such mind-reading sessions he is correct seven times. Is this sufficient evidence to accept his claim?

3 Test, at the 5% level, the hypothesis that the variable x is Normally distributed with mean 80 and standard deviation 10, if we obtain:
(*i*) a value of x equal to 100;
(*ii*) a value of x equal to 65;
(*iii*) three values of x, all less than 70;
(*iv*) three values of x, whose average is equal to 70.

4 Over a number of years an average of 40 out of 100 patients who underwent a difficult operation survived. Last year new medical techniques were introduced and 73 out of 150 patients survived the operation. Explain whether or not the maintaining of these techniques is statistically justified. (S.M.P.)

5 In a certain constituency, in the last election, $\frac{5}{8}$ of the electorate voted for the Nationalist Party. In a sample of 200 voters it is now found that 107 support the Nationalists. Test the hypothesis that the support for the Nationalist Party has not decreased since the last election.

6 A tyre company claims that the lifetimes of its tyres are Normally distributed with mean 34 000 km and standard deviation 6 000 km:
(*i*) A motorist finds that one of their tyres lasts only 26 000 km. Does this lead you to doubt the manufacturers' claims?
(*ii*) A car-hire firm using the tyres finds that the average life-time of 100 such tyres is 32 200 km. Does this evidence lead you to doubt the manufacturers' claims?

7 The masses of a certain species of rabbit are known to be Normally distributed with mean 1.68 kg and standard deviation 0.24 kg. Nine rabbits are fed on specially enriched foodstuffs, and their average mass after two months is found to be 1.85 kg. Test at
(*i*) the 5% level,
(*ii*) the 1% level, the hypothesis that the foodstuff does not increase the rabbits' masses.

8 The following experiment was carried out 100 times by each of 16 pairs of children in a class. 'Each pair has an ordinary pack of cards from which one child selects a card and replaces it; the other child does likewise, and they score a point if the suit is the same for both'. The average number of points scored per pair was 27.2. Test at the 5% significance level, the hypothesis that all the cards were drawn at random. (S.M.P.)

9 A random sample of 200 persons from a large population is examined for eye colour. In this sample 50 persons are found to have blue eyes. Is this result consistent with the hypothesis that the proportion of blue-eyed persons in this population is 0.2? What would have been your conclusion if there had been 200 blue-eyed persons in a random sample of 800 persons? (M.E.I.)

10 A grower sows 200 Carlton lettuce seeds under conditions such that the average germination rate is 75%. By using a suitable approximation, find

(*i*) the probability that more than 170 seeds germinate,

(*ii*) the probability that less than 140 seeds germinate.

The grower also sows 120 Alberni lettuce seeds under the same conditions as the Carlton seeds, and finds that 82 seeds germinate. Test whether the Alberni seeds have a germination rate less than 75%. (C.)

11 The protein intakes for a sample of six unemployed men was observed to be 86, 82, 66, 66, 94, 104 g day^{-1}. A large number of employed men had an average intake of 97.6 g day^{-1} and the variance of their intake was 225.

Do these data support the view that at the time of the study unemployed men were underfed by comparison with men in employment. (M.E.I.)

12 You are engaged as an expert witness for the prosecution in a court case in which a gaming club is accused of running an unfair roulette wheel. The evidence is that out of 3 700 trial spins, zero (on which the club wins) turned up 140 times. There are 37 possible scores on a trial spin, labelled 0 to 36, and these should have equal probability. Test whether there is evidence that the wheel is biased. (O.)

13 According to a certain hypothesis the variable x is uniformly distributed in (0, 1). Determine the probability that the smallest member of a sample of n observations of x has value not less than X where $0 \leqslant X \leqslant 1$.

Ten observations were taken of x; the values of the smallest and largest were 0.2 and 0.7. Calculate the probability that ten observations:

(*i*) are each not less than 0.2;

(*ii*) are each not greater than 0.7;

(*iii*) lie within an interval of length 0.5.

State whether your results would lead you to reject the hypothesis. (C.)

14 In a sample of 1 000 voters, an opinion pollster finds that 520 intend to vote for Jones and 480 for Smith in the forthcoming election. What is the probability of a result such as this, or one even more favourable for Jones, if in fact Jones has only the support of 50% of the electorate? How many people in the sample would need to support Jones to make it significant

(*i*) at the 5% level,

(*ii*) at the 1% level;

that he has more support than Smith.

15 A children's test which had been standardised some years ago to have a mean of 100 and a standard deviation of 15 was applied recently to a random group of children, and their mean score was recorded as 101.6, but no record was made of the number of children tested. What is the least number of children tested if the result shows that the population mean is unlikely still to be 100? (You should assume that the mean value will certainly not have decreased. Work at the 1% significance level.) (S.M.P.)

Confidence intervals

The links in lengths of chain produced in a factory are known to have breaking tensions which are Normally distributed with standard deviation 0.24 kN. Eight links are tested to destruction, and are found to have breaking tensions of 3.61, 3.58, 3.66, 3.65, 3.89, 3.48, 3.50, 3.59 kN. What can we say, if anything, about the average breaking tension of all links produced in the factory?

Let us denote the average breaking tension (of the background population) by μ.

Now the sample mean (which for this sample is 3.62) is distributed Normally, with the same μ as the background population, and $\sigma = 0.24/\sqrt{8} = 0.085$.

We know that $(m - \mu)/\sigma$ has a standard Normal distribution, and from the properties of the standard Normal curve, it is 95% certain that $(m - \mu)/\sigma$ lies between -1.96 and 1.96.

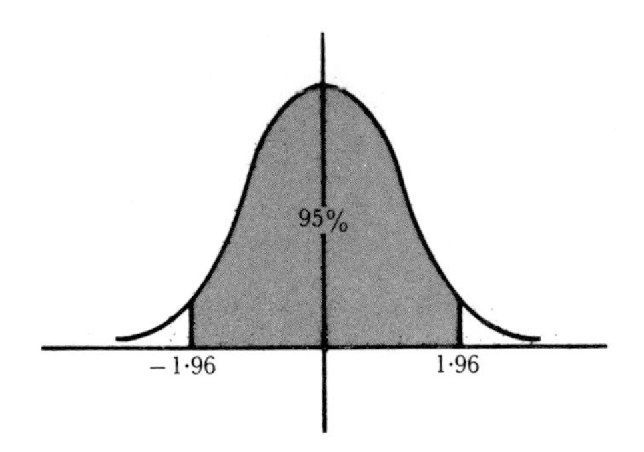

$$\text{So} \quad p\left(-1.96 < \frac{m - \mu}{\sigma} < 1.96\right) = 0.95$$

$$\Rightarrow \quad p(-1.96\sigma < m - \mu < 1.96\sigma) = 0.95.$$

Rewriting the orderings,

$$p(m - 1.96\sigma < \mu < m + 1.96\sigma) = 0.95.$$

Therefore for the chain links, it is 95% certain that

$$3.62 - 1.96 \times 0.085 < \mu < 3.62 + 1.96 \times 0.085$$

$$\Rightarrow \quad 3.62 - 0.17 < \mu < 3.62 + 0.17$$

$$\Rightarrow \quad 3.45 < \mu < 3.79.$$

The interval 3.45 to 3.79 is referred to as the 95% confidence interval for μ, and 3.45 and 3.79 are the 95% confidence limits.

Similarly the 99% confidence limits for μ are

$$3.62 \pm 2.58 \times 0.085$$

$$= 3.62 \pm 0.22$$

(i.e., 3.40 and 3.84).

The 95% and 99% confidence intervals are the standard intervals used, but there is no reason why others should not be employed. Quite naturally, the more certain we need to be of the interval within which μ lies, the larger that interval becomes. So for any experimental results, we have to balance the usefulness of the result with the confidence we can put in it, in choosing a suitable interval. Also, of course, the larger the sample used, the narrower will be the confidence interval so obtained.

The general procedure for establishing confidence limits for μ (assuming a Normal distribution, with known standard deviation) is as follows:

(*i*) Obtain a representative value — this may be a single reading or (better) the mean of a sample — and determine its standard deviation.
(*ii*) Decide on a confidence coefficient: 0.95 is for example the confidence coefficient corresponding to a 95% confidence interval. Denote this coefficient by $1 - 2\alpha$ (so for a 0.95 coefficient, $\alpha = 0.025$).
(*iii*) Define c_α by

$$\Phi(-c_\alpha) = \alpha$$

$$\Rightarrow \quad \Phi(c_\alpha) = 1 - \alpha,$$

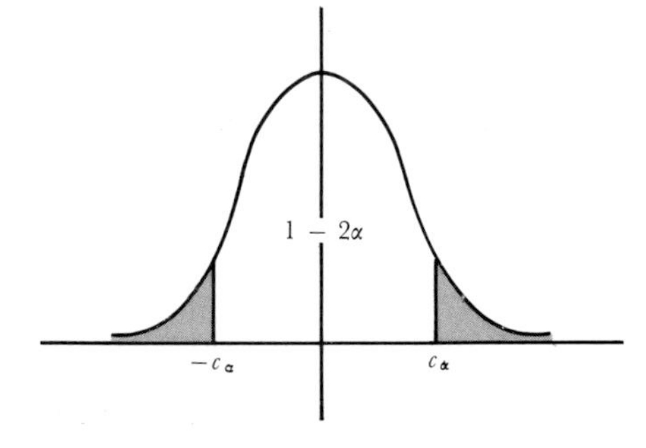

therefore the unshaded area is $1 - 2\alpha$, and the shaded areas are each α.

(*iv*) Then, in the case of the sample mean.

$$p\left(m - \frac{\sigma}{\sqrt{n}}c_\alpha < \mu < m + \frac{\sigma}{\sqrt{n}}c_\alpha\right) = 1 - 2\alpha$$

and the confidence limits are $m \pm \sigma c_\alpha/\sqrt{n}$.

Similarly, for a single observation x, the confidence limits for μ are $x \pm \sigma c_\alpha$. However the confidence intervals obtained from *single* observations are usually so wide as to be of little real use.

This technique of obtaining confidence intervals is very useful. But it does have limitations, the most obvious of which is that the standard deviation of the population has to be known; in practice, however, σ is often not known, and has to be estimated from the sample. A more complicated technique has then to be employed (using what is known as the 't' distribution), and the effect of using an estimated value of σ is to widen the equivalent confidence intervals for a known value of σ. However, for *large* samples the process described above can given good approximations to the confidence limits even when the standard deviation has to be estimated. (See exercise 12.6b, no. **7**.)

Exercise 12.6b

1 It is known that the variable x is Normally distributed with standard deviation 8. The average size of a sample of 10 values of x is found to be 63. Find (*i*) 90%, (*ii*) 95%, and (*iii*) 99% confidence limits for the mean value of x.

2 The melting points, in °C, of 10 samples of a certain metal were found to be 1 154, 1 151, 1 154, 1 150, 1 148, 1 152, 1 155, 1 153, 1 149, 1 154. Past experience indicates that these observations will be Normally distributed, with standard deviation equal to 3 °C. Find (*i*) 95%, and (*ii*) 99% confidence limits for the mean melting point of the metal.

3 It is known that the heights of 14-year old schoolboys are Normally distributed, with a standard deviation of 9.2 cm. The average height of 16 such schoolboys, chosen at random in a certain school, is found to be 159.1 cm. Find (*i*) 95%, and (*ii*) 99% confidence limits for the average height of 14-year old schoolboys.

4 The average length of 200 nails produced by a certain machine was found to be 7.42 cm. If the standard deviation of nail lengths is known to be 0.53 cm, find (*i*) 95% and (*ii*) 99% confidence limits for the mean length of all nails produced by the machine.

5 It is known that an examination paper is marked in such a way that the standard deviation of the marks is 15.1. In a certain school, 80 candidates take the examination, and they have an average mark of 57.4. Find (*i*) 95% and (*ii*) 99% confidence limits for the mean mark in the examination.

6 A factory manufacturing ammeters tests them for zero errors in their calibration. From past routine tests, it is known that the standard deviation of these errors is 0.3.

A batch of 9 ammeters, taken from one worker's production has zero errors of 1.0, −0.1, −0.3, 1.6, 0.5, 0.4, 0.5, 0.2, −0.2. Test whether there is evidence of bias in the ammeters produced by this worker, and establish a 95% confidence interval for the mean zero error of his ammeters. (o.)

7 A random sample of 100 capacitors, each of nominal capacitance 2 μF was taken from a very large batch. The capacitance of each capacitor in the sample was measured. The results of these measurements are summarised in the following table:

Capacitance/μF (mid-interval value)	1.85	1.90	1.95	2.00	2.05	2.10	2.15
No. of capacitors	2	12	23	31	20	10	2

Calculate the mean and the standard deviation of these capacitances. Calculate also the standard error of the mean, and use it to determine 95% confidence limits for the mean capacitance of capacitors in the batch. (Use a Normal approximation.) (M.E.I.)

8 In a random sample of 1 000 housewives from a large population 300 stated that they used a certain detergent. Show that 95% confidence limits for the proportion of the population using this detergent are (approximately) 0.272 to 0.328. (Use a Normal approximation to the binomial distribution, and an estimated value of σ.)

Following an advertising campaign a second sample of 800 housewives was taken, and of these 260 stated that they used the detergent. Is this evidence of the success of the campaign?

Subsequently it was decided to give a free gift with each packet of the detergent, and later in a random sample of 600 housewives 216 stated that they used the detergent. Does this indicate the success of the free gift scheme? (M.E.I.)

9 The probability density function of a variable x takes the form shown in the diagram, the constant a being unknown. A single observation x_1 is made of x. Give an unbiased estimate of a, and calculate 95% confidence limits for a. (Use first principles.)

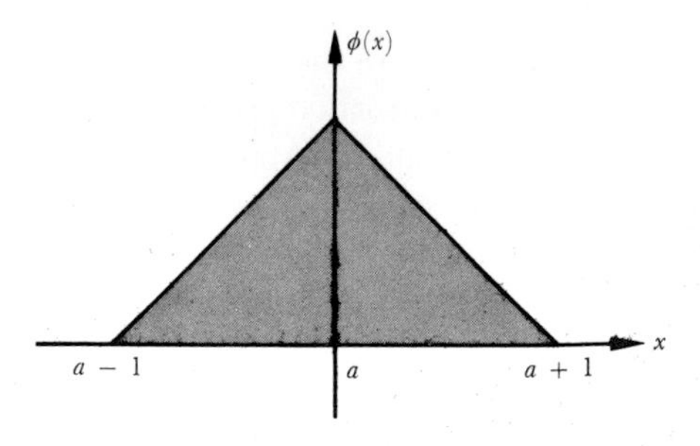

After a total of 24 observations have been made, the mean $\bar{x}$ is calculated. Use this mean to obtain an unbiased estimate of a, and calculate approximate 95% confidence limits for a. (C.)

*12.7 Correlation and regression

Bivariate data

The marks of a class of 20 pupils in a mathematics examination, and a French examination are given in the table below.

Mathematics	10	13	17	19	21	22	22	22	23	24
French	10	16	11	16	17	17	18	20	22	18

Mathematics	24	26	27	28	31	31	31	34	35	37
French	25	20	21	18	18	24	26	29	22	25

Since the marks are paired for each pupil, such data is referred to as bivariate data (as opposed to univariate data, where each item is treated individually; all the data so far dealt with in this chapter has been univariate, though we rarely use this term). It is generally convenient to represent bivariate data in the form of ordered pairs. Thus the data above would be (10, 10), (13, 16), ...

Scatter diagrams

To display bivariate data pictorially, we use a scatter diagram. This is obtained by regarding the ordered pairs of bivariate data as Cartesian coordinates, and marking the associated points on a graph:

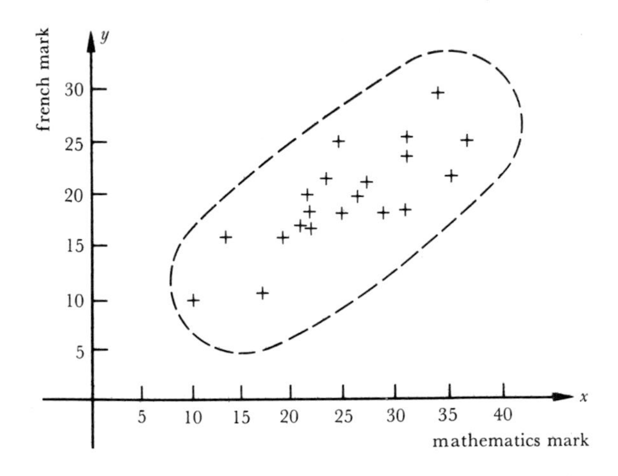

The points in the diagram all lie within a relatively narrow band, and this is emphasised by the boundary which has been superimposed around the points. This suggests some association between the x and y coordinates (i.e., between the mathematics and French marks). Such association is generally referred to as *correlation*.

The scatter diagram illustrates bivariate data which is closely correlated (also referred to as *positive* or *direct* correlation), i.e., small values of x correspond to small values of y and large x to large y. An example of such data might be the number of pupils and number of staff in secondary schools.

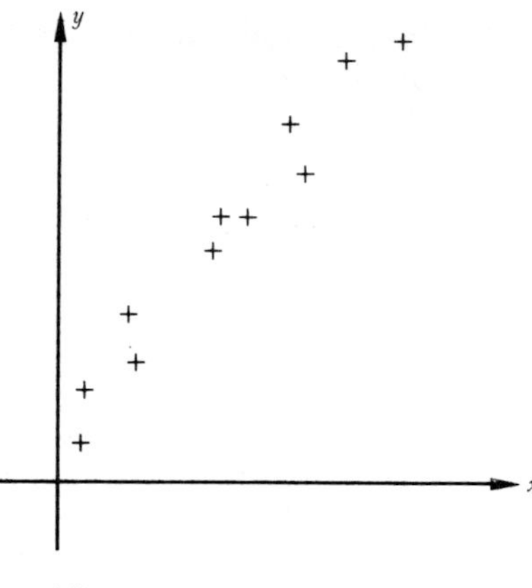

The second diagram shows data for which there is little or no correlation, for example the number of pupils, and number of trees planted in the grounds of a group of schools.

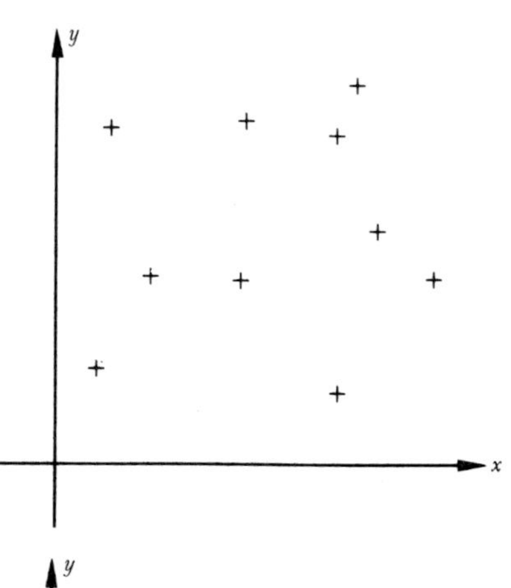

The third diagram illustrates *negative* (or inverse) correlation — the data here for example might be the numbers of pupils in schools, and the average percentages of the school's pupils per class.

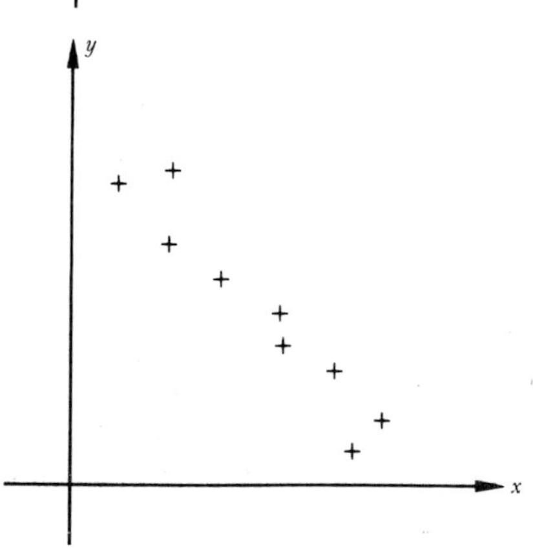

Regression

A precise measure of correlation follows naturally from investigation of the related topic of regression. Whilst in the study of correlation both variates are regarded as having equal status, in regression one (say x, though either may be chosen) is regarded as the independent variate, and the dependence of the other variate, y, on x is investigated. To be precise, we need to find a relationship between each fixed value of x, and the expected (or mean) value of y for that particular x. Let us illustrate this by a theoretical example:

Two dice, one red and the other blue, are thrown together; let x denote the score on the red die and y the total score on the two dice. Then the possible (equally likely) values of y for each value of x are shown in the diagram:

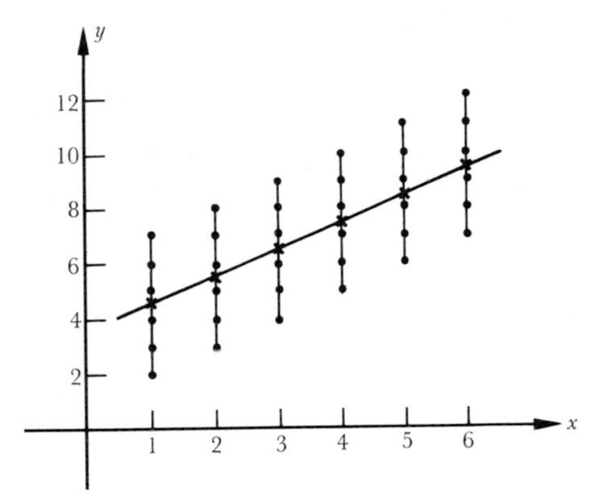

Now when $x = 1$, the expected value of y is equal to

$\frac{1}{6}(2 + 3 + 4 + 5 + 6 + 7) = 4.5.$

This point is marked on the diagram by a cross, and equivalent points for the other values of x are similarly marked. The straight line upon which all these points lie is then referred to as the *regression line of y on x*, and its equation in this case is $y = x + 3.5$.

Similarly in the diagram below we have the regression line of x on y, which we see has equation $y = 2x$.

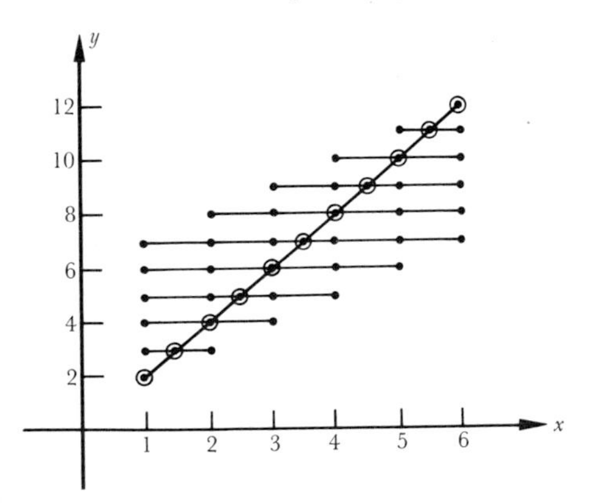

In this case y has been regarded as the independent variable, and for a particular total score, we are considering the expected value of the score on the red die.

This example, of course, is an idealised probability model. In general, the relationship between x and the equivalent expected value of y will not necessarily be linear. But whatever the relationship, linear or otherwise, we shall restrict our task (at this stage) to finding a straight line approximation. Such an approximation is generally referred to as the line of regression of y on x, though it must be emphasised that such a line, derived from a bivariate sample, will only provide an approximation to the true regression relationship in the background population from which the sample is drawn.

There is of course no single criterion for determining the 'best' line of regression for a particular set of data, but in practice the most useful one is known as the *least squares criterion*, and the line so obtained is called the *least squares line of regression of y on x*:

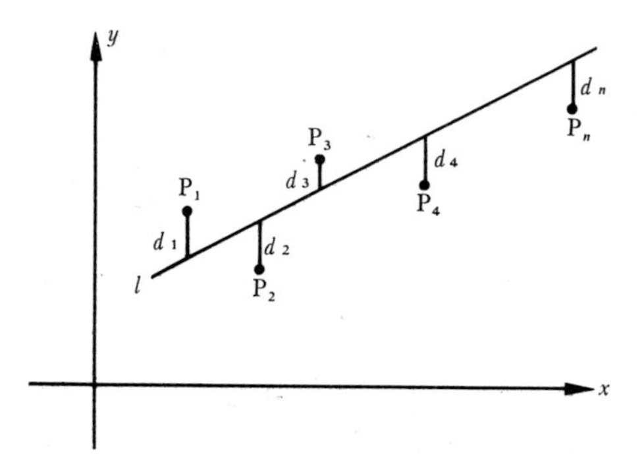

If d_i is the distance of the point P_i from the line l in the y-direction, and l is chosen in such a way that $\sum d_i^2$ is a minimum, then l is the least squares line of regression of y on x.

If instead of the line of regression of y on x, we need that of x on y, the least squares criterion is still used, but the deviations from the regression line are in this case measured in the x-direction:

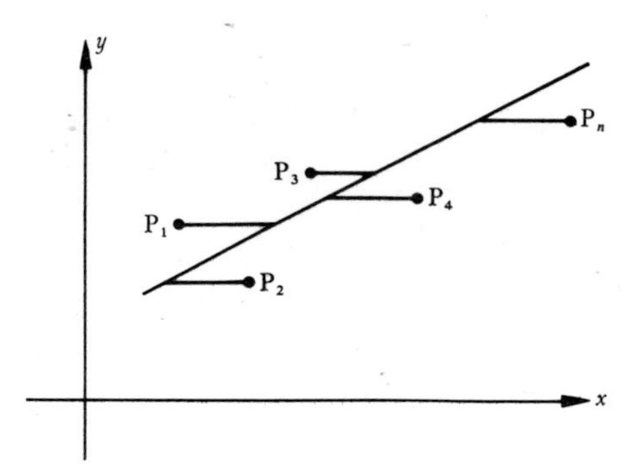

The obvious question arises: why should we choose this particular criterion? Why not the more obvious one of minimising the sum of perpendicular distances from the line? Our original definition of the regression relationship throws some light on the choice of deviation parallel to the axes, but the main justification for the criterion is on more advanced mathematical grounds. On a practical level, however, the algebra following from the least squares criterion is simpler than that arising, for example, from considerations of perpendicular distance, and the very neat and convenient results are in themselves some justification for the method.

Derivation of the line of regression of y on x

Let the points $P_1, P_2, \ldots, P_i, \ldots, P_n$ of the scatter diagram have coordinates $(x_1, y_1), (x_2, y_2), \ldots (x_i, y_i), \ldots (x_n, y_n)$. Now if we take the equation of the regression line to be $y = a + bx$, then

$$\text{deviation} \quad d_i = y_i - (a + bx_i)$$
$$= y_i - a - bx_i.$$

We therefore need to find values of a, b such that

$$S = \sum_{i=1}^{n} (y_i - a - bx_i)^2 \text{ is a minimum.}$$

We minimise S by regarding it as a function of a, b and then equating the *partial derivatives* of S with respect to a, b (written $\partial S/\partial a$, $\partial S/\partial b$) to zero.

($\partial S/\partial a$ denotes the derivative of S with respect to a, *while regarding b as a constant.* Similarly $\partial S/\partial b$ is the derivative of S with respect to b, *keeping a constant.* A thorough treatment of partial differentiation can be found in most books on advanced calculus, but a detailed knowledge of the subject is not essential for the purposes of this section.)

$$\text{Now} \quad \frac{\partial S}{\partial a} = -2\sum (y_i - a - bx_i)$$

$$\text{and} \quad \frac{\partial S}{\partial b} = -2\sum x_i(y_i - a - bx_i).$$

So if $\partial S/\partial a = 0$ and $\partial S/\partial b = 0$,

we obtain the equations:

$$\sum (y_i - a - bx_i) = 0 \Rightarrow \sum y_i - na - b\sum x_i = 0 \quad (1)$$
$$\sum x_i(y_i - a - bx_i) = 0 \Rightarrow \sum x_i y_i - a\sum x_i - b\sum x_i^2 = 0 \quad (2)$$

and in their most convenient form, these equations can be written:

$$\sum y_i = na + b\sum x_i \quad (1)$$
$$\sum x_i y_i = a\sum x_i + b\sum x_i^2. \quad (2)$$

Dividing equation (1) by n, we obtain

$$\frac{1}{n}\sum y_i = a + b\frac{1}{n}\sum x_i$$

and writing the means of x and y as $\bar{x}$ and $\bar{y}$, the equation becomes

$$\bar{y} = a + b\bar{x}.$$

This tells us that the regression line passes through the 'mean point', $(\bar{x}, \bar{y})$, so that the equation of the regression line can be rewritten as

$$y - \bar{y} = b(x - \bar{x}).$$

Hence we only need find the value of b from equations (1) and (2). This is done by multiplying equation (1) by $\sum x_i$ and equation (2) by n, and then subtracting (1) from (2). This gives:

$$n \sum x_i y_i - (\sum x_i)(\sum y_i) = b[n \sum x_i^2 - (\sum x_i)^2]$$

and dividing through by n^2,

$$\frac{\sum x_i y_i}{n} - \left(\frac{\sum x_i}{n}\right)\left(\frac{\sum y_i}{n}\right) = b\left[\frac{\sum x_i^2}{n} - \left(\frac{\sum x_i}{n}\right)^2\right].$$

Now $\dfrac{\sum x_i^2}{n} - \left(\dfrac{\sum x_i}{n}\right)^2 = s_x^2$, the *variance* of x,

and we also write

$$\frac{\sum x_i y_i}{n} - \left(\frac{\sum x_i}{n}\right)\left(\frac{\sum y_i}{n}\right) = s_{xy},$$ which is called the *covariance* of x, y.

(s_{xy} is a simplified statistical version of the covariance of x, y which arose in 12.5.)

So $s_{xy} = bs_x^2 \quad \Rightarrow \quad b = \dfrac{s_{xy}}{s_x^2}.$

This quantity, s_{xy}/s_x^2, known as the *coefficient of regression* of y on x, is therefore the gradient of the line of regression.

The equation of the line of regression can therefore be written in its most convenient form as:

$$\boxed{y - \bar{y} = \frac{s_{xy}}{s_x^2}(x - \bar{x})}$$

If, alternatively, y is taken as the independent variable, it is clear by symmetry that the line of regression of x on y is given by

$$x - \bar{x} = \frac{s_{xy}}{s_y^2}(y - \bar{y}).$$

Now let us return to our original bivariate data, where the two variates were mathematics (x) and French (y) marks. We will see how the French mark depends linearly on the mathematics mark by obtaining the line of regression of y on x.

x	y	x^2	xy
10	10	100	100
13	16	169	208
17	11	289	187
19	16	361	304
21	17	441	357
22	17	484	374
22	18	484	396
22	20	484	440
23	22	529	506
24	18	576	432
24	25	576	600
26	20	676	520
27	21	729	567
28	18	784	504
31	18	961	558
31	24	961	744
31	26	961	806
34	29	1 156	986
35	22	1 225	770
37	25	1 369	925

$$\sum x_i = 497 \qquad \sum y_i = 393 \qquad \sum x_i^2 = 13\,315 \qquad \sum x_i y_i = 10\,284$$

$$\Rightarrow \bar{x} = 24.85 \qquad \bar{y} = 19.65 \qquad \frac{\sum x_i^2}{20} = 665.75 \qquad \frac{\sum x_i y_i}{20} = 514.2.$$

Hence $s_x^2 = 665.75 - (24.85)^2$; $s_{xy} = 514.20 - 24.85 \times 19.65$

$$= 665.75 - 617.52 \qquad = 514.20 - 488.30$$

$$= 48.23 \qquad = 25.90$$

$$\Rightarrow \quad \frac{s_{xy}}{s_x^2} = \frac{25.90}{48.23} = 0.537.$$

Therefore the line of regression of y on x is

$$y - 19.65 = 0.537(x - 24.85)$$

$$= 0.537x - 13.35$$

$$\Rightarrow \quad y = 0.54x + 6.30, \quad \text{to 2 decimal places.}$$

In a similar way, we can calculate the regression line of x on y to be

$$y = 0.84x - 1.30.$$

The scatter diagram for the data is reproduced below with the two regression lines superimposed. The intersection of the two lines is the point $(\bar{x}, \bar{y})$, since of course we have shown that this point must lie on both lines.

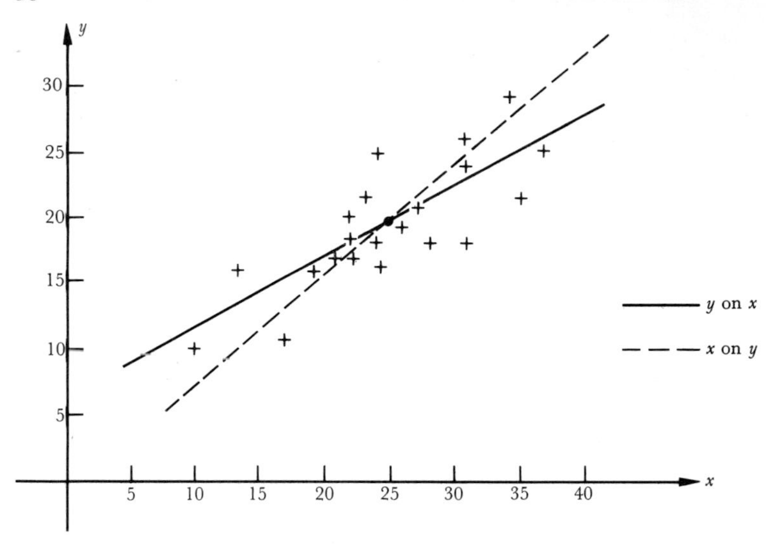

Example 1 (Grouped data)

In a sample of 150 married couples, the relationship between age of husband and wife was investigated. The results are illustrated in the table below, where the numbers in the table are frequencies (e.g., 5 couples with both husband and wife in the age-range 16–19).

Age of wife (Y)	Age of husband (X)						
	16–19	20–29	30–39	40–49	50–59	60–69	70–79
16–19	5	3					
20–29	1	25	7	2			
30–39		1	28	5	1	1	
40–49		1	2	24	4		
50–59				3	18		
60–69					1	11	2
70–79						1	4

Find, as conveniently as possible, the two lines of regression for the data.

We first of all label each interval by its average value, so that for instance the interval 20–29 (which includes ages actually up to 30), is labelled as 25. We then choose 45 as a working zero for both X and Y and scale down the values of X, Y by 10 to obtain the transformed (or coded) variables x, y. Working in terms of x, y will simplify the necessary arithmetic. The table then takes the form shown opposite, and we add the columns to obtain the frequencies f_x, and the rows to obtain f_y. The columns and rows signifying xf_x, x^2f_x, yf_y and y^2f_y can then be easily completed and their totals calculated.

$y \backslash x$	-2.7	-2	-1	0	1	2	3	f_y	yf_y	y^2f_y
-2.7	5	3						8	-21.6	58.3
-2	1	25	7	2				35	-70	140
-1		1	28	5	1	1		36	-36	36
0		1	2	24	4			31	0	0
1				3	18			21	21	21
2					1	11	2	14	28	56
3						1	4	5	15	45
f_x	6	30	37	34	24	13	6	150	-63.6	356.3
xf_x	-16.2	-60	-37	0	24	26	18	-45.2		
x^2f_x	43.7	120	37	0	24	52	54	330.7		

So $\sum f_x = \sum f_y = \sum f = 150.$

and $\sum xf_x = -45.2 \quad \Rightarrow \quad \bar{x} = \dfrac{\sum xf_x}{\sum f} = -0.301$

$$\sum x^2 f_x = 330.7 \quad \Rightarrow \quad \frac{\sum x^2 f_x}{\sum f} = 2.205$$

$$\sum yf_y = -63.6 \quad \Rightarrow \quad \bar{y} = \frac{\sum yf_y}{\sum f} = -0.424$$

$$\sum y^2 f_y = 356.3 \quad \Rightarrow \quad \frac{\sum y^2 f_y}{\sum f} = 2.375$$

It remains to calculate $\sum xyf$. This is done simply by drawing out the table again, but with the entries this time representing xyf rather than just the frequency f. Then the numbers in each row are added to obtain a further column, and the total for this column gives $\sum xyf$.

$y \backslash x$	-2.7	-2	-1	0	1	2	3	
-2.7	36.4	16.2						52.6
-2	5.4	100	14	0				119.4
-1		2	28	0	-1	-2		27
0		0	0	0	0			0
1				0	18			18
2					2	44	12	58
3						6	36	42
								317

So $\sum xyf = 317 \quad \Rightarrow \quad \dfrac{\sum xyf}{\sum f} = 2.133.$

$$\begin{aligned} \text{Hence} \quad s_x^2 &= 2.205 - (-0.301)^2 = 2.114 \\ s_y^2 &= 2.375 - (-0.424)^2 = 2.195 \\ s_{xy} &= 2.133 - (-0.301) \times (-0.424) = 2.005. \end{aligned}$$

Now transforming back to the original variables X and Y,

$$\begin{aligned} \bar{X} &= 10\bar{x} + 45 = 41.99 \\ \bar{Y} &= 10\bar{y} + 45 = 40.76 \\ s_X^2 &= 100 s_x^2 = 211.4 \\ s_Y^2 &= 100 s_y^2 = 219.5 \\ s_{XY} &= 100 s_{xy} = 200.5. \end{aligned}$$

So the regression coefficient of Y on $X = \dfrac{200.5}{211.4} = 0.948$

and the regression coefficient of X on $Y = \dfrac{200.5}{219.5} = 0.913.$

Therefore the equation of the regression line of Y on X is

$Y - 40.76 = 0.948\,(X - 41.99)$

$\Rightarrow \qquad Y = 0.9X + 0.9,$ to 1 decimal place,

and the equation of the regression line of X on Y is

$X - 41.99 = 0.913\,(Y - 40.76)$

$\Rightarrow \qquad Y = 1.1X - 5.2,$ to 1 decimal place.

Exercise 12.7a

1 (*i*) Obtain the heights (x) and masses (y) of 10 people of about the same age. Represent this data on a scatter diagram, and draw by eye through the points of your diagram what you consider to be the line of 'best fit'. Calculate the equations of the lines of regression of y on x and of x on y, and draw them on the same diagram.
(*ii*) Obtain the heights and masses of a larger number of people, and group this data in a suitable way. Then calculate the equations of the two regression lines, and compare with the results of part (*i*).

2 An experiment was conducted to discover the effect of adding a certain compound to the diet of mice. The results of the experiment are:

No. of units of compound added (x)	1	2	3	4	5	6	7	8
Mean gain in mass of mice (y)/mg	8.7	10.9	9.2	10.9	11.6	11.1	12.3	13.8

Find the equation of the line of regression of y on x.

3 In the following table W g is the mass of a certain chemical substance which dissolved in water at T °C:

T/°C	10	20	30	40	50	60	70	80	90
W/g	45	46	50	56	59	63	64	67	74

Calculate the equation of the line of regression of W on T. Use this equation to obtain a tentative value for W when $T = 56$. (M.E.I.)

4 A sample of eight families of a certain monkey species was chosen in which each family had just a single fully grown male offspring, and the heights of father and son for each family are given below.

Height of father/cm	(x)	67	64	70	71	65	67	68	69
Height of son/cm	(y)	68	62	70	73	64	69	71	69

Plot the data on a scatter diagram. Find the equation of the line of regression of y on x, and superimpose this line on the diagram.

5 The breadth (x) and length (y) of 12 leaves from a certain tree are given below:

Breadth/mm	39	36	35	35	30	31	34	28	33	30	19	29
Length/mm	81	80	74	89	71	73	76	75	84	81	77	82

Plot the data on a scatter diagram. Find the equations of the regression lines of (*i*) y on x, and (*ii*) x on y, and draw these lines on the scatter diagram.

6 A certain machine processes raw material. The purities (x) of 10 batches of raw material are given below along with the yield (y) from each batch.

Purity (x)	0.49	0.41	0.53	0.46	0.47	0.39	0.51	0.42	0.49	0.52
Yield (y)	24	21	26	25	20	22	25	23	25	27

Plot the data on a scatter diagram. Find the equation of the line of regression of y on x, and superimpose this line on the diagram.

7 The chronological ages and the reading ages (in years and months) of 10 children from a primary school are given on next page:

Chronological age	7.5	7.11	8.1	8.2	8.8	9.0	9.4	9.10	10.2	10.6
Reading age	7.8	7.10	8.0	7.9	8.9	9.1	8.8	9.10	12.1	10.7

Find the equation of the line of regression of reading age (y) on chronological age (x), if x and y are measured in months.

8 To test the effect of a new drug, twelve patients were examined before the drug was administered and given an initial score (I) depending on the severity of various symptoms. After taking the drug they were examined again and given a final score (F). A decrease of score represented an improvement. The scores for the twelve patients are given in the table.

	Patient	1	2	3	4	5	6	7	8	9	10	11	12
Score	Initial (I)	61	23	8	14	42	34	32	31	41	25	20	50
	Final (F)	49	12	3	4	28	27	20	20	34	15	16	40

Calculate the equation of the line of regression of F on I.

On the average, what improvement would you expect for a patient whose initial score was 30? (M.E.I.)

9 The following table gives the results of 4 estimations of a variable y at each of four values of a variable x. Calculate the equation of the regression line of y on x.

x	y			
2	17	15	20	18
4	15	14	16	15
6	12	14	10	12
8	8	10	10	12

Estimate the mean value of y when $x = 3$. (M.E.I.)

10 The marks of 100 pupils in English language and English literature examinations were graded from 1 to 9 and their results are given in the frequency table below:

	English language (x)									
		1	2	3	4	5	6	7	8	9
English literature (y)	1	6	4	1						
	2	2	1	1		1	1			
	3	3	3	3	7	2	1			
	4	1	1	4	1	1	1		1	
	5		2	1	3	1		2		
	6			2	11	6	4			
	7				2	2	3	1		
	8				1		3	2		1
	9				2	3	1		1	

Find the equations of the lines of regression of (*i*) y on x, and (*ii*) x on y.

11 The following table gives, in coded form, the breadths and lengths of 1 306 human heads.

Frequencies of heads

		X (length)										
		−4	−3	−2	−1	0	1	2	3	4	5	Total
	−4		2	2	1	1	2		1			9
	−3	1	5	5	15	8	4	1	1			40
	−2	2	3	21	34	48	41	18	3		1	171
Y	−1		4	21	57	92	110	57	19	5	1	366
(breadth)	0	1	2	17	53	93	116	62	27	12	1	384
	1		2	3	12	39	65	57	40	12	2	232
	2			1	4	9	17	24	12	9	1	77
	3				1	1	5	10	6		2	25
	4							2				2
Total		4	18	70	177	291	360	231	109	38	8	1 306

The length X is coded with 18.85 cm as origin and the breadth Y is coded with 15.15 cm as origin. The interval of grouping is 0.4 cm in both cases. Calculate the average length and the average breadth of the heads, and the equation of the line of regression of breadth on length. (M.E.I.)

12 The assets of a certain company have grown over the period 1850 to 1966 as shown in the table:

Year X	1850	1870	1900	1930	1960	1964	1965	1966
Assets/£10^5 Y	5	56	161	464	1 498	2 250	2 527	2 905

Calculate the line of regression of $\log_{10} Y$ on X. (M.E.I.)

13 Six pairs of values of x and y are given in the following table:

x	1	2	3	4	5	6
y	1.5	3.1	4.9	7.0	9.4	12.2

It is thought that these values are connected by a relation of the form $y = ax + bx^3$. By writing $Y = y/x$, and $X = x^2$, reduce the relation to a linear one and calculate:

(*i*) the means of X and Y;
(*ii*) the variance of X;
(*iii*) the covariance of X and Y;
(*iv*) estimated values of a and b on the assumption that there is a linear relationship between X and Y. (O.C.)

The correlation coefficient

Let us consider further the general line of regression of y on x:

$$y - \bar{y} = \frac{s_{xy}}{s_x^2}(x - \bar{x}).$$

If we divide through by s_y, the standard deviation of the y values, the equation can be written

$$\frac{y - \bar{y}}{s_y} = \frac{s_{xy}}{s_x s_y} \times \frac{x - \bar{x}}{s_x}.$$

Now in this equation $(y - \bar{y})/s_y$ and $(x - \bar{x})/s_x$ are standardised versions of the variates x and y, and the quantity $s_{xy}/s_x s_y$, which is symmetrical in x and y, is known as the *coefficient of correlation* of x and y; or, more exactly, the *product-moment correlation coefficient.*

The correlation coefficient is normally denoted by the letter r, and if the standardised variates are written as x' and y', the regression lines of y on x and x on y become:

$$y' = rx' \quad \text{and} \quad x' = ry' \quad \text{respectively.}$$

$$\Rightarrow \qquad y' = \frac{1}{r}x'$$

The limits on the value of r can now be determined, using a result known as *Cauchy's inequality*:

If a_i, b_i are real numbers (for $i = 1, 2, \ldots, n$)

then $\quad (\sum a_i b_i)^2 \leqslant (\sum a_i^2)(\sum b_i^2)$.

[For consider the expression $\sum (\lambda a_i + b_i)^2$, where λ is any real number.

Since each term of the summation is squared, and hence positive, it follows that

$$\sum (\lambda a_i + b_i)^2 \geqslant 0 \quad \text{for all } \lambda$$

$$\Rightarrow \quad \sum (\lambda^2 a_i^2 + 2\lambda a_i b_i + b_i^2) \geqslant 0$$

$$\Rightarrow \quad \lambda^2 \sum a_i^2 + 2\lambda \sum a_i b_i + \sum b_i^2 \geqslant 0 \quad \text{for all } \lambda.$$

Now we know that in general if $ax^2 + 2bx + c \geqslant 0$ for all x, then $b^2 \leqslant ac$.

So if we regard $\lambda^2 \sum a_i^2 + 2\lambda \sum a_i b_i + \sum b_i^2$ as a quadratic function in λ, it follows that

$$(\sum a_i b_i)^2 \leqslant (\sum a_i^2)(\sum b_i^2).]$$

Now for the correlation coefficient, r,

$$r^2 = \frac{s_{xy}^2}{s_x^2 s_y^2} = \frac{[\sum (x_i - \bar{x})(y_i - \bar{y})]^2}{\sum (x_i - \bar{x})^2 \sum (y_i - \bar{y})^2}.$$

So, using Cauchy's inequality with $(x_i - \bar{x}) = a_i$ and $(y_i - \bar{y}) = b_i$, we see that $r^2 \leqslant 1$.

Hence $-1 \leqslant r \leqslant 1$.

Summarising,

$$r = \frac{s_{xy}}{s_x s_y} \quad \text{and} \quad -1 \leqslant r \leqslant 1$$

Let us now look at three particular values of r:

(*i*) $r = 1$: In this case, the regression lines of y on x and x on y coincide, and if standardised they have gradient 1; this means that all the points of the scatter diagram are collinear, and we say that there is exact *positive* (*or direct*) *correlation* between the variates.

(*ii*) $r = 0$: The regression line of y on x is horizontal, and that of x on y is vertical. This indicates a complete lack of linear association between the variates, which are said to be *uncorrelated.*

(*iii*) $r = -1$: The regression lines coincide, and if standardised have gradient -1; small values of x correspond to large values of y and vice-versa, all points being collinear. In this case we say that there is exact *negative* (*or inverse*) *correlation* between the variates.

Significance of the correlation coefficient

If, in a particular example we obtain a value of r close to 1 or a value which is nearly zero, then little further analysis is necessary; but how do we interpret a correlation coefficient of (say) 0.5? We wish to know its significance, that is the probability of obtaining such a high value if the variates are uncorrelated (i.e., with the null hypothesis that $r = 0$ for the background population). This will depend on the size of the sample from which r is calculated: the more data we use, the more confident we can feel about the value of r. The mathematics involved in the relevant significance test is too involved for this text, but we can (without at present being concerned about the mathematical justification) use a graph like that overleaf. This shows the size of samples required to give 5%, 1%, and 0.1% significance levels for values of the correlation coefficient r (using a one-tailed significance test).

So, using the graph, we can say that a correlation coefficient of 0.5 obtained from a bivariate sample of 15 pairs is significantly different from zero at the 5% level, but if obtained from a sample of only 10 pairs is not significant. Similarly a value of $r = 0.8$, obtained from a sample of size 9, is significant at the 1% level, but not at the 0.1% level.

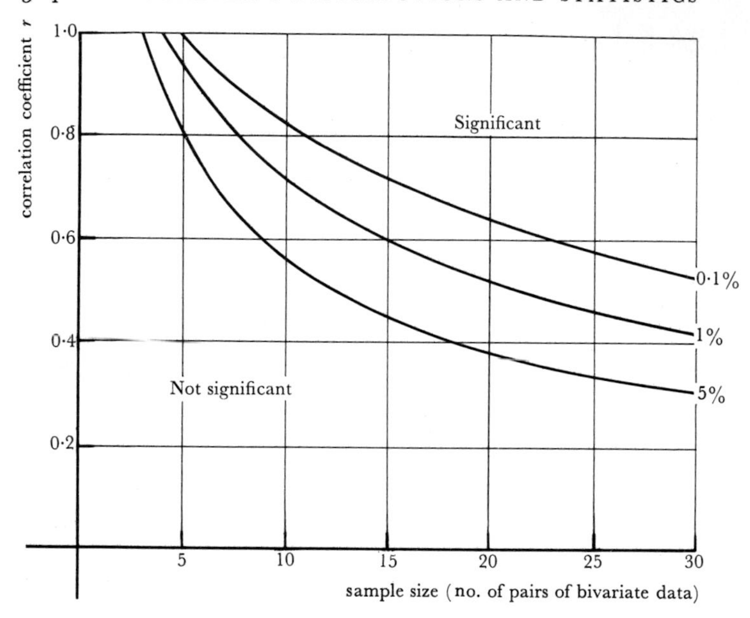

Warning

A significant correlation coefficient does not necessarily imply causal connection between two quantities. There is, for instance, a correlation coefficient of 0.75, which is significant at the 5% level, between the number of indictable crimes committed and life expectation in Great Britain between 1900 and 1960 (measured at 10-yearly intervals). But we would hardly accept that an extended life-span increased a person's criminal tendencies; nor even vice-versa!

Example 2

Calculate the product-moment correlation coefficient for the mathematics and French marks given at the beginning of the section, and investigate the significance of this coefficient.

We have already calculated $\quad s_{xy} = 25.90$

and $\quad s_x^2 = 48.23 \Rightarrow s_x = 6.94.$

Now for the data, $\quad \sum y_i^2 = 8\,159$

$$\Rightarrow \quad \frac{\sum y_i^2}{n} = 407.95$$

$$\Rightarrow \quad s_y^2 = 407.95 - (19.65)^2$$

$$= 407.95 - 386.12$$

$$= 21.83$$

$$\Rightarrow \quad s_y = 4.67.$$

Therefore $r = \dfrac{s_{xy}}{s_x s_y} = \dfrac{25.90}{6.94 \times 4.67} = 0.80.$

Now this value of r was calculated from 20 pairs of data, so using the significance graph we can see that this correlation coefficient is definitely significant at the 0.1% level. This is therefore extremely strong evidence of correlation between the mathematics and French marks in the population from which the sample was drawn.

Example 3

The heights and masses of a random sample of 10 first-formers from Watford Grammar School are given below.

Height/cm	156	151	152	160	146	157	149	142	158	141
Mass/kg	47	38	44	55	46	39	45	30	45	32

On the basis of this sample, how significant is the correlation between height and mass?

Calculation will be considerably simplified if we subtract 150 cm from each height and 40 kg from each mass. This is equivalent to translating the axes of the scatter diagram, so does not affect the value of the correlation coefficient. The data therefore becomes:

Height (x)	6	1	2	10	−4	7	−1	−8	8	−9
Weight (y)	7	−2	4	15	6	−1	5	−10	5	−8

and the calculations are set out in tabular form:

x	y	x^2	y^2	xy
6	7	36	49	42
1	−2	1	4	−2
2	4	4	16	8
10	15	100	225	150
−4	6	16	36	−24
7	−1	49	1	−7
−1	5	1	25	−5
−8	−10	64	100	80
8	5	64	25	40
−9	−8	81	64	72
12	21	416	545	354

Therefore $\bar{x} = 1.2$ and $\bar{y} = 2.1.$

So, working correct to 3 significant figures

$$s_x^2 = 41.6 - 1.2^2, \quad s_y^2 = 54.5 - 2.1^2, \quad s_{xy} = 35.4 - 1.2 \times 2.1$$
$$= 40.2 \quad\quad = 50.1 \quad\quad = 32.9.$$
$$\Rightarrow \quad s_x = 6.34, \quad s_y = 7.08,$$

Hence $r = \dfrac{32.9}{6.34 \times 7.08} = 0.73,$

and using the significance graph, we find that this correlation coefficient is significant at the 1% level.

Exercise 12.7b

1–6 Calculate the correlation coefficients for the data in exercise 12.7a, nos. **1–6**, and comment on the significance of the coefficient in each case.

7 Calculate the coefficient of correlation between the mass of the heart and mass of the liver in mice, using the following data.

	Mass of organ							
Heart/10^{-2} g	20	16	20	21	26	24	18	18
Liver/10^{-2} g	230	126	203	241	159	230	140	242

What do you conclude from this analysis? (M.E.I.)

8 The following table gives the number of goals scored at home and away by ten teams in the First Division of the Football League, part way through the 1972–3 season:

Home	25	20	12	15	14	24	20	13	14	10
Away	10	14	12	13	6	10	8	5	6	7

Calculate the product-moment correlation coefficient for home and away goals.

9 The intelligence quotients of 8 students were assessed, and the time (in minutes) it took them to complete a certain piece of work was measured. Their I.Qs. and times are given below:

I.Q.	121	118	132	141	140	137	124	130
Time	28	26	23	20	16	17	22	19

Is the negative correlation between these significant?

10 Calculate the correlation coefficient for the data of exercise 12.7a, no. **10**.

11 Calculate the correlation coefficient for the data of exercise 12.7a, no. **11**.

12 Calculate the product-moment correlation coefficient for the bivariate data given in the following frequency table:

y \ x	-3	-2	-1	0	1	2	3	4
-2	4	6	2	1				
-1	1	8	4	3		2		
0		2	6	7	2	3	2	1
1		1	4	3	9	4	1	1
2		2		1	6	6	3	5

13 Ten sets of readings for the variates x, y, and z are given below.

x	2	3	5	7	8	11	13	16	18	19
y	8	5	12	11	22	9	24	17	12	25
z	8	10	10	7	12	14	15	21	28	26

Calculate the coefficients of correlation between (*i*) x and y, (*ii*) y and z, and (*iii*) z and x, and investigate the significance of each.

Rank correlation: Spearman's rank correlation coefficient

Six athletes competed in both a 1 500 m race and a 400 m race, and finished in the following positions:

Athlete	A	B	C	D	E	F
1 500 m	1	2	3	4	5	6
400 m	2	3	5	1	4	6

Can anything be said about correlation of performance at these two distances?

If the times for the athletes were given, then the product-moment correlation coefficient for these times could be calculated, as in previous examples. Here, however, the only information is about position or rank. Therefore the product-moment correlation coefficient of the *ranks* is calculated, and is referred to as the *coefficient of rank correlation.*

Since ranks (or positions) are less informative than actual measurements, the rank coefficient is correspondingly less informative than the product-moment coefficient. However, since only integers are involved, the rank coefficient does have the advantage of being easily calculated. It is evaluated by means of the formula:

$$\rho = 1 - \frac{6 \sum d_i^2}{n(n^2 - 1)}$$

where n is the total number of ranks involved, and d_i the ith rank difference.

The derivation of this formula follows the worked examples.

This rank correlation coefficient was first defined by Spearman in 1906, and until recently has been the most widely used. But there are others, so it is usually called *Spearman's rank correlation coefficient* and its significance graph is given below:

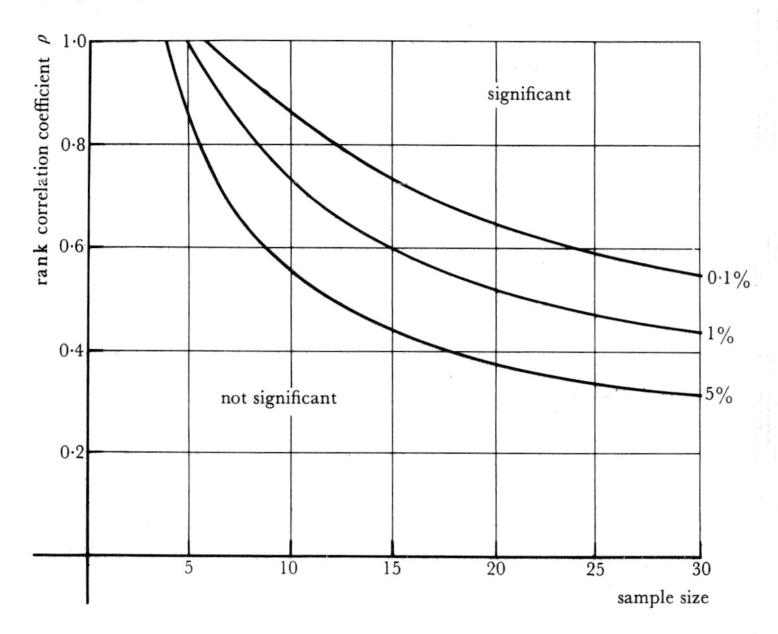

Returning to the data for the two races, ρ is evaluated as follows:

Athlete	A	B	C	D	E	F	$n = 6$
1 500 m	1	2	3	4	5	6	
400 m	2	3	5	1	4	6	
difference d	−1	−1	−2	3	1	0	
d^2	1	1	4	9	1	0	$\sum d^2 = 16.$

Hence $\rho = 1 - \dfrac{6 \times 16}{6(36 - 1)}$

$= 1 - \frac{16}{35}$

$= \frac{19}{35}$

$= 0.54;$

which, using the significance graph, is not significantly different from zero at the 5% level.

Example 4

What is the rank correlation coefficient for the heights and masses of schoolboys given in example 3, and how significant is this coefficient?

The first step is to rank (in ascending order) the heights and masses. But the problem arises of how to deal with the two equal masses of 45 kg; between them, they must occupy the 6th and 7th positions, so by convention each is given the rank 6.5. (It is not difficult to envisage how this is extended to more than two equal ranks, each being given the average of the ranks they replace.) The heights and masses are now ranked in the following table:

Height/cm	Mass/kg	Height rank	Mass rank	d	d^2
156	47	7	9	2	4
151	38	5	3	2	4
152	44	6	5	1	1
160	55	10	10	0	0
146	46	3	8	5	25
157	39	8	4	4	16
149	45	4	6.5	2.5	6.25
142	30	2	1	1	1
158	45	9	6.5	2.5	6.25
141	32	1	2	1	1
					64.5

So $n = 10$ and $\sum d^2 = 64.5$

$$\Rightarrow \quad \rho = 1 - \frac{6 \times 64.5}{10 \times 99}$$

$$= 1 - 0.39$$

$$= 0.61;$$

and using the significance graph, we find that this coefficient is significant at the 5% level. (As compared with the product-moment coefficient which was significant at the 1% level.)

Derivation of the formula for Spearman's rank correlation coefficient

The product-moment correlation coefficient is required where the two variates, x and y, are ranks, each taking values $1, 2, \ldots, n$ in some order or other.

We shall assume the standard results (see chapter 6)

$$\sum_1^n i = \tfrac{1}{2}n(n+1)$$

$$\sum_1^n i^2 = \tfrac{1}{6}n(n+1)(2n+1).$$

Since we need to calculate $\rho = s_{xy}/s_x s_y$, it is necessary first of all to evaluate s_x, s_y and s_{xy}.

Now $\sum_1^n x_i = \sum_1^n y_i = \sum_1^n i = \frac{1}{2}n(n+1)$

$$\Rightarrow \quad \bar{x} = \bar{y} = \tfrac{1}{2}(n+1).$$

Also $\sum_1^n x_i^2 = \sum_1^n y_i^2 = \sum_1^n i^2 = \frac{1}{6}n(n+1)(2n+1)$

$$\begin{aligned}
\Rightarrow \quad s_x^2 &= \frac{1}{n}\sum_1^n x_i^2 - \bar{x}^2 \\
&= \tfrac{1}{6}(n+1)(2n+1) - \tfrac{1}{4}(n+1)^2 \\
&= \tfrac{1}{12}(n+1)(4n+2-3n-3) \\
&= \tfrac{1}{12}(n+1)(n-1) = \tfrac{1}{12}(n^2-1)
\end{aligned}$$

$$\Rightarrow \quad s_x = \sqrt{[\tfrac{1}{12}(n^2-1)]}, \quad \text{and similarly} \quad s_y = \sqrt{[\tfrac{1}{12}(n^2-1)]}$$

$$\Rightarrow \quad s_x s_y = \tfrac{1}{12}(n^2-1).$$

We now use the identity:

$$\begin{aligned}
x_i y_i &= \tfrac{1}{2}x_i^2 + \tfrac{1}{2}y_i^2 - \tfrac{1}{2}(x_i - y_i)^2 \\
\Rightarrow \quad \sum x_i y_i &= \tfrac{1}{2}\sum x_i^2 + \tfrac{1}{2}\sum y_i^2 - \tfrac{1}{2}\sum (x_i - y_i)^2 \\
&= \tfrac{1}{6}n(n+1)(2n+1) - \tfrac{1}{2}\sum d_i^2 \\
\Rightarrow \quad \frac{1}{n}\sum x_i y_i &= \tfrac{1}{6}(n+1)(2n+1) - \frac{\sum d_i^2}{2n}.
\end{aligned}$$

Now
$$s_{xy} = \frac{1}{n}\sum x_i y_i - \bar{x}\bar{y}$$

$$\begin{aligned}
\Rightarrow \quad s_{xy} &= \tfrac{1}{6}(n+1)(2n+1) - \frac{\sum d_i^2}{2n} - \tfrac{1}{4}(n+1)^2 \\
&= \tfrac{1}{12}(n^2-1) - \frac{\sum d_i^2}{2n}.
\end{aligned}$$

Therefore
$$\begin{aligned}
\rho &= \frac{s_{xy}}{s_x s_y} \\
&= \frac{\frac{1}{12}(n^2-1) - \sum d_i^2/2n}{\frac{1}{12}(n^2-1)} = 1 - \frac{6\sum d_i^2}{n(n^2-1)}.
\end{aligned}$$

It should be noted that we have assumed in these calculations that the ranks are all different. If, however, equal ranks occur, $\sum x_i^2$ and/or $\sum y_i^2$ are no longer equal to $\frac{1}{6}n(n+1)(2n+1)$, and the above formula is no longer strictly valid; but the error involved is generally very small.

Kendall's rank correlation coefficient

Another coefficient of rank correlation, known as Kendall's rank correlation coefficient (τ), is derived quite differently, but has the same range of values (-1 to 1) as the other correlation coefficients, and values of τ equal to -1, 0, and 1 have the same meaning as r and ρ. It is a measure based on a simple principle of agreements and disagreements between the orders in which the objects are placed.

If we have a total of n objects (A, B, ...) ranked in two different ways, then we can choose $\binom{n}{2} = \frac{1}{2}n(n-1)$ different pairs, such as A and B. Now for the two rankings, the relative positions (A before B or B before A) can agree or disagree. So the maximum number of agreements, and the maximum number of disagreements of these pairs are both equal to $\frac{1}{2}n(n-1)$. Kendall's coefficient is calculated by subtracting the total number of such disagreements from the total number of agreements, and then dividing by the maximum possible number of agreements, i.e.,

$$\tau = \frac{\text{no. of agreements} - \text{no. of disagreements}}{\frac{1}{2}n(n-1)} = \frac{\delta}{\frac{1}{2}n(n-1)}.$$

Clearly if both rankings are exactly the same, all pairs will agree in relative position, so that the number of agreements will be $\frac{1}{2}n(n-1)$ and the number of disagreements will be zero; hence τ will equal 1. Similarly for total disagreement τ will equal -1.

The significance graph for τ is given below:

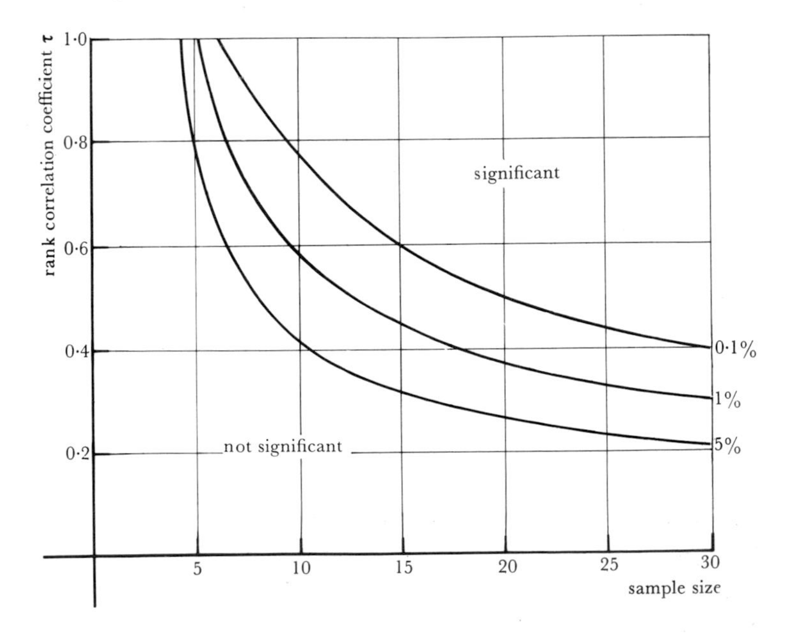

At first glance it would seem a very lengthy procedure to calculate τ, but in practice it can be calculated very quickly and conveniently, by either of the two methods in the following example.

Example 5

In a brass-band competition, two judges place the eight bands in the following order:

Band	A	B	C	D	E	F	G	H
1st judge	3	6	7	2	4	8	1	5
2nd judge	5	7	4	1	3	8	2	6

Calculate Kendall's rank correlation coefficient for the two judgements.

Method 1

Re-order the bands so that they are in the correct order of merit according to the first judge. Take the rank given by the second judge to the first band, G, that is 2. Now count $+1$ for every number to the right which is greater than 2 and -1 for each number which is less than 2; these totals are given below the rank. Repeat this procedure for the next rank in the second row, remembering to count only numbers to the *right* of it. And so on for each rank in the second row. The positive numbers so obtained represent agreements, and the negative numbers disagreements, so that their sum gives δ, the total number of agreements less the disagreements.

G	D	A	E	H	B	C	F	$n = 8$
1	2	3	4	5	6	7	8	
2	1	5	3	6	7	4	8	
6	6	3	4	2	1	1		
-1	0	-2	0	-1	-1	0		
5	6	1	4	1	0	1		$\delta = 18$

So $\tau = \dfrac{\delta}{\frac{1}{2}n(n-1)} = \dfrac{18}{28} = \dfrac{9}{14}$.

Method 2

Set out the bands in the order decided by each judge. Then join the same letter in each ranking by a straight line. Let c denote the number of crossings.

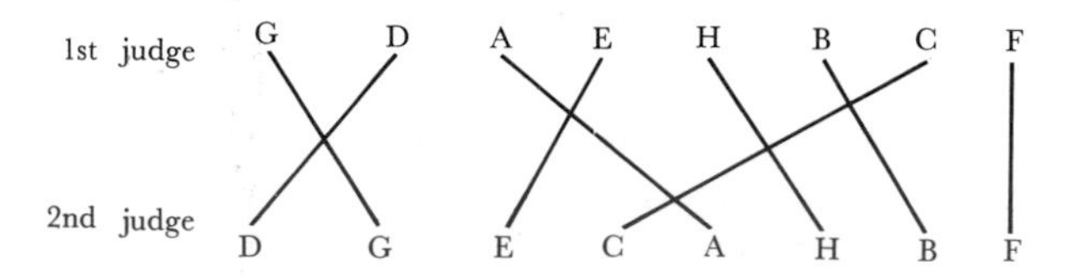

Then $\delta = \frac{1}{2}n(n-1) - 2c$

$$\Rightarrow \qquad \tau = 1 - \frac{4c}{n(n-1)}.$$

In this case, $c = 5$.

So
$$\tau = 1 - \frac{20}{8 \times 7}$$
$$= 1 - \tfrac{5}{14}$$
$$= \tfrac{9}{14} = 0.64.$$

(The justification for this second method is left as an exercise for the reader.)

So Kendall's rank correlation coefficient is in this case equal to 0.64; which, using the graph, is found to be significant at the 5% level.

Example 6

Calculate τ for the race positions given earlier in this section.

	A	B	C	D	E	F	$n = 6$
1 500 m	1	2	3	4	5	6	
400 m	2	3	5	1	4	6	
	4	3	1	2	1		
	−1	−1	−2	0	0		
	3	2	−1	2	1		$\delta = 7$

Hence $\tau = \dfrac{7}{\frac{1}{2} \times 6 \times 5} = \dfrac{7}{15} = 0.47,$

and this value of τ compares with the value (already calculated) of Spearman's coefficient, ρ, equal to 0.54.

Exercise 12.7c

1 Six varieties of chocolate, denoted by A, B, C, D, E, F are ranked for taste in the following order (best first): F C A D B E. The ranking on price (most expensive first) is C F B A E D. Calculate a rank correlation coefficient, and comment on the result. (C.)

2 At a village fête the fruit cake competition was judged by the vicar and the local squire. They placed the entries as follows:

Cake	A	B	C	D	E	F	G	H	I	J
Vicar	5	1	9	2	3	8	7	4	10	6
Squire	3	5	4	7	10	8	2	1	6	9

Calculate a rank correlation coefficient. Can you deduce anything about the vicar's and squire's tastes in fruit cake?

3 Twelve chemistry students were each given a theory and a practical examination. Their positions in these two examinations were as follows:

Theory	1	2	3	4	5	6	7	8	9	10	11	12
Practical	1	4	7	3	5	2	9	8	10	6	12	11

Calculate (*i*) Spearman's and (*ii*) Kendall's rank correlation coefficient for the data, and test the significance of each coefficient.

4 Rank the data of exercise 12.7a, no. **4**, and calculate Spearman's rank correlation coefficient.

5 Rank the data of exercise 12.7a, no. **5**, and calculate Spearman's rank correlation coefficient for the lengths and breadths of leaves.

6 Rank the data of exercise 12.7a, no. **6**, and calculate Spearman's coefficient. Is the rank correlation coefficient between parity and yield significant?

7 Rank the data of exercise 12.7a, no. **7**, and calculate (*i*) Spearman's and (*ii*) Kendall's rank correlation coefficient for the chronological and reading ages of the children.

8 The batting and bowling averages of seven of the members of a cricket team are put in order as follows:

Player	A	B	C	D	E	F	G
Batting order	1	2	3	4	5	6	7
Bowling order	7	4	6	5	3	1	2

(*i*) Calculate Kendall's rank correlation coefficient between the two orders.
(*ii*) Is this significant at the 1% level?
(*iii*) If the positions of A and C in the bowling order were reversed, show that Kendall's r.c.c. would be significant at the 5% level. (S.M.P.)

9 A, B, C, D, E was the order of merit for the five marrows entered in a local vegetable show. One admirer put them in the order B, A, C, D, E; show that he got 0.8 as a Kendall coefficient of rank correlation on comparing his order with the official result. List all other possible orders which would give the same coefficient when compared with A B C D E. How many different orders could there be for the five marrows? How likely is he to obtain a coefficient as high, or higher than he did, purely by chance? (S.M.P.)

10 At the end of a particular season, the goals scored by the ten teams forming a football league, as compared with their league positions were as follows:

Position	1	2	3	4	5	6	7	8	9	10
Goals scored	34	19	31	24	26	20	22	18	19	15

By ranking the number of goals scored (in descending order), calculate a rank correlation coefficient between league position and goal score. Is this coefficient significant?

11 Eleven boys from the same class took part in a sponsored walk for a charity. The distance each walked, and the amount per kilometre that he earned are given below. Calculate the Kendall rank correlation coefficient for these two sets of figures.

Distance/km	25	23	22	21	19	18	12	10	9	5	3
Rate/p km^{-1}	14	16	12	13	11	7	9	6	10	4	8

Comment briefly on the result. (S.M.P.)

Miscellaneous problems 12

1 In a game of Ludo a six has to be thrown with a die before each of a player's four counters can be moved. Find an expression for the probability that a player will be able to move his fourth counter with his tenth throw.

Find the probability, for a binomial distribution with parameters n, p and q, that r trials are required to obtain k successes.

2 A certain device, used for measuring Earth tremors in a seismological station, can only maintain the required degree of sensitivity for a total of ten tremors, and after this is disposed of. If such tremors occur randomly, but averaging 4 per day, find the probability density function for the useful lifetime (in days) of such a device, and use this to show that the mean lifetime is $2\frac{1}{2}$ days. What is the variance of the lifetime?

Find the p.d.f. if the device can be used for k tremors and tremors average λ per day; the probability distribution with this p.d.f. is referred to as the *gamma distribution*.

3 For a certain type of bacterium, the time x from birth to death is a random variable with p.d.f. $(1 + x)^{-2}$, $(x \geqslant 0)$. A culture of such bacteria is routinely inspected at regular intervals of time.

One bacterium, inspected at time t after birth, is found to be dead already. Find the probability that this bacterium has been dead for at least time $kt(0 < k < 1)$.

Another bacterium of the same type is alive at time t after birth, but is found to be dead by the end of a further time t. Find the probability that this bacterium had been dead for at least time $kt(0 < k < 1)$. (M.E.I.)

4 When a patient A arrives at a doctor's consulting room he will be seen

at once, and will occupy the doctor for 30 minutes; he is equally likely to arrive at any time between 2.00 and 3.30 p.m. Patient B, who comes independently, is equally likely to arrive at any time between 2.00 and 4.00 p.m.
(*i*) What is the probability that B will arrive while the doctor is seeing A?
(*ii*) What is B's expected waiting time? (O.S.)

5 In a certain large population of men, heights are distributed Normally about a mean of 180 cm with standard deviation 5 cm. Random samples are taken with three men in each sample and their heights are arranged in increasing order. In 1 000 such samples, approximately how many will have:
(*i*) the middle height under 175 cm?
(*ii*) the least height less than 175 cm?
(*iii*) the least height between 175 and 180 cm? (S.M.P.)

6 The number of eggs laid by a farmyard hen in a week is a Poisson variable with mean λ. The chance that an egg is fertile is p. Write down the probability that in one week n eggs are laid, and r of these are fertile.

By summing this probability over appropriate values of n find the probability that r fertile eggs are collected in one week.

Show that if this number r is given, then the distribution of $(n - r)$ conditional on it is of Poisson form, and that the mean of n is $r + \lambda(1 - p)$. (C.)

7 By means of probability generators, calculate the expected value and variance of the number of throws of a normal unbiased die required to obtain the sequence 1 2 3 4 5 6 (not necessarily consecutively).

8 Matches are put into a box five at a time until the mass of the box and matches combined reaches M g when the box is said to be full. The mass of an individual match is Normally distributed with mean m g and standard deviation σ g. The mass of an empty box is Normally distributed with mean $5m$ g and standard deviation 2σ g. Find the value of M such that there is only one chance in a hundred that a full match-box contains fewer than 50 matches (C.S.)

9 Each of four players is dealt 13 cards from a pack of 52 which contains 4 aces. Player A looks at his hand and winks at his partner, Player B which is a pre-arranged signal that his hand has at least one ace. Player B winks back to show that he has at least one ace as well. Player C looks at his hand and sees that he has just one ace. From Player C's point of view what is the probability that his partner, Player D has at least one ace if:
(*i*) he saw the winks and understood their meaning;
(*ii*) he knows nothing about his opponents' signals. (C.S.)

10 R is the number of successes in a binomial distribution with parameters n, p; S is the number of failures.
(*i*) Find the covariance of R and S.
(*ii*) Use the expectation and variance of $(R - S)$, together with a suit-

able approximation to the distribution of $(R - S)$ when $p = \frac{1}{2}$, to find λ such that

$$p(|R - S| \geqslant \lambda) = 0.05.$$

(*iii*) Calculate the probability $p(|R - S| \geqslant 8)$ when $n = 16$ and $p = \frac{1}{2}$, to two significant figures, using the exact distribution. (M.E.I.)

11 (*i*) If $A_1, A_2, A_3, \ldots, A_n$ are certain events, we know (from chapter 7) that

$$p(A_1 + A_2) = p(A_1) + p(A_2) - p(A_1A_2).$$

Find corresponding expressions for $p(A_1 + A_2 + A_3)$

and $p(A_1 + A_2 + \cdots + A_n)$.

(*ii*) n boys go to a riotous party, each with a girl; and though the party ends in complete confusion, each boy also leaves with a girl. What is the probability that no boy leaves with the girl he brought, and to what value does this probability tend for a large party?

12 Initially a machine is in good running order but it is subsequently liable to break down. As soon as a breakdown occurs repairs begin. If the machine is in good order at time t then the probability that a breakdown occurs in a small interval $(t, t + \delta t)$ is $\alpha\, \delta t$, and if it is under repair at time t the probability that the repair is completed in time $(t, t + \delta t)$ is $\beta\, \delta t$. Let $p(t)$ be the probability that the machine is under repair at time t.

Write down an equation relating $p(t + \delta t)$ to $p(t)$ and hence show that $p(t)$ is $\alpha/(\alpha + \beta)\,\{1 - e^{-(\alpha+\beta)t}\}$. (C.S.)

13 Two jars, one white and the other black contain $a + b$ balls each; in the white jar there are a white and b black balls and in the black jar b white and a black. Single draws are made as follows: at the rth draw a ball is drawn from the white or black jar according as the $(r - 1)$th ball drawn was white or black, the colour of the ball noted, and then returned. If p_n is the probability that the nth ball is white, show that

$$(a + b)p_n = b + (a - b)p_{n-1}.$$

By means of the substitution

$$p_n = \tfrac{1}{2} + q_n,$$

or otherwise, determine p_n, when the jar from which the first draw is made is (*i*) chosen at random, (*ii*) white. (O.S.)

14 Three deadly enemies A, B, and C take part in a three-cornered duel with pistols. The probability of A hitting his target is 0.6, that of B 0.8, and C 1. They draw lots for the order of firing and each uses his best strategy (they all know one another's capabilities). If they take it in turns to keep on firing until just one is left, who is most likely to survive, and what is his probability of survival?

15 In a game between two players both players have an equal chance of winning each point. The game continues until one player has scored N points. Find the probability p_r that the winning player has a lead of exactly r points when the game is complete. Deduce that

$$(2N - r - 1)p_{r+1} = 2(N - r)p_r \quad (r = 1, 2, \ldots, N)$$

and hence find the expected value of the lead at the end of the game.

(C.S.)

13

Matrices, determinants, and linear algebra

13.1 Introduction: matrices and matrix addition

Matrices as stores of numerical information

Suppose the boys in a school are placed in 'sets' for mathematics according to ability, and that there are four sets in the first and second forms and five sets thereafter. We use the notation III_2, for example, to denote the second set of the third forms. If conditions were 'ideal' or extremely rigid!, all the boys in I_1 would move up the following year into set II_1; all those in IV_3 would go into V_3, and so on. But of course, in practice, things are not like that: if they were, life would be very dull! Some make unexpected progress and earn promotion to higher sets; others do the reverse. Some develop a dislike of the subject, whilst others acquire a fascination for it. So it may be that, of the 31 boys in IV_2, 4 go into V_1 the next year, 20 into V_2, 5 into V_3 and 2 into V_4, none into V_5. Suppose we wished to give an account of the flow of boys through the school. One way would be to describe the composition of the various sets in words: 'of the 31 boys in II_1, 18 went into III_1, 6 into II_2, 4 into III_3, none into III_4 and 1 into the lowest set, III_5, while 2 boys left to join other schools, and so on...'.

One diagrammatic way of representing this data might be as follows:

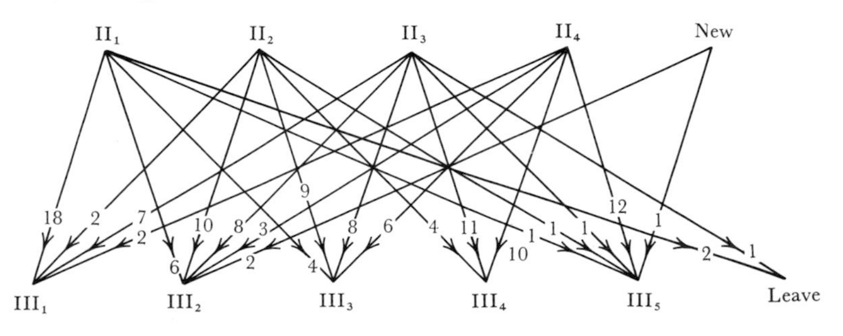

A similar diagram showing the 'flow' between I's and II's, III's and IV's, and between IV's and V's would give some sort of a picture of the 'flow' of boys through the school. Though this may be an improvement on the dry prose of the descriptive paragraph, in that one can 'see' the movement by following the various arrows, it may be objected that the representation is somewhat lacking in clarity!

What is the best way of depicting this information? Surely by means of a two-dimensional array, or table, or *matrix*, as follows:

		From				
		II_1	II_2	II_3	II_4	New
To	III_1	18	2	7	2	0
	III_2	6	10	8	3	2
	III_3	4	9	8	6	0
	III_4	0	4	11	10	0
	III_5	1	1	1	12	1
	Leave	2	0	1	0	–

This is usually written

$$\begin{pmatrix} 18 & 2 & 7 & 2 & 0 \\ 6 & 10 & 8 & 3 & 2 \\ 4 & 9 & 8 & 6 & 0 \\ 0 & 4 & 11 & 10 & 0 \\ 1 & 1 & 1 & 12 & 1 \\ 2 & 0 & 1 & 0 & 0 \end{pmatrix}$$

and, as it has six rows and five columns, is called a 6×5 matrix.

This method of displaying the data can scarcely be improved. Each *row* of the matrix shows clearly the composition of the new III's sets, e.g., that set III_4 consists of 4 boys who were in last year's II_2, 11 from II_3, 10 from II_4, and 1 new boy. The row, or 'list', (0, 4, 11, 10, 1) may, in fact be regarded as a *row vector*, and the 6×5 matrix as a collection of six such row vectors. Similarly the columns show the distribution of the boys in the old II's, the final *column vector*

$$\begin{pmatrix} 0 \\ 2 \\ 0 \\ 1 \\ 1 \\ 0 \end{pmatrix}$$

is a list indicating the disposition of the four new boys between the five sets (the final entry may be deemed to be zero, since a new boy is hardly likely to leave before even arriving — though it has been known!).

A large school with an intake of, say, 300 pupils per year would require larger matrices (say 12 × 12) to describe the composition of the new sets. A small school with 3 sets per year may possibly have a policy of moving boys only in exceptional circumstances, and it could be that a 'promotion matrix' for one level would read:

$$\begin{pmatrix} 28 & 1 & 0 \\ 2 & 25 & 1 \\ 0 & 0 & 27 \end{pmatrix}.$$

Notice that all such matrices are nothing more than *stores of numerical information* (or just tables, if you like) and you can imagine how such systems of matrices are useful for describing the stages in an industrial process.

Matrix addition

We can now show how *matrix addition* may be defined. Suppose five boys, Don, Joe, Ken, Tom and Vic tabulated their marks in maths (M), physics (P), chemistry (C), and biology (B) during each of two exams, as follows:

	1st exam				2nd exam			
	M	P	C	B	M	P	C	B
Boy D	55	42	71	60	70	56	82	57
J	35	62	29	73	46	58	34	68
K	69	48	40	57	73	55	52	78
T	24	38	47	49	29	32	47	58
V	70	68	67	85	76	63	61	83

Then the following table would give their aggregate scores (out of a total 200)

	M	P	C	B
Boy D	125	98	153	117
J	81	120	63	141
K	142	103	92	135
T	53	70	94	107
V	146	131	128	168

We may write this in matrix form as the addition of the two 5 × 4 matrices to give a single 5 × 4 matrix, thus:

$$\begin{pmatrix} 55 & 42 & 71 & 60 \\ 35 & 62 & 29 & 73 \\ 69 & 48 & 40 & 57 \\ 24 & 38 & 47 & 49 \\ 70 & 68 & 67 & 85 \end{pmatrix} + \begin{pmatrix} 70 & 56 & 82 & 57 \\ 46 & 58 & 34 & 68 \\ 73 & 55 & 52 & 78 \\ 29 & 32 & 47 & 58 \\ 76 & 63 & 61 & 83 \end{pmatrix} = \begin{pmatrix} 125 & 98 & 153 & 117 \\ 81 & 120 & 63 & 141 \\ 142 & 103 & 92 & 135 \\ 53 & 70 & 94 & 107 \\ 146 & 131 & 128 & 168 \end{pmatrix}$$

More generally, we may define the *sum* of two $p \times q$ matrices, where essentially each has the same number of rows (p) and the same number of columns (q), by asserting that the elements in the ith row and jth column of the two given matrices are added to give the element of the ith row and jth column of the sum matrix. Again, for example, when $p = 2$, $q = 5$, we have

$$\begin{pmatrix} a_1 & a_2 & a_3 & a_4 & a_5 \\ b_1 & b_2 & b_3 & b_4 & b_5 \end{pmatrix} + \begin{pmatrix} c_1 & c_2 & c_3 & c_4 & c_5 \\ d_1 & d_2 & d_3 & d_4 & d_5 \end{pmatrix}$$

$$= \begin{pmatrix} a_1 + c_1 & a_2 + c_2 & a_3 + c_3 & a_4 + c_4 & a_5 + c_5 \\ b_1 + d_1 & b_2 + d_2 & b_3 + d_3 & b_4 + d_4 & b_5 + d_5 \end{pmatrix}$$

Similarly, as in the previous example, one may possibly wish to add a number of 'promotion matrices'; if, for instance, one wished to find the aggregate movement of pupils from II's to III's over a period of ten years, one could add those ten 6 × 5 matrices to give a single 6 × 5 matrix.

An ordinary football league table may also be regarded as a matrix, the rows corresponding to the various teams A, B, C, . . . , and the columns giving the number of games played (P), won (W), drawn (D), lost (L), goals for (F), against (A), and total points (T):

$$\begin{array}{c} \\ A \\ B \\ C \\ \vdots \end{array} \begin{array}{c} \begin{array}{ccccccc} P & W & D & L & F & A & T \end{array} \\ \begin{pmatrix} 12 & 9 & 2 & 1 & 31 & 10 & 20 \\ 11 & 8 & 3 & 0 & 28 & 12 & 19 \\ 12 & 8 & 1 & 3 & 30 & 15 & 17 \\ \vdots & \vdots & \vdots & \vdots & \vdots & \vdots & \vdots \end{pmatrix} \end{array}$$

The result of a particular week's play might be represented as follows:

$$\begin{pmatrix} 1 & 0 & 1 & 0 & 3 & 3 & 1 \\ 1 & 0 & 0 & 1 & 1 & 4 & 0 \\ 0 & 0 & 0 & 0 & 0 & 0 & 0 \\ \vdots & \vdots & \vdots & \vdots & \vdots & \vdots & \vdots \end{pmatrix}$$

and the table
$$\begin{pmatrix} 13 & 9 & 3 & 1 & 34 & 13 & 21 \\ 12 & 8 & 3 & 1 & 29 & 16 & 19 \\ 12 & 8 & 1 & 3 & 30 & 15 & 17 \\ \vdots & \vdots & \vdots & \vdots & \vdots & \vdots & \vdots \end{pmatrix}^{\dagger}$$

informing us of the final state of the teams, is derived from the first two by matrix addition.

Note that each row of the matrix is a list representing the performance of one particular team; while the column vectors are lists providing statistics about one aspect of all the teams' performances, e.g., column (D) gives the whole story of the league from the point of view of matches drawn.

Exercise 13.1a

1 The following is a matrix showing the expenditure in £ of a house on types of fuel during consecutive quarters in 1973:

Quarter	Oil	Electric	Solid fuel
1	40	15	12
2	28	9	7
3	21	11	2
4	37	8	9

The corresponding matrices for 1974 and 1975 are:

$$\begin{pmatrix} 36 & 13 & 14 \\ 23 & 18 & 5 \\ 20 & 21 & 2 \\ 41 & 15 & 3 \end{pmatrix} \quad \text{and} \quad \begin{pmatrix} 45 & 18 & 7 \\ 15 & 14 & 0 \\ 18 & 19 & 0 \\ 30 & 22 & 0 \end{pmatrix}$$

Sum the three matrices and interpret.

2 Five bowlers had the following figures at the end of three test matches:

† Of course, if the final table is also to show the teams in order of merit, there will usually have to be a reshuffling of the rows, and this could be effected by 'pre-multiplying' by a so-called '*permutation matrix*'.

Bowler	Overs	Maidens	Runs	Wickets
A	110	26	213	15
B	89	11	305	18
C	50	8	110	7
D	25	2	64	5
E	26	10	33	3

The performance at the end of the series of five matches was given by the 5×4 matrix:

$$\begin{pmatrix} 213 & 41 & 418 & 32 \\ 89 & 11 & 305 & 18 \\ 80 & 12 & 198 & 11 \\ 25 & 2 & 64 & 5 \\ 37 & 14 & 90 & 8 \end{pmatrix}$$

Set up a matrix showing the performance of the five bowlers over the last two games.

3 The matrix below shows the readings of barometer, thermometer, rainfall, windspeed and wind direction at three stations A, B, C:

	Barometer	Thermometer	Rainfall	Wind Speed	Direction
A	1 021	24	13	14	280
B	1 053	29	0	29	350
C	998	18	35	40	085

Changes in the next 24 hours are recorded in the following matrix:

$$\begin{pmatrix} +20 & -2 & +6 & +12 & -15 \\ +8 & -7 & +21 & -11 & +20 \\ -11 & 0 & 0 & -22 & +5 \end{pmatrix}$$

Record the final state of the weather at A, B, C as a 3×5 matrix.

4 If $\boldsymbol{A}$, $\boldsymbol{B}$, $\boldsymbol{C}$ are $p \times q$ matrices, and $+$ denotes matrix addition, prove that

$$(\boldsymbol{A} + \boldsymbol{B}) + \boldsymbol{C} = \boldsymbol{A} + (\boldsymbol{B} + \boldsymbol{C})$$

(the associative law of addition for matrices).

5 $\boldsymbol{A}$ is a $p \times q$ matrix. Define $-\boldsymbol{A}$, the negative of $\boldsymbol{A}$, as the matrix $\boldsymbol{A}$ with every element multiplied by -1. Proceed to define matrix subtraction, and evaluate

$$\begin{pmatrix} 3 & -4 & 7 \\ -2 & 0 & -1 \end{pmatrix} - \begin{pmatrix} 2 & -3 & -1 \\ -5 & 1 & 4 \end{pmatrix}$$

Linear dependence and independence: preliminary discussion

It is worth remarking here that some of these column vectors are *not independent* of each other. For example, in the above table of football results:

col (P) = col. (W) + col. (D) + col. (L)

$$\begin{pmatrix} 12 \\ 11 \\ 12 \\ \vdots \end{pmatrix} = \begin{pmatrix} 9 \\ 8 \\ 8 \\ \vdots \end{pmatrix} + \begin{pmatrix} 2 \\ 3 \\ 1 \\ \vdots \end{pmatrix} + \begin{pmatrix} 1 \\ 0 \\ 3 \\ \vdots \end{pmatrix}$$

Using the notation $\boldsymbol{p}$ to represent column vector (P), we might write:

$$\boldsymbol{p} = \boldsymbol{w} + \boldsymbol{d} + \boldsymbol{l} \quad \text{or} \quad \boldsymbol{l} = \boldsymbol{p} - \boldsymbol{w} - \boldsymbol{d}, \text{ etc.} \tag{1}$$

The fact that the column vectors are not independent means that there is a sense in which one of them is redundant — if we know any three of $\boldsymbol{p}$, $\boldsymbol{w}$, $\boldsymbol{d}$, $\boldsymbol{l}$, then the other may be found from the vector equation (1). Again, we have $\boldsymbol{t} = 2\boldsymbol{w} + \boldsymbol{d}$, the three columns (T), (W), (D) are said to be 'linearly dependent'.

It may be observed that columns (F) and (A) are not independent, since their *totals* must be equal. Even so, it would not be correct to describe them as a pair of *linearly* dependent columns. This would only happen when a *linear* relation, of similar form to (1), connected the columns, and this must mean, in the case of two columns, that one would have to be a multiple of the other. For example, if there were an extra column for 'points from matches won':

$$\boldsymbol{x} = \begin{pmatrix} 18 \\ 16 \\ 16 \\ \vdots \end{pmatrix},$$

then clearly $\boldsymbol{x} = 2\boldsymbol{w}$, the two columns having corresponding entries in the fixed ratio 2:1.

Again, suppose there was a column (G) for 'goal average'

= goals for/goals against:

$$\boldsymbol{g} = \begin{pmatrix} 3.1 \\ 2.33 \\ 2.0 \\ \vdots \end{pmatrix}$$

This column *depends* very much on columns (F) and (A), but it is by no means *linearly* dependent upon them: we cannot find values of α and β such that $\boldsymbol{g} = \alpha\boldsymbol{f} + \beta\boldsymbol{a}$.

Indeed, even the first three equations

$$\begin{aligned} 3.1 &= 31\alpha + 10\beta \\ 2\tfrac{1}{3} &= 28\alpha + 12\beta \\ 2 &= 30\alpha + 15\beta \end{aligned}$$

are inconsistent.

The relations that do exist are $g_1 = f_1/a_1$, $g_2 = f_2/a_2$, etc., but these are not what is needed for *linear* dependence.

Note that 'vector addition' such as we have encountered in this paragraph occurs in many other contexts. For example, in adding polynomials, we are carrying out a similar process in adding like terms:

$$\begin{array}{lrrrr} & 3x^3 & -\ 2x^2 & & -\ 11 \\ \text{Add} & & 5x^2 & -\ 7x & -\ \ 4 \\ \hline & 3x^3 & +\ 3x^2 & -\ 7x & -\ 15 \\ \hline \end{array}$$

This might be written

$$(3 \quad -2 \quad 0 \quad -11) + (0 \quad 5 \quad -7 \quad -4) = (3 \quad 3 \quad -7 \quad -15),$$

each polynomial being represented by a row vector or 'list' of coefficients.

The idea of linear independence and linear dependence also arises when we consider the solution of linear equations. The pair of equations:

$$\begin{aligned} 2x - 5y - 4 &= 0 \\ -10x + 25y + 20 &= 0 \end{aligned}$$

is a perfectly good pair of simultaneous equations, but they do not have a unique solution. They have an infinity of solutions for the simple reason that the second is simply a multiple of the first, and is in effect redundant, the matrix of coefficients

$$\begin{pmatrix} 2 & -5 & -4 \\ -10 & 25 & 20 \end{pmatrix}$$ consisting of two *linearly dependent rows*.

On the other hand, the equally respectable pair of linear equations

$$\begin{aligned} 2x - 5y - 4 &= 0 \\ -10x + 25y + 9 &= 0 \end{aligned}$$

do not have any solution for the unknowns x, y. Here the first two columns $\begin{pmatrix} 2 \\ -10 \end{pmatrix}$ and $\begin{pmatrix} -5 \\ 25 \end{pmatrix}$ are linearly dependent, but not the rows. We shall return to a fuller consideration of this (see pp. 442, 445).

Exercise 13.1b

1 Three batsmen have the following analyses. Fill in the blanks, and state which rows and which columns are linearly dependent.

	Innings	Not out	Runs	100s	Completed innings	Average
A	24	4	1 050	3	20	51.0
B	32	0	1 507	5	...	...
C	...	9	967	5	19	...

2 A boy records the number of ice creams, buns and packets of crisps in columns a, c, and e during four successive weeks, and the money spent on them (prices 3p, 5p, and 2p respectively) in columns b, d, f. The total expenditure in p is shown in g, while h shows the change from £1.

Week	Ice-cream		Buns		Crisps		Total	Change
	No.	Cost	No.	Cost	No.	Cost		
	a	b	c	d	e	f	g	h
1	4	12	8	40	11	22	74	26
2	0	0	10	50	9	18	68	32
3	5	15	3	15	5	10	40	60
4	3	9	4	20	0	0	29	71
Totals	12	36	25	125	25	50	211	189

State which of the columns and which of the rows are linearly dependent (taking them in pairs, in threes, etc.).

3 Decide which of the following row vectors are linearly dependent (in twos and in threes):

$\boldsymbol{a} = (2, 5, -3)$ $\quad \boldsymbol{b} = (6, 15, -9)$ $\quad \boldsymbol{c} = (4, -10, -6)$

$\boldsymbol{d} = (-2, 5, 3)$ $\quad \boldsymbol{e} = (6, -5, -9)$ $\quad \boldsymbol{f} = (0, 30, 0)$

4 In the matrix $\begin{pmatrix} 2 & -3 & 1 \\ 5 & 6 & -2 \end{pmatrix}$ decide which columns are linearly dependent. Replace one element in row 2 so that the rows may be linearly dependent.

5 Examine the matrix $\begin{pmatrix} 2 & 3 & 1 & -4 \\ -1 & 2 & 0 & 2 \\ 3 & 1 & 1 & -6 \end{pmatrix}$

for linear dependence of rows and of columns in pairs and in threes.

6 Find x so that the three rows may be linearly dependent:

$$\begin{pmatrix} -3 & 4 & 0 \\ 8 & 5 & -2 \\ x & -7 & 6 \end{pmatrix}$$

Are the columns then linearly dependent or not?

7 The atomic weights of C, O, and H are 12, 16 and 1 respectively. The matrix shows the numbers of atoms of each element in various compounds, and the molecular weight in the final column.

	C	H	O	
Water	0	2	1	18
Methane	1	4	0	16
Benzene	6	6	0	78
Alcohol	2	6	1	46
Sugar	12	22	11	352
Acetic acid	2	4	2	60

Show that the row vectors for water, sugar and acetic acid are linearly dependent.

13.2 Linear transformations and their matrices

Transformations in two dimensions

It will be remembered from our discussion in the prologue to Book 1 that linear transformations† in the plane were governed by equations of the form

$$x' = ax + by$$
$$y' = cx + dy$$

† Confining ourselves to those in which the origin is invariant (i.e., excluding translations).

which we expressed as a matrix equation

$$\begin{pmatrix} x' \\ y' \end{pmatrix} = \begin{pmatrix} a & b \\ c & d \end{pmatrix} \begin{pmatrix} x \\ y \end{pmatrix}.$$

These may be written more briefly as $\boldsymbol{x}' = \boldsymbol{Mx}$, where $\boldsymbol{x}$, $\boldsymbol{x}'$ are the vectors

$\begin{pmatrix} x \\ y \end{pmatrix}$, $\begin{pmatrix} x' \\ y' \end{pmatrix}$ (or 2×1 matrices, with 2 rows and 1 column),

and $\boldsymbol{M}$ is the 2×2 matrix $\begin{pmatrix} a & b \\ c & d \end{pmatrix}$.

The effect of this transformation is to take the grid of unit squares in the x, y plane into a grid of parallelograms in the x', y' plane, so that the unit square OUKV transforms into the parallelogram OU′K′V′,

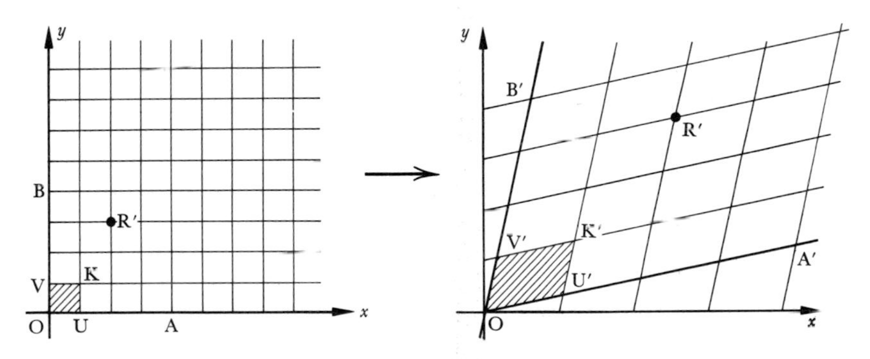

where U′ is $\begin{pmatrix} a \\ c \end{pmatrix}$, V′ is $\begin{pmatrix} b \\ d \end{pmatrix}$ and K′ is $\begin{pmatrix} a+b \\ c+d \end{pmatrix}$,

the area of the basic parallelogram cell being $ad - bc$. (In the 'singular case', when $ad - bc = 0$, the network of parallelograms collapses into a straight line.) The salient results are that the first column of the 2×2 matrix,

$\begin{pmatrix} a \\ c \end{pmatrix}$, gives the transform U′ of the point U $\begin{pmatrix} 1 \\ 0 \end{pmatrix}$, for $\begin{pmatrix} a & b \\ c & d \end{pmatrix} \begin{pmatrix} 1 \\ 0 \end{pmatrix} = \begin{pmatrix} a \\ c \end{pmatrix}$;

while the second column $\begin{pmatrix} b \\ d \end{pmatrix}$ gives the coordinates of V′, the transform of

V $\begin{pmatrix} 0 \\ 1 \end{pmatrix}$, since $\begin{pmatrix} a & b \\ c & d \end{pmatrix} \begin{pmatrix} 0 \\ 1 \end{pmatrix} = \begin{pmatrix} b \\ d \end{pmatrix}$.

Using this important property we are able to write down *at sight* the 2×2 matrices which represent such transformations as rotations about the origin, reflections in a line through O, shears, enlargements, and so on.

For example, suppose we wish to reflect in the line $y = x \tan \theta$. We see that the image of U is the point with coordinates $(\cos 2\theta, \sin 2\theta)$. It is not quite so easy to 'read off' the coordinates of V′, but remembering that OU′ ⊥ OV′, it is clear that V′ is the point $(\sin 2\theta, -\cos 2\theta)$.

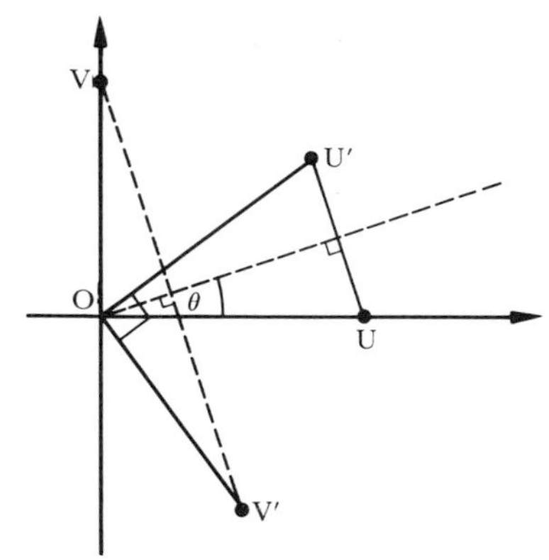

Thus the required matrix is $\begin{pmatrix} \cos 2\theta & \sin 2\theta \\ \sin 2\theta & -\cos 2\theta \end{pmatrix}$.

Notice that its determinant $(ac - bd)$ is $-\cos^2 2\theta - \sin^2 2\theta$, and so has the value -1: this numerical value shows that areas are preserved, while the minus sign signifies that the transformation is 'opposite', i.e., the reflection has turned the triangle OUV over, the anti-clockwise sense of the lettering OUV being replaced by the clockwise sense of OU′V′.

Exercise 13.2a

1 Write down the matrices representing the following transformations:
(*i*) Rotation about O through π, $\frac{1}{2}\pi$, $-\frac{1}{2}\pi$, $\frac{1}{3}\pi$, $-\frac{1}{3}\pi$, $2\pi/3$, $-2\pi/3$, $3\pi/4$, $-3\pi/4$, $\tan^{-1}(3/4)$, etc.
(*ii*) Reflections in the lines $x = 0$, $y = 0$, $x = y$, $x = -y$, $x = 2y$.
(*iii*) An enlargement in O with ratio 3:1; an x-stretch, ratio 2:1; a y-stretch, ratio 1:3; a y-shear which moves the point (1, 1) to the point (1, 0).

2 What transformations are represented by the following matrices:

(*i*) $\begin{pmatrix} 2 & 0 \\ 0 & 3 \end{pmatrix}$, (*ii*) $\begin{pmatrix} -3 & 0 \\ 0 & -3 \end{pmatrix}$, (*iii*) $\begin{pmatrix} 3 & 4 \\ -4 & 3 \end{pmatrix}$, (*iv*) $\begin{pmatrix} 1 & 2 \\ 0 & 1 \end{pmatrix}$,

(*v*) $\begin{pmatrix} 1 & 0 \\ -3 & 1 \end{pmatrix}$, (*vi*) $\begin{pmatrix} 0 & -2 \\ -2 & 0 \end{pmatrix}$, (*vii*) $\begin{pmatrix} 2 & -1 \\ 0 & 2 \end{pmatrix}$,

(*viii*) $\begin{pmatrix} \cos\alpha & \sin\alpha \\ \sin\alpha & -\cos\alpha \end{pmatrix}$; (*ix*) $\begin{pmatrix} -\sin\alpha & \cos\alpha \\ \cos\alpha & \sin\alpha \end{pmatrix}$,

(*x*) $\begin{pmatrix} \cos 40^\circ & -\sin 40^\circ \\ \sin 40^\circ & \cos 40^\circ \end{pmatrix}$, (*xi*) $\begin{pmatrix} 2 & 4 \\ 1 & 2 \end{pmatrix}$, (*xii*) $\begin{pmatrix} 1 & 0 \\ 1 & 0 \end{pmatrix}$?

3 Describe the transformations represented by the eight matrices

$\begin{pmatrix} \pm 1 & 0 \\ 0 & \pm 1 \end{pmatrix}$, $\begin{pmatrix} 0 & \pm 1 \\ \pm 1 & 0 \end{pmatrix}$.

4 In nos. **2** and **3** above, check the effects of the transformations on areas, and compare the values of the determinant $ad - bc$ in each case.

5 Obtain the matrix for rotation about the origin through θ
(*i*) by obtaining x' and y' in terms of x and y by projection;
(*ii*) by using complex numbers, starting from the equation $z' = z \operatorname{cis} \alpha$;
(*iii*) by the method shown in the text above.

6 Obtain the matrix for reflection in the line $y = x \tan \theta$, by first rotating the coordinate axes so that the x-axis coincides with the reflection axis (operation R), then performing the reflection in this axis, and finally restoring the axes to their original position (R^{-1}).

7 A triangle has vertices $A\begin{pmatrix} 2 \\ 1 \end{pmatrix}$, $B\begin{pmatrix} -1 \\ 2 \end{pmatrix}$, $C\begin{pmatrix} 0 \\ 3 \end{pmatrix}$.

What is its image under the transformations represented by the matrices:
(*i*) $\begin{pmatrix} 2 & 3 \\ -1 & -2 \end{pmatrix}$; (*ii*) $\begin{pmatrix} 3 & 1 \\ 2 & 0 \end{pmatrix}$; (*iii*) $\begin{pmatrix} 4 & -6 \\ -2 & 3 \end{pmatrix}$.

8 Investigate the effect of the transformation represented by $\begin{pmatrix} 1 & 0 \\ 1 & 1 \end{pmatrix}$ on the rectangle with vertices $\begin{pmatrix} 0 \\ -1 \end{pmatrix}, \begin{pmatrix} 0 \\ 1 \end{pmatrix}, \begin{pmatrix} 1 \\ 1 \end{pmatrix}, \begin{pmatrix} 1 \\ -1 \end{pmatrix}$.

9 What is the image of the point $\begin{pmatrix} 3 \\ -2 \end{pmatrix}$ under the matrix which moves $\begin{pmatrix} 1 \\ 0 \end{pmatrix}$ to $\begin{pmatrix} -2 \\ 2 \end{pmatrix}$ and $\begin{pmatrix} 1 \\ 1 \end{pmatrix}$ to $\begin{pmatrix} 0 \\ 3 \end{pmatrix}$?

10 Show that $\boldsymbol{M} = \begin{pmatrix} 2 & -1 \\ -3 & 4 \end{pmatrix}$ operating on *any* vector in the direction $\begin{pmatrix} 1 \\ 1 \end{pmatrix}$ leaves it invariant. Show that vectors in the direction $\begin{pmatrix} -3 \\ 1 \end{pmatrix}$ are also unchanged in direction, but are enlarged by a factor 5. Repeat, using the matrix $\boldsymbol{M}^2$.

11 Find a 2×2 matrix which transforms the vector $\begin{pmatrix} 0 \\ 2 \end{pmatrix}$ to the vector $\begin{pmatrix} 1 \\ 3 \end{pmatrix}$ and leaves $\begin{pmatrix} 1 \\ 1 \end{pmatrix}$ invariant. What type of transformation is it?

12 Show that, if a 2×2 matrix represents an *isometry*, i.e., a distance-preserving transformation, then it must be either of the form

$$\begin{pmatrix} \cos \theta & \sin \theta \\ -\sin \theta & \cos \theta \end{pmatrix} \text{ or else of the form } \begin{pmatrix} \cos \theta & \sin \theta \\ \sin \theta & -\cos \theta \end{pmatrix},$$

i.e., a rotation or a reflection.

Transformations in three dimensions

All the above can be extended to a higher number of dimensions. In three dimensions, for instance, we can consider the equations

$$x' = a_1x + b_1y + c_1z$$
$$y' = a_2x + b_2y + c_2z$$
$$z' = a_3x + b_3y + c_3z$$

as representing a transformation which maps

the vector $\begin{pmatrix} x \\ y \\ z \end{pmatrix}$ into the vector $\begin{pmatrix} x' \\ y' \\ z' \end{pmatrix}$.

Just as in the two dimensional case, this linear transformation which preserves the origin may be completely described by a matrix, in this case the 3×3 matrix

$$\boldsymbol{M} = \begin{pmatrix} a_1 & b_1 & c_1 \\ a_2 & b_2 & c_2 \\ a_3 & b_3 & c_3 \end{pmatrix};$$

so that the transformation may be written

$$\begin{pmatrix} x' \\ y' \\ z' \end{pmatrix} = \begin{pmatrix} a_1 & b_1 & c_1 \\ a_2 & b_2 & c_2 \\ a_3 & b_3 & c_3 \end{pmatrix}\begin{pmatrix} x \\ y \\ z \end{pmatrix},$$

or more briefly as $\boldsymbol{x}' = \boldsymbol{Mx}$.

If we label the points (1, 0, 0), (0, 1, 0) and (0, 0, 1) as U, V, W respectively, the transform of the point U will be given by

$$\begin{pmatrix} a_1 & b_1 & c_1 \\ a_2 & b_2 & c_2 \\ a_3 & b_3 & c_3 \end{pmatrix}\begin{pmatrix} 1 \\ 0 \\ 0 \end{pmatrix} = \begin{pmatrix} a_1 \\ a_2 \\ a_3 \end{pmatrix},$$

i.e., the coordinates of the point U′, the image of U, are the *first column*† of the matrix. Similarly, the second and third columns of the matrix give the coordinates of V′ and W′:

$$\begin{pmatrix} a_1 & b_1 & c_1 \\ a_2 & b_2 & c_2 \\ a_3 & b_3 & c_3 \end{pmatrix}\begin{pmatrix} 0 \\ 1 \\ 0 \end{pmatrix} = \begin{pmatrix} b_1 \\ b_2 \\ b_3 \end{pmatrix}; \quad \begin{pmatrix} a_1 & b_1 & c_1 \\ a_2 & b_2 & c_2 \\ a_3 & b_3 & c_3 \end{pmatrix}\begin{pmatrix} 0 \\ 0 \\ 1 \end{pmatrix} = \begin{pmatrix} c_1 \\ c_2 \\ c_3 \end{pmatrix}.$$

† It might therefore seem that columns are more important than rows. But this is only because we have chosen to represent vectors by columns. Had we used *row* vectors, and *post*-multiplied by the transforming matrix, its *rows* would have given the images of U, V, W.

Thus the images of the vectors

$$\begin{pmatrix}1\\0\\0\end{pmatrix}, \begin{pmatrix}0\\1\\0\end{pmatrix}, \begin{pmatrix}0\\0\\1\end{pmatrix} \text{ are given by the columns of the matrix } \begin{pmatrix}a_1\\a_2\\a_3\end{pmatrix}, \begin{pmatrix}b_1\\b_2\\b_3\end{pmatrix}, \begin{pmatrix}c_1\\c_2\\c_3\end{pmatrix}.$$

In this sense, the 3×3 matrix may be thought of as a collection of three column vectors $\boldsymbol{a}$, $\boldsymbol{b}$, $\boldsymbol{c}$. Conversely, if we wish to write down the nine elements of the matrix which represents a particular three-dimensional transformation of the type described, all we have to do is to find the images of the three points U, V, W, and their coordinates will give the three columns of the required matrix.

The cube having OU, OV and OW as adjacent edges is taken by the above transformation into a *parallelepiped* having OU′, OV′, OW′ as adjacent edges, and the effect of the transformation on the lattice (i.e., solid network) of unit cubes will be to take them into a lattice of parallelepipeds.

Effect of linear transformations on volumes

We have seen that the linear transformation whose matrix is

$$\boldsymbol{M} = \begin{pmatrix} a_1 & b_1 & c_1 \\ a_2 & b_2 & c_2 \\ a_3 & b_3 & c_3 \end{pmatrix}$$

transforms the unit vectors

$$\boldsymbol{i} = \begin{pmatrix}1\\0\\0\end{pmatrix}, \quad \boldsymbol{j} = \begin{pmatrix}0\\1\\0\end{pmatrix}, \quad \boldsymbol{k} = \begin{pmatrix}0\\0\\1\end{pmatrix}$$

into $\boldsymbol{a} = \begin{pmatrix}a_1\\a_2\\a_3\end{pmatrix}, \quad \boldsymbol{b} = \begin{pmatrix}b_1\\b_2\\b_3\end{pmatrix}, \quad \boldsymbol{c} = \begin{pmatrix}c_1\\c_2\\c_3\end{pmatrix}.$

But the volume of the parallelepiped formed by $\boldsymbol{a}$, $\boldsymbol{b}$, $\boldsymbol{c}$ has been shown (in 11.5) to be the scalar triple product

$$\boldsymbol{a}.(\boldsymbol{b} \times \boldsymbol{c}) = a_1b_2c_3 - a_1b_3c_2 + b_1c_2a_3 - b_1c_3a_2 + c_1a_2b_3 - c_1a_3b_2$$

which is called the *determinant*† of the Matrix $\boldsymbol{M}$ and written

$\Delta = |\boldsymbol{M}|$ (or det $\boldsymbol{M}$).

† Just as the area $ad - bc$ of the basic parallelogram in the two-dimensional case was the determinant of $\begin{pmatrix} a & b \\ c & d \end{pmatrix}$.

So the transformation whose matrix is $\boldsymbol{M}$ leads to volumes being altered by a factor $|\boldsymbol{M}|$.

Determinants will be discussed more fully in 13.5 and 13.6, but at this stage we simply note that when the determinant is zero the parallelepiped has zero volume, i.e., the three vectors are coplanar:

$$|\boldsymbol{M}| = 0 \quad \Leftrightarrow \quad \boldsymbol{a}, \boldsymbol{b}, \boldsymbol{c} \text{ are coplanar.}$$

In this case, a vector $\begin{pmatrix} x \\ y \\ z \end{pmatrix}$ in the original space will now be transformed into a vector

$$x\begin{pmatrix} a_1 \\ a_2 \\ a_3 \end{pmatrix} + y\begin{pmatrix} b_1 \\ b_2 \\ b_3 \end{pmatrix} + z\begin{pmatrix} c_1 \\ c_2 \\ c_3 \end{pmatrix} = x\boldsymbol{a} + y\boldsymbol{b} + z\boldsymbol{c},$$

which lies in the plane of the vectors $\boldsymbol{a}$, $\boldsymbol{b}$, $\boldsymbol{c}$. So the entire original three-dimensional space is mapped onto a plane, a two-dimensional space; or, in exceptional cases, onto a line or even a point.

Examples of three-dimensional transformations

Let us set up the 3×3 matrices for transformations in a number of simple cases.

First consider the reflection in the plane UOV ($z = 0$). The points U and V are invariant, and the point W clearly goes to W′ $(0, 0, -1)$. Hence the matrix is

$$\begin{pmatrix} 1 & 0 & 0 \\ 0 & 1 & 0 \\ 0 & 0 & -1 \end{pmatrix}.$$

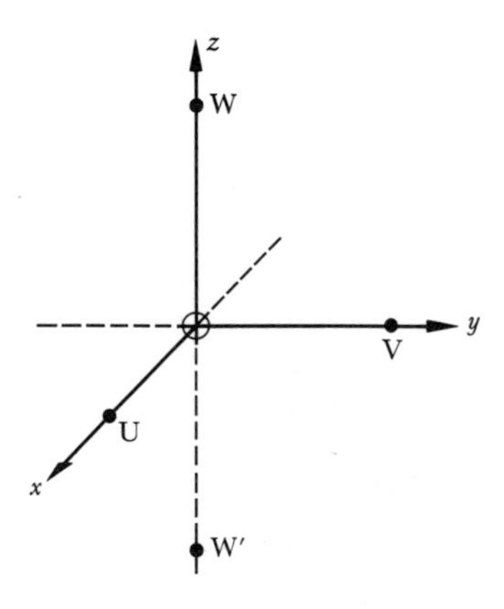

Next we consider a quarter-turn about the y-axis, the sense being such as to take the point W into the point U, i.e., W′ is $(1, 0, 0)$. We see that U′ has coordinates $(0, 0, -1)$ and since V is invariant, the required matrix is

$$\begin{pmatrix} 0 & 0 & 1 \\ 0 & 1 & 0 \\ -1 & 0 & 0 \end{pmatrix}.$$

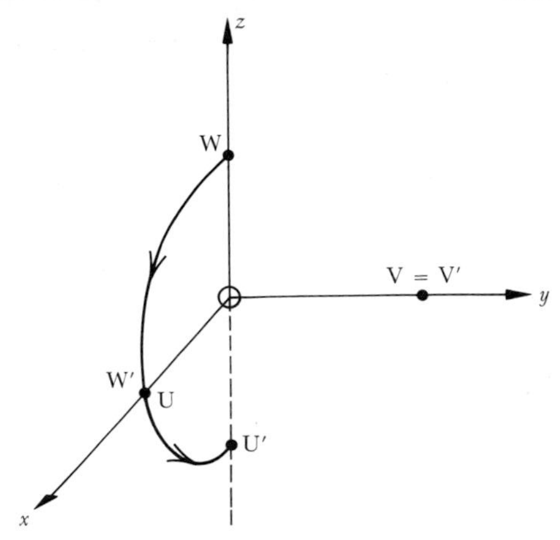

We now ask what transformation is represented by the matrix

$$\begin{pmatrix} 0 & 0 & 1 \\ 1 & 0 & 0 \\ 0 & 1 & 0 \end{pmatrix}.$$

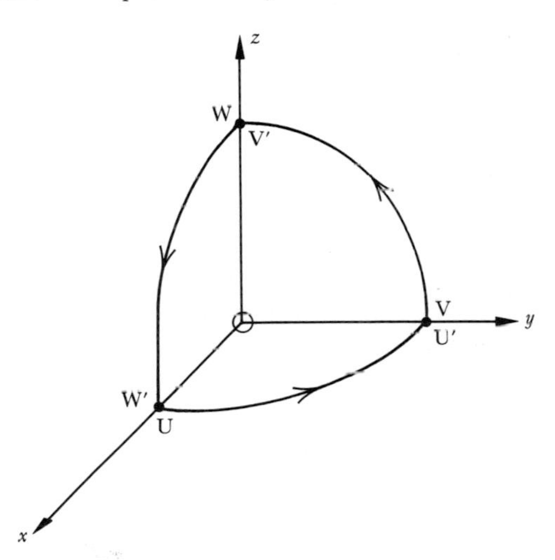

From the columns, we can read off the positions of U′ (0, 1, 0) = V, V′ (0, 0, 1) = W, and W′ (1, 0, 0) = U. The diagram makes it evident that we are concerned with a rotation through 120° about an axis passing through (1, 1, 1). We may check that the vector $\begin{pmatrix} 1 \\ 1 \\ 1 \end{pmatrix}$ is invariant under this rotation, for

$$\begin{pmatrix} 0 & 0 & 1 \\ 1 & 0 & 0 \\ 0 & 1 & 0 \end{pmatrix}\begin{pmatrix} 1 \\ 1 \\ 1 \end{pmatrix} = \begin{pmatrix} 1 \\ 1 \\ 1 \end{pmatrix}.$$

[An important idea is suggested here, that of an *eigenvector*, whose direction is unchanged in a linear transformation.]

All the matrices so far considered have included only 0 and 1, but

$$\begin{pmatrix} 3 & 0 & 0 \\ 0 & 1 & 0 \\ 0 & 0 & 1 \end{pmatrix}$$

represents an x-stretch in the ratio 3:1, while

$$\begin{pmatrix} 3 & 0 & 0 \\ 0 & 3 & 0 \\ 0 & 0 & 3 \end{pmatrix}$$

is an enlargement in the ratio 3:1, the vector $\boldsymbol{OP}$ being replaced by $\boldsymbol{OP}' = 3\boldsymbol{OP}$. This matrix is known as a 'scalar' matrix, since its effect on every vector $\boldsymbol{OP}$ is the same as the effect of multiplication by the scalar number 3. The general scalar matrix

$$\begin{pmatrix} k & 0 & 0 \\ 0 & k & 0 \\ 0 & 0 & k \end{pmatrix}$$

represents a pure enlargement by a factor k when $k > 0$, but when $k < 0$ it involves 'reflection through the origin', or 'central inversion', exemplified by the matrix

$$\begin{pmatrix} -1 & 0 & 0 \\ 0 & -1 & 0 \\ 0 & 0 & -1 \end{pmatrix},$$

whereby every vector is *reversed in direction.*

Lastly, we consider a transformation in which the origin is not fixed. Under a transformation of the form $\boldsymbol{x}' = \boldsymbol{M}\boldsymbol{x}$, the origin maps into itself. How can one express in matrix form a transformation which does not possess this property? The simplest example might be a *translation* represented by $\boldsymbol{x}' = \boldsymbol{x} + \boldsymbol{a}$, where $\boldsymbol{a}$ is a fixed vector. In the two-dimensional case, we have

$$\left.\begin{aligned} x' &= x + a \\ y' &= y + b \end{aligned}\right\}$$

and this may be written

$$\begin{pmatrix} x' \\ y' \\ 1 \end{pmatrix} = \begin{pmatrix} 1 & 0 & a \\ 0 & 1 & b \\ 0 & 0 & 1 \end{pmatrix} \begin{pmatrix} x \\ y \\ 1 \end{pmatrix}, \quad \text{or as} \quad \begin{pmatrix} x' \\ y' \end{pmatrix} = \begin{pmatrix} 1 & 0 & a \\ 0 & 1 & b \end{pmatrix} \begin{pmatrix} x \\ y \\ 1 \end{pmatrix}.$$

Exercise 13.2b

1 Write down the 3×3 matrices representing the following:

(*i*) reflection in the plane $y = 0$;
(*ii*) reflection in the plane $y = x$;
(*iii*) reflection in the plane $y = x \tan \alpha$;
(*iv*) reflection in the plane $x + y + z = 0$;
(*v*) half-turn about the x-axis;
(*vi*) half-turn about the line $x = y, z = 0$;
(*vii*) half-turn about the line $x = y = z$;
(*viii*) quarter-turns about z-axis;
(*ix*) quarter-turns about the line $x = 0 = y + z$;
(*x*) rotation through θ about the y-axis;
(*xi*) an enlargement from the origin in the ratio $5:1$;
(*xii*) an orthogonal projection on to the plane $y = 0$.

2 What transformations are represented by the following matrices:

$$\begin{pmatrix} 1 & 0 & 1 \\ 0 & -1 & 0 \\ 0 & 0 & 1 \end{pmatrix}, \begin{pmatrix} 1 & 0 & 0 \\ 0 & 1 & 0 \\ 0 & 0 & 0 \end{pmatrix}, \begin{pmatrix} 0 & 1 & 0 \\ 0 & 0 & 1 \\ 1 & 0 & 0 \end{pmatrix}, \begin{pmatrix} 1 & 0 & 0 \\ 0 & 0 & -1 \\ 0 & 1 & 0 \end{pmatrix}, \begin{pmatrix} 2 & 0 & 0 \\ 0 & 3 & 0 \\ 0 & 0 & -1 \end{pmatrix},$$

$$\begin{pmatrix} -3 & 0 & 0 \\ 0 & -3 & 0 \\ 0 & 0 & -3 \end{pmatrix}, \begin{pmatrix} 0.8 & -0.6 & 0 \\ 0.6 & 0.8 & 0 \\ 0 & 0 & 1 \end{pmatrix}, \begin{pmatrix} 1 & 0 & 0 \\ 0 & 0.6 & 0.8 \\ 0 & 0.8 & -0.6 \end{pmatrix}?$$

3 Consider the effects of the above transformations on volumes and check by evaluating the determinants in each case.

4 Show that the matrix $\begin{pmatrix} 6 & -4 & 3 \\ 29 & -20 & 13 \\ 33 & -23 & 14 \end{pmatrix}$ transforms the vector $\begin{pmatrix} 1 \\ 2 \\ 1 \end{pmatrix}$ into itself.

5 Check that the vectors $\begin{pmatrix} 2 \\ 1 \\ 0 \end{pmatrix}, \begin{pmatrix} 1 \\ 0 \\ 1 \end{pmatrix}, \begin{pmatrix} 0 \\ -1 \\ 2 \end{pmatrix}$ are coplanar, and investigate the effect of the matrix

$$\begin{pmatrix} 2 & 1 & 0 \\ 1 & 0 & -1 \\ 0 & 1 & 2 \end{pmatrix}.$$

6 Find the effect on the cube with vertices $\begin{pmatrix} \pm 1 \\ \pm 1 \\ \pm 1 \end{pmatrix}$ of the matrix

$$\begin{pmatrix} 0 & -1 & 0 \\ 1 & 0 & 0 \\ 0 & 0 & -1 \end{pmatrix}.$$

Repeat for some of the matrices mentioned in nos. 1, 2, 5 above.

7 Check that, if a 3×3 matrix $\boldsymbol{M}$ transforms the vectors $\boldsymbol{x}$ and $\boldsymbol{y}$ into the vectors $\boldsymbol{x}'$ and $\boldsymbol{y}'$ (i.e., $\boldsymbol{x}' = \boldsymbol{Mx}$ and $\boldsymbol{y}' = \boldsymbol{My}$), then the vector $\boldsymbol{z} = \lambda\boldsymbol{x} + \mu\boldsymbol{y}$ will be transformed into the vector $\boldsymbol{z}' = \lambda\boldsymbol{x}' + \mu\boldsymbol{y}'$.

8 Show that a glide reflection in the x-axis, in two dimensions, whereby each point is reflected in the x-axis and then undergoes a translation parallel to the x-axis, may be represented by a 3×3 matrix.

What does the equation $\begin{pmatrix} x' \\ y' \\ 1 \end{pmatrix} = \begin{pmatrix} 0 & 1 & a \\ 1 & 0 & a \\ 0 & 0 & 1 \end{pmatrix}\begin{pmatrix} x \\ y \\ 1 \end{pmatrix}$ represent?

Describe the effect of the matrices $\begin{pmatrix} \pm 1 & 0 & a \\ 0 & \pm 1 & b \\ 0 & 0 & 1 \end{pmatrix}$

in the four possible cases of selection of signs.

13.3 Products of linear transformations and matrices

Suppose that the general 3-dimensional vector $\boldsymbol{x}$ is transformed to give a vector $\boldsymbol{x}'$, and that this in turn is transformed to give $\boldsymbol{x}''$. Let us further suppose that these transformations can be represented as

$$\boldsymbol{x}'' = \boldsymbol{Ax}' \quad \text{and} \quad \boldsymbol{x}' = \boldsymbol{Bx}$$

where $\boldsymbol{A} = \begin{pmatrix} \alpha_1 & \beta_1 & \gamma_1 \\ \alpha_2 & \beta_2 & \gamma_2 \\ \alpha_3 & \beta_3 & \gamma_3 \end{pmatrix}$ and $\boldsymbol{B} = \begin{pmatrix} a_1 & b_1 & c_1 \\ a_2 & b_2 & c_2 \\ a_3 & b_3 & c_3 \end{pmatrix}$.

Hence
$$\left.\begin{aligned} x'' &= \alpha_1 x' + \beta_1 y' + \gamma_1 z' \\ y'' &= \alpha_2 x' + \beta_2 y' + \gamma_2 z' \\ z'' &= \alpha_3 x' + \beta_3 y' + \gamma_3 z' \end{aligned}\right\} \qquad \left.\begin{aligned} x' &= a_1 x + b_1 y + c_1 z \\ y' &= a_2 x + b_2 y + c_2 z \\ z' &= a_3 x + b_3 y + c_3 z \end{aligned}\right\}.$$

Substitution gives

$$\begin{aligned} x'' &= \alpha_1(a_1 x + b_1 y + c_1 z) + \beta_1(a_2 x + b_2 y + c_2 z) + \gamma_1(a_3 x + b_3 y + c_3 z) \\ &= (\alpha_1 a_1 + \beta_1 a_2 + \gamma_1 a_3)x + (\alpha_1 b_1 + \beta_1 b_2 + \gamma_1 b_3)y \\ &\quad + (\alpha_1 c_1 + \beta_1 c_2 + \gamma_1 c_3)z. \end{aligned}$$

Similarly we can write down expressions for y'' and z'', and the three results can be combined in matrix form:

$$\begin{pmatrix} x'' \\ y'' \\ z'' \end{pmatrix} = \begin{pmatrix} \alpha_1 a_1 + \beta_1 a_2 + \gamma_1 a_3 & \alpha_1 b_1 + \beta_1 b_2 + \gamma_1 b_3 & \alpha_1 c_1 + \beta_1 c_2 + \gamma_1 c_3 \\ \alpha_2 a_1 + \beta_2 a_2 + \gamma_2 a_3 & \alpha_2 b_1 + \beta_2 b_2 + \gamma_2 b_3 & \alpha_2 c_1 + \beta_2 c_2 + \gamma_2 c_3 \\ \alpha_3 a_1 + \beta_3 a_2 + \gamma_3 a_3 & \alpha_3 b_1 + \beta_3 b_2 + \gamma_3 b_3 & \alpha_3 c_1 + \beta_3 c_2 + \gamma_3 c_3 \end{pmatrix} \begin{pmatrix} x \\ y \\ z \end{pmatrix}$$

or $\boldsymbol{x}'' = \boldsymbol{Cx}$,

where $\boldsymbol{C}$ represents the combined effect, or *product*, of the two transformations.

But $\boldsymbol{x}'' = \boldsymbol{Ax}'$ and $\boldsymbol{x}' = \boldsymbol{Bx}$

can also be combined to give

$\boldsymbol{x}'' = \boldsymbol{A}(\boldsymbol{Bx})$,

so we now call $\boldsymbol{C}$ the *product* of $\boldsymbol{A}$ and $\boldsymbol{B}$ and write

$\boldsymbol{AB} = \boldsymbol{C}$ and $\boldsymbol{x}'' = (\boldsymbol{AB})\boldsymbol{x}$:

$$\begin{pmatrix} \alpha_1 & \beta_1 & \gamma_1 \\ \alpha_2 & \beta_2 & \gamma_2 \\ \alpha_3 & \beta_3 & \gamma_3 \end{pmatrix} \begin{pmatrix} a_1 & b_1 & c_1 \\ a_2 & b_2 & c_2 \\ a_3 & b_3 & c_3 \end{pmatrix}$$

$$= \begin{pmatrix} \alpha_1 a_1 + \beta_1 a_2 + \gamma_1 a_3 & \alpha_1 b_1 + \beta_1 b_2 + \gamma_1 b_3 & \alpha_1 c_1 + \beta_1 c_2 + \gamma_1 c_3 \\ \alpha_2 a_1 + \beta_2 a_2 + \gamma_2 a_3 & \alpha_2 b_1 + \beta_2 b_2 + \gamma_2 b_3 & \alpha_2 c_1 + \beta_2 c_2 + \gamma_2 c_3 \\ \alpha_3 a_1 + \beta_3 a_2 + \gamma_3 a_3 & \alpha_3 b_1 + \beta_3 b_2 + \gamma_3 b_3 & \alpha_3 c_1 + \beta_3 c_2 + \gamma_3 c_3 \end{pmatrix}$$

Hence we see that the products of 3×3 matrices are formed just like those of 2×2 matrices (see 0.7) by 'row and column multiplication'. More generally, we define the product $\boldsymbol{AB}$ of two matrices $\boldsymbol{A}$ and $\boldsymbol{B}$ in just the same way, the element of $\boldsymbol{AB}$ in the ith row and jth column being called the *inner product* of the ith row of $\boldsymbol{A}$ with the jth column of $\boldsymbol{B}$. This, of course, necessitates that these two have the same number of elements, i.e., that the number of columns of $\boldsymbol{A}$ is equal to the number of rows of $\boldsymbol{B}$, such matrices being called *conformable for multiplication*.

Hence if $\boldsymbol{A}$ is an $m \times n$ matrix
and $\boldsymbol{B}$ is an $n \times p$ matrix,
$\boldsymbol{AB}$ will be an $m \times p$ matrix.

Example

Find $\boldsymbol{AB}$ and $\boldsymbol{BA}$ when

$$(i)\quad \boldsymbol{A} = \begin{pmatrix} 0 & 1 & 2 \\ 0 & -1 & -2 \end{pmatrix} \qquad \boldsymbol{B} = \begin{pmatrix} 3 & -3 \\ 4 & -4 \\ 5 & -5 \end{pmatrix};$$

(*ii*) $A = \begin{pmatrix} 1 & 2 & 3 \end{pmatrix} \qquad B = \begin{pmatrix} 4 & 5 \\ 6 & 7 \\ 8 & 9 \end{pmatrix}.$

(*i*) $AB = \begin{pmatrix} 0 & 1 & 2 \\ 0 & -1 & -2 \end{pmatrix}\begin{pmatrix} 3 & -3 \\ 4 & -4 \\ 5 & -5 \end{pmatrix} = \begin{pmatrix} 14 & -14 \\ -14 & 14 \end{pmatrix};$

$BA = \begin{pmatrix} 3 & -3 \\ 4 & -4 \\ 5 & -5 \end{pmatrix}\begin{pmatrix} 0 & 1 & 2 \\ 0 & -1 & -2 \end{pmatrix} = \begin{pmatrix} 0 & 6 & 12 \\ 0 & 8 & 16 \\ 0 & 10 & 20 \end{pmatrix}.$

(*ii*) $AB = \begin{pmatrix} 1 & 2 & 3 \end{pmatrix}\begin{pmatrix} 4 & 5 \\ 6 & 7 \\ 8 & 9 \end{pmatrix} = \begin{pmatrix} 40 & 46 \end{pmatrix};$

whilst $BA = \begin{pmatrix} 4 & 5 \\ 6 & 7 \\ 8 & 9 \end{pmatrix}\begin{pmatrix} 1 & 2 & 3 \end{pmatrix}$ does not exist.

Exercise 13.3a

1 $A = \begin{pmatrix} 4 & 1 \\ 3 & 2 \end{pmatrix} \quad B = \begin{pmatrix} 1 & 0 \\ 2 & 1 \end{pmatrix} \quad C = \begin{pmatrix} 1 & 1 \\ 2 & 0 \\ 0 & 4 \end{pmatrix} \quad D = \begin{pmatrix} -1 & 0 & 2 \\ 2 & -3 & 1 \end{pmatrix}.$

Find where possible ***AB***, ***BA***, ***CD***, ***DC***, ***DB***, ***BD***, ***DA***, ***AD***, ***CA***, ***AC***, ***BC***, ***CB***.

2 Find the missing elements:

$$\begin{pmatrix} 1 & -3 \\ 1 & . \\ 1 & . \\ . & 1 \end{pmatrix}\begin{pmatrix} -1 & . & 7 & . \\ . & 1 & . & 0 \end{pmatrix} = \begin{pmatrix} -25 & -1 & 1 & 3 \\ -1 & . & . & . \\ . & . & 5 & . \\ . & . & . & 0 \end{pmatrix}.$$

3 Show that $I = \begin{pmatrix} 1 & 0 & 0 \\ 0 & 1 & 0 \\ 0 & 0 & 1 \end{pmatrix}$

(unit matrix) is an identity for post-multiplication of matrices with 3 columns, and also for pre-multiplication of matrices with 3 rows, while ***MI*** = ***IM*** = ***M*** where ***M*** is any 3 × 3 matrix. Generalise.

4 If $A = \begin{pmatrix} 6 & -3 \\ -4 & 2 \end{pmatrix}, \quad B = \begin{pmatrix} 4 & 3 \\ 8 & 6 \end{pmatrix},$

find ***AB*** and ***BA***. Find other counter-examples to illustrate that $AB = 0 \nRightarrow A = 0$ or $B = 0$, where ***0*** is the null matrix. Deduce a counter-example to show that $AC = BC \nRightarrow A = B$. Obtain examples with 3×3 matrices, taking

$$A = \begin{pmatrix} -2 & 1 & 0 \\ 1 & -3 & -1 \\ 1 & 7 & 3 \end{pmatrix}.$$

5 Represent the following five quadratic equations by a single matrix equation:
$2x^2 + 5x - 8 = 0$; $x^2 - 7x + 11 = 0$; $3x^2 = 4x + 1$;
$5 - x - x^2 = 0$; $4x^2 + 3x = 0$.

6 If A is a 2×2 matrix such that $A^2 = 0$, find the general form of A. Repeat when $A^2 = I$.

7 Find all the matrices that can be obtained by forming products (e.g., ABA^2) from

$$A = \begin{pmatrix} 0 & i \\ i & 0 \end{pmatrix}, \quad B = \begin{pmatrix} 0 & -1 \\ 1 & 0 \end{pmatrix}.$$

Repeat with $C = \begin{pmatrix} \omega & 0 \\ 0 & \omega^2 \end{pmatrix}$, $D = \begin{pmatrix} 0 & \omega \\ -\omega^2 & 0 \end{pmatrix}$.

(Note that in this question $i^2 = -1$, $\omega^3 = 1$.)

8 Show that the only 2×2 matrices which commute with *every* 2×2 matrix are those of the form $\begin{pmatrix} k & 0 \\ 0 & k \end{pmatrix}$. Extend the result to 3×3 matrices.

9 If $A = \begin{pmatrix} a_1 & a_2 \\ -a_2 & a_1 \end{pmatrix}$, $B = \begin{pmatrix} b_1 & b_2 \\ -b_2 & b_1 \end{pmatrix}$,
find AB and BA. What does this result remind you of?

10 Find the products:

(*i*) $\begin{pmatrix} 2 & -1 & 0 \\ 3 & 1 & -2 \\ 0 & 4 & -3 \end{pmatrix} \begin{pmatrix} 5 & -3 & 2 \\ 9 & -6 & 4 \\ 12 & -8 & 5 \end{pmatrix}$;

(*ii*) $\begin{pmatrix} 3 - \sqrt{2} & 4 \\ -4 & 3 + \sqrt{2} \end{pmatrix} \begin{pmatrix} 2 & 5 - 2\sqrt{2} \\ -5 - 2\sqrt{2} & 2 \end{pmatrix}$;

(*iii*) $\begin{pmatrix} 3 - i & 4 \\ -4 & 3 + i \end{pmatrix} \begin{pmatrix} 2 & 5 - 2i \\ -5 - 2i & 2 \end{pmatrix}$.

11 If $A = \begin{pmatrix} 2 & -3 \\ -3 & 5 \end{pmatrix}$, find $A^2, A^3, A^4, \ldots$.
If $AX = XA$ where X is a 2×2 matrix, show that X may be expressed in the form $\begin{pmatrix} p & q \\ q & p - q \end{pmatrix}$.

Find the general form of matrices which commute with a given matrix $\begin{pmatrix} a & b \\ c & d \end{pmatrix}$.

12 If $\boldsymbol{M} = \begin{pmatrix} 1 & 0 & 0 \\ 0 & 0 & -1 \\ 0 & -1 & 0 \end{pmatrix}$, prove that $\boldsymbol{M}^{2n+1} = \boldsymbol{M}$.

13 If $\boldsymbol{C} = \begin{pmatrix} 0 & i \\ -i & i \end{pmatrix}$, find $\boldsymbol{C}^2, \boldsymbol{C}^3, \ldots$ and show that $\boldsymbol{C}^{12} = \boldsymbol{I}$.

14 Express $\begin{pmatrix} 1 & 2 & -1 \\ 1 & 1 & 0 \\ 4 & 8 & -1 \end{pmatrix}$ in the form $\begin{pmatrix} a & 0 & 0 \\ b & c & 0 \\ d & e & f \end{pmatrix}\begin{pmatrix} 1 & g & h \\ 0 & 1 & k \\ 0 & 0 & 1 \end{pmatrix}$.

15 If $\boldsymbol{A}$, $\boldsymbol{B}$, $\boldsymbol{C}$ are 2×2 matrices, prove that $(\boldsymbol{A} \times \boldsymbol{B}) \times \boldsymbol{C} = \boldsymbol{A} \times (\boldsymbol{B} \times \boldsymbol{C})$ (associative law of multiplication). Extend the proof to apply to 3×3 matrices. How can the proof be generalised to apply to any three matrices which are conformable for multiplication?

16 Interpret row by column multiplication of two vectors $\boldsymbol{r}_1, \boldsymbol{r}_2$ in 3-space, expressed by the equation

$$(x_1 \ y_1 \ z_1)\begin{pmatrix} x_2 \\ y_2 \\ z_2 \end{pmatrix} = x_1x_2 + y_1y_2 + z_1z_2.$$

Show that this 'inner product' of the two vectors $\boldsymbol{r}_1, \boldsymbol{r}_2$ has the value $r_1r_2 \cos \theta$, where r_1, r_2 are the lengths of the two vectors, and θ is the angle between them. How may the concept of 'angle' be extended into more than three dimensions?

17 Express $(x \ \ y \ \ z)\begin{pmatrix} a & h & g \\ h & b & f \\ g & f & c \end{pmatrix}\begin{pmatrix} x \\ y \\ z \end{pmatrix}$

as a single expression, and show how $x^2 + y^2 + 2gx + 2fy + c$ may be similarly expressed.
Repeat with the expression $2x^2 + 4y^2 - 6xz + 3yz$.

18 Investigate the effect of pre-multiplication of a column vector of dimension n by an $n \times n$ matrix containing n 1s and the rest 0s, each row and each column of which contains a single 1. Repeat for post-multiplication of a row-vector (such matrices are known as 'permutation matrices').

19 Interpret the matrix equation

$$\begin{pmatrix} 0 & 1 \\ -1 & 0 \end{pmatrix}\begin{pmatrix} 2 & 3 & -1 & 0 \\ -1 & 5 & 0 & -4 \end{pmatrix} = \begin{pmatrix} -1 & 5 & 0 & -4 \\ -2 & -3 & 1 & 0 \end{pmatrix}$$

as a geometrical transformation applied to a certain quadrilateral.

20 The equation $\begin{pmatrix} x'_1 & x'_2 \ldots x'_n \\ y'_1 & y'_2 \ldots y'_n \\ z'_1 & z'_2 \ldots z'_n \end{pmatrix} = \begin{pmatrix} a_1 & b_1 & c_1 \\ a_2 & b_2 & c_2 \\ a_3 & b_3 & c_3 \end{pmatrix} \begin{pmatrix} x_1 & x_2 \ldots x_n \\ y_1 & y_2 \ldots y_n \\ z_1 & z_2 \ldots z_n \end{pmatrix}$

may be interpreted as follows: the n points $\begin{pmatrix} x_r \\ y_r \\ z_r \end{pmatrix}$ $(r = 1, 2, \ldots, n)$ are transformed by the matrix $(\boldsymbol{a} \quad \boldsymbol{b} \quad \boldsymbol{c})$ into the n points $\begin{pmatrix} x'_r \\ y'_r \\ z'_r \end{pmatrix}$, so that we are thinking of each of the first and third matrices as collections of column vectors. Interpret in this way the equation:

$$\begin{pmatrix} 0 & 0 & 1 \\ 1 & 0 & 0 \\ 0 & 1 & 0 \end{pmatrix} \begin{pmatrix} 1 & 4 & 0 & 7 & -1 \\ 2 & -3 & 2 & 0 & -1 \\ 3 & 0 & -5 & 1 & 2 \end{pmatrix} = \begin{pmatrix} 3 & 0 & -5 & 1 & 2 \\ 1 & 4 & 0 & 7 & -1 \\ 2 & -3 & 2 & 0 & -1 \end{pmatrix}.$$

Successive transformations

We may now consider the combined effect of certain successive three-dimensional transformations (which would be difficult to draw or even to visualise) by simply applying the multiplication rule for the corresponding matrices. For example, the products

$$\begin{pmatrix} 1 & 0 & 0 \\ 0 & 0 & 1 \\ 0 & 1 & 0 \end{pmatrix} \begin{pmatrix} 0 & 1 & 0 \\ -1 & 0 & 0 \\ 0 & 0 & 1 \end{pmatrix} = \begin{pmatrix} 0 & 1 & 0 \\ 0 & 0 & 1 \\ -1 & 0 & 0 \end{pmatrix} \qquad (1)$$

and
$$\begin{pmatrix} 0 & 1 & 0 \\ -1 & 0 & 0 \\ 0 & 0 & 1 \end{pmatrix} \begin{pmatrix} 1 & 0 & 0 \\ 0 & 0 & 1 \\ 0 & 1 & 0 \end{pmatrix} = \begin{pmatrix} 0 & 0 & 1 \\ -1 & 0 & 0 \\ 0 & 1 & 0 \end{pmatrix} \qquad (2)$$

illustrate (1) that the quarter-turn about the z-axis, represented by the matrix

$$\begin{pmatrix} 0 & 1 & 0 \\ -1 & 0 & 0 \\ 0 & 0 & 1 \end{pmatrix},$$

followed by the reflection in the plane $y = z$, represented by

$$\begin{pmatrix} 1 & 0 & 0 \\ 0 & 0 & 1 \\ 0 & 1 & 0 \end{pmatrix},$$

combine to give a transformation represented by

$$\begin{pmatrix} 0 & 1 & 0 \\ 0 & 0 & 1 \\ -1 & 0 & 0 \end{pmatrix};$$

(2) that if these operations are commuted, a different transformation results. (In both cases, the composition transformation is not a 'simple' one, being known as a *rotatory reflection* or *rotatory inversion* (see Coxeter, H., *Introduction to geometry*, Wiley, 1961, p. 99.)

On the other hand, the scalar matrix

$$\begin{pmatrix} k & 0 & 0 \\ 0 & k & 0 \\ 0 & 0 & k \end{pmatrix}$$

does commute with other 3×3 matrices; so that, for example, when the quarter-turn

$$\begin{pmatrix} 1 & 0 & 0 \\ 0 & 0 & 1 \\ 0 & -1 & 0 \end{pmatrix} \text{ is combined with the enlargement } \begin{pmatrix} 3 & 0 & 0 \\ 0 & 3 & 0 \\ 0 & 0 & 3 \end{pmatrix},$$

the result is the same when the order of operations is reversed, being represented by

$$\begin{pmatrix} 3 & 0 & 0 \\ 0 & 0 & 3 \\ 0 & -3 & 0 \end{pmatrix} \text{ in each case.}$$

Suppose next we consider the transformations represented by the matrix product

$$\begin{pmatrix} 0 & 0 & 1 \\ 0 & 1 & 0 \\ -1 & 0 & 0 \end{pmatrix}\begin{pmatrix} 1 & 0 & 0 \\ 0 & 0 & -1 \\ 0 & 1 & 0 \end{pmatrix} = \begin{pmatrix} 0 & 1 & 0 \\ 0 & 0 & -1 \\ -1 & 0 & 0 \end{pmatrix}.$$

Each matrix on the L.H.S. represents a quarter-turn, but it is not obvious what the product represents. We consider its effect on the general vector

$$\begin{pmatrix} x \\ y \\ z \end{pmatrix}, \text{ namely } \begin{pmatrix} 0 & 1 & 0 \\ 0 & 0 & -1 \\ -1 & 0 & 0 \end{pmatrix}\begin{pmatrix} x \\ y \\ z \end{pmatrix} = \begin{pmatrix} y \\ -z \\ -x \end{pmatrix}$$

and ask under what circumstances such a vector would be unmoved by the transformation. This would require $x = y$, $y = -z$, $z = -x$, i.e., any vector of the form $\begin{pmatrix} k \\ k \\ -k \end{pmatrix}$, or $k\begin{pmatrix} 1 \\ 1 \\ -1 \end{pmatrix}$, would be invariant under this transformation – another example of an eigenvector. We suspect, for this reason, that it is a rotation. Surprisingly, the rotation is through 120°, for if the matrix is applied again to $\begin{pmatrix} y \\ -z \\ -x \end{pmatrix}$ it becomes $\begin{pmatrix} -z \\ x \\ -y \end{pmatrix}$, and after another application it finally becomes $\begin{pmatrix} x \\ y \\ z \end{pmatrix}$. As an exercise, you should find the effect of the same two quarter-turns when they are (*a*) reversed in sense (one at a time), (*b*) commuted, (*c*) both. Finally, observe that,

since $$\begin{pmatrix} 0 & 1 & 0 \\ 0 & 0 & -1 \\ -1 & 0 & 0 \end{pmatrix} = \begin{pmatrix} 1 & 0 & 0 \\ 0 & -1 & 0 \\ 0 & 0 & -1 \end{pmatrix}\begin{pmatrix} 0 & 1 & 0 \\ 0 & 0 & 1 \\ 1 & 0 & 0 \end{pmatrix},$$

the transformation may be analysed into a 120° turn about the line $x = y = z$, represented by

$$\begin{pmatrix} 0 & 1 & 0 \\ 0 & 0 & 1 \\ 1 & 0 & 0 \end{pmatrix},$$

followed by a half-turn about the x-axis represented by

$$\begin{pmatrix} 1 & 0 & 0 \\ 0 & -1 & 0 \\ 0 & 0 & -1 \end{pmatrix}.$$

Exercise 13.3b

1 If $\boldsymbol{A} = \begin{pmatrix} 0 & -1 \\ 1 & 0 \end{pmatrix}$, $\boldsymbol{B} = \begin{pmatrix} -1 & 0 \\ 0 & 1 \end{pmatrix}$, $\boldsymbol{C} = \begin{pmatrix} 2 & 0 \\ 0 & 2 \end{pmatrix}$, $\boldsymbol{D} = \begin{pmatrix} 1 & 1 \\ 0 & 1 \end{pmatrix}$

evaluate the products $\boldsymbol{AB}$, $\boldsymbol{BA}$, $\boldsymbol{AC}$, $\boldsymbol{CA}$, etc., also $\boldsymbol{C}^2$, $\boldsymbol{D}^2$, $\boldsymbol{ABC}$, $\boldsymbol{CBA}$, $\boldsymbol{BAD}$, etc. and interpret geometrically in each case.

2 Repeat the above question using the matrices

$$\boldsymbol{P} = \begin{pmatrix} 0 & 1 & 0 \\ 0 & 0 & 1 \\ 1 & 0 & 0 \end{pmatrix}; \quad \boldsymbol{Q} = \begin{pmatrix} -1 & 0 & 0 \\ 0 & -1 & 0 \\ 0 & 0 & 1 \end{pmatrix}; \quad \boldsymbol{R} = \begin{pmatrix} 0 & 0 & 1 \\ 0 & 1 & 0 \\ 1 & 0 & 0 \end{pmatrix};$$

$$\boldsymbol{S} = \begin{pmatrix} 2 & 0 & 0 \\ 0 & 2 & 0 \\ 0 & 0 & 2 \end{pmatrix}.$$

3 If $\boldsymbol{R}_1 = \begin{pmatrix} \cos\alpha & \sin\alpha \\ -\sin\alpha & \cos\alpha \end{pmatrix}$ and $\boldsymbol{R}_2 = \begin{pmatrix} \cos\beta & \sin\beta \\ -\sin\beta & \cos\beta \end{pmatrix}$, find $\boldsymbol{R}_1\boldsymbol{R}_2$ and $\boldsymbol{R}_2\boldsymbol{R}_1$, and deduce certain trigonometrical formulae.

4 Decompose the following as the products of two matrices representing 'simple' two-dimensional transformations such as rotations, reflections, enlargements, etc.:

(*i*) $\begin{pmatrix} -3 & 0 \\ 0 & -3 \end{pmatrix}$; (*ii*) $\begin{pmatrix} 0 & -3 \\ -3 & 0 \end{pmatrix}$; (*iii*) $\begin{pmatrix} 1 & \sqrt{3} \\ \sqrt{3} & -1 \end{pmatrix}$;

(*iv*) $\begin{pmatrix} 5 & -12 \\ 12 & 5 \end{pmatrix}$; (*v*) $\begin{pmatrix} 2 & -3 \\ 0 & 2 \end{pmatrix}$; (*vi*) $\begin{pmatrix} 0 & -2 \\ 3 & 0 \end{pmatrix}$;

(*vii*) $\begin{pmatrix} 1 & 1 \\ -1 & 3 \end{pmatrix}$.

5 Repeat the above question with the 3×3 matrices:

(*i*) $\begin{pmatrix} 0 & 0 & -2 \\ 0 & -2 & 0 \\ -2 & 0 & 0 \end{pmatrix}$; (*ii*) $\begin{pmatrix} -1 & 0 & 0 \\ 0 & 0 & 1 \\ 0 & 3 & 0 \end{pmatrix}$; (*iii*) $\begin{pmatrix} 0 & 2 & 0 \\ 2 & 0 & 0 \\ 0 & 0 & 2 \end{pmatrix}$;

(*iv*) $\begin{pmatrix} 0 & 1 & 0 \\ 1 & 0 & 0 \\ 0 & 0 & 0 \end{pmatrix}$.

6 Show that if $\boldsymbol{M} = \begin{pmatrix} 6 & -4 & 3 \\ 29 & -20 & 13 \\ 33 & -23 & 14 \end{pmatrix}$,

then $\boldsymbol{M}\begin{pmatrix} 1 \\ 2 \\ 1 \end{pmatrix} = \begin{pmatrix} 1 \\ 2 \\ 1 \end{pmatrix}$ and $\boldsymbol{M}^3 = \boldsymbol{I}$.

What do you suspect?

7 Find vectors ('eigenvectors') whose directions are unaltered by

$$N = \begin{pmatrix} 1 & -2 & 0 \\ 1 & -1 & 0 \\ -1 & 0 & 1 \end{pmatrix}.$$

Show that $N^4 = I$, but that N does *not* represent a quarter-turn.

8 Experiment with the composition of rotations of period 2 (half-turns), 3 and 4 about various axes as was done in the last paragraph of the text, and also with reflections in various planes.

(A matrix M is said to have 'period n' if $M^r = I$ when $r = n$, but $M^r \neq I$ when $r < n$.)

9 Construct 2×2 and 3×3 matrices with period 2, 3, 4, 5, ...

10 If a 3×3 matrix has the property $M^2 = I$ and $MX = X$, show that it may well *not* represent a half-turn about the vector X. What conditions are needed in addition to the two given in order that it may represent a half-turn?

11 M is a certain 3×3 matrix whose elements are 0, 1 or -1, and each row and each column of M contains exactly one non-zero element. Prove that $M^2, M^3, \ldots, M^n$ are all of the same form, and deduce that $M^h = I$ for some positive integer $h \leqslant 48$. Interpret the action of M on a vector (x, y, z) geometrically. (M.E.I.)

12 If the matrix of a plane translation is

$$T = \begin{pmatrix} 1 & 0 & a \\ 0 & 1 & b \\ 0 & 0 & 1 \end{pmatrix}, \quad \text{show that} \quad T^n = \begin{pmatrix} 1 & 0 & an \\ 0 & 1 & bn \\ 0 & 0 & 1 \end{pmatrix}.$$

Investigate, by using 3×3 matrices the composition of translations with other types of plane transformations.

13 Interpret the plane transformations

$$\begin{pmatrix} x' \\ y' \\ 1 \end{pmatrix} = \begin{pmatrix} 1 & 0 & 0 \\ 0 & 1 & k \\ 0 & 0 & 1 \end{pmatrix} \begin{pmatrix} \cos\theta & -\sin\theta & 0 \\ \sin\theta & \cos\theta & 0 \\ 0 & 0 & 1 \end{pmatrix} \begin{pmatrix} 1 & 0 & 0 \\ 0 & -1 & 0 \\ 0 & 0 & 1 \end{pmatrix} \begin{pmatrix} x \\ y \\ 1 \end{pmatrix},$$

and expand the product of the three 3×3 matrices.

13.4 Inverse transformations and matrices

So far we have discussed the effect of a particular transformation on a given vector: 'What becomes of it?' We now ask the reverse question: 'Under such and such a transformation, from where does a given vector originate?'

Or, in matrix notation,

if $\boldsymbol{x}' = \boldsymbol{M}\boldsymbol{x}$, can we find a matrix $\boldsymbol{N}$ such that $\boldsymbol{x} = \boldsymbol{N}\boldsymbol{x}'$?

If so, it will clearly follow that

$$\boldsymbol{x} = \boldsymbol{N}\boldsymbol{x}' = \boldsymbol{N}\boldsymbol{M}\boldsymbol{x}$$

and $\boldsymbol{x}' = \boldsymbol{M}\boldsymbol{x} = \boldsymbol{M}\boldsymbol{N}\boldsymbol{x}'$.

So, if these are true for all vectors $\boldsymbol{x}$, $\boldsymbol{x}'$, then $\boldsymbol{MN} = \boldsymbol{NM} = \boldsymbol{I}$†

and our problem is to find $\boldsymbol{N}$ which is *inverse* of $\boldsymbol{M}$ (and which we shall write $\boldsymbol{M}^{-1}$).

Two dimensional case

It will be recalled (from 0.8) that in the two-dimensional case the transformation

$$\begin{pmatrix} x' \\ y' \end{pmatrix} = \begin{pmatrix} a & b \\ c & d \end{pmatrix} \begin{pmatrix} x \\ y \end{pmatrix}$$

had such an inverse, provided that its determinant $\Delta = ad - bc$ was non-zero.

If $\boldsymbol{M} = \begin{pmatrix} a & b \\ c & d \end{pmatrix}$, then $\boldsymbol{M}^{-1} = \dfrac{1}{ad - bc}\begin{pmatrix} d & -b \\ -c & a \end{pmatrix}$,

$$\text{so that}\quad \boldsymbol{M}^{-1}\boldsymbol{M} = \frac{1}{ad - bc}\begin{pmatrix} d & -b \\ -c & a \end{pmatrix}\begin{pmatrix} a & b \\ c & d \end{pmatrix}$$

$$= \frac{1}{ad - bc}\begin{pmatrix} ad - bc & 0 \\ 0 & ad - bc \end{pmatrix}$$

$$= \begin{pmatrix} 1 & 0 \\ 0 & 1 \end{pmatrix} = \boldsymbol{I} \quad (\text{and similarly } \boldsymbol{M}\boldsymbol{M}^{-1} = \boldsymbol{I}).$$

If, however, $\Delta = 0$, there was no such inverse, and the transformation together with its matrix $\boldsymbol{M}$ were said to be *singular*.

Three-dimensional case

Let us now consider the transformation

$$\left.\begin{aligned} x' &= a_1 x + b_1 y + c_1 z \\ y' &= a_2 x + b_2 y + c_2 z \\ z' &= a_3 x + b_3 y + c_3 z \end{aligned}\right\} \quad \text{or} \quad \begin{pmatrix} x' \\ y' \\ z' \end{pmatrix} = \begin{pmatrix} a_1 & b_1 & c_1 \\ a_2 & b_2 & c_2 \\ a_3 & b_3 & c_3 \end{pmatrix}\begin{pmatrix} x \\ y \\ z \end{pmatrix}.$$

† We shall use $\boldsymbol{I}$ to denote the identity (or unit) matrix of any order. So here, for instance,

$\boldsymbol{I} = \begin{pmatrix} 1 & 0 & 0 \\ 0 & 1 & 0 \\ 0 & 0 & 1 \end{pmatrix}$; but in the two-dimensional case which follows, $\boldsymbol{I} = \begin{pmatrix} 1 & 0 \\ 0 & 1 \end{pmatrix}$.

Our problem is to find the vector $\begin{pmatrix} x \\ y \\ z \end{pmatrix}$ from which the vector $\begin{pmatrix} x' \\ y' \\ z' \end{pmatrix}$ originated.

We start by asking the simpler question of where the vector $\begin{pmatrix} 1 \\ 0 \\ 0 \end{pmatrix}$ originated,

i.e., we try to find $\begin{pmatrix} x \\ y \\ z \end{pmatrix}$ such that

$$a_1x + b_1y + c_1z = 1 \tag{1}$$

$$a_2x + b_2y + c_2z = 0 \tag{2}$$

$$a_3x + b_3y + c_3z = 0. \tag{3}$$

Looking at the last two of these, we see that they are requiring us to find a vector $\begin{pmatrix} x \\ y \\ z \end{pmatrix}$ which is perpendicular to both $\begin{pmatrix} a_2 \\ b_2 \\ c_2 \end{pmatrix}$ and $\begin{pmatrix} a_3 \\ b_3 \\ c_3 \end{pmatrix}$.

Now we know (see 11.5) that their *vector product* is precisely such a vector, and this we abbreviate by writing the vector product thus:

$$\begin{pmatrix} b_2c_3 - b_3c_2 \\ c_2a_3 - c_3a_2 \\ a_2b_3 - a_3b_2 \end{pmatrix} = \begin{pmatrix} A_1 \\ B_1 \\ C_1 \end{pmatrix}.$$

Similarly, we write

$$\begin{pmatrix} b_3c_1 - b_1c_3 \\ c_3a_1 - c_1a_3 \\ a_3b_1 - a_1b_3 \end{pmatrix} = \begin{pmatrix} A_2 \\ B_2 \\ C_2 \end{pmatrix} \quad \text{and} \quad \begin{pmatrix} b_1c_2 - b_2c_1 \\ c_1a_2 - c_2a_1 \\ a_1b_2 - a_2b_1 \end{pmatrix} = \begin{pmatrix} A_3 \\ B_3 \\ C_3 \end{pmatrix}.$$

We now consider the matrix $\boldsymbol{M}^*$ formed from these capital letters, with the A's forming the first row:

$$\boldsymbol{M}^* = \begin{pmatrix} A_1 & A_2 & A_3 \\ B_1 & B_2 & B_3 \\ C_1 & C_2 & C_3 \end{pmatrix}.$$

This is called the *adjoint matrix*, and we now consider the product

$$\boldsymbol{MM}^* = \begin{pmatrix} a_1 & b_1 & c_1 \\ a_2 & b_2 & c_2 \\ a_3 & b_3 & c_3 \end{pmatrix}\begin{pmatrix} A_1 & A_2 & A_3 \\ B_1 & B_2 & B_3 \\ C_1 & C_2 & C_3 \end{pmatrix}$$

$$= \begin{pmatrix} a_1A_1+b_1B_1+c_1C_1 & a_1A_2+b_1B_2+c_1C_2 & a_1A_3+b_1B_3+c_1C_3 \\ a_2A_1+b_2B_1+c_2C_1 & a_2A_2+b_2B_2+c_2C_2 & a_2A_3+b_2B_3+c_2C_3 \\ a_3A_1+b_3B_1+c_3C_1 & a_3A_2+b_3B_2+c_3C_2 & a_3A_3+b_3B_3+c_3C_3 \end{pmatrix}.$$

Now $\begin{pmatrix} A_1 \\ B_1 \\ C_1 \end{pmatrix}$ is, by definition, perpendicular to $\begin{pmatrix} a_2 \\ b_2 \\ c_2 \end{pmatrix}$ and $\begin{pmatrix} a_3 \\ b_3 \\ c_3 \end{pmatrix}$.

So $a_2A_1 + b_2B_1 + c_2C_1 = 0$ and $a_3A_1 + b_3B_1 + c_3C_1 = 0$ (as may be verified by direct substitution). Indeed, the only terms in $\boldsymbol{MM}^*$ which do not vanish are those on the leading diagonal:

$$a_1A_1 + b_1B_1 + c_1C_1, \quad a_2A_2 + b_2B_2 + c_2C_2, \quad a_3A_3 + b_3B_3 + c_3C_3.$$

But each of these is quickly seen to be equal to the expression

$$\Delta = a_1b_2c_3 - a_1b_3c_2 + b_1c_2a_3 - b_1c_3a_2 + c_1a_2b_3 - c_1a_3b_2,$$

which we have already called the *determinant* of $\boldsymbol{M}$.

$$\text{So} \quad \boldsymbol{MM}^* = \begin{pmatrix} \Delta & 0 & 0 \\ 0 & \Delta & 0 \\ 0 & 0 & \Delta \end{pmatrix} = \Delta\boldsymbol{I}, \text{ where } \Delta = \det(\boldsymbol{M})$$

and it can similarly be proved that

$$\boldsymbol{M}^*\boldsymbol{M} = \Delta\boldsymbol{I}.$$

Now if $\Delta \neq 0$, it follows that

$$\frac{\boldsymbol{M}^*}{\Delta}\boldsymbol{M} = \boldsymbol{M}\frac{\boldsymbol{M}^*}{\Delta} = \boldsymbol{I},$$

so that $\boldsymbol{M}^*/\Delta$ is an inverse of $\boldsymbol{M}$ for both pre- and post-multiplication and we can write

$$\boldsymbol{M}^{-1} = \frac{\boldsymbol{M}^*}{\Delta} \quad (\text{provided } \Delta \neq 0).$$

It is immediately clear that $\boldsymbol{M}^{-1}$ is unique, for if we suppose that $\boldsymbol{L}$ is another inverse for pre-multiplication (or *left inverse*) such that $\boldsymbol{LM} = \boldsymbol{I}$,

then $\boldsymbol{M}^{-1} = \boldsymbol{I}\boldsymbol{M}^{-1} = (\boldsymbol{L}\boldsymbol{M})\boldsymbol{M}^{-1} = \boldsymbol{L}(\boldsymbol{M}\boldsymbol{M}^{-1})\dagger = \boldsymbol{L}\boldsymbol{I} = \boldsymbol{L}$
and $\boldsymbol{L}$ is identical with $\boldsymbol{M}^{-1}$.

Similarly, $\boldsymbol{M}^{-1}$ is the unique inverse for post-multiplication.
If, however, $\Delta = 0$, then $\boldsymbol{M}$ has no such inverse and is said to be *singular*.

Example 1

Find the adjoints and the inverses of the matrices:

$$(i)\ \boldsymbol{A} = \begin{pmatrix} 1 & 0 & 0 \\ 0 & 0 & 2 \\ 0 & -1 & 0 \end{pmatrix}; \qquad (ii)\ \boldsymbol{B} = \begin{pmatrix} 0 & 1 & 2 \\ 1 & 2 & 3 \\ 2 & 3 & 4 \end{pmatrix}.$$

$$(i)\ \text{Since}\quad \boldsymbol{A} = \begin{pmatrix} 1 & 0 & 0 \\ 0 & 0 & 2 \\ 0 & -1 & 0 \end{pmatrix},\quad \text{we see that}\quad \boldsymbol{A}^* = \begin{pmatrix} 2 & 0 & 0 \\ 0 & 0 & -2 \\ 0 & 1 & 0 \end{pmatrix}$$

$$\Rightarrow \boldsymbol{A}\boldsymbol{A}^* = \begin{pmatrix} 1 & 0 & 0 \\ 0 & 0 & 2 \\ 0 & -1 & 0 \end{pmatrix}\begin{pmatrix} 2 & 0 & 0 \\ 0 & 0 & -2 \\ 0 & 1 & 0 \end{pmatrix} = \begin{pmatrix} 2 & 0 & 0 \\ 0 & 2 & 0 \\ 0 & 0 & 2 \end{pmatrix} = 2\boldsymbol{I}$$

$$\Rightarrow \quad \boldsymbol{A}^{-1} = \tfrac{1}{2}\boldsymbol{A}^* = \begin{pmatrix} 1 & 0 & 0 \\ 0 & 0 & -1 \\ 0 & \frac{1}{2} & 0 \end{pmatrix}.$$

$$(ii)\ \text{Since}\quad \boldsymbol{B} = \begin{pmatrix} 0 & 1 & 2 \\ 1 & 2 & 3 \\ 2 & 3 & 4 \end{pmatrix},\quad \text{we see that}\quad \boldsymbol{B}^* = \begin{pmatrix} -1 & 2 & -1 \\ 2 & -4 & 2 \\ -1 & 2 & -1 \end{pmatrix}$$

$$\Rightarrow \quad \boldsymbol{B}\boldsymbol{B}^* = \begin{pmatrix} 0 & 1 & 2 \\ 1 & 2 & 3 \\ 2 & 3 & 4 \end{pmatrix}\begin{pmatrix} -1 & 2 & -1 \\ 2 & -4 & 2 \\ -1 & 2 & -1 \end{pmatrix} = \begin{pmatrix} 0 & 0 & 0 \\ 0 & 0 & 0 \\ 0 & 0 & 0 \end{pmatrix} = \boldsymbol{0}.$$

Hence $\Delta = 0$, so that $\boldsymbol{B}$ is singular and does not possess an inverse.

Example 2

If $\boldsymbol{A}$ and $\boldsymbol{B}$ are both non-singular matrices, show that:
$(i)\ (\boldsymbol{A}^{-1})^{-1} = \boldsymbol{A}$; $\qquad (ii)\ (\boldsymbol{A}\boldsymbol{B})^{-1} = \boldsymbol{B}^{-1}\boldsymbol{A}^{-1}$.

(i) The inverse of $\boldsymbol{A}$ is $\boldsymbol{A}^{-1}$,

so $\boldsymbol{A}\boldsymbol{A}^{-1} = \boldsymbol{A}^{-1}\boldsymbol{A} = \boldsymbol{I}$.

† Using the associativity property of matrix multiplication, (see Exercise 13.3a).

Hence the inverse of $\boldsymbol{A}^{-1}$ is $\boldsymbol{A}$

and $(\boldsymbol{A}^{-1})^{-1} = \boldsymbol{A}.$

(ii) $(\boldsymbol{AB})(\boldsymbol{B}^{-1}\boldsymbol{A}^{-1}) = \boldsymbol{A}(\boldsymbol{BB}^{-1})\boldsymbol{A}^{-1} = \boldsymbol{AIA}^{-1} = \boldsymbol{AA}^{-1} = \boldsymbol{I}$

and $(\boldsymbol{B}^{-1}\boldsymbol{A}^{-1})(\boldsymbol{AB}) = \boldsymbol{B}^{-1}(\boldsymbol{A}^{-1}\boldsymbol{A})\boldsymbol{B} = \boldsymbol{B}^{-1}\boldsymbol{IB} = \boldsymbol{B}^{-1}\boldsymbol{B} = \boldsymbol{I}.$

So the inverse of $\boldsymbol{AB}$ is $\boldsymbol{B}^{-1}\boldsymbol{A}^{-1}$

or $(\boldsymbol{AB})^{-1} = \boldsymbol{B}^{-1}\boldsymbol{A}^{-1}.$

Example 3

Given the transformation

$$\begin{aligned} x' &= x + 2y + 3z \\ y' &= 3x + y + 2z \\ z' &= 2x + 3y + z, \end{aligned}$$

find: *(i)* its inverse transformation;
(ii) the point from which $(1, 2, 3)$ arises under the given transformation.

The matrix of the original transformation is

$$\boldsymbol{A} = \begin{pmatrix} 1 & 2 & 3 \\ 3 & 1 & 2 \\ 2 & 3 & 1 \end{pmatrix},$$

and we can easily find its adjoint matrix

$$\boldsymbol{A}^* = \begin{pmatrix} -5 & 7 & 1 \\ 1 & -5 & 7 \\ 7 & 1 & -5 \end{pmatrix}.$$

Furthermore, it is seen that

$$\boldsymbol{AA}^* = \begin{pmatrix} 1 & 2 & 3 \\ 3 & 1 & 2 \\ 2 & 3 & 1 \end{pmatrix}\begin{pmatrix} -5 & 7 & 1 \\ 1 & -5 & 7 \\ 7 & 1 & -5 \end{pmatrix} = \begin{pmatrix} 18 & 0 & 0 \\ 0 & 18 & 0 \\ 0 & 0 & 18 \end{pmatrix} = 18\boldsymbol{I}.$$

So the determinant Δ of the transformation is 18, the transformation is non-singular and

$$\boldsymbol{A}^{-1} = \frac{\boldsymbol{A}^*}{\Delta} = \frac{1}{18}\begin{pmatrix} -5 & 7 & 1 \\ 1 & -5 & 7 \\ 7 & 1 & -5 \end{pmatrix}.$$

Hence the inverse transformation is

$$x = \tfrac{1}{18}(-5x' + 7y' + z')$$
$$y = \tfrac{1}{18}(x' - 5y' + 7z')$$
$$z = \tfrac{1}{18}(7x' + y' - 5z').$$

Finally, putting $x' = 1$, $y' = 2$, $z' = 3$ we see that the point $(1, 2, 3)$ arises from $x = \frac{2}{3}$, $y = \frac{2}{3}$, $z = -\frac{1}{3}$, i.e., from the point $(\frac{2}{3}, \frac{2}{3}, -\frac{1}{3})$.

Example 4

Solve the equations

$$a_1x + b_1y + c_1z = d_1$$
$$a_2x + b_2y + c_2z = d_2$$
$$a_3x + b_3y + c_3z = d_3\,.$$

These may be written as

$$\boldsymbol{Mx} = \boldsymbol{d},$$

where $\boldsymbol{M} = \begin{pmatrix} a_1 & b_1 & c_1 \\ a_2 & b_2 & c_2 \\ a_3 & b_3 & c_3 \end{pmatrix}$ and $\boldsymbol{d} = \begin{pmatrix} d_1 \\ d_2 \\ d_3 \end{pmatrix}$.

Case (i) If $\boldsymbol{M}$ is non-singular, then it possesses an inverse $\boldsymbol{M}^{-1}$ and

$$\begin{aligned} & \boldsymbol{Mx} = \boldsymbol{d} \\ \Rightarrow\quad & \boldsymbol{M}^{-1}(\boldsymbol{Mx}) = \boldsymbol{M}^{-1}\boldsymbol{d} \\ \Rightarrow\quad & (\boldsymbol{M}^{-1}\boldsymbol{M})\boldsymbol{x} = \boldsymbol{M}^{-1}\boldsymbol{d} \\ \Rightarrow\quad & \boldsymbol{Ix} = \boldsymbol{M}^{-1}\boldsymbol{d} \\ \Rightarrow\quad & \boldsymbol{x} = \boldsymbol{M}^{-1}\boldsymbol{d}. \end{aligned}$$

In a particular case, with numerical values for all the a's b's and c's, we could find the matrix $\boldsymbol{M}^{-1}$ and hence solve the equations. But this is a tedious process and, though we shall investigate it further (in 13.6), we shall also (in 13.7) learn how more practical methods can be employed.

Case (ii) If $\boldsymbol{M}$ is singular, the above method breaks down as $\boldsymbol{M}^{-1}$ does not exist. But we shall see (in 13.6) that a useful result can still be obtained simply by using the adjoint matrix $\boldsymbol{M}^*$ rather than $\boldsymbol{M}^*/\Delta$.

Example 5

Given $\quad l_1^2 + m_1^2 = l_2^2 + m_2^2 = 1 \quad$ and $\quad l_1l_2 + m_1m_2 = 0$;
prove that $\quad l_1^2 + l_2^2 = m_1^2 + m_2^2 = 1 \quad$ and $\quad l_1m_1 + l_2m_2 = 0$.

First of all the reader should attempt to prove this by ordinary manipulation of the equations. Even if successful, it will be agreed that the task is vexing. But how much simpler it is to use matrices. For if we let

$$A = \begin{pmatrix} l_1 & m_1 \\ l_2 & m_2 \end{pmatrix} \quad \text{and} \quad B = \begin{pmatrix} l_1 & l_2 \\ m_1 & m_2 \end{pmatrix},$$

we immediately see that

$$AB = \begin{pmatrix} l_1 & m_1 \\ l_2 & m_2 \end{pmatrix}\begin{pmatrix} l_1 & l_2 \\ m_1 & m_2 \end{pmatrix} = \begin{pmatrix} l_1^2 + m_1^2 & l_1 l_2 + m_1 m_2 \\ l_1 l_2 + m_1 m_2 & l_2^2 + m_2^2 \end{pmatrix}$$

$$= \begin{pmatrix} 1 & 0 \\ 0 & 1 \end{pmatrix} = I.$$

Hence A and B are inverses, so that

$$BA = \begin{pmatrix} l_1 & l_2 \\ m_1 & m_2 \end{pmatrix}\begin{pmatrix} l_1 & m_1 \\ l_2 & m_2 \end{pmatrix} = \begin{pmatrix} l_1^2 + l_2^2 & l_1 m_1 + l_2 m_2 \\ l_1 m_1 + l_2 m_2 & m_1^2 + m_2^2 \end{pmatrix} = \begin{pmatrix} 1 & 0 \\ 0 & 1 \end{pmatrix}$$

$$\Rightarrow \qquad l_1^2 + l_2^2 = m_1^2 + m_2^2 = 1 \quad \text{and} \quad l_1 m_1 + l_2 m_2 = 0.$$

Should it be protested that the saving in work is not very great, consider an extension of the given problem into three dimensions:

Given the equations

$$l_1^2 + m_1^2 + n_1^2 = 1 \qquad (1) \qquad\qquad l_2 l_3 + m_2 m_3 + n_2 n_3 = 0 \qquad (4)$$

$$l_2^2 + m_2^2 + n_2^2 = 1 \qquad (2) \qquad\qquad l_3 l_1 + m_3 m_1 + n_3 n_1 = 0 \qquad (5)$$

$$l_3^2 + m_3^2 + n_3^2 = 1 \qquad (3) \qquad\qquad l_1 l_2 + m_1 m_2 + n_1 n_2 = 0 \qquad (6)$$

prove that

$$l_1^2 + l_2^2 + l_3^2 = 1; \qquad m_1^2 + m_2^2 + m_3^2 = 1; \qquad n_1^2 + n_2^2 + n_3^2 = 1;$$

$$\text{and} \quad m_1 n_1 + m_2 n_2 + m_3 n_3 = n_1 l_1 + n_2 l_2 + n_3 l_3 = l_1 m_1 + l_2 m_2 + l_3 m_3 = 0.$$

To attempt this by 'ordinary algebra' would be prohibitive. By matrices, it is scarcely any more difficult than the 2 by 2 case above. For we may take

$$A = \begin{pmatrix} l_1 & m_1 & n_1 \\ l_2 & m_2 & n_2 \\ l_3 & m_3 & n_3 \end{pmatrix},$$

and B to be its 'transpose' (i.e., with rows and columns interchanged), and the reader will have no difficulty in completing the proof on the lines of the 2×2 case. Note that equations (1)–(6) are represented by the *single* matrix equation $AB = I$.

Exercise 13.4a

1 Obtain the inverse of $\begin{pmatrix} a & b \\ c & d \end{pmatrix}$:

(*i*) by letting it be $\begin{pmatrix} p & q \\ r & s \end{pmatrix}$;

(*ii*) by solving the equations $\begin{aligned} ax + by &= x' \\ cx + dy &= y' \end{aligned}$

to obtain x and y in terms of x' and y'.

2 Find the inverse of $\begin{pmatrix} 3 & -2 \\ -2 & 2 \end{pmatrix}$ and hence find the quadrilateral which is mapped by this matrix into the unit square $\begin{pmatrix} \pm 1 \\ \pm 1 \end{pmatrix}$.

3 Show that the inverse of the product $\boldsymbol{ABCD}\ldots$ is $\ldots\boldsymbol{D}^{-1}\boldsymbol{C}^{-1}\boldsymbol{B}^{-1}\boldsymbol{A}^{-1}$, where $\boldsymbol{A}, \boldsymbol{B}, \boldsymbol{C}, \boldsymbol{D}, \ldots$ are square matrices of the same order.

4 A 2×2 matrix $\boldsymbol{M}$ has the property that

$$\boldsymbol{M}\begin{pmatrix} a \\ b \end{pmatrix} = \begin{pmatrix} p \\ q \end{pmatrix} \quad \text{and} \quad \boldsymbol{M}\begin{pmatrix} c \\ d \end{pmatrix} = \begin{pmatrix} r \\ s \end{pmatrix}.$$

Prove that, provided a, b, c, d satisfy a certain restriction, then $\boldsymbol{M} = \boldsymbol{V}\boldsymbol{U}^{-1}$,

where $\boldsymbol{U} = \begin{pmatrix} a & c \\ b & d \end{pmatrix}$ and $\boldsymbol{V} = \begin{pmatrix} p & r \\ q & s \end{pmatrix}$.

State the restriction on a, b, c, d. (S.M.P.)

5 $\boldsymbol{A}$ and $\boldsymbol{B}$ are matrices such that $\boldsymbol{AB} = \boldsymbol{A}$, $\boldsymbol{BA} = \boldsymbol{B}$. If $\boldsymbol{A}$ is non-singular, prove that $\boldsymbol{A} = \boldsymbol{B} = \boldsymbol{I}$. If $\boldsymbol{A}$ is singular, prove $\boldsymbol{B}$ is also singular, and that $(\boldsymbol{A} - \boldsymbol{B})^2 = \boldsymbol{0}$. (M.E.I.)

6 Find the adjoint and the inverse of $\begin{pmatrix} 2 & -1 & 4 \\ 4 & 0 & 2 \\ 3 & -2 & 7 \end{pmatrix}$ and hence solve the equations

$$\begin{aligned} 2x - y + 4z &= 1 \\ 4x + 2z &= -1 \\ 3x - 2y + 7z &= -3. \end{aligned}$$

7 Find the adjoint and inverse of $\boldsymbol{M} = \begin{pmatrix} 1 & 2 & 0 \\ -1 & 3 & 1 \\ 0 & 4 & 2 \end{pmatrix}$ and use to find the column vector which, when transformed by $\boldsymbol{M}$, becomes $\begin{pmatrix} -2 \\ 1 \\ 1 \end{pmatrix}$.

8 Find the inverse of the triangular matrix $\begin{pmatrix} 1 & a & b \\ 0 & 1 & c \\ 0 & 0 & 1 \end{pmatrix}$.

9 Find the inverse of $\begin{pmatrix} a & b & c \\ c & a & b \\ b & c & a \end{pmatrix}$ and state under what circumstances it is singular.

10 Solve $\begin{matrix} a + b = 3 \\ b + c = 4 \\ c + a = 5 \end{matrix}$ by first inverting the matrix $\begin{pmatrix} 1 & 1 & 0 \\ 0 & 1 & 1 \\ 1 & 0 & 1 \end{pmatrix}$.

11 Under what conditions does $\boldsymbol{M} = \boldsymbol{M}^{-1}$, where $\boldsymbol{M}$ is:

(i) $\begin{pmatrix} a & b \\ c & d \end{pmatrix}$; (ii) $\begin{pmatrix} a_1 & b_1 & c_1 \\ a_2 & b_2 & c_2 \\ a_3 & b_3 & c_3 \end{pmatrix}$?

Which types of geometrical transformations may such matrices represent?

12 Invert the matrix $\boldsymbol{M} = \begin{pmatrix} 1 & 1 & 0 \\ 5 & 1 & -3 \\ 2 & 7 & 4 \end{pmatrix}$ and also the transposed matrix $\boldsymbol{M}' = \begin{pmatrix} 1 & 5 & 2 \\ 1 & 1 & 7 \\ 0 & -3 & 4 \end{pmatrix}$.

Can you draw any general conclusion about the inverse of the transpose of a 3×3 matrix? Deduce that the inverse of a symmetric matrix (i.e., one for which $\boldsymbol{M}' = \boldsymbol{M}$) is itself symmetric.

13 A matrix which satisfies $\boldsymbol{MM}' = \boldsymbol{M}'\boldsymbol{M} = \boldsymbol{I}$, i.e., whose inverse is its transpose, is called an *orthogonal* matrix. Investigate such matrices (i) in the 2×2 case (ii) in the 3×3 case, and give examples of each.

14 Show that if M is an orthogonal 3×3 matrix, and $\boldsymbol{MX} = \boldsymbol{X}'$, where $\boldsymbol{X}$, $\boldsymbol{X}'$ are column 3-vectors, then these vectors have the same *length*, i.e.,

if $\boldsymbol{X} = \begin{pmatrix} x \\ y \\ z \end{pmatrix}$ and $\boldsymbol{X}' = \begin{pmatrix} x' \\ y' \\ z' \end{pmatrix}$, then $x^2 + y^2 + z^2 = x'^2 + y'^2 + z'^2$.

15 Find the inverse of $\begin{pmatrix} \cos\alpha\cos\beta & \sin\alpha\cos\beta & -\sin\beta \\ \cos\alpha\sin\beta & \sin\alpha\sin\beta & \cos\beta \\ \sin\alpha & -\cos\alpha & 0 \end{pmatrix}$.

How do you account for the result geometrically?

16 If $\boldsymbol{A}$ is a matrix of period n (i.e., $\boldsymbol{A}^r = \boldsymbol{I}$ when $r = n$, but $\boldsymbol{A}^r \neq \boldsymbol{I}$ when $r < n$), show that $\boldsymbol{BAB}^{-1}$ is also of period n, where $\boldsymbol{B}$ is another square matrix of the same order as $\boldsymbol{A}$.

17 Further practice for finding inverses:

$$(i)\ \begin{pmatrix} 1 & 4 & -3 \\ -2 & -5 & 7 \\ 3 & 3 & -10 \end{pmatrix},\quad (ii)\ \begin{pmatrix} 1 & 2 & 3 \\ 0 & 1 & 2 \\ 0 & 0 & 1 \end{pmatrix},\quad (iii)\ \begin{pmatrix} -15 & 11 & 7 \\ 11 & -8 & -5 \\ -1 & 2 & 1 \end{pmatrix},$$

$$(iv)\ \begin{pmatrix} 6 & 0 & -1 \\ 3 & -3 & -2 \\ -3 & 6 & 3 \end{pmatrix},\quad (v)\ \begin{pmatrix} 2 & -1 & 0 \\ -1 & 2 & -1 \\ 0 & -1 & 1 \end{pmatrix},\quad (vi)\ \begin{pmatrix} 1 & a & a \\ a & 1 & a \\ a & a & 1 \end{pmatrix},$$

$$(vii)\ \begin{pmatrix} 1 & a & b \\ a & 1 & c \\ b & c & 1 \end{pmatrix},\quad (viii)\ \begin{pmatrix} i & 1 \\ -1 & -i \end{pmatrix},\quad (ix)\ \begin{pmatrix} \omega & i \\ i & \omega^2 \end{pmatrix},$$

$$(x)\ \begin{pmatrix} 1 & \omega & \omega^2 \\ \omega^2 & 1 & \omega \\ \omega & \omega^2 & 1 \end{pmatrix},\quad \text{where}\quad \begin{matrix} i^2 = -1 \\ \omega^3 = 1. \end{matrix}$$

2×2 linear equations: three interpretations

At the beginning of this section, we saw how the inverse of a 2×2 matrix helped in the solution of two simultaneous equations. Following this, we considered briefly the solution of three simultaneous linear equations and how the inverse of a 3×3 matrix might be found and used. The whole question of the solution of linear equations will be dealt with in more detail in 13.6 and 13.7, but this is a convenient point at which to make some preliminary observations in the case of two equations.

Consider the equations

$$\left.\begin{aligned} a_1x + b_1y &= c_1 \\ a_2x + b_2y &= c_2 \end{aligned}\right\}. \tag{1}$$

If $\boldsymbol{M} = \begin{pmatrix} a_1 & b_1 \\ a_2 & b_2 \end{pmatrix}$, $\quad \boldsymbol{x} = \begin{pmatrix} x \\ y \end{pmatrix}$, $\quad \boldsymbol{c} = \begin{pmatrix} c_1 \\ c_2 \end{pmatrix}$,

these may be written as

$$\begin{pmatrix} a_1 & b_1 \\ a_2 & b_2 \end{pmatrix}\begin{pmatrix} x \\ y \end{pmatrix} = \begin{pmatrix} c_1 \\ c_2 \end{pmatrix}, \quad \text{or} \quad \boldsymbol{Mx} = \boldsymbol{c}. \tag{2}$$

And if $\boldsymbol{a} = \begin{pmatrix} a_1 \\ a_2 \end{pmatrix}$, $\quad \boldsymbol{b} = \begin{pmatrix} b_1 \\ b_2 \end{pmatrix}$,

they can be written as

$$x\begin{pmatrix} a_1 \\ a_2 \end{pmatrix} + y\begin{pmatrix} b_1 \\ b_2 \end{pmatrix} = \begin{pmatrix} c_1 \\ c_2 \end{pmatrix}, \quad \text{or} \quad x\boldsymbol{a} + y\boldsymbol{b} = \boldsymbol{c}. \tag{3}$$

Each of these alternatives carries a different interpretation. The two separate equations (1) suggest two straight lines in the x, y plane, whose intersection one is trying to find. Equation (2) suggests that one is looking for an unknown vector $\boldsymbol{x} = \begin{pmatrix} x \\ y \end{pmatrix}$ which is transformed by the matrix $\boldsymbol{M} = \begin{pmatrix} a_1 & b_1 \\ a_2 & b_2 \end{pmatrix}$ into the known vector $\boldsymbol{c} = \begin{pmatrix} c_1 \\ c_2 \end{pmatrix}$: where did this particular vector originate under the given transformation? The final version (3) starts with three given vectors,

$$\boldsymbol{a} = \begin{pmatrix} a_1 \\ a_2 \end{pmatrix}, \quad \boldsymbol{b} = \begin{pmatrix} b_1 \\ b_2 \end{pmatrix}, \quad \boldsymbol{c} = \begin{pmatrix} c_1 \\ c_2 \end{pmatrix},$$

and we ask what multiples of the first two must be combined to give the third, or how can we 'mix' the $\boldsymbol{a}$ and $\boldsymbol{b}$ vectors to give the $\boldsymbol{c}$ vector?

In general, one expects to get a *unique solution* to each of these questions. Thus, in the case $\left.\begin{aligned} 3x + 2y &= 1 \\ 4x - y &= -6 \end{aligned}\right\}$,

we obtain $x = -1$, $y = 2$. This means

(1) that the point common to the two lines with given equations has coordinates $(-1, 2)$;

(2) that under the transformations represented by the matrix $\begin{pmatrix} 3 & 2 \\ 4 & -1 \end{pmatrix}$, the vector $\begin{pmatrix} 1 \\ -6 \end{pmatrix}$ arises from the vector $\begin{pmatrix} -1 \\ 2 \end{pmatrix}$;

(3) that in order to obtain the vector $\begin{pmatrix} 1 \\ -6 \end{pmatrix}$, one should take $-1\begin{pmatrix} 3 \\ 4 \end{pmatrix} + 2\begin{pmatrix} 2 \\ -1 \end{pmatrix}$. This is illustrated in the figure.

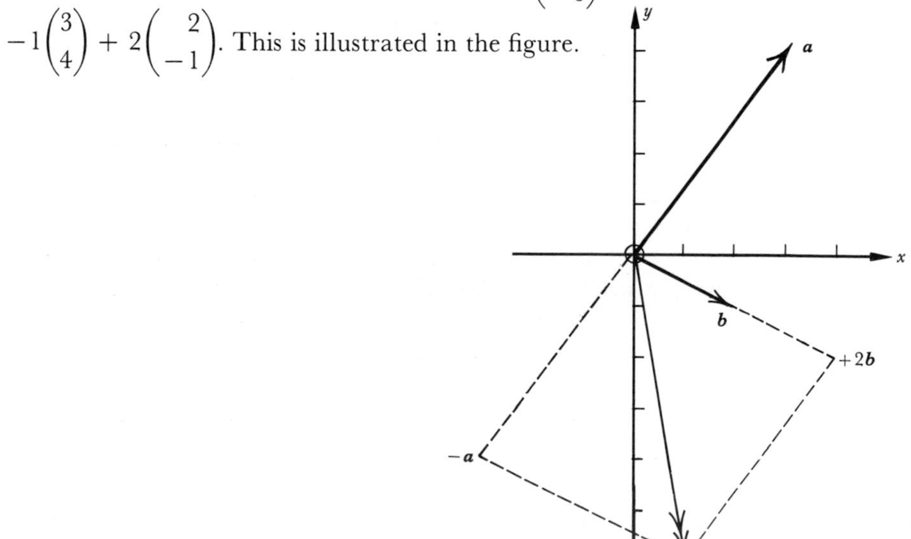

More generally, such a unique solution for x, y may always be found by the use of the inverse matrix *when this exists*, (i.e., when $\Delta = a_1b_2 - a_2b_1 \neq 0$). For then

$$\begin{pmatrix} x \\ y \end{pmatrix} = \frac{1}{a_1b_2 - a_2b_1}\begin{pmatrix} b_2 & -b_1 \\ -a_2 & a_1 \end{pmatrix}\begin{pmatrix} c_1 \\ c_2 \end{pmatrix}$$

$$\Rightarrow \quad x = \frac{b_2c_1 - b_1c_2}{a_1b_2 - a_2b_1}; \qquad y = \frac{a_1c_2 - a_2c_1}{a_1b_2 - a_2b_1}.$$

2 × 2 linear equations: singular case

The exceptional case gives two possibilities, illustrated by the examples:

$$(i)\ \begin{aligned} 3x - 2y &= 1 \\ 6x - 4y &= -1; \end{aligned} \qquad (ii)\ \begin{aligned} 3x - 2y &= 1 \\ 6x - 4y &= 2. \end{aligned}$$

Equations (*i*) have no finite solutions in x, y. The first of our three interpretations shows this by the fact that the lines are parallel, and so fail to intersect. The second interpretation is that the transformation represented by the matrix $\begin{pmatrix} 3 & -2 \\ 6 & -4 \end{pmatrix}$ is singular, and the effect on all vectors in the x, y plane is to take them into vectors having the direction $\begin{pmatrix} 1 \\ 2 \end{pmatrix}$, for

$$\begin{pmatrix} 3 & -2 \\ 6 & -4 \end{pmatrix}\begin{pmatrix} x \\ y \end{pmatrix} = \begin{pmatrix} 3x - 2y \\ 6x - 4y \end{pmatrix} = (3x - 2y)\begin{pmatrix} 1 \\ 2 \end{pmatrix}.$$

No conceivable values of x, y can cause this vector to be $\begin{pmatrix} 1 \\ -1 \end{pmatrix}$, which has a *different direction*. The third interpretation is that we cannot possibly take a certain multiple of $\boldsymbol{a} = \begin{pmatrix} 3 \\ 6 \end{pmatrix}$ and a certain multiple of $\boldsymbol{b} = \begin{pmatrix} -2 \\ -4 \end{pmatrix}$ and combine them to obtain $\boldsymbol{c} = \begin{pmatrix} 1 \\ -1 \end{pmatrix}$, simply because $\boldsymbol{a}$ and $\boldsymbol{b}$ have the same direction, i.e., lie in the same one-dimensional space (line), whereas $\boldsymbol{c}$ is outside that space.

Equations (*ii*), on the other hand, have an infinity of solutions for x, y. The lines, taking the first interpretation, are coincident, and so have an infinity of points in common. (One of the equations, you may prefer to think, is superfluous.) You should think out the second and third interpretations in case (*ii*) as well.

Note that the exceptional cases in the solution of $\boldsymbol{Mx} = \boldsymbol{c}$ arise precisely in those cases when the equations

$$\begin{aligned} a_1x + b_1y &= 0 \\ a_2x + b_2y &= 0 \end{aligned} \quad \text{or} \quad \boldsymbol{Mx} = \boldsymbol{0}$$

do possess solutions other than $\boldsymbol{x} = \boldsymbol{0}$. For if $\Delta \neq 0$, the only possible solution to the homogeneous equations

$$\begin{aligned} a_1x + b_1y &= 0 \\ a_2x + b_2y &= 0 \end{aligned} \quad \text{is} \quad x = 0,\ y = 0.$$

Exercise 13.4b

1 Give the three interpretations suggested in the text in the case of the equations:

$$(i)\quad \begin{aligned} x + 2y &= 2 \\ 3x - y &= 13; \end{aligned} \qquad (ii)\quad \begin{aligned} 5x - 6y &= 10\tfrac{1}{2} \\ -3x + 4y &= -6\tfrac{1}{2}; \end{aligned}$$

and draw diagrams to illustrate.

2 Apply the three interpretations to the equations

$$\begin{aligned} 2x - y &= a \\ 6x - 3y &= b \end{aligned} \quad \text{in the cases } \begin{pmatrix} a \\ b \end{pmatrix} = \begin{pmatrix} 4 \\ -3 \end{pmatrix} \text{ and } \begin{pmatrix} a \\ b \end{pmatrix} = \begin{pmatrix} -1 \\ -3 \end{pmatrix}$$

and give a parametric representation of the infinity of solutions in the latter case.

3 What multiples of the vectors $\begin{pmatrix} -3 \\ 2 \end{pmatrix}$ and $\begin{pmatrix} 8 \\ -1 \end{pmatrix}$ must be combined to give the vector $\begin{pmatrix} 46 \\ -9 \end{pmatrix}$?

4 Prove that the equations $a_1x + b_1y = 0$, $a_2x + b_2y = 0$ have an infinity of solutions if and only if $a_1b_2 - a_2b_1 = 0$. Prove however, that in this case, the equations $a_1x + b_1y = c_1$ and $a_2x + b_2y = c_2$ do not in general have a solution, and find under what conditions they do.

5 Show that the system of equations

$$\begin{aligned} 5x - 6y + 3 &= 0 \\ -3x + 4y - 2 &= 0 \end{aligned}$$

has a unique solution even though the matrix $\begin{pmatrix} 5 & -6 & 3 \\ -3 & 4 & -2 \end{pmatrix}$ has the final two columns linearly dependent.

Codes

Having established the idea of an inverse, we are now in a position to consider a possible application of matrices to the coding and decoding of messages. One obvious way of constructing a code is to use a number to stand for each letter of the alphabet, thus:

	A	B	C	D	E	F	G	H	I	J	K	L	M
Row 1	1	2	3	4	5	6	7	8	9	10	11	12	13
Row 2	3	6	9	12	15	18	21	24	1	4	7	10	13

	N	O	P	Q	R	S	T	U	V	W	X	Y	Z
Row 1	14	15	16	17	18	19	20	21	22	23	24	25	26
Row 2	16	19	22	25	2	5	8	11	14	17	20	23	26

Row 1 provides too transparent a method. Row 2 is an improvement. It might have been a *random* permutation of the numbers 1, 2, 3, . . ., 25, 26, but it will be seen that there is a 'system', whereby each number in row 1 has been multiplied by 3 and reduced modulo 26 where necessary, e.g., $21 \times 3 = 63 = 11 \pmod{26}$. The idea of this system is that it makes it easy to decode a message without having to keep referring to the crib. For, as coding is performed by multiplication of row 1 by 3, so decoding is carried out by division by 3, e.g., $11 \div 3 = 63 \div 3 \pmod{26} = 21$, giving U. We may write:

$$3 \times (\text{WORD}) \longrightarrow \text{CODED WORD}$$

$$3^{-1} \times (\text{CODED WORD}) \longrightarrow \text{WORD}$$

This method has the obvious disadvantage that it may be easily broken down on account of each letter of the alphabet being invariably represented by the same number — it is a simple 'replacement' code. It also suffers from the inconvenience of having to reduce modulo 26, a rather awkward number. The latter objection may be removed by adding four characters as shown:

.	,	?	□ (space)
27	28	29	0

Then it becomes necessary to discard multiples of 30, which is easier. To remove the former objection, we abandon row 2,† and re-arrange the message, say: 'COME QUICKLY TO MY ASSISTANCE' as follows:

C M □ U C L □ O M □ S I T N E □
O E Q I K Y T □ Y A S S A C . □

(□ means 'space'). Coding by row 1,† we set up a two-row matrix:

$$\boldsymbol{W} = \begin{pmatrix} 3 & 13 & 0 & 21 & 3 & 12 & 0 & 15 & 13 & 0 & 19 & 9 & 20 & 14 & 5 & 0 \\ 15 & 5 & 17 & 9 & 11 & 25 & 20 & 0 & 25 & 1 & 19 & 19 & 1 & 3 & 27 & 0 \end{pmatrix}$$

which is then to be 'coded' by a *coding matrix*,

† A *random* permutation of the set 0, 1, 2, . . ., 29 would be marginally better.

$C = \begin{pmatrix} 3 & 1 \\ 5 & 2 \end{pmatrix}$, pre-multiplying as follows:

$$\begin{aligned} CW &= \begin{pmatrix} 3 & 1 \\ 5 & 2 \end{pmatrix}\begin{pmatrix} 3 & 13 & 0 & 21 & 3 & 12 & 0 & 15 & 13 & 0 & 19 & 9 & 20 & 14 & 5 & 0 \\ 15 & 5 & 17 & 9 & 11 & 25 & 20 & 0 & 25 & 1 & 19 & 19 & 1 & 3 & 27 & 0 \end{pmatrix} \\ &= \begin{pmatrix} 24 & 44 & 17 & 72 & 20 & 61 & 20 & 45 & 64 & 1 & 76 & 46 & 61 & 45 & 42 & 0 \\ 45 & 75 & 34 & 123 & 37 & 110 & 40 & 75 & 115 & 2 & 133 & 83 & 102 & 76 & 79 & 0 \end{pmatrix} \\ &= \begin{pmatrix} 24 & 14 & 17 & 12 & 20 & 1 & 20 & 15 & 4 & 1 & 16 & 16 & 1 & 15 & 12 & 0 \\ 15 & 15 & 4 & 3 & 7 & 20 & 10 & 15 & 25 & 2 & 13 & 23 & 12 & 16 & 19 & 0 \end{pmatrix}, \\ &\quad (\text{mod. } 30), \end{aligned}$$

giving the coded message:

XONOQDLCTGATTJOODYABPMPWALOLS.

In order to decode, we need the decoder matrix, $\boldsymbol{D}$, which will take us from $\boldsymbol{CW}$ to $\boldsymbol{W}$: $\boldsymbol{D}(\boldsymbol{CW}) = \boldsymbol{W}$. Thus $\boldsymbol{DC} = \boldsymbol{I}$, so that $\boldsymbol{D}$ (as expected) should be the *inverse* of $\boldsymbol{C}$, i.e., $\boldsymbol{D} = \boldsymbol{C}^{-1} = \begin{pmatrix} 2 & -1 \\ -5 & 3 \end{pmatrix}$.

Going in the reverse direction then, we have

$$\begin{aligned} &\begin{pmatrix} 2 & -1 \\ -5 & 3 \end{pmatrix}\begin{pmatrix} 24 & 14 & 17 & 12 & 20 & 1 & 20 & 15 & 4 & 1 & 16 & 16 & 1 & 15 & 12 & 0 \\ 15 & 15 & 4 & 3 & 7 & 20 & 10 & 15 & 25 & 2 & 13 & 23 & 12 & 16 & 19 & 0 \end{pmatrix} \\ &= \begin{pmatrix} 33 & 13 & 30 & 21 & 33 & -18 & 30 & 15 & -17 & 0 & 19 & 9 & -10 & 14 & 5 & 0 \\ -75 & -25 & -73 & -51 & -79 & 55 & -70 & -30 & 55 & 1 & -41 & -11 & 31 & -27 & -3 & 0 \end{pmatrix} \\ &= \begin{pmatrix} 3 & 13 & 0 & 21 & 3 & 12 & 0 & 15 & 13 & 0 & 19 & 9 & 20 & 14 & 5 & 0 \\ 15 & 5 & 17 & 9 & 11 & 25 & 20 & 0 & 25 & 1 & 19 & 19 & 1 & 3 & 27 & 0 \end{pmatrix} \quad (\text{mod. } 30) \end{aligned}$$

and the decoded message can then be read off. Notice that both the word spacing and the repetition of letters are completely obscured by this method: the three Ss in 'assistance' come out as P, M, and W respectively.

Of course, the coder $\boldsymbol{C}$ may be chosen in many ways, but in order that the decoder $\boldsymbol{D}$ may contain integers only, we must arrange for the *determinant* of $\boldsymbol{C}$ ($= ac - bd$) to be ± 1. In the above case, $|\boldsymbol{C}| = 3 \times 2 - 1 \times 5 = 1$.

For example,

$$\text{if} \quad \boldsymbol{C} = \begin{pmatrix} 2 & 1 \\ 1 & 0 \end{pmatrix}, \quad |\boldsymbol{C}| = -1, \quad \text{and} \quad \boldsymbol{D} = \boldsymbol{C}^{-1} = \begin{pmatrix} 0 & 1 \\ 1 & -2 \end{pmatrix}.$$

The presence of a zero makes computation easy:

$$\begin{pmatrix} 2 & 1 \\ 1 & 0 \end{pmatrix}\begin{pmatrix} 3 & 13 & 0 & 21 & \cdots \\ 15 & 5 & 17 & 9 & \cdots \end{pmatrix} = \begin{pmatrix} 21 & 1 & 17 & 21 & \cdots \\ 3 & 13 & 0 & 21 & \cdots \end{pmatrix}$$

and the fact that the second row of $\boldsymbol{C}$, (1, 0) simply moves the top row of $\boldsymbol{W}$ to the bottom scarcely gives the game away; for the zero moving to the bottom row causes the first 'word' of the coded message to have 5 letters instead of 4.

Evidently the method could be made more subtle by using a 3 × 3 coder, the message being written out in three rows. The computation might be heavy, but use of a computer would avoid the tedium. One of the difficulties would be to find a 3 × 3 matrix with integral coefficients whose inverse has the same property. However, you might try with the following coder:

$$\boldsymbol{C} = \begin{pmatrix} 2 & 1 & 3 \\ 1 & 0 & -1 \\ 4 & 2 & 5 \end{pmatrix}, \quad \text{the corresponding decoder being} \quad \boldsymbol{D} = \begin{pmatrix} 2 & 1 & -1 \\ -9 & -2 & 5 \\ 2 & 0 & -1 \end{pmatrix}$$

(first checking for yourself that $\boldsymbol{CD} = \boldsymbol{I}$).

Exercise 13.4c

1 Using row (1) at the beginning of this section (A = 1, B = 2, etc., with . , ? □ represented by 27, 28, 29, 30, and the encoder

$$\boldsymbol{C} = \begin{pmatrix} 2 & 1 \\ 1 & 0 \end{pmatrix} \quad \text{and decoder} \quad \boldsymbol{D} = \begin{pmatrix} 0 & 1 \\ 1 & -2 \end{pmatrix} \quad \text{in the text,}$$

construct and decode various messages. (This should be done in pairs, each pupil constructing a message and exchanging with his partner.)

2 Find other suitable encoders and decoders, avoiding fractions by arranging for $ad - bc = \pm 1$.

3 Code the message, 'Come quickly to my assistance', using

$$\begin{pmatrix} 2 & 1 & 3 \\ 1 & 0 & -1 \\ 4 & 2 & 5 \end{pmatrix} \quad \text{as encoder, and} \quad \begin{pmatrix} 2 & 1 & -1 \\ -9 & -2 & 5 \\ 2 & 0 & -1 \end{pmatrix} \quad \text{as decoder,}$$

with the modulo 30 system established in the text. Repeat with the message 'The quick brown dog jumps over the lazy fox.'

4 Using row (2) on p. 421 (A = 3, B = 6, . . . , Z = 26) . = 27, , = 28, ? = 29, □ = 30), the word ADAM was coded as CDNE. What was the encoding matrix? Using this matrix, how would the message 'BEEF ESSENCE' be coded?

13.5 Determinants

We have already met instances of both 2 × 2 and 3 × 3 determinants. Unlike a matrix, a determinant is not just a table of numerical data, but (quite differently) a shorthand way of writing an algebraic expression, whose *value* can be computed when the values of the 'terms' or 'elements' are known.

There is little to be said about 2×2 determinants:

$$\begin{vmatrix} a & b \\ c & d \end{vmatrix} = ad - bc$$

and this method of writing such a two-termed expression does not carry many advantages. However, there is some point in doing a little work on them in the following exercise since the rules for manipulating 2×2 determinants are imitated by those for larger determinants. For example,

$$\text{if} \quad \Delta = \begin{vmatrix} a & b \\ c & d \end{vmatrix}, \quad \text{then} \quad \begin{vmatrix} ak & b \\ ck & d \end{vmatrix} = k\Delta, \quad \text{and} \quad \begin{vmatrix} ak & bk \\ ck & dk \end{vmatrix} = k^2\Delta$$

(not $k\Delta$, as with matrices).

So with 3×3 determinants,

$$\text{if} \quad \Delta = \begin{vmatrix} a_1 & b_1 & c_1 \\ a_2 & b_2 & c_2 \\ a_3 & b_3 & c_3 \end{vmatrix}, \quad \text{then} \quad \begin{vmatrix} a_1k & b_1k & c_1k \\ a_2 & b_2 & c_2 \\ a_3 & b_3 & c_3 \end{vmatrix} = k\Delta;$$

while if every element were multiplied by k, the value would be $k^3\Delta$.

Exercise 13.5a

1 Evaluate:

(*i*) $\begin{vmatrix} 3 & -4 \\ -2 & 1 \end{vmatrix}$; (*ii*) $\begin{vmatrix} 8 & -3 \\ 2 & 6 \end{vmatrix}$; (*iii*) $\begin{vmatrix} 8 & 5 \\ 80 & 50 \end{vmatrix}$;

(*iv*) $\begin{vmatrix} 173 & 163 \\ 94 & 104 \end{vmatrix}$; (*v*) $\begin{vmatrix} \cdot 42 & 6 \\ -36 & 48 \end{vmatrix}$; (*vi*) $\begin{vmatrix} 13 & 1 \\ 1\,001 & 77 \end{vmatrix}$;

(*vii*) $\begin{vmatrix} 300 & -400 \\ -240 & 320 \end{vmatrix}$.

2 Show that $\begin{vmatrix} a & b \\ a^2 & b^2 \end{vmatrix}$ has a factor $(a - b)$. Generalise the result.

3 Simplify:

(*i*) $\begin{vmatrix} a+b & a-b \\ a-b & a+b \end{vmatrix}$; (*ii*) $\begin{vmatrix} \cos\theta & \sin\theta \\ -\sin\theta & \cos\theta \end{vmatrix}$;

(*iii*) $\begin{vmatrix} \cos\theta & \sin\theta \\ \sin\theta & \cos\theta \end{vmatrix}$; (*iv*) $\begin{vmatrix} \cosh u & \sinh u \\ \sinh u & \cosh u \end{vmatrix}$;

(*v*) $\begin{vmatrix} 2t & 1-t^2 \\ 1-t^2 & 2t \end{vmatrix}$; (*vi*) $\begin{vmatrix} px+qz & rx+sz \\ py+qz & ry+sz \end{vmatrix}$.

4 If $\Delta = \begin{vmatrix} a & b \\ c & d \end{vmatrix}$, show that $\begin{vmatrix} a+c & b+d \\ c & d \end{vmatrix}$ and $\begin{vmatrix} a & b+ka \\ c & d+kc \end{vmatrix}$ both have the value Δ, and write down some other determinants which have the same value.

5 Express $\begin{vmatrix} a+x & b+y \\ c+p & d+q \end{vmatrix}$ as the sum of determinants.

6 Find the value of λ so that the equations

$$2x + 5y = \lambda x$$
$$3x + 4y = \lambda y$$

may have non-trivial solution.

7 Solve the equation for x: $\begin{vmatrix} 2-x & -1 \\ -6 & 3+x \end{vmatrix} = 0.$

8 Show that, in general, the solutions of the simultaneous equations

$$a_1x + b_1y + c_1 = 0$$
$$a_2x + b_2y + c_2 = 0$$

may be written $\dfrac{x}{\begin{vmatrix} b_1 & c_1 \\ b_2 & c_2 \end{vmatrix}} = \dfrac{y}{\begin{vmatrix} c_1 & a_1 \\ c_2 & a_2 \end{vmatrix}} = \dfrac{1}{\begin{vmatrix} a_1 & b_1 \\ a_2 & b_2 \end{vmatrix}}$.

Use this:

(*i*) to solve $3x - y = -8$, $x + 7y = 1$;

(*ii*) to eliminate x from the equations $ax^2 + bx + c = 0$, $px^2 + qx + r = 0$.

3 × 3 determinants

We now consider the 3 straight lines with equations:

$$a_1x + b_1y + c_1 = 0 \quad (1)$$
$$a_2x + b_2y + c_2 = 0 \quad (2)$$
$$a_3x + b_3y + c_3 = 0 \quad (3)$$

and the condition for them to be *concurrent*. Solving (2) and (3), gives (in the case of non-zero denominators):

$$\frac{x}{b_2c_3 - b_3c_2} = \frac{y}{c_2a_3 - c_3a_2} = \frac{1}{a_2b_3 - a_3b_2},$$

the point of intersection of lines (2) and (3). If this point is to lie also on the first line, its coordinates must satisfy (1), and so by substitution we obtain

$$a_1(b_2c_3 - b_3c_2) + b_1(c_2a_3 - c_3a_2) + c_1(a_2b_3 - a_3b_2) = 0 \quad (4)$$

The expression on the left-hand side is precisely what we have already called the determinant of the coefficients in the three equations, written

$$\Delta = \begin{vmatrix} a_1 & b_1 & c_1 \\ a_2 & b_2 & c_2 \\ a_3 & b_3 & c_3 \end{vmatrix}.$$

Such expressions occur so frequently in mathematics that the determinant shorthand notation was devised for them. Just as expressions like $x \times x \times x \times x \times x$ have the abbreviation x^5, and with this abbreviated notation comes manipulation using the *laws of indices,* so with determinants we shall want to devise corresponding methods of manipulation.

Another instance of the occurrence of 3×3 determinants in coordinate geometry is in the area of a triangle formed by 3 points (x_1, y_1), (x_2, y_2), (x_3, y_3) which is given by the expression

$$\Delta = \tfrac{1}{2}(x_1y_2 - x_2y_1 + x_2y_3 - x_3y_2 + x_3y_1 - x_1y_3) = \tfrac{1}{2}\begin{vmatrix} x_1 & y_1 & 1 \\ x_2 & y_2 & 1 \\ x_3 & y_3 & 1 \end{vmatrix}.$$

The condition for collinearity of 3 points is of course that this determinant should *vanish.*

In this section we shall deal principally with 3×3 determinants† and then indicate briefly how their behaviour is imitated in the case of determinants of higher order.

Expansion of determinants: minors and cofactors

The expansion of the general 3×3 determinant has already been given:

$$\Delta = \begin{vmatrix} a_1 & b_1 & c_1 \\ a_2 & b_2 & c_2 \\ a_3 & b_3 & c_3 \end{vmatrix} = a_1b_2c_3 - a_1b_3c_2 + b_1c_2a_3 - b_1c_3a_2 + c_1a_2b_3 - c_1a_3b_2.$$

Each term is the product of 3 elements of the determinant, no two in the same row or column (i.e., there are no terms like $b_1a_2b_3$, or $a_3b_1c_3$), so it is easily seen that the number of terms must be $3 \times 2 \times 1 = 6$. In the case of 4×4 determinants, the number of terms (such as $a_2b_1c_4d_3$) would be $4 \times 3 \times 2 \times 1 = 4! = 24$, and in the general case of a $n \times n$ determinant, there will be $n!$ terms in the expansion. Thus the notation gains in brevity as the order of Δ increases: a 5×5 determinant is a very efficient abbreviation for an algebraic expression with 120 terms.

In dealing with determinants of higher order, one needs to know the rule for deciding which of the $n!$ terms carry a $+$ and which a $-$ sign.

† Determinants must be square: there is no such thing as a rectangular determinant.

This rule is bound up with the theory of odd and even permutations,† and we do not need to concern ourselves with it here.

The value of the 3 × 3 determinant may be written in many alternative ways, each of which we abbreviate using the notation of 13.4:

$$a_1(b_2c_3 - b_3c_2) + b_1(c_2a_3 - c_3a_2) + c_1(a_2b_3 - a_3b_2)$$
$$= a_1A_1 + b_1B_1 + c_1C_1 \qquad (1)$$

$$a_2(b_3c_1 - b_1c_3) + b_2(c_3a_1 - c_1a_3) + c_2(a_3b_1 - a_1b_3)$$
$$= a_2A_2 + b_2B_2 + c_2C_2 \qquad (2)$$

$$a_3(b_1c_2 - b_2c_1) + b_3(c_1a_2 - c_2a_1) + c_3(a_1b_2 - a_2b_1)$$
$$= a_3A_3 + b_3B_3 + c_3C_3 \qquad (3)$$

$$a_1(b_2c_3 - b_3c_2) + a_2(b_3c_1 - b_1c_3) + a_3(b_1c_2 - b_2c_1)$$
$$= a_1A_1 + a_2A_2 + a_3A_3 \qquad (4)$$

$$b_1(c_2a_3 - c_3a_2) + b_2(c_3a_1 - c_1a_3) + b_3(c_1a_2 - c_2a_1)$$
$$= b_1B_1 + b_2B_2 + b_3B_3 \qquad (5)$$

$$c_1(a_2b_3 - a_3b_2) + c_2(a_3b_1 - a_1b_3) + c_3(a_1b_2 - a_2b_1)$$
$$= c_1C_1 + c_2C_2 + c_3C_3 \qquad (6)$$

(1) is called the 'expansion of the determinant by the first row': the elements a_1, b_1, c_1 of the first row are multiplied by their corresponding *cofactors* A_1, B_1, C_1 to form the *inner product* $a_1A_1 + b_1B_1 + c_1C_1$. (2) and (3) are expansions by the second and third rows respectively, while in (4), (5), (6) we have written the expansion by the first, second and third columns.

We now need to recognise at sight the cofactors of the several terms of the determinant

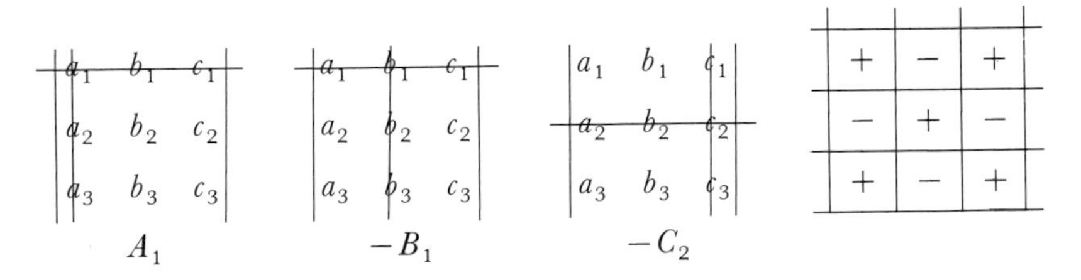

It will be seen from the examples given above that each cofactor is either plus or minus the 2 × 2 determinant obtained by striking out the row and column through the element concerned, the appropriate sign being given by the chessboard scheme on the right. For example, if we want the cofactor of c_2, we strike out the second row and third column, i.e., those which intersect in the element c_2 as in the third part of the diagram, thereby obtaining the determinant $\begin{vmatrix} a_1 & b_1 \\ a_3 & b_3 \end{vmatrix}$, which is known as the *minor* of c_2.

† See Budden, F. J., *The Fascination of Groups*, C.U.P., 1972, p. 312; Moakes, A. J. and Neill, H., *Vectors, Matrices and Linear Equations*, Oliver and Boyd, 1967.

The scheme of signs shows that C_2 has a value which is the negative of this minor, i.e.,

$$C_2 = -\begin{vmatrix} a_1 & b_1 \\ a_3 & b_3 \end{vmatrix} = -(a_1b_3 - a_3b_1) = a_3b_1 - a_1b_3.$$

The rule is true for determinants of any order (and enables the general determinant to be defined recursively):

$$\begin{vmatrix} a_1 & b_1 & c_1 & d_1 & e_1 \\ a_2 & b_2 & c_2 & d_2 & e_2 \\ a_3 & b_3 & c_3 & d_3 & e_3 \\ a_4 & b_4 & c_4 & d_4 & e_4 \\ a_5 & b_5 & c_5 & d_5 & e_5 \end{vmatrix} \qquad \text{Signs: } \begin{matrix} + & - & + & - & + \\ - & + & - & + & - \\ + & - & + & - & + \\ - & + & - & + & - \\ + & - & + & \ominus & + \end{matrix}$$

$$\text{The minor of } d_5 = \begin{vmatrix} a_1 & b_1 & c_1 & e_1 \\ a_2 & b_2 & c_2 & e_2 \\ a_3 & b_3 & c_3 & e_3 \\ a_4 & b_4 & c_4 & e_4 \end{vmatrix} \quad \text{so that} \quad D_5 = -\begin{vmatrix} a_1 & b_1 & c_1 & e_1 \\ a_2 & b_2 & c_2 & e_2 \\ a_3 & b_3 & c_3 & e_3 \\ a_4 & b_4 & c_4 & e_4 \end{vmatrix}$$

In theoretical work we prefer to work with cofactors rather than minors, since the expansions as in (1)...(6) do not contain any $-$ signs, the cofactors themselves incorporating the appropriate signs. The latter are therefore sometimes referred to as 'rectified' minors.

Exercise 13.5b

1 Evaluate the following determinants and check your answers by expanding them in several different ways:

$$(i)\ \begin{vmatrix} 6 & 3 & -7 \\ 5 & 0 & 2 \\ 2 & -1 & 4 \end{vmatrix}; \qquad (ii)\ \begin{vmatrix} 1 & 3 & -1 \\ 2 & 3 & 1 \\ 3 & 3 & 4 \end{vmatrix}; \qquad (iii)\ \begin{vmatrix} -1 & 3 & 1 \\ -6 & 12 & 3 \\ 6 & 1 & 0 \end{vmatrix};$$

$$(iv)\ \begin{vmatrix} 60 & 18 & 49 \\ 50 & 0 & -14 \\ 20 & -6 & -28 \end{vmatrix}; \qquad (v)\ \begin{vmatrix} 1 & -2 & -3 & 4 \\ -2 & 3 & 4 & -5 \\ 3 & -4 & -5 & 6 \\ -4 & 5 & 6 & -7 \end{vmatrix};$$

2 Expand the determinants

$$\begin{vmatrix} a & b & c \\ c & a & b \\ b & c & a \end{vmatrix}, \quad \begin{vmatrix} a & c & b \\ b & a & c \\ c & b & a \end{vmatrix}, \quad \begin{vmatrix} a & b & c \\ b & c & a \\ c & a & b \end{vmatrix},$$

and deduce the values of

$$\begin{vmatrix} a & b & 0 \\ b & 0 & a \\ 0 & a & b \end{vmatrix} \quad \text{and} \quad \begin{vmatrix} 1 & x^2 & x \\ x & 1 & x^2 \\ x^2 & x & 1 \end{vmatrix}.$$

3 Evaluate the cofactors of some of the elements in the above determinants.

4 Write down the cofactors in $\begin{vmatrix} a & h & g \\ h & b & f \\ g & f & c \end{vmatrix}$ beginning $A = bc - f^2$.

Show that $hA + bH + fG = 0$, and obtain similar results.

5 Solve the equation for x:

$$\begin{vmatrix} 2-x & -3 & 5 \\ 0 & 5-x & 2 \\ 8 & -6 & 8-x \end{vmatrix} = 0.$$

6 If $\boldsymbol{A} = \begin{vmatrix} 2 & -3 & 1 \\ 1 & 4 & 6 \\ 0 & -1 & -1 \end{vmatrix}$, show that $|\boldsymbol{A}| = 0$,

and that $|\text{adj. } \boldsymbol{A}| \begin{vmatrix} 2 & -4 & -22 \\ 1 & -2 & -11 \\ -1 & 2 & 11 \end{vmatrix}$.

What do you notice? (Compare exercise 13.5*d*, no. 8).

7 Expand the determinants:

$$(i) \begin{vmatrix} a & a & 0 \\ 0 & b & b \\ c & 0 & c \end{vmatrix}; \quad (ii) \begin{vmatrix} 1 & 1 & 1 \\ x & y & z \\ y+z & z+x & x+y \end{vmatrix}; \quad (iii) \begin{vmatrix} bc^2+b^2c & bc & 1 \\ ca^2+c^2a & ca & 1 \\ ab^2+a^2b & ab & 1 \end{vmatrix}.$$

8 Eliminate x, y, and z from:

$$(i)\quad \begin{aligned} x + ay - bz &= 0 \\ -ax + y + cz &= 0; \\ bx - cy + z &= 0 \end{aligned} \qquad (ii)\quad \begin{aligned} a(y - z) &= x \\ b(z - x) &= y. \\ c(x - y) &= z \end{aligned}$$

9 What is the volume of a parallelepiped, three adjacent edges of which are the vectors $\boldsymbol{OP_r}$, where P_r is (x_r, y_r, z_r), $r = 1, 2, 3$. Find the volume of a tetrahedron having vertices O $(0, 0, 0)$, P_1 $(-2, 1, 0)$, P_2 $(3, 1, 1)$, P_3 $(1, 0, -4)$.

10 Show that the result of eliminating x from the equations

$$\begin{aligned} ax^2 + bx + c &= 0 \\ px^2 + qx + r &= 0 \end{aligned} \quad \text{can be expressed} \quad \begin{vmatrix} a & b & c & 0 \\ 0 & a & b & c \\ p & q & r & 0 \\ 0 & p & q & r \end{vmatrix} = 0.$$

(Compare exercise 13.5*a*, no. 8).

11 Eliminate x from the equations

$$ax^2 + bx + c = 0 \quad \text{and} \quad px^3 + qx^2 + rx + s = 0,$$

giving the result as a 5×5 determinant equated to zero.

Row and column operations

Before dealing with the rules for the manipulation of determinants, we shall introduce a notation which will be found convenient throughout the remainder of this chapter for both matrices and determinants.

We may regard a matrix as a collection of row vectors which we shall call $\boldsymbol{r}_1, \boldsymbol{r}_2, \boldsymbol{r}_3$, etc., and similarly the columns will be denoted $\boldsymbol{c}_1, \boldsymbol{c}_2, \boldsymbol{c}_3, \ldots$.

So, in the determinant $\begin{vmatrix} 0 & 3 & -1 \\ 2 & 5 & 1 \\ -1 & 0 & 2 \end{vmatrix}$, $\boldsymbol{c}_3$ denotes the vector $\begin{pmatrix} -1 \\ 1 \\ 2 \end{pmatrix}$.

Furthermore, we shall use a notation similar to that of transformations to indicate operations of replacing individual rows or columns. For example, $\boldsymbol{r}_2' = \boldsymbol{r}_2 - 2\boldsymbol{r}_1$ simply means 'create a new second row by subtracting from its elements twice the corresponding elements of the first row'; so that the above determinant becomes

$$\begin{vmatrix} 0 & 3 & -1 \\ 2 - 2\times 0 & 5 - 2\times 3 & 1 - 2x-1 \\ -1 & 0 & 2 \end{vmatrix}, \quad \text{or} \quad \begin{vmatrix} 0 & 3 & -1 \\ 2 & -1 & 3 \\ -1 & 0 & 2 \end{vmatrix}.$$

Similarly, the operation $\boldsymbol{c}_2' = 2\boldsymbol{c}_2$ has the effect of creating a new second column by doubling the elements of its predecessor.

It may be noticed that row operations on matrices can themselves be performed by the use of matrices. For example, $\boldsymbol{r}_2' = \boldsymbol{r}_2 - 2\boldsymbol{r}_1$ can be achieved by *pre-multiplying* the given matrix by

$$\begin{pmatrix} 1 & 0 & 0 \\ -2 & 1 & 0 \\ 0 & 0 & 1 \end{pmatrix}.$$

Similarly, the operation $\boldsymbol{r}_1' = \alpha\boldsymbol{r}_2 + \beta\boldsymbol{r}_3$ is achieved by pre-multiplying by

$$\begin{pmatrix} 0 & \alpha & \beta \\ 0 & 1 & 0 \\ 0 & 0 & 1 \end{pmatrix},$$

for
$$\begin{pmatrix} 0 & \alpha & \beta \\ 0 & 1 & 0 \\ 0 & 0 & 1 \end{pmatrix}\begin{pmatrix} a_1 & b_1 & c_1 \\ a_2 & b_2 & c_2 \\ a_3 & b_3 & c_3 \end{pmatrix} = \begin{pmatrix} \alpha a_2 + \beta a_3 & \alpha b_2 + \beta b_3 & \alpha c_2 + \beta c_3 \\ a_2 & b_2 & c_2 \\ a_3 & b_3 & c_3 \end{pmatrix};$$

whilst the effect of pre-multiplication by

$$\begin{pmatrix} 1 & 0 & 0 \\ 0 & 1 & 0 \\ 0 & 0 & \alpha \end{pmatrix}$$

would be to carry out the operation $\boldsymbol{r}_3' = \alpha\boldsymbol{r}_3$. Matrices which perform these row operations are known as *elementary matrices* and we shall use them extensively in the solution of systems of linear equations (13.7).

Lastly, the reader might like to investigate how *column* operations can be achieved by means of *post-multiplication* by similar elementary matrices.

Rules for manipulation of determinants

We shall not go into the tedious process of proving these rules, but will simply illustrate them, letting them speak for themselves.

(*i*) $|\boldsymbol{M}| = |\boldsymbol{M}'|$, i.e., *the determinant of a matrix is unchanged when the matrix is transposed.*

$$\Delta' = \begin{vmatrix} a_1 & a_2 & a_3 \\ b_1 & b_2 & b_3 \\ c_1 & c_2 & c_3 \end{vmatrix} = \Delta$$ (the reader should verify that the expansion of Δ' agrees term by term with that of Δ).

(*ii*) *Multiple of row* (*or column*),

e.g., $$\Delta' = \begin{vmatrix} ka_1 & b_1 & c_1 \\ ka_2 & b_2 & c_2 \\ ka_3 & b_3 & c_3 \end{vmatrix} = k\Delta; \qquad \Delta' = \begin{vmatrix} a_1 & b_1 & c_1 \\ a_2 & b_2 & c_2 \\ \lambda a_3 & \lambda b_3 & \lambda c_3 \end{vmatrix} = \lambda\Delta,$$ etc.

(*iii*) *When two rows (or columns) are interchanged, the determinant changes sign,*

e.g., $$\Delta' = \begin{vmatrix} c_1 & b_1 & a_1 \\ c_2 & b_2 & a_2 \\ c_3 & b_3 & a_3 \end{vmatrix} = -\Delta:$$ this may be verified by actual expansion.

Corollaries:

(*a*) *If two rows (or columns) are identical, the determinant has the value zero* (for in this case: $\Delta = -\Delta \Rightarrow \Delta = 0$).

(*b*) *If two rows (or columns) are proportional, then the determinant is zero,*

The two proportional rows (or columns) are, in fact, linearly dependent (see 13.1) and the determinant is also zero if its three rows (or columns) are linearly dependent, but this cannot usually be spotted at a glance.

(*iv*) *A determinant is unchanged when to the elements of any row (or column) are added a constant multiple of the elements of any other row (or column).*

For example,

if $$\Delta = \begin{vmatrix} a_1 & b_1 & c_1 \\ a_2 & b_2 & c_2 \\ a_3 & b_3 & c_3 \end{vmatrix},$$

let $\boldsymbol{r}_1' = \boldsymbol{r}_1 + \lambda \boldsymbol{r}_2$.

Then $$\Delta' = \begin{vmatrix} a_1 + \lambda a_2 & b_1 + \lambda b_2 & c_1 + \lambda c_2 \\ a_2 & b_2 & c_2 \\ a_3 & b_3 & c_3 \end{vmatrix}$$

$$= \begin{vmatrix} a_1 & b_1 & c_1 \\ a_2 & b_2 & c_2 \\ a_3 & b_3 & c_3 \end{vmatrix} + \lambda \begin{vmatrix} a_2 & b_2 & c_2 \\ a_2 & b_2 & c_2 \\ a_3 & b_3 & c_3 \end{vmatrix}$$

$$= \Delta + \lambda \times 0$$

$$= \Delta.$$

Finally, it is also worth expressing corollary *iii* (*a*) in terms of cofactors. Suppose, for instance, that the second row of Δ is replaced by its first row (i.e., $\boldsymbol{r}_2' = \boldsymbol{r}_1$).

Then $$\begin{vmatrix} a_1 & b_1 & c_1 \\ a_2 & b_2 & c_2 \\ a_3 & b_3 & c_3 \end{vmatrix} = \Delta$$ is replaced by

$$\begin{vmatrix} a_1 & b_1 & c_1 \\ a_1 & b_1 & c_1 \\ a_3 & b_3 & c_3 \end{vmatrix} = 0 \quad \text{(since two rows are identical).}$$

Now the cofactors of the second row have been unaffected by the replacement, so the equation

$$a_2A_2 + b_2B_2 + c_2C_2 = \Delta$$

has been replaced by

$$a_1A_2 + b_1B_2 + c_1C_2 = 0.$$

This expression consists of the elements of row 1 of the original determinant taken with the cofactors of a *different* row (row 2), or *alien cofactors* as they are usually called.

Similarly $a_1A_3 + b_1B_3 + c_1C_3 = 0$;

and more generally we may say, calling any such expression an *inner product*, that

> The inner product of the elements of a row (or column) with the cofactors of the *same* row (or column) gives the *value of the determinant*; but the inner product of the elements of a row (or column) with the *alien* cofactors of a *different* row (or column) gives *zero*.

Example 1

Evaluate $\Delta = \begin{vmatrix} 2 & 4 & 18 \\ -3 & 10 & -27 \\ 1 & 3 & 11 \end{vmatrix}$.

We try to use row and column operations to obtain as many zeros as possible:

$$\Delta = 2\begin{vmatrix} 1 & 2 & 9 \\ -3 & 10 & -27 \\ 1 & 3 & 11 \end{vmatrix}$$

$$= 2\begin{vmatrix} 1 & 2 & 9 \\ -3 & 10 & -27 \\ 0 & 1 & 2 \end{vmatrix} \quad (\boldsymbol{r}_3' = \boldsymbol{r}_3 - \boldsymbol{r}_1)$$

$$= 2\begin{vmatrix} 1 & 2 & 5 \\ -3 & 10 & -47 \\ 0 & 1 & 0 \end{vmatrix} \quad (\boldsymbol{c}_3' = \boldsymbol{c}_3 - 2\boldsymbol{c}_2)$$

$$= -2\begin{vmatrix} 1 & 5 \\ -3 & -47 \end{vmatrix}$$

$$= -2(-47 + 15)$$

$$= 64.$$

Alternatively,

$$\Delta = 2\begin{vmatrix} 1 & 2 & 9 \\ -3 & 10 & -27 \\ 1 & 3 & 11 \end{vmatrix} = 2\begin{vmatrix} 1 & 2 & 9 \\ 0 & 16 & 0 \\ 1 & 3 & 11 \end{vmatrix} \quad (\boldsymbol{r}_2' = \boldsymbol{r}_2 + 3\boldsymbol{r}_1)$$

$$= 32\begin{vmatrix} 1 & 9 \\ 1 & 11 \end{vmatrix} = 32 \times 2 = 64.$$

Example 2

Factorise $\begin{vmatrix} 1 & a & a^3 \\ 1 & b & b^3 \\ 1 & c & c^3 \end{vmatrix}$.

Putting $b = c$ gives two identical rows, thus causing the determinant to vanish. Therefore, by the remainder theorem, $(b - c)$ is a factor. Similarly $(c - a)$ and $(a - b)$ are also factors. Hence $\Delta = (b - c)(c - a)(a - b)F$, and consideration of the degree of the determinant enables us to see that the remaining factor F must have degree 1. Clearly it is symmetric in a, b and c, so must have the form $k(a + b + c)$, where k is a constant. The 'leading term' $1bc^3$ (the product of the elements in the leading diagonal) has coefficient 1, and this shows us that $k = 1$. Hence

$$\Delta = (b - c)(c - a)(a - b)(a + b + c)$$

Example 3

Evaluate $\Delta = \begin{vmatrix} a & b & c & d \\ d & a & b & c \\ c & d & a & b \\ b & c & d & a \end{vmatrix}$.

Now $\Delta = \begin{vmatrix} a+d+c+b & b+a+d+c & c+b+a+d & d+c+b+a \\ d & a & b & c \\ c & d & a & b \\ b & c & d & a \end{vmatrix}$

$$(\boldsymbol{r}_1' = \boldsymbol{r}_1 + \boldsymbol{r}_2 + \boldsymbol{r}_3 + \boldsymbol{r}_4)$$

$$= (a+b+c+d)\begin{vmatrix} 1 & 1 & 1 & 1 \\ d & a & b & c \\ c & d & a & b \\ b & c & d & a \end{vmatrix}.$$

It is, in fact, unnecessary to go so far: all we need is to obtain the factor $(a+b+c+d)$. Similarly, $\boldsymbol{r}_1' = \boldsymbol{r}_1 - \boldsymbol{r}_2 + \boldsymbol{r}_3 - \boldsymbol{r}_4$ detects the factor $(a-d+c-b)$, and so $\Delta = (a+b+c+d)(a-b+c-d)F$. The remaining factor F, being homogeneous of degree 2, may contain terms in $a^2, b^2, c^2, d^2, ab, ac, ad, bc, bd, cd$. But it is easier to find it by using complex numbers. For $\boldsymbol{r}_1' = \boldsymbol{r}_1 + i\boldsymbol{r}_2 - \boldsymbol{r}_3 - i\boldsymbol{r}_4$ replaces $\boldsymbol{r}_1$ by

$$\begin{aligned} &\big(a+id-c-ib \quad b+ia-d-ic \quad c+ib-a-id \quad d+ic-b-ia\big) \\ &= \big(a-ib-c+id \quad i(a-ib-c+id) \quad -(a-ib-c+id) \\ &\qquad -i(a-ib-c+id)\big), \end{aligned}$$

showing that $(a-ib-c+id)$ is a factor. Similarly, the conjugate $(a+ib-c-id)$ is also a factor, and we get

$$\Delta = (a+b+c+d)(a-b+c-d)(a+ib-c-id)(a-ib-c+id).$$

Finally, combining the conjugate factors gives

$$\Delta = (a+b+c+d)(a-b+c-d)(a^2+b^2+c^2+d^2-2ac-2bd).$$

Exercise 13.5c

1 Perform the operations specified on the given determinants, and hence evaluate:

(i) $\begin{vmatrix} 4 & -18 & 3 \\ 2 & 12 & -4 \\ -4 & 0 & 5 \end{vmatrix}$ $\quad \begin{aligned} \boldsymbol{r}_1' &= \boldsymbol{r}_1 + \boldsymbol{r}_3 \\ \boldsymbol{r}_3' &= \boldsymbol{r}_3 + 2\boldsymbol{r}_2; \end{aligned}$

(ii) $\begin{vmatrix} 1 & a & b \\ 1 & a^2 & b^2 \\ 1 & a^3 & b^3 \end{vmatrix}$ $\quad \begin{aligned} \boldsymbol{c}_2' &= \boldsymbol{c}_2 - \boldsymbol{c}_3 \\ \boldsymbol{r}_1'' &= \boldsymbol{r}_1' - \boldsymbol{r}_2'; \end{aligned}$

(iii) $\begin{vmatrix} a-b-c & 2a & 2a \\ 2b & b-c-a & 2b \\ 2c & 2c & c-a-b \end{vmatrix}$ $\quad \begin{aligned} \boldsymbol{c}_2' &= \boldsymbol{c}_2 - \boldsymbol{c}_3 \\ \boldsymbol{r}_2'' &= \boldsymbol{r}_2' + \boldsymbol{r}_3'; \end{aligned}$

(iv) $\begin{vmatrix} a & b & c \\ b & c & a \\ c & a & b \end{vmatrix}$ $\quad \boldsymbol{c}_1' = \boldsymbol{c}_1 + \omega\boldsymbol{c}_2 + \omega^2\boldsymbol{c}_3$, where $\omega^3 = 1$.

2 Use the cube roots of unity to factorise $\begin{vmatrix} a & b & c \\ c & a & b \\ b & c & a \end{vmatrix}$

($\boldsymbol{r}_1' = \boldsymbol{r}_1 + \boldsymbol{r}_2 + \boldsymbol{r}_3$; compare the text above, and exercise 10.4b, no **9**).

3 Use the result of no. **2** above to write down the factors of

$$\begin{vmatrix} a & -1 & 0 \\ 0 & a & -1 \\ -1 & 0 & a \end{vmatrix} \text{ and of } \begin{vmatrix} 1 & a & a^2 \\ a^2 & 1 & a \\ a & a^2 & 1 \end{vmatrix}.$$

What is the value of the latter determinant when a is a complex cube root of unity?

4 Factorise the following determinants:

$$(i) \begin{vmatrix} 1 & bc & a \\ 1 & ca & b \\ 1 & ab & c \end{vmatrix}; \qquad (ii) \begin{vmatrix} b+c & c+a & a+b \\ bc & ca & ab \\ a & b & c \end{vmatrix};$$

$$(iii) \begin{vmatrix} b+c & c+a & a+b \\ c & a & b \\ 1 & 1 & 1 \end{vmatrix}; \qquad (iv) \begin{vmatrix} a & bc & a^2 \\ b & ca & b^2 \\ c & ab & c^2 \end{vmatrix}; \qquad (v) \begin{vmatrix} 1 & a & b+c \\ 1 & b & c+a \\ 1 & c & a+b \end{vmatrix}.$$

5 Show that:

$$\begin{vmatrix} a & b & c \\ a^2 & b^2 & c^2 \\ b+c & c+a & a+b \end{vmatrix} = k(a+b+c)(b-c)(c-a)(a-b),$$

and find the value of the constant k.

6 Evaluate $\begin{vmatrix} b & a & a \\ a & b & a \\ a & a & b \end{vmatrix}$,

and find other possible values of a 3×3 determinant which contains six a's and three b's.

7 Express in factorised form the determinants:

$$(i) \begin{vmatrix} b^2+c^2 & ab & ac \\ ab & c^2+a^2 & bc \\ ca & cb & a^2+b^2 \end{vmatrix}; \qquad (ii) \begin{vmatrix} 1 & a & a & a \\ 1 & b & a & a \\ 1 & a & b & a \\ 1 & a & a & b \end{vmatrix};$$

$(iii)\ \begin{vmatrix} 1 & 1 & 1 & 1 \\ 1 & -1 & 1 & 1 \\ 1 & -1 & -1 & 1 \\ 1 & -1 & -1 & -1 \end{vmatrix}$; $(iv)\ \begin{vmatrix} a & b & c & d \\ b & a & d & c \\ c & d & a & b \\ d & c & b & a \end{vmatrix}$; $(v)\ \begin{vmatrix} a & b & c & d \\ b & a & c & d \\ a & b & d & c \\ b & a & d & c \end{vmatrix}$.

8 Evaluate all the cofactors in the determinant $\begin{vmatrix} 1 & 2 & 4 \\ 2 & 4 & 8 \\ 4 & 8 & 16 \end{vmatrix}$.

Write down several other examples of determinants which have all the 2×2 minors equal to zero.
Compose a 4×4 determinant which has every one of its 3×3 minors zero.

Multiplication rule for determinants

Suppose that $\boldsymbol{A}$ and $\boldsymbol{B}$ are two 3×3 matrices and that their determinants are denoted by $|\boldsymbol{A}|$ and $|\boldsymbol{B}|$. Now we have seen (in 13.3) how to form, by row and column multiplication, their product matrix $\boldsymbol{AB}$, which in turn will have a determinant denoted by $|\boldsymbol{AB}|$. Is it, we wonder, possible to state the value of $|\boldsymbol{AB}|$ simply from our knowledge of $|\boldsymbol{A}|$ and $|\boldsymbol{B}|$ without having to find the product $\boldsymbol{AB}$ and then go through the tedious process of calculating its determinant?

A little reflection will show that this is easy. For if the successive transformations

$$\boldsymbol{x}'' = \boldsymbol{A}\boldsymbol{x}' \qquad \boldsymbol{x}' = \boldsymbol{B}\boldsymbol{x}$$

are combined to give

$$\boldsymbol{x}'' = \boldsymbol{AB}\boldsymbol{x},$$

volumes are multiplied by the factors $|\boldsymbol{A}|$ and $|\boldsymbol{B}|$ under the separate transformations, and by $|\boldsymbol{AB}|$ under the combined transformation $\boldsymbol{AB}$ (see 13.2).

But the final effects in both cases must be identical,

so $$\boxed{|\boldsymbol{AB}| = |\boldsymbol{A}| \times |\boldsymbol{B}|}$$

Similarly $|\boldsymbol{BA}| = |\boldsymbol{B}| \times |\boldsymbol{A}|$, so $\boldsymbol{AB}$ and $\boldsymbol{BA}$ have the same determinant, *even though the matrices themselves may well be different.*

A similar proof, using areas, applies in the case of 2×2 determinants and the result is also generally true for determinants of any order.

Exercise 13.5d

1 (*i*) Verify, by expansion of the determinants, that

$$\begin{vmatrix} \alpha & \beta \\ \gamma & \delta \end{vmatrix} \times \begin{vmatrix} a & b \\ c & d \end{vmatrix} = \begin{vmatrix} \alpha a + \beta c & \alpha b + \beta d \\ \gamma a + \delta c & \gamma b + \delta d \end{vmatrix}.$$

(*ii*) Express the square of $D = \begin{vmatrix} a & b \\ c & d \end{vmatrix}$ in four different ways as a 2×2 determinant.

(Note that if $D' = \begin{vmatrix} a & c \\ b & d \end{vmatrix}$, then we may have DD', $D'D$, DD and $D'D'$.)

2 Find a counter-example to show that, if $\boldsymbol{X}$, $\boldsymbol{Y}$, $\boldsymbol{Z}$ are square matrices, such that $\boldsymbol{Z} = \boldsymbol{X} + \boldsymbol{Y}$, then $\det \boldsymbol{X} + \det \boldsymbol{Y} \neq \det \boldsymbol{Z}$.

3 If $D = \begin{vmatrix} a & c & 0 \\ b & 0 & c \\ 0 & a & b \end{vmatrix}$, express D^2 as a 3×3 determinant in two different ways.

4 Express $\begin{vmatrix} a^2 + p^2 & ab + pq & ac + pr \\ ab + pq & b^2 + q^2 & bc + qr \\ ac + pr & bc + qr & c^2 + r^2 \end{vmatrix}$ as the square of a determinant, and hence prove that it is zero.

5 Using the notation $|\boldsymbol{a} \;\; \boldsymbol{b} \;\; \boldsymbol{c}|$ to denote $\begin{vmatrix} a_1 & b_1 & c_1 \\ a_2 & b_2 & c_2 \\ a_3 & b_3 & c_3 \end{vmatrix}$,
where $\boldsymbol{a}$, $\boldsymbol{b}$, $\boldsymbol{c}$ are column vectors, prove that

$$|\boldsymbol{a} + \boldsymbol{x} \;\; \boldsymbol{b} \;\; \boldsymbol{c}| = |\boldsymbol{a} \;\; \boldsymbol{b} \;\; \boldsymbol{c}| + |\boldsymbol{x} \;\; \boldsymbol{b} \;\; \boldsymbol{c}|,$$

and that $$|\boldsymbol{a}+\boldsymbol{x}+\boldsymbol{y} \;\; \boldsymbol{b} \;\; \boldsymbol{c}| = |\boldsymbol{a} \;\; \boldsymbol{b} \;\; \boldsymbol{c}| + |\boldsymbol{x} \;\; \boldsymbol{b} \;\; \boldsymbol{c}| + |\boldsymbol{y} \;\; \boldsymbol{b} \;\; \boldsymbol{c}|,$$

and express $|\boldsymbol{a} + \boldsymbol{x} \quad \boldsymbol{b} + \boldsymbol{y} \quad \boldsymbol{c} + \boldsymbol{z}|$ as the sum of eight determinants.

6 Prove the row-by-column rule for multiplication of 3×3 determinants by expressing the determinant of the product matrix in 13.3 (see p. 399) as the sum of 27 determinants, 21 of which may be proved to be zero.

7 If $\Delta = \begin{vmatrix} a_1 & b_1 & c_1 \\ a_2 & b_2 & c_2 \\ a_3 & b_3 & c_3 \end{vmatrix}$ and $\Delta^* = \begin{vmatrix} A_1 & A_2 & A_3 \\ B_1 & B_2 & B_3 \\ C_1 & C_2 & C_3 \end{vmatrix}$, (see 13.4, p. 410),

show that $\Delta^* = \Delta^2$, and that $\begin{vmatrix} B_2 & B_3 \\ C_2 & C_3 \end{vmatrix} = a_1 \Delta$,

with similar results for the other minors of Δ^*.

8 If a 3×3 determinant is zero, prove that any pair of rows or of columns of the adjoint determinant are linearly dependent,

e.g. that $\dfrac{A_1}{B_1} = \dfrac{A_2}{B_2} = \dfrac{A_3}{B_3}$, assuming $B_1, B_2, B_3 \neq 0$

(use the result of the previous question).

9 Show how to construct 2×2 and 3×3 matrices with elements which are integers so that their inverses may also have elements which are integers. (*cf.* 'codes', pp. 420–423).

*13.6 Systems of linear equations

Use of determinants in the solution of linear equations

For practical purposes, particularly in the case of large numbers of equations with large numbers of unknowns (as are frequently encountered in commercial, industrial and scientific situations), the method of reducing to échelon form (see 13.7) is the most convenient. But it is also instructive to approach the problem of the solution of a set of linear equations by using determinants.

We shall merely hint at the case of a 2×2 system:

$$\left.\begin{aligned} a_1x + b_1y &= c_1 \\ a_2x + b_2y &= c_2 \end{aligned}\right\} \quad \text{by noting that, if} \quad \Delta = \begin{vmatrix} a_1 & b_1 \\ a_2 & b_2 \end{vmatrix},$$

the solutions may be written

$$\Delta x = \begin{vmatrix} c_1 & b_1 \\ c_2 & b_2 \end{vmatrix}, \qquad \Delta y = \begin{vmatrix} a_1 & c_1 \\ a_2 & c_2 \end{vmatrix},$$

these being true in the non-regular case ($\Delta = 0$) as well as in the regular case. (*cf.* pp. 417–419).

Proceeding immediately to the 3×3 case,

$$a_1x + b_1y + c_1z = d_1 \tag{1}$$

$$a_2x + b_2y + c_2z = d_2 \tag{2}$$

$$a_3x + b_3y + c_3z = d_3, \tag{3}$$

a possible elementary way of obtaining solutions would be to eliminate z from (1) and (2) to obtain a new equation (4) in x and y; then to eliminate z from (2) and (3) to obtain a second equation (5) in x and y; and finally to eliminate y from (4) and (5) thereby enabling x to be found. How much simpler if this multi-step solution could be telescoped into a single step! In fact, this can be done by multiplying (1) by A_1, (2) by A_2, (3) by A_3, and adding, leading to

$$(a_1A_1 + a_2A_2 + a_3A_3)x + (b_1A_1 + b_2A_2 + b_3A_3)y \\ + (c_1A_1 + c_2A_2 + c_3A_3)z = d_1A_1 + d_2A_2 + d_3A_3.$$

In view of the properties of cofactors (see 13.5), we know

$$a_1A_1 + a_2A_2 + a_3A_3 = \Delta,$$

while the coefficients of y and z are zero because the elements of columns 2 and 3 have been combined with 'alien' cofactors, those of column 1. Thus

y and z have been eliminated in a single *tour de force*! The expression on the R.H.S. has the value

$$d_1A_1 + d_2A_2 + d_3A_3 = \begin{vmatrix} d_1 & b_1 & c_1 \\ d_2 & b_2 & c_2 \\ d_3 & b_3 & c_3 \end{vmatrix} = \det(\boldsymbol{d}, \boldsymbol{b}, \boldsymbol{c}),$$

where $\det(\boldsymbol{d}, \boldsymbol{b}, \boldsymbol{c})$, which we will call Δ_1, is simply the determinant Δ with column 1 (the col. vector $\boldsymbol{a}$), replaced by the col. vector $\boldsymbol{d}$.

Thus $\Delta x = \Delta_1$; and similarly we obtain

$\Delta y = \Delta_2$,

where Δ_2 is Δ with its second column replaced by the column vector $\boldsymbol{d}$; and finally,

$\Delta z = \Delta_3$.

So $$\boxed{\Delta x = \Delta_1, \qquad \Delta y = \Delta_2, \qquad \Delta z = \Delta_3}$$

If we make use of the matrix $\boldsymbol{M} = \begin{pmatrix} a_1 & b_1 & c_1 \\ a_2 & b_2 & c_2 \\ a_3 & b_3 & c_3 \end{pmatrix}$

and its adjoint matrix $\boldsymbol{M}^*$ (see 13.4), the foregoing proof can be considerably shortened. For the original equations (1), (2), (3) can be written in matrix form as

$$\begin{aligned} \boldsymbol{Mx} &= \boldsymbol{d} \\ \Rightarrow \quad \boldsymbol{M^*Mx} &= \boldsymbol{M^*d} \\ \Rightarrow \quad \Delta\boldsymbol{Ix} &= \boldsymbol{M^*d} \\ \Rightarrow \quad \Delta\boldsymbol{x} &= \boldsymbol{M^*d} \end{aligned}$$

$$\Rightarrow \quad \begin{pmatrix} \Delta x \\ \Delta y \\ \Delta z \end{pmatrix} = \begin{pmatrix} A_1 & A_2 & A_3 \\ B_1 & B_2 & B_3 \\ C_1 & C_2 & C_3 \end{pmatrix} \begin{pmatrix} d_1 \\ d_2 \\ d_3 \end{pmatrix} \quad \Rightarrow \quad \begin{aligned} \Delta x &= A_1d_1 + A_2d_2 + A_3d_3 \\ \Delta y &= B_1d_1 + B_2d_2 + B_3d_3 \\ \Delta z &= C_1d_1 + C_2d_2 + C_3d_3 \end{aligned}$$

So, again $\Delta x = \Delta_1, \quad \Delta y = \Delta_2, \quad \Delta z = \Delta_3$.

This solution in determinant form is known as *Cramer's rule* and is clearly capable of generalisation to any number of equations. But it suffers from the very serious disadvantage that evaluation of determinants is usually extremely tedious.

Example 1

Solve the equations

$$x + y - z = 1$$
$$x - y + z = 2$$
$$2x \quad - z = 4.$$

Here $\Delta = \begin{vmatrix} 1 & 1 & -1 \\ 1 & -1 & 1 \\ 2 & 0 & -1 \end{vmatrix} = 1(1) - 1(-3) - 1(2) = 2$

$$\Delta_1 = \begin{vmatrix} 1 & 1 & -1 \\ 2 & -1 & 1 \\ 4 & 0 & -1 \end{vmatrix} = 1(1) - 1(-6) - 1(4) = 3$$

$$\Delta_2 = \begin{vmatrix} 1 & 1 & -1 \\ 1 & 2 & 1 \\ 2 & 4 & -1 \end{vmatrix} = 1(-6) - 1(-3) - 1(0) = -3$$

$$\Delta_3 = \begin{vmatrix} 1 & 1 & 1 \\ 1 & -1 & 2 \\ 2 & 0 & 4 \end{vmatrix} = 1(-4) - 1(0) + 1(2) = -2.$$

So $2x = 3, \quad 2y = -3, \quad 2z = -2$

$\Rightarrow \quad x = \frac{3}{2}, \quad y = -\frac{3}{2}, \quad z = -1$

(as we could have found more simply by elimination!).

Example 2

Solve the *homogeneous* equations

$$a_1x + b_1y + c_1z = 0$$
$$a_2x + b_2y + c_2z = 0$$
$$a_3x + b_3y + c_3z = 0.$$

Here it is immediately clear that

$$\Delta_1 = \Delta_2 = \Delta_3 = 0 \quad \Rightarrow \quad \Delta x = \Delta y = \Delta z = 0.$$

So if $\Delta \neq 0$, it follows that the equations have a unique set of solutions $x = y = z = 0$; in other words, if there should exist a set of solutions which are *not* all zero, then we may be sure that $\Delta = 0$.

Conversely (see exercise 13.6a, no. **8**), it can be shown that if $\Delta = 0$, then it is possible to find values of x, y, z *which are not all zero* satisfying the original equations.

So

The existence of values x, y, z *not all zero* satisfying the *homogeneous* equations	$\Leftrightarrow$	$\Delta = 0$

Regarding these three equations as representing three planes through the origin O, we see that they also have some other point P in common if, and only if, they intersect in at least a line (OP), in which case their equations are linearly dependent and $\Delta = 0$.

But this important result may also be reconsidered in several other ways:

First, $\Delta = 0$ is the condition for the three straight lines $a_i x + b_i y + c_i = 0$ $(i = 1, 2, 3)$ to be concurrent (see 13.5, p. 425).

Second, $\Delta = |\boldsymbol{M}| = 0$ is also the condition for the matrix $\boldsymbol{M}$ to be singular (see 13.4), that is, for the non-existence of an inverse.

Third, in the context of the system of non-homogeneous equations

$$\boldsymbol{Mx} = \boldsymbol{d} \quad (\boldsymbol{d} \neq \boldsymbol{0})$$

is the condition for the system to be non-regular, and this we shall deal with in 13.7.

Fourth, $|\boldsymbol{M}|$ is the value of the scalar triple product $\boldsymbol{a}.(\boldsymbol{b} \times \boldsymbol{c})$ which represents the volume of the parallelepiped formed by the vectors $\boldsymbol{a}$, $\boldsymbol{b}$, $\boldsymbol{c}$ as adjacent edges (see 13.2). Hence $|\boldsymbol{M}| = 0$ is the condition for its volume to vanish, and so for the vectors $\boldsymbol{a}$, $\boldsymbol{b}$, $\boldsymbol{c}$ to be linearly dependent. And, similarly, it is also the condition for the three row-vectors

$$(a_i, b_i, c_i) \quad (i = 1, 2, 3) \quad \text{to be linearly dependent.}$$

Example 3

It is also useful to see how Cramer's rule can be obtained by vector methods. For equations (1), (2), and (3) above may be written in vector form as

$$x\boldsymbol{a} + y\boldsymbol{b} + z\boldsymbol{c} = \boldsymbol{d}.$$

Remembering that vector product distributes over addition, we take the *vector* product of both sides $\times\, \boldsymbol{b}$, giving

$$x\boldsymbol{a} \times \boldsymbol{b} + y\boldsymbol{b} \times \boldsymbol{b} + z\boldsymbol{c} \times \boldsymbol{b} = \boldsymbol{d} \times \boldsymbol{b}.$$

But $\boldsymbol{b} \times \boldsymbol{b} = \boldsymbol{0}$.

Hence $x\boldsymbol{a} \times \boldsymbol{b} + z\boldsymbol{c} \times \boldsymbol{b} = \boldsymbol{d} \times \boldsymbol{b}$.

Now take the *scalar* product of both sides with $\boldsymbol{c}$, and again remember that scalar product distributes over addition.

Then $x(\boldsymbol{a} \times \boldsymbol{b}).\boldsymbol{c} + z(\boldsymbol{c} \times \boldsymbol{b}).\boldsymbol{c} = (\boldsymbol{d} \times \boldsymbol{b}).\boldsymbol{c}$.

But $(\boldsymbol{c} \times \boldsymbol{b}).\boldsymbol{c} = 0$.

Hence $x(\boldsymbol{a} \times \boldsymbol{b}).\boldsymbol{c} = (\boldsymbol{d} \times \boldsymbol{b}).\boldsymbol{c}$.

But $\boldsymbol{a} \times \boldsymbol{b} = (a_2b_3 - a_3b_2)\boldsymbol{i} + (a_3b_1 - a_1b_3)\boldsymbol{j} + (a_1b_2 - a_2b_1)\boldsymbol{k}$

and $\boldsymbol{c} = c_1\boldsymbol{i} + c_2\boldsymbol{j} + c_3\boldsymbol{k}$.

So $(\boldsymbol{a} \times \boldsymbol{b}).\boldsymbol{c} = (a_2b_3 - a_3b_2)c_1 + (a_3b_1 - a_1b_3)c_2 + (a_1b_2 - a_2b_1)c_3$

$$= \begin{vmatrix} a_1 & b_1 & c_1 \\ a_2 & b_2 & c_2 \\ a_3 & b_3 & c_3 \end{vmatrix} = \Delta;$$

and similarly, $(\boldsymbol{d} \times \boldsymbol{b}).\boldsymbol{c} = \begin{vmatrix} d_1 & b_1 & c_1 \\ d_2 & b_2 & c_2 \\ d_3 & b_3 & c_3 \end{vmatrix} = \Delta_1.$

So $\Delta x = \Delta_1$, and similarly $\Delta y = \Delta_2$, $\Delta z = \Delta_3$.

Exercise 13.6a

1 Use Cramer's rule to solve:

(*i*) $2x - 5y + 3 = 0$
$3x + y - 11 = 0$;

(*ii*) $2x - 3y - 1 = 0$
$4x - 6y + 5 = 0$;

(*iii*) $lx + my + nz = 0$
$mx - ly + pz = 0$.

2 Use Cramer's rule to solve the simultaneous equations:

(*i*) $x - z = 4$
$2x + y + 3z = -2$
$4x + 2y + 5z = -3$;

(*ii*) $4x - y - 2z = -7$
$3x + 5y - 6z = 4$
$6x - 13y + 6z = -29$.

3 Use Cramer's rule to solve for $\cos\theta$ and $\sin\theta$ in terms of a, b, c, p, q, and r, and hence eliminate θ:

$a\cos\theta + b\sin\theta = c$
$p\cos\theta + q\sin\theta = r$.

4 Obtain the condition for the quadratic equation $ax^2 + bx + c = 0$, $px^2 + qx + r = 0$ to have a common root by first solving these as a pair of simultaneous equations in x^2 and x, using Cramer's rule, and then eliminating x. (Compare exercises 13.5a, no. **8** and 13.5b, no. **10**.)

5 Find the condition that the equations

$ax + by + cz = 0$
$bx + cy + az = 0$
$cx + ay + bz = 0$

have a solution other than $(x, y, z) = (0, 0, 0)$.

6 Find the values of λ for which the equations

$$\begin{aligned} 4x - 6y - z &= \lambda x \\ x - 4y - z &= \lambda y \\ 2x + 3y + z &= \lambda z \end{aligned}$$

have a solution other than $x = y = z = 0$, and find the ratios $x:y:z$ for each of these values of λ. (O.C.)

7 If ω denotes one of the cube roots of 1 other than 1 itself, solve the equations

$$\begin{aligned} x + y + z &= a \\ x + \omega y + \omega^2 z &= b \\ x + \omega^2 y + \omega z &= c \end{aligned}$$

for x, y, z in terms of a, b, c. Give your answer in as simple terms as possible. (S.M.P.)

8 Consider the three homogeneous equations

$$\begin{pmatrix} a_1 & b_1 & c_1 \\ a_2 & b_2 & c_2 \\ a_3 & b_3 & c_3 \end{pmatrix} \begin{pmatrix} x \\ y \\ z \end{pmatrix} = \begin{pmatrix} 0 \\ 0 \\ 0 \end{pmatrix}.$$

Prove that, if solutions other than $x = y = z = 0$ exist, then $\Delta = 0$ (the 3×3 determinant of the coefficients).

Prove conversely, that if $\Delta = 0$, then solutions other than the trivial ones may be found.

(Note that $\Delta x = \Delta_1 = 0$, etc., and that if $\Delta = 0$, then any arbitrary value of x will satisfy. Note also that

$$\begin{pmatrix} a_2 & b_2 & c_2 \\ a_3 & b_3 & c_3 \end{pmatrix} \begin{pmatrix} x \\ y \\ z \end{pmatrix} = \begin{pmatrix} 0 \\ 0 \end{pmatrix} \quad \text{are satisfied by} \quad \begin{pmatrix} x \\ y \\ z \end{pmatrix} = \begin{pmatrix} A_1 \\ B_1 \\ C_1 \end{pmatrix} \quad \text{in any case.)}$$

9 Prove that if α, β are real numbers and $\alpha \neq 0$, then the equations

$$\begin{aligned} x + y + 2z &= 5 \\ 2x + y + z &= 2 \\ 3x + y + \alpha z &= \beta \end{aligned}$$

have a unique solution. Prove further, that if $\alpha = 0$, then these equations have a solution only if β takes one particular value, which should be found. Find all the solutions of the equations in this case. (O.S.)

10 Given that k is a real constant, solve completely, when possible, the simultaneous equations

$$\begin{aligned} kx + 2y + 8z &= 0 \\ x - y + 2kz &= 0 \\ x + y + 6z &= 2k \end{aligned}$$

discussing carefully any special cases. (O.S.)

11 Three planes through the origin have equations

$$\begin{aligned} \Pi_1 \quad & ax + hy + gz = 0 \\ \Pi_2 \quad & hx + by + fz = 0 \\ \Pi_3 \quad & gx + fy + cz = 0. \end{aligned}$$

Let Γ_r, $r = 1, 2, 3$, be the three planes defined as follows: Γ_1 is the plane through the intersection of Π_2 and Π_3, and also containing the x-axis. Γ_2 and Γ_3 are similarly defined. Prove that Γ_1, Γ_2, Γ_3 have a common point.

Regular System — 'Ill-conditioned' case

We know that in the case $\Delta \neq 0$ our original system of equations will have a unique solution. Note, however, that if Δ is very small compared with Δ_1, Δ_2, and Δ_3, the values of x, y, z may be large: the point of intersection of the three planes is a long way off, if we think in terms of the first interpretation of p. 418; or the matrix is nearly singular, in the second interpretation; or the 3 vectors $\boldsymbol{a}$, $\boldsymbol{b}$, $\boldsymbol{c}$ are very nearly coplanar in the case of the third.

In such a situation, the equations (and the matrix $\boldsymbol{M}$) are described as *ill-conditioned*. This means that the solution is very sensitive to changes in the data: move one of our three planes by a very slight alteration of perhaps just one of its coefficients, and the common point of the three planes may be moved a long way.

The singular case, in terms of Cramer's rule

In a non-regular set of equations, where $\Delta = 0$, we know that no finite solutions are possible in x, y, z unless *all three* of the determinants Δ_1, Δ_2, Δ_3 are zero. This may happen, of course by having $\boldsymbol{d} = \boldsymbol{0}$, i.e., $d_1 = 0$, $d_2 = 0$, and $d_3 = 0$.

In the more general case when $\boldsymbol{d} \neq \boldsymbol{0}$, it may be wondered whether it is possible for Δ_1 to be zero without Δ_2 or Δ_3 necessarily being zero. At first it seems that this is not possible, for one may argue as follows:

Firstly, $\Delta = 0$ implies that $\boldsymbol{a}$, $\boldsymbol{b}$, $\boldsymbol{c}$ are linearly dependent,

$\Delta_1 = 0$ implies that $\boldsymbol{d}$, $\boldsymbol{b}$, $\boldsymbol{c}$ are linearly dependent.

Hence it appears that $\boldsymbol{a}$, $\boldsymbol{b}$, $\boldsymbol{d}$ are linearly dependent, so that Δ_3 is also zero, and similarly $\Delta_2 = 0$.

Alternatively, we may think of the three planes represented by the three linear equations. Now since in any case $\Delta = 0$, we know the three planes either have a common line or else form a triangular prism, including the case when some pairs are parallel. Consider now the lines in which the planes intersect the plane $z = 0$.

They have equations

$$\left.\begin{aligned} a_1x + b_1y &= d_1 \\ a_2x + b_2y &= d_2 \\ a_3x + b_3y &= d_3 \end{aligned}\right\} \text{ and the vanishing of the determinant } \Delta_3 = \begin{vmatrix} a_1 & b_1 & d_1 \\ a_2 & b_2 & d_2 \\ a_3 & b_3 & d_3 \end{vmatrix}$$

ensures that these three lines have a common point (taking the more general case when they are not all three parallel). In such conditions, the three planes do not form a triangular prism, but will have a common line. This means that they also cut the plane $y = 0$ in three concurrent lines, whose equations are

$$a_1x + c_1z = d_1$$

$$a_2x + c_2z = d_2$$

$$a_3x + c_3z = d_3$$

and whose determinant should therefore vanish. This would give the result $\Delta_2 = 0$, and similarly we should have $\Delta_3 = 0$.

However, the above arguments overlook the possibility that Δ and Δ_1 vanish as a result of the *two* vectors $\boldsymbol{b}$ and $\boldsymbol{c}$ being linearly dependent. In such a case, $\boldsymbol{a}$ and $\boldsymbol{d}$ can be freely chosen as *any* two vectors, so it would clearly be possible to avoid the vanishing of either Δ_2 or Δ_3, or of them both.

Exercise 13.6b

1 Consider the lines

$$2x + 3y - 4 = 0$$
$$6x + ky - 2 = 0$$

in the case when $k = 9$, and when $k = 9 + \delta$, where δ is very small.

2 Find k so that the planes

$$2x - 3y + z = c$$
$$3x - y = -4$$
$$x + ky + 2z = 3$$

may form a triangular prism in general.

When k has this value, find c so that the planes may have a common line, and check that, in this case, all four of the 3×3 determinants of the 3×4 matrix (the 'augmented matrix' of the set of equations) are zero.

In this case, find constants p, q, and r, such that:

$$p(2 \;\; -3 \;\; 1 \;\; c) + q(3 \;\; -1 \;\; 0 \;\; -4) + r(1 \;\; k \;\; 2 \;\; 3) = (0 \;\; 0 \;\; 0 \;\; 0).$$

In the case $c = 1$, find the effect of changing k slightly from its critical value.

3 (*i*) Find the values of λ and μ for which the equations

$$\begin{aligned} \lambda x + y &= 0 \\ x + y - z &= 0 \\ 2x + 3y - 2z &= 0 \end{aligned}$$

have a solution other than the trivial solution $x = y = z = 0$.

(*ii*) For what values of λ do the equations

$$\begin{aligned} \lambda x + y &= 1 \\ x + \lambda y - z &= 1 \\ 2x + 2y - 2z &= 1 \end{aligned}$$

have a solution? (O.S.)

4 Find a vector which is perpendicular to each of the following pairs:

(*i*) $(2 \quad 1 \quad 0)$ and $(-2 \quad 0 \quad 3)$;

(*ii*) $(1 \quad 4 \quad -1)$ and $(-1 \quad 2 \quad 0)$;

(*iii*) $(\cos \alpha \quad \sin \alpha \quad 0)$ and $(0, \quad \cos \beta \quad \sin \beta)$.

*13.7 Systematic reduction

Three linear equations with three unknowns

We now deal with the case of three simultaneous linear equations in the form

(*i*)
$$\left.\begin{aligned} a_1x + b_1y + c_1z &= d_1 \\ a_2x + b_2y + c_2z &= d_2 \\ a_3x + b_3y + c_3z &= d_3 \end{aligned}\right\}$$

or (*ii*)
$$\begin{pmatrix} a_1 & b_1 & c_1 \\ a_2 & b_2 & c_2 \\ a_3 & b_3 & c_3 \end{pmatrix}\begin{pmatrix} x \\ y \\ z \end{pmatrix} = \begin{pmatrix} d_1 \\ d_2 \\ d_3 \end{pmatrix}; \quad \boldsymbol{Mx} = \boldsymbol{d}$$

or (*iii*)
$$x\begin{pmatrix} a_1 \\ a_2 \\ a_3 \end{pmatrix} + y\begin{pmatrix} b_1 \\ b_2 \\ b_3 \end{pmatrix} + z\begin{pmatrix} c_1 \\ c_2 \\ c_3 \end{pmatrix} = \begin{pmatrix} d_1 \\ d_2 \\ d_3 \end{pmatrix}; \quad x\boldsymbol{a} + y\boldsymbol{b} + z\boldsymbol{c} = \boldsymbol{d}$$

Let us consider now the threefold† interpretation in this case, just as we did with two equations (see 13.4).

In the first place, as an equation such as $2x - y + 5z = -3$ connecting the coordinates (x, y, z) of a point in three-dimensional space represents a *plane*, our first interpretation concerns three planes in space; and in the general case (when the equations are regular), we expect there to exist a

† In due course it will become apparent that the second and third interpretations are really the same, but expressed differently. The similarity of these to the first interpretation is not however so apparent, *cf.* 13.4, p. 417.

unique point common to the three planes, as for example the three planes which intersect at the ('trihedral') vertex of a cube. The non-regular cases, when there is something unusual, will be considered presently.

In the second interpretation, we require to find which vector is transformed by the matrix $\boldsymbol{M}$ into the given vector $\boldsymbol{d}$, and in the general case this problem again has a unique solution.

Thirdly, we are attempting to find what multiples of the given vectors $\boldsymbol{a}$, $\boldsymbol{b}$, $\boldsymbol{c}$ must be 'mixed' together in order to obtain the vector $\boldsymbol{d}$; and if all is well, there will again be a unique answer.

Whenever the matrix $\boldsymbol{M}$ is non-singular, i.e., possesses an inverse, the unique solution is applicable. For then

$$\boldsymbol{Mx} = \boldsymbol{d} \quad \Rightarrow \quad \boldsymbol{M}^{-1}\boldsymbol{Mx} = \boldsymbol{M}^{-1}\boldsymbol{d},$$

so that $\boldsymbol{x} = \boldsymbol{M}^{-1}\boldsymbol{d}$ is the unique solution.

The question of actually *finding the inverse* of a 3×3 matrix was discussed in 13.4, and a second method will emerge in this section (p. 455). But in numerical cases, even with the facility of a computer, it is usually simpler to use another method, rather than to compute the inverse. Similarly, a problem in mechanics might lead to the following set of equations, in which T, S represent unknown tensions, while a, b, and c represent unknown accelerations:

$$m_1 g - S = m_1 a; \qquad m_2 g - T = m_2 c; \qquad c = 2a - 2b;$$
$$m_3 g - 2T = m_3 b; \qquad S - 2T = m_4 a.$$

In such a case, the values of the unknowns are best found by 'common-sense' elimination; it would be absurd to force the problem into the format of a 5×5 matrix with a very large number of zeros.

Echelon form

Suppose we are asked to express

$2r^3 - 5r^2 - 3r + 11$ in the form

$Ar(r+1)(r+2) + Br(r+1) + Cr + D.$

$$\begin{array}{lrll}
\text{Put } r = 0, & 11 = & D & (1) \\
r = -1, & 7 = & -C + D & (2) \\
r = -2, & -19 = & 2B - 2C + D & (3) \\
r = 1, & 5 = & 6A + 2B + C + D & (4)
\end{array}$$

Equation (1) gives immediately $D = 11$; substitution into (2) gives $C = 4$. Knowing C and D, (3) gives $B = -11$, and finally (4) enables us to find $A = 2$.

Thus $2r^3 - 5r^2 - 3r + 11 \equiv 2r(r+1)(r+2) - 11r(r+1) + 4r + 11.$

Now these equations were particularly easy to solve because of the form of the 4×4 matrix

$$\begin{pmatrix} 0 & 0 & 0 & 1 \\ 0 & 0 & -1 & 1 \\ 0 & 2 & -2 & 1 \\ 6 & 2 & 1 & 1 \end{pmatrix} \begin{pmatrix} A \\ B \\ C \\ D \end{pmatrix} = \begin{pmatrix} 11 \\ 7 \\ -19 \\ 5 \end{pmatrix}$$

where it will be noticed that there is a triangle of zeros above the N.E.–S.W. diagonal. Such a matrix is said to be in *triangular*, or *échelon*, form. Had we written the four equations in reverse order, we should have had

$$\begin{pmatrix} 6 & 2 & 1 & 1 \\ 0 & 2 & -1 & 1 \\ 0 & 0 & -1 & 1 \\ 0 & 0 & 0 & 1 \end{pmatrix} \begin{pmatrix} A \\ B \\ C \\ D \end{pmatrix} = \begin{pmatrix} 5 \\ -19 \\ 7 \\ 11 \end{pmatrix}.$$

The latter is a more usual form for triangular matrices — we prefer to have all the zeros either (as in this case) below the *leading* diagonal (N.W. to S.E.) or above it, thus:

$$\begin{pmatrix} \cdot & \cdot & \cdot & \cdot \\ 0 & \cdot & \cdot & \cdot \\ 0 & 0 & \cdot & \cdot \\ 0 & 0 & 0 & \cdot \end{pmatrix} \quad \text{or} \quad \begin{pmatrix} \cdot & 0 & 0 & 0 \\ \cdot & \cdot & 0 & 0 \\ \cdot & \cdot & \cdot & 0 \\ \cdot & \cdot & \cdot & \cdot \end{pmatrix}$$

(leading diagonal marked in each case)

(*lower diagonal form*) (*upper diagonal form*)

Reduction to échelon form

If we start with a set of linear equations, we may attempt their solution by trying to obtain from them an equivalent set of equations *whose matrix is triangular*. The evaluation of the several unknowns will then be a simple matter, as in the example above. The method of reduction to such an *échelon form*, resembles the method used in elementary algebra:

Example I

Solve the equations

$$x - z = 4 \qquad (1)$$

$$2x + y + 3z = -2 \qquad (2)$$

$$4x + 2y + 5z = -3 \qquad (3)$$

Multiply equation (1) by 2, and subtract from equation (2).

Also, multiply equation (1) by 4, and subtract from equation (3).

These operations we shall abbreviate respectively: (2) $-2\times(1)$,
and: (3) $-4\times(1)$.

$$\text{Then}\quad \begin{cases} x \qquad\; - z = 4 & (1)\\ y + 5z = -10 & (5) = (2) - 2\times(1)\\ 2y + 9z = -19 & (6) = (3) - 4\times(1)\end{cases}$$

$$\Rightarrow\quad \begin{cases} x \qquad\; - z = 4 & (1)\\ y + 5z = -10 & (5)\\ \qquad\;\; - z = 1 & (7) = (6) - 2\times(5)\end{cases}$$

The equations are now in échelon form, and by 'going up the ladder' (*échelon* = ladder), we obtain successively $z = -1$, $y = -5$, $x = 3$. In this particular example, there was a short cut, which we have ignored, namely (3) $-$ $2\times(2)$, which leads to $z = -1$ straight away, but we must be prepared for less fortunate cases!

Reminding ourselves of the notation introduced in 13.5, the working can now be condensed:

$$\begin{matrix} \\ \boldsymbol{r}_2' = \boldsymbol{r}_2 - 2\boldsymbol{r}_1 \\ \boldsymbol{r}_3' = \boldsymbol{r}_3 - 4\boldsymbol{r}_1 \end{matrix} \qquad \begin{pmatrix} 1 & 0 & -1 & 4\\ 0 & 1 & 5 & -10\\ 0 & 2 & 9 & -19 \end{pmatrix}$$

$$\begin{matrix} \\ \\ \boldsymbol{r}_3' = \boldsymbol{r}_3 - 2\boldsymbol{r}_2 \end{matrix} \qquad \begin{pmatrix} 1 & 0 & -1 & 4\\ 0 & 1 & 5 & -10\\ 0 & 0 & -1 & 1 \end{pmatrix}$$

corresponding to equations (1), (6) and (7) above.

We note the three interpretations of the solution $(x, y, z) = (3, -5, -1)$:

(*i*) the three planes intersect at the point with coordinates $(3, -5, -1)$;

or (*ii*) in the linear transformation of three-dimensional space represented by the matrix

$$\begin{pmatrix} 1 & 0 & -1\\ 2 & 1 & 3\\ 4 & 2 & 5 \end{pmatrix}, \quad \text{the vector} \quad \begin{pmatrix} 4\\ -2\\ -3 \end{pmatrix}$$

arises from the vector $\begin{pmatrix} 3\\ -5\\ -1 \end{pmatrix}$.

or (*iii*) Letting $\boldsymbol{a} = \begin{pmatrix} 1\\ 2\\ 4 \end{pmatrix}$, $\boldsymbol{b} = \begin{pmatrix} 0\\ 1\\ 2 \end{pmatrix}$, $\boldsymbol{c} = \begin{pmatrix} -1\\ 3\\ 5 \end{pmatrix}$, $\boldsymbol{d} = \begin{pmatrix} 4\\ -2\\ -3 \end{pmatrix}$;

we have found that

$$d = 3a - 5b - c,$$

so that $\boldsymbol{d}$ can be expressed as a linear combination of $\boldsymbol{a}$, $\boldsymbol{b}$, and $\boldsymbol{c}$.

Example 2

$$\begin{aligned} 4x - y - 2z &= -7 \\ 3x + 5y - 6z &= 4 \\ x - 3y - 4z &= -9. \end{aligned}$$

We show the matrix of coefficients at successive stages, and on the right give the values of the row sums (see below)

$$\begin{pmatrix} 4 & -1 & -2 & -7 \\ 3 & 5 & -6 & 4 \\ 1 & -3 & -4 & -9 \end{pmatrix} \quad \begin{matrix} -6 \\ 6 \\ -15 \end{matrix}$$

$$\boldsymbol{r}_1' = 2\boldsymbol{r}_1 - \boldsymbol{r}_3 \qquad \begin{pmatrix} 7 & 1 & 0 & -5 \\ 3 & 5 & -6 & 4 \\ 1 & -3 & -4 & -9 \end{pmatrix} \quad \begin{matrix} 3 \\ 6 \\ -15 \end{matrix}$$

This time we are going to obtain an 'upper' triangular form.

$$\boldsymbol{r}_2' = 2\boldsymbol{r}_2 - 3\boldsymbol{r}_3 \qquad \begin{pmatrix} 7 & 1 & 0 & -5 \\ 3 & 19 & 0 & 35 \\ 1 & -3 & -4 & -9 \end{pmatrix} \quad \begin{matrix} 3 \\ 57 \\ -15 \end{matrix}$$

It remains to get a zero as the second element in the first row without disturbing the two zeros already obtained.

$$\boldsymbol{r}_1' = 19\boldsymbol{r}_1 - \boldsymbol{r}_2 \qquad \begin{pmatrix} 130 & 0 & 0 & -130 \\ 3 & 19 & 0 & 35 \\ 1 & -3 & -4 & -9 \end{pmatrix} \quad \begin{matrix} 0 \\ 57 \\ -15 \end{matrix}$$

The equations are now in échelon form:

$$\begin{aligned} 130x \qquad\qquad &= -130 \\ 3x + 19y \qquad &= 35 \\ x - 3y - 4z &= -9 \end{aligned}$$

and so $x = -1, \quad y = 2, \quad z = \frac{1}{2}.$

Note that $\boldsymbol{r}_1' = 2\boldsymbol{r}_1 - \boldsymbol{r}_3$ is equivalent to the two steps $\boldsymbol{r}_1' = 2\boldsymbol{r}_1$ (i.e., multiply throughout equation (1) by 2), followed by $\boldsymbol{r}_1' = \boldsymbol{r}_1 - \boldsymbol{r}_3$. We might have used $\boldsymbol{r}_1' = \boldsymbol{r}_1 - \frac{1}{2}\boldsymbol{r}_3$, but this would have caused the work to become encumbered by fractions. On the right, a tally is shown of row sums, providing a useful check on accuracy, which is always used in practice. For example, when we applied $\boldsymbol{r}_2' = 2\boldsymbol{r}_2 - 3\boldsymbol{r}_3$, the sum of the newly formed second row (57) does satisfy the relation:

$$57 = 2 \times 6 - 3(-15).$$

Exercise 13.7a

1 Solve the following sets of equations using the method of the above section, showing the row (or column) operations, and keeping a check of row sums:

(*i*)
$$\begin{aligned} 2x + 2y - 3z &= 0 \\ 3x + 2y - 10z &= 7 \\ 3x + 4y + z &= 8; \end{aligned}$$

(*ii*)
$$\begin{aligned} 2x - 7y + 4z &= 2 \\ 4x + 2z &= -3 \\ 3x - 2y + 7z &= 1; \end{aligned}$$

(*iii*)
$$\begin{aligned} x + y + z &= 1 \\ x - 2y + z &= -2 \\ 2x + 3y - 2z &= -1; \end{aligned}$$

(*iv*)
$$\begin{pmatrix} 1 & 1 & -1 \\ 1 & 2 & 3 \\ 1 & 3 & -1 \end{pmatrix} \begin{pmatrix} x \\ y \\ z \end{pmatrix} = \begin{pmatrix} 1 \\ -1 \\ 1 \end{pmatrix};$$

(*v*)
$$\begin{pmatrix} 4 & 6 & 5 \\ -3 & 8 & 14 \\ 9 & -4 & 2 \end{pmatrix} \begin{pmatrix} x \\ y \\ z \end{pmatrix} = \begin{pmatrix} 17 \\ 67 \\ -8 \end{pmatrix}.$$

2 Find numbers p, q, r such that:

$$p\begin{pmatrix} 1 \\ 1 \\ 2 \end{pmatrix} + q\begin{pmatrix} -1 \\ 1 \\ -1 \end{pmatrix} + r\begin{pmatrix} 1 \\ 2 \\ 3 \end{pmatrix} = \begin{pmatrix} 1 \\ 0 \\ 2 \end{pmatrix}.$$

3 Solve the homogeneous equations for the ratios $x:y:z:w$:

$$\begin{aligned} x + y - z + w &= 0 \\ 2x - 3y + 4z - 8w &= 0 \\ x + 2y + 5z + 3w &= 0. \end{aligned}$$

4 Let $\boldsymbol{A}$ stand for the matrix

$$\begin{pmatrix} 17 & -4 & 0 \\ 23 & 49 & 3 \\ 18 & 13 & 1 \end{pmatrix}$$

Find two matrices $\boldsymbol{M}_1$, $\boldsymbol{M}_2$ such that:

(*i*) each has only one non-zero entry off the leading diagonal;

(*ii*) the product $\boldsymbol{M}_2\boldsymbol{M}_1\boldsymbol{A}$ has zeros above and to the right of the leading diagonal, so that its final form is

$$\begin{pmatrix} . & 0 & 0 \\ . & . & 0 \\ . & . & . \end{pmatrix}$$

where the places indicated by the dots are occupied by numbers.
From your result, deduce the value of the determinant of $\boldsymbol{A}$. (S.M.P.)

5 Describe a method of finding the solution of three simultaneous linear equations and illustrate your answer by solving the system:

$$\begin{aligned} 3.9x + 2.5y - 0.5z &= 4.3 \\ 5.2x + 2.9y - 1.6z &= 10.9 \\ 2.6x + 1.8y - 0.7z &= 2.8 \end{aligned}$$

working to three significant figures. (M.E.I.)

6 Use the method of pivotal condensation,† with a sum check, or any other suitable method, to find the values of x, y, z from the following simultaneous equations. Tabulate the details of your working, and give results to three significant figures (you may use a slide rule).

$$\begin{aligned} 9.37x + 3.04y - 2.44z &= 9.23 \\ 3.04x + 6.18y + 1.22z &= 8.20 \\ -2.44x + 1.22y + 8.44z &= 3.93. \end{aligned}$$

Find the residual when your results are substituted in the second equation. Comment on the value of this residual as a guide to the accuracy of your results. (M.E.I.)

Identity matrix by using row operations

It is possible to proceed still further in this process. Reverting to Example 1, and resuming from the point we reached when triangular form was achieved:

$$\begin{pmatrix} 1 & 0 & -1 & 4 \\ 0 & 1 & 5 & -10 \\ 0 & 0 & -1 & 1 \end{pmatrix} \quad \begin{matrix} 4 \\ -4 \\ 0 \end{matrix}$$

$\boldsymbol{r}_1' = \boldsymbol{r}_1 - \boldsymbol{r}_3$

$$\begin{pmatrix} 1 & 0 & 0 & 3 \\ 0 & 1 & 5 & -10 \\ 0 & 0 & -1 & 1 \end{pmatrix} \quad \begin{matrix} 4 \\ -4 \\ 0 \end{matrix}$$

$\boldsymbol{r}_2' = \boldsymbol{r}_2 + 5\boldsymbol{r}_3$

$$\begin{pmatrix} 1 & 0 & 0 & 3 \\ 0 & 1 & 0 & -5 \\ 0 & 0 & -1 & 1 \end{pmatrix} \quad \begin{matrix} 4 \\ -4 \\ 0 \end{matrix}$$

† i.e. by reduction to échelon form.

Finally

$$\boldsymbol{r}_3' = -\boldsymbol{r}_3 \quad \text{gives} \quad \begin{pmatrix} 1 & 0 & 0 & 3 \\ 0 & 1 & 0 & -5 \\ 0 & 0 & 1 & -1 \end{pmatrix} \quad \begin{matrix} 4 \\ -4 \\ 0 \end{matrix}$$

and it will be seen that we have effectively succeeded in setting up the identity matrix on the left-hand side of the equation. However, it is not necessary to go beyond the point where the échelon form is obtained.

It is instructive to follow through the above working using elementary matrices. In Example 1, the successive operations

$$\boldsymbol{r}_2' = \boldsymbol{r}_2 - 2\boldsymbol{r}_1, \qquad \boldsymbol{r}_3' = \boldsymbol{r}_3 - 4\boldsymbol{r}_1, \quad \text{and} \quad \boldsymbol{r}_3' = \boldsymbol{r}_3 - 2\boldsymbol{r}_2$$

are carried out by using

$$\begin{pmatrix} 1 & 0 & 0 \\ -2 & 1 & 0 \\ 0 & 0 & 1 \end{pmatrix}, \quad \begin{pmatrix} 1 & 0 & 0 \\ 0 & 1 & 0 \\ -4 & 0 & 1 \end{pmatrix} \quad \text{and} \quad \begin{pmatrix} 1 & 0 & 0 \\ 0 & 1 & 0 \\ 0 & -2 & 1 \end{pmatrix}$$

as pre-multipliers. Their combined effect is the product

$$\begin{pmatrix} 1 & 0 & 0 \\ 0 & 1 & 0 \\ 0 & -2 & 1 \end{pmatrix} \begin{pmatrix} 1 & 0 & 0 \\ 0 & 1 & 0 \\ -4 & 0 & 1 \end{pmatrix} \begin{pmatrix} 1 & 0 & 0 \\ -2 & 1 & 0 \\ 0 & 0 & 1 \end{pmatrix} = \begin{pmatrix} 1 & 0 & 0 \\ -2 & 1 & 0 \\ 0 & -2 & 1 \end{pmatrix};$$

and when this pre-multiplies the original matrix, we obtain

$$\begin{pmatrix} 1 & 0 & 1 \\ -2 & 1 & 0 \\ 0 & -2 & 1 \end{pmatrix} \begin{pmatrix} 1 & 0 & -1 & 4 \\ 2 & 1 & 3 & -2 \\ 4 & 2 & 5 & -3 \end{pmatrix} = \begin{pmatrix} 1 & 0 & -1 & 4 \\ 0 & 1 & 5 & -10 \\ 0 & 0 & -1 & 1 \end{pmatrix}$$

corresponding to the result above.

$$\text{Thus the matrix} \quad \begin{pmatrix} 1 & 0 & 0 \\ -2 & 1 & 0 \\ 0 & -2 & 1 \end{pmatrix} \quad \text{pre-multiplying} \quad \begin{pmatrix} 1 & 0 & -1 \\ 2 & 1 & 3 \\ 4 & 2 & 5 \end{pmatrix}$$

changes it into *lower triangular form*.

Proceeding with the three succeeding transformations,

$$\boldsymbol{r}_1' = \boldsymbol{r}_1 - \boldsymbol{r}_3, \qquad \boldsymbol{r}_2' = \boldsymbol{r}_2 + 5\boldsymbol{r}_3, \qquad \boldsymbol{r}_3' = -\boldsymbol{r}_3,$$

their combined effect is given in a similar way by the product of the elementary matrices:

$$\begin{pmatrix} 1 & 0 & 0 \\ 0 & 1 & 0 \\ 0 & 0 & -1 \end{pmatrix}\begin{pmatrix} 1 & 0 & 0 \\ 0 & 1 & 5 \\ 0 & 0 & 1 \end{pmatrix}\begin{pmatrix} 1 & 0 & -1 \\ 0 & 1 & 0 \\ 0 & 0 & 1 \end{pmatrix} = \begin{pmatrix} 1 & 0 & -1 \\ 0 & 1 & 5 \\ 0 & 0 & -1 \end{pmatrix}.$$

The combined effect of all six matrices is given by

$$\begin{pmatrix} 1 & 0 & -1 \\ 0 & 1 & 5 \\ 0 & 0 & -1 \end{pmatrix}\begin{pmatrix} 1 & 0 & 0 \\ -2 & 1 & 0 \\ 0 & -2 & 1 \end{pmatrix} = \begin{pmatrix} 1 & 2 & -1 \\ -2 & -9 & 5 \\ 0 & 2 & -1 \end{pmatrix}.$$

This, pre-multiplying the original equation, gives:

$$\begin{pmatrix} 1 & 2 & -1 \\ -2 & -9 & 5 \\ 0 & 2 & -1 \end{pmatrix}\begin{pmatrix} 1 & 0 & -1 & 4 \\ 2 & 1 & 3 & -2 \\ 4 & 2 & 5 & -3 \end{pmatrix} = \begin{pmatrix} 1 & 0 & 0 & 3 \\ 0 & 1 & 0 & -5 \\ 0 & 0 & 1 & -1 \end{pmatrix}.$$

It will therefore be seen that, by making use of the six elementary matrices we have succeeded in showing that

$$\begin{pmatrix} 1 & 2 & -1 \\ -2 & -9 & 5 \\ 0 & 2 & -1 \end{pmatrix} \text{ is the inverse of } \begin{pmatrix} 1 & 0 & -1 \\ 2 & 1 & 3 \\ 0 & 2 & 5 \end{pmatrix}.$$

Formation of inverse matrix by row operations

The above process gives a clue to the formation of the inverse of any non-singular 3×3 matrix, starting from the unit matrix and applying the appropriate row operations in turn. These are shown in the centre, with the corresponding matrices on each side:

$$\text{original matrix} \rightarrow \begin{pmatrix} 1 & 0 & -1 \\ 2 & 1 & 3 \\ 4 & 2 & 5 \end{pmatrix} \qquad\qquad \begin{pmatrix} 1 & 0 & 0 \\ 0 & 1 & 0 \\ 0 & 0 & 1 \end{pmatrix} \leftarrow \text{unit matrix}$$

$$\begin{pmatrix} 1 & 0 & -1 \\ 0 & 1 & 5 \\ 4 & 2 & 5 \end{pmatrix} \qquad \boldsymbol{r}_2' = \boldsymbol{r}_2 - 2\boldsymbol{r}_1 \qquad \begin{pmatrix} 1 & 0 & 0 \\ -2 & 1 & 0 \\ 0 & 0 & 1 \end{pmatrix}$$

$$\begin{pmatrix} 1 & 0 & -1 \\ 0 & 1 & 5 \\ 0 & 2 & 9 \end{pmatrix} \qquad \boldsymbol{r}_3' = \boldsymbol{r}_3 - 4\boldsymbol{r}_1 \qquad \begin{pmatrix} 1 & 0 & 0 \\ -2 & 1 & 0 \\ -4 & 0 & 1 \end{pmatrix}$$

échelon form → $\begin{pmatrix} 1 & 0 & -1 \\ 0 & 1 & 5 \\ 0 & 0 & -1 \end{pmatrix}$ $\boldsymbol{r}_3' = \boldsymbol{r}_3 - 2\boldsymbol{r}_2$ $\begin{pmatrix} 1 & 0 & 0 \\ -2 & 1 & 0 \\ 0 & -2 & 1 \end{pmatrix}$

$\begin{pmatrix} 1 & 0 & 0 \\ 0 & 1 & 5 \\ 0 & 0 & -1 \end{pmatrix}$ $\boldsymbol{r}_1' = \boldsymbol{r}_1 - \boldsymbol{r}_3$ $\begin{pmatrix} 1 & 2 & -1 \\ -2 & 1 & 0 \\ 0 & -2 & 1 \end{pmatrix}$

$\begin{pmatrix} 1 & 0 & 0 \\ 0 & 1 & 0 \\ 0 & 0 & -1 \end{pmatrix}$ $\boldsymbol{r}_2' = \boldsymbol{r}_2 + 5\boldsymbol{r}_3$ $\begin{pmatrix} 1 & 2 & -1 \\ -2 & -9 & 5 \\ 0 & -2 & 1 \end{pmatrix}$

unit matrix → $\begin{pmatrix} 1 & 0 & 0 \\ 0 & 1 & 0 \\ 0 & 0 & 1 \end{pmatrix}$ $\boldsymbol{r}_3' = -\boldsymbol{r}_3$ $\begin{pmatrix} 1 & 2 & -1 \\ -2 & -9 & 5 \\ 0 & 2 & -1 \end{pmatrix}$ ← inverse matrix

In the process of forming the inverse, it is not necessary to pass through the échelon stage, so long as one contrives to produce zeros in the correct positions. Moreover, the process involving row operations could be replaced by one using *column operations*, but row and column operations must not be mixed, for column operations are equivalent to *post*-multiplication by elementary matrices.

This method of inverting matrices, by using *either* row operations *or* column operations, is basically the one used in practice, e.g., by computers in the case of larger matrices, the elements of which are possibly awkward numbers. Even using a computer, the method is a great deal shorter† than that of forming the adjoint matrix, which is chiefly of theoretical interest.

Exercise 13.7b

1 Find the inverses of the following square matrices, using the method of the preceding section:

(*i*) $\begin{pmatrix} 4 & 3 \\ 2 & 2 \end{pmatrix}$ (*ii*) $\begin{pmatrix} 7 & -5 \\ 5 & -4 \end{pmatrix}$ (*iii*) $\begin{pmatrix} a & b \\ c & d \end{pmatrix}$

(*iv*) $\begin{pmatrix} 2 & 2 & -3 \\ 3 & 2 & -10 \\ 3 & 4 & 1 \end{pmatrix}$; (*v*) $\begin{pmatrix} 2 & -1 & 4 \\ 4 & 0 & 2 \\ 3 & -2 & 7 \end{pmatrix}$; (*vi*) $\begin{pmatrix} 1 & 1 & 1 \\ 1 & -2 & 1 \\ 2 & 3 & -2 \end{pmatrix}$;

(*vii*) $\begin{pmatrix} 1 & 1 & -1 \\ 1 & 2 & 3 \\ 1 & 3 & -1 \end{pmatrix}$; (*viii*) $\begin{pmatrix} 4 & 6 & 5 \\ -3 & 8 & 14 \\ 9 & -4 & 2 \end{pmatrix}$.

† See Neill, H. and Moakes, A. J., *Vectors, Matrices and Linear Equations*, Oliver and Boyd, 1967, p. 157.

2 Use the results of no. **1** above to solve the equations in exercise 13.7a, no. **1**.
(Use row operations in most cases, but try using column operations in others.)

3 Find the inverse of the matrix $\boldsymbol{M} = \begin{pmatrix} 1 & -1 & 1 \\ 3 & -9 & 5 \\ 1 & -3 & 3 \end{pmatrix}$.

Rewrite the set of equations
$$\begin{aligned} x - y + z &= 3 \\ 3x - 9y + 5z &= 6 \\ x - 3y + 3z &= 13 \end{aligned}$$
in a form using the matrix $\boldsymbol{M}$, and hence solve the equations. (M.E.I.)

4 Use the method of this section to find the inverses of some of the matrices in exercise 13.4a.

5 Construct some examples of singular 3×3 matrices:
(*i*) by arranging for the 3×3 determinant to have the value zero.
(*ii*) by arranging the third row to be linearly dependent on the first two rows (i.e., $\boldsymbol{r}_3 = a\boldsymbol{r}_1 + b\boldsymbol{r}_2$, see p. 430).

6 Express the following simultaneous linear equations in matrix form
$$\begin{aligned} x + y + 2z &= 4 \\ y - 3z &= -2 \\ 2x + 5y - 5z &= k \end{aligned} \qquad \begin{aligned} X + Y + Z &= 3x \\ X - 2Y + Z &= 3y \\ Y - Z &= z \end{aligned}$$
Find the value of k for which they are consistent, and obtain the general solution for X, Y, Z in that case. Give a geometrical illustration. (M.E.I.)

7 Solve where possible the equations
$$\begin{pmatrix} 1 & 0 & -5 \\ 2 & 1 & 2 \\ -4 & -3 & -16 \end{pmatrix} \begin{pmatrix} x \\ y \\ z \end{pmatrix} = \begin{pmatrix} a \\ b \\ c \end{pmatrix}$$
in the cases $\begin{pmatrix} a \\ b \\ c \end{pmatrix} = (i) \begin{pmatrix} 0 \\ 0 \\ 0 \end{pmatrix}$, $(ii) \begin{pmatrix} 1 \\ 1 \\ 1 \end{pmatrix}$, $(iii) \begin{pmatrix} 5 \\ 0 \\ 10 \end{pmatrix}$, $(iv) \begin{pmatrix} 5 \\ 1 \\ 10 \end{pmatrix}$.

8 What is the effect of the singular matrix $\begin{pmatrix} 1 & -2 & 5 \\ 3 & 1 & -1 \\ 5 & -3 & 9 \end{pmatrix}$ on the points $\begin{pmatrix} x \\ y \\ z \end{pmatrix}$ of 3-D space?

9 Find a 3×3 matrix which takes all the points of 3-D space into the plane $2x - 4y + z = 0$.

10 Find a 3×3 matrix which takes all the points of 3-D space into the line of intersection of the planes $2x - 4y + z = 0$ and $x + y - 4z = 0$.

11 Show that the matrix $\begin{pmatrix} 2 & -3 & 1 \\ 4 & -6 & 2 \\ -3 & 4\frac{1}{2} & -1\frac{1}{2} \end{pmatrix}$

takes all the points of 3-D space into the line $x:y:z = 2:4:-3$. What do you notice about the (*i*) minors, (*ii*) rows and columns?

*13.8 Further applications of matrices

There are many applications of the use of matrices throughout mechanics and other branches of applied mathematics but we shall concentrate our attention on a group of problems involving the impact and collisions of a particle.

Example 1

The bouncing of a particle on an inclined plane

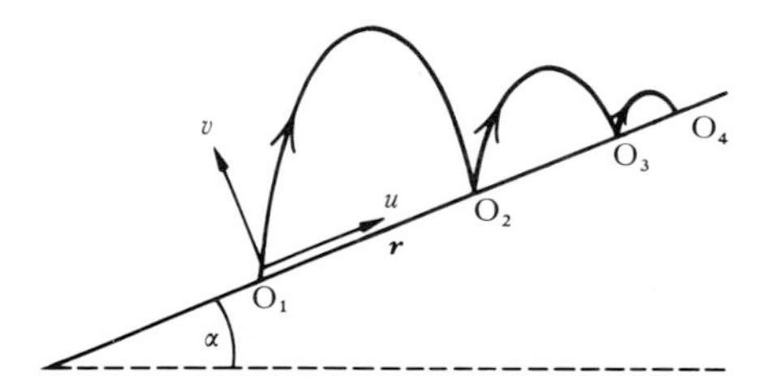

Consider a projectile launched from a point O_1 so as to land on a plane through O_1 inclined to the horizontal at an angle α. If the magnitude and direction of the initial velocity are known, it is possible to calculate the range r, that is the distance between the point O_1 and the point O_2 where the projectile hits the plane. It is convenient to work with the components of the initial velocity along the plane and normal to it. Calling these u and v, it can be shown that

$$r = \frac{2v}{g \cos \alpha}(u - v \tan \alpha).$$

Of course, if $u < v \tan \alpha$, r would be negative, and you should have no difficulty in appreciating the meaning in this case.

Next suppose that the ball bounces after landing at O_2 after which it proceeds along another parabolic trajectory under gravity, and again hits the plane at a point O_3, whereupon it rebounds and the process continues.

We may wish to compute the position of the successive bounces, and to know just what happens in the subsequent motion.

The key to the problem is that we should like to be able to obtain the components (along and perpendicular to the plane) of the velocity of the ball as it sets out on each of its respective trajectories. For, once we know these, the above formula enables us to find where it lands at the end of that trajectory.

If the velocity just after the first bounce has components u_1, v_1, it is easy to show that

$$u_1 = u - 2v \tan\alpha$$
$$v_1 = ev,$$

where it is assumed that the plane is smooth, and that there is a coefficient of restitution e $(0 < e \leqslant 1)$ governed by the elastic properties of the materials. These equations may be written

$$\begin{pmatrix} u_1 \\ v_1 \end{pmatrix} = \begin{pmatrix} 1 & -2\tan\alpha \\ 0 & e \end{pmatrix} \begin{pmatrix} u \\ v \end{pmatrix},$$

so that we see the matrix $\boldsymbol{M} = \begin{pmatrix} 1 & -2\tan\alpha \\ 0 & e \end{pmatrix}$ transforms the initial velocity represented by a column vector into the velocity vector following the bounce. This we may write $\boldsymbol{v}_1 = \boldsymbol{Mv}$. Similarly, if $\boldsymbol{v}_2 = \begin{pmatrix} u_2 \\ v_2 \end{pmatrix}$ is the velocity immediately after the second bounce, we have $\boldsymbol{v}_2 = \boldsymbol{Mv}_1$. Combining these, $\boldsymbol{v}_2 = \boldsymbol{MMv} = \boldsymbol{M}^2\boldsymbol{v}$, giving the velocity after the second bounce in terms of the initial velocity. Similarly, $\boldsymbol{v}_3 = \boldsymbol{M}^3\boldsymbol{v}$ is the velocity after the third bounce, and $\boldsymbol{v}_n = \boldsymbol{M}^n\boldsymbol{v}$ is the velocity after the nth rebound.

Thus, to find out what happens to the ball after repeated rebounds from the plane, we need to be able to multiply the matrix $\begin{pmatrix} 1 & -2\tan\alpha \\ 0 & e \end{pmatrix}$ by itself repeatedly, i.e., we need to find $\boldsymbol{M}^2, \boldsymbol{M}^3, \boldsymbol{M}^4, \ldots$, or 'powers' of the matrix $\boldsymbol{M}$. As an example, consider first the case when $\tan\alpha = \frac{1}{2}$, $e = 1$ (perfect elasticity).

$$\text{Then} \quad \boldsymbol{M} = \begin{pmatrix} 1 & -1 \\ 0 & 1 \end{pmatrix}, \quad \boldsymbol{M}^2 = \begin{pmatrix} 1 & -2 \\ 0 & 1 \end{pmatrix}; \quad \boldsymbol{M}^3 = \begin{pmatrix} 1 & -3 \\ 0 & 1 \end{pmatrix}, \ldots,$$

and it is easy to see inductively that $\boldsymbol{M}^n = \begin{pmatrix} 1 & -n \\ 0 & 1 \end{pmatrix}$.

Thus if the initial velocity were given by the vector $\boldsymbol{v} = \begin{pmatrix} 15 \\ 4 \end{pmatrix}$, we have

$$\boldsymbol{v}_1 = \boldsymbol{Mv} = \begin{pmatrix} 11 \\ 4 \end{pmatrix}; \quad \boldsymbol{v}_2 = \boldsymbol{M}^2\boldsymbol{v} = \begin{pmatrix} 7 \\ 4 \end{pmatrix}; \quad \boldsymbol{v}_3 = \boldsymbol{M}^3\boldsymbol{v} = \begin{pmatrix} 3 \\ 4 \end{pmatrix};$$

$$\boldsymbol{v}_4 = \boldsymbol{M}^4\boldsymbol{v} = \begin{pmatrix} -1 \\ 4 \end{pmatrix}.$$

The range formula gives successive values of r to be:

$r_1 = 13k; \quad r_2 = 9k; \quad r_3 = 5k; \quad r_4 = k$ and so on, where

$$k = \frac{2v}{g \cos \alpha} = \frac{10}{9.8 \times 2/\sqrt{5}}.$$

Evidently, after the third bounce, the ball starts coming down the plane. Had the coefficient of restitution been smaller (say $\frac{1}{2}$), our matrix would have been

$$\boldsymbol{M} = \begin{pmatrix} 1 & -1 \\ 0 & \frac{1}{2} \end{pmatrix}, \text{ with } \boldsymbol{M}^2 = \begin{pmatrix} 1 & -1\frac{1}{2} \\ 0 & \frac{1}{4} \end{pmatrix}, \quad \boldsymbol{M}^3 = \begin{pmatrix} 1 & -1\frac{3}{4} \\ 0 & \frac{1}{8} \end{pmatrix}, \text{ and so on.}$$

In this case, with initial velocity the same, we obtain

$$\boldsymbol{v}_1 = \begin{pmatrix} 1 & -1 \\ 0 & \frac{1}{2} \end{pmatrix}\begin{pmatrix} 15 \\ 4 \end{pmatrix} = \begin{pmatrix} 11 \\ 2 \end{pmatrix}; \qquad \boldsymbol{v}_2 = \begin{pmatrix} 1 & -1\frac{1}{2} \\ 0 & \frac{1}{4} \end{pmatrix}\begin{pmatrix} 15 \\ 4 \end{pmatrix} = \begin{pmatrix} 9 \\ 1 \end{pmatrix};$$

$$\boldsymbol{v}_3 = \begin{pmatrix} 1 & -1\frac{3}{4} \\ 0 & \frac{1}{8} \end{pmatrix}\begin{pmatrix} 15 \\ 4 \end{pmatrix} = \begin{pmatrix} 8 \\ \frac{1}{2} \end{pmatrix}; \qquad \boldsymbol{v}_4 = \begin{pmatrix} 1 & -1\frac{7}{8} \\ 0 & \frac{1}{16} \end{pmatrix}\begin{pmatrix} 15 \\ 4 \end{pmatrix} = \begin{pmatrix} 7\frac{1}{2} \\ \frac{1}{4} \end{pmatrix}, \ldots,$$

and successive values of r:

$r_1 = 13k; \quad r_2 = 10k; \quad r_3 = 8\frac{1}{2}k; \quad r_4 = 7\frac{3}{4}k, \ldots.$

Here we may well ask whether the value of r ever becomes negative — whether the ball ever leaps *down* the plane — before bouncing ceases for ever.

Example 2

Collision of elastic particles moving in a line

We consider next another situation in mechanics, that of a number of billiard balls moving along a straight line, with impacts occurring between pairs of adjacent balls. We shall suppose that the balls are all of equal mass m, and that the coefficient of restitution between any pair is e. Suppose the two balls have speeds u_1, u_2 before impact and v_1, v_2 after impact:

Then, by conservation of momentum:

$$mv_1 + mv_2 = mu_1 + mu_2$$

and by Newton's law of restitution:

$$v_2 - v_1 = -e(u_2 - u_1).$$

Thus $v_1 + v_2 = u_1 + u_2$

$$v_1 - v_2 = -eu_1 + eu_2$$

or $$\begin{pmatrix} 1 & 1 \\ 1 & -1 \end{pmatrix}\begin{pmatrix} v_1 \\ v_2 \end{pmatrix} = \begin{pmatrix} 1 & 1 \\ -e & e \end{pmatrix}\begin{pmatrix} u_1 \\ u_2 \end{pmatrix}.$$

Pre-multiplying by the inverse of $\begin{pmatrix} 1 & 1 \\ 1 & -1 \end{pmatrix}$, namely $\begin{pmatrix} \frac{1}{2} & \frac{1}{2} \\ \frac{1}{2} & -\frac{1}{2} \end{pmatrix}$,

$$\begin{pmatrix} v_1 \\ v_2 \end{pmatrix} = \begin{pmatrix} \frac{1}{2} & \frac{1}{2} \\ \frac{1}{2} & -\frac{1}{2} \end{pmatrix}\begin{pmatrix} 1 & 1 \\ -e & e \end{pmatrix}\begin{pmatrix} u_1 \\ u_2 \end{pmatrix}$$

$$= \begin{pmatrix} \frac{1}{2}(1-e) & \frac{1}{2}(1+e) \\ \frac{1}{2}(1+e) & \frac{1}{2}(1-e) \end{pmatrix}\begin{pmatrix} u_1 \\ u_2 \end{pmatrix} = \begin{pmatrix} p & q \\ q & p \end{pmatrix}\begin{pmatrix} u_1 \\ u_2 \end{pmatrix}$$

where $p = \frac{1}{2}(1-e)$, $q = \frac{1}{2}(1+e)$, and $p + q = 1$, or more briefly as $\boldsymbol{v} = \boldsymbol{Mu}$, where $\boldsymbol{v}$ is the column vector giving the speeds after impact, $\boldsymbol{u}$ is the column vector giving the speeds before impact, and $\boldsymbol{M}$ is the matrix $\begin{pmatrix} p & q \\ q & p \end{pmatrix}$ which converts velocities before into velocities after impact. (Note that $\boldsymbol{v}$ is not a vector in the same sense that $\boldsymbol{v}$ was in Example 1. There, the two constituents of the column vector were the *components* of $\boldsymbol{v}$, along and normal to the plane, so that $\boldsymbol{v}$ was an 'ordinary' vector, namely a velocity. Here, $\boldsymbol{v}$ is simply a 'list' of the (signed) *speeds* of the two balls in the system. If there had been more than two balls, then a column with more than two components would have been necessary.)

Now the two balls, having collided, will not do so again. But one could contrive to make them do so, for example, by arranging for them to rebound at right angles from two cushions as in the figure, further impacts

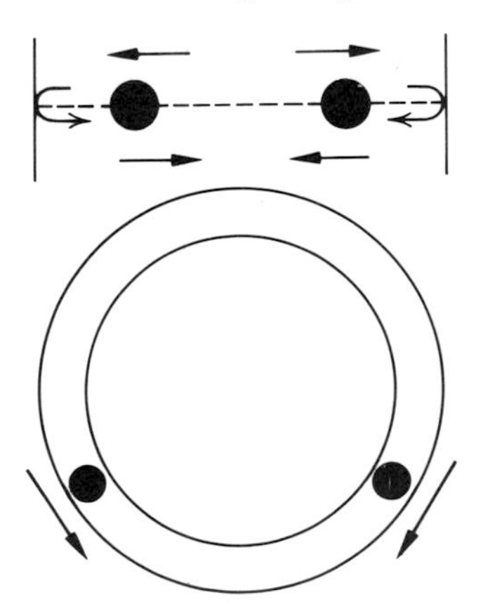

occurring in the line of motion. Or the system may consist of tiny elastic particles inside a smooth circular tube, so that after colliding they rebound and collide repeatedly by moving round the circumference of the tube. The result of repeated collisions would be determined by computing successive powers of the matrix $\boldsymbol{M}$: $\boldsymbol{M}^2$, $\boldsymbol{M}^3$, and so on.

Example 3

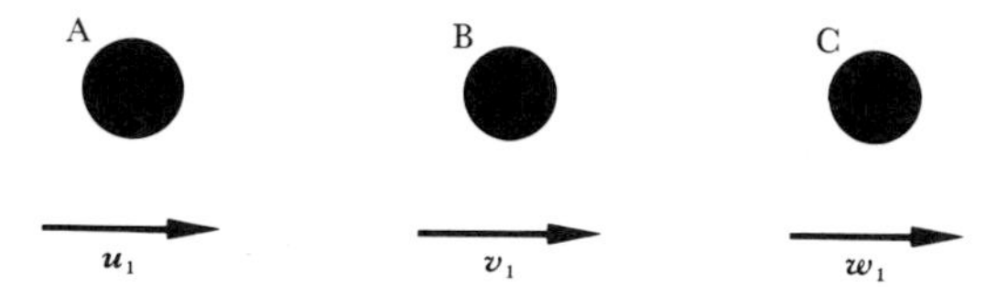

Repeated impacts of three balls moving in a line

Suppose now we have three balls, A, B, and C of equal mass moving in a straight line, and that A first strikes B, after which B strikes C. We shall suppose the speeds to be u_1, v_1, w_1 initially, u_2, v_2, w_2 after the first impact between A and B, and finally u_3, v_3, w_3 after the second impact. We have, as in the previous section:

$$\left.\begin{aligned} u_2 &= \tfrac{1}{2}(1-e)u_1 + \tfrac{1}{2}(1+e)v_1 = pu_1 + qv_1 \\ v_2 &= \tfrac{1}{2}(1+e)u_1 + \tfrac{1}{2}(1-e)v_1 = qu_1 + pv_1 \\ w_2 &= w_1 \end{aligned}\right\} \Rightarrow \begin{pmatrix} u_2 \\ v_2 \\ w_2 \end{pmatrix} = \begin{pmatrix} p & q & 0 \\ q & p & 0 \\ 0 & 0 & 1 \end{pmatrix} \begin{pmatrix} u_1 \\ v_1 \\ w_1 \end{pmatrix}.$$

For the impact between B and C we have similarly,

$$\begin{aligned} u_3 &= u_2 \\ v_3 &= \tfrac{1}{2}(1-e)v_2 + \tfrac{1}{2}(1+e)w_2 = pv_2 + qw_2 \\ w_3 &= \tfrac{1}{2}(1+e)v_2 + \tfrac{1}{2}(1-e)w_2 = qv_2 + pw_2 \end{aligned} \Rightarrow \begin{pmatrix} u_3 \\ v_3 \\ w_3 \end{pmatrix} = \begin{pmatrix} 1 & 0 & 0 \\ 0 & p & q \\ 0 & q & p \end{pmatrix} \begin{pmatrix} u_2 \\ v_2 \\ w_2 \end{pmatrix}.$$

Abbreviating the notation:

$$\boldsymbol{v}_2 = \boldsymbol{M}_1 \boldsymbol{v}_1$$
$$\boldsymbol{v}_3 = \boldsymbol{M}_2 \boldsymbol{v}_2$$

and these when combined give the final velocities in terms of the initial ones: $\boldsymbol{v}_3 = \boldsymbol{M}_2 \boldsymbol{M}_1 \boldsymbol{v}_1$, where the product matrix $\boldsymbol{M}_2 \boldsymbol{M}_1$ will be found to be

$$\begin{pmatrix} p & q & 0 \\ pq & p^2 & q \\ q^2 & pq & p \end{pmatrix}.$$

This matrix describes the final result after the two successive collisions. As expected, $\boldsymbol{M}_2 \boldsymbol{M}_1 \neq \boldsymbol{M}_1 \boldsymbol{M}_2$, so the final result would be different if the collision between B and C had preceded that between A and B.

Example 4

Perfectly elastic collisions

The case $e = 1$ deserves special attention. Here we have

$$M_1 = \begin{pmatrix} 0 & 1 & 0 \\ 1 & 0 & 0 \\ 0 & 0 & 1 \end{pmatrix}, \qquad M_2 = \begin{pmatrix} 1 & 0 & 0 \\ 0 & 0 & 1 \\ 0 & 1 & 0 \end{pmatrix}$$

and these are *permutation* matrices, confirming the fact that the pairs of colliding balls exchange velocities upon impact. The diagram

A	B	C
u_1	v_1	w_1
v_1	u_1	w_1
v_1	w_1	u_1
w_1	v_1	u_1

shows the velocities after successive collisions between A and B, B and C, and a further collision between A and B, in the case when $u_1 > v_1 > w_1$. Evidently no further collisions can occur because after the third impact both pairs of adjacent balls are separating, with $u_1 > v_1$ and $v_1 > w_1$. It is interesting to extend to the case when there are 4 (or more) perfectly elastic balls moving in a line. The matrices will then be

$$\begin{pmatrix} 0 & 1 & 0 & 0 \\ 1 & 0 & 0 & 0 \\ 0 & 0 & 1 & 0 \\ 0 & 0 & 0 & 1 \end{pmatrix} \quad \text{for an impact between A and B;}$$

$$\begin{pmatrix} 1 & 0 & 0 & 0 \\ 0 & 0 & 1 & 0 \\ 0 & 1 & 0 & 0 \\ 0 & 0 & 0 & 1 \end{pmatrix} \quad \text{for an impact between B and C;}$$

$$\text{and} \quad \begin{pmatrix} 1 & 0 & 0 & 0 \\ 0 & 1 & 0 & 0 \\ 0 & 0 & 0 & 1 \\ 0 & 0 & 1 & 0 \end{pmatrix} \quad \text{for an impact between C and D.}$$

The order in which impacts occur will depend upon the initial position and initial velocities of the 4 balls. If 6 impacts occur in the sequenc AB, BC, CD, BC, AB, BC, the velocities follow the sequence:

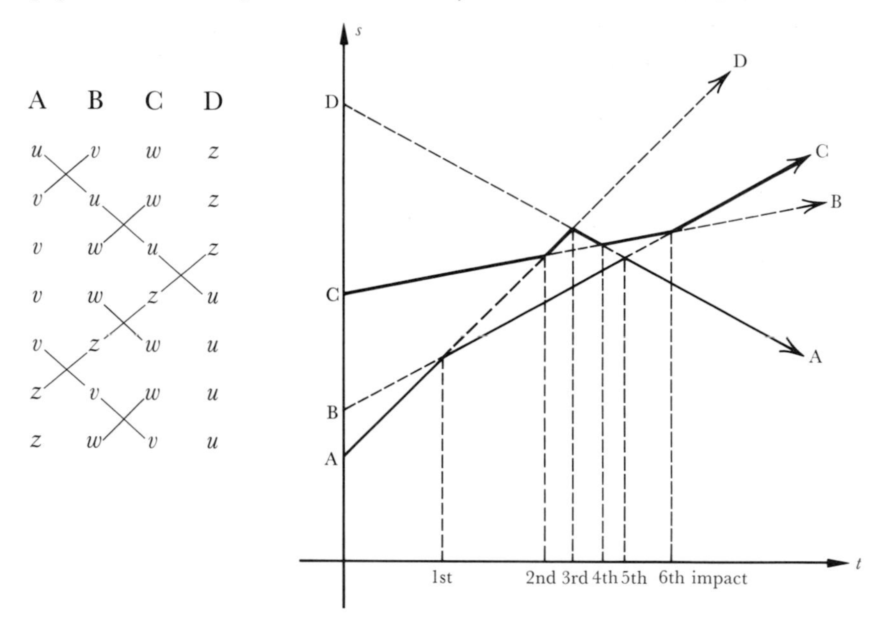

the final velocities z, w, v, u being the original ones in the reverse order. This is shown graphically in the figure where a displacement time graph is given for each ball. At each impact, the two bodies 'swop graphs'. Clearly, since four lines intersect in at most six points, there can be no more than six impacts under the most favourable conditions, with $u > v > w > z$. This may obviously be extended to the case of five balls moving in a line, when the maximum number of possible collisions will be 10 $(= {}^5C_2)$.

Example 5

As a final example, which may be answered by the use of matrices and which would be very difficult by any other means, suppose one considers the following problem.

A body is first given a rotation through 180° about an axis, and then a quarter-turn (through 90°) about a second axis, the two axes intersecting at a point O and making an angle θ with each other. It seems a reasonable assumption that these successive rotations may be replaced by a *single* rotation about some other axis through O. We now ask whether it is possible for the angle θ between the axes of rotation to be chosen in such a way that the single rotation will be through $\pm 120°$. In other words, can a half-turn, followed by a quarter-turn, ever lead to a one-third turn?

There are probably not very many people who can visualise the solution of such a problem, as it would require much more three-dimensional insight than any but the most gifted possess. Even if one were allowed to use apparatus — rods, matches, pieces of string, cardboard, etc. — one would still be hard put to it to obtain a solution. The reader, trusting his intuition, may well wonder whether the problem is soluble, and may like to perform experiments to discover whether his intuition has misled him before reading on.

Using matrices, however, the solution is comparatively simple. We are concerned with rotations about two axes intersecting at an angle θ. We therefore choose our coordinate system so that the origin is the point of intersection of the lines, one of them is the axis of x, and the other lies in the plane of the x- and y-axes, i.e., the plane $z = 0$, so that it has equations $z = 0$, $y = x \tan \theta$.

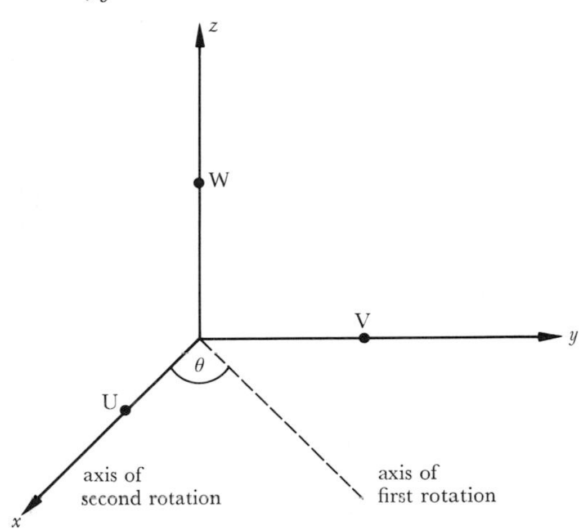

Our first rotation, through 180°, will be about this axis, and in order to set up the matrix, we must enquire the destinations of the points U, V, W under this half-turn. Clearly the point W′ has coordinates $(0, 0, -1)$, so the third column of the matrix is determined.

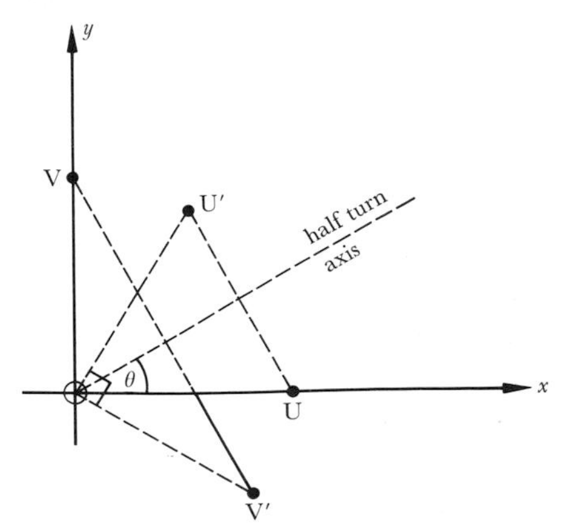

A 'plan' view on to the xOy plane shows that the half-turn takes the point U to U′ with coordinates $(\cos 2\theta, \sin 2\theta, 0)$, while V′ (remembering that $\angle U'OV' = 90°$) will be found to have coordinates $(\sin 2\theta, -\cos 2\theta, 0)$.†

† Compare the matrix for the two-dimensional reflection in a line through the origin; see 13.2.

Thus the matrix representing the half-turn is

$$A = \begin{pmatrix} \cos 2\theta & \sin 2\theta & 0 \\ \sin 2\theta & -\cos 2\theta & 0 \\ 0 & 0 & -1 \end{pmatrix}.$$

Since our second rotation is about the x-axis, it will be easier to find the matrix, but we must specify in which sense the quarter-turn is to be carried out (this did not matter in the case of the half-turn, of course). If the sense of thc rotation is that of a right-hand screw in the positive direction of the x-axis, i.e., such as to carry the point V to move into coincidence with the point W, then we can easily see that U′ is (1, 0, 0), this point being *invariant* under the rotation, V′ is (0, 0, 1), and W′ is (0, −1, 0). Thus our second matrix is

$$B = \begin{pmatrix} 1 & 0 & 0 \\ 0 & 0 & -1 \\ 0 & 1 & 0 \end{pmatrix}.$$

The combined effect of the two rotations will be represented by the product matrix BA (first A, then B, since the matrices are pre-multiplying column vectors), and we have

$$BA = \begin{pmatrix} 1 & 0 & 0 \\ 0 & 0 & -1 \\ 0 & 1 & 0 \end{pmatrix}\begin{pmatrix} \cos 2\theta & \sin 2\theta & 0 \\ \sin 2\theta & -\cos 2\theta & 0 \\ 0 & 0 & -1 \end{pmatrix} = \begin{pmatrix} \cos 2\theta & \sin 2\theta & 0 \\ 0 & 0 & 1 \\ \sin 2\theta & -\cos 2\theta & 0 \end{pmatrix} = C.$$

The question is: can C conceivably represent a third-turn, i.e., a rotation through $\pm 120°$? In order to answer this, we first consider the matrix C^2, which should represent a rotation about $\pm 240°$: Writing $c = \cos 2\theta$, $s = \sin 2\theta$,

$$C^2 = \begin{pmatrix} c & s & 0 \\ 0 & 0 & 1 \\ s & -c & 0 \end{pmatrix}\begin{pmatrix} c & s & 0 \\ 0 & 0 & 1 \\ s & -c & 0 \end{pmatrix} = \begin{pmatrix} c & cs & s \\ s & -c & 0 \\ cs & s^2 & -c \end{pmatrix}$$

and $$C^3 = \begin{pmatrix} c^2 & cs & s \\ s & -c & 0 \\ cs & s^2 & -c \end{pmatrix}\begin{pmatrix} c & s & 0 \\ 0 & 0 & 1 \\ s & -c & 0 \end{pmatrix} = \begin{pmatrix} c^3 + s^2 & c^2 s - cs & cs \\ cs & s^2 & -c \\ c^2 s - cs & cs^2 + c^2 & s^2 \end{pmatrix}.$$

Now if C is to represent a 120° rotation, C^3 *must be the identity matrix*, and the equation $C^3 = I$ would mean that the nine equations:

$$c^3 + s^2 = 1 \quad (i) \qquad c^2s - cs = 0 \quad (ii) \qquad cs = 0 \quad (iii)$$

$$cs = 0 \quad (iv) \qquad s^2 = 1 \quad (v) \qquad -c = 0 \quad (vi)$$

$$c^2s - cs = 0 \quad (vii) \qquad cs^2 + c^2 = 0 \quad (viii) \qquad s^2 = 1 \quad (ix)$$

ought all to be satisfied. It may well be that there does not exist an angle θ which satisfies all these equations. However, our fears on this score are soon dispelled. The key to the solution is equation (vi), requiring that $c = 0$, in which case (ii), (iii), (iv), (vii) and $(viii)$ are automatically true. The remaining three equations (i), (v), and (ix) then all reduce to $s^2 = 1$, which is consistent with $c = 0$, and we summarise our findings:

$$\cos 2\theta = 0; \qquad \sin 2\theta = \pm 1.$$

These are satisfied by $\theta = 45°$ (and by $135°$, as well as other values of θ which need not concern us). Thus we have verified that a rotation through 180°, followed by a rotation through 90° about an axis *inclined to the first at* 45°, may be together equivalent to a single rotation through 120°.

We may go further and use matrices to *locate the axis* of this single rotation. This is not difficult, for points *on* the axis of rotation are unchanged, or *invariant*, under the rotation. Thus if $\boldsymbol{x} = \begin{pmatrix} x \\ y \\ z \end{pmatrix}$ is such a point, we have

$$\boldsymbol{Cx} = \boldsymbol{x}.$$

But $\boldsymbol{C} = \begin{pmatrix} 0 & 1 & 0 \\ 0 & 0 & 1 \\ 1 & 0 & 0 \end{pmatrix}$ (taking $\sin 2\theta = +1$),

and so $\begin{pmatrix} 0 & 1 & 0 \\ 0 & 0 & 1 \\ 1 & 0 & 0 \end{pmatrix}\begin{pmatrix} x \\ y \\ z \end{pmatrix} = \begin{pmatrix} x \\ y \\ z \end{pmatrix} \quad \Rightarrow \quad \begin{pmatrix} y \\ z \\ x \end{pmatrix} = \begin{pmatrix} x \\ y \\ z \end{pmatrix}$

and such a point therefore satisfies the equations $x = y = z$. Clearly the axis of rotation is a line making equal angles with the coordinate axes, and in fact passes through the point (1, 1, 1). Note that the matrix $\boldsymbol{C}$ is a permutation matrix, its effect on the coordinates of any point being to 'move them round one place' — a cyclic permutation — whereby x is replaced by y, y by z, and z by x. Evidently this is geometrically a rotation through 120°.

If your experiments with matches and string failed, you may now care to demonstrate the geometry of the above by using a cube as shown overleaf:

The diagram should be self explanatory, and you may of course perform the experiment with an ordinary die.

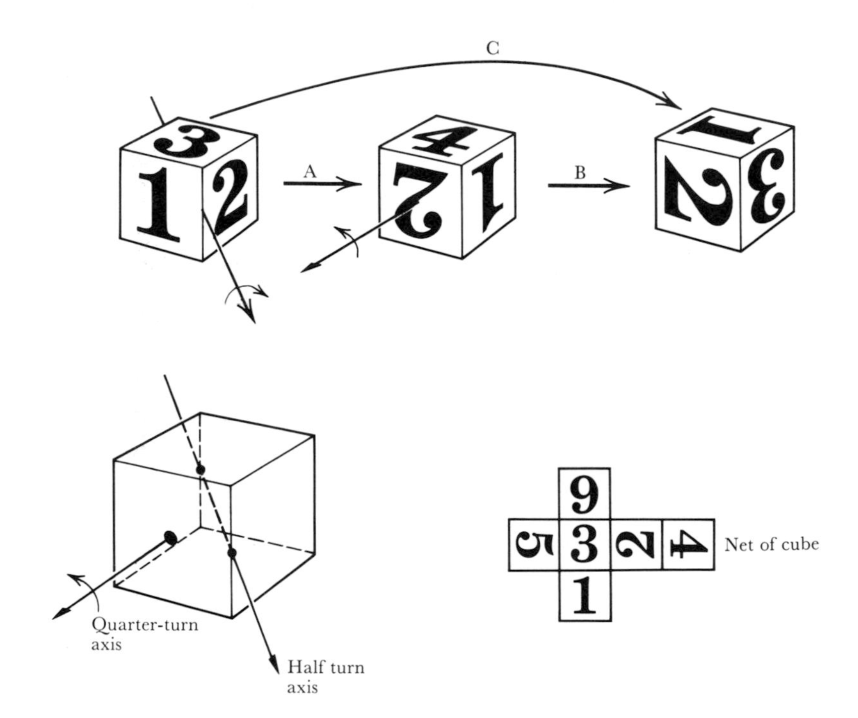

Exercise 13.8

1 Find the condition on e for three particles of equal masses moving in a line as in example 3 to collide more than three times. Is there any limit to the number of collisions that are possible when e is small?

2 When a rotation through $2\pi/p$ about one axis through O is followed by a second rotation through $2\pi/q$ about another axis through O, the effect is that of a single rotation through $2\pi/r$. In example 5 we saw that if $p = 2$, $q = 4$, it is possible for r to have the value 3. Use matrices to find other possible sets of values of p and q for which r is rational.

Miscellaneous Problems 13

1 $Oxyz$ make a set of mutually perpendicular right-handed axes in three-dimensional space. The transformation Q_1 is a quarter-turn about the ray Ox, its sense being clockwise if one looks out from the origin along the ray; and Q_2 is a quarter-turn about a ray from $(0, 0, 1)$ parallel to Oy, clockwise if one looks out from $(0, 0, 1)$ along the ray. Find the images of a general point (x, y, z) under Q_1 and under Q_2, and hence find the image of (x, y, z) under Q_2Q_1 (i.e., Q_1 followed by Q_2). Prove that the transformation Q_2Q_1 has no fixed point.

Give the equations of four planes with the property that $\boldsymbol{M}_4\boldsymbol{M}_3\boldsymbol{M}_2\boldsymbol{M}_1 =$

Q_2Q_1, where M_i denotes the reflection in the ith plane. Explain why Q_2Q_1 cannot be expressed as the product of fewer than four reflections. (S.M.P.)

2 Square matrices A, B of the same order have the property that $A^TA = I$, $B^TB = I$ (where A^T, B^T denote the transposes of A, B, and I is the unit matrix). Prove that:

(*i*) If $C = A^{-1}$, then $C^TC = I$; (*ii*) If $D = AB$, then $D^TD = I$.

What can you deduce about the set of matrices of the given order such that $X^TX = I$? (S.M.P.)

3 A is a non-singular 3×3 matrix; A^T denotes the transpose and A^{-1} is the inverse. Prove that:

$(AB)^T = B^TA^T$ and that $(A^T)^{-1} = (A^{-1})^T$.

Denote by A^* the matrix $(A^T)^{-1}$. Operations e, r, s, t are defined on the set of all non-singular 3×3 matrices by

$e(A) = A$, $r(A) = A^*$, $s(A) = A^{-1}$, $t(A) = A^T$.

If tr denotes the composition 'first r, then t', prove that $tr = s$, justify the other entries in the composition table, and complete the table.

2nd \ 1st	e	r	s	t
e				
r				s
s				
t	t	s	r	e

(M.E.I.)

4 (A statistical application of the transpose of a matrix.)

Suppose we have two variables x, y, and a number of observed pairs of values have been recorded: (x_i, y_i), $i = 1, 2, \ldots, n$. It is supposed theoretically that the points ought to lie on a straight line, whose equations we will take to be $y = mx + c$. We cannot of course satisfy all the n equations,

$$y_i = mx_i + c, \quad \text{or} \quad \begin{pmatrix} x_1 & 1 \\ x_2 & 1 \\ \vdots & \vdots \\ x_n & 1 \end{pmatrix} \begin{pmatrix} m \\ c \end{pmatrix} = \begin{pmatrix} y_1 \\ y_2 \\ \vdots \\ y_n \end{pmatrix}$$

for the two unknowns m and c. However the best values of m and c according to the 'principle of least squares' may be found by *pre-multiplying by the transpose* of the $n \times 2$ matrix, thus:

This gives

$$\begin{pmatrix} x_1 & x_2 & \dots & x_n \\ 1 & 1 & \dots & 1 \end{pmatrix}\begin{pmatrix} x_1 & 1 \\ x_2 & 1 \\ \vdots & \vdots \\ x_n & 1 \end{pmatrix}\begin{pmatrix} m \\ c \end{pmatrix} = \begin{pmatrix} x_1 & x_2 & \dots & x_n \\ 1 & 1 & \dots & 1 \end{pmatrix}\begin{pmatrix} y_1 \\ y_2 \\ \vdots \\ y_n \end{pmatrix}.$$

Proceed to show that the values of m and c determined by this equation are those which make the line pass through the mean centre of the n points, and have gradient m, where

$$m = \frac{s_{xy}}{s_x^2}$$

(see gradient of regression line of y on x, p. 354).
Find the equation of the regression line of y on x through the five points $(-3, -1)$, $(-2, 3)$, $(2, 2)$, $(5, 6)$, $(8, 5)$.

5 Solve for x, y, z the equations:

$$\begin{aligned} x + 2y - z &= 4 \\ 3x - 4y + 5z &= -2 \\ 5x - 5y + 7z &= -1. \end{aligned}$$

Describe geometrically the intersection of the three planes given by these equations. (S.M.P.)

6 Write a short essay on the solution of three simultaneous linear equations using the method of pivotal condensation. Include in your essay:
(*i*) at least one example to illustrate the method,
(*ii*) a brief description of the nature of 'ill-conditioning'. (M.E.I.)

7 What condition must be satisfied by the constants a, b, c, l, m, and n in order for the system of equations

$$\begin{aligned} -ny + mz &= a \\ nx \qquad - lz &= b \\ -mx + ly \qquad &= c \end{aligned}$$

to have solutions?
If this condition is satisfied, show that any solution of the equations will then also satisfy the additional equation $ax + by + cz = 0$, and write down the most general solution in this case.
Do any of these solutions also satisfy the equation $lx + my + nz = 0$?

8 A 4×4 determinant contains 24 terms. When there is a single zero among its 16 elements, there will be $24 - 6 = 18$ non-zero terms. Find the greatest and the least possible number of non-zero terms when *three* zero elements are present, showing where they may be placed.

9 If n is a positive integer, show that $b - c$, $c - a$, $a - b$ are factors of the determinant

$$\begin{vmatrix} 1 & 1 & 1 \\ a & b & c \\ a^{n+2} & b^{n+2} & c^{n+2} \end{vmatrix}$$

and that the remaining factor is the sum of all products of a, b, c of degree n (with repetitions of a, b, c included).
Obtain the corresponding result when the last row of the determinant is replaced by a^{-n}, b^{-n}, c^{-n}. (O.C.)

10 Given that $\begin{vmatrix} x^3 & a^2+1 & 1 \\ a^3 & x^2+1 & 1 \\ 1 & x^2+a^2 & 1 \end{vmatrix} = 0,$

show that $x = 1$, or $x = a$, or $x = -a/(a+1)$ will satisfy the equation. Are there any other roots?

11 Prove that the determinant Δ_n of order n

$$\begin{vmatrix} 2\cos\theta & 1 & 0 & 0 & \cdots & & \cdots & 0 \\ 1 & 2\cos\theta & 1 & 0 & & & & 0 \\ 0 & 1 & 2\cos\theta & 1 & & & & 0 \\ \cdot & \cdot & \cdot & \cdot & & & & \\ 0 & \cdot & \cdot & \cdot & \cdots & 1 & 2\cos\theta & 1 \\ 0 & \cdot & \cdot & \cdot & & 0 & 1 & 2\cos\theta \end{vmatrix}$$

satisfies the recurrence relation $\Delta_{n+1} = 2\Delta_n \cos\theta - \Delta_{n-1}$.

Prove that, if $\theta \neq k\pi$, where k is an integer, then $\Delta_n = \dfrac{\sin(n+1)\theta}{\sin\theta}$.

If the term in the top left-hand corner is changed to $\cos\theta$, what is the value of the determinant? (O.C.)

12 Expand the determinants $\begin{vmatrix} a & b & 0 \\ b & a & b \\ 0 & b & a \end{vmatrix}$, $\begin{vmatrix} a & b & 0 & 0 \\ b & a & b & 0 \\ 0 & b & a & b \\ 0 & 0 & b & a \end{vmatrix}$,

and continue with similar determinants of order 5, 6, . . ., etc.

Show that the determinant

$$\begin{vmatrix} 1 & 1 & 0 & 0 & 0 & . & . & . & . \\ 1 & 1 & 1 & 0 & 0 & . & . & . & . \\ 0 & 1 & 1 & 1 & 0 & . & . & . & . \\ 0 & 0 & 1 & 1 & 1 & . & . & . & . \\ . & . & . & . & . & & & & \\ . & . & . & . & . & & & & \\ . & . & . & . & . & & & & \end{vmatrix}.$$

must have the value 0, 1 or -1.

14

Introduction to algebraic structures

14.1 The idea of a structure

In 9.5, we proved that, if A, B, C, D are four points in space such that $AB \perp CD$ and $AC \perp BD$, then $AD \perp BC$. This was done by using the scalar product of vectors, and during the course of the proof we made use of the *distributive* property of scalar product,

$$\boldsymbol{a}.(\boldsymbol{b} - \boldsymbol{c}) = \boldsymbol{a}.\boldsymbol{b} - \boldsymbol{a}.\boldsymbol{c},$$

whereby vectors behave like ordinary numbers in the relationship

$$a(b - c) = ab - ac.$$

The property of distributivity of scalar product over subtraction was not something we could take for granted, but had to be established, and it was essential that our dependence on the result should be acknowledged in the course of the proof concerning the orthocentric tetrahedron. Once distributivity *has* been proved, however, we have available this powerful vector tool, enabling us to produce a very concise and elegant proof.

This draws attention to the importance of what has come to be called *structure*, which (roughly speaking) refers to the various rules which the elements of a set — the set of three-dimensional vectors in this case — have to obey in order that the system may behave itself in a manner which avoids inconsistency or contradiction. We know of two methods of 'multiplying' vectors — the scalar, and the vector, products.† These are good definitions precisely because they lead to an orderly structure. One might try to invent other ways of 'multiplying' vectors; e.g., $\boldsymbol{AB} \times \boldsymbol{CD}$ = the product of the lengths AD and BC. Such a definition will lead to chaos!

† Note the structure of grammar here: 'the scalar and the vector products' is short for 'the scalar product and the vector product', so that the word 'products' *distributes* over 'and' in this case. Again, in 'W. D. and H. O. Wills', the surname distributes over the conjunction 'and'; so too in 'Mr and Mrs Wills'.

For a start, if $\boldsymbol{AB}$ and $\boldsymbol{A'B'}$ were 'equivalent' vectors (having equal length and the same direction), the 'product' $\boldsymbol{A'B'} \times \boldsymbol{CD}$ would be different from the 'product' $\boldsymbol{AB} \times \boldsymbol{CD}$, when we feel we have a right for it to be the same. In the second place, $\boldsymbol{AB} \times \boldsymbol{DC}$ would be $AC \times BD$ when one would rather expect it to give the *negative* of $\boldsymbol{AB} \times \boldsymbol{CD}$. It is true that $\boldsymbol{AB} \times \boldsymbol{CD} = \boldsymbol{CD} \times \boldsymbol{AB}$, i.e., our 'product' would be commutative. But this is by no means essential, since such a useful definition as vector product (see 11.5) is in fact *not* commutative. Much more important, we should no longer be able to claim distributivity of $\times$ over $+$ and $-$, for $\boldsymbol{AB} \times (\boldsymbol{CD} + \boldsymbol{DE}) = \boldsymbol{AB} \times \boldsymbol{CE} = AE \times BC$; whereas $\boldsymbol{AB} \times \boldsymbol{CD} + \boldsymbol{AB} \times \boldsymbol{DE} = AD \times BC + AE \times BD$, and these are by no means equal.

The simpler features of structure such as commutativity and associativity may appear to be of little interest — the smallest child can quickly learn to appreciate that $8 + 3 = 3 + 8$ (i.e., that addition is *commutative*), that $2 \times (3 \times 5) = (2 \times 3) \times 5$ (i.e., that multiplication is *associative*) and so on, and the realisation of these fundamental laws quickly becomes a matter of habit. Even so, the question of commutativity is not trivial and loses its aura of inevitability when one considers non-commutative operations, such as subtraction and division of numbers and the multiplication of matrices. In the case of subtraction, the small child may become confused if asked 'By how much does 253 fall short of 527?' and commonly enquires about which number to put at the top.

The idea of distributivity, involving both the operations of $+$ and $\times$, is a little more complicated. A young child may appreciate it when dealing with a problem like, 'How much profit is made by a man who buys 50 articles at 28p each and sells them at 40p each?:

$$\text{Profit} = 50 \times 40 - 50 \times 28 = 50(40 - 28) = 50 \times 12 = 600\text{p} = £6,$$

and repeated practice in this sort of situation eventually imprints on his mind the idea of distributivity of multiplication over subtraction as being normal and inevitable. It is possible that he may in some other situation make a false assumption about distributivity, as for example if he says $(ab)c = ac \times bc$ (making the assumption that multiplication distributes over multiplication); or that $(a + b)^2 = a^2 + b^2$, which wrongly presumes that 'squaring' distributes over addition. Indeed a very common error is what may humorously be called the 'universal law of distributivity', $c \circ (a + b) = c \circ a + c \circ b$, where $\circ$ is some operation; e.g., 'exponentiation': $2^{x+y} = 2^x + 2^y$(!) or 'sine': $\sin(x + y) = \sin x + \sin y$(!), heresies not unknown in school work. One of the virtues of studying structure is, hopefully, that this type of 'structural' error is less likely to be perpetrated.

It is interesting to trace the structural implications in a simple arithmetical calculation, such as the multiplication 24×3:

For the formal layout

$$\begin{array}{r} 24 \\ \times \; 3 \\ \hline 72 \\ \hline \end{array}$$

is essentially an abbreviation of

$$\begin{aligned}
24 \times 3 &= (2\times 10 + 4) \times 3 \\
&= (2\times 10)\times 3 + 4\times 3 \\
&= (2\times 10)\times 3 + 12 \\
&= (2\times 10)\times 3 + (1\times 10 + \mathbf{2}) \\
&= 2\times (10\times 3) + (1\times 10 + \mathbf{2}) \\
&= 2\times (3\times 10) + (1\times 10 + \mathbf{2}) \\
&= (2\times 3)\times 10 + (1\times 10 + \mathbf{2}) \\
&= 6\times 10 + (1\times 10 + \mathbf{2}) \\
&= (6\times 10 + 1\times 10) + \mathbf{2} \\
&= (6+1)\times 10 + \mathbf{2} \\
&= \mathbf{7}\times 10 + \mathbf{2} \\
&= \mathbf{72}.
\end{aligned}$$

You should check in this analysis the stages at which you have used the associative, commutative and distributive laws of addition and multiplication, and then be grateful that we do not usually need to take such a scrupulous view of elementary arithmetic!

Instead of building up our idea of structure from primitive beginnings, let us now approach it from the starting point of a very highly organised structure — the *field*. The most familiar example is perhaps the set of rational numbers, Q, with the operations† of addition $(+)$ and multiplication $(\times)$, obeying the following basic requirements, which constitute the axioms for a field, (where it is to be understood that x, y, z denote *any* choice of elements from our set).

A1	$(x + y) + z = x + (y + z)$	Associative law of addition
A2	$x + y = y + x$	Commutative law of addition
A3	$x + 0 = x = 0 + x$	Existence of identity element (zero) for addition
A4	$x + (-x) = 0$	Existence of inverse under addition (the 'negative' of x)
M1	$(x \times y) \times z = x \times (y \times z)$	Associative law of multiplication
M2	$x \times y = y \times x$	Commutative law of multiplication
M3	$x \times 1 = x = 1 \times x$	Existence of identity element (unit) for multiplication
M4	$x \times x^{-1} = 1 \quad (x \neq 0)$	Existence of inverse under multiplication (the 'reciprocal' of x)
D	$x \times (y + z) = x \times y + x \times z$	Distributive law

† i.e., closed operations (see p. 478).

It will be seen that the first four rules (A1–A4) are concerned with addition, M1–M4 with multiplication; D is the distributive law of multiplication over addition (to which we have chiefly referred) and which relates addition with multiplication. These laws govern the behaviour of the rationals under the ordinary manipulations of algebra: subtraction is taken care of by the application of A1, A2, A3 and A4. For if we ask 'What must be added to x to obtain y?', we can easily show that

$$\begin{aligned} x + z = y \quad &\Rightarrow \quad -x + (x + z) = -x + y \\ &\Rightarrow \quad (-x + x) + z = -x + y \\ &\Rightarrow \quad 0 + z = -x + y \\ &\Rightarrow \quad z = -x + y = y + (-x). \end{aligned}$$

In the same way, if $x \times z = y$, we may show that $z = x^{-1}y = yx^{-1}$, and this is a formal way of writing the division $y \div x$. We may loosely describe a field as a number system in which addition, subtraction, multiplication and division may all be carried out, with the exception of division by zero. Of course, it may be that some of the laws for a field as given above may be redundant: it is possible, for instance, to dispense with the second half of A3 and derive it from the first half of A2. When there are no such redundancies, we should have what may be called a *minimum set* of axioms.

The above was chosen as an example, first because the field is one of the fullest and most satisfying structures, and also because the rationals are so familiar. You will readily check that the reals R form a larger field (of which the rationals Q are a sub-field), and the complex numbers C again form a field which contains the reals as a sub-field. Do not imagine, however, that fields are necessarily infinite: it is common to have finite fields, and you may care to show that the two numbers $\{0, 1\}$ form a field when addition and multiplication are defined, modulo 2, thus:

$+$	0	1
0	0	1
1	1	0

$\times$	0	1
0	0	0
1	0	1

Clearly, however, the integers Z do not form a field, since multiplicative inverses do not exist for numbers other than $+1$ or -1. So M4 is not satisfied and the integers with the operations $+$ and $\times$ form a 'lower' structure, known as a *ring*. By contrast, there do exist even more highly organised structures than fields, e.g., *vector spaces* and *modules*, but these need not concern us here, and we shall concentrate upon the most important of all mathematical structures, known as a *group*. So central is this that its study forms the backbone of the present chapter, but first we must discuss more fully the idea of a binary operation.

Exercise 14.1

1 Consider the long multiplication

```
  2 7 5
    7 9
  -----
2 4 7 5
1 9 2 5
---------
2 1 7 2 5
---------
```

and analyse the uses of the field axioms in the course of the work in the same way as was done in the case $24 \times 3 = 72$ above.

2 Revise the proofs of $\boldsymbol{a}.(\boldsymbol{b} + \boldsymbol{c}) = \boldsymbol{a}.\boldsymbol{b} + \boldsymbol{a}.\boldsymbol{c}$

and $\boldsymbol{a} \times (\boldsymbol{b} + \boldsymbol{c}) = \boldsymbol{a} \times \boldsymbol{b} + \boldsymbol{a} \times \boldsymbol{c}$,

where $\boldsymbol{a}$, $\boldsymbol{b}$, $\boldsymbol{c}$ are vectors in three dimensions.

3 Show that $\{0, 1, 2, 3, 4\}$ is a field under addition and multiplication modulo 5, zero being excluded in the case of multiplication, but that $\{0, 1, 2, 3, 4, 5\}$ cannot be made into a field with addition and multiplication modulo 6, even when zero is excluded.

14.2 Binary operations

In the set of axioms for a field given in 14.1, we were dealing with a familiar example in which x, y, z represented *numbers*. One of the deficiencies of traditional school mathematics was the impression given that letters always stood for numbers. But here you have already encountered many examples of the use of letters for other entities — sets, matrices, rotations, points, vectors, permutations, lines, curves, ordered pairs, and so on. We may now ask whether sets of such objects may be organised into structures.

That the answer to this question must be 'yes' you cannot have doubted. But whatever the nature of the elements of the set which is being provided with a structure, the fundamental step to such provision is the idea of a *binary operation* or *law of composition*, by which we can take any two elements of the set and *combine* them in some way, according to some rule. Thus, for example, multiplication is a binary operation on the set of integers, for when one takes two integers such as -3 and 2 and combines them by this particular binary operation, one obtains the integer -6, which is also in Z, and we write $(-3) \times (+2) = -6$, the binary operation of multiplication being represented by the familiar symbol $\times$. More generally, if the binary operation is represented by $*$, we may describe the result of combining members x and y of the background set S formally as follows:

$$z = x * y$$

and this states that the result of applying the binary operation $*$ to the two elements x and y of S is to produce the element z. If (as in the multiplication of integers) the result of an operation is always in the original set S, then we say that our operation is *closed*; but if there is some z which is not in S, then the operation is not closed, as example, $S = Z$ (the integers) and $*$ is the operation of division. Then $(-3) * (+2) = -1\frac{1}{2}$. But $-1\frac{1}{2}$ is not in the set Z, so the operation $\div$ on the integers is *not* closed. Again, suppose $**$ denotes 'exponentiation' (as in the computer programming language Fortran), then $(-3) ** (+2) = (-3)^2 = +9$. However, $(+2) ** (-3) = 2^{-3} = \frac{1}{8}$, and this operation would be closed if our set had been the reals R, but is not closed when we confine our attention to the integers Z. (Note, in this case, that we must be careful to exclude the case $x = y = 0$, for no value can be assigned to 0^0).

One may also describe a closed binary operation on a set S as a mapping from *ordered pairs* of the set into members of the set itself: $(x, y) \mapsto z$ where $x, y, z \in S$. If however, the binary operation is not closed, the image may be outside the set, as for example in the case of the operation . on the set of vectors, which is a mapping from the ordered pairs of vectors into the reals, R.

It is important to realise that by the use of a binary operation, an element may be combined *with itself*, and that $x * x$ must be defined for all x in S: e.g., if $S = R$ and $*$ is multiplication, then $x * x = x^2$; again, if $S = Z^+$ and $*$ is the operation of finding the h.c.f. of x and y, then in the case $x = y$, we have $x * x = x$ for all x in Z^+ (since the h.c.f. of a positive integer and itself is that same number). Moreover, if $z = x * y$, it is perfectly possible for z to be equal to x or y. For, if $**$ is exponentiation, $3 ** 1 = 3$, and $1 ** 3 = 1$.

Linear transformations

We now consider an example where the elements of the set are not numbers but two-dimensional linear transformations, such as rotations, translations, reflections, shears, or enlargements. Given two elements in the set of all such transformations, how do we define a binary operation $\circ$ on them? Simply by considering the *composition* of the movements, one after the other. So if T_1 and T_2 are two transformations, we shall denote the result of first performing the transformation T_1, and then performing T_2 by $T_2 \circ T_1$, the reason for this order being our similar practice when linear transformations were represented by matrices (see 13.2). Suppose now that R_1 represents a $+90°$ turn about O_1, and R_2 represents a $+60°$ turn about point O_2 in the plane of the rotations. Then in the case $O_1 = O_2$, $R_2 \circ R_1$ evidently represents a $+150°$ turn; and also, clearly $R_2 \circ R_1 = R_1 \circ R_2$. In the case when O_1 and O_2 are distinct, intuition tells us that $R_2 \circ R_1$ is a $+150°$ turn, but it is not immediately clear about which point the rotation is centred. A little experiment should however convince you that $R_2 \circ R_1 \neq R_1 \circ R_2$.

Again, suppose our set of transformations consists of all possible rotations in the plane, through any possible angle about all possible points. In view of the above informal discussion, it may seem that this set is closed under the operation $\circ$. However, this is not so, and it is not difficult to produce a counter-example: consider a positive 90° turn about A followed by a negative 90° turn about a different point B:

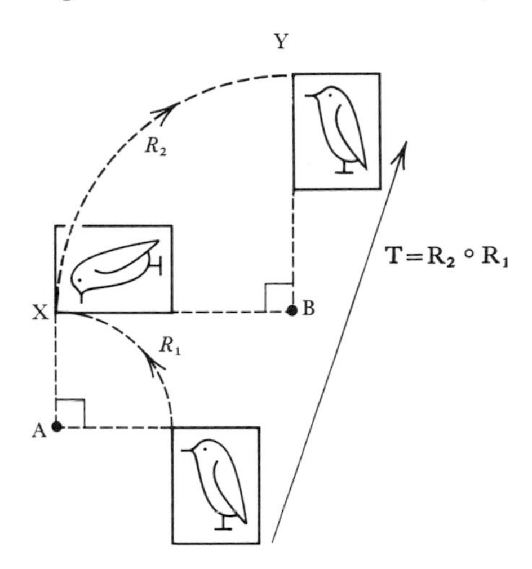

It is clear from the figure that the composition of the two rotations is not a rotation, but a translation T, for after rotation through two equal and opposite angles, the orientation of the figure is restored — the sort of manoeuvre might be executed in the course of a dance, at the end of which the dancer would be facing the same way. Evidently, the composition of rotations about different points through equal and opposite angles will always be a translation, so that the set of plane rotations is not closed under composition. Closure may be achieved, however, by adding the plane *translations* to give a set of transformations comprising both rotations and translations. Notice, however, that the subset of translations alone *is* closed within itself.

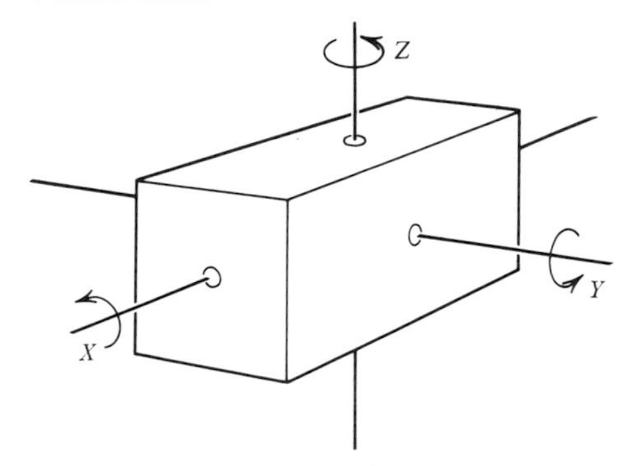

As another illustration, consider the half-turns (180° rotations) of a cuboid about its three perpendicular central axes, and let us call these X, Y, and Z. Here is a set containing only three objects, and we consider whether we have closure. A simple experiment with a matchbox will convince you that $X \circ Y = Y \circ X = Z$, with four other similar relations. But our hopes of closure are dashed when we remember that $X \circ X$ is neither X nor Y nor Z, but has the effect of restoring the original position of the box. This is the *identity transformation*, which may be called I, and it must be included in our set if we are to have a closed system. The table below shows the way in which our four transformations combine:

$\circ$	I	X	Y	Z
I	I	X	Y	Z
X	X	I	Z	Y
Y	Y	Z	I	X
Z	Z	Y	X	I

Permutations

As another illustration of a set with a binary operation, we consider the composition of permutations, a most important idea in mathematics. A finite permutation is strictly a mapping from a finite set on to itself, such as

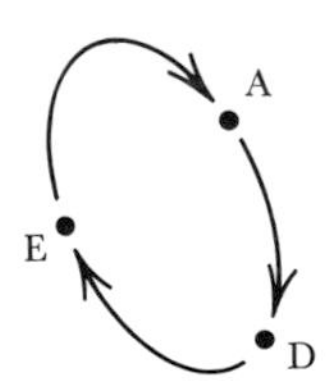

Here there are five elements A, B, C, D, E, and the arrows indicate that the element A is 'mapped' into D, D is mapped into E, and E into A, while B and C are interchanged. It is perhaps convenient to think of a permutation as a 'rearrangement', and to represent it as $\begin{pmatrix} A & B & C & D & E \\ D & C & B & E & A \end{pmatrix}$. Calling this permutation p, suppose now we wish to combine it with another permutation q which replaces the letters according to the scheme $\begin{pmatrix} A & B & C & D & E \\ C & E & B & D & A \end{pmatrix}$. Consider the effect of first performing the permutation p, and then following this by the permutation q. The effect will be that A will be replaced first by D, and finally by D; B will be replaced by C, and subsequently by B, and so on. The final effect of the composite permutation arising from the successive application of first p then q will be thus: $\begin{pmatrix} A & B & C & D & E \\ D & B & E & A & C \end{pmatrix}$. Calling this permutation r, and adhering to the rule of writing the first operation on the right, we may write

$$r = q \circ p = \begin{pmatrix} A & B & C & D & E \\ D & B & E & A & C \end{pmatrix}$$

where $\circ$ represents the binary operation of composition of the permutations. You should verify that

$$s = p \circ q = \begin{pmatrix} A & B & C & D & E \\ B & A & C & E & D \end{pmatrix},$$

showing that composition of permutations is not commutative.

Composition of functions

As another example, we may consider types of binary operations on the set of functions.

If $f(x) = 3 - x$ and $g(x) = \dfrac{x}{2 - x}$

are two functions, then one simple way of combining them is by straightforward addition:

$$f(x) + g(x) = 3 - x + \frac{x}{2 - x} = \frac{6 - 4x + x^2}{2 - x},$$

this being denoted $(f + g)(x)$.

Again, if $f(x) = \ln x$ and $g(x) = \sin x$, the function $f + g$ would be $\ln x + \sin x$. However, we are perhaps more familiar with the composition of functions by successive substitution, $(f \circ g)(x)$ being defined as $f[g(x)]$ so that in the latter example above we get the familiar $\ln(\sin x)$, whilst in the former case,

$$(f \circ g)(x) = 3 - \frac{x}{2 - x} = \frac{6 - 4x}{2 - x}.$$

Note that this operation $\circ$ is not commutative, for

$$(g \circ f)(x) = \frac{3 - x}{2 - (3 - x)} = \frac{3 - x}{2 - x} \neq (f \circ g)(x).$$

Exercise 14.2a

1 Consider the set of permutations of five letters which interchange two pairs, such as (AB)(CD), which changes ABCDE into BADCE. Is this set closed under composition of permutations?

2 Is the following a valid binary operation on the set of rationals?

$$\frac{a}{b} * \frac{c}{d} = \frac{a + c}{b + d} \quad \text{where } a, b, c, d \in Z.$$

Is the following a valid binary operation on the set of points in a plane? Given P, Q, P * Q is defined as the mirror image of P in OQ, where O is a given fixed point.

3 Find which permutation(s) must be added to the following set in order for the set to be closed under composition of permutations:

1 2 3 4 5 6

4 1 2 3 5 6

3 4 1 2 6 5

2 3 4 1 6 5

4 Show that the set of 3 × 3 matrices which have positive determinants are closed under matrix multiplication. Are they closed under matrix addition? Consider also the set of matrices whose determinants are (*i*) negative, (*ii*) zero.

5 Consider the set of 120° rotations of the regular tetrahedron about its altitudes. Is this closed under successive composition?

6 Is the set $\{1, 3, 5, 9, 13\}$ closed under multiplication mod 14? Is there a subset which is closed?

7 Consider the closure or otherwise of the following number systems under the operations $+, -, \times, \div$:

(*i*) N; (*ii*) Z; (*iii*) Z^-; (*iv*) Q; (*v*) Q^+; (*vi*) R; (*vii*) R^+; (*viii*) C; (*ix*) the irrationals; (*x*) $\cos\theta + i\sin\theta$ $(0 \leqslant \theta < 2\pi)$; (*xi*) $x + iy$ $(x, y \in Z)$.

8 Is the set of 2 × 2 *singular* matrices closed under matrix addition, under matrix multiplication?

9 In the following cases, find $(f + g), f \circ g, g \circ f, f^2$ and g^2:

(*i*) $f(x) = \dfrac{1}{x}, \quad g(x) = \dfrac{1}{1 - x}$; (*ii*) $f(x) = \ln x, \quad g(x) = \sin x$;

(*iii*) $f(x) = \dfrac{x - 3}{2x - 1}, \quad g(x) = -x$;

(*iv*) $f(x) = 2x - 1, \quad g(x) = 5 - 2x$.

10 Is the operation $x * y = x^y$ closed in (*i*) Z^+, (*ii*) Z; (*iii*) Q; (*iv*) R^+?

11 Show that functions of the type $f(x) = (ax + b)/(cx + d)$ are closed under function composition.

12 Construct a finite set of 2×2 matrices which is closed under matrix multiplication.

13 Consider the operations on vectors: (*i*) vector addition; (*ii*) scalar product; (*iii*) vector product. Which of these are closed and which not closed (*a*) in two-dimensional space; (*b*) in three-dimensional space?

14 S denotes the set of matrices of the form $\begin{pmatrix} x & y \\ y & x \end{pmatrix}$ where x and y are real. Investigate whether, for the operation of matrix multiplication, (*i*) S is closed; (*ii*) the identity belongs to S; (*iii*) every element of S has an inverse. (S.M.P.)

Associativity and commutativity

Once we have a set with a closed binary operation (or possibly more than one operation), we have what may be called an *algebra*. If the existence of the operation is the only structural feature, then the algebra is very primitive. But usually the operation satisfies a number of axioms which make the structure more interesting.

Generally speaking, the most desirable property, apart from closure, of a binary operation is that it should be *associative*, that is

$$\forall a, b, c \in S, \qquad (a * b) * c = a * (b * c).$$

If this feature is lacking, the structure is unlikely to lead to fruitful mathematical results (though the operation of vector product is a notable exception). On the other hand, *commutativity*,

$$\forall a, b \in S, \qquad a * b = b * a$$

can often be dispensed with, and many examples of non-commutative operations which are nonetheless mathematically powerful have already been encountered.

It is usually a simple matter to see whether an operation is commutative by inspection, but often a laborious process to check for associativity. For example, consider the binary operation on R defined by the formula

$$v_1 * v_2 = \frac{v_1 + v_2}{1 + v_1 v_2}, \quad \text{with } |v_1| < 1, \quad |v_2| < 1$$

(a formula used in special relativity theory). We wish to show that $*$ is associative, i.e., that

$$(v_1 * v_2) * v_3 = v_1 * (v_2 * v_3).$$

$$\text{Now} \quad (v_1 * v_2) * v_3 = \frac{v_1 + v_2}{1 + v_1 v_2} * v_3 = \frac{\dfrac{v_1 + v_2}{1 + v_1 v_2} + v_3}{1 + \dfrac{(v_1 + v_2) v_3}{1 + v_1 v_2}}$$

$$= \frac{v_1 + v_2 + v_3 + v_1 v_2 v_3}{1 + v_1 v_2 + v_1 v_3 + v_2 v_3}$$

and $$v_1 * (v_2 * v_3) = v_1 * \frac{v_2 + v_3}{1 + v_2v_3} = \frac{v_1 + \dfrac{v_2 + v_3}{1 + v_2v_3}}{1 + \dfrac{v_1(v_2 + v_3)}{1 + v_2v_3}}$$

$$= \frac{v_1 + v_2 + v_3 + v_1v_2v_3}{1 + v_1v_2 + v_1v_3 + v_2v_3}$$

so the operation is associative, as well as being obviously commutative. On the other hand, the operation $x \sim y = |x - y|$ is commutative but not associative, as may be confirmed by a single counter-example: e.g., if

$x = 2, \quad y = 1, \quad z = 4,$

then $(x \sim y) \sim z = |2 - 1| \sim 4 = 1 \sim 4 = |4 - 1| = 3,$

but $x \sim (y \sim z) = 2 \sim |1 - 4| = 2 \sim 3 = |2 - 3| = 1.$

Usually there is no short cut to discovering whether or not a particular operation is associative and it cannot be ascertained by inspection. However, one should remember that associativity is always guaranteed in the following cases:

(*i*) multiplication and addition in a number system, including finite arithmetics (see 14.3);

(*ii*) composition of mappings (geometrical transformations, permutations, matrices, vectors, etc.).

Exercise 14.2b

1 State whether each of the operations in exercises 14.2a, nos. 2, 10 are: (*a*) commutative; (*b*) associative.

2 Discover which of the operations on the reals described by the following formulae are (*a*) commutative (*b*) associative:

(*i*) $x * y = \dfrac{xy - 1}{x + y} \qquad x * y = \dfrac{1 - xy}{x + y} \quad (x + y \neq 0);$

(*ii*) $x * y = k(x + y);$ (*iii*) $x * y = |x - y|;$ (*iv*) $x * y = x;$

(*v*) $x * y = \dfrac{x + y}{xy} \quad (x, y \neq 0);$ (*vi*) $x * y = x \times |y|;$

(*vii*) $x * y = x^y.$

3 Show that the non-commutative operation shown in the table is not associative:

$*$	a	b	c
a	a	b	c
b	c	a	b
c	b	c	a

4 Show that the operation $*$ on the set $\{a, b, c, d, e\}$ where products are given in the following table, is not associative:

$*$	a	b	c	d	e
a	a	b	c	d	e
b	b	c	e	a	d
c	c	a	d	e	b
d	d	e	a	b	c
e	e	d	b	c	a

5 Show that, for vectors in three-dimensional space, vector addition is associative, but that vector product is not.

6 Prove the associativity of the following law of composition for pairs of reals:

$$(x_1, y_1) * (x_2, y_2) = (x_1x_2 - y_1y_2, x_1y_2 + x_2y_1).$$

7 Show that the operation on ordered pairs of reals defined by

$$(x_1, y_1) * (x_2, y_2) = (x_1x_2, x_2y_1 + y_2)$$

is associative, and find whether the operation on ordered triples defined by

$$(x_1, y_1, z_1) * (x_2, y_2, z_2) = (x_1 + x_2 + y_2z_1, y_1 + y_2, z_1 + z_2)$$

is or is not associative.

8 Show that the operation on R expressed by the formula

$$x * y = \frac{xy + k}{x + y} \quad (x + y \neq 0)$$

is associative for all k, and suggest some interpretations for special values of k.

9 The operation $\circ$ is defined on the number pairs $A = (a_1, a_2)$ $B = (b_1, b_2)$ so that $A \circ B = (a_1 + b_2, a_2 + b_1)$ and $A = B$ if and only if $a_1 = b_1$ and $a_2 = b_2$. Find whether the operation is associative.

Find P such that $A \circ P = A$, and Q such that $Q \circ A = A$, and determine whether I, J can be found such that, for all A, $A \circ I = A$ and $J \circ A = A$.

Given that $A^* = (-a_2, -a_1)$, discuss the following:

$$\begin{aligned} & B \circ A = C \circ A \\ \Rightarrow\quad & B \circ A \circ A^* = C \circ A \circ A^* \\ \Rightarrow\quad & B \circ (0, 0) = C \circ (0, 0) \\ \Rightarrow\quad & B = C. \end{aligned}$$

(M.E.I.)

Identity and inverse

Before we are ready to talk about groups, we must consider two other ideas, those of *identity* and *inverse*. If a set S is provided with an operation $*$ and there exists a *fixed* element e of S which combines with *any* element of S to leave it unchanged, i.e.,

$$\forall x \in S, \quad \underset{\text{(right identity)}}{x * e = x} \quad \text{and} \quad \underset{\text{(left identity)}}{e * x = x},$$

then e is described as an identity element for the set. For example, when the operation is addition in the set of integers, the identity is 0; while for multiplication, it is 1. These are the two numbers, 0 and 1, which play special parts in the axioms for a field (see 14.1). For addition in the set of 2×2 matrices, the identity is $\begin{pmatrix} 0 & 0 \\ 0 & 0 \end{pmatrix}$, the 'null' matrix; for multiplication it is $\begin{pmatrix} 1 & 0 \\ 0 & 1 \end{pmatrix}$, the 'unit' matrix. Note that $\begin{pmatrix} 1 & 0 & 0 \\ 0 & 1 & 0 \\ 0 & 0 & 1 \end{pmatrix}$ is a left identity for the set of all $3 \times n$ matrices, but a right identity for the set of all $n \times 3$ matrices, and both a left and right identity only for square 3×3 matrices. Similarly, we have already encountered the *identity transformation* in a set of transformations, the *identity permutation* and the *identity function*. In the case of subtraction on the set Z (or Q or R), we have a *right* identity $x - 0 = x$ (for all x), but no left identity; similarly there is a right- but no left-identity in the case of division and also exponentiation. When we speak of 'the identity element' in future, we shall be concerned with associative closed operations having an identity which is both a left- and right-identity.

The concept of *inverse* is one that has already arisen in several situations during the course of this book. The general idea is that of 'undoing' an operation, i.e., of finding the element y which combines with x to give the identity, thus

$$\underset{(y\text{ is a right inverse})}{x * y = e} \qquad \underset{(y\text{ is a left inverse}).}{y * x = e}$$

For example, in the set Z with the operation $+$ (the identity being 0) we have

$$x + (-x) = 0 \quad \text{and} \quad (-x) + x = 0,$$

so that $-x$, or the *negative* of x, is the inverse of x under addition. In the set Q with the operation $\times$, (the identity being 1) we have $x \times 1/x = 1$ $(x \neq 0)$, and the multiplicative inverse is the reciprocal, and written x^{-1}. We have considered the inverse of a matrix under matrix multiplication in some detail, and we use the same notation, $\boldsymbol{M}^{-1}$, to denote the inverse of a non-singular matrix $\boldsymbol{M}$, satisfying $\boldsymbol{M}\boldsymbol{M}^{-1} = \boldsymbol{I} = \boldsymbol{M}^{-1}\boldsymbol{M}$ (see 13.4).

A similar notation is used in the case of inverses of geometrical transformations, permutations and functions. For example, if R is an anti-clockwise rotation through 40° about O, then the inverse of this, a clockwise 40° rotation about 0, would be denoted R^{-1}. Again, if x is the permutation

$$\begin{pmatrix} 1 & 2 & 3 & 4 & 5 & 6 \\ 5 & 1 & 6 & 4 & 2 & 3 \end{pmatrix},$$

then the inverse permutation may be written

$$\begin{pmatrix} 5 & 1 & 6 & 4 & 2 & 3 \\ 1 & 2 & 3 & 4 & 5 & 6 \end{pmatrix} \quad \text{or more simply} \quad \begin{pmatrix} 1 & 2 & 3 & 4 & 5 & 6 \\ 2 & 5 & 6 & 4 & 1 & 3 \end{pmatrix},$$

and is denoted x^{-1}.

You should check that the result of combining x with x^{-1} either way round is to produce the identity permutation.

Lastly, we have also considered (see 1.1, Book 1, *Advanced Mathematics*) the inverse of a function or mapping. Taking a particular example,

if $f{:}x \mapsto \dfrac{x+1}{2-x}$, we wish to specify the inverse $f^{-1}(x)$.

$$\text{Calling} \quad y = \frac{x+1}{2-x}$$

$$\Rightarrow \quad 2y - xy = x + 1$$

$$\Rightarrow \quad x = \frac{2y+1}{y+1},$$

so the required inverse function is $f^{-1}(x) = \dfrac{2x+1}{x+1}$.

You should check that the 'composition' of the functions f and f^{-1}, in either order, does in fact give the identity function $i{:}x \mapsto x$.

Finally we notice that if x and y are two elements of a set which combine under an associative binary operation $*$, then the inverse of $x * y$ is $y^{-1} * x^{-1}$; this may be abbreviated by dropping the $*$ and using the 'juxtaposition', or multiplicative notation of ordinary algebra:

$$(xy)^{-1} = y^{-1}x^{-1}.$$

The proof is straightforward:

$$(xy)\,(y^{-1}x^{-1}) = x\,(y\,(y^{-1}x^{-1})) = x\,((yy^{-1})x^{-1}) = x\,(ex^{-1}) = xx^{-1} = e.$$

(Note the repeated use of associativity.)

Similarly, $(y^{-1}x^{-1})(xy) = e$, so that xy and $y^{-1}x^{-1}$ are inverse elements (just as the inverse of the combined operation of putting on one's socks and then one's shoes is the operation of first removing the shoes, then removing the socks). Evidently the rule may be extended to

$$(xyz)^{-1} = z^{-1}y^{-1}x^{-1}, \text{ etc.}$$

Exercise 14.2c

1 Find the identity elements for sets under the operations $\cap$ and under $\cup$ (see 0.2 and 14.6), and show that it is impossible to find inverses for these operations. Find an operation which does admit inverses.

2 (*i*) What must be included in the set $\{4, 8\}$ in order to achieve closure under multiplication mod 14. What is the identity element, and what is the inverse of 4?
(*ii*) Which is the identity element in the system $\{2, 4, 8, 10, 14, 16\}$ under multiplication mod 18?
(*iii*) Show that $\{2, 4, 6, 8, 12, 14, 16, 18\}$ is closed under multiplication mod 20, but has no identity.

3 Find:
(*i*) the inverse of 27 and of 12 under addition mod 40 of $\{0, 1, 2, \ldots, 39\}$;
(*ii*) the inverse of 17 under multiplication mod 20 of $\{0, 1, 2, \ldots, 19\}$;
(*iii*) the inverse of 53 under multiplication mod 100 of $\{0, 1, 2, \ldots, 99\}$.

4 Find identity elements where they exist in operations described in 14.2b, **2**, **3**, **4**, **5**, **6**, **7**, **8**, **9**.

5 Find the inverses of the matrices under matrix multiplication, and interpret geometrically where possible:

(*i*) $\begin{pmatrix} 3 & 4 \\ -4 & 3 \end{pmatrix}$; (*ii*) $\begin{pmatrix} 0.6 & 0.8 \\ 0.8 & -0.6 \end{pmatrix}$; (*iii*) $\begin{pmatrix} 1 & 1 \\ -1 & 0 \end{pmatrix}$;

(*iv*) $\begin{pmatrix} 0 & 0 & 1 \\ 1 & 0 & 0 \\ 0 & -1 & 0 \end{pmatrix}$; (*v*) $\begin{pmatrix} -5 & -1 & 3 \\ 4 & 1 & -2 \\ 1 & 2 & 2 \end{pmatrix}$;

(*vi*) $\begin{pmatrix} \cos 80^\circ & -\sin 80^\circ \\ \sin 80^\circ & \cos 80^\circ \end{pmatrix}$.

6 Find the inverses of the following functions:

(*i*) $1 - \dfrac{1}{x}$; (*ii*) $\dfrac{2 - x}{x}$; (*iii*) $\dfrac{2 + x}{2 - x}$; (*iv*) x^3;

(*v*) a^x; (*vi*) $\dfrac{3x - 7}{5x - 3}$; (*vii*) $\log_a (\log_a x)$.

7 Find the inverses of the permutations

(*i*) $\begin{pmatrix} 1 & 2 & 3 & 4 & 5 \\ 2 & 4 & 5 & 1 & 3 \end{pmatrix}$; (*ii*) $\begin{pmatrix} 1 & 2 & 3 & 4 & 5 & 6 \\ 4 & 6 & 5 & 1 & 2 & 3 \end{pmatrix}$;

(*iii*) $\begin{pmatrix} 1 & 2 & 3 & 4 & 5 & 6 & 7 \\ 4 & 7 & 5 & 1 & 3 & 6 & 2 \end{pmatrix}$.

Verify in each case that $x \circ x^{-1}$ and $x^{-1} \circ x$ are both the identity permutation.

8 Verify the rule $(xy)^{-1} = y^{-1}x^{-1}$ in the case when x and y are the permutations

$$x = \begin{pmatrix} 1 & 2 & 3 & 4 & 5 \\ 4 & 1 & 2 & 3 & 5 \end{pmatrix} \qquad y = \begin{pmatrix} 1 & 2 & 3 & 4 & 5 \\ 3 & 5 & 1 & 4 & 2 \end{pmatrix}.$$

9 Let S be the set of 2×2 matrices with elements in $\{0, 1, 2, 3, 4, 5\}$. If $A \in S$ and A has an inverse in S, show that $\det A = 1$ or 5. (J.M.B.)

10 Show that the set of matrices of the form

$$\begin{pmatrix} a & b & c \\ c & a & b \\ b & c & a \end{pmatrix} \quad (a, b, c \in R)$$

are closed under matrix multiplication, and that the identity is in the set. Under what circumstances do we have inverses?

11 Let $\oplus$ and $\otimes$ be two binary operations (rules of composition) defined on the field Q of rational numbers, as follows:

$$a \oplus b = a + b - 1 \qquad a \otimes b = a + b - ab.$$

Find the identity element, and the inverse of any number a, for each of these operations.
Find all functions $f\colon Q \mapsto Q$ which simultaneously satisfy the equations

$$f(a + b) = f(a) \oplus f(b) \quad \text{and} \quad f(ab) = f(a) \otimes f(b).$$ (M.E.I.)

14.3 Groups

We are now ready to describe the vitally important structure known as a group, but before giving a general definition we shall first take two examples.

Example 1

Let us consider the set of integers $\{1, 3, 7, 9\}$ and their products modulo 10 (i.e., the remainders, or *residues*, when their ordinary products are divided by the modulus 10).

As $3 \times 7 = 21$, $3 \times 9 = 27$, $9 \times 7 = 63$ etc., we may write $3 \times 7 = 1$, $3 \times 9 = 7$, $9 \times 7 = 3$ (mod 10), and these products can be summarised:

Table 1

$\times_{10}$	1	3	7	9
1	1	3	7	9
3	3	9	1	7
7	7	1	9	3
9	9	7	3	1

The system has the following features. First, it is closed. Had we taken the set $\{3, 5, 7\}$, this would have failed for closure, for products such as 3×7 would not have been contained inside the set. Second, the operation of multiplication is undoubtedly associative. Third, we have an identity element, 1. And fourth, every element has an inverse; e.g. since $3 \times 7 = 1$, the residues 3 and 7 are a pair of inverses. (This would not have been so had the number 5 been included, for the element 5 would have no inverse though closure would still have held good.) A system which passes these four tests — of *closure*, *associativity*, *identity* and *inverses* — is called a *group*, and as the present example has four elements, it is said to be a *group of order* 4.

Example 2

We now give an example of an infinite group, provided by the set of all *non-singular* 2×2 matrices with real elements under the operation of matrix multiplication.

It is immediately clear that

(*i*) The product of two such matrices with real elements is another such matrix, so closure is established.

(*ii*) Multiplication of matrices is known to be associative.

(*iii*) The set includes the unit matrix $\begin{pmatrix} 1 & 0 \\ 0 & 1 \end{pmatrix}$.

(*iv*) Every non-singular 2×2 matrix has an inverse which is also in the set.

So the fourfold requirements for a group are satisfied.

This time we have not only an infinite group, but one in which the binary operation is non-commutative, since matrices do not commute under multiplication. By contrast, our first example:† $\{1, 3, 7, 9\}$, $\times$ (mod 10) was a *finite commutative* group. Such a group, in which *every* product is commutative is called *Abelian*, after the Norwegian pioneer of group theory, N. H. Abel (1802–29). Note that, within non-Abelian groups, it is possible for *some* products to commute: the identity will always commute with every element; and in the group of non singular 2×2 matrices under multiplication, we note that,

$$\begin{pmatrix} 4 & 1 \\ 2 & -1 \end{pmatrix}\begin{pmatrix} 0 & -1 \\ -2 & 5 \end{pmatrix} = \begin{pmatrix} -2 & 1 \\ 2 & -7 \end{pmatrix} = \begin{pmatrix} 0 & -1 \\ -2 & 5 \end{pmatrix}\begin{pmatrix} 4 & 1 \\ 2 & -1 \end{pmatrix}.$$

Having provided these two very dissimilar examples, we are now ready to give our general definition:

† We shall specify a group by stating first the set, then the binary operation, these being separated by a comma.

A *group* is a set G with a binary operation $*$ subject to the following requirements:

C	$*$ is *closed* in G, i.e., for all $x, y \in G$, $x * y \in G$
A	$*$ is *associative*, i.e., for all $x, y, z \in G$, $(x * y) * z = x * (y * z)$
N	There is an identity or neutral element e satisfying $x * e = e * x = x$, for all $x \in G$
I	Each element x has an *inverse*, x^{-1}, satisfying $x * x^{-1} = x^{-1} * x = e$

If we now use the symbols $\forall$ to denote 'for all', $\exists$ to denote 'there exists' and : to denote 'such that', these four axioms of a group $G, *$ may be conviently summarised:

C	$\forall x, y \in G,$	$x * y \in G$
A	$\forall x, y, z \in G,$	$(x * y) * z = x * (y * z)$
N	$\exists e \in G\colon\ \forall x \in G,$	$x * e = e * x = x$
I	$\forall x \in G, \exists x^{-1}\colon x * x^{-1} = x^{-1} * x = e$	

If you refer to the axioms for a field (see 14.1), you will notice that $A1$–4 and $M1$–4 are equivalent to stating that we should have a group under addition and under multiplication, always remembering that the 'zero' element (the identity for addition) must be removed in the case of the multiplicative group, since it has no multiplicative inverse. Thus the rationals, Q, are a group under addition, and also group under multiplication, provided 0 is excluded (just as all singular matrices had to be expunged from the set before group structure of 2×2 matrices could be claimed).

Exercise 14.3a

1 Show that the following are finite groups and state the order of the group in each case

(*i*) integers with a specified number of binary digits under the operation of addition mod 2 with no carrying (e.g., $011 * 110 = 101$);

(*ii*) the functions $x, \dfrac{x-1}{x+1}, \dfrac{-1}{x}, \dfrac{1+x}{1-x}$ under function composition;

(*iii*) all the 120 permutations of five letters;

(*iv*) the set $\{0, 1, 2, \ldots, n-1\}$ under addition mod n;

(*v*) the set $\{1, 2, 4, 7, 8, 11, 13, 14\}$ under multiplication mod 15;

(*vi*) all the symmetries (rotations and reflections) of the rectangle; the square; the equilateral triangle, the cuboid, etc.

2 Show that the following† are infinite groups

(*i*) $Z, +$; (*ii*) $R, +$; (*iii*) $R^*, \times$; (*iv*) $Q^*, \times$; (*v*) $p \times q$ matrices, under matrix addition (suggest numbers systems from which the elements might be drawn);
(*vi*) $n \times n$ non-singular matrices under matrix multiplication;
(*vii*) all isometries in the plane (i.e., transformations which preserve distance) under successive composition;
(*viii*) vectors in 2D space under vector addition;
(*ix*) all integral powers of 3 under multiplication.

3 Find subsets of the examples in no. **1** and no. **2** above which are themselves groups (called *subgroups*) under the operation mentioned.

4 Say why the following fail to be groups:
(*i*) odd integers under addition;
(*ii*) integers under multiplication;
(*iii*) sets under the binary operation 'intersection';
(*iv*) the negative rationals, under addition;
(*v*) $\{a + b\sqrt{2} + c\sqrt{3}\}$ $(a, b, c \in Z)$, under multiplication;
(*vi*) $\{a + b\sqrt{2} + c\sqrt{3}\}$ $(a, b, c \in Q)$, under multiplication;
(*vii*) the permutations $\begin{pmatrix} 1 & 2 & 3 & 4 \\ 2 & 1 & 4 & 3 \end{pmatrix}$ $\begin{pmatrix} 1 & 2 & 3 & 4 \\ 3 & 4 & 1 & 2 \end{pmatrix}$ $\begin{pmatrix} 1 & 2 & 3 & 4 \\ 4 & 3 & 2 & 1 \end{pmatrix}$;
(*viii*) $\{1, -1, i\}$ under multiplication;
(*ix*) all irrationals together with zero, under addition.

5 Show that the first table is a group table, but the second is not. In the third, fill in the blanks in the group table, given that it is Abelian, and that every element (say g) satisfies $g^3 = 1$ (identity)

	a	*b*	*c*	*d*	*e*	*f*
a	*e*	*c*	*b*	*f*	*a*	*d*
b	*d*	*e*	*f*	*a*	*b*	*c*
c	*f*	*a*	*d*	*e*	*c*	*b*
d	*b*	*f*	*e*	*c*	*d*	*a*
e	*a*	*b*	*c*	*d*	*e*	*f*
f	*c*	*d*	*a*	*b*	*f*	*e*

	a	*b*	*c*	*d*	*e*
a	*b*	*d*	*e*	*c*	*a*
b	*c*	*a*	*d*	*e*	*b*
c	*d*	*e*	*a*	*b*	*c*
d	*e*	*c*	*b*	*a*	*d*
e	*a*	*b*	*c*	*d*	*e*

	1	*p*	*q*	*r*	*s*	*t*	*u*	*v*	*w*
1	1	*p*	*q*	*r*	*s*	*t*	*u*	*v*	*w*
p	*p*	*q*	1	*t*	.	.	.	.	.
q	*q*	1	*p*	.	.	.	.	.	.
r	*r*	.	.	.	1	.	.	.	.
s	*s*	.	.	.	.	.	.	.	.
t	*t*	.	.	.	.	.	.	.	.
u	*u*	.	.	.	.	1	.	.	.
v	*v*	.	.	.	.	.	.	.	.
w	*w*	.	.	.	.	.	.	.	.

6 Show that the set $\{x, 1, -x, -1\}$ of linear polynomials in x, when multiplied modulo $(x^2 + 1)$, are a group, and write out the group table.

7 Show that the matrices $\begin{pmatrix} \cos\theta & -\sin\theta \\ \sin\theta & \cos\theta \end{pmatrix}$ are a group under matrix multiplication. What is the order of the group when $\theta = 15°$?

† See Notation (p. xiii).

8 The set S consists of all transformations T of the real numbers given by
$T(x) = ax + b \quad (a \neq 0,\ a,\ b \text{ real})$.
Show that if $T_1, T_2 \in S$, then $T_1 T_2 \in S$. Prove that the set S forms a group under composition of functions. Find the elements T of S of order 2 (i.e., $TT = I$ and $T \neq I$ where I is the identity transformation). Prove that these elements are all reflections. Show that there are no elements of order 3 in S. (J.M.B.)

9 Let G be the set of all matrices of the form $\begin{pmatrix} 1 & x \\ 0 & 1 \end{pmatrix}$ with x a real number.
(*i*) Show that G does not form a group under matrix addition.
(*ii*) Show that G does form a group under matrix multiplication.
(You may assume associativity of matrix addition and multiplication.) (J.M.B.)

10 G is a *finite* group, and g is an element other than the identity. A set S consists of all positive integral powers of g (so that $S = \{g, g^2, g^3, \ldots\}$).
Prove that: (*i*) the identity belongs to S; (*ii*) g^{-1} belongs to S.
Give an example to show that neither of these need be true if the group is infinite. (S.M.P.)

We now resume the illustration of groups in a variety of contexts.

Infinite groups

Other examples of infinite groups include the set of plane rotations and translations (see p. 478), the law of composition being as described there; in three dimensions, the set of transformations (called *isometries*), such as rotations, translations and reflections, which preserve distance between all pairs of points; and also the set of vectors in three-dimensional space under vector addition. In each case, it should be confirmed that the four requirements for a group are satisfied.

Finite groups

Turning to groups with a finite number of elements, we saw at the beginning of this section that there is a useful way of displaying all the products of elements in the group in the form of a table, and this is valuable in the case of small groups. For example, the matrices

$$I = \begin{pmatrix} 1 & 0 \\ 0 & 1 \end{pmatrix}, \quad X = \begin{pmatrix} 1 & 0 \\ 0 & -1 \end{pmatrix}, \quad Y = \begin{pmatrix} -1 & 0 \\ 0 & 1 \end{pmatrix}, \quad Z = \begin{pmatrix} -1 & 0 \\ 0 & -1 \end{pmatrix}$$

under multiplication combine according to the entries of

Table 2

$\times$	I	X	Y	Z
I	I	X	Y	Z
X	X	I	Z	Y
Y	Y	Z	I	X
Z	Z	Y	X	I

and this describes the group completely. It will be recalled that precisely the same table arose (p. 480) when we combined the half-turns of the rectangular box, so that the table serves to describe the composition of the various rotational symmetries of the cuboid.

By contrast, a quite different table arises when the four matrices

$$I = \begin{pmatrix} 1 & 0 \\ 0 & 1 \end{pmatrix}, \quad R = \begin{pmatrix} 0 & -1 \\ 1 & 0 \end{pmatrix}, \quad S = \begin{pmatrix} 0 & 1 \\ -1 & 0 \end{pmatrix}, \quad H = \begin{pmatrix} -1 & 0 \\ 0 & -1 \end{pmatrix},$$

combine under multiplication:

Table 3

$\times$	I	R	S	H
I	I	R	S	H
R	R	H	I	S
S	S	I	H	R
H	H	S	R	I

These matrices correspond to rotations about the origin through angles 0, $\frac{1}{2}\pi$, π, $\frac{3}{2}\pi$, and so the table describes the rotational symmetries of a figure like

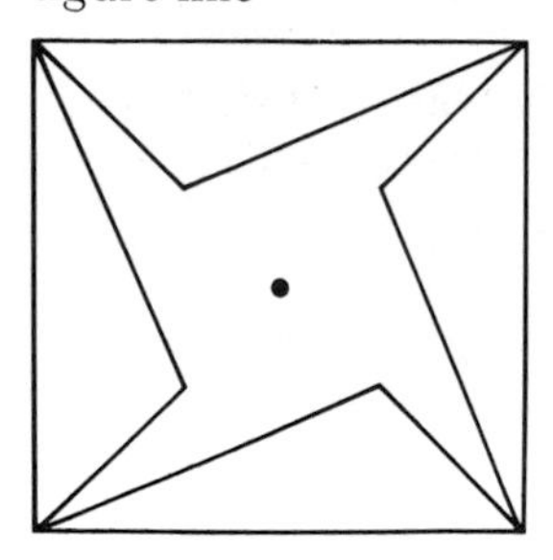

Permutation groups

Next we consider a group of permutations.

Let x be the permutation $\begin{pmatrix} 1 & 2 & 3 & 4 & 5 \\ 2 & 3 & 4 & 5 & 1 \end{pmatrix}$, i.e., x replaces 1 by 2, 2 by 3, . . ., 5 by 1.

It is usual to denote such a 'cycle' by (1 2 3 4 5), each digit in the cycle enclosed in brackets being replaced by the succeeding one, the cycle being completed by the replacement of the final digit by the first. (So that this particular cycle might also be denoted as (4 5 1 2 3), and in various other ways). Similarly, the permutation

$$y = \begin{pmatrix} 1 & 2 & 3 & 4 & 5 & 6 \\ 4 & 6 & 5 & 2 & 1 & 3 \end{pmatrix}$$

can be shown in cycle notation as $y = (1\ 4\ 2\ 6\ 3\ 5)$; while the permutation

$$z = \begin{pmatrix} 1 & 2 & 3 & 4 & 5 & 6 \\ 5 & 3 & 6 & 4 & 1 & 2 \end{pmatrix}$$

contains two cycles, $a = (1\ 5)$, and $p = (2\ 3\ 6)$, the digit 4 being unchanged by the permutation. We also see that the cycle $x = (1\ 2\ 3\ 4\ 5)$, when applied repeatedly to the five digits, changes the line successively as follows:

$$1\ 2\ 3\ 4\ 5 \xrightarrow{x} 2\ 3\ 4\ 5\ 1 \xrightarrow{x} 3\ 4\ 5\ 1\ 2 \xrightarrow{x} 4\ 5\ 1\ 2\ 3$$

$$\xrightarrow{x} 5\ 1\ 2\ 3\ 4 \xrightarrow{x} 1\ 2\ 3\ 4\ 5,$$

so five applications of the permutation x lead to the *identity permutation*, and we write $x * x * x * x * x = e$.

From now on we shall usually drop the symbol $*$, or whatever sign is used for the binary operation, and use *juxtaposition* to denote the composition of two elements: thus $x * y$ will be abbreviated to xy (first y, then x). By analogy with algebra, we also use x^5 to abbreviate $xxxxx$ and, since we are now committed to a *multiplicative* notation, we shall henceforward use 1 for the identity element. Thus, in the above, $x^5 = 1$ describes the behaviour of the cycle $x = (1\ 2\ 3\ 4\ 5)$.

Evidently, since $x^2 = \begin{pmatrix} 1 & 2 & 3 & 4 & 5 \\ 3 & 4 & 5 & 1 & 2 \end{pmatrix}$,

we may write x^2 in cycle notation as $(1\ 3\ 5\ 2\ 4)$, though it is easy enough to do this without first writing down the cumbersome replacement notation. The permutations $1, x, x^2, x^3, x^4$ combine according to

Table 4

		1	x	x^2	x^3	x^4
	1	1	x	x^2	x^3	x^4
	x	x	x^2	x^3	x^4	1
C_5	x^2	x^2	x^3	x^4	1	x
	x^3	x^3	x^4	1	x	x^2
	x^4	x^4	1	x	x^2	x^3

and this particular type of group, generated by a single cycle, is called a *cyclic group* and denoted C_5. In a similar way, the permutation

$$y = \begin{pmatrix} 1 & 2 & 3 & 4 & 5 & 6 \\ 4 & 6 & 5 & 2 & 1 & 3 \end{pmatrix} = (1 \quad 4 \quad 2 \quad 6 \quad 3 \quad 5)$$

generates a cyclic group, this time containing 6 elements, or a group of order 6, called C_6. More generally, the cyclic group of order n is abbreviated C_n.

Finite arithmetic

Suppose in the case of the permutation x in the previous paragraph we had tabulated not 1, x, x^2, x^3, and x^4, but the 'indices' (i.e., the *number of times* the permutation x had been applied, namely 0, 1, 2, 3, and 4). Then we obtain

Table 5

	0	1	2	3	4
0	0	1	2	3	4
1	1	2	3	4	0
2	2	3	4	0	1
3	3	4	0	1	2
4	4	0	1	2	3

For example, when 4 is combined with 3, we get 2; because, in this system, $x^4 \times x^3 = x^7 = x^2$ (as $x^5 = 1$). So far as the indices are concerned, we may say $4 + 3 = 7$, but since any superfluous 5's are of no interest, we may work *modulo* 5, and declare that $4 + 3 = 2 \pmod 5$. Thus in Table 5 above, the operation is *addition mod* 5 and we have an example of a *finite arithmetic* providing us with a group.

More interesting groups arise in finite arithmetics when the operation is multiplication to some modulus: for example, the group of order 4 at the beginning of this section. Another simple example is provided by the set $\{0, 1, 2, 3, 4\}$ under multiplication mod 5, when we obtain

Table 6

$\times_5$	0	1	2	3	4
0	0	0	0	0	0
1	0	1	2	3	4
2	0	2	4	1	3
3	0	3	1	4	2
4	0	4	3	2	1

Here there is an identity element 1, but the element 0 has no inverse, so the structure is not a group. If, however, 0 is deleted from the set, we do have group structure, described by

Table 7

$\times_5$	1	2	3	4
1	1	2	3	4
2	2	4	1	3
3	3	1	4	2
4	4	3	2	1

Sometimes with finite arithmetic under multiplication it is necessary to delete more than merely 0 from the set. For example, working mod 6 on the set 1, 2, 3, 4, 5, we arrive at

Table 8

$\times_6$	1	2	3	4	5
1	1	2	3	4	5
2	2	4	0	2	4
3	3	0	3	0	3
4	4	2	0	4	2
5	5	4	3	2	1

and we see that, with identity 1, the only elements which are inverses are 1 and 5. So we must omit 2, 3 and 4 to obtain the set $\{1, 5\}$ with only two elements forming a group under multiplication mod 6. Nevertheless, the set $\{2, 4\}$ also forms a group under $\times_6$, only here the 'identity' is 4:

Table 9

$\times_6$	4	2
4	4	2
2	2	4

Again, working mod 15, the set {1, 2, 4, 8} is closed under multiplication:

Table 10

$\times_{15}$	1	2	4	8
1	1	2	4	8
2	2	4	8	1
4	4	8	1	2
8	8	1	2	4

But it is more surprising to discover the set {3, 6, 9, 12} which has 6 as its identity under multiplication mod 15:

Table 11

$\times_{15}$	6	3	9	12
6	6	3	9	12
3	3	9	12	6
9	9	12	6	3
12	12	6	3	9

Note that in Table 10, the elements 1, 2, 4, 8 can be written as 1, 2, 2^2, 2^3, with $2^4 = 1$ (mod 15), so there is a certain resemblance to the situation arising from $R^4 = I$ in Table 3. Similarly in Table 11, the elements may be written 3, 3^2, 3^3 ($= 12$), $3^4 = 81 = 6$ (mod 15).

Symmetry groups

The finite groups illustrated above have been drawn mostly from groups of permutations and from finite arithmetics—groups of residues to some modulus under addition and multiplication. But there have also been cases of groups of matrices and groups of symmetries, and we shall now have a little more to say about the latter.

It was clear on p. 494 that a plane figure possessing rotational symmetry gave rise to a group of order 4. Similarly, a figure with rotational symmetry of order 9, i.e., which is unchanged by rotations though multiples of 40°, as in the case of the 'plate' illustrated, will have a symmetry group whose elements are rotations through 0°, 40°, 80°, . . ., 280°, 320°.

Calling r an anti-clockwise rotation through 40°, the elements of the group (C_9) will be 1, r, r^2, r^3, r^4, r^5, r^6, r^7, r^8; where, for example, $r^7 r^5 = r^3$, indices being added modulo 9, since $r^9 = 1$. It is clear that such rotations may be represented by permutations of the vertices; for example, rotation r

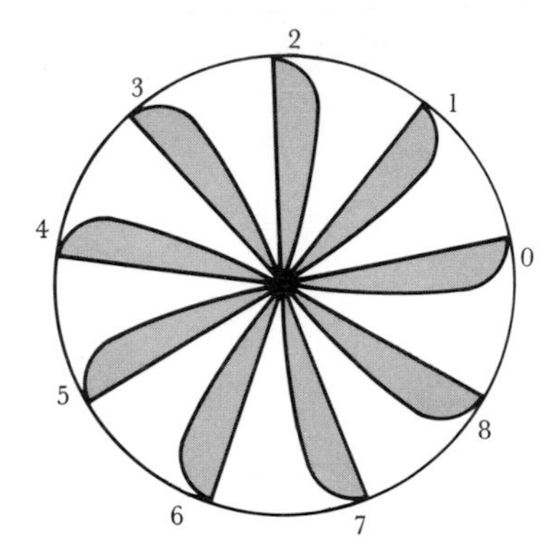

replaces the vertices 0, 1, 2, . . ., 8 by 8, 1, 2, . . ., 7 respectively, and so corresponds to the cycle (0 8 7 6 5 4 3 2 1). Moreover, the same set of rotations may be represented by matrices; r, for example, being represented by

$$\begin{pmatrix} \cos 40^\circ & -\sin 40^\circ \\ \sin 40^\circ & \cos 40^\circ \end{pmatrix}.$$

So we have a set of nine matrices forming the group of order 9 under matrix multiplication. Again, in the Argand diagram, rotation through 40° is the geometrical equivalent of multiplication by the complex number cis 40°. Calling this complex number ω, we have the nine complex numbers $\{1, \omega, \omega^2, \omega^3, \omega^4, \omega^5, \omega^6, \omega^7, \omega^8\}$ forming the group of order 9 under multiplication, e.g., $\omega^7 \times \omega^5 = \omega^3$ (remembering $\omega^9 = 1$). These are the nine complex ninth roots of unity (cf. p. 135).

Groups also appear in the case of sets of functions under function composition (see 481). For example, if

$$f(x) = \frac{x-1}{x+1},$$

then $$f(f(x)) = f^2(x) = \left(\frac{x-1}{x+1} - 1\right)\bigg/\left(\frac{x-1}{x+1} + 1\right)$$

$$= \frac{x-1-x-1}{x-1+x+1} = -\frac{1}{x};$$

$$f^3(x) = \frac{-\dfrac{1}{x} - 1}{-\dfrac{1}{x} + 1} = \frac{1+x}{1-x},$$

and finally, $f^4(x) = x = i(x)$, where i is the identity function.

So these four functions: i, f, f^2, f^3 combine according to

Table 12

	i	f	f^2	f^3
i	i	f	f^2	f^3
f	f	f^2	f^3	i
f^2	f^2	f^3	i	f
f^3	f^3	i	f	f^2

Cyclic groups

Apart from the group of Table 2, all the finite groups mentioned so far have been *cyclic* groups. This means that all the elements of the groups *may be expressed in terms of a single element*, or *generator* (as it is called). Thus in the case of the functions in Table 12 above, the elements were i, f, f^2, f^3, where i might be expressed as f^0, f is of course f^1, and f^3 might have been written f^{-1}, the group being governed by the over-riding relation $f^4 = i$. Indeed, whenever (as in Table 3) we have a situation of the type $x^4 = 1$, there we must have the cyclic group of order 4, usually called C_4; and the same group would arise in the multiplication of the set $\{1, i, -1, -i\}$, where $i^2 = -1$.

Exercise 14.3b

1 Write out the table for multiplication of the set $\{1, 3, 5, 9, 11, 13\}$ under multiplication mod 14, and show that it is a group.

2 Show that, if x is the permutation

$$\begin{pmatrix} 1 & 2 & 3 & 4 & 5 \\ 4 & 1 & 5 & 2 & 3 \end{pmatrix},$$

and e is the identity permutation, then e, x, x^2, x^3, x^4 and x^5, form a group.

3 What is the order of the group generated by the permutation

$$y = \begin{pmatrix} 1 & 2 & 3 & 4 & 5 & 6 & 7 & 8 \\ 2 & 7 & 4 & 8 & 5 & 1 & 6 & 3 \end{pmatrix}?$$

4 Show that $\{1, 4, 7, 13\}$, $\times_{15}$ is a group, and find a group under $\times_{15}$ which contains the number 3.

5 Find a group under $\times_{20}$, containing the number 8.

6 Show that the functions $x \mapsto x$, $1 - x$, $1 - 1/x$, together with their reciprocals form a group under function composition. Verify that it is not commutative.

7 What is the order of the symmetry group taking both rotations and reflections in all cases, of:
(*i*) a playing card (say Jack of Hearts);
(*ii*) an isosceles triangle;
(*iii*) a rhombus;
(*iv*) a pyramid on a regular pentagonal base;
(*v*) a regular hexagon;
(*vi*) a square.

8 Draw examples of plane figures whose symmetry groups have order 3, 4, etc.

9 Show that the permutations

$$\begin{pmatrix} 1 & 2 & 3 & 4 & 5 & 6 \\ 4 & 6 & 5 & 1 & 3 & 2 \end{pmatrix} \text{ and } \begin{pmatrix} 1 & 2 & 3 & 4 & 5 & 6 \\ 6 & 5 & 4 & 3 & 2 & 1 \end{pmatrix}$$

when combined in every possible way give rise to a group of order six.

10 Take the matrices

$$A = \begin{pmatrix} 1 & 0 \\ 0 & -1 \end{pmatrix} \quad \text{and} \quad B = \begin{pmatrix} 0 & 1 \\ 1 & 0 \end{pmatrix}$$

and consider all possible matrices obtained by combining them by multiplication (e.g., AB, BA, ABA, etc.). Show that these form a group of order 8.

11 Show that the complex 12th roots of 1 form the group C_{12} under multiplication. Is it possible to obtain the group C_{12} under *addition* of complex numbers?

12 Obtain a set of permutations forming the group C_7. Also obtain a set of permutations of seven objects forming the group C_{10}.

13 What are the periods of each element in the groups C_9, C_{12}?

14 By writing each of the set $\{1, 2, 3, \ldots, 12\}$ as a power of 2 (mod 13), e.g., $12 = 2^6$ (mod 13), show that this set forms the group C_{12}.

15 A rectangular cuboid has the following symmetries:
(*i*) half-turns about three mutually perpendicular axes (call these a, b, c);
(*ii*) reflections in three planes of symmetry. Call these p, q, r, where p is the reflection in the plane containing the axes of b and c, etc.
If the identity is called e, write out the table showing the composition of e, a, b, c and show that this is a group. What is its structure? Write out the table showing the composition of the seven symmetries listed, showing that a group is obtained (the 'full group' of the cuboid), only when an eighth symmetry is included, and specify the nature of this eighth element.

16 In the set of ordered pairs of reals under the operation $*$ defined by the formula

$$(x_1, y_1) * (x_2, y_2) = (x_1 x_2, x_1 y_2 + y_1),$$

what is the identity, and what is the inverse of (x, y)? What must be done to arrive at group structure?

14.4 Isomorphisms

Saying that Table 12 is the 'same' group as Table 10 and Table 11 is rather like saying that $\frac{3}{4}$ is the 'same' fraction as $\frac{6}{8}$ and $\frac{9}{12}$. It is more correct to describe these fractions which are equal to $\frac{3}{4}$ as belonging to an 'equivalence class' of fractions, of which it is convenient to take $\frac{3}{4}$ as the simplest representative. The tables for the groups discussed are the 'same' in the sense that they have identical patterns, as one can readily see by the 'stripes' running down from N.E. to S.W. The stripes are not apparent in the case of Table 3 till a rearrangement is made by interchanging R and S:

Table 13

	I	R	H	S
I	I	R	H	S
R	R	H	S	I
H	H	S	I	R
S	S	I	R	H

The correspondence with Table 14 (below) for $\{1, i, -1, -i\}$, $\times$ is now clear when the elements are paired off:

$$I \leftrightarrow 1; \quad R \leftrightarrow i; \quad H \leftrightarrow -1; \quad S \leftrightarrow -i.$$

Table 14

$\times$	1	i	-1	$-i$
1	1	i	-1	$-i$
i	i	-1	$-i$	1
-1	-1	$-i$	1	i
$-i$	$-i$	1	i	-1

This conforms with our knowledge that multiplication by i is represented geometrically by a quarter turn (R). However, there would be nothing to stop us setting up the correspondence

$$I \leftrightarrow 1, \quad R \leftrightarrow -i, \quad H \leftrightarrow -1, \quad S \leftrightarrow i,$$

with R and S interchanged. The reader should do the same in the case of the group whose Table 7 is shown on p. 497.

Groups which have identical structures are said to be *isomorphic*, and the idea of isomorphism is one of the most important in mathematics. Indeed, the fact that the same pattern may arise in a wide variety of situations is the main justification for a close study of structure as it means that techniques devised for one situation may well be applicable in a seemingly different context, because the two situations may, for all their apparent diversity, really be examples of the same structure. The relation between two groups of being isomorphic is usually denoted '$\cong$'. It will emerge later that this is a so-called *equivalence relation* between groups, and that isomorphic groups form an *equivalence class*, which may be called the *abstract group* of that structure.

Groups of order 4

We have already seen, however, that there are two quite distinct groups of order 4:

	1	x	y	z
1	1	x	y	z
x	x	y	z	1
y	y	z	1	x
z	z	1	x	y

and

	1	x	y	z
1	1	x	y	z
x	x	1	z	y
y	y	z	1	x
z	z	y	x	1

The first is the cyclic group C_4, and we have had several examples ranging from the rotations of the figure on p. 494 to the set $\{1, 2, 3, 4\}$ under multiplication mod 5. Other examples of this group are to be found in Tables 1, 3, 7, 10, 11, 12, 13 and 14.

The second, although so far provided with only two illustrations, (p. 480 and Table 2), is no less important. This is known as the '*Klein 4-Group*', and contains three elements of period 2. Other representations of this group which the reader may check are $\{1, 3, 5, 7\} \times$ mod 8, and the combination of the four functions x, $-x$, $1/x$, $-1/x$. For representations of the Klein 4-Group in a large number of contexts, see Budden, F. J., *The Fascination of Groups*, C.U.P., 1972, pp. 131–8.

The above two are, in fact, the only possible distinct groups of order 4. As for groups of higher orders, it is easy to show that there is only one possible group of order n whenever n is prime; whilst for composite values of n,

$n = 6$: 2 groups (including one non-Abelian)
$n = 8$: 5 groups (two non-Abelian)
$n = 9$: 2 groups
$n = 10$: 2 groups (one non-Abelian)
$n = 12$: 5 groups (three non-Abelian)

Groups of order 6

Consider the three sets of permutations:

(*i*)

1	1	2	3	4	5	6
x	2	3	4	5	6	1
x^2	3	4	5	6	1	2
x^3	4	5	6	1	2	3
x^4	5	6	1	2	3	4
x^5	6	1	2	3	4	5

(*ii*)

1	1	2	3	4	5
r	3	1	2	4	5
r^2	2	3	1	4	5
d	1	2	3	5	4
y	3	1	2	5	4
z	2	3	1	5	4

(*iii*)

1	1	2	3
p	3	1	2
p^2	2	3	1
a	1	3	2
b	3	2	1
c	2	1	3

It may easily be verified that each of these sets is closed, and that in each case we get a group of order 6:

Table 15

(*i*)

	1	x	x^2	x^3	x^4	x^5
1	1	x	x^2	x^3	x^4	x^5
x	x	x^2	x^3	x^4	x^5	1
x^2	x^2	x^3	x^4	x^5	1	x
x^3	x^3	x^4	x^5	1	x	x^2
x^4	x^4	x^5	1	x	x^2	x^3
x^5	x^5	1	x	x^2	x^3	x^4

(*ii*)

1	1	r	r^2	d	y	z
1	1	r	r^2	d	y	z
r	r	r^2	1	y	z	d
r^2	r^2	1	r	z	d	y
d	d	y	z	1	r	r^2
y	y	z	d	r	r^2	1
z	z	d	y	r^2	1	r

(*iii*)

1	1	p	p^2	a	b	c
1	1	p	p^2	a	b	c
p	p	p^2	1	b	c	a
p^2	p^2	1	p	c	a	b
a	a	c	b	1	p^2	p
b	b	a	c	p	1	p^2
c	c	b	a	p^2	p	1

(It should be noted that the permutation which is performed first is shown across the top of the table, the second permutation being shown down the side, so that in Table (*iii*), for example, $ac = p$ signifies that permutation c

is performed first, so that the a is read off on the left and the c at the top). We shall now show that the first two are isomorphic, but that their structure is different from that of the third group.

First note that x may be denoted in cycle notation as $(1\ 2\ 3\ 4\ 5\ 6)$, and clearly $x^6 = 1$, whereas $x^n \neq 1$ for $n = 1, 2, 3, 4, 5$. We say that the permutation x has *period* 6 or *order* 6, meaning that 6 repetitions of the permutation are needed in order to get back to the identity. The 6 permutations of the group are thus $\{1, x, x^2, x^3, x^4, x^5\}$, and the group is C_6.

Looking at the second set of permutations, we see that r and r^2 are each of period 3, that d is of period 2, but when we come to y $(= dr = rd)$, we have a permutation of period 6. This encourages us to begin to set up a correspondence between the first two groups by matching x with y. Indeed, it turns out that $y^2 = r^2$, $y^3 = d$, $y^4 = r$, $y^5 = z$, so that we have the cyclic group $\{1, y, y^2, y^3, y^4, y^5\}$ governed by $y^6 = 1$. The tables for the first two groups would thus correspond exactly; for wherever x occurs in the first, we should have y in the second. This same group would arise in other situations, in fact whenever an equation of the type $x^6 = 1$ appears, e.g., if $\omega = \text{cis } 60°$, $\omega^6 = 1$; or if R were a rotation through $60°$, we should have $R^6 = I$ (the identity); if $M = \begin{pmatrix} 0 & -1 \\ 1 & 1 \end{pmatrix}$, then $M^6 = I$; again, since $3^6 \equiv 1 \pmod{14}$, we have the group C_6 appearing as a multiplicative group in a finite arithmetic.

The group of the third set of permutations is different from the first two in the sense that, while we can easily enough set up a 1,1 correspondence, we *cannot do so in such a way as to preserve products*. That this must be so is immediately apparent from the fact that the third group contains no permutation of period 6, the periods of permutations being respectively 1, 3, 3, 2, 2, 2. Another easy way of seeing that the third group is different is because, unlike the first two, it is not Abelian; for $b = (1\ 3)$, $c = (1\ 2)$, then remembering that bc means 'first c, then b', we have $bc = (1\ 2\ 3) = p^2$, whereas $cb = (1\ 3\ 2) = p$. This non-Abelian group of order 6 will be discussed in more detail on p. 509 when we consider the symmetries of the equilateral triangle.

Isomorphism of infinite groups

Now it is all very well saying that two groups are isomorphic if their tables may be rearranged to correspond, but this is only meaningful in the case of finite groups. How may one describe isomorphism in the case of infinite groups? To answer this, we must give a more precise definition for two groups G, G' with different operations $\circ$ and $*$: the groups $G, \circ$ and $G', *$ are said to be *isomorphic* provided there is a one-to-one correspondence between the elements

$$x, y, \ldots \in G \quad \text{and} \quad x', y', \ldots \in G'$$

such that for all $x, y \in G$ and $x', y' \in G'$, it is also true that $x \circ y$ corresponds to $x' * y'$.

Now take the group $R^+, \times$ of positive reals under multiplication and the group $R, +$ of all reals under addition.

These can easily be shown to be isomorphic by the simple process of mapping the first set onto the second by the logarithm function $x \mapsto \ln x$. Then if $z = xy$ in the first group, we have $\ln z = \ln x + \ln y$, so that $z' = x' + y'$ in the second group and the structural properties of the first group are reproduced in the second (where the operation is, of course, addition).

Again, consider the groups $Z, +$; $10Z, +$; and also the geometrical group generated by a translation t:

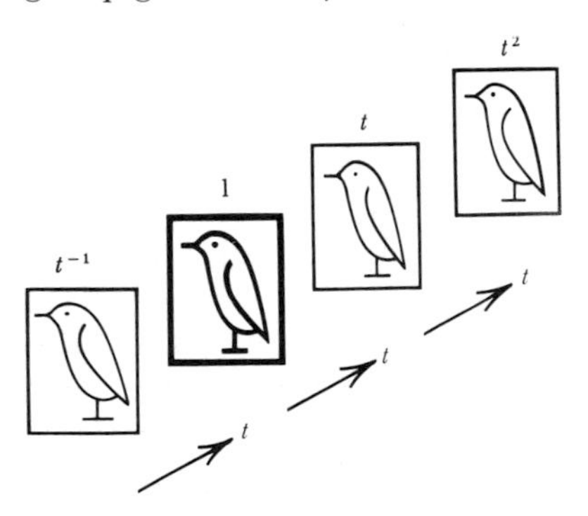

We may show that these three groups are in fact isomorphic by setting up a correspondence where x in $Z, +$ corresponds with $10x$ in $10Z, +$ and to t^x in the group of translations. Then 'products' (as we shall continue to call them, though the operation is addition) are preserved.

For example, $\quad -5 + 2 = -3 \quad$ in the group $Z, +$

becomes $\quad -50 + 20 = -30 \quad$ in the second group,

and $\quad t^{-5} \times t^2 = t^{-3} \quad$ in the case of the third group.

Or, more generally, $\quad x + y = z \quad$ in $Z, +$

corresponds to $\quad 10x + 10y = 10z \quad$ in $10Z, +$,

and to $\quad t^x t^y = t^z \quad$ in the third group.

Consider next the operation on the integers given by the formula

$$m * n = m + (-1)^m n.$$

This is clearly non-commutative, but is in fact associative. Closure is self-evident, while 0 is a left and right identity. Finally, the inverse of an *even* integer is its *negative*, e.g.,

$$18 * (-18) = 18 + (-1)^{18}(-18) = 0,$$

$$(-18) * 18 = -18 + (-1)^{-18}18 = 0,$$

but the inverse of an *odd* integer is itself, e.g.,

$$19 * 19 = 19 + (-1)^{19}19 = 0$$
$$(-19) * (-19) = -19 + (-1)^{-19}(-19) = 0.$$

Thus $Z, *$ is a group.

For a second group we take the series of images set up by reflection in two parallel mirrors.

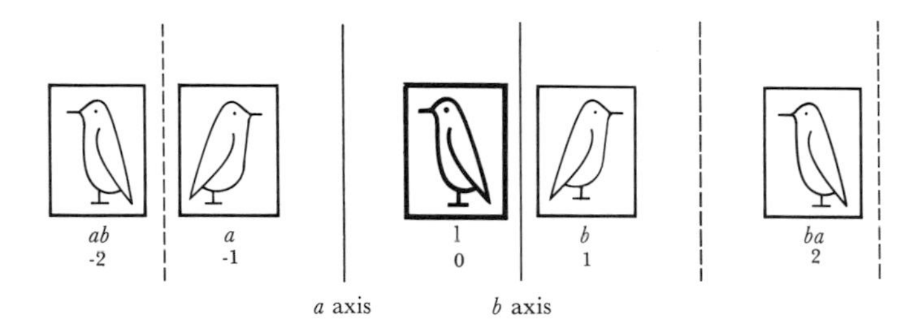

In the figure, a represents the operation of reflecting in the first mirror, b represents a reflection in the second mirror, and following the usual convention, *baba* (for example) represents the result of two pairs of alternate reflections in the two mirrors, starting with the first. In the figure, each image of the bird motif is labelled according to the reflections leading to it. These two reflections generate an infinite group, the elements of which are every possible 'word' formed from the two letters a and b, e.g., *abababababa*, *bababa*, etc. (since $a^2 = b^2 = 1$, there will be no powers of a and b higher than 1).

We now show that these two infinite non-Abelian groups are isomorphic. The method of setting up the correspondence between $Z, *$ and the group generated by reflections a and b is that shown in the figure whereby each image is labelled not only with the 'word' describing the reflections, but is also allotted an integer in the bottom row. The way the correspondence works out is that the number in the bottom row, regardless of sign, says how many letters there are in the 'word', e.g., -3 for *aba*; $+6$ for *bababa*; while the sign is $+$ if the initial letter of the word is b, $-$ if it is a. The way in which the products are 'imitated' between the two groups is illustrated in two cases:

$Z, *$	Reflections
$5 * 2 = 3$	$(babab)(ba) = babab^2a = bab$
$-5 * 3 = -8$	$(ababa)(bab) = abababab$
$6 * -3 = 3$	$(bababa)(aba) = bababa^2ba = bab$ (using $a^2 = b^2 = 1$).

Thus the two groups are isomorphic, and in giving this example we have incidentally illustrated how groups may arise in the symmetries of an *infinite* geometrical pattern, here a one-dimensional 'frieze' pattern formed by the bird motifs.

Exercise 14.4a

1 Show that $\{1, i, -1, -i\}, \times$ is a group isomorphic to the group $\{1, 3, 7, 9\}, \times$ mod 20, and to the group of rotations of the figure on p. 494, and also to the permutations $\{i, x, x^2, x^3\}$ where

$$x = \begin{pmatrix} 1 & 2 & 3 & 4 & 5 & 6 \\ 2 & 1 & 4 & 5 & 6 & 3 \end{pmatrix}.$$

2 Invent other examples of the Klein four group other than those in the text:

(*i*) as a finite arithmetic under multiplication;
(*ii*) as a set of permutations (of four objects);
(*iii*) as a set of four 2×2 matrices;
(*iv*) as a symmetry group of a plane figure other than the rectangle.

3 Why is it not possible to set up an isomorphism between
(*i*) the symmetries of the equilateral triangle and the rotations of a regular hexagon;
(*ii*) the symmetries of the ellipse and $\{1, 3, 7, 9\}, \times_{10}$
(*iii*) the complex numbers under addition and the vectors in three-dimensional space under addition;
(*iv*) the positive reals under multiplication, and the rationals under addition.

4 If a group is isomorphic to a proper subgroup of itself, what must necessarily be true?

5 Show that $\{2^n, n \in Z\}, \times$ is isomorphic to the set of translations in a line generated by a single given translation.

6 Let S be the set $\{x : 0 \leqslant x < 1, x \in \boldsymbol{R}\}$. Show that S with the operation addition mod 1 (e.g., $0.7 + 0.84 = 0.54$), is a group, and that this is isomorphic to the group of rotations of the circle.

7 Establish the isomorphism between the groups shown in these two tables:

	a	b	c	d	f	g
a	a	b	c	d	f	g
b	b	c	a	f	g	d
c	c	a	b	g	d	f
d	d	f	g	a	b	c
f	f	g	d	b	c	a
g	g	d	f	c	a	b

	p	q	r	s	t	u
p	p	q	r	s	t	u
q	q	p	u	t	s	r
r	r	u	s	q	p	t
s	s	t	q	u	r	p
t	t	s	p	r	u	q
u	u	r	t	p	q	s

8 In the group C_9 $\{0, 1, 2, \ldots, 8\}$ $+_9$, which elements are generators? Which are generators in similar representations of C_8, C_{10}, C_{12}? If every element of the cyclic group C_n is a generator (except the identity), what can you say?

9 Consider the group $\{1, p, p^2, p^3\}$ with $p^4 = 1$. Show that the permutation of the elements which interchanges p and p^3 preserves the structure, but no other permutation of the four elements has this property. (Such a permutation is known as an 'automorphism', in this case of C_4).

10 Find the automorphisms of the group $\{1, a, b, c\}$ with $a^2 = b^2 = 1$, $ab = ba = c$.

11 Show that the correspondence $x \leftrightarrow -x$ is an automorphism of $Z, +$.

12 Show that $\{1, \alpha, \alpha^2, \alpha^3, \alpha^4, \alpha^5, \alpha^6\}, \times$ where $\alpha^7 = 1$ has six automorphisms, one of which is $\alpha \rightarrow \alpha^2$, and find the others. Find the automorphisms of the group C_8 formed by the eighth complex roots of 1.

The group of equilateral triangle, D_3

So far the only non-Abelian finite group has been the one of order 6 consisting of every possible permutation of 3 digits. This is the smallest case of a non-Abelian group, and we now look at it in more detail. In setting up the table showing the products, we must (since many of the products are non-commutative) specify that when we read the product xy from row x and column y, this is to be the value of xy; so y, the first operation, is shown across the top of the table, and the second operation is down the left-hand side. The table showing the composition of the permutations will be found to be

Table 16

1st operation

2nd operation	1	p	q	a	b	c	
1	1	p	q	a	b	c	
p	p	q	1	b	c	a	(1 3 2)
q	q	1	p	c	a	b	(1 2 3)
a	a	c	b	1	q	p	(2 3)
b	b	a	c	p	1	q	(1 3)
c	c	b	a	q	p	1	(2 1)

To complete the table, it is not necessary to find every single product from the permutations as we did on p. 505. For, once having obtained $bc = q$, we may immediately deduce (for example):

$bq = bbc = b^2c = c,\quad$ since $b^2 = 1$

$qc = bcc = bc^2 = b,\quad$ since $c^2 = 1$

$cp = bqp = b1 = b,\quad$ since $qp = 1$, and so on.

We next show that the same group arises as a group of symmetrics. Consider an equilateral triangle 1 2 3 which is subjected to the following rotations:

$p = (132)$.

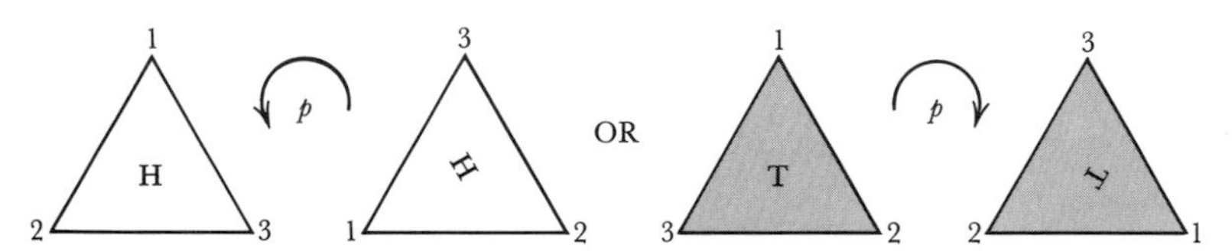

It will be seen that p is an anti-clockwise or a clockwise rotation according to whether the triangle has or has not been turned over (is showing 'heads' or 'tails'); q is simply the inverse of this, so $q = p^{-1} = p^2$; whilst a is a rotation about the median through 1, so interchanges the vertices 2 and 3. Similarly, $b = (1\ 3)$ and $c = (1\ 2)$. These five rotations together with the identity constitute the symmetry group of the equilateral triangle. It is clear, not only that every rotation has been matched with a corresponding permutation of the vertices, but also that the rotations will combine in a way which exactly mimics the composition of the permutations. So we have an isomorphism. For example, corresponding to the equation $pc = a$ expressing the composition of the permutations $(1\ 3\ 2)\ (1\ 2) = (2\ 3)$, we observe that if the triangle is subjected first to a rotation about the median through 3 (thereby turning it over) and then to a (now clockwise) $120°$ turn p, we obtain the final position which would arise from a single rotation about the median through 1, i.e., $a = (2\ 3)$:

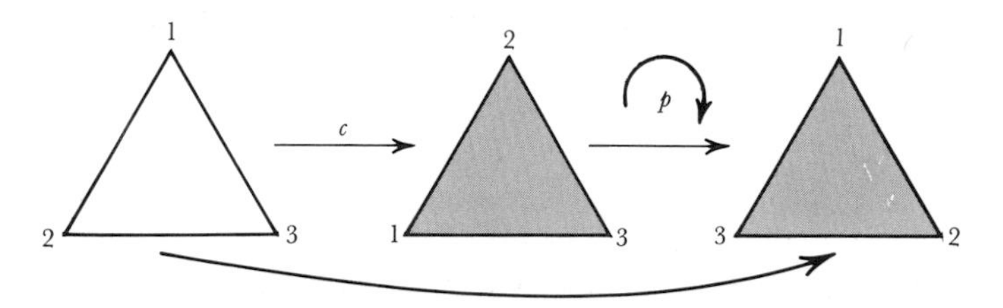

Note that

(*i*) a, b, c were described as rotations, taking the triangle into three dimensions, but could equally well have been regarded as *reflections*; and

(*ii*) the triangle could just as well have been a uniform triangular *prism*, the rotations a, b, c being about symmetry axes situated midway between the triangular faces.

In the previous paragraph we have been considering the group of symmetries of the equilateral triangle which includes not only the plane rotational symmetries through 0, $120°$, $240°$ $(1, p, q)$ which would be possessed by a figure like , but also the bilateral symmetries about its three medians. Similarly, if we look at a square instead of an equilateral triangle, it would possess not only the rotational symmetries, as in Table 3, but also the bilateral symmetries about four axes,

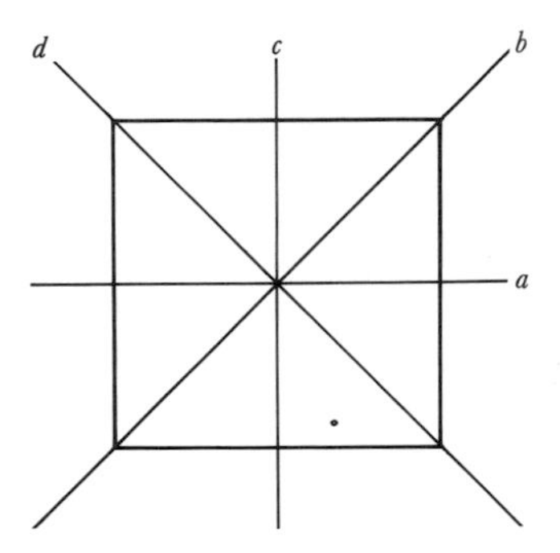

and these will together give a group of order 8, the symmetry group of the square. The full group of rotations and reflections of a regular n-sided polygon is called the *dihedral group of order* $2n$, and is denoted D_n. A useful exercise for the reader would be to construct the group table for D_4, using the notation of this figure where (for a change) we use a, b, c, d to be *fixed* symmetry axes (unlike the axes which moved with the triangle in p. 510) while r is an anti-clockwise quarter-turn. The table should be set out with the elements in the order $1, r, r^2, r^3, a, b, c, d$.

Infinite symmetry groups

Usually the symmetry group of a geometrical figure is finite, but we have already met (p. 507) one case of an infinite group, when the geometrical figure itself extended to infinity. The groups arising from repeating patterns are of great interest, but there is not space to deal with them here, and the reader is referred to Budden, F. J., *The Fascination of Groups*, C.U.P., 1972, Chapter 26. Another case when we obtain an infinite symmetry group is that of the *circle*, which is not only symmetrical by rotation through any angle, but also possesses bilateral symmetry about any diameter.

Symmetry groups of solids: the tetrahedral group

We have already mentioned the symmetry groups of the cuboid (Table 2), and more recently the six rotational symmetries of the equilateral triangular prism. When dealing with the symmetries of three-dimensional geometrical figures, one has to specify whether one is concerned with rotations only (as above), or whether one is going to include reflective, or 'opposite' symmetries, in which case the cuboid would have a group of order 8, and the triangular prism one of order 12. A good illustration of this is provided by the regular tetrahedron, which has eight 120° turn symmetries about its four altitudes, but also three half-turns about its joins of mid-points of opposite edges which, with the identity, make a group of order 12. If the vertices were labelled 1, 2, 3, 4, the 120° turn would correspond to such permutations as $(1\ 2\ 3)$, while the half-turns correspond to $(1\ 2)\ (3\ 4)$, etc. When reflections are included we should get all the other 12 permutations of the four vertices, such as $(1\ 2)$, $(1\ 2\ 3\ 4)$ etc., which cannot be obtained as a result of direct geometrical movements, but only with the aid of mirrors! The group of all 24 $(= 4!)$ permutations of 4 digits is called the *symmetric group*, S_4. More generally, the group of permutations of n symbols is the symmetric group S_n, and is of order $n!$. On p. 509, we saw that the group of the equilateral triangle was in fact isomorphic to S_3, so we may write $D_3 \cong S_3$. As an exercise, the reader should construct the group of order 12 of the rotations of the regular tetrahedron. This should be done using the permutations $p = (1\ 2\ 3)$ and $a = (1\ 2)\ (3\ 4)$ described above, and it will be found that the whole group is generated by using these two permutations only, i.e., by forming all possible 'words' from a and p. The group is known as the tetrahedral group, or the *alternating group of order* 12.

Exercise 14.4b

1 What are the symmetry groups of the octagons:

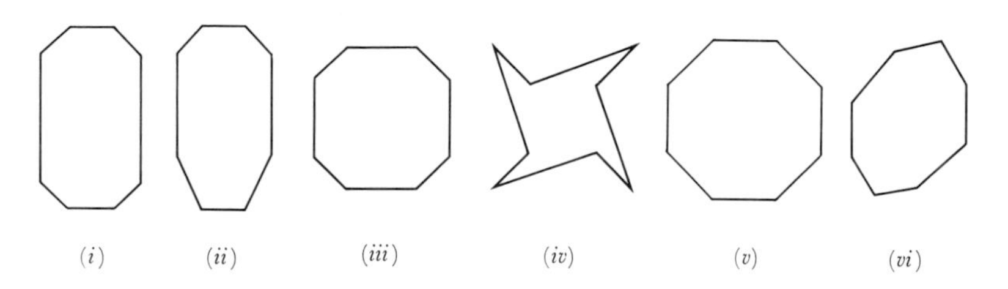

2 Draw hexagons whose symmetry groups are C_3, C_2, D_2, D_3.

3 Show that two mirrors at an angle of 90° will produce a pattern with the symmetry group D_2; and that if the angle were 60°, the group would be D_3. Generalise.

4 Show that the matrices

$$A = \begin{pmatrix} 0 & 1 \\ 1 & 0 \end{pmatrix} \quad \text{and} \quad B = \begin{pmatrix} -1 & 0 \\ 0 & 1 \end{pmatrix}$$

generate a group of order 8, and that this group is isomorphic with the group of symmetries of the square.

5 If $f(x) = 2 - x$ and $g(x) = 2/x$, find $gf(x)$ and $fg(x)$, and the other functions generated by f and g. What group do they form?

6 Note that the group of the equilateral triangle has the following relations connecting the elements:

$$ap = p^2a; \quad ap^2 = pa; \quad pap = a; \quad (ab)^3 = 1; \quad aba = bab,$$

and interpret these in geometric terms. Find other relations connecting: (*i*) b and q; (*ii*) b and c.
Show that the relations $a^2 = p^3 = 1$, $ap = p^2a$ are sufficient to determine the group completely, and find another set of 'defining relations' based on the 'generators' a and b.

7 For a regular hexagon, let r denote a $60°$ rotation about its centre, and a, b, c, d, f, g be reflections in the six lines of symmetry. Show that, $a = rar$, and obtain b, c, d, f, g each in terms of the generators a and r, it being supposed that the six axes are labelled consecutively in the direction of the rotation r. Draw up the group table of the twelve symmetries of the regular hexagon (D_6).

8 Discuss various tetrahedra with rotation groups smaller than the group of the regular tetrahedron (e.g., a tetrahedron with one face equilateral and the other three congruent isosceles triangles has the group C_3).

9 Consider the rotations of the cube and classify them according to their period (e.g., a rotation through $90°$ has period 4).

10 Show that the regular octahedron has symmetry group isomorphic to that of the cube. How many of the symmetries are rotations, and how many are opposite?

11 Investigate (infinite) symmetry groups generated by:
(*i*) a single glide reflection;
(*ii*) reflections in two parallel mirrors;
(*iii*) reflections in two mirrors at various angles;
(*iv*) a half-turn and a translation;
(*v*) a half-turn and a mirror reflection.
Invent other examples.

Epilogue

How many numbers are there?

We have come a long way since first learning to count and now retrace our steps to this apparently childish question. By now, however, we are aware of a variety of sets of numbers. In arithmetic modulo 2, for instance, there are just two elements, 0 and 1, whilst in the system of complex numbers there is a far richer variety including such diverse numbers as

$0, \quad 1, \quad -3, \quad \frac{2}{3}i, \quad \sqrt{2}, \quad \sqrt{3} - i, \quad$ etc.

In this epilogue we shall confine ourselves to the real numbers R and shall start by looking again at their subset of positive integers, Z^+.

The positive integers, Z^+

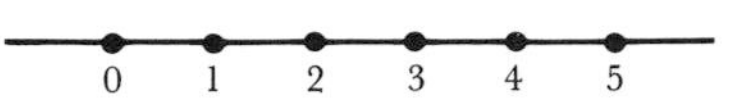

A child very soon realises that there is no end to the sequence of natural numbers, or positive integers. Even so, they are capable of being *counted*, or set in order, because this is precisely what is done when we say ‘one, two, three, four, five, . . . : we name them in sequence, so that — given life and breath — any given number would ultimately be reached. We therefore say that the positive integers Z^+ are countable.

The integers, Z

When, however, we consider *all* the integers (positive, negative and zero)

an immediate difference is apparent, that there is no obvious starting-point for such an operation of counting. Even so, they can clearly be set down in the order

$$0, \quad +1, \quad -1, \quad +2, \quad -2, \quad +3, \quad -3, \quad \ldots.$$

Here again, every integer — positive, zero or negative, and however large — has its place in this sequence, which therefore provides us with a means of threading our way through them. So, just as for natural numbers, *the set Z of all integers is countable.*

The rational numbers, Q

There are, of course, many gaps between the integers and into these gaps we can place other quotients of integers like $\frac{1}{2}$, $\frac{5}{3}$, $-\frac{7}{6}$, so obtaining the set Q of rational numbers:

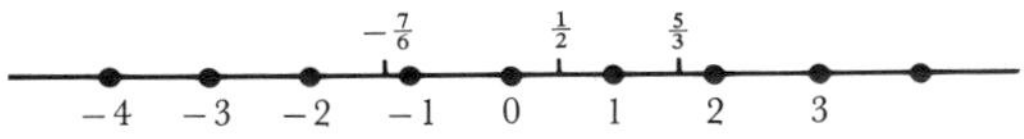

$-\frac{7}{6}$ $\frac{1}{2}$ $\frac{5}{3}$

-4 -3 -2 -1 0 1 2 3

It is clear that there is no end to the number of such rationals that can be placed between two integers. The interval between 0 and 1, for instance, could be divided into tenths, hundredths, thousandths or millionths, and each of the points of subdivision would correspond to a rational number. So the number of rationals between 0 and 1 is clearly infinite.

This is, of course, just as true of the number of rationals that can be placed between any two given rationals, however close they are. For if p/q and r/s are the given rational numbers (p, q, r, s being integers), their mean is

$$\frac{1}{2}\left(\frac{p}{q} + \frac{r}{s}\right) = \frac{ps + qr}{2qs}.$$

Now $ps + qr$ and $2qs$ are integers, so this mean is itself a rational number and we see that between any two rational numbers there is always another, and, therefore, by continued bisection, an infinity.

There is, therefore, certainly no shortage of rational numbers and the former question again arises. Is it still possible to count our way through this multitude of rationals, or are they too numerous? Can we construct a sequence in which every rational number has its place?

This question was first raised in 1873 by the German mathematician George Cantor and did not prove difficult for him to answer. Rather, as we shall see, did his genius lie in *asking* the question.

Let us start by considering the positive integers, together with the number 0. First we construct the pattern

$$0$$

$$\frac{1}{1}$$

$$\frac{1}{2} \quad \frac{2}{1}$$

$$\frac{1}{3} \quad \cancel{\frac{2}{2}} \quad \frac{3}{1}$$

$$\frac{1}{4} \quad \frac{2}{3} \quad \frac{3}{2} \quad \frac{4}{1}$$

$$\frac{1}{5} \quad \cancel{\frac{2}{4}} \quad \cancel{\frac{3}{3}} \quad \cancel{\frac{4}{2}} \quad \frac{5}{1}$$

$$\cdot \quad \cdot \quad \cdot \quad \cdot \quad \cdot \quad \cdot \quad \cdot \quad \cdot \quad \cdot$$

and delete any number which is not in its lowest terms. Now if this pattern is continued indefinitely, every positive rational number will have its place. We notice, for instance, that the fifth row consists of all positive rationals whose numerator and denominator add up to five and the sixth row consists of those whose numerator and denominator (when expressed in their lowest terms) add up to six. Similarly the rational number $\frac{9}{11}$ would occur in the 20th row; and, more generally, the number p/q (if expressed in its lowest terms) in the $(p + q)$th row.

So, although the positive (and zero) rationals are extremely numerous, we have invented a way of counting them:

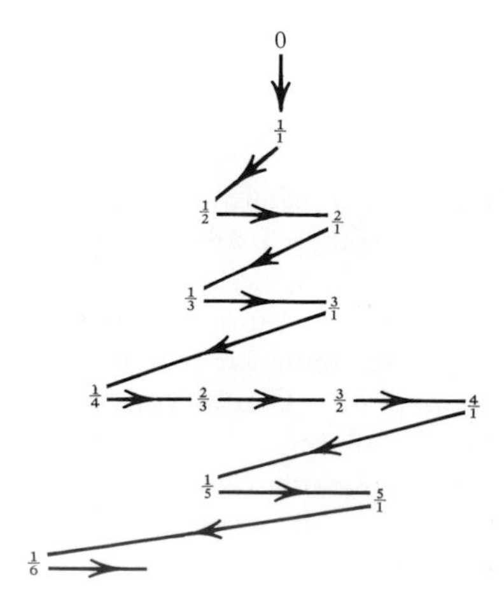

Now we can just as easily count *all* the rationals (positive, zero and negative) simply by placing each negative rational immediately after the corresponding positive rational:

$0; \quad +1, -1; \quad +\frac{1}{2}, -\frac{1}{2}, +2, -2; \quad +\frac{1}{3}, -\frac{1}{3}, \quad \ldots.$

Hence the set Q of all rationals is countable.

The continuum of real numbers, *R*

−3 −2 −1 0 1 2 3

There is such a multitude of rational numbers, with an infinity of them inside any interval, that the question naturally arises whether there is room for any other kind of number. Now we have seen that Euclid had effectively answered this question about 300 BC when showed that $\sqrt{2}$ is not a rational number and we naturally ask whether such numbers are rare or common. Are most numbers rational or irrational? Certainly we have seen that the rational numbers are countable. Can we similarly count all the numbers on the straight line, rational and irrational?
This question too was first asked by Cantor and at the beginning of December in 1873 — just a hundred years ago as this epilogue is being written — he answered it by proving one of the most famous results of mathematics, that of the *non-enumerability of the continuum: the real numbers R, rational and irrational, are too numerous to count.*

Let us consider the numbers between 0 and 1 and *suppose* that they *can* be arranged in a sequence

$a_1, a_2, a_3, a_4, \ldots,$

which includes every such number.

Now each of these numbers can be expressed as a decimal. It is true that there is a slight difficulty about decimals which terminate as they can be represented either with a recurring 0 or a recurring 9: $\frac{3}{20}$, for instance, can be written

either as 0.150 000...

or as 0.149 999....

But in such cases we simply stipulate that a recurring 9 is never to be used. So each number can be represented uniquely in decimal form, and we can write

$$\begin{aligned} a_1 &= 0.\alpha_{11}\alpha_{12}\alpha_{13}\alpha_{14}\ldots \\ a_2 &= 0.\alpha_{21}\alpha_{22}\alpha_{23}\alpha_{24}\ldots \\ a_3 &= 0.\alpha_{31}\alpha_{32}\alpha_{33}\alpha_{34}\ldots \\ a_4 &= 0.\alpha_{41}\alpha_{42}\alpha_{43}\alpha_{44}\ldots \\ &\quad \cdot \quad \cdot \quad \cdot \quad \cdot \quad \cdot \quad \cdot \end{aligned}$$

as a sequence containing *every* number between 0 and 1.

Having done this, we now attempt to construct another number

$$b = 0.\beta_1\beta_2\beta_3\beta_4\ldots$$

which is different from every a. Deciding to be as perverse as we possibly can, we see that this could be achieved if b were

different from a_1 in its first decimal place,
different from a_2 in its second decimal place,
different from a_3 in its third decimal place, and so on.

We therefore *choose* $\beta_1, \beta_2, \beta_3 \ldots$ by the following rule:

$$\text{if } a_{nn} \neq 5, \text{ choose } \beta_n = 5,$$
$$\text{and if } a_{nn} = 5, \text{ choose } \beta_n = 4.$$

Then $\beta_n \neq a_{nn}$, for all values of n.

Hence $b = 0.\beta_1\beta_2\beta_3\ldots$

is different from every single member of the sequence $a_1, a_2, a_3 \ldots$. By this means, therefore, we have succeeded in constructing a number between 0 and 1 which is not a member of this sequence. Hence our initial assumption is contradicted, and we see that it is *not* possible to count even the real numbers between 0 and 1. It is, therefore, certainly impossible to count all the real numbers R, so that *the continuum is not enumerable.*

This, of course, provides us with another proof, over two thousand years later than that of Euclid, that there are other numbers besides the rationals. For the rational numbers were countable, but the continuum is certainly not countable, so must be far more numerous; and whereas Euclid's proof produces a single irrational number, Cantor produced an uncountable infinity all at once.

Transcendental numbers

It is now interesting to speculate about the numbers which constitute this uncountable multitude. Which are they? A first guess would be that they are irrational numbers like $\sqrt{2}$, $\sqrt[3]{5}$ and $\sqrt[4]{\frac{7}{11}}$. But this is mistaken. For the positive rational numbers Q^+, as they are countable, can be written

$$r_1,\ r_2,\ r_3 \ldots$$

so their square roots are $\sqrt{r_1}, \sqrt{r_2}, \sqrt{r_3} \ldots$

their cube roots are $\sqrt[3]{r_1}, \sqrt[3]{r_2}, \sqrt[3]{r_3} \ldots$

their fourth roots are $\sqrt[4]{r_1}, \sqrt[4]{r_2}, \sqrt[4]{r_3} \ldots$ etc.

Now we can count through this set of numbers by this route:

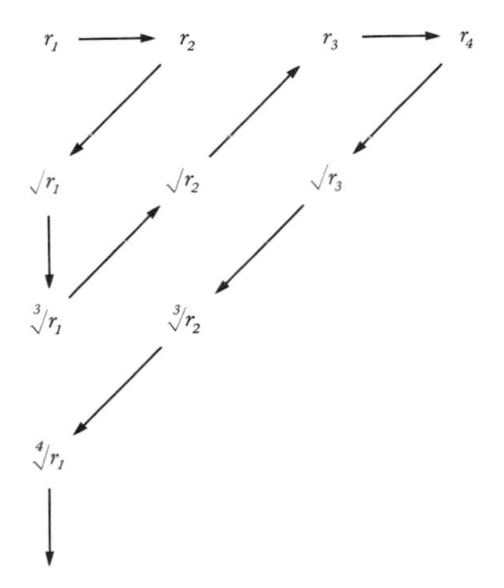

and so the set of *all positive rational numbers and all their roots is countable.*

It can similarly be shown that the set of all polynomial equations with rational coefficients have roots (known as the set of *algebraic* numbers) which are also countable. So it is clear that there must exist a vast uncountable multitude of numbers which are not the roots of any such equation. These numbers, which are not algebraic, are called *transcendental.*

Which numbers are transcendental? Certainly it is clear that no rational number or root of a rational number is transcendental, and we might well wonder what comprises this vast galaxy. Now by coincidence it was also proved in 1873 that e cannot be the root of any algebraic equation with rational coefficients, so must be just such a transcendental number; and in 1882 it was shown that π is another. Even today, however, we know very few kinds of naturally occurring transcendental numbers: $\ln 2$, e^{π}, $2^{\sqrt{2}}$, $\sin 1$ have been proved transcendental, but they are rarities.

What is known is simply that when all rationals, together with all roots of all rationals, and together with all the roots of all algebraic equations with rational coefficients, have been removed from the continuum, it has hardly been diminished, and that the infinitely vaster multitude of transcendental numbers has remained secretly in hiding. And in this, it might be said, these numbers are typical of the whole of mathematics, where much is known but much more is not.

Answers to exercises

Exercise 9.1a

1 (*i*) 0.69 (*ii*) 1.10 (*iii*) 1.79 (*iv*) 2.20
(*v*) 2.48 (*vi*) −0.69 (*vii*) −1.10

Exercise 9.1b†

1 (*i*) 1.386 30 (*ii*) 1.791 76 (*iii*) 2.302 59 (*iv*) 2.484 91
(*v*) −0.693 15 (*vi*) −1.791 76 (*vii*) −2.484 91
(*viii*) 13.815 54 (*ix*) 0.346 57(5) (*x*) 1.535 06

2 (*i*) 3 (*ii*) 0 (*iii*) −1 (*iv*) 0.5

3 (*i*) e^2 (*ii*) e^{100} (*iii*) $\sqrt[10]{e}$ (*iv*) $\dfrac{1}{e^2}$

4 (*i*) $\dfrac{1}{x}$ (*ii*) $\dfrac{3}{x}$ (*iii*) $\dfrac{4}{x}$ (*iv*) $\dfrac{1}{x}$ (*v*) $-\dfrac{1}{x}$ (*vi*) $\dfrac{1}{2x}$ (*vii*) $\dfrac{2x}{x^2+1}$
(*viii*) $-\dfrac{1}{x(x+1)}$ (*ix*) $\dfrac{x}{x^2-1}$ (*x*) $\dfrac{3x+2}{2x(x+1)}$ (*xi*) $1 + \ln x$
(*xii*) $\dfrac{1-\ln x}{x^2}$

5 (*i*) $\cot x$ (*ii*) $-\tan x$ (*iii*) $\tan x$ (*iv*) $\sec x$ (*v*) $\cot x$
(*vi*) $3 \cot 3x$ (*vii*) $2 \cot x$ (*viii*) $2 \cot x$ (*ix*) $\dfrac{2 \ln x}{x}$ (*x*) $\dfrac{1}{x \ln x}$

6 (*i*) $\frac{1}{3} \ln x$ (*ii*) $\ln (x+2)$ (*iii*) $\frac{1}{2} \ln (2x+3)$ (*iv*) $x + \ln x$
(*v*) $-\ln (1-x)$ (*vi*) $\ln (x^2+1)$
(*vii*) $-\frac{1}{3} \ln (1-x^3)$ (*viii*) $\frac{1}{2} \ln (x^2 - 2x + 3)$ (*ix*) $\frac{1}{3} \ln (x^3 + 3x)$

† In this and subsequent exercises, the arbitrary constant has been omitted from all indefinite integrals.

(*x*) $\ln \sin x$ (*xi*) $\frac{1}{3} \ln \sec 3x$ (*xii*) $2 \ln \sin \dfrac{x}{2}$

(*xiii*) $\ln(\sin x - \cos x)$ (*xiv*) $\ln(1 + \tan x)$

7 (*i*) 0.916 3 (*ii*) 0.336 5 (*iii*) 0.125 6 (*iv*) −1.098 6
(*v*) Impossible (*vi*) 2.5 (*vii*) −0.804 7 (*viii*) Impossible
(*ix*) 0.693 1 (*x*) 0.346 5

8 (*i*) 0.693 65 (*ii*) 0.693 10

9 (*i*) $\pi \ln 2 \approx 2.18$ (*ii*) $\pi \ln 4 \approx 4.35$

10 2.21

11 (*i*) 1.295, 12.71 (*ii*) $\dfrac{1}{e}$

Exercise 9.2

1 2.208×10^4, 2.692×10^{43}, 1.105, $0.452\,9 \times 10^{-5}$

2 (*i*) 2.303, 6.908, 12.82 (*ii*) 4.605, 9.210, 12.82

3 (*i*) $3e^{3x}$ (*ii*) $-\frac{1}{2}e^{-\frac{1}{2}x}$ (*iii*) $(1 - x)e^{-x}$ (*iv*) $(x^2 + 2x)e^x$

(*v*) $-xe^{-\frac{1}{2}x^2}$ (*vi*) $\dfrac{xe^x - e^x}{x^2}$ (*vii*) $e^{-x}(\sin x - \cos x)$

(*viii*) $e^{2x}(2\cos 3x - 3\sin 3x)$ (*ix*) $3^x \ln 3$ (*x*) $(\ln x + 1)x^x$

4 (*i*) $-\frac{1}{2}e^{-2x}$ (*ii*) $3e^{\frac{1}{3}x}$ (*iii*) $\frac{1}{2}e^{x^2}$ (*iv*) $e^{\sin x}$ (*v*) $-e^{-\frac{1}{2}x^2}$

(*vi*) $\dfrac{10^x}{\ln 10}$

5 (*i*) $2\left(1 - \dfrac{1}{e}\right)$ (*ii*) $\frac{1}{2}e^2(e^4 - 1)$ (*iii*) $1 - \dfrac{1}{\sqrt{e}}$ (*iv*) $\dfrac{99}{\ln 10}$

6 (*i*) $\dfrac{1}{e}$ (*ii*) $\dfrac{4}{e^2}$ (*iii*) $\dfrac{n^n}{e^n}$ (*iv*) 1

7 $1 - e^{-X}$; 1

8 (*i*) $n\pi$ (*ii*) $n\pi + \frac{1}{4}\pi$

9 (*iii*) 10.52% (*iv*) 6.93 days

10 (*iii*) $\frac{1}{1600} \ln 2 \approx 4.33 \times 10^{-4}$ (*iv*) 4.3%

11 (*i*) 565 m (*ii*) 13.8 min (*iii*) Never

Exercise 9.4a

2 (*i*) $1 - x + \frac{1}{2}x^2 - \frac{1}{6}x^3 + \frac{1}{24}x^4$ (*ii*) $1 + 2x + 2x^2 + \frac{4}{3}x^3 + \frac{2}{3}x^4$
(*iii*) $1 + \frac{1}{2}x^2 + \frac{1}{24}x^4$ (*iv*) $x + \frac{1}{6}x^3$

3 (*i*) 1.648 72 (*ii*) 0.367 88 (*iii*) 1.105 17

Exercise 9.4b

2 (*i*) $\frac{1}{2}x - \frac{1}{8}x^2 + \frac{1}{24}x^3 - \frac{1}{64}x^4$ (*ii*) $2x - 2x^2 + \frac{8}{3}x^3 - 4x^4$
(*iii*) $x^2 + \frac{1}{2}x^4$

3 (*i*) 1.099 (*ii*) 1.511 (*iii*) 2.610

Exercise 9.4c

1 (*i*) $1 + \frac{1}{2}x - \frac{1}{8}x^2 + \frac{1}{16}x^3$ $|x| < 1$
(*ii*) $1 - x + x^2 - x^3$ $|x| < 1$
(*iii*) $1 - 3x + 6x^2 - 10x^3$ $|x| < 1$
(*iv*) $1 + x + \frac{3}{2}x^2 + \frac{5}{2}x^3$ $|x| < \frac{1}{2}$
(*v*) $\frac{1}{2} - \frac{1}{4}x + \frac{1}{8}x^2 - \frac{1}{16}x^3$ $|x| < 2$
(*vi*) $2 - \frac{1}{2}x - \frac{1}{16}x^2 - \frac{1}{64}x^3$ $|x| < 2$

2 (*i*) 1.004 99 (*ii*) 0.997 49 (*iii*) 2.024 84 (*iv*) 3.009 97
(*v*) 0.247 512 (*vi*) 0.126 89

3 4

4 (*ii*) $\dfrac{1}{x^2} + \dfrac{2}{x^3} + \dfrac{3}{x^4}$, $x + \dfrac{2}{3x^2} - \dfrac{4}{9x^5}$

Exercise 9.4d

1 (*i*) $x - \frac{1}{6}x^3$ (*ii*) $1 - \frac{1}{2}x^2 + \frac{1}{24}x^4$ (*iii*) $x + \frac{1}{3}x^3$ (*iv*) $1 + \frac{1}{2}x^2 + \frac{5}{24}x^4$
(*v*) $x + x^2 + \frac{1}{3}x^3$ (*vi*) $\frac{1}{2}x^2 + \frac{1}{12}x^4$

2 0.100 33, 0.017 45, 0.000 29

3 1.035 53, 1.000 58

Exercise 9.5a

1 (*i*) $x\,e^x - e^x$ (*ii*) $\sin x - x\cos x$ (*iii*) $\frac{1}{4}(2x - 1)\,e^{2x}$
(*iv*) $\frac{1}{2}x^2 \ln x - \frac{1}{4}x^2$ (*v*) $2x \sin\dfrac{x}{2} + 4\cos\dfrac{x}{2}$ (*vi*) $-\dfrac{1}{x}(\ln x + 1)$

2 (*i*) $-(x^2 + 2x + 2)\,e^{-x}$ (*ii*) $2x\sin x + (2 - x^2)\cos x$
(*iii*) $\theta \tan\theta - \log\sec\theta$ (*iv*) $\dfrac{u10^u}{\ln 10} - \dfrac{10^u}{(\ln 10)^2}$

3 (*i*) $\frac{1}{2}e^x(\sin x - \cos x)$ (*ii*) $\frac{1}{2}e^x(\sin x + \cos x)$
(*iii*) $\frac{1}{2}e^{-x}(\sin x - \cos x)$ (*iv*) $\frac{1}{5}e^x(\sin 2x - 2\cos 2x)$

4 (*i*) $x\{(\ln x)^2 - 2\ln x + 2\}$ (*ii*) $x\tan^{-1}x - \frac{1}{2}\ln(1 + x^2)$
(*iii*) $u\sin^{-1}u + \sqrt{(1 - u^2)}$ (*iv*) $\frac{1}{2}(x^2 + 1)\tan^{-1}x - \frac{1}{2}x$

5 (*i*) $1 - \dfrac{2}{e}$ (*ii*) $\frac{1}{4}\pi$ (*iii*) $2\pi^2 - 16$ (*iv*) $\frac{8}{3}\ln 2 - \frac{7}{9}$

6 £25

Exercise 9.5b

1 (*i*) $\dfrac{3}{8}$ (*ii*) $\dfrac{\pi}{3}$ (*iii*) $\dfrac{5\pi}{32}$ (*iv*) $\dfrac{16}{35}$ (*v*) $\dfrac{35\pi}{256}$

2 (*i*) $\dfrac{16}{15}$ (*ii*) $\dfrac{2\pi}{3}$ (*iii*) 0 (*iv*) $\dfrac{5\pi}{16}$ (*v*) 0

3 $x(\ln x)^3 + 3x(\ln x)^2 - 6x\ln x + 6x$

4 $I_n = \dfrac{1}{n - 1}\tan^{n-1}\theta - I_{n-2}$
$\frac{1}{3}\tan^3\theta - \tan\theta + \theta$
$\frac{1}{5}\tan^5\theta - \frac{1}{3}\tan^3\theta + \tan\theta - \theta$

Exercise 9.6a

1 (*i*) $\dfrac{-2}{x-1}+\dfrac{3}{x-2}$ (*ii*) $\dfrac{\frac{1}{2}}{x-2}+\dfrac{\frac{1}{2}}{x+2}$ (*iii*) $\dfrac{1}{x}-\dfrac{2}{2x+1}$

(*iv*) $\dfrac{\frac{1}{2}}{x-1}-\dfrac{4}{x-2}+\dfrac{\frac{9}{2}}{x-3}$ (*v*) $-\dfrac{1}{x}+\dfrac{2}{x+1}$

(*vi*) $1-\dfrac{\frac{1}{2}}{x}+\dfrac{\frac{5}{2}}{x-2}$ (*vii*) $1+\dfrac{1}{x-2}-\dfrac{1}{x+2}$

(*viii*) $x+\dfrac{\frac{1}{2}}{x-1}+\dfrac{\frac{1}{2}}{x+1}$

2 (*i*) $\dfrac{1}{x}-\dfrac{1}{x-1}+\dfrac{1}{(x-1)^2}$ (*ii*) $\dfrac{2}{x}-\dfrac{1}{x^2}-\dfrac{2}{x+1}$

(*iii*) $\dfrac{\frac{1}{4}}{x-1}+\dfrac{\frac{1}{2}}{(x-1)^2}-\dfrac{\frac{1}{4}}{x+1}$ (*iv*) $\dfrac{\frac{5}{4}}{x-1}+\dfrac{\frac{1}{2}}{(x-1)^2}-\dfrac{\frac{1}{4}}{x+1}$

3 (*i*) $\dfrac{1}{x}-\dfrac{x}{x^2+1}$ (*ii*) $\dfrac{\frac{1}{4}}{x-1}-\dfrac{\frac{1}{4}}{x+1}-\dfrac{\frac{1}{2}}{x^2+1}$

(*iii*) $\dfrac{\frac{1}{3}}{x-1}-\dfrac{\frac{1}{3}(x+2)}{x^2+x+1}$ (*iv*) $\dfrac{\frac{1}{3}}{x+1}-\dfrac{\frac{1}{3}(x-2)}{x^2-x+1}$

4 $1-3x+7x^2-15x^3$

5 (*i*) $\dfrac{1}{r(r+1)}$, $\dfrac{n}{n+1}$, 1

(*ii*) $\dfrac{1}{r(r+2)}$, $\dfrac{n+1}{2(n+2)}$, $\dfrac{1}{2}$

(*iii*) $\dfrac{1}{r(r+1)(r+2)}$, $\dfrac{1}{4}-\dfrac{1}{2(n+1)(n+2)}$, $\dfrac{1}{4}$

Exercise 9.6b

1 (*i*) $\frac{1}{2}\ln\dfrac{x-1}{x+1}$ (*ii*) $\ln\dfrac{x}{x+1}$ (*iii*) $\frac{1}{2}\ln\dfrac{x}{2-x}$

(*iv*) $x+\ln\dfrac{x-2}{x+2}$ (*v*) $\frac{1}{4}\ln\dfrac{2x-1}{2x+1}$ (*vi*) $\frac{1}{3}\ln\dfrac{2x-1}{x+1}$

2 (*i*) $-\dfrac{1}{x+1}$ (*ii*) $-\dfrac{1}{2x}+\dfrac{1}{4}\ln\dfrac{x+2}{x}$ (*iii*) $-\dfrac{2}{x-1}+\ln\dfrac{x}{x-1}$

(*iv*) $x-\dfrac{1}{x-1}+2\ln(x-1)$

3 (*i*) $\ln(x^2+9)$; $\frac{1}{3}\tan^{-1}\dfrac{x}{3}$; $\frac{1}{2}\ln(x^2+9)+\tan^{-1}\dfrac{x}{3}$

(*ii*) $\ln(4x^2+9)$; $\frac{1}{6}\tan^{-1}\dfrac{2x}{3}$; $\frac{1}{4}\ln(4x^2+9)-\frac{1}{2}\tan^{-1}\dfrac{2x}{3}$

(*iii*) $\ln(x^2 - 6x + 10)$; $\tan^{-1}(x - 3)$;
$\frac{1}{2}\ln(x^2 - 6x + 10) + 3\tan^{-1}(x - 3)$

4 (*i*) $x + \ln(x - 1)$ (*ii*) $x + \ln\dfrac{x - 1}{x + 1}$ (*iii*) $\frac{1}{2}x^2 + \ln(x - 1)$

(*iv*) $-\dfrac{1}{x - 1} + \ln(x - 1)$ (*v*) $\frac{1}{3}\ln(x^3 + 1)$

(*vi*) $\frac{1}{6}\ln\dfrac{(x - 1)^2}{x^2 + x + 1} + \dfrac{1}{\sqrt{3}}\tan^{-1}\dfrac{2x + 1}{\sqrt{3}}$

Exercise 9.7

1 (*i*) x^5 (*ii*) $\frac{1}{7}t^7$ (*iii*) $-\dfrac{3}{u}$ (*iv*) $\sqrt{x}$ (*v*) $\frac{2}{3}v^{3/2}$ (*vi*) $\frac{1}{5}(x + 1)^5$

(*vii*) $-\dfrac{1}{4(2u + 1)^2}$ (*viii*) $\frac{2}{3}\sqrt{(3v - 2)}$ (*ix*) $\frac{1}{12}(x^2 - 1)^6$

(*x*) $-\dfrac{1}{3(1 - x^3)}$ (*xi*) $\frac{1}{3}(x^2 + 1)^{3/2}$ (*xii*) $-\sqrt{(1 - x^2)}$

2 (*i*) $2\ln x$ (*ii*) $3\ln(x - 3)$ (*iii*) $-\ln(1 - u)$ (*iv*) $\frac{1}{2}\ln(v^2 - 1)$
(*v*) $-\frac{1}{3}\ln(1 - x^3)$ (*vi*) $\ln(x^2 + x)$ (*vii*) $\frac{1}{2}e^{2x}$ (*viii*) $\frac{1}{6}e^{3x}$

(*ix*) $-\frac{1}{2}e^{-x^2}$ (*x*) $\ln(1 + e^u)$ (*xi*) $\dfrac{10^t}{\ln 10}$ (*xii*) $\ln(\ln x)$

3 (*i*) $\frac{1}{2}\sin 2x$ (*ii*) $-2\cos\dfrac{x}{2}$ (*iii*) $\frac{1}{3}\sin^3\theta$ (*iv*) $-\frac{1}{3}\cos^3\theta$

(*v*) $\frac{1}{12}\sin 3\theta + \frac{3}{4}\sin\theta$ (*vi*) $-\frac{3}{4}\cos\theta + \frac{1}{12}\cos 3\theta$ (*vii*) $\frac{1}{2}\sin^2 x$
(*viii*) $\frac{1}{2}(x - \sin x)$ (*ix*) $\frac{1}{4}\sin 2x + \frac{1}{2}x$ (*x*) $2\sqrt{(\sin x)}$ (*xi*) $\ln\sin\theta$
(*xii*) $-\ln\cos\theta$

4 (*i*) $\tan x$ (*ii*) $\tan x - x$ (*iii*) $\sec\theta$ (*iv*) $-\operatorname{cosec}\theta$ (*v*) $-\cot\theta$
(*vi*) $-\cot\theta - \theta$ (*vii*) $\frac{1}{2}\tan^2\theta$ (*viii*) $\frac{1}{3}\sec^3\theta$ (*ix*) $-\frac{1}{2}\ln\cos 2x$
(*x*) $\frac{1}{2}\tan^2 x + \ln\cos x$ (*xi*) $2\ln\sin\frac{1}{2}x$ (*xii*) $-\frac{1}{2}\cot^2 x - \ln\sin x$

5 (*i*) $\sin^{-1}\dfrac{x}{2}$ (*ii*) $\frac{1}{2}\sin^{-1}\dfrac{2x}{3}$ (*iii*) $\frac{1}{2}\tan^{-1}2x$ (*iv*) $\frac{1}{6}\tan^{-1}\dfrac{3x}{2}$

(*v*) $\ln[u + \sqrt{(1 + u^2)}]$ (*vi*) $\ln[u + \sqrt{(u^2 - 1)}]$
(*vii*) $\frac{1}{2}[\sin^{-1}x + x\sqrt{(1 - x^2)}]$ (*viii*) $-\frac{1}{3}(1 - x^2)^{3/2}$

(*ix*) $\frac{1}{6}\tan^{-1}\dfrac{2x}{3}$ (*x*) $\frac{1}{8}\ln(4x^2 + 9)$

6 (*i*) $x\sin x + \cos x$ (*ii*) $-x\,e^{-x} - e^{-x}$ (*iii*) $-\frac{1}{3}t\cos 3t + \frac{1}{9}\sin 3t$
(*iv*) $2x\sin x - (x^2 - 2)\cos x$ (*v*) $x\sin^{-1}x + \sqrt{(1 - x^2)}$
(*vi*) $x\ln x - x$ (*vii*) $\theta\tan\theta - \ln\cos\theta$ (*viii*) $(\frac{1}{2}x^2 - \frac{1}{2}x + \frac{1}{4})\,e^{2x}$
(*ix*) $-\frac{1}{2}e^{-t}(\sin t + \cos t)$ (*x*) $\frac{1}{13}e^{2t}(2\cos 3t + 3\sin 3t)$

(*xi*) $\frac{1}{9}x^3(3\ln x - 1)$ (*xii*) $\dfrac{10^x(x\ln 10 - 1)}{(\ln 10)^2}$

7 (*i*) 1 (*ii*) $\ln 2$ (*iii*) $\dfrac{1}{\sqrt{2}}\ln(3 + 2\sqrt{2})$

8 (*i*) $\frac{1}{2}\pi$ (*ii*) $\frac{1}{4}\pi \ln 2$ (*iii*) $\dfrac{\pi}{4\sqrt{2}} \ln (3 + 2\sqrt{2})$

9 (*i*) $\ln \dfrac{x}{x+1}$ (*ii*) $\ln \dfrac{x-1}{x}$ (*iii*) $\ln \dfrac{x-1}{x} + \dfrac{1}{x}$ (*iv*) $\ln \dfrac{x}{\sqrt{(x^2+1)}}$
(*v*) $\ln (1 + e^x)$ (*vi*) $x - \ln (1 + e^x)$ (*vii*) $\frac{1}{2} \tan^{-1} x^2$
(*viii*) $x (\ln x)^2 - 2x \ln x + 2x$
(*ix*) $\frac{1}{3}x^3 \tan^{-1} x - \frac{1}{6}x^2 + \frac{1}{6} \ln (1 + x^2)$
(*x*) $\dfrac{2}{\sqrt{3}} \tan^{-1} \left(\dfrac{1}{\sqrt{3}} \tan \dfrac{x}{2} \right)$ (*xi*) $\sec^{-1} x$ (*xii*) $\sin^{-1} x + \sqrt{(1 - x^2)}$

10 (*i*) $\frac{1}{4} \ln 2 + \frac{1}{8}\pi$ (*ii*) $\frac{1}{4}$ (*iii*) $\frac{1}{8}(\pi + 2)$ (*iv*) $\dfrac{1}{\sqrt{2}}$ (*v*) $\dfrac{8}{105}$
(*vi*) $\dfrac{5\pi}{256}$ (*vii*) $\frac{1}{9}(2 e^3 + 1)$ (*viii*) $\frac{1}{3}\pi + \frac{1}{2}\sqrt{3}$ (*ix*) $\dfrac{\pi}{3\sqrt{3}}$ (*x*) $\frac{1}{2} \ln 2$
(*xi*) $\dfrac{2\pi}{3\sqrt{3}}$ (*xii*) $\frac{1}{2}\pi - 1$

Exercise 9.8a

2 (*i*) $x \dfrac{dy}{dx} + y = 0$ (*ii*) $\dfrac{dy}{dx} = y$ (*iii*) $2x \dfrac{dy}{dx} = y$ (*iv*) $y \dfrac{dy}{dx} = x$
(*v*) $x^2 \dfrac{d^2y}{dx^2} - 2x \dfrac{dy}{dx} + 2y = 0$ (*vi*) $\dfrac{d^2x}{dt^2} = x$

Exercise 9.8b

1 (*i*) $y = A e^x$ (*ii*) $y = A e^{-\frac{1}{2}x^2}$ (*iii*) $y = x^2 + A$ (*iv*) $y = \dfrac{1}{x + c}$
(*v*) $y = Ax$ (*vi*) $2x^2 + y^2 = A$

2 (*i*) $\dfrac{dy}{dx} = \dfrac{y}{x}$, $\dfrac{dy}{dx} = -\dfrac{x}{y}$, $x^2 + y^2 = A$
(*ii*) $\dfrac{dy}{dx} = 1$, $\dfrac{dy}{dx} = -1$, $y = -x + A$
(*iii*) $\dfrac{dy}{dx} = \dfrac{2y}{x}$, $\dfrac{dy}{dx} = -\dfrac{x}{2y}$, $x^2 + 2y^2 = A$
(*iv*) $\dfrac{dy}{dx} = \dfrac{1}{2y}$, $\dfrac{dy}{dx} = -2y$, $y = A e^{-2x}$
(*v*) $\dfrac{dy}{dx} = -y$, $\dfrac{dy}{dx} = \dfrac{1}{y}$, $y^2 = 2x + A$
(*vi*) $\dfrac{dy}{dx} = -\dfrac{2y}{x}$, $\dfrac{dy}{dx} = \dfrac{x}{2y}$, $x^2 - 2y^2 = A$

3 (*i*) $\sin x = 2 \sin y \Rightarrow y = \sin^{-1} (\frac{1}{2} \sin x)$
(*ii*) $y = 2 \tan^{-1} \frac{1}{2}x$

4 $S = A e^{kt}$, 57 000 **6** 100 years

7 160 000 km

8 $x = \dfrac{1}{1 + 9(\frac{2}{3})^{2t/3}}$, 26%

9 171 years, 141 years

10 $\dfrac{d\theta}{dt} = a - b\theta$; 3.76 min ($a = 21.3$, $b = 0.170$)

11 $a\dfrac{dM}{dt} + bM = f(t)$, 21.2 days

12 (*i*) $v\dfrac{dv}{dx} = -g - kv^2$ (*ii*) $x = \dfrac{1}{2k}\ln\dfrac{g + ku^2}{g + kv^2}$

(*iii*) $\dfrac{1}{2k}\ln\left(1 + \dfrac{ku^2}{g}\right)$ (*iv*) $\sqrt{\left(\dfrac{g}{k}\right)}$

13 $x = \left(\dfrac{10 - t}{10}\right)^3$, $\frac{1}{8}$ kg **14** 132 min

Exercise 9.8c

1 (*i*) $y = A\,e^{-x} + x - 1$ (*ii*) $y = x(\ln x + A)$
(*iii*) $y = e^x + Ax^2$ (*iv*) $y = A\,e^{-x} + x^2 - 2x + 2$

2 (*i*) $y = A(1 - x^2)^{-1/2} - 1$ (*ii*) $(1 + x)y = A\,e^x - x + 1$

3 $A\,e^{-t} + \frac{1}{2}(\sin t - \cos t)$; $\frac{1}{2}(\sin t - \cos t)$

4 $2\,e^{-t}(1 - e^{-t})$

5 (*i*) $e^{Rt/L}$; $A\,e^{-(Rt/L)} + \dfrac{E_0}{R^2 + \omega^2L^2}(R\sin\omega t - \omega L\cos\omega t)$;

$\dfrac{LE_0}{R^2 + \omega^2L^2}(R\sin\omega t + \omega L\cos\omega t)$

(*ii*) $e^{t/RC}$; $A\,e^{-t/RC} + \dfrac{\omega RCE_0}{1 + \omega^2R^2C^2}(\omega RC\sin\omega t - \cos\omega t)$

$\dfrac{\omega RCE_0}{1 + \omega^2R^2C^2}(\omega RC\sin\omega t - \cos\omega t)$

Exercise 9.8d

1 (*i*) $y = A\,e^{3x} + B\,e^x$ (*ii*) $y = A\,e^{5x} + B\,e^{-x}$
(*iii*) $y = A\,e^{2x} + B\,e^{-2x}$ (*iv*) $y = A + B\,e^{4x}$
(*v*) $y = (A + Bx)\,e^{2x}$

2 (*i*) $y = A\,e^x + B\,e^{2x} + 3$
(*ii*) $y = A\,e^x + B\,e^{2x} + \frac{1}{4}(2x^2 + 6x + 7)$
(*iii*) $x = A\,e^{-t} + B\,e^{-2t} + \frac{1}{6}e^t$
(*iv*) $x = A\,e^{-t} + B\,e^{-2t} + \frac{1}{10}(\sin t - 3\cos t)$

3 (*i*) $y = A\,e^{3x} + B\,e^{-x} - \frac{1}{3}x + \frac{2}{9}$
(*ii*) $y = A\,e^{-3x} + B\,e^x - \frac{1}{4}e^{-x}$
(*iii*) $x = A\,e^t + B\,e^{2t} + \frac{1}{10}(\sin t + 3\cos t)$
(*iv*) $\theta = A\,e^{-2t} + B\,e^{-t} + \frac{1}{10}(3\sin t + \cos t)$

4 (*i*) $y = e^{2x} + 2\,e^{x}$ (*ii*) $x = e^{2t} - e^{-2t}$ (*iii*) $\theta = e^{t} + e^{-2t} - 2$

5 $2\,(e^{-2t} - e^{-3t})$; $\frac{8}{27}$ rad ($\approx 17°$) when $t = \ln 1.5$ (≈ 0.4 s)

6 $q = A\,e^{-200t} + B\,e^{-1\,000t} + \frac{1}{10}$

Exercise 9.9a

1 $\dfrac{b}{a}\coth\theta$

3 (*i*) $2\cosh 2x$ (*ii*) $\frac{1}{2}\sinh\dfrac{x}{2}$ (*iii*) $x\sinh x + \cosh x$ (*iv*) $\coth x$
(*v*) $\tanh x$ (*vi*) $-\operatorname{sech} x \tanh x$ (*vii*) $-\operatorname{cosech} x \coth x$
(*viii*) $\operatorname{cosech} x$

4 (*i*) $\cosh x$ (*ii*) $\frac{1}{2}\sinh 2x$ (*iii*) $\tanh x$ (*iv*) $\ln\cosh x$
(*v*) $x\sinh x - \cosh x$ (*vi*) $x\tanh x - \ln\cosh x$

5 (*i*) $2\sinh 1 = 2.35$ (*ii*) $\frac{1}{2}\pi\,(\sinh 2 + 1) = 8.837$

6 (*i*) $x + \dfrac{x^3}{3!} + \dfrac{x^5}{5!} + \cdots$ (*ii*) $1 + \dfrac{x^2}{2!} + \dfrac{x^4}{4!} + \cdots$

Exercise 9.9b

1 (*i*) $\ln 2$ (*ii*) $\ln(2 + \sqrt{3})$ (*iii*) $\frac{1}{2}\ln\dfrac{1+x}{1-x}$

2 (*i*) $\sinh^{-1}\dfrac{x}{2}$ (*ii*) $\cosh^{-1}\dfrac{x}{3}$ (*iii*) $\frac{1}{2}\sinh^{-1}\dfrac{2x}{3}$ (*iv*) $\frac{1}{3}\cosh^{-1}\dfrac{3x}{2}$
(*v*) $\cosh^{-1}(x+1)$ (*vi*) $\sinh^{-1}(x+1)$
(*vii*) $x\sinh^{-1}x - \sqrt{(1+x^2)}$ (*viii*) $x\cosh^{-1}x - \sqrt{(x^2-1)}$
(*ix*) $\frac{1}{2}\{x\sqrt{(x^2+1)} + \sinh^{-1}x\}$ (*x*) $\frac{1}{2}x\sqrt{(x^2-4)} - 2\cosh^{-1}\dfrac{x}{2}$

3 (*i*) $\sinh^{-1}2 = \ln(2+\sqrt{5}) = 1.44$
(*ii*) $\cosh^{-1}\frac{3}{2} - \cosh^{-1}1 = \ln\left(\dfrac{3}{2} + \dfrac{\sqrt{5}}{2}\right) = 0.96$
(*iii*) $\sqrt{3} - \frac{1}{2}\cosh^{-1}2 = \sqrt{3} - \frac{1}{2}\ln(2+\sqrt{3}) = 1.07$
(*iv*) $\frac{1}{2}\sqrt{5} + \frac{1}{4}\sinh^{-1}2 = \frac{1}{2}\sqrt{5} + \frac{1}{4}\ln(2+\sqrt{5}) = 1.48$

4 (*i*) $\dfrac{1}{a}\tanh^{-1}\dfrac{x}{a}$
(*ii*) $\left\{\dfrac{g}{k}(1 - e^{-2kx})\right\}^{1/2}$; $\sqrt{\dfrac{g}{k}}\tanh\sqrt{(gk)}\,t$; $\sqrt{\left(\dfrac{g}{k}\right)}$

Exercise 9.10a

1 (*i*) $\frac{8}{27}(10\sqrt{10} - 1)$ (*ii*) $\ln(1+\sqrt{2})$ (*iii*) $2(\sqrt{2}-1)a$
(*iv*) $a(\sqrt{2} + \ln(1+\sqrt{2}))$

2 $6a$

Exercise 9.10b

1 (*i*) $\frac{1}{4}\pi c^2(e^2 - e^{-2} + 4)$ (*ii*) $\frac{8}{3}\pi a^2(2\sqrt{2} - 1)$ (*iii*) $\frac{12}{5}\pi a^2$
(*iv*) $\frac{64}{3}\pi a^2$

Exercise 9.11

1 (*i*) $\pi a, \frac{1}{4}\pi a^2, \phi = \frac{1}{2}\pi + \theta$ (*ii*) $8a, \frac{3}{2}\pi a^2, \phi = \frac{1}{2}(\pi + \theta)$
2 $a^2, \phi = 2\theta$
3 a^2
4 $\frac{1}{6}a^2, \frac{1}{2}a(\sqrt{2} + \ln(1 + \sqrt{2}))$

Miscellaneous problems 9

1 (*i*) 2, 2.25, 2.37, 2.49, 2.59, 2.69
8 (*iv*) $\sqrt{\pi}$

Exercise 10.1

1 (*i*) $-3 - 6i$ (*ii*) $-2 - 8i$ (*iii*) $10 + 5i$ (*iv*) $11 + 2i$
(*v*) $-13 - 19i$ (*vi*) $-3 - 4i, -7 + 24i, -527 - 336i$
(*vii*) $-20 - 20i$ (*viii*) $-11 - 2i$ (*ix*) 61 (*x*) 50
(*xi*) $(x^5 - 10x^3 + 5x) + i(-5x^4 + 10x^2 - 1)$
(*xii*) $(x^6 - 15x^4y + 15x^2y^4 - y^6) + i(6x^5y - 20x^3y^3 + 6xy^5)$
(*xiii*) $\dfrac{7 + i}{50}, \dfrac{7 - i}{50}$

2 (*i*) $\dfrac{2 - i}{5}$ (*ii*) $0.2(3 + i)$ (*iii*) $0.02(1 - 7i)$ (*iv*) $1 + 2i$
(*v*) $-i$ (*vi*) $0.008(-11 + 2i)$ (*vii*) $1 + 2i$ (*viii*) $\cos\theta + i\sin\theta$

3 $p = 7k, q = 3k \ (k \in Z)$

4 (*i*) $(x^3 - 3xy^2) + i(3x^2y - y^3)$ (*ii*) $(x^2 - y^2 - 2ixy)(x^2 + y^2)^{-2}$
(*iii*) $\left(x + \dfrac{x}{x^2 + y^2}\right) + i\left(y - \dfrac{y}{x^2 + y^2}\right)$

5 $(4 + 5i)(2 - 3i)$; $83^2 + 1^2 = 71^2 + 43^2$

6 $-1 + 2i$ 7 $-\frac{4}{11}, \frac{2}{11}$

8 $x^2 = \frac{1}{2}(1 + \sqrt{2}), y^2 = \frac{1}{2}(-1 + \sqrt{2})$; $\pm(3 - 2i)$

9 $2 - i$ 10 1, 2

11 (*i*) $-2 \pm 5i$ (*ii*) $1\frac{1}{2} \pm 2i$ (*iii*) $i(-1 \pm \sqrt{2})$
(*iv*) $\frac{1}{2}(\pm\sqrt{7} - i)$ (*v*) $-2, -4 \pm \sqrt{5}i$
(*vi*) $0.1(1 \pm \sqrt{19}i)(2 - i)$

12 (*i*) $z^2 - 6z + 4 = 0$ (*ii*) $z^2 + 6z + 10 = 0$
(*iii*) $z^2 - 4z + 7 = 0$ (*iv*) $z^2 - (1 + i)z + (2 - i) = 0$

13 (*i*) 5, 7, +11
(*ii*) $-1, 7, -13$; $z_1^2 + z_2^2 < 0 \Rightarrow z_1, z_2$ complex, but not conversely

14 (*i*) $\frac{1}{2}(1 - i^{n+1})(1 + i)$ (*ii*) $0.2(1 - 2i)[1 - (-2i)^n]$
(*iii*) $\frac{1}{2}i[1 - (n + 1)i^n + ni^{n+1}]$

15 $A = \frac{1}{2}i,\ B = -\frac{1}{2}i$

Exercise 10.2a

3 $z_1 + z_2 = 3 + 2i;\quad z_1 - z_2 = 1 - 4i;\quad z_1z_2 = 5 + 5i;$
$\dfrac{z_1}{z_2} = 0.1(-1 - 7i);\quad \dfrac{z_2}{z_1} = 0.2(-1 + 7i);\quad z_1^2 = 3 - 4i$

4 $iz = -12 - 5i;\quad z - 1 = -6 + 12i;\quad 3z = -15 + 36i;$
$-3z = 15 - 36i;\quad z^{-1} = \dfrac{-5 - 12i}{169};\quad z^2 = -119 - 120i,$
$zz' = 169$ (on real axis)

6 Line segments AB and CD are equal and perpendicular

8 $-2 + i,\ 6 - 3i$ **9** $-4 + 4i,\ -2i,\ -5 - i$

10 $6 + 2i,\ 7 - i$; or $0,\ 1 - 3i$

11 $7\frac{1}{2} + 1\frac{1}{2}i;\quad -5\frac{1}{2} - 4\frac{1}{2}i$

12 $(3 - 2i) \pm \sqrt{3}(1 + i)$

Exercise 10.2b

1 $-2 - 2i$ **5** $c = -a\omega^2 - b\omega;\quad c = -a\omega - b\omega^2$

Exercise 10.2c

1 (*i*) $1 \operatorname{cis} \pi$ (*ii*) $\sqrt{2} \operatorname{cis}\left(-\dfrac{\pi}{4}\right)$ (*iii*) $2 \operatorname{cis} 60°$ (*iv*) $1 \operatorname{cis}(-90°)$
(*v*) $13 \operatorname{cis}(\pi + \tan^{-1}\frac{5}{12})$ (*vi*) $\sqrt{13} \operatorname{cis}(\pi - \tan^{-1} 1.5)$
(*vii*) $1 \operatorname{cis}(\frac{1}{2}\pi - \alpha)$ (*viii*) $1 \operatorname{cis}(\alpha - \frac{1}{2}\pi)$

2 (*i*) $2\sqrt{3} + 2i$ (*ii*) -5 (*iii*) $5\sqrt{2}(1 - i)$
(*iv*) $-8 \cos 15° + 8i \sin 15°$ (*v*) $2 \cos 80° + 2i \sin 80°$
(*vi*) $3\sqrt{2}(1 - i)$

3 (*i*) $ab = 5\sqrt{2} \operatorname{cis}(\pi - \tan^{-1} 7)$ (*ii*) $a/b = \frac{1}{2}\sqrt{2} \operatorname{cis}(-\frac{1}{4}\pi)$
(*iii*) $b/a = \sqrt{2} \operatorname{cis} \frac{1}{4}\pi$ (*iv*) $a^2 = 5 \operatorname{cis}(\tan^{-1}\frac{4}{3})$
(*v*) $b^2 = 10 \operatorname{cis}(\pi - \tan^{-1}\frac{3}{4})$ (*vi*) $a^2 + b^2 = 5\sqrt{5} \operatorname{cis}(\pi - \tan^{-1} 2)$

4 $a^n = \operatorname{cis} \frac{2}{3}\pi n;\quad b^n = 2^n \operatorname{cis}(-\frac{1}{4}n\pi);\quad ab = 2 \operatorname{cis} \dfrac{5\pi}{12};$
$\dfrac{a}{b} = \frac{1}{2} \operatorname{cis} \dfrac{11\pi}{12}$

5 $\operatorname{cis}(\theta + \phi);\quad \operatorname{cis} \dfrac{3\pi}{2} = -i;\quad \operatorname{cis} 240° = -\frac{1}{2} - \frac{1}{2}\sqrt{3}i$

6 (*i*) $\operatorname{cis}(\pm\frac{1}{3}\pi)$ (*ii*) $\operatorname{cis}(\pm\frac{2}{3}\pi)$ (*iii*) $\frac{1}{3}\sqrt{5} \operatorname{cis}(\pi \pm \tan^{-1}\frac{1}{2})$
(*iv*) $\sqrt{2} \operatorname{cis} \frac{1}{4}\pi$ or $\operatorname{cis}(-\frac{1}{2}\pi)$

7 (*i*) $z^2 + z + 1 = 0$ (*ii*) $z^2 - 2(\cos 36°)z + 1 = 0$
(*iii*) $z^2 - 2(\cos \alpha)z + 1 = 0$

11 Apollonius' median theorem

12 (*i*) circle centre i, radius 2 (*ii*) perp. bisector of -1, $2i$
(*iii*) interior of circle centre $-3i$, radius 2 (*iv*) portion of plane inside circle centre O, radius 4 and outside circle centre 2, radius 1
(*v*) ellipse foci ± 3, major axis length 10

13 (*i*) region below real axis (*ii*) region outside circle centre O, radius 3 and inside circle centre 3, radius 3

14 First circle entirely inside region between concentric circles

Exercise 10.2d

1 (*i*) $20 \operatorname{cis} 230°$ (*ii*) $\operatorname{cis}(-\theta - \phi)$ (*iii*) 1 (*iv*) $\operatorname{cis}(-1\tfrac{1}{2}\pi) = -i$
(*v*) $2\sqrt{2} \operatorname{cis} \dfrac{5\pi}{12}$ (*vi*) $\operatorname{cis} \dfrac{7\pi}{2} = -i$ (*vii*) $2(1 + \cos \alpha) = 4 \cos^2 \tfrac{1}{2}\alpha$
(*viii*) $4 \sin^2 \tfrac{1}{2}\beta\,(-\cos \beta + i \sin \beta)$ (*ix*) $\operatorname{cis}(-3\theta)$ (*x*) $2 \operatorname{cis} \theta$
(*xi*) $\operatorname{cis}(\alpha + \beta - 2\gamma)$ (*xii*) $\tan \tfrac{1}{2}\theta\,(\sin \theta + i \cos \theta)$

3 (*i*) $2 \cos \tfrac{1}{2}\theta \operatorname{cis} \tfrac{1}{2}\theta$ (*ii*) $i \tan \tfrac{1}{2}\theta$ (*iii*) $\tfrac{1}{2} \sec \theta$
(*iv*) $\dfrac{i \cos \theta}{1 - \sin \theta} = i \tan(\tfrac{1}{2}\theta + \tfrac{1}{4}\pi)$

5 Rewrite $k = \dfrac{c - a}{b - a}$

6 $\dfrac{a - c}{b - c} = \dfrac{a' \quad c'}{b' - c'} \Leftrightarrow \angle C = \angle C'$ and $\dfrac{AC}{BC} = \dfrac{A'C'}{B'C'} \Leftrightarrow \triangle$s ABC, A'B'C' similar in same sense

7 Use $\triangle$ABC similar to $\triangle$BCA

8 See no. 5: $l = (1 - k)b + kc$, etc. Add three equations

Exercise 10.3

3 $z_1^* z_2 = r_1 r_2 \operatorname{cis}(-\theta_1 + \theta_2)$

4 $z = z^* \operatorname{cis} 2\theta \Rightarrow \arg z = \theta$ or $\theta + \pi$

5 (*i*) $8zz^* + z + z^* - 1 = 0$
(*ii*) $\dfrac{z - 1}{z + 1} = \dfrac{z^* - 1}{z^* + 1} \operatorname{cis} \dfrac{4\pi}{3} \Leftrightarrow (zz^* - 1)(3 + \sqrt{3}i)$
$(z - z^*)(1 - \sqrt{3}i)$

8 $1 + i$, $1\tfrac{1}{2}$ **9** $2 - i$, $\tfrac{1}{2}$

10 $(z + 2)(z^2 + 4z + 8)$; $2 \operatorname{cis} \pi$, $\sqrt{8} \operatorname{cis}\left(\pm\dfrac{3\pi}{4}\right)$

11 $\operatorname{cis}(\pm 60°)$, 1 (repeated)

12 $2 \pm i$, $-2 \pm i$ **13** $1 \pm i$, $2 \pm i$

15 OA $\perp$ BC $\Rightarrow b - c = \lambda i a$ $(\lambda \in R)$: conjugate and eliminate λ

17 $\begin{vmatrix} 1 & 1 & 1 \\ a & b & c \\ a^* & b^* & c^* \end{vmatrix} = 0$

Exercise 10.4a

1 (*i*) $\text{cis}\,(-72°)$ (*ii*) $\text{cis}\,220°$ (*iii*) $10^8\,\text{cis}\,(4\tan^{-1}\frac{24}{7})$
(*iv*) $\text{cis}\,80°$ (*v*) $2^{7/4}\,\text{cis}\,(-52\frac{1}{2}°)$ (*vi*) $32\,\text{cis}\,2\frac{1}{2}\pi = 32i$
(*vii*) $2^{20}\,\text{cis}\,120°$ (*viii*) $2^{1/6}\,\text{cis}\,15°$ (*ix*) $\sqrt{2}\,\text{cis}\,(-30°)$
(*x*) $\text{cis}\,(5\theta - \frac{1}{2}\pi)$ (*xi*) $\sec^n\theta\,\text{cis}\,n(\frac{1}{2}\pi - \theta)$ (*xii*) $\text{cis}\,\alpha$

2 $\pm 2^{-\frac{1}{2}}(1 + i) = \pm\text{cis}\,\frac{1}{4}\pi$; $\sqrt[4]{i} = \text{cis}\left(\frac{\pi}{8} + k\frac{1}{2}\pi\right)$;
$\sqrt[5]{(-i)} = \text{cis}\left(-\frac{\pi}{10} + \frac{k2\pi}{5}\right)$

4 $-110°, -110°, -110°$; $\text{cis}\,(-94°), \text{cis}\,(-22°)$

5 $2^{1/12}\,\text{cis}\,(7\frac{1}{2}° + 60k°)$, $k = 0, 1, 2, 3, 4, 5$.

6 (*i*) $1, \text{cis}\,(\pm\frac{2}{3}\pi)$ (*ii*) $-1, \text{cis}\,(\pm\frac{1}{3}\pi)$
(*iii*) $\text{cis}\,50°, \text{cis}\,170°, \text{cis}\,(-70°)$

7 $32c^6 - 48c^4 + 18c^2 - 1$; $64c^6 - 80c^4 + 24c^2 - 1$;
$\frac{5t - 10t^3 + t^5}{1 - 10t^2 + 5t^4}$; $32c^5 - 32c^3 + 6c$

8 $z + z^{-1} = -1 \Rightarrow z = \text{cis}\,(\pm 120°)$; $z^n + z^{-n} = 2\,\text{cis}\,(n120°)$

10 $2^n\cos^n\frac{1}{2}\theta\,(\cos\frac{1}{2}n\theta + i\sin\frac{1}{2}n\theta)$; sum $= 2^{n+1}\cos^n\frac{1}{2}\theta\cos\frac{1}{2}n\theta$

11 $\tan 15° = 2 - \sqrt{3}$

12 $\tan\,(12°\,41' + k36°)$ $(k = 0, 1, 2, 3, 4)$

13 $\frac{1}{2}k\pi$ or $k\pi \pm \tan^{-1}\frac{1}{\sqrt{2}}$

15 $\cot^2\frac{k\pi}{7}$ $k = 0, 1, 2$

Exercise 10.4b

1 $1 + 3i, -3 + i, -1 - 3i, 3 - i$

2 (*i*) $2\,\text{cis}\,\theta(\theta = 20°, 140°, 260°)$
(*ii*) $\sqrt{2}\,\text{cis}\,\theta(\theta = -16°, 56°, 128°, 200°, 272°)$
(*iii*) $\text{cis}\,(-20° + k45°)$ $k = 0, 1, 2, \ldots, 7$
(*iv*) $\text{cis}\,(10° + k40°)$ $k = 0, 1, 2, \ldots, 8$
(*v*) $2^{3/4}\,\text{cis}\,(-45° + k90°)$ $k = 0, 1, 2, 3$
(*vi*) $2^{1/6}\,\text{cis}\,(25° + k60°)$, $k = 0, 1, 2, 3, 4, 5$
(*vii*) $\text{cis}\,(22\frac{1}{2}° + 90°k)$ $k = 0, 1, 2, 3$
(*viii*) $2^{1/8}\,\text{cis}\,(8k + 3)\frac{\pi}{16}$, $k = 0, 1, 2, 3$

3 (*i*) $1 \pm \sqrt{3}i, -2$ (*ii*) $z = \text{cis}\,\frac{2\pi k}{7}, k = 1, 2, 3, 4, 5, 6$

(*iii*) $\text{cis}\left(\pm\frac{\pi}{5}\right), \text{cis}\left(\pm\frac{3\pi}{5}\right)$ (*iv*) $\text{cis}\left(\frac{2\pi k}{5}\right), k = 0, 1, 2, 3, 4$

(*v*) $\sqrt{5}\,\text{cis}\,\theta, \theta = \frac{1}{3}(\pi \pm \tan^{-1}\frac{2}{11} + \frac{2}{3}\pi k), k = 0, 1, 2$

4 (*i*) $\dfrac{4 - 5\cos\alpha - 3i\sin\alpha}{5 - 4\cos\alpha}, \alpha = \dfrac{2\pi k}{5}, k = 0, 1, 2, 3, 4$

(*ii*) $z = \frac{1}{2}i, 0.1(3 - i), -\frac{1}{4}i, 0.1(-3 - i)$

(*iii*) $2^{1/6}\,\text{cis}\,(\pm 15^\circ + k120^\circ), k = 0, 1, 2$

(*iv*) $\text{cis}\,\frac{2\pi k}{5}\ (k = 0, 1, 2, 3, 4);\quad \text{cis}\,(\frac{1}{4}\pi + \frac{1}{2}\pi k), k = 0, 1, 2, 3$

7 $z^2 - 5z + 7 = 0;\quad z^2 + z + 7 = 0$

9 Condition for equilateral triangle (cf. Ex. 10.2b, no. **5**, Ex. 10.2d, no. **7**)

10 (*i*) $\dfrac{\cos 5\theta + 5\cos 3\theta + 10\cos\theta}{16}$

(*ii*) $\dfrac{10 - 15\cos 2\theta + 6\cos 4\theta - \cos 6\theta}{32}$

(*iii*) $\dfrac{35\sin\theta - 21\sin 3\theta + 7\sin 5\theta - \sin 7\theta}{64}$

(*iv*) $\sin^4\theta\cos^6\theta = 2^{-9}(\cos 10\theta + 2\cos 8\theta - 3\cos 6\theta - 8\cos 4\theta + 2\cos 2\theta + 6)$

11 $\dfrac{60\theta - 45\sin 2\theta + 9\sin 4\theta - \sin 6\theta}{192};\quad 3\pi \times 2^{-9}$

12 $\frac{1}{2}\sec\alpha[1 - (-1)^n\cos 2n\alpha];\quad \frac{1}{2}\sec\alpha(-1)^{n+1}\sin 2n\alpha$

13 $\dfrac{\sin(n+1)\frac{1}{2}\alpha\cos n\frac{1}{2}\alpha}{2\sin\frac{1}{2}\alpha}[n\tan\frac{1}{2}n\alpha + \cot\frac{1}{2}\alpha - (n+1)\cot(n+1)\frac{1}{2}\alpha]$

14 $\cot\theta\left[= I\left(\dfrac{\cos\theta\,\text{cis}\,\theta}{1 - \cos\theta\,\text{cis}\,\theta}\right)\right]$

Exercise 10.5

1 (*i*) i (*ii*) -1 (*iii*) $e^{-1}i$ (*iv*) $2i\sin 1$

(*v*) $\text{sh}\,2\cos 5 - i\,\text{ch}\,2\sin 5$ (*vi*) $\text{cis}\,3\,\text{ch}\,1 - i\sin 3\,\text{sh}\,1$

(*vii*) $\ln 8 + i(2k+1)\pi\ (k \in Z)$ (*viii*) $\ln 8 + i(2k + \frac{1}{2})\pi\ (k \in Z)$

(*ix*) $\ln 2 + i(2k + \frac{1}{6})\pi$ (*x*) $\ln 2 + i(2k - \frac{5}{6})\pi$

4 $e^{-2x}\dfrac{-2\cos 3x + 3\sin 3x}{13};\quad e^{-2x}\dfrac{-3\cos 3x - 2\sin 3x}{13}$

6 (*i*) $\sin x\,\text{ch}\,y + i\cos x\,\text{sh}\,y;\quad r^2 = \frac{1}{2}(\text{ch}\,2y - \cos 2x)$

(*ii*) $\cos x\,\text{ch}\,y - i\sin x\,\text{sh}\,y;\quad r^2 = \frac{1}{2}(\text{ch}\,2y + \cos 2x)$

(*iii*) $\dfrac{\sin 2x + i\,\text{sh}\,2y}{\cos 2x + \text{ch}\,2y};\quad r^2 = \dfrac{\text{ch}\,2y - \cos 2x}{\text{ch}\,2y + \cos 2x}$

7 $-\frac{1}{2}\pi \pm i \ln(3 + 2\sqrt{2})(+2\pi k)$

8 $\sinh^2 y = \frac{1}{2}\sqrt{3}$; $\cos^2 x = \frac{1}{2}\sqrt{3}$

9 $w = e^z$; $\cosh z$

10 $\tan 2y = \dfrac{2v}{1 - u^2 - v^2}$; $\tanh 2x = \dfrac{e^{4x} - 1}{e^{4x} + 1}$,

where $e^{4x} = \dfrac{(1 - u^2 - v^2)^2 + 4v^2}{[(1 - u)^2 + v^2]^2}$

12 (*i*) $e^{4x}(A \cos 3x + B \sin 3x)$ (*ii*) $e^x (A \cos x + B \sin x)$
(*iii*) $A \cos 2t + B \sin 2t + 1\frac{1}{2}$ (*iv*) $A \cos 3t + B \sin 3t - \frac{1}{7} \sin 4t$

13 (*i*) $3e^{3t} \sin t$ (*ii*) $\frac{4}{9}(1 - \cos 3t) + \frac{2}{3} \sin 3t$

Exercise 10.6

1 (*i*) translation
(*ii*) half-turn about $1 - 2i$
(*iii*) enlargement $\times 5$ in point $-\frac{1}{2}i$
(*iv*) enlargement $\times 2$ and rotation $+\frac{1}{2}\pi$ about O
(*v*) spiral similarity about O, $\times 5$ through $(-\tan^{-1} 1\frac{1}{3})$
(*vi*) spiral similarity about O, $\times 5$ through $100°$
(*vii*) enlargement $\times \frac{1}{2}$ in point 2
(*viii*) spiral similarity $\times \sqrt{2}$ through $-\frac{1}{4}\pi$ about O
(*ix*) rotation through $+\frac{1}{2}\pi$ about $-\frac{1}{2} - \frac{1}{2}i$

2 (*i*) $z' = z + 3i$ (*ii*) $z' = 3z \operatorname{cis} 135°$
(*iii*) $z' - i = 3\sqrt{2}(-1 + i)(z - i)$ (*iv*) $z' = -iz + 2$
(*v*) $z' = 2 + \frac{1}{2}i - z$ (*vi*) $z' = 4z - 6 - 3i$ (*vii*) $z' = 3z - 2i$
(*viii*) $z' = 3z - 6i$ (*ix*) $z' = -2iz + 1 + 2i$
(*x*) $z' = -2iz + 4 + 3i$

3 (*i*) Rotation $+\frac{1}{2}\pi$ about point 1 (*ii*) reflection in line $y = x - 1$

5 (*i*) Reflection in real axis (*ii*) reflection in $y = x$
(*iii*) reflection in $y = x \tan \alpha$ (*iv*) glide reflection in line $y = x + 1$
(*v*) glide reflection in $y = x - 1$

Exercise 10.7

1 (*i*) Circle centre O, radius 2 (*ii*) imaginary axis (*iii*) $y + x = 0$
(*iv*) real axis
(*v*) portion of line $y = 1 - \sqrt{3}x$ in first and fourth quadrants
(*vi*) ellipse, foci O and $1 - i$ (*vii*) circle, diameter $1, -1$
(*viii*) circle, diameter $\frac{3}{4}, 1\frac{1}{2}$ (*ix*) line segment joining 1 and $-i$
(*x*) ellipse $\dfrac{x^2}{4} + \dfrac{y^2}{3} = 1$

2 (*i*) Interior of circle (*ii*) exterior of circle
(*iii*) circumference and exterior of circle, centre $-4i$, radius 3
(*iv*) exterior of circle, centre $3 - i$, radius 2
(*v*) interior of ellipse in no. 1 (*x*) (*vi*) all points for which $y > 0$

3 The locus of points whose distance from -1 and from -16 are $4:1$ has centre O, radius 4

4 $|z| = 1$

5 $3zz^* + 4iz - 4iz^* + 4 = 0$;
$zz^*(1 - \lambda^2) + z(-a^* + \lambda^2 b^*) + z^*(-a + \lambda^2 b) + aa^* - \lambda^2 bb^* = 0$

6 (*i*) half-circle $|z| = 3$
(*ii*) semicircle $|z| = 2$ from $-2i$, -2, $2i$
(*iii*) semicircle $|z| = 1$ through -1, i, 1
(*iv*) the same moved $+2$ in direction of real axis
(*v*) semicircle $|z| - 2$ through $-2i$, 2, $2i$
(*vi*) whole circle $|z| = 1$, beginning and ending at -1
(*vii*) whole circle $|z + 1| = 1$
(*viii*) quadrant of circle $|z| = 1$

7 Semicircle $|z| = 1$ through 1, i, -1

9 (*i*) the same translated $+1$ in direction of imaginary axis
(*ii*) the same enlarged $\times 2$ in origin
(*iii*) the same rotated $+\frac{1}{2}\pi$ about origin
(*iv*) the circle $|z| = 1$ anti-clockwise from 1 to 1, thence along real axis to O
(*v*) the semicircle $|z| = 1$ clockwise through 1, $-i$, -1, thence along real axis to $-\infty$, and from $+\infty$ to $+1$

10 (*i*) From 0 to 1, thence along parabola $y^2 = 4 - 4x$ to i, thence along $y^2 = 4 + 4x$ to -1, thence to 0
(*ii*) from 1 to e, thence round circle $|z| = e$ to e cis 1, thence along radius to cis 1, and finally along circle $|z| = 1$ to 1
(*iii*) from 1 to cosh 1, thence along ellipse $\dfrac{x^2}{\text{ch}^2\,1} + \dfrac{y^2}{\text{sh}^2\,1} = 1$ to ch $(1 + i)$, thence along hyperbola $\dfrac{x^2}{\cos^2 1} - \dfrac{y^2}{\sin^2 1} = 1$ to $\cos 1$, finally along real axis to 1

Exercise 10.8

1 Imaginary axis with $y > 0$ **2** Anti-clockwise along $|w| = 1$

3 $|z + 4 - 3i| = 5$; $3x - 4y = -24$

4 Coaxal system of Apollonius circles based on the points 1 and $-2i$; the origin side of the line $2x + 4y + 3 = 0$

8 $x =$ constant $\mapsto$ circles centre O, radius e^x
$y =$ constant $\mapsto$ half-lines, $\arg w = y$

10 $2 - i$, $-1 + 2i$

11 Three quarters of the circle $x^2 + y^2 - x + y = 0$, starting at O, passing through $1 - i$, and finishing at 1

12 Confocal hyperbolae and ellipses

Miscellaneous problems 10

1 $a = \pm 2,\ b = \pm 3$; $(z^2 - 4z + 13)(z^2 + 4z + 13)$

3 Rotation through $+\theta$ about c; half-turn about $\sqrt{3}i$

4 (*i*) $2 + 3i,\ \pm 1$
(*ii*) the points of $|z| = 2$ in the third and fourth quadrants

5 (*i*) True (*ii*) false, e.g. $a = 0,\ b = 1,\ c = i$
(*iii*) true, cf. Ex. 10.2a, no. 7

6 (*i*) $r = 1$; $f(z)$ describes line segment $[-1, 1]$
(*iii*) $\theta = 0$, real axis, $x \geqslant 1$
$\theta = \pi$, real axis, $x \leqslant -1$
$\theta = \pm\frac{1}{2}\pi$, imaginary axis

7 If U, P, Q represent 1, z, z^2, and if $\angle \mathrm{UOP} = \theta$, then UP and PQ are inclined to real axis at angles 2θ, 3θ (Δs OUP, OPQ are isosceles)

8 $\cos^2 \dfrac{\pi}{5}$, $\cot^2 \dfrac{2\pi}{5}$ roots of $5z^4 - 10z^2 + 1 = 0$; $2, \dfrac{1}{\sqrt{5}}$

11 $e^{-\frac{1}{2}\pi} \times e^{2\pi k}$

12 $x =$ constant $\mapsto$ hyperbolae
$y =$ constant $\mapsto$ ellipses

13 $\begin{pmatrix} \cos 2\alpha & \sin 2\alpha \\ \sin 2\alpha & -\cos 2\alpha \end{pmatrix}$

Exercise 11.1a

1 (*i*) $\frac{1}{2}(\boldsymbol{a} + \boldsymbol{b})$ (*ii*) $\frac{3}{4}\boldsymbol{a} + \frac{1}{4}\boldsymbol{b}$ (*iii*) $\frac{1}{4}\boldsymbol{a} + \frac{3}{4}\boldsymbol{b}$ (*iv*) $\frac{3}{2}\boldsymbol{a} - \frac{1}{2}\boldsymbol{b}$
(*v*) $-\frac{1}{2}\boldsymbol{a} + \frac{3}{2}\boldsymbol{b}$

Exercise 11.1b

1 (*i*) $\frac{1}{2}(\boldsymbol{b} + \boldsymbol{c})$, $\frac{1}{2}(\boldsymbol{c} + \boldsymbol{a})$, $\frac{1}{2}(\boldsymbol{a} + \boldsymbol{b})$ (*ii*) $\frac{1}{3}(\boldsymbol{a} + \boldsymbol{b} + \boldsymbol{c})$

2 (*i*) $\frac{2}{5}\boldsymbol{b} + \frac{3}{5}\boldsymbol{c}$ (*ii*) $\frac{1}{4}\boldsymbol{a} + \frac{3}{4}\boldsymbol{c}$ (*iii*) $\frac{1}{3}\boldsymbol{a} + \frac{2}{3}\boldsymbol{b}$
(*iv*) (*v*) (*vi*) $\frac{1}{6}\boldsymbol{a} + \frac{1}{3}\boldsymbol{b} + \frac{1}{2}\boldsymbol{c}$

3 (*i*) $-2\boldsymbol{b} + 3\boldsymbol{c}$ (*ii*) $\frac{3}{2}\boldsymbol{c} - \frac{1}{2}\boldsymbol{a}$ (*iii*) $-\boldsymbol{a} + 2\boldsymbol{b}$

4 $\boldsymbol{r} = \frac{1}{2}(1 - t)\boldsymbol{a} + \frac{1}{2}t\boldsymbol{b} + \frac{1}{2}t\boldsymbol{c}$,
$\frac{1}{4}(\boldsymbol{a} + \boldsymbol{b} + \boldsymbol{c})$

5 $\lambda = \frac{4}{5},\ \mu = \frac{2}{5}$

Exercise 11.2a

1 (*i*) $\frac{1}{3}(\boldsymbol{b} + \boldsymbol{c} + \boldsymbol{d})$ (*ii*) $\frac{1}{4}(\boldsymbol{a} + \boldsymbol{b} + \boldsymbol{c} + \boldsymbol{d})$
(*iii*) $\frac{1}{4}(\boldsymbol{a} + \boldsymbol{b} + \boldsymbol{c} + \boldsymbol{d})$

3 (*i*) $\frac{1}{6}\boldsymbol{a} + \frac{1}{3}\boldsymbol{b} + \frac{1}{2}\boldsymbol{c}$ (*ii*) $(\frac{1}{2}, \frac{3}{2}, \frac{7}{5})$ (*iii*) $-0.4\boldsymbol{i} + 2.1\boldsymbol{j} - 0.3\boldsymbol{k}$

4 12 cm 5 $\frac{4}{7}, \frac{26}{7}$

Exercise 11.2b

1 (*i*) $(1\frac{3}{5}, 2\frac{2}{7})$ (*ii*) $(1\frac{17}{28}, 1\frac{23}{70})$ (*iii*) $(5\frac{2}{5}, 0)$ (*iv*) $(\frac{2}{3}h, \frac{1}{3}a)$

2 (*i*) $(1\frac{3}{4}, 0)$ (*ii*) $(1\frac{43}{62}, 0)$ (*iii*) $(6, 0)$ (*iv*) $(\frac{3}{4}h, 0)$

3 $\bar{x} = \frac{3}{8}a$ **5** $\frac{5}{24}\pi a^3, \frac{27}{40}a$ **6** $x = 2.71$

Exercise 11.3b

1 (*i*) 3, 18° 27′ (*ii*) 1, 70° 32′ (*iii*) −4, 161° 31′

2 (*i*) $2\boldsymbol{i} + 2\boldsymbol{k}, 3\boldsymbol{j} + 2\boldsymbol{k}$, 66° 55′
(*ii*) $2\boldsymbol{j} + 2\boldsymbol{k}, -4\boldsymbol{i} + 3\boldsymbol{j}$, 64° 54′
(*iii*) $-2\boldsymbol{i} + 3\boldsymbol{j} - \boldsymbol{k}, -2\boldsymbol{i} - 3\boldsymbol{j} + \boldsymbol{k}$, 115° 23′
(*iv*) $3\boldsymbol{j} - \boldsymbol{k}, \boldsymbol{j} - 2\boldsymbol{k}$, 45°

3 Lines meet at (2, 1, 1), 70° 30′

4 $(\frac{3}{2}, \frac{3}{2})$

5 $\begin{pmatrix} 1 - 2u_1 - 2t_1 \\ u_1 - t_1 \\ -4 + u_1 + t_1 \end{pmatrix}, \; t_1 = u_1 = \frac{3}{5}, \; \frac{7}{\sqrt{5}}$

Exercise 11.4a

1 (*i*) $\boldsymbol{r} = 2\lambda\boldsymbol{i} + \boldsymbol{j} + 2(1 - \lambda)\boldsymbol{k}; \quad x = \frac{y - 1}{0} = \frac{z - 2}{2}$

(*ii*) $\boldsymbol{r} = \lambda\boldsymbol{i} + (1 + \lambda)\boldsymbol{j} + (2 + \lambda)\boldsymbol{k}; \quad x = y - 1 = z - 2$

2 (*i*) $2x + y + 3z = 13$ (*ii*) $x + 2y + 3z = 8$

(*iii*) $\frac{x - 1}{2} = \frac{y - 1}{1} = \frac{z - 2}{1}$

3 (*i*) 61° 53′ (*ii*) 38° 13′

Exercise 11.4b

1 (*i*) $\boldsymbol{i} + 2\boldsymbol{j} + 3\boldsymbol{k}, \boldsymbol{i} - \boldsymbol{j} - \boldsymbol{k}$, 128° 6′ (*ii*) $\boldsymbol{i} - \boldsymbol{j}, \boldsymbol{j} + \boldsymbol{k}$, 120°

2 (*i*) 45° 35′ (*ii*) 38° 34′

3 (*i*) 3, 1 (*ii*) $7x + 56y - 5 = 0$ and $16x - 2y + 35 = 0$

4 (*i*) 6, 3 (*ii*) $3x + 3y + 4z + 6 = 0$ and $x - y - 8 = 0$

5 (*i*) 35° 17′ (*ii*) 54° 43′ (*iii*) $\frac{1}{47}$

6 $\sqrt{5}$ **7** 3

8 $2x + 2y - z = 9$, (4, 1, 1)

Exercise 11.5

1 $\boldsymbol{i} - \boldsymbol{j}, \boldsymbol{i} + \boldsymbol{j}, -\boldsymbol{i} - \boldsymbol{j} + 2\boldsymbol{k}$

2 (*i*) $\boldsymbol{i} - \boldsymbol{j}, \boldsymbol{i} + \boldsymbol{j}, \boldsymbol{i} - \boldsymbol{j} + 2\boldsymbol{k}$ (*ii*) 90° (*iii*) 54° 44′, 90°
(*iv*) 39° 14′, 39° 14′

Exercise 11.6

1 (*i*) $r = e^{\theta}$ (*ii*) $2\,e^{2t}, 2\,e^{2t}; 0, 4\,e^{2t}$

4 $-20t^2, 10; \quad -20t^2, 2k, 10$

Exercise 11.7a

1 (*i*) 3×10^4 N s (*ii*) 10^5 N s (*iii*) 0.75 N s (*iv*) 2.7×10^{-23} N s

2 (*i*) 300 N s (*ii*) 12 N s (*iii*) 80 N s (*iv*) 2 N s

3 (*i*) 2 000 N s (*ii*) 2 000 N s (*iii*) 2 m s^{-1}

4 (*i*) 1 000 N s (*ii*) 1 000 N s (*iii*) 1 m s^{-1}

5 (*i*) 667 N s (*ii*) 667 N s (*iii*) 0.667 m s^{-1}

6 (*i*) −40 000 N s (*ii*) 260 000 N s (*iii*) 26 m s^{-1}

7 (*i*) −20 000 N s (*ii*) 280 000 N s (*iii*) 28 m s^{-1}

8 (*i*) −20 000 N s (*ii*) 280 000 N s (*iii*) 28 m s^{-1}

9 4 N s, 200 N

10 400 N

11 (*i*) $0.0016\pi \approx 0.005$ N s (*ii*) 0.005 N

12 50 N

13 1 500 N

14 50 N, increase

15 7.2×10^6 N s, 7.2×10^6 N s, 7.2 m s^{-1}

Exercise 11.7b

1 (*i*) $12\boldsymbol{i} + 4\boldsymbol{j} + 2\boldsymbol{k}$ (*ii*) $-3\boldsymbol{i} - 9\boldsymbol{j} + 6\boldsymbol{k}$ (*iii*) $2\boldsymbol{i} + 8\boldsymbol{k}$
(*iv*) $\boldsymbol{i} + 1.23\boldsymbol{j}$

2 (*i*) $10\boldsymbol{i} + 5\boldsymbol{j} + 3\boldsymbol{k}$ (*ii*) $\frac{5}{2}\boldsymbol{i} - \frac{3}{2}\boldsymbol{j} + 5\boldsymbol{k}$ (*iii*) $5\boldsymbol{i} + 3\boldsymbol{j} + 4\boldsymbol{k}$
(*iv*) $4.5\boldsymbol{i} + 3.6\boldsymbol{j} + 2\boldsymbol{k}$

3 (*i*) $-2\boldsymbol{i} + 10\boldsymbol{k}$ (*ii*) $-2.4\boldsymbol{i} - 1.6\boldsymbol{j} + 1.6\boldsymbol{k}$

4 $\dfrac{pAr}{36\,bv}$ m, $\dfrac{pArv\rho\sqrt{2}}{36} \times 10^3$ N

Exercise 11.7c

1 (*i*) $2\boldsymbol{i} + 3\boldsymbol{j} + \boldsymbol{k}$ (*ii*) $\frac{5}{3}\boldsymbol{i} + \frac{11}{6}\boldsymbol{j} - \boldsymbol{k}$

2 $\begin{pmatrix} 1.8 \\ 0.8 \end{pmatrix}, \pm \begin{pmatrix} 2.4 \\ 8.4 \end{pmatrix}$

3 (600, 400, 20)

4 1 600 N

5 $\frac{2}{49}$ km s^{-1}, $\frac{9}{245}$ km s^{-1}, 5 s

6 (*i*) 30°, $8v$ (*ii*) $25.2\,mv$

Exercise 11.7d

1 $1\frac{5}{6}, 3\frac{1}{3}$ m s^{-1}

2 3.6 m s^{-1}, 0.4

3 $-\frac{1}{2}, +2$ m s^{-1}

4 1 m s^{-1}, $\frac{1}{8}$

5 10, 7.5 m s^{-1}

8 0.952, 40.7 m

Exercise 11.8a

1 (*i*) 300 J (*ii*) 5.48 m s^{-1}, 4.27 m s^{-1}

2 8 000 N, 32 000 N

3 3.2×10^4 N

4 100 J, 0.064

5 (*i*) 4×10^5 J (*ii*) 5 000 N (*iii*) 10 000 N (*iv*) 20 000 N

6 $\dfrac{mv^2}{2F}$

7 2.4 m s^{-1}; 0.6 J

8 2.1, 2.6 m s^{-1}; 0.45 J

9 1.8, 2.8 m s^{-1}; 0

10 (*i*) 18.2 N (*ii*) 6.6 N

Exercise 11.8b

1 (*i*) 8 N (*ii*) 1 N (*iii*) 0.2 m (*iv*) 0.05 m

2 (*i*) 0.16 J (*ii*) 2.5×10^{-3} J (*iii*) 4 J (*iv*) 0.25 J

3 (*i*) 2×10^3 N m^{-1} (*ii*) 0.004 J

4 (*i*) 5×10^4 N m^{-1} (*ii*) 2.5 J

5 (*i*) 4 m (*ii*) 400 J (*iii*) 800 J

6 (*i*) $200(2 + x)$ (*ii*) $25x^2$ (*iii*) $400 + 200x - 25x^2$, 9.7 m

7 4.06×10^4 J, 15 m, 32.8 m s^{-1}, 53.9 m s^{-2}

8 (*i*) gR (*ii*) $\frac{1}{2}gR$

9 $\sqrt{2gR}$ ($\approx$ 11.3 km s^{-1}), $\sqrt{gR}$ ($\approx$ 8 km s^{-1})

Exercise 11.8c

1 (*i*) 8 J (*ii*) −4 J (*iii*) 0

2 (*i*) 3.6×10^5 J (*ii*) 45.9 m

3 (*i*) 9.9 m s^{-1} (*ii*) 10.8 m s^{-1}

4 (*i*) 10 m s^{-1} (*ii*) 500 J (*iii*) 5.1 m

5 (*i*) 3.13 m s^{-1}, 0.39 N (*ii*) 4.43 m s^{-1}, 0.98 N
(*iii*) 3.83 m s^{-1}, 0.73 N

6 (*i*) 784 J (*ii*) 6.26 m s^{-1} (*iii*) 1 180 N

7 11 m s^{-1}

8 (*i*) $v^2 = 2\,ag(1 - \cos\theta)$ (*ii*) $mg(3\cos\theta - 2)$ (*iii*) $\cos^{-1}\frac{2}{3} \approx 48°$

Exercise 11.8d

1 2.26 kW

2 9.5 kW

3 0.039 m s^{-2}

4 (*i*) 1.47 m s^{-2} (*ii*) 1.47 m s^{-2}

5 4.9×10^4 W

6 2 160 N, 1 700 N, 80 m

7 25 N s, 25 N, 0.85 m s^{-1}

8 $\dfrac{Mg}{v}$ kg s^{-1}, $\frac{1}{2}\,Mgv$ W

9 (*i*) $a = \dfrac{P}{mv}$ (*ii*) $v = \sqrt{\dfrac{2Pt}{m}}$ (*iii*) $v = \sqrt{\dfrac{3Px}{m}}$ (*iv*) $x = \dfrac{2}{3}\sqrt{\dfrac{2Pt^3}{m}}$

10 $m\dot{v} + kv^2 = F\ (v \leqslant U),\ m\dot{v} = \dfrac{Fu}{v} - kv^2\ (v > U)$;

$\sqrt{\dfrac{F}{k}}\ (F < kU^2),\ \sqrt[3]{\dfrac{FU}{k}}\ (F \leqslant kU^2)$; $k = 0.5$;

4.98, 4.78, 4.38, 2.35 m s^{-2}; 10.6 s

Exercise 11.9

1 (*i*) $\ddot{x} = -4x$ (*ii*) 3.14 s (*iii*) $x = 3\cos 2t$ (*iv*) 6 m s^{-1}

3 (*i*) $\frac{3}{2}$ (*ii*) 6.28 s (*iii*) $\frac{1}{3}$ (*iv*) 0.13

4 (*i*) 0.63 s (*ii*) 0.68 s

7 $(\frac{1}{2}\pi + \frac{1}{2})\sqrt{\dfrac{a}{2g}}$

8 $mg(1 - \cos nt)$

9 $\dfrac{v}{\sqrt{(gl)}},\ \frac{1}{4}\pi\sqrt{\dfrac{l}{g}}$

10 $\ddot{\theta} = -\dfrac{4T}{ml}\theta,\ \pi\sqrt{\dfrac{ml}{T}}$

Exercise 11.10a

1 (*i*) $\boldsymbol{i} + \boldsymbol{k}$ (*ii*) $-5\boldsymbol{i} + 3\boldsymbol{j} - \boldsymbol{k}$ (*iii*) $2\boldsymbol{i} - \boldsymbol{j}$
(*iv*) $-11\boldsymbol{i} + 9\boldsymbol{j} - 5\boldsymbol{k}$ (*v*) $-4\boldsymbol{i} - 3\boldsymbol{j} + \boldsymbol{k}$ (*vi*) $-12\boldsymbol{i} - 6\boldsymbol{j} - 12\boldsymbol{k}$

3 (*i*) Force $6\boldsymbol{i} + 3\boldsymbol{j}$ along $3x - 6y + 4 = 0$ (*ii*) Couple $4\boldsymbol{k}$
(*iii*) Equilibrium (*iv*) Force $2\boldsymbol{i} + 15\boldsymbol{j}$ along $15x - 2y - 2 = 0$

4 $(30P, 30P)$ along $x - y = \frac{1}{3}a$; force $(30P, 30P)$, couple $10aP$

Exercise 11.10b

1 700 N, 800 N; 75 kg

2 (*i*) 0.213 N (*ii*) 0.106 N

3 $\frac{1}{3}W(4x - y + 6),\ \frac{1}{3}W(-x + 4y + 6)$

4 $11\frac{1}{4}$ kg, $8\frac{1}{4}$ kg **5** $\frac{2}{3}W\cot\beta$ **6** 40°

8 (*i*) $W\tan 15°$, $\tan 15° \approx 0.27$ (*ii*) $\frac{1}{3}W$, $\tan 30° \approx 0.58$
(*iii*) $\frac{1}{4}W,\ \dfrac{1}{2\sqrt{3}} \approx 0.29$

10 $\frac{5}{9}l$ from A, $\dfrac{5\,mgl^2}{6a},\ \dfrac{mgl(18a - 5l)}{12a}$

11 $\dfrac{2d - l\cos\theta}{l\sin\theta}$

12 (*i*) $2\sqrt{13}P$, $6x - 8y = 5a$ (*ii*) Pa (*iii*) $6Pa$

Exercise 11.11a

2 $\frac{5}{3}\boldsymbol{i} + \boldsymbol{j}$, towards O

Exercise 11.11b

1 $\frac{4}{3}M(a^2 + b^2)$

2 $\frac{1}{2}Ma^2$, $\frac{3}{2}Ma^2$, $2Ma^2$

3 $\frac{5}{4}Ma^2$, $\frac{3}{2}Ma^2$

4 $\frac{2}{5}Ma^2$, $\frac{7}{5}Ma^2$

5 $\frac{2}{3}Ma^2$, $\frac{5}{3}Ma^2$

6 (*i*) $\frac{3}{10}Ma^2$ (*ii*) $\dfrac{3M}{20}(a^2 + 4h^2)$ (*iii*) $\dfrac{M}{20}(3a^2 + 2h^2)$

Exercise 11.11c

1 (*i*) 12 kg m^2 s^{-1}; 36 J
(*ii*) 0.48 kg m^2 s^{-1}; 0.96 J
(*iii*) 0.24 kg m^2 s^{-1}; 0.48 J
(*iv*) 96 kg m^2 s^{-1}; 288 J
(*v*) 168 kg m^2 s^{-1}; 877 J
(*vi*) 7.15×10^{33} kg m^2 s^{-1}; 2.60×10^{29} J

2 (*i*) $2F/ma$; (*ii*) $ma\omega/2F$; (*iii*) $ma\omega^2/4F$

3 5.82 N, 1.8 s, 58.2 N

4 62.8 rad s^{-2}, 126 rad s^{-1}, 78.5 N

5 3.65 kg m^2

6 (*i*) mga, $\sqrt{3g/2a}$, $3g/4a$; (*ii*) $2\,mga$, $\sqrt{3g/a}$, 0

7 38.7 rad s^{-1}

8 (*i*) $\frac{1}{2}mgl\sin\theta$; (*ii*) $\ddot{\theta} = -\dfrac{3g}{2l}\sin\theta$; (*iii*) $\ddot{\theta} = -\dfrac{3g\theta}{2l}$;
(*iv*) $2\pi\sqrt{2l/3g}$

9 (*i*) $2\sqrt{\dfrac{Mgx}{2M + m}}$; (*ii*) $\dfrac{2Mg}{2M + m}$

10 $2\pi\sqrt{5a/3g}$

Exercise 11.12

1 (*i*) $\mathbf{ML^{-1}T^{-2}}$ (*ii*) $\mathbf{ML^{-1}}$ (*iii*) $\mathbf{ML^{-2}}$ (*iv*) $\mathbf{T^{-1}}$ (*v*) $\mathbf{T^{-1}}$
(*vi*) $\mathbf{T^{-2}}$ (*vii*) 1 (*viii*) $\mathbf{ML^2}$ (*ix*) $\mathbf{MT^{-1}}$ (*x*) $\mathbf{M^{-1}L^3T^{-2}}$
(*xi*) $\mathbf{MT^{-2}}$ (*xii*) $\mathbf{MT^{-2}}$

2 (*i*) $T \propto \dfrac{mv^2}{r}$ (*ii*) $h \propto \dfrac{E}{mg}$ (*iii*) $v \propto \sqrt{\dfrac{p}{\rho}}$ (*iv*) $f \propto \sqrt{\dfrac{k}{m}}$
(*v*) $v \propto \sqrt{g\lambda}$

3 $T = kv^2\rho A$; $\mathbf{M^{-1}LT^2}$

Miscellaneous problems 11

7 Distance $\dfrac{3(a^2 - 2l^2)}{4(2a + 3l)}$ from common base

9 $\dfrac{\cos^2 \beta + \cos^2 \gamma - 2 \cos \alpha \cos \beta \cos \gamma}{\sin^2 \alpha}$

10 $\dfrac{d\boldsymbol{v}}{dt} = -c\boldsymbol{v} + \boldsymbol{g}$

$$\boldsymbol{r} = \frac{1}{c}(ct + e^{-ct} - 1)\boldsymbol{g} + \frac{1}{c}(1 - e^{-ct})\boldsymbol{u}$$

11 $\boldsymbol{A} = 18\boldsymbol{i} + 3\boldsymbol{j}$; $\boldsymbol{B} = -2\boldsymbol{i} - 3\boldsymbol{j}$

12 $y = \left(\dfrac{V \sin \alpha}{k} + \dfrac{g}{k^2}\right)(1 - e^{-kt}) - \dfrac{gt}{k}$

13 $A\rho(1 - k)u^2$; 4.5 m s^{-1}

15 3 h 35 min, 6.19×10^6 N s, 1.03×10^3 N

16 $\pi\sqrt{\dfrac{a}{g}}$

17 $l + \dfrac{mgl}{\lambda}\left[1 + \sqrt{\left(1 + \dfrac{2\lambda}{mg}\right)}\right]$;

$$\sqrt{\frac{2l}{g}} + \sqrt{\frac{ml}{\lambda}}\left[\tfrac{1}{2}\pi + \sin^{-1}\frac{1}{\sqrt{(1 + 2\lambda/mg)}}\right]$$

18 G moves with constant velocity $\dfrac{P}{3m}$; $\dfrac{P^2}{3m}$

19 $P = \frac{1}{2}Av(v^2 + 2gh) \times 10^{-4}$

$$\ddot{\theta} = \frac{Av^2c}{10I};\quad 0$$

20 $\mathbf{M}^{-1}\mathbf{L}^0\mathbf{T}$; $\dfrac{dV}{dt} = -\lambda A\rho g V$

Exercise 12.1a

1 1.5, 0.75
2 (*i*) 2.92 (*ii*) 5.83 (*iii*) 1.46
3 4.5, 8.25
4 5.96, 1.52
5 $4\frac{17}{36}$
6 1.94, 1.44
7 1.73, 3.13
8 5.58, 0.60
9 3.61, 6.69
10 £2.78
11 29.75, 230; 0.5
12 $p(T = 30) = p, p(T = 45) = 1 - p$; $45 - 15p, 225p(1 - p)$
$p(T = 30) = \frac{1}{2}p, p(T = 35) = \frac{1}{2}, p(T = 45) = \frac{1}{2}(1 - p)$;
$40 - \frac{15}{2}p$

Exercise 12.1b

1 $\dfrac{t}{2-t}$; $2, 2$

2 (*i*) 2, 1 (*ii*) 18, 3 (*iii*) 50, 5

3 $36, \dfrac{48}{\sqrt{n}}$

4 $\dfrac{3}{2}, \dfrac{3\sqrt{2}}{4}$

5 0.59; 7, 6.3; 8.26

6 4.49, 2.47

7 (*i*) 0.96, 0.88 (*ii*) 1.58

9 8, 7.92; approximation np; 1.01%

10 (*i*) $\frac{8}{27}$ (*ii*) $\frac{11}{27}$; $\frac{20}{3}, \frac{40}{9}$

11 $(\frac{1}{6}t + \frac{1}{3}t^2 + \frac{1}{2}t^3)^5$; $-\frac{1}{243}, \frac{121}{243}$

12 $\dfrac{t^2}{(6-5t)^2}$; 12, 60

13 (*i*) $(\frac{1}{5}t^4 + \frac{4}{5}t^{-1})^{25}$, 0, 10 (*ii*) 16.67, 11.78

Exercise 12.2a

1 $a = 2$; $\dfrac{x^2}{8} + \dfrac{x}{4}$; (*i*) $\frac{5}{32}$ (*ii*) $\frac{5}{8}$

2 $X = \sqrt[3]{2}$; (*i*) 0.001 (*ii*) $\frac{7}{64}$

3 (*i*) $\frac{1}{4}$ (*ii*) $\frac{2}{3}$;

$$\Phi(x) = \begin{cases} 0 & x < 0 \\ \dfrac{x^2}{12} & 0 \leqslant x \leqslant 3 \\ 2x - \dfrac{x^2}{4} - 3 & 3 \leqslant x \leqslant 4 \\ 1 & 4 < x \end{cases}$$

$\sqrt{6}$

4 $k = \frac{3}{4}$; (*i*) $\frac{11}{16}$ (*ii*) $\frac{7}{27}$; $\Phi(x) = \begin{cases} 0 & x \leqslant -1 \\ \frac{1}{4}(2 + 3x - x^3) & -1 \leqslant x \leqslant 1 \\ 1 & 1 \leqslant x \end{cases}$

5 $a = \dfrac{\pi}{2}$; $\Phi(x) = \begin{cases} 0 & x < 0 \\ \frac{1}{2}(1 - \cos \pi x) & 0 \leqslant x \leqslant 1; \\ 1 & 1 < x \end{cases}$ (*i*) $\frac{1}{4}$ (*ii*) $\frac{1}{4}$

6 $\phi(t) = \dfrac{1}{2\,000} e^{-t/2000}$; (*i*) 0.135 (*ii*) 0.145 (*iii*) 0.018

Exercise 12.2b

1 $\frac{7}{6}, \frac{11}{36}$

2 $\frac{3}{2}, \frac{3}{4}$

3 $\frac{7}{3}, \frac{13}{18}$

4 $0, \frac{1}{5}$

5 $\dfrac{1}{2}, \dfrac{1}{4} - \dfrac{2}{\pi^2}$

6 $\phi(x) = \dfrac{1}{b-a}$ for $a \leqslant x \leqslant b$, and zero elsewhere; $\dfrac{a+b}{2}, \dfrac{(b-a)^2}{12}$

7 $\alpha = \frac{3}{32}$; $2, \frac{4}{5}$; $\frac{5}{32}$

8 $\frac{3}{2}, \sqrt[3]{4}, \frac{3}{20}$

9 $2 + \dfrac{1}{\lambda}, 2 + \dfrac{\ln 2}{\lambda}, \dfrac{1}{\lambda}$

10 $\phi(t) = \dfrac{1}{\mu} e^{-t/\mu}$; 0.135; 5 000

11 $C = \dfrac{1}{2a}$; $1.53a$; (*i*) 0.53 (*ii*) 1; $k = 1.28$

12 (*i*) $k = \dfrac{1}{\pi}$ (*ii*) $\frac{1}{2}$ (*ii*) $\frac{1}{3}$

13 λ, λ

Exercise 12.3a

1 (*i*) 16 (*ii*) 92 (*iii*) 33 (*iv*) 23
2 (*i*) 6.7% (*ii*) 10.6% (*iii*) 0.6%
3 (*i*) 0.319 (*ii*) 0.145 (*iii*) 0.030 (*iv*) 0.537
4 0.753   5 (*i*) 6 (*ii*) 210 (*iii*) 34
6 66.77, 2.28; (*i*) 7.9% (*ii*) 66.2 cm
7 3.63, 1.71; 2 (since x takes only integer values)
8 (*i*) 0.675 (*ii*) 1.65
9 (*i*) 9.1% (*ii*) 37% (*iii*) 530.8
10 63 g
11 (*i*) 54.6 min (*ii*) 44.7 and 49.3 min (*iii*) 39.4 and 44.7 min
12 (*i*) 5.1% (*ii*) 103.26 mm (*iii*) less than 1% exceed 107 mm
13 1.004 cm, 0.020 cm
14 14.6, 5.2 months
15 74.35, 3.23
16 7.52 a.m.
17 (*i*) 8.48 a.m. (*ii*) 5 min, 118, 4 entrances
18 7.9%, 15.5%; 24.45 mm, 6.1%

Exercise 12.3b

1 (*i*) 0.018 (*ii*) 0.729 (*iii*) 0.067
2 (*i*) effectively zero (*ii*) 0.999 (*iii*) 0.89
3 (*i*) 0.052 (*ii*) 0.198   4 (*i*) 0.131 (*ii*) 0.063
5 (*i*) 0.222 (*ii*) 0.018 (*iii*) effectively zero
6 (*i*) 0.827 (*ii*) 0.062 (*iii*) 0.021
7 (*i*) 0.228 (*ii*) 0.051

8 (*i*) 7.8% (*ii*) 95.8% (*iii*) 0.8%

9 $n = 610$; 0.09

Exercise 12.3c

2 (*i*) 0.5, 0.1 (*ii*) 0.5, 0.05

3 1, 1; 1, $\frac{1}{\sqrt{10}}$

4 2.5, 0.112; 0.037

Exercise 12.4

1 (*i*) 0.368 (*ii*) 0.368 (*iii*) 0.080

2 (*i*) 0.330 (*ii*) 0.865

3 (*i*) 0.018 (*ii*) 0.195 (*iii*) 0.156 (*iv*) 0.0003

4 (*i*) 0.165 (*ii*) 0.269 (*iii*) 0.373

5 (*i*) 0.003 (*ii*) 0.037

6 52

7 228, 211, 98, 30, 7, 1, 0 (to the nearest integers)

8 Theoretical frequencies: 3, 9, 16, 17, 14, 9, 5, 2.5, 1, 0.4, 0—close correspondence with observed frequencies, suggesting lucky and unlucky tenants

9 2, 1.97; so $\mu \approx \sigma^2$, as in case of Poisson distribution

10 50

11 1.05; 0.350, 0.367, 0.193, 0.0675, 0.0177; 105; 27

13 (*i*) 0.997 5 (*ii*) 0.849

14 $1 - e^{-2\mu}$; $\frac{1}{1 + e^{\mu}}$

15 20, 60, 90, 90, 67, 40, 33; (*i*) £157.50 (*ii*) £167.50 (*iii*) £160.75

16 (*i*) 0.406 (*ii*) 0.184 (*iii*) 0.37

Exercise 12.5a

2 (*i*) $aE[x] + bE[y]$, $a^2V[x] + b^2V[y]$
(*ii*) $aE[x] + bE[y]$, $a^2V[x] + 2ab \operatorname{Cov}[x, y] + b^2V[y]$

3 (*i*) $aE[x] + bE[y] + cE[z]$ (*ii*) $a^2V[x] + b^2V[y] + c^2V[z]$

4 10.5, 8.75 5 10.5, 14.58 6 0, 2.42

7 Bob has the advantage; Alan's variance = 11.67,
Bob's variance = 11.25

8 12.25

9 (*i*) $6\frac{2}{3}$ (*ii*) $13\frac{1}{3}$ (*iii*) $26\frac{2}{3}$ (*iv*) $33\frac{1}{3}$

10 60 minutes, 3.16 minutes; (*i*) 0.005 7 (*ii*) 0.057 (*iii*) 0.565

11 (*i*) 0.168 (*ii*) 0.027

12 $\sum_i \lambda_i \mu_i, \sqrt{(\sum_i \lambda_i^2 \sigma_i^2)}$; 150 s, 8.124 s (*i*) 0.7% (*ii*) 11%

13 0.638

14 $V[z] = [(\lambda_1 a_1 + \lambda_2 b_1)^2 + (\lambda_1 a_2 + \lambda_2 b_2)^2]\sigma^2$

Exercise 12.5b

1 (*i*) 0.057 (*ii*) 0.57

2 (*i*) 0.072 (*ii*) 0.165

3 0.048

4 (*i*) 0.115 (*ii*) 0.008

5 (*i*) effectively zero (*ii*) 0.998 7

6 0.001, suspicions justified

7 136

8 96

Exercise 12.5c

1 58.5, 31.7

2 5.08, 0.55

3 (*i*) 1.79 (*ii*) 1.64 (*iii*) 1.61

4 (*ii*) $\sum_1^n \lambda_i = 1$

5 $V[\bar{x}] = \frac{\sigma^2}{n}$; $E[s^2] = \sigma^2$; $\lambda = \frac{n-1}{m+n-2}$

Exercise 12.6a

1 Significant at the 1% level

2 Significant at the 5% level

3 (*i*) Significant (*ii*) Not significant (*iii*) Significant
(*iv*) Not significant evidence against the hypothesis

4 Significant at 5% level that techniques are effective

5 Significant at 1% level that support has decreased

6 (*i*) Not a significant result
(*ii*) Significant at 1% level that claim is not justified

7 (*i*) Significant (*ii*) Not significant evidence against the hypothesis

8 Significant at 5% level that cards were not drawn at random

9 Consistent with hypothesis; significant at 0.1% level that hypothesis is incorrect

10 (*i*) 0.000 4 (*ii*) 0.043; not significant at 5% level that germination rate is less than 75%

11 Significant at 1% level

12 Highly significant, almost conclusive evidence

13 $(1 - X)^n$; (*i*) 0.107 (*ii*) 0.028 (*iii*) 0.001
Significant at 0.1% level that hypothesis is incorrect

14 0.109; (*i*) 527 (*ii*) 537

15 475

Exercise 12.6b

1 (*i*) 58.8, 67.2 (*ii*) 58.0, 68.0 (*iii*) 56.5, 69.5
2 (*i*) 1 150.1, 1 153.9 (*ii*) 1 149.6, 1 154.4
3 (*i*) 154.6, 163.6 (*ii*) 153.2, 165.0
4 (*i*) 7.35, 7.49 (*ii*) 7.32, 7.52
5 (*i*) 54.1, 60.7 (*ii*) 53.05, 61.75
6 Highly significant that there is bias; 0.2 to 0.6
7 1.996 5, 0.064 5; 0.006 5; 1.98 to 2.01
8 No; yes
9 $x_1, x_1 \pm 0.776$; $\bar{x}, \bar{x} \pm 0.163$

Exercise 12.7a

2 $y = 0.58x + 8.4$
3 $W = 0.36T + 40.4$; $W = 58$
4 $y = 1.4x - 26.3$
5 (*i*) $y = 0.28x + 69.8$ (*ii*) $y = 3.6x - 36.4$
6 $y = 33.6x + 8.1$
7 $y = 1.24x - 24.4$
8 $F = 0.9I - 6.3$; 9
9 $y = -1.28x + 20$; 16.2
10 (*i*) $y = 0.8x + 1.5$ (*ii*) $y = 1.9x - 3.1$
11 19.14, 15.03; $y = 0.33x + 8.8$
12 $\log_{10} Y = 0.021X - 37.2$
13 (*i*) 15.2, 1.72 (*ii*) 149 (*iii*) 2.3 (*iv*) $a = 1.5$, $b = 0.015$

Exercise 12.7b

2 0.88 just significant at the 0.1% level
3 0.99 highly significant
4 0.92 highly significant
5 0.28 not significant
6 0.72 just significant at the 1% level
7 0.23 no significant correlation
8 0.35
9 correlation coefficient of −0.84 significant at the 1% level
10 0.65 11 0.43 12 0.65
13 (*i*) 0.61 significant at 5% level (*ii*) 0.45 not significant
(*iii*) 0.93 significant at 0.1% level

Exercise 12.7c

1 $\rho = 0.66, \tau = 0.47$ not significant

2 $\rho = -0.08, \tau = -0.02$ tastes quite dissimilar

3 (*i*) 0.79 (*ii*) 0.61; both significant at the 1% level

4 0.90 5 0.22

6 $\rho = 0.79$ significant at the 1% level

7 (*i*) $\rho = 0.92$ (*ii*) $\tau = 0.78$

8 (*i*) $\tau = -0.71$ (*ii*) not quite significant

9 ACBDE, ABDCE, ABCED; 120; $\frac{1}{24}$

10 $\rho = 0.74, \tau = 0.62$ significant at the 1% level

11 $\tau = 0.67$ significant at the 1% level

Miscellaneous problems 12

1 $\binom{9}{3}\left(\frac{1}{6}\right)^4\left(\frac{5}{6}\right)^6$; $\binom{r-1}{k-1}p^k q^{r-k}$

2 $\dfrac{4(4t)^9 e^{-4t}}{9!}$; $\dfrac{5}{8}$; $\dfrac{\lambda(\lambda t)^{k-1} e^{-\lambda t}}{(k-1)!}$

3 (*i*) $\dfrac{(1-k)(1+t)}{1+(1-k)t}$ (*ii*) $\dfrac{(1-k)(1+2t)}{1+(2-k)t}$

4 (*i*) $\frac{1}{4}$ (*ii*) $3\frac{3}{4}$ min

5 (*i*) 68 (*ii*) 404 (*iii*) 471

6 $\binom{n}{r}p^r(1-p)^{n-r}\dfrac{\lambda^n}{n!}e^{-\lambda}$; $\dfrac{(\lambda p)^r e^{-\lambda p}}{r!}$

7 36, 180

8 $50m + 16.3\sigma$

9 (*i*) $\frac{1}{3}$ (*ii*) $\frac{19}{27}$

10 (*i*) $-np(1-p)$ (*ii*) $\lambda = 1.96\sqrt{n}$ (*iii*) 0.077

11 (*i*) $p(\boldsymbol{A}_1) + p(\boldsymbol{A}_2) + p(\boldsymbol{A}_3) - p(\boldsymbol{A}_1\boldsymbol{A}_2) - p(\boldsymbol{A}_2\boldsymbol{A}_3)$
$- p(\boldsymbol{A}_3\boldsymbol{A}_1) + p(\boldsymbol{A}_1\boldsymbol{A}_2\boldsymbol{A}_3)$
$\sum_i p(\boldsymbol{A}_i) - \sum\sum_{i \neq j} p(\boldsymbol{A}_i\boldsymbol{A}_j) + \cdots + (-1)^{n-1} p(\boldsymbol{A}_1\boldsymbol{A}_2\ldots\boldsymbol{A}_n)$

(*ii*) $1 - \dfrac{1}{1!} + \dfrac{1}{2!} - \dfrac{1}{3!} + \cdots + (-1)^n\dfrac{1}{n!}$, which tends to e^{-1}

12 $p(t+\delta t) = (1-\beta\delta t)p(t) + \alpha\delta t(1-p(t))$

13 (*i*) $p_n = \frac{1}{2}$ (*ii*) $p_n = \dfrac{1}{2}\left[1 + \left(\dfrac{a-b}{a+b}\right)^{n-1}\right]$

14 A, 0.621

15 $p_r = \binom{2N-r-1}{N-1}\dfrac{1}{2^{2N-r}}$; $\binom{2N-2}{N-1}\dfrac{2N-1}{2^{2N-1}}$

Exercise 13.1a

1 $\begin{pmatrix} 121 & 46 & 33 \\ 66 & 41 & 12 \\ 59 & 51 & 4 \\ 108 & 45 & 12 \end{pmatrix}$ **2** $\begin{pmatrix} 103 & 15 & 205 & 17 \\ 0 & 0 & 0 & 0 \\ 50 & 4 & 88 & 4 \\ 0 & 0 & 0 & 0 \\ 11 & 4 & 37 & 5 \end{pmatrix}$

3 $\begin{pmatrix} 1\,041 & 22 & 19 & 26 & 265 \\ 1\,061 & 22 & 21 & 18 & 010 \\ 987 & 18 & 35 & 18 & 090 \end{pmatrix}$ **5** $\begin{pmatrix} 1 & -1 & 8 \\ 3 & -1 & -5 \end{pmatrix}$

Exercise 13.1b

1 $\begin{pmatrix} 24 & 4 & 1\,050 & 3 & 20 & 51.0 \\ 32 & 0 & 1\,507 & 5 & 32 & 47.1 \\ 28 & 9 & 967 & 5 & 19 & 50.9 \end{pmatrix}$ columns 1, 2, 5 are linearly dependent

2 $\boldsymbol{b} = 3\boldsymbol{a}$; $\boldsymbol{d} = 5\boldsymbol{c}$; $\boldsymbol{f} = 2\boldsymbol{e}$; $\boldsymbol{g} = \boldsymbol{b} + \boldsymbol{d} + \boldsymbol{f} = 3\boldsymbol{a} + 5\boldsymbol{c} + 2\boldsymbol{e}$, etc. $\boldsymbol{h}$ is independent of the others, including $\boldsymbol{g}$

3 $\boldsymbol{b} = 3\boldsymbol{a}$; $\boldsymbol{c} = -2\boldsymbol{d}$; $3\boldsymbol{a} + 3\boldsymbol{d} - \boldsymbol{f} = \boldsymbol{0}$; $\boldsymbol{a} - 2\boldsymbol{d} - \boldsymbol{e} = \boldsymbol{0}$, etc.

4 col. 2 = −3 col. 3; replace 5 by −4

5 row 1 − row 2 − row 3 = 0; 2 col. 1 + col. 2 − 7 col. 3 = 0; col. 2 − 7 col. 3 − col. 4 = 0, etc.

6 $x = -30$; yes: 8 col. 1 + 6 col. 2 − 47 col. 3 = 0

7 $H_2O + C_{12}H_{22}O_{11}$ contains same as $6C_2H_4O_2$

Exercise 13.2a

1 (*i*) $\begin{pmatrix} -1 & 0 \\ 0 & -1 \end{pmatrix}$, $\begin{pmatrix} 0 & -1 \\ 1 & 0 \end{pmatrix}$, $\begin{pmatrix} 0 & 1 \\ -1 & 0 \end{pmatrix}$,

$\begin{pmatrix} \frac{1}{2} & -\frac{\sqrt{3}}{2} \\ \frac{\sqrt{3}}{2} & \frac{1}{2} \end{pmatrix}$, $\begin{pmatrix} \frac{1}{2} & \frac{\sqrt{3}}{2} \\ -\frac{\sqrt{3}}{2} & \frac{1}{2} \end{pmatrix}$, $\begin{pmatrix} -\frac{1}{2} & \frac{\sqrt{3}}{2} \\ -\frac{\sqrt{3}}{2} & -\frac{1}{2} \end{pmatrix}$,

$\begin{pmatrix} -\frac{1}{2} & -\frac{\sqrt{3}}{2} \\ \frac{\sqrt{3}}{2} & -\frac{1}{2} \end{pmatrix}$, $\begin{pmatrix} -\frac{1}{\sqrt{2}} & -\frac{1}{\sqrt{2}} \\ \frac{1}{\sqrt{2}} & -\frac{1}{\sqrt{2}} \end{pmatrix}$, $\begin{pmatrix} -\frac{1}{\sqrt{2}} & \frac{1}{\sqrt{2}} \\ -\frac{1}{\sqrt{2}} & -\frac{1}{\sqrt{2}} \end{pmatrix}$,

$\begin{pmatrix} 0.8 & -0.6 \\ 0.6 & 0.8 \end{pmatrix}$

(*ii*) $\begin{pmatrix}1 & 0\\0 & -1\end{pmatrix}$, $\begin{pmatrix}-1 & 0\\0 & 1\end{pmatrix}$, $\begin{pmatrix}0 & 1\\1 & 0\end{pmatrix}$, $\begin{pmatrix}0 & -1\\-1 & 0\end{pmatrix}$, $\begin{pmatrix}-0.6 & 0.8\\0.8 & 0.6\end{pmatrix}$
$\begin{pmatrix}3 & 0\\0 & 3\end{pmatrix}$, $\begin{pmatrix}2 & 0\\0 & 1\end{pmatrix}$, $\begin{pmatrix}1 & 0\\0 & 3\end{pmatrix}$, $\begin{pmatrix}1 & 0\\-1 & 1\end{pmatrix}$

2 x and y stretches; enlargement and half-turn; enlargement $\times 5$ and rotation; x-shear; y-shear; reflection in $y = -x$ and enlargement; $\begin{pmatrix}2 & -1\\1 & 2\end{pmatrix} = \begin{pmatrix}1 & -\frac{1}{2}\\0 & 1\end{pmatrix}\begin{pmatrix}2 & 0\\0 & 2\end{pmatrix}$ (x-shear and enlargement); reflection in $y = x \tan \frac{1}{2}\alpha$; reflection in line $y = x \tan (45° + \frac{1}{2}\alpha)$; rotation $+40°$; all points map into line $y = \frac{1}{2}x$; $(x, y) \mapsto (x, x)$ lying on $y = x$

3 Reflections in $y = 0$, $x = 0$, $y = \pm x$; rotation through $\frac{1}{2}n\pi$, $(n \in Z)$

7 $\begin{pmatrix}2 & 3\\-1 & -2\end{pmatrix}\begin{pmatrix}2 & -1 & 0\\1 & 2 & 3\end{pmatrix} = \begin{pmatrix}7 & 4 & 9\\-4 & -3 & -6\end{pmatrix}$;
$\begin{pmatrix}3 & 1\\2 & 0\end{pmatrix}\begin{pmatrix}2 & -1 & 0\\1 & 2 & 3\end{pmatrix} = \begin{pmatrix}7 & -1 & 3\\4 & -2 & 0\end{pmatrix}$;
$\begin{pmatrix}4 & -6\\-2 & 3\end{pmatrix}\begin{pmatrix}2 & -1 & 0\\1 & 2 & 3\end{pmatrix} = \begin{pmatrix}2 & -16 & -18\\1 & 8 & 9\end{pmatrix}$,
all lying on line $2y + x = 0$

8 $\begin{pmatrix}1 & 0\\1 & 1\end{pmatrix}\begin{pmatrix}0 & 0 & 1 & 1\\-1 & 1 & 1 & -1\end{pmatrix} = \begin{pmatrix}0 & 0 & 1 & 1\\-1 & 1 & 2 & 0\end{pmatrix}$ (parallelogram)

9 $\begin{pmatrix}-2 & 2\\2 & 1\end{pmatrix}\begin{pmatrix}3\\-2\end{pmatrix} = \begin{pmatrix}-10\\4\end{pmatrix}$

10 $M^2 = \begin{pmatrix}7 & -6\\-18 & 19\end{pmatrix}$ vectors in direction $(1, 1)$ unchanged; vectors in direction $(1, -3) \times 25$

11 $\begin{pmatrix}\frac{1}{2} & \frac{1}{2}\\-\frac{1}{2} & 1\frac{1}{2}\end{pmatrix}$ shear in line $y = x$

Exercise 13.2b

1 (*i*) $\begin{pmatrix}1 & 0 & 0\\0 & -1 & 0\\0 & 0 & 1\end{pmatrix}$ (*ii*) $\begin{pmatrix}0 & 1 & 0\\1 & 0 & 0\\0 & 0 & 1\end{pmatrix}$ (*iii*) $\begin{pmatrix}\cos 2\alpha & \sin 2\alpha & 0\\\sin 2\alpha & -\cos 2\alpha & 0\\0 & 0 & 1\end{pmatrix}$

(*iv*) $\frac{1}{3}\begin{pmatrix}1 & -2 & -2\\-2 & 1 & -2\\-2 & -2 & 1\end{pmatrix}$ (*v*) $\begin{pmatrix}1 & 0 & 0\\0 & -1 & 0\\0 & 0 & -1\end{pmatrix}$ (*vi*) $\begin{pmatrix}0 & 1 & 0\\1 & 0 & 0\\0 & 0 & -1\end{pmatrix}$

(*vii*) $\frac{1}{3}\begin{pmatrix}-1 & 2 & 2\\2 & -1 & 2\\2 & 2 & -1\end{pmatrix}$ (*viii*) $\begin{pmatrix}0 & \pm 1 & 0\\\pm 1 & 0 & 0\\0 & 0 & 1\end{pmatrix}$

(*ix*) $\frac{1}{2}\begin{pmatrix}0 & -\sqrt{2} & -\sqrt{2}\\\sqrt{2} & 1 & -1\\\sqrt{2} & -1 & 1\end{pmatrix}$ and inverse (*x*) $\begin{pmatrix}\cos\theta & 0 & \sin\theta\\0 & 1 & 0\\-\sin\theta & 0 & \cos\theta\end{pmatrix}$

$(xi)\ \begin{pmatrix} 5 & 0 & 0 \\ 0 & 5 & 0 \\ 0 & 0 & 5 \end{pmatrix}$ $(xii)\ \begin{pmatrix} 1 & 0 & 0 \\ 0 & 0 & 0 \\ 0 & 0 & 1 \end{pmatrix}$

2 (*i*) Reflection in $y = 0$ (*ii*) orthogonal projection on to $z = 0$
(*iii*) 120° rotation about $x = y = z$ (*iv*) quarter turn about x-axis
(*v*) x and y stretches, z reflection
(*vi*) enlargement $\times 3$ and central inversion
(*vii*) rotation through $\tan^{-1}\frac{3}{4}$ about z-axis
(*viii*) rotation through $\tan^{-1}\frac{4}{3}$ about x-axis

3 $-1, 0, 1, 1, -6, -27, 1, 1$

5 All points have images in the plane $x - 2y - z = 0$

6 $$\begin{pmatrix} 0 & -1 & 0 \\ 1 & 0 & 0 \\ 0 & 0 & -1 \end{pmatrix}\begin{pmatrix} 1 & -1 & 1 & 1 & 1 & -1 & -1 & -1 \\ 1 & 1 & -1 & 1 & -1 & 1 & -1 & -1 \\ 1 & 1 & 1 & -1 & -1 & -1 & 1 & -1 \end{pmatrix}$$
$$= \begin{pmatrix} -1 & -1 & 1 & -1 & 1 & -1 & 1 & 1 \\ 1 & -1 & 1 & 1 & 1 & -1 & -1 & -1 \\ -1 & -1 & -1 & 1 & 1 & 1 & -1 & 1 \end{pmatrix}$$

8 Glide reflection in line $y = x$; $+ +$ translation; $+ -$ and $- +$ glide reflection; $- -$ half-turn

Exercise 13.3a

1 $AB = \begin{pmatrix} 6 & 1 \\ 7 & 2 \end{pmatrix}$ $BA = \begin{pmatrix} 4 & 1 \\ 11 & 4 \end{pmatrix}$ $CD = \begin{pmatrix} 1 & -3 & 3 \\ -2 & 0 & 4 \\ 8 & -12 & 4 \end{pmatrix}$

$DC = \begin{pmatrix} -1 & 7 \\ -4 & 6 \end{pmatrix}$ $BD = \begin{pmatrix} -1 & 0 & 2 \\ 0 & -3 & 5 \end{pmatrix}$ $AD = \begin{pmatrix} -2 & 3 & 9 \\ 1 & -6 & 8 \end{pmatrix}$

$CA = \begin{pmatrix} 7 & 3 \\ 8 & 2 \\ 12 & 8 \end{pmatrix}$; $CB = \begin{pmatrix} 3 & 1 \\ 2 & 0 \\ 8 & 4 \end{pmatrix}$ **DB**, **DA**, **AC**, **BC** impossible

2 $$\begin{pmatrix} 1 & -3 \\ 1 & 0 \\ 1 & -1 \\ 0 & 1 \end{pmatrix}\begin{pmatrix} -1 & 2 & 7 & 3 \\ 8 & 1 & 2 & 0 \end{pmatrix} = \begin{pmatrix} -25 & -1 & 1 & 3 \\ -1 & 2 & 7 & 3 \\ -9 & 1 & 5 & 3 \\ 8 & 1 & 2 & 0 \end{pmatrix}$$

4 $AB = BA = \begin{pmatrix} 0 & 0 \\ 0 & 0 \end{pmatrix}$; $\begin{pmatrix} a & b \\ ka & kb \end{pmatrix}\begin{pmatrix} lb & mb \\ -la & -ma \end{pmatrix} = \begin{pmatrix} 0 & 0 \\ 0 & 0 \end{pmatrix}$

5 $(x^2 \quad x \quad 1)\begin{pmatrix} 2 & 1 & 3 & -1 & 4 \\ 5 & -7 & -4 & -1 & 3 \\ -8 & -11 & -1 & 5 & 0 \end{pmatrix} = (0 \quad 0 \quad 0)$ (or transposed)

6 $\begin{pmatrix} a & ka \\ -\frac{a}{k} & -a \end{pmatrix}$; $\begin{pmatrix} a & b \\ c & -a \end{pmatrix}$ where $bc = 1 - a^2$

7 $\boldsymbol{I}, \boldsymbol{A}, \boldsymbol{A}^2 = \boldsymbol{B}^2, \boldsymbol{A}^3, \boldsymbol{AB}, \boldsymbol{BA}, \boldsymbol{A}^3\boldsymbol{B}, \boldsymbol{B}^3\boldsymbol{A}$; twelve in second group

8 Only scalar matrices: $\begin{pmatrix} k & 0 & 0 \\ 0 & k & 0 \\ 0 & 0 & k \end{pmatrix}$ commute with all square matrices

9 Multiplication of complex numbers: $(a_1 + ia_2)(b_1 + ib_2)$

10 $\begin{pmatrix} 1 & 0 & 0 \\ 0 & 1 & 0 \\ 0 & 0 & 1 \end{pmatrix}$; $\begin{pmatrix} -14 - 10\sqrt{2} & 27 - 11\sqrt{2} \\ -27 - 11\sqrt{2} & -14 + 10\sqrt{2} \end{pmatrix}$;

$\begin{pmatrix} -14 - 10i & 21 - 11i \\ -21 - 11i & -14 + 10i \end{pmatrix}$

11 $\begin{pmatrix} 13 & -21 \\ -21 & 34 \end{pmatrix}$, $\begin{pmatrix} 90 & -144 \\ -144 & 233 \end{pmatrix}$, etc. (Fibonacci)

$\begin{pmatrix} p & q \\ r & s \end{pmatrix}$ where $r = \dfrac{cq}{b}$, $s = \left(p - \dfrac{q}{b}\right)(a - d)$ $(b \neq 0)$

13 $\boldsymbol{C}^2 = \begin{pmatrix} 1 & -1 \\ 1 & 0 \end{pmatrix}$ $\boldsymbol{C}^6 = \begin{pmatrix} -1 & 0 \\ 0 & -1 \end{pmatrix}$

14 $a = 1, b = 1, c = -1, d = 4, e = 0, f = 3, g = 2, h - -1, k = -1$

16 $\cos\theta = \sum x_1 x_2 / \sqrt{(\sum x_1^2 \sum x_2^2)}$

17 $(x \quad y \quad 1)\begin{pmatrix} 1 & 0 & g \\ 0 & 1 & f \\ g & f & c \end{pmatrix}\begin{pmatrix} x \\ y \\ 1 \end{pmatrix}$; $(x \quad y \quad z)\begin{pmatrix} 2 & 0 & -3 \\ 0 & 4 & 1\frac{1}{2} \\ -3 & 1\frac{1}{2} & 0 \end{pmatrix}\begin{pmatrix} x \\ y \\ z \end{pmatrix}$

19 Clockwise rotation of quadrilateral about O through $\frac{1}{2}\pi$

20 Five points in three dimensions rotated through 120° about line $x = y = z$

Exercise 13.3b

1 $\boldsymbol{AB} = \begin{pmatrix} 0 & -1 \\ -1 & 0 \end{pmatrix}$ $\boldsymbol{BA} = \begin{pmatrix} 0 & 1 \\ 1 & 0 \end{pmatrix}$ $\boldsymbol{BC} = \boldsymbol{CB} = \begin{pmatrix} -2 & 0 \\ 0 & -2 \end{pmatrix}$

$\boldsymbol{AC} = \boldsymbol{CA} = \begin{pmatrix} 0 & -2 \\ -2 & 0 \end{pmatrix}$ $\boldsymbol{C}^2 = \begin{pmatrix} 4 & 0 \\ 0 & 4 \end{pmatrix}$ $\boldsymbol{D}^2 = \begin{pmatrix} 1 & 2 \\ 0 & 1 \end{pmatrix}$;

$\boldsymbol{A}$, quarter-turn; $\boldsymbol{B}$, reflection; $\boldsymbol{C}$, enlargement; $\boldsymbol{D}$, shear

2 $\boldsymbol{P}$, 120° rotation about $x = y = z$; $\boldsymbol{Q}$, half-turn about z-axis; $\boldsymbol{R}$, reflection in plane $x = z$; $\boldsymbol{S}$, enlargement $\times 2$

3 $\boldsymbol{R}_1\boldsymbol{R}_2 = \boldsymbol{R}_2\boldsymbol{R}_1$, formulae for $\cos(\alpha + \beta)$, $\sin(\alpha + \beta)$

4 (*i*) $\begin{pmatrix}3 & 0\\0 & 3\end{pmatrix}\begin{pmatrix}-1 & 0\\0 & -1\end{pmatrix}$ enlargement and half-turn

(*ii*) $\begin{pmatrix}0 & -3\\-3 & 0\end{pmatrix}\begin{pmatrix}0 & -1\\-1 & 0\end{pmatrix}$ enlargement and reflection

(*iii*) $2\begin{pmatrix}\cos\frac{1}{3}\pi & \sin\frac{1}{3}\pi\\ \sin\frac{1}{3}\pi & -\cos\frac{1}{3}\pi\end{pmatrix}$ enlargement and reflection

(*iv*) $13\begin{pmatrix}\cos\alpha & -\sin\alpha\\ \sin\alpha & \cos\alpha\end{pmatrix}$, $\tan\alpha = \frac{12}{5}$

(*v*) $\begin{pmatrix}2 & 0\\0 & 2\end{pmatrix}\begin{pmatrix}1 & -1\frac{1}{2}\\0 & 1\end{pmatrix}$; shear and enlargement

(*vi*) $\begin{pmatrix}2 & 0\\0 & 3\end{pmatrix}\begin{pmatrix}0 & -1\\1 & 0\end{pmatrix}$; x- and y-stretches and quarter-turn

(*vii*) $\begin{pmatrix}1 & 1\\-1 & 3\end{pmatrix} = \begin{pmatrix}1 & 0\\-1 & 1\end{pmatrix}\begin{pmatrix}1 & 0\\0 & 4\end{pmatrix}\begin{pmatrix}1 & 1\\0 & 1\end{pmatrix}$; shears and y-stretch

5 (*i*) $\begin{pmatrix}2 & 0 & 0\\0 & 2 & 0\\0 & 0 & 2\end{pmatrix}\begin{pmatrix}-1 & 0 & 0\\0 & -1 & 0\\0 & 0 & -1\end{pmatrix}\begin{pmatrix}0 & 0 & 1\\0 & 1 & 0\\1 & 0 & 0\end{pmatrix}$

(*ii*) $\begin{pmatrix}1 & 0 & 0\\0 & 1 & 0\\0 & 0 & 3\end{pmatrix}\begin{pmatrix}-1 & 0 & 0\\0 & 0 & 1\\0 & 1 & 0\end{pmatrix}$ (*iii*) $\begin{pmatrix}2 & 0 & 0\\0 & 2 & 0\\0 & 0 & 2\end{pmatrix}\begin{pmatrix}0 & 1 & 0\\1 & 0 & 0\\0 & 0 & 1\end{pmatrix}$

(*iv*) $\begin{pmatrix}1 & 0 & 0\\0 & 1 & 0\\0 & 0 & 0\end{pmatrix}\begin{pmatrix}0 & 1 & 0\\1 & 0 & 0\\0 & 0 & 1\end{pmatrix}$

6 Suspect 120° rotation about line $2x = y = 2z$

7 Vectors $(0\ 0\ z)$

9 e.g. permutation matrices, rotations, etc.

10 e.g. from no. **7**, $\boldsymbol{M} = \boldsymbol{N}^2$, $\boldsymbol{X} = \begin{pmatrix}0\\0\\1\end{pmatrix}$.

Must be an orthogonal matrix, with unit determinant (see Ex. 13.4a, no. **13**)

11 Composition of a permutation matrix with one of $\begin{pmatrix}\pm 1 & 0 & 0\\0 & \pm 1 & 0\\0 & 0 & \pm 1\end{pmatrix}$

13 Reflection, rotation and translation $+k$ parallel to y-axis

Exercise 13.4a

2 $\begin{pmatrix}1 & 1\\1 & 1\frac{1}{2}\end{pmatrix}\begin{pmatrix}1 & 1 & -1 & -1\\1 & -1 & 1 & -1\end{pmatrix} = \begin{pmatrix}2 & 0 & 0 & -2\\2\frac{1}{2} & -\frac{1}{2} & \frac{1}{2} & -2\frac{1}{2}\end{pmatrix}$

4 $ad \neq bc$

6 $\begin{pmatrix} 4 & -1 & -2 \\ -22 & 2 & 12 \\ -8 & 1 & 4 \end{pmatrix}$; $\begin{pmatrix} -2 & \frac{1}{2} & 1 \\ 11 & -1 & -6 \\ 4 & -\frac{1}{2} & -2 \end{pmatrix}$; $\begin{pmatrix} -5\frac{1}{2} \\ 30 \\ 10\frac{1}{2} \end{pmatrix}$

7 $\begin{pmatrix} 2 & -4 & 2 \\ 2 & 2 & -1 \\ -4 & -4 & 5 \end{pmatrix} = \text{inverse} \times 6$; $(x \ y \ z) = (-1 \ -\frac{1}{2} \ 1\frac{1}{2})$

8 $\begin{pmatrix} 1 & -a & ac-b \\ 0 & 1 & -c \\ 0 & 0 & 1 \end{pmatrix}$

9 $\dfrac{1}{\Delta}\begin{pmatrix} A & C & B \\ B & A & C \\ C & B & A \end{pmatrix}$,

where $A = a^2 - bc$, etc.; $\Delta = \Sigma a^3 - 3abc = 0$ when $a + b + c = 0$, or $a = b = c$

10 $\frac{1}{2}\begin{pmatrix} 1 & -1 & 1 \\ 1 & 1 & -1 \\ -1 & 1 & 1 \end{pmatrix}$; $\begin{pmatrix} a \\ b \\ c \end{pmatrix} = \begin{pmatrix} 2 \\ 1 \\ 3 \end{pmatrix}$

11 $\boldsymbol{M}^2 = \boldsymbol{I}$, see Ex. 13.3a, no. **6**; e.g. half-turn, reflection, etc.

12 $\boldsymbol{M}^{-1} = \begin{pmatrix} -25 & 4 & 3 \\ 26 & -4 & -3 \\ -33 & 5 & 4 \end{pmatrix} (\boldsymbol{M}')^{-1} = (\boldsymbol{M}^{-1})'$

15 $\begin{pmatrix} \cos\alpha\cos\beta & -\cos\alpha\sin\beta & \sin\alpha \\ \sin\alpha\cos\beta & \sin\alpha\sin\beta & -\cos\alpha \\ -\sin\beta & \cos\beta & 0 \end{pmatrix}$ matrix orthogonal; rotation

17 $\frac{1}{6}\begin{pmatrix} 29 & 31 & 13 \\ 1 & -1 & -1 \\ 9 & 9 & 3 \end{pmatrix}$; $\begin{pmatrix} 1 & -2 & 1 \\ 0 & 1 & -1 \\ 0 & 0 & 1 \end{pmatrix}$; $\frac{1}{2}\begin{pmatrix} 2 & 3 & 1 \\ -6 & -8 & 2 \\ 14 & 19 & -1 \end{pmatrix}$;

$\frac{1}{3}\begin{pmatrix} 1 & -2 & -1 \\ -1 & 5 & 3 \\ 3 & -12 & -6 \end{pmatrix}$; $\begin{pmatrix} 1 & 1 & 1 \\ 1 & 2 & 2 \\ 1 & 2 & 3 \end{pmatrix}$;

$(1-a)^{-1}(1+2a)^{-1}\begin{pmatrix} 1+a & -a & -a \\ -a & 1+a & -a \\ -a & -a & 1+a \end{pmatrix}$;

$(a^2+b^2+c^2-1-2abc)^{-1}\begin{pmatrix} c^2-1 & a-bc & b-ac \\ a-cb & b^2-1 & c-ab \\ b-ac & c-ab & a^2-1 \end{pmatrix}$

$\frac{1}{2}\begin{pmatrix} -i & -1 \\ 1 & i \end{pmatrix}$; $\frac{1}{2}\begin{pmatrix} \omega^2 & -i \\ -i & \omega \end{pmatrix}$; $\begin{pmatrix} 1 & \omega & \omega^2 \\ \omega^2 & 1 & \omega \\ \omega & \omega^2 & 1 \end{pmatrix}$ is singular

Exercise 13.4b

2 t, $2t + 1$ **3** -2 and $+5$

4 Non-trivial solution when $a_1 b_2 - a_2 b_1 \neq 0$

Exercise 13.4c

3 Using row 1 on p. 421

(*i*) □TQARO□R.SPWEOY,EA,KGN□IWFBBFG

(*ii*) COATISX,GZLDEAZSOWN.RMEJWDAKCDNVTP WTMFAUO.RXF

4 $\begin{pmatrix} 5 & 7 \\ 2 & 3 \end{pmatrix}$; E.GHEE QKF A

Exercise 13.5a

1 (*i*) -5 (*ii*) 54 (*iii*) 0 (*iv*) 2 670 (*v*) 2 232 (*vi*) 0 (*vii*) 0

2 $\begin{vmatrix} a & b \\ a^n & b^n \end{vmatrix}$ has factor $a - b$

3 (*i*) $4ab$ (*ii*) 1 (*iii*) $\cos 2\theta$ (*iv*) 1 (*v*) $-1 - t^4$
(*vi*) $(ps - rq)(x - y)z$

4 e.g. $\begin{vmatrix} pa + qb & ra + sb \\ pc + qd & rc + sd \end{vmatrix} = \Delta \times (ps - qr)$

5 $\begin{vmatrix} a & b \\ c & d \end{vmatrix} + \begin{vmatrix} a & y \\ c & q \end{vmatrix} + \begin{vmatrix} x & b \\ p & d \end{vmatrix} + \begin{vmatrix} x & y \\ p & q \end{vmatrix}$

6 $\lambda = -1, 7$ **7** $x = 0, -1$

8 $x = -2\frac{1}{2}, y = \frac{1}{2}$ $(ar - cp)^2 = (br - cp)(aq - bp)$

Exercise 13.5b

1 (*i*) -1 (*ii*) -3 (*iii*) -21 (*iv*) $-19\,320$ (*v*) 0

2 $\Delta = a^3 + b^3 + c^3 - 3abc, \Delta, \Delta, -\Delta$; $a^3 + b^3$; $1 - 2x^3 + x^6$

4 See alien cofactors, 13.5, p. 000 **5** $-2, 8, 9$

6 Any pair of rows or columns are linearly dependent

7 (*i*) $2abc$ (*ii*) 0, (*iii*) 0

8 (*i*) $1 + a^2 + b^2 + c^2 + 2abc = 0$ (*ii*) $bc + ca + ab + 1 = 0$

9 $3\frac{1}{2}$

11 $\begin{vmatrix} p & q & r & s & 0 \\ 0 & p & q & r & s \\ a & b & c & 0 & 0 \\ 0 & a & b & c & 0 \\ 0 & 0 & a & b & c \end{vmatrix} = 0$

Exercise 13.5c

1 (*i*) 276 (*ii*) $-ab(a-b)(1-a)(1-b)$
(*iii*) $(a+b+c)^3$
(*iv*) $-(a+b+c)(a+b\omega+c\omega^2)(a+b\omega^2+c\omega)$

2 $(a+b+c)(a+b\omega+c\omega^2)(a+b\omega^2+c\omega)$
$= (a+b+c)(\sum a^2 - \sum bc)$

3 $(a-1)(a^2+a+1)$; $(1+a+a^2)^2(1-a)^2$; 0

4 (*i*) $(b-a)(a-c)(c-b)$ (*ii*) $(a+b+c)(b-a)(a-c)(c-b)$
(*iii*) $\sum bc - \sum a^2$ (*iv*) $(b-a)(a-c)(c-b)(bc+ca+ab)$
(*v*) 0

5 -1 **6** $(2a+b)(a-b)^2$; $-(2a+b)(a-b)^2$; 0

7 (*i*) $\begin{vmatrix} 0 & c & b \\ c & 0 & a \\ b & a & 0 \end{vmatrix}^2 = 4a^2b^2c^2$; (*ii*) $(b-a)^3$ (*iii*) -8
(*iv*) $(a+b+c+d)(a-b+c-d)(a+b-c-d)(a-b-c+d)$
(*v*) 0

8 e.g. $\begin{vmatrix} 1 & 2 & 3 & 4 \\ 2 & 3 & 4 & 5 \\ 3 & 4 & 5 & 6 \\ 1 & 1 & 1 & 1 \end{vmatrix}$

Exercise 13.5d

1 e.g. $\boldsymbol{D}'\boldsymbol{D} = \begin{vmatrix} a^2+c^2 & ab+cd \\ ab+cd & b^2+d^2 \end{vmatrix}$

2 e.g. $\begin{vmatrix} 1 & 0 \\ 0 & 1 \end{vmatrix} + \begin{vmatrix} 1 & 1 \\ 1 & 1 \end{vmatrix} = \begin{vmatrix} 2 & 1 \\ 1 & 2 \end{vmatrix}$

3 $\begin{vmatrix} a^2+bc & ac & c^2 \\ ab & bc+ac & bc \\ ab & ab & ac+b^2 \end{vmatrix} = \begin{vmatrix} a^2+c^2 & ab & ac \\ ab & b^2+c^2 & bc \\ ac & bc & a^2+b^2 \end{vmatrix}$

4 $\begin{vmatrix} a & p & 0 \\ b & q & 0 \\ c & r & 0 \end{vmatrix}^2 = 0$

5 $|\boldsymbol{a}+\boldsymbol{x}\ \ \boldsymbol{b}+\boldsymbol{y}\ \ \boldsymbol{c}+\boldsymbol{z}| = |\boldsymbol{a}\ \boldsymbol{b}\ \boldsymbol{c}| + |\boldsymbol{a}\ \boldsymbol{b}\ \boldsymbol{z}| + |\boldsymbol{x}\ \boldsymbol{b}\ \boldsymbol{c}| + |\boldsymbol{x}\ \boldsymbol{b}\ \boldsymbol{z}| +$
$+ |\boldsymbol{a}\ \boldsymbol{y}\ \boldsymbol{z}| + |\boldsymbol{x}\ \boldsymbol{y}\ \boldsymbol{c}| + |\boldsymbol{x}\ \boldsymbol{y}\ \boldsymbol{z}|$

9 Must have value of determinant = 1

Exercise 13.6a

1 (*i*) $\frac{52}{17}, \frac{31}{17}$ (*ii*) $\dfrac{x}{-21} = \dfrac{y}{-14} = \dfrac{1}{0}$; no solutions

(*iii*) $\dfrac{x}{ln+mp} = \dfrac{y}{mn-lp} = \dfrac{-1}{l^2+m^2}$

2 (*i*) 3, −5, −1 (see 13.6, Example 1)
(*ii*) infinity of solutions: 3 × eqn. 1 − 2 × eqn. 2 = eqn. 3 (see 13.7, pp. 445–6)

3 $(br - cq)^2 + (ar - pc)^2 = (aq - bp)^2$

5 $a + b + c = 0 \Rightarrow x = y = z$; or $a = b = c \Rightarrow x + y + z = 0$

6 $\lambda = 3, -1, -1$; $\lambda = 3, x:y:z = 1:0:1$;
$\lambda = -1, x:y:z = 3:4:-9$

7 $3x = a + b + c$; $3y = a + b\omega^2 + c\omega$; $3z = a + b\omega + c\omega^2$

9 $\beta = -1\ (\lambda, -3\lambda, \lambda + 3)$

10 $k = 1$, no solutions; $k = -2(\lambda, 5\lambda + 8, -\lambda - 2)$

Exercise 13.6b

1 $x = 2 + \dfrac{15}{\delta}, \ y = -\dfrac{10}{\delta}$ **2** $k = -5, c = -\frac{1}{2}$

3 (*i*) $\lambda = 0, x:y:z = 1:0:1$
(*ii*) values other than 0 or 1:
$$x = \frac{2\lambda - 3}{2\lambda(\lambda - 1)}, \quad y = \frac{1}{2(\lambda - 1)}, \quad z = \frac{-\lambda^2 + 4\lambda - 3}{2\lambda(\lambda - 1)}$$

4 $(-3, 6, 2), (2, 1, 6)$; $(\sin\alpha \sin\beta, -\cos\alpha \sin\beta, \cos\alpha \cos\beta)$

Exercise 13.7a

1 (*i*) No solutions (*ii*) $-4\frac{1}{2}, 19, 7\frac{1}{2}$ (*iii*) $-1, 1, 1$ (*iv*) $\frac{1}{2}, 0, -\frac{1}{2}$
(*v*) $\dfrac{-106}{65}, \dfrac{111}{260}, \dfrac{109}{26}$

2 $-1, -1, 1$ **3** $x:y:z:w = 1:-2:0:1$

4 e.g. $\boldsymbol{M}_2 = \begin{pmatrix} 1 & 2 & 0 \\ 0 & 1 & 0 \\ 0 & 0 & 1 \end{pmatrix}$ $\boldsymbol{M}_1 = \begin{pmatrix} 5 & 0 & 0 \\ 0 & 1 & -3 \\ 0 & 0 & 1 \end{pmatrix}$ $\det \boldsymbol{A} = 46$

5 $x = 5.23, y = -6.91, z = -2.33$

6 $x = 0.898, y = 0.763, z = 0.618$

Exercise 13.7b

1 (*i*) $\begin{pmatrix} 1 & -1\frac{1}{2} \\ -1 & 2 \end{pmatrix}$ (*ii*) $\frac{1}{3}\begin{pmatrix} 4 & -5 \\ 5 & -7 \end{pmatrix}$ (*iii*) $\dfrac{1}{ad - bc}\begin{pmatrix} d & -b \\ -c & a \end{pmatrix}$

(*iv*) singular (*v*) $\begin{pmatrix} -2 & \frac{1}{2} & 1 \\ 11 & -1 & -6 \\ 4 & -\frac{1}{2} & -2 \end{pmatrix}$ (*vi*) $\frac{1}{12}\begin{pmatrix} 1 & 5 & 3 \\ 4 & -4 & 0 \\ 7 & -1 & -3 \end{pmatrix}$

(*vii*) $\frac{1}{8}\begin{pmatrix} 11 & 2 & -5 \\ -4 & 0 & 4 \\ -1 & 2 & -1 \end{pmatrix}$ (*viii*) $\frac{1}{780}\begin{pmatrix} 72 & -32 & 44 \\ 132 & -37 & -71 \\ -60 & 70 & 50 \end{pmatrix}$

2 (*i*) No solutions (*ii*) $-4\frac{1}{2}, 19, 7\frac{1}{2}$ (*iii*) $-1, 1, 1$
(*iv*) $-\frac{106}{65}, \frac{111}{260}, \frac{109}{26}$

3 $\frac{1}{4}\begin{pmatrix} 6 & 0 & -2 \\ 2 & -1 & 1 \\ 0 & -1 & 3 \end{pmatrix}$; $-2, 3\frac{1}{4}, 8\frac{1}{4}$

6 $k = 2$; $x = 6 - 5z, y = 3z - 2$;
$X = 2 + 2z, Y = 8 - 8z, Z = 8 - 9z$

7 (*i*) $5t, -12t, t$ (*ii*) no solutions (*iii*) $5t + 5, -12t - 10, t$
(*iv*) no solutions

8 All points map into the plane $2x + y - z = 0$

9 $\begin{pmatrix} a_1 & b_1 & c_1 \\ a_2 & b_2 & c_2 \\ a_3 & b_3 & c_3 \end{pmatrix}$ where $\begin{matrix} 2a_1 - 4a_2 + a_3 = 0 \\ 2b_1 - 4b_2 + b_3 = 0, \\ 2c_1 - 4c_2 + c_3 = 0 \end{matrix}$ e.g. $\begin{pmatrix} 0 & -2 & -1 \\ 1 & 0 & 0 \\ 4 & 4 & 2 \end{pmatrix}$

10 $\begin{pmatrix} 5a & 5b & 5c \\ 3a & 3b & 3c \\ 2a & 2b & 2c \end{pmatrix}$

11 All 2×2 minors zero; rows and columns linearly dependent in pairs

Exercise 13.8

1 $e < 3 - 2\sqrt{2}$;

no, let $M_1 = \frac{1}{2}\begin{pmatrix} 1-e & 1+e & 0 \\ 1+e & 1-e & 0 \\ 0 & 0 & 2 \end{pmatrix} = \frac{1}{2}\left[\begin{pmatrix} 1 & 1 & 1 \\ 1 & 1 & 0 \\ 0 & 0 & 2 \end{pmatrix} + e\begin{pmatrix} -1 & 1 & 0 \\ 1 & -1 & 0 \\ 0 & 0 & 0 \end{pmatrix}\right]$

$= \frac{1}{2}(\boldsymbol{A} + e\boldsymbol{B})$, and similarly $\boldsymbol{M}_2 = \frac{1}{2}(\boldsymbol{C} + e\boldsymbol{D})$. Work out successive products... $\boldsymbol{M}_1\boldsymbol{M}_2\boldsymbol{M}_1\boldsymbol{M}_2\boldsymbol{M}_1$, neglecting terms in $e^2, e^3, \ldots$

Miscellaneous problems 13

1 $(x, y, z) \xmapsto{Q_1} (x, -z, y)$; $(x, y, z) \xmapsto{Q_2} (z - 1, y, -x + 1)$;
$(x, y, z) \xmapsto{Q_2Q_1} (y - 1, -z, -x + 1)$;
$x = y, x = z - 1, y = 0, z = 1$.
Reflections in three planes would have a fixed point

2 Closure **3** $r^2 = s^2 = e$; $\boldsymbol{D}_2$, as on p. 502

4 $86y = 41x + 176$

5 $(3t, 3 - 4t, 2 - 5t)$ through points $(0, 3, 2)$ and $(3, -1, -3)$

7 Let $\boldsymbol{l} = (l\,m\,n), \boldsymbol{x} = (x\,y\,z), \boldsymbol{a} = (a\,b\,c)$. Equations state that $\boldsymbol{a} = \boldsymbol{l} \times \boldsymbol{x}$, so we must have $\boldsymbol{a}, \boldsymbol{l}$ perpendicular, i.e. $al + bm + cn = 0$; also $\boldsymbol{a}, \boldsymbol{x}$ perpendicular, so $ax + by + cz = 0$. A vector parallel to $\boldsymbol{a} \times \boldsymbol{l}$ satisfies $lx + my + nz = 0$, i.e. $(bn - cm, cl - an, am - bl)\,(l^2 + m^2 + n^2)^{-1}$

8 Least 6 (3 zeros in same row or column); greatest 11 (no two zeros in same row or column)

9 $(abc)^{-1}(b-c)(c-a)(a-b)\sum a^{-r}b^{-s}c^{-t}\ (r+s+t=n-1)$

10 No **11** $\cos n\theta$

Exercise 14.2a

1 No **2** (*i*) Exclude $b+d=0$ (*ii*) exclude $Q=0$

3 1 2 3 4 6 5, 4 1 2 3 6 5, 3 4 1 2 5 6, 2 3 4 1 5 6

4 No, e.g. $\begin{pmatrix}1 & 4\\0 & 1\end{pmatrix}+\begin{pmatrix}0 & -1\\1 & 0\end{pmatrix}=\begin{pmatrix}1 & 3\\1 & 1\end{pmatrix}$;

$\{\boldsymbol{M}: \det \boldsymbol{M}<0\}$, $\times$ not closed

$\{\boldsymbol{M}: \det \boldsymbol{M}=0\}$, $\times$ closed

5 No. 120° turns about two altitudes may combine to give a half-turn (see 14.4, p. 512)

6 Include 11 for closure. $\{1, 13\}$

7

	N	Z	Z^-	Q	Q^+	R	R^+	C	$R\backslash Q$	cis θ	$x+iy: x, y \in Z$
$+$	1	1	1	1	1	1	1	1	0	0	1
$-$	0	1	0	1	0	1	0	1	0	0	1
$\times$	0	1	0	1	1	1	1	1	0	1	1
$\div$	0	0	0	1	1	1	1	1	0	1	0

(zero excluded for $\times$ and $\div$)

8 Multiplication closed, addition not closed

9

	$f+g$	$f\circ g$	$g\circ f$	f^2	g^2
(*i*)	$\dfrac{1}{x(1-x)}$	$1-x$	$\dfrac{x}{1-x}$	x	$\dfrac{x-1}{x}$
(*ii*)	$\ln x+\sin x$	$\ln(\sin x)$	$\sin(\ln x)$	$\ln(\ln x)$	$\sin(\sin x)$
(*iii*)	$\dfrac{(-3+2x-2x^2)}{2x-1}$	$\dfrac{x+3}{2x+1}$	$\dfrac{-x+3}{2x-1}$	x	x
(*iv*)	4	$9-4x$	$7-4x$	$4x-3$	$4x-5$

10 (*i*) Yes (*ii*) no (*iii*) no (*iv*) yes

13

	$+$	$\cdot$	$\times$
2D	1	0	0
3D	1	0	1

14 (*i*) Closed (*ii*) yes (*iii*) yes provided $x\neq y$

Exercise 14.2b

2

	(*i*)		(*ii*)	(*iii*)	(*iv*)	(*v*)	(*vi*)	(*vii*)
C	1	1	1	1	0	1	0	0
A	1	0	0	0	1	0	1	0

3 $(ba)c \neq b(ac)$ **4** $(be)d \neq b(ed)$ **7** Associative

8 e.g. $k = -1$, put $x = \cot \alpha$, $y = \cot \beta$

9 Not associative; $(0, 0)$, $(a_1 - a_2, a_2 - a_1)$; no. Invalid because associativity assumed. $(B \circ A) \circ A^* \neq B \circ (A \circ A^*)$

Exercise 14.2c

1 $\mathscr{E}, \varnothing$; symmetric difference: $A \Delta B = (A \cap B') \cup (B \cap A')$

2 (*i*) $\{2, 4, 8\} \times \pmod{14}$ closed (*ii*) 10

3 (*i*) $27 + 13 = 0$; $12 + 28 = 0 \pmod{40}$
(*ii*) $17 \times 13 = 1 \pmod{20}$ (*iii*) $53 \times 17 = 1 \pmod{100}$

5 (*i*) $0.04\begin{pmatrix} 3 & -4 \\ 4 & 3 \end{pmatrix}$ (rotation) (*ii*) $\begin{pmatrix} 0.6 & 0.8 \\ 0.8 & -0.6 \end{pmatrix}$ (reflection)

(*iii*) $\begin{pmatrix} 0 & -1 \\ 1 & 1 \end{pmatrix}$ (*iv*) $\begin{pmatrix} 0 & -1 & 0 \\ 0 & 0 & 1 \\ -1 & 0 & 0 \end{pmatrix}$ (*v*) $\begin{pmatrix} 6 & 8 & -1 \\ -10 & -13 & 2 \\ 7 & 9 & -1 \end{pmatrix}$

(*vi*) $\begin{pmatrix} \cos 80^\circ & \sin 80^\circ \\ -\sin 80^\circ & \cos 80^\circ \end{pmatrix}$ (rotation)

6 (*i*) $\dfrac{1}{1 - x}$ (*ii*) $\dfrac{2}{x + 1}$ (*iii*) $\dfrac{2(x - 1)}{x + 1}$ (*iv*) $x^{1/3}$ (*v*) $\log_a x$

(*vi*) $\dfrac{3x - 7}{5x - 3}$ (*vii*) a^{a^x}

7 (*i*) $\begin{pmatrix} 1 & 2 & 3 & 4 & 5 \\ 4 & 1 & 5 & 2 & 3 \end{pmatrix}$ (*ii*) $\begin{pmatrix} 1 & 2 & 3 & 4 & 5 & 6 \\ 4 & 5 & 6 & 1 & 3 & 2 \end{pmatrix}$ (*iii*) $\begin{pmatrix} 1 & 2 & 3 & 4 & 5 & 6 & 7 \\ 4 & 7 & 5 & 1 & 3 & 6 & 2 \end{pmatrix}$

10 Inverse requires $a^3 + b^3 + c^3 - 3abc \neq 0$, i.e. $a + b + c \neq 0$ and a, b, c not all the same

Exercise 14.3a

1 (*i*) 8 (*ii*) 4 (*iii*) 120 (*iv*) n (*v*) 8 (*vi*) 4, 8, 6, 8

4 (*i*) Fails for closure, identity, inverse (*ii*) fails for inverses
(*iii*) no inverses (*iv*) neither inverses nor identity (*v*) not closed
(*vi*) not closed (*vii*) include identity (*viii*) include $-i$
(*ix*) not closed

5 (*i*) $b^2 = f^2 = e = (bf)^3$ (*ii*) $a(bc) \neq (ab)c$
(*iii*) table (with v and w interchangeable):

	1	p	q	r	s	t	u	v	w
1	1	p	q	r	s	t	u	v	w
p	p	q	1	t	w	v	s	r	u
q	q	1	p	v	u	r	w	t	s
r	r	t	v	s	1	w	q	u	p
s	s	w	u	1	r	p	v	q	t
t	t	v	r	w	p	y	1	s	q
u	u	s	w	q	v	1	t	p	r
v	v	r	t	u	q	s	p	w	1
w	w	u	s	p	t	q	r	1	v

6

	1	x	-1	$-x$
1	1	x	-1	$-x$
x	x	-1	$-x$	1
-1	-1	$-x$	1	x
$-x$	$-x$	1	x	-1

7 24
8 $T(x) = -x + b$ (reflection in $\frac{1}{2}b$)
10 e.g. $G = R^+, \times, g = 2$

Exercise 14.3b

1 $3^2 = 9, 3^3 = 13, 3^4 = 11, 3^5 = 5, 3^6 = 1$ (mod 14)
3 12 **4** $7^4 = 1$ (mod 15); $\{3, 6, 9, 12\} \times$ mod 15 (identity 6).
5 $\{4, 8, 12, 16\} \times$ mod 20 (identity 16)
6 $\left\{x,\ 1-x,\ 1-\frac{1}{x},\ \frac{1}{1-x},\ \frac{x}{x-1},\ \frac{1}{x}\right\}$ (D_3, as on p. 509)
7 (*i*) 2 (*ii*) 2 (*iii*) 4 (*iv*) 10 (*v*) 12
9 D_3, as on p. 509
10 ***I*, *A*, *B*, *AB*, *BA*, *BAB*, *ABA*, *ABAB* = *BABA*** (D_4, as on p. 511)
11 No finite additive subgroups of C, except $\{0\}$
12 e.g. generated by (1 2 3 4 5 6 7); generated by (1 2 3 4 5) (6 7)
13 C_9; 2 of period 3, 6 of period 9;
C_{12}, 4 of period 12, 2 of period 6, 4 and 3; 1 of period 2
14 Generated by 2, 6, 11 or 7
15 (*i*) D_2
(*ii*) eighth symmetry is central inversion, $(x, y, z) \mapsto (-x, -y, -z)$;
group is $C_2 \times C_2 \times C_2$.

16 $(1, 0), \left(\frac{1}{x}, -\frac{y}{x}\right)$. Remove elements having $x = 0$

Exercise 14.4a

1 $\left.\begin{array}{r} 1 \leftrightarrow 1 \leftrightarrow i \\ -1 \leftrightarrow 9 \leftrightarrow x^2 \\ \{i, -i\} \leftrightarrow \{3, 7\} \leftrightarrow \{x, x^3\} \end{array}\right\} C_4$

3 (*i*) non-Abelian $\ncong$ Abelian (*ii*) D_2, C_4
(*iii*) $C, + \cong V_2, + \ncong V_3, +$
(*iv*) impossible to set up one-to-one correspondence between R^+ and Q

4 Must be infinite

6 $x\ (\in S) \leftrightarrow 2\pi x$ ($\in$ rotations of circle)

7 C_6; $\{f, g\} \leftrightarrow \{r, t\}$ of period 6

8 $\{1, 2, 4, 5, 7, 8\}$; $\{1, 3, 5, 7\}$; $\{1, 3, 7, 9\}$; $\{1, 5, 7, 11\}$; n prime

10 All six permutations of a, b, c are automorphisms

12 $\alpha \leftrightarrow \alpha^r$ $(r = 1, 2, 3, 4, 5, 6)$; cis $\frac{1}{4}\pi \leftrightarrow$ cis $r\frac{1}{4}\pi$ $(r = 1, 3, 5, 7)$

Exercise 14.4b

1 (*i*) D_2 (*ii*) D_1 (*iii*) D_4 (*iv*) C_4 (*v*) D_8 (*vi*) C_2

2

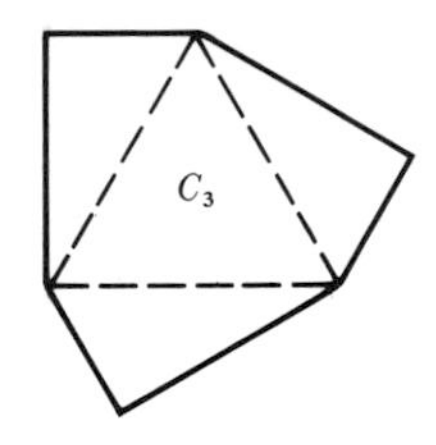

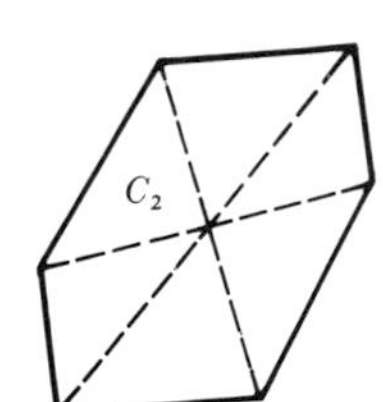

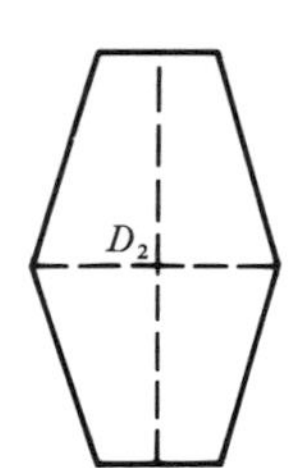

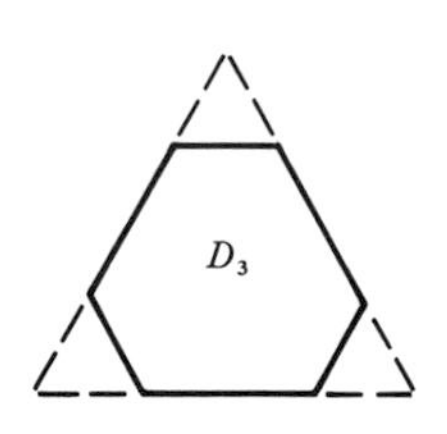

3 Mirrors at $\frac{\pi}{n}$, group is D_n

4 See Ex. 14.3b, no. **10**

5 $gf(x) = \frac{2}{2 - x}$; $fg(x) = 2 - \frac{2}{x}$ $(gf)^4 = i$; D_4

6 $a^2 = b^2 = (ab)^3 = 1$

7 $ar = r^5 a = g$; $ar^2 = r^4 a = f$; $ar^3 = r^3 a = d$;
$ar^4 = r^2 a = c$; $ar^5 = ra = b$

9 Nine of period 2, eight of period 3, six of period 4

10 24 of each

11 (*i*) C_∞ (*ii*) D_∞ (*iii*) see no. **3** (*iv*) D_∞ (*v*) D_∞